MEP 804A/B AND 814A/B
15 KW GENERATOR SET
MANUAL TM 9-6115-643-24
PMCS, GENERAL MAINTENANCE
AND
DIRECT SUPPORT MAINTENANCE

GENERATOR SET
SKID MOUNTED
TACTICAL QUIET
15 KW, 50/60 HZ, MEP-804A
(NSN: 6115-01-274-7388) (EIC: VG4)
15 KW, 50/60 HZ, MEP-804B
(NSN: 6115-01-530-1458) (EIC: N/A)
AND
GENERATOR SET
SKID MOUNTED
TACTICAL QUIET
15 KW, 400 HZ, MEP-814A
(NSN: 6115-01-274-7393) (EIC: VN4)
15 KW, 400 HZ, MEP-814B
(NSN: 6115-01-529-9494) (EIC: N/A)

edited by
Brian Greul

The MEP series of Military Generators are reknowned for their quiet, durable operation and conservative power ratings. This is the PMCS, General Maintenance, and Direct Support Maintenance manual for the 15KW version of the generator issued under models 804 and 814. The A series has analog controls and the B series has digital controls. Various units are manufactured for the US Government by different contractors with different power plants. This book is a reprint of the operator manual published by the US Army. It is printed as a courtesy to enthusiasts and owners of these generator sets. Other important manuals for this generator are also printed by this publisher.

An 8.5x11 3 hole punched loose leaf copy may be purchased for your 3 ring binder. Email books@ocotillopress.com for current information.

Should you have suggestions or feedback on ways to improve this book please send email to Books@OcotilloPress.com We also welcome requests

---

Edited 2021 Ocotillo Press
ISBN 978-1-954285-17-0

Printed in the United States of America

Ocotillo Press
Houston, TX 77017
Books@OcotilloPress.com

ARMY TM 9-6115-643-24

TECHNICAL MANUAL

UNIT, DIRECT SUPPORT, AND GENERAL SUPPORT MAINTENANCE MANUAL

FOR

GENERATOR SET, SKID MOUNTED, TACTICAL QUIET, 15 kW, 50/60 Hz,
MEP-804A (NSN: 6115-01-274-7388)
MEP-804B (NSN: 6115-01-530-1458)

GENERATOR SET, SKID MOUNTED, TACTICAL QUIET, 15 kW, 400 Hz,
MEP-814A (NSN: 6115-01-274-7393)
MEP-814B (NSN: 6115-01-529-9494)

This manual supersedes TM 9-6115-643-24 dated 1 September 1993, including all changes.

HEADQUARTERS, DEPARTMENTS OF THE ARMY AND AIR FORCE

1 MAY 2008

WARNING SUMMARY

This Warning Summary provides a summary of all critical safety information in this manual. The Summary contains all warnings used throughout this manual.

Prior to starting any procedure, the WARNINGS included in the text and at the beginning of each maintenance procedure must be reviewed and understood.

The WARNINGS located in the generator set technical manuals and the trailer technical manuals must also be considered.

This manual describes physical and chemical processes that may require the use of chemicals, solvents, paints, or other commercially available material. Users of the manual should obtain the material safety data sheets (Occupational Safety and Health Act (OSHA) Form 20 or equivalent) from the manufacturers or suppliers of materials to be used. Users must be completely familiar with manufacturer/supplier information and adhere to their procedures, recommendations, warnings, and cautions for safe use, handling, storage, and disposal of these materials.

WARNING

All metal jewelry can conduct electricity and become entangled in generator set components. Remove all jewelry when working on generator set. Failure to comply with this warning can cause injury or death to personnel.

WARNING

DO NOT wear loose clothing when performing checks, services and maintenance. Failure to comply with this warning can cause injury or death to personnel.

WARNING

High voltage is produced when this generator set is in operation. Make sure generator set is completely shut down and free of any power source before attempting any repair or maintenance on the set, or when connecting or disconnecting load cables. Failure to comply with this warning can cause injury or death to personnel.

WARNING

High voltage is produced when this generator set is in operation. SHUT DOWN generator set and make sure it is free of any power source before attempting any repair or maintenance on the set, or when connecting or disconnecting load cables. Failure to comply with this warning can cause injury or death to personnel.

WARNING

High voltage is produced when the generator set is in operation. Never attempt to start the generator set unless it is properly grounded. Failure to comply with this warning can cause injury or death to personnel.

WARNING SUMMARY – Cont'd

WARNING

Shut down generator set before performing inspection of wiring. Failure to comply with this warning can cause injury or death to personnel.

WARNING

Ensure nuts on ground terminals are properly secured creating a good ground. Failure to comply with this warning can cause injury or death to personnel.

WARNING

High voltage is produced when the generator set is in operation. DO NOT touch live voltage connections. Never attempt to connect or disconnect load cables or paralleling cables while the generator set is running. Failure to comply with this warning can cause injury or death to personnel.

WARNING

Dangerous voltage exists on live circuits. Always observe precautions and never work alone. Failure to comply with this warning can cause injury or death to personnel.

WARNING

DC voltages are present at generator set electrical components even with generator set shut down. Avoid shorting any positive with ground/negative. Failure to comply with this warning can cause injury to personnel, and damage to equipment.

WARNING

Prior to making any connections for parallel operation, ensure that there is no input to the load and that the generator sets are shut down. Failure to comply with this warning can cause injury or death to personnel.

WARNING

If it is necessary to move a generator set which has been operating in parallel with another generator set, shut down remaining generator set connected to the load, prior to removing load and ground cables. Failure to comply with this warning can cause injury or death to personnel.

WARNING

Power is available when the main contactor is open. Avoid accidental contact. Failure to comply with this warning can cause injury or death to personnel.

WARNING SUMMARY – Cont'd

WARNING

High voltage power is available when the main contactor is closed. Avoid accidental contact with live components. Ensure load cables are properly connected and the load cable door is shut before closing main contactor. Ensure load is turned off before closing main contactor. Ensure that soldiers working with/on loads connected to the generator set are aware that main contactor is about to be closed before closing main contactor. Failure to comply with this warning can cause injury or death to personnel.

WARNING

Slave receptacle (NATO connector) is electrically live at all times and is unfused. The Battery Disconnect Switch does not remove power from the slave receptacle. NATO slave receptacle has 24 VDC even when Battery Disconnect Switch is set to OFF. This circuit is only dead when the batteries are fully disconnected. Disconnect the batteries before performing maintenance on the slave receptacle. Failure to comply with this warning can cause injury or death to personnel.

WARNING

Diesel fuel is flammable and toxic to eyes, skin, and respiratory tract. Skin and eye protection are required when working in contact with diesel fuel. Avoid repeated or prolonged contact. Provide adequate ventilation. Operators are to wash exposed skin and change chemical soaked clothing promptly if exposed to fuel. Failure to comply with this warning can cause injury or death to personnel.

WARNING

Fuels used in the generator set are flammable. Do not smoke or use open flames when performing maintenance. Failure to comply with this warning can cause injury or death to personnel, and damage to the generator set.

WARNING

Fuels used in the generator set are flammable. When filling the fuel tank, maintain metal-to-metal contact between filler nozzle and fuel tank opening to eliminate static electrical discharge. Failure to comply with this warning can cause injury or death to personnel, and damage to the generator set.

WARNING

Hot engine surfaces from the engine and generator circuitry are possible sources of ignition. When hot refueling with DF-1, DF-2, JP5 or JP8, avoid fuel splash and fuel spill. Do not smoke or use open flame when performing refueling. Remember PMCS is still required. Failure to comply with this warning can cause injury or death to personnel, and damage to the generator set.

WARNING SUMMARY – Cont'd

WARNING

When running, generator set engine has hot metal surfaces that will burn flesh on contact. Shut down generator set and allow engine to cool before performing checks, services, and maintenance. Wear gloves and additional protective clothing as required. Failure to comply with this warning can cause injury or death to personnel.

WARNING

Cooling system operates at high temperature and pressure. Contact with high pressure steam and/or liquids can result in burns and scalding. Shut down generator set, and allow system to cool before performing checks, services and maintenance, or wear gloves and additional protective clothing and goggles as required. Failure to comply with this warning can cause injury or death to personnel.

WARNING

In extreme cold weather, skin can stick to metal. Avoid contacting metal items with bare skin in extreme cold weather. Failure to comply with this warning can cause injury to personnel.

WARNING

Operating the generator set exposes personnel to a high noise level. Hearing protection must be worn when operating or working near the generator set when the generator set is running. Failure to comply with this warning can cause hearing damage to personnel.

WARNING

Exhaust discharge contains deadly gases including carbon monoxide. DO NOT operate generator set in enclosed areas unless exhaust discharge is properly vented outside. Failure to comply with this warning can cause injury or death to personnel.

WARNING

Hot exhaust gases can ignite flammable materials. Allow room for safe discharge of hot gases and sparks. Failure to comply with this warning can cause injury or death to personnel.

WARNING

Top housing panels and exhaust system can get very hot. Shut down generator set, and allow system to cool before performing checks, services, and maintenance. Failure to comply with this warning can cause severe burns and injury to personnel.

WARNING SUMMARY – Cont'd

WARNING

Top housing panels and exhaust system can get very hot. When performing DURING PMCS, wear gloves and additional protective clothing as required. Failure to comply with this warning can cause severe burns and injury to personnel.

WARNING

Exercise extreme caution when performing DURING PMCS checks inside engine compartment. Avoid contact with moving or hot engine parts. Failure to comply with this warning can cause injury or death to personnel.

WARNING

When running, winterization heater has hot metal surfaces that will burn flesh on contact. Shut down generator set and allow heater to cool before performing maintenance. Wear gloves and additional protective clothing as required. Failure to comply with this warning can cause injury or death to personnel.

WARNING

Each battery weighs more than 70 pounds (32 kg) and requires a two-person lift. Lifting batteries can cause back strain. Ensure proper lifting techniques are used when lifting batteries. Failure to comply with this warning can cause injury to personnel.

WARNING

Batteries give off a flammable gas. Do not smoke or use open flame when performing maintenance. Failure to comply with this warning can cause injury or death to personnel, and damage to the generator set.

WARNING

Battery acid can cause burns to unprotected skin. Wear safety goggles and chemical gloves and avoid acid splash while working on batteries. Failure to comply with this warning can cause injury to personnel.

WARNING

When disconnecting or removing batteries, disconnect the negative lead that connects directly to the grounding stud first; disconnect the negative end of the interconnection cable next. When installing batteries, reverse the connection sequence. Failure to comply with this warning can cause injury to personnel.

WARNING

The connection of any electrical equipment and the disconnection of any electrical equipment may cause an explosion hazard. Do not connect any electrical equipment or disconnect any electrical equipment in an explosive atmosphere. Failure to comply with this warning can cause injury or death to personnel.

WARNING SUMMARY – Cont'd

WARNING

Many components require a two-person lift. Lifting heavy components can cause back strain. Ensure proper lifting techniques are used when lifting heavy components. Failure to comply with this warning can cause injury to personnel.

WARNING

Solvent used to clean parts is potentially dangerous to personnel and property. Clean parts in a well-ventilated area. Avoid inhalation of solvent fumes. Wear goggles and rubber gloves to protect eyes and skin. Wash exposed skin thoroughly. Do not smoke or use near open flame or excessive heat. Failure to comply with this warning can cause injury to personnel, and damage to the equipment.

WARNING

Do not operate generator set while servicing radiator. Failure to comply with this warning can cause injury to personnel and damage to the equipment.

WARNING

Fan has sharp blades. Use caution and wear gloves when removing or installing belts. Failure to comply with this warning can cause injury to personnel.

WARNING

Oil filter base and housing springs are under tension and can act as projectiles when being removed. Use eye protection when removing springs. Failure to comply with this warning can cause injury to personnel.

WARNING

The high pressure oil system operates at high temperature and pressure. Contact with hot oil can result in burns and scalding. Shut down generator set, and allow system to cool before performing checks, services, and maintenance. Wear heat resistant gloves and avoid contacting hot surfaces. Do not allow hot oil or components to contact skin or hands. Failure to comply with this warning can cause injury or death to personnel.

WARNING

Wear heat resistant gloves and avoid contacting hot metal surfaces with your hands after components have been heated. Wear additional protective clothing as required. Failure to comply with this warning can cause injury to personnel.

WARNING SUMMARY – Cont'd

WARNING

Ensure that the engine cannot be started while maintenance is being performed. (ENGINE CONTROL switch set to OFF/RESET; Battery Disconnect Switch is OFF; DEAD CRANK SWITCH is OFF). Failure to comply with this warning can cause injury or death to personnel.

WARNING

If not shielded, hot exhaust pipe can ignite flammable wall materials. Failure to

WARNING

An unwrapped exhaust pipe can cause injury if touched. Failure to comply with this

WARNING

Always remove radiator cap slow ly to permit any pressure to escape. Failure to

WARNING

Avoid breathing fumes generated by soldering. Eye protection is required. Good general ventilation is normally adequate. Failure to comply with this warning can cause injury to personnel.

WARNING

CARC paint is a health hazard, and is irritating to eyes, skin, and respiratory system. Wear protective eyewear, mask, and gloves when applying or removing CARC paint. Failure to comply with this warning can cause injury to personnel.

WARNING

Eye protection is required when working with compressed air. Compressed air can propel particles at high velocity and injure eyes. Do not exceed 15 psi pressure when using compressed air. Failure to comply with this warning can cause injury to personnel.

WARNING

High pressure steam can blow particles or chemicals into eyes, can cause severe burns, and creates hazardous noise levels. Wear protective eye, skin, and hearing protection when using high pressure steam. Failure to comply with this warning can cause injury to personnel.

WARNING SUMMARY – Cont'd

WARNING

The generator set, engine, and generator are extremely heavy and require an assistant and a lifting device (forklift, overhead lifting device) with sufficient capacity. Failure to comply with this warning can cause injury or death to personnel.

WARNING

Rated capacity of overhead hoist should be at least 1,500 pounds (680 kg). Do not use a hoist with less capacity. Failure to comply with this warning can cause injury or death to personnel, and damage to equipment.

WARNING

Keep hands and feet from underside of engine and generator while using lifting device to remove them from the skid base. Failure to comply with this warning can cause injury or death to personnel.

WARNING

Support components when removing attaching hardware or component may fall. Failure to comply with this warning can cause injury to personnel, and damage to equipment.

WARNING

A qualified technician must make the power connections and perform all continuity checks. The power source may be a generator or commercial power. Failure to comply with this warning can cause injury or death to personnel.

WARNING

Cleaning solvent is flammable and toxic to eyes, skin, and respiratory tract. Skin and eye protection are required when working in contact with cleaning solvent. Avoid repeated or prolonged contact. Work in ventilated area only. Failure to comply with this warning can cause injury or death to personnel.

WARNING

Cleaning compound is toxic. Avoid prolonged breathing of vapors. Use only in a well-ventilated area. Failure to comply with this warning can cause injury to personnel.

WARNING

Conversion coating material is toxic to eyes, skin, and respiratory tract. Skin and eye protection are required when working in contact with conversion coating material. Avoid repeated or prolonged contact. Work in ventilated area only. Failure to comply with this warning can cause injury or death to personnel.

WARNING SUMMARY – Cont'd

WARNING

Hot metal surfaces can cause burns to skin. Wear protective gloves and eye protection when applying heat to generator housing. Failure to comply with this warning can cause injury to personnel.

WARNING

Use protective gloves when handling heated rectifier hub. Failure to comply with

WARNING

Do not use the engine starter to turn the flywheel. Failure to comply with this warn-

WARNING

Catch fuel in suitable container. Keep spilled fuel away from hot engine and all fires.

WARNING

Do not attempt to perform any maintenance tasks on the Winterization Kit while the generator set is operating. Failure to comply with this warning can cause injury or death to personnel.

WARNING

Disconnect negative battery cable from right battery and the positive battery cable from the left battery before doing the following procedures. Reconnect cables in reverse order. Failure to comply with this warning can cause injury or death to personnel.

WARNING

Muffler and flex hoses can get very hot. Allow them to cool before touching them. Failure to comply with this warning can cause severe burns and injury to personnel.

WARNING

Engine exhaust fumes contain deadly poisonous gases.

Severe exposure can cause death or permanent brain damage.

Exhaust gases are most dangerous in places with poor airflow. Best defense against exhaust gas poisoning is very good airflow.

To protect yourself and your partners, always obey the following rules:

WARNING SUMMARY – Cont'd

- DO NOT run engine indoors unless you have VERY GOOD AIRFLOW.
- DO NOT idle engine for a long time unless there is VERY GOOD AIRFLOW.
- Be alert at all times. Check for smell of exhaust fumes.
- REMEMBER: Best defense against exhaust gas poisoning is VERY GOOD AIRFLOW.
- Exhaust gas poisoning causes dizziness, headache, loss of muscle control, sleepiness, coma, and death. If anyone shows signs of exhaust gas poisoning, get ALL PERSONNEL clear of exhaust area. Make sure they have lots of fresh air. KEEP THEM WARM, CALM, AND INACTIVE. GET MEDICAL HELP. If anyone stops breathing, give artificial respiration. See FM 4-25.11 for first aid.

FOR FIRST AID, REFER TO FM 4-25.11.

LIST OF EFFECTIVE PAGES

INSERT LATEST CHANGED PAGES, DESTROY SUPERSEDED PAGES

Note: a vertical line in the outer margins of the page indicates the portion of the text affected by the changes. Changes to illustrations are indicated by miniature pointing hands. Shaded areas indicate changes to diagrams.

Dates of issue for original and changed pages are:

Original ... 0 ... 1 May 2008

TOTAL NUMBER OF PAGES IN THIS PUBLICATION IS 830, CONSISTING OF THE FOLLOW-

Page No.	*Change No.
Cover	0
Blank	0
a – j	0
A	0
B Blank	0
i – xvii	0
xviii Blank	0
1-1 – 1-30	0
2-1 – 2-320	0
3-1 – 3-52	0
4-1 – 4-109	0
4-110 Blank	0
5-1 – 5-112	0
6-1 – 6-84	0
FP-1	0
FP-2 Blank	0
FP-3	0
FP-4 Blank	0
FP-5	0
FP-6 Blank	0
FP-7	0
FP-8 Blank	0
FP-9	0
FP-10 Blank	0
FP-11	0
FP-12 Blank	0
FP-13	0
FP-14 Blank	0
FP-15	0
FP-16 Blank	0
FP-17	0
FP-18 Blank	0
FP-19	0
FP-20 Blank	0
FP-21	0
FP-22 Blank	0
Metric Cover	0
A-1 – A-2	0
B-1 – B-11	0
Back Cover	0

* Zero in this column indicates an original

TECHNICAL MANUAL

NO. 9-6115-643-24

HEADQUARTERS, DEPARTMENTS
OF THE ARMY AND AIR FORCE
Washington, DC, 1 May 2008

UNIT, DIRECT SUPPORT, AND GENERAL SUPPORT MAINTENANCE MANUAL

FOR

GENERATOR SET, SKID MOUNTED, TACTICAL QUIET, 15 kW, 50/60 Hz, MEP-804A (NSN: 6115-01-274-7388) 15 kW, 50/60 Hz, MEP-804B (NSN: 6115-01-530-1458)

GENERATOR SET, SKID MOUNTED, TACTICAL QUIET, 15 kW, 400 Hz, MEP-814A (NSN: 6115-01-274-7393) 15 kW, 400 Hz, MEP-814B (NSN: 6115-01-529-9494)

REPORTING ERRORS AND RECOMMENDING IMPROVEMENTS

You can help improve this manual. If you find any mistakes or if you know of a way to improve the procedures, please let us know. Reports, as applicable by the requiring Service, should be submitted as follows:

(a) (A) Army - Mail your letter or DA Form 2028 (Recommended Changes to Publications and Blank Forms) located in the back of this manual directly to: Commander, U.S. Army Communications-Electronics Life Cycle Management Command (C-E LCMC) and Fort Monmouth, ATTN: AMSEL-LC-LEO-E-ED, Fort Monmouth, NJ 07703-5006. You may also send in your recommended changes via electronic mail or by fax. Our fax number is 732-532-1556, DSN 992-1556. Our e-mail address is MONM-MSELLEOPUBSCHG@conus.army.mil. Our online web address for entering and submitting DA Form 2028s is http://edm.monmouth.army.mil/pubs/2028.html.

(b) (F) Air Force - By Air Force AFTO Form 22 (Technical Manual (TM) Change Recommendation and Reply) in accordance with paragraph 6-5, Section VI, TO 00-5-1 directly to prime ALC/MST.

A reply will be furnished to you.

DISTRIBUTION STATEMENT A. Approved for public release; distribution is unlimited.

***This manual supersedes TM 9-6115-643-24 dated 1 September 1993, including all changes.**

TABLE OF CONTENTS

TABLE OF CONTENTS - Continued

Page

TABLE OF CONTENTS - Continued

TABLE OF CONTENTS – Continued

LIST OF ILLUSTRATIONS

Figure Title Page

TABLE OF CONTENTS

LIST OF ILLUSTRATIONS – Continued

Figure Title Page

TABLE OF CONTENTS

LIST OF ILLUSTRATIONS – Continued

Figure Title Page

TABLE OF CONTENTS

LIST OF ILLUSTRATIONS – Continued

Figure Title Page

TABLE OF CONTENTS

LIST OF ILLUSTRATIONS – Continued

Figure Title Page

TABLE OF CONTENTS – Continued

LIST OF TABLES

HOW TO USE THIS MANUAL

DESCRIPTION OF THE MANUAL

Manual Organization. This manual is designed to help you operate and maintain the MEP-804A, MEP-804B, MEP-814A and MEP-814B Tactical Quiet Generator (TQG) Sets. Warning pages are located in the front of this manual. Read the warnings before operating or doing maintenance on the equipment.

The major elements of this manual are its chapters and appendices. Each chapter has one or more sections. The Table of Contents, beginning on page ii, is provided for quick reference to the subjects covered by each chapter, section, and appendix. Each chapter also has a chapter index. The chapter index lists the chapter sections and paragraphs.

The front cover of this manual has an index that lists the most important areas of the manual.

A glossary follows the last appendix. The glossary lists and explains the special or unique abbreviations and the unusual terms used in this manual.

An alphabetical index follows the glossary. That index is for use in locating specific items of information.

Chapters. This manual has six chapters and four appendices. Each chapter is divided into sections. Each section is divided into descriptive paragraphs. The paragraphs have specific information about the generator sets and their major components.

Paragraph Numbering. All paragraphs are numbered. This helps you find what you need when you need it. USE THE TABLE OF CONTENTS OR ALPHABETICAL INDEX TO FIND THE SECTION OR PARAGRAPH YOU NEED. Some paragraphs have a related illustration, to show the items discussed in the paragraph. Also, some paragraphs have a related table that provides a detailed list of items introduced by the paragraph. Each primary paragraph, illustration, and table is identified by the number of the chapter in which it appears, followed by a dash and another number. The number after the dash indicates the sequence in which the paragraph, illustration, or table appears in the chapter. Some paragraphs are further divided into subparagraphs. Subparagraphs are identified by the number of the primary paragraph followed by a decimal number.

Appendices. Each appendix covers a specific subject.

CHAPTER 1 - INTRODUCTION

Chapter 1 provides an introduction to the generator sets. It is divided into three sections, as follows:

Section I - General Information. This section provides general information about this manual and the related forms and records. Instructions are provided for making equipment improvement recommendations. Coverage includes a reference to the TM that contains instructions on destruction of materiel to prevent enemy use.

Section II - Equipment Description and Data. This section describes generator set capabilities, characteristics, and features. It provides basic equipment data and shows the locations of major generator set components. Descriptions of the major components are also provided.

Section III - Principles of Operation. This section provides functional descriptions of the generator sets.

CHAPTER 2 - UNIT MAINTENANCE INSTRUCTIONS

Chapter 2 provides information on servicing the generator sets and components upon receipt, Unit Level

Preventive Maintenance Checks and Services (PMCS), troubleshooting procedures used to recognize and correct generator set malfunctions, and all maintenance procedures authorized at Unit Level. The chapter is divided into seventeen sections, as follows:

Section I - Service Upon Receipt of Equipment. This section provides information and guidance for inspecting, servicing, fabrication/assembly of parts, and installing the generator set under normal conditions.

Section II - Repair Parts; Special Tools; Test, Measurement, and Diagnostic Equipment (TMDE); and Special Support Equipment. This section contains information concerning the RPSTL, support equipment, and maintenance repair parts at Unit Level.

Section III - Special Lubrication Instructions. This section provides reference to the applicable lubrication order.

Section IV - Preventive Maintenance Checks and Services. This section provides the inspections and care of the generator set required to keep it in good operating condition.

Section V - Troubleshooting. This section contains troubleshooting information for locating and correcting operating problems which may occur in the generator set.

Section VI - Radio Interference Suppression. This section provides general methods used to attain proper suppression and discusses primary and secondary suppression components.

Section VII - Special Instructions. This section provides information concerning Nuclear, Biological, Chemical (NBC) contamination and decontamination procedures for the generator set.

Section VIII - Maintenance of DC Electrical System. This section provides unit maintenance procedures for the DC electrical system.

Section IX - Maintenance of Housing. This section provides unit maintenance procedures for generator set access doors, control box top panel, and housing sections.

Section X - Maintenance of Control Box Assembly. This section provides unit maintenance procedures for the control box assembly components and structures.

Section XI - Maintenance of Air Intake and Exhaust System. This section provides unit maintenance procedures for the muffler, exhaust, and air cleaner components.

Section XII - Maintenance of Coolant System. This section provides unit maintenance procedures for the radiator, fan guards, fan, fan belt, hoses, and coolant recovery system.

Section XIII - Maintenance of Fuel System. This section provides unit maintenance procedures for the fuel lines, fittings and components of the fuel system.

Section XIV - Maintenance of Output Box Assembly. This section provides unit maintenance procedures for the terminal boards (output and reconnection), wiring harness, transformers, relays, load terminals and varistors.

Section XV - Maintenance of Engine Accessories. This section provides unit maintenance procedures for the switches, senders, magnetic pickup, heating system contactors and filters (fuel and CCV).

Section XVI - Maintenance of Lubrication System. This section provides unit maintenance procedures for the oil drain line.

Section XVII - Preparation for Shipment and Storage. This section provides information concerning preservation, packing, marking, storage and use of corrosion-preventive materials.

CHAPTER 3 - GENERAL MAINTENANCE INSTRUCTIONS

Chapter 3 contains Direct Support Level troubleshooting procedures used to recognize and correct generator set malfunctions, and procedures for the removal and installation of major components. The chapter is divided into three sections, as follows:

Section I - Repair Parts; Special Tools; Test, Measurement, and Diagnostic Equipment (TMDE); and Special Support Equipment. This section contains information concerning the RPSTL, support equipment, and maintenance repair parts at the Direct Support Level.

Section II - Troubleshooting. This section provides Direct Support troubleshooting procedures for locating and correcting operating problems which may develop in the generator set.

Section III - Removal and Installation of Major Components. This section contains information for the removal and installation of the engine and generator. The engine and generator may be removed as an assembly or individually.

CHAPTER 4 - DIRECT SUPPORT MAINTENANCE INSTRUCTIONS

Chapter 4 contains all maintenance procedures authorized to be performed on the generator set at the Direct Support Level. The chapter is divided into seven sections, as follows:

Section I - Maintenance of Control Box Assembly. This section provides direct support maintenance procedures for the AC voltage regulator, governor control unit, wiring harness and load measuring unit.

Section II - Maintenance of Coolant System. This section provides direct support information concerning repair of the radiator.

Section III - Maintenance of Fuel System. This section provides direct support maintenance procedures for the fuel tank.

Section IV - Maintenance of Output Box Assembly. This section provides direct support maintenance procedures for the output box assembly, voltage reconnection terminal board, wiring harness, transformers, and output box panels.

Section V - Maintenance of Engine Accessories. This section provides direct support maintenance procedures for the governor actuator

Section VI - Maintenance of Generator Assembly. This section provides direct support maintenance procedures for the generator assembly components and generator housing.

Section VII - Maintenance of Skid Base. This section provides direct support maintenance procedures for the skid base.

CHAPTER 5 - RE-ENGINE INSTRUCTIONS

Chapter 5 contains detailed instructions for how to remove an Isuzu diesel engine from the generator set and replace it with a Yanmar turbocharged diesel engine. The process converts the generator set from an MEP-

804A to an MEP-804B or an MEP-814A to an MEP-814B. The chapter is divided into five sections, as follows:

Section I - Introduction. This section provides general information concerning the scope of the re-engine task.

Section II - Engine Removal. This section provides information and guidance for removal of the Isuzu diesel engine from the MEP-804A/MEP-814A generator set.

Section III - Pre-Engine Installation Modification. This section provides information concerning new engine preparation and pre-engine installation modification requirements.

Section IV - Engine Installation. This section provides new engine installation procedures; coolant, oil and fuel service; and new engine rehost procedures.

Section V - Final Adjustments. This section contains engine adjustment procedures, generator set adjustment procedures and instructions for placing the generator set into service.

CHAPTER 6 - WINTERIZATION KIT

Chapter 6 provides Unit and Direct Support Level information on the operation, troubleshooting and maintenance of the winterization kit designed to be mounted in generator sets where extreme cold temperatures are anticipated. The chapter is divided into thirteen sections, as follows:

Section I - Introduction. This section provides general information concerning the scope of this chapter in the discussion of the operation, maintenance and troubleshooting procedures for the winterization kit.

Section II - Equipment Description and Data. This section describes winterization kit capabilities, characteristics, and features. It provides descriptions of the major components and tabulated data for the heater.

Section III - Principles of Operation. This section provides a functional description of the winterization kit.

Section IV - Description and Use of Controls and Indicators. This section describes and illustrates winterization kit controls and indicators to ensure proper operations.

Section V - Preventive Maintenance Checks and Services (PMCS). This section contains detailed instructions that must be performed before, during, and after preventive maintenance checks and services of the winterization kit.

Section VI - Repair Parts: Special Tools: Test, Measurement, and Diagnostic Equipment (TMDE): and Special Support Equipment. This section contains information concerning the RPSTL, support equipment, and maintenance repair parts for the winterization kit at the Unit and Direct Support Levels.

Section VII - Service Upon Receipt of Materiel. This section provides information concerning the service upon receipt required by unit for the winterization kit already installed on the generator set.

Section VIII - Unit Lubrication. This section provides the information that no lubrication is required on the winterization kit.

Section IX - Installation Instructions. This section provides the procedures for installing the winterization kit on the 15kW generator set.

Section X - Troubleshooting Procedures. This section lists diagnostic and symptom related malfunctions that may occur during operation of the generator set with the winterization kit installed.

Section XI - Symptom Index, Winterization Kit. This section provides a fault symptom index for the winterization kit and corrective actions to be taken.

Section XII - Maintenance Procedures. This section provides the maintenance procedures for the winterization kit components.

Section XIII - Installation Instructions. This section provides instructions for removing the winterization kit from the 15kW generator set.

APPENDICES

APPENDIX A – REFERENCES

This appendix lists all publications referenced in the various chapters of the technical manual. The listing includes the title and document number of each publication.

APPENDIX B - COMPONENTS OF END ITEM (COEI) AND BASIC ISSUE ITEMS (BII) LISTS

This appendix lists the items usually packaged separately but needed for installation and operation of the generator sets. The appendix has three sections, as follows:

Section I - Introduction. This section explains what is covered in Sections II and III.

Section II - Components of End Item. The generator sets are normally shipped fully assembled, so this section is not applicable.

Section III - Basic Issue Items. This section contains a list of the accessories needed for installation and operation of the generator sets.

APPENDIX C - EXPENDABLE AND DURABLE ITEMS LIST

This appendix lists expendable/durable supplies and materials needed to operate and maintain the generator sets. The appendix contains two sections, as follows:

Section I - Introduction. This section explains the entries in Section II.

Section II - Expendable and Durable Items List. The list indicates the maintenance level that needs each item and identifies the items by National Stock Number (NSN), description, and unit of measure.

APPENDIX D - FABRICATION/ASSEMBLY OF PARTS

This appendix includes complete instructions for fabricating or assembling parts as required on this generator set.

ALPHABETICAL INDEX

An alphabetical index at the back of this technical manual provides a listing of subjects covered, cross-referenced to the applicable Paragraph/Figure (F)/Table (T) Number.

HOW TO FIX A GENERATOR SET MALFUNCTION

Determining the Cause. Finding the cause of a malfunction, troubleshooting, is the first step in fixing the generator set and returning it to operation. Follow these simple steps to determine the root of the problem:

a. Turn to the Table of Contents in this manual (page ii).

b. Locate "Troubleshooting Procedures" under Chapter 2 (Unit), Chapter 3 (GS/DS) or Chapter 6 (Winterization Kit). Turn to the page indicated.

c. Follow the instructions in the references listed in Chapter 2, Chapter 3 or Chapter 6.

Preparing for a Task. Be sure that you understand the entire maintenance procedure before beginning any maintenance task. Make sure that all parts, materials, and tools are handy. Read all steps before beginning. Prepare to do the task as follows:

a. Carefully read the entire task before starting. It tells you what you will need and what you have to know to start the task. DO NOT START THE TASK UNTIL:

(1) You know what is needed

(2) You have everything you need

(3) You understand what to do

b. If parts are listed, they can be drawn from technical supply. Before you start the task, check to make sure you can get the needed parts.

c. If expendable/durable supplies or materials are needed, get them before starting the task. Refer to Appendix C for the correct nomenclature and NSN.

How to Do the Task. Before starting, read the entire task. Be sure that you understand the entire procedure before you begin the task. As you read, remember the following:

a. PAY ATTENTION TO WARNINGS, CAUTIONS, AND NOTES.

b. Use the GLOSSARY if you do not understand the special abbreviations or unusual terms used in this manual.

c. The following are standard maintenance practices. Instructions about these practices are usually not included in task steps. When standard maintenance practices do not apply, the task steps will tell you.

(1) Tag electrical wiring before disconnecting it.

(2) Discard used preformed packing, retainers, gaskets, cotter pins, lockwashers, and similar items. Install new parts to replace the discarded items.

(3) Coat packing before installation, in accordance with the task instructions.

(4) Disassembly procedures describe the disassembly needed for total authorized repair. You may not need to disassemble an item as far as described in the task. Follow the disassembly steps only as far as needed to repair/replace worn or damaged parts.

(5) Clean the assembly, subassembly, or part before inspecting it.

(6) Before installing components having mating surfaces, inspect the mating surfaces to make sure they are in serviceable condition.

(7) Hold the bolt (or screw) head with a wrench (or screwdriver) while tightening or loosening a nut on the bolt (or screw).

(8) Torque to the special torque cited when the task instructions include the words “torque to.” Use standard torques at all other times.

(9) When a cotter pin is required, align the cotter pin holes within the allowable torque range.

(10) Inspect for foreign objects after performing maintenance.

CHAPTER 1

INTRODUCTION

Subject Index Page

Section I. GENERAL INFORMATION

1-1 SCOPE

1-1.1 Type of Manual. This manual contains Unit, Direct Support, and General Support maintenance instructions for the 15 kW 50/60 Hz and 400 Hz Tactical Quiet (TQ) Generator Sets (Figure 1-1), herein referred to as generator sets. Included are descriptions of major components and their functions in relation to other components.

1-1.2 Model Numbers and Equipment Names.

Model Number	Equipment Name
MEP-804A/ MEP- 804B	Generator Set, Skid Mounted, Tactical Quiet 15 kW 50/60 Hz
MEP-814A/ MEP- 814B	Generator Set, Skid Mounted, Tactical Quiet 15 kW 400 Hz

1-1.3 Purpose of Equipment. The generator set provides tactical quiet AC power. The generator set is easily transported, operated, and maintained.

1-2 LIMITED APPLICABILITY

Some portions of this publication are not applicable to all services. These portions are prefixed to indicate the service(s) to which they pertain: (A) for Army, (F) for Air Force, and (N) for Navy.

1-3 MAINTENANCE FORMS AND RECORDS

1-3.1 (A) Department of the Army forms and procedures used for equipment maintenance will be those prescribed by DA PAM 738-750, The Army Maintenance Management System (TAMMS) Maintenance Management UPDATE).

1-3.2 (F) Maintenance Forms and Records maintained by the Air Force are prescribed in AFR 66-1 and the applicable TO 00-20 Series Technical Orders.

1-3.3 (N) Navy users should refer to their service peculiar directives to determine the applicable maintenance forms and records to be used.

1-4 REPORTING OF ERRORS

Reporting of errors, omissions, and recommendations for improvement of this publication by the individual user is encouraged. Reports should be submitted as follows:

1-4.1 (A) Mail your letter, DA Form 2028 (Recommended Changes to Publications and Blank Forms), or DA Form 2028-2 located in back of this manual directly to: Commander, U.S. Army CECOM Life Cycle Management Command (LCMC) and Fort Monmouth, ATTN: AMSEL-LC-LEO-E-ED, Fort Monmouth, NJ 07703-5006. You may also send in your recommended changes via electronic mail or by fax. Our fax number is 732-532-1556, DSN 992-1556. Our e-mail address is MONM-AMSELLEOPUBSCHG@conus.army.mil. Our online web address for entering and submitting DA Form 2028s is http://edm.monmouth.army.mil/pubs/2028.html.

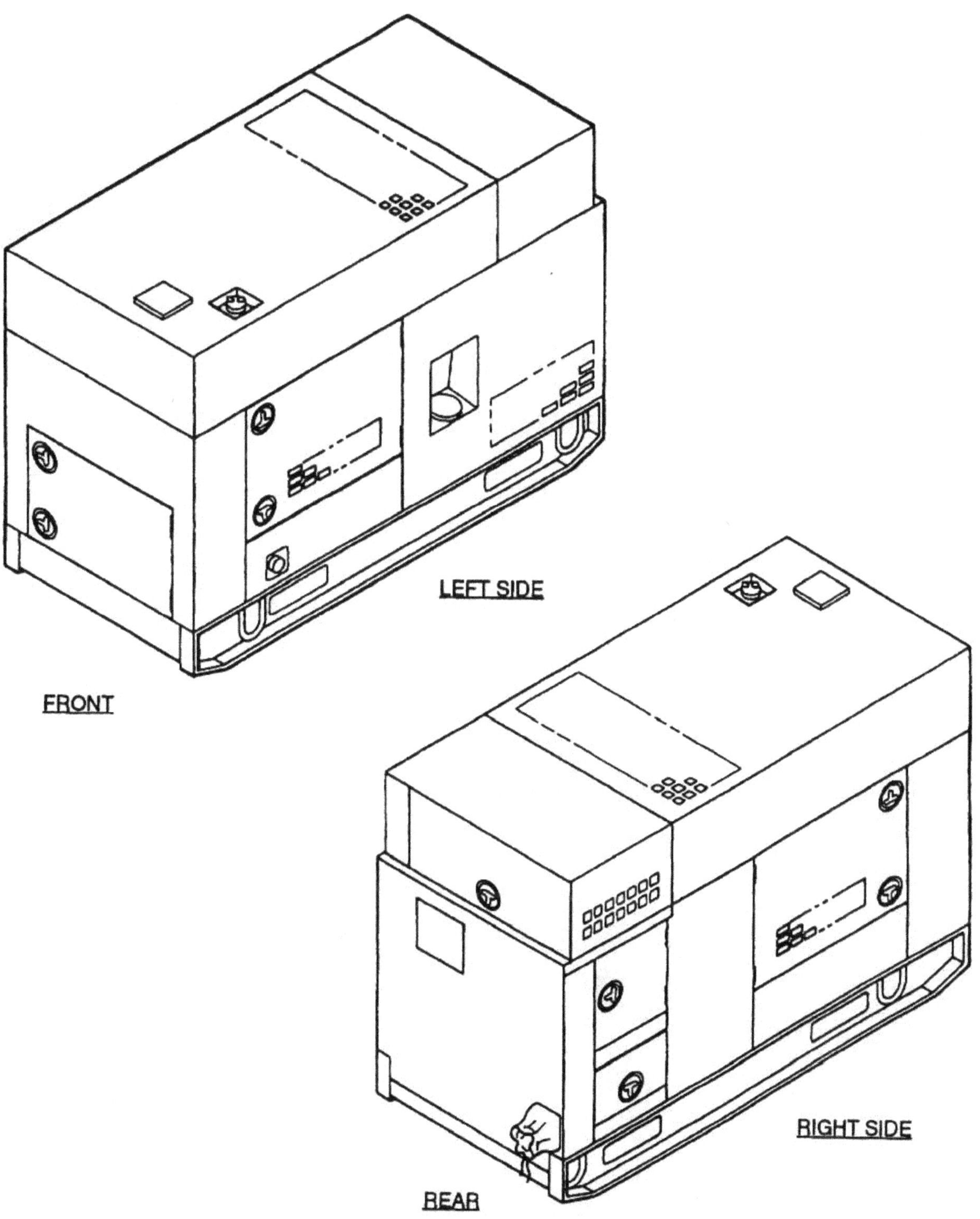

Figure 1-1. 15 kW 50/60 Hz and 400 Hz Tactical Quiet Generator Sets.

1-4.3 (N) Navy – By letter directly to Commander, Space and Naval Warfare Systems Command, ATTN: SPAWAR 8122, Washington, DC 20363-5100.

1-5 EQUIPMENT IMPROVEMENT RECOMMENDATION (EIR)

1-5.1 If your generator set needs improvement, let us know. Send us an EIR. You, the user, are the only one who can tell us what you don't like about your equipment. Let us know why you don't like the design or performance. We will send you a reply.

1-5.2 (A) If you have Internet access, the easiest and fastest way to report problems or suggestions is to go to https://aeps.ria.army.mil/aepspublic.cfm (scroll down and choose the "Submit Quality Deficiency Report" bar). The Internet form lets you choose to submit an Equipment Improvement Recommendation (EIR), a Product Quality Deficiency Report
(PQDR or a Warranty Claim Action (WCA). You may also submit your information using an SF 368 (Product Quality
Deficiency Report). You can send your SF 368 via e-mail, regular mail, or facsimile using the addresses/facsimile numbers specified in DA PAM 750-8, The Army Maintenance Management System (TAMMS) Users Manual.

1-5.3 (N) Navy personnel are encouraged to submit EIR's through their local Beneficial Suggestion Program.

1-5.4 (F) Air Force personnel are encouraged to submit EIR's in accordance with AFR 900-4.

1-6 LEVELS OF MAINTENANCE

1-6.1 (A) Army users shall refer to the Maintenance Allocation Chart (MAC) for tasks and levels of maintenance to be performed.

1-6.2 (F) Refer to the Source Maintenance Recoverability (SMR) Codes for maintenance to be performed.

1-6.3 (N) Navy users shall determine their maintenance levels in accordance with their service directives.

1-7 DESTRUCTION OF ARMY MATERIEL TO PREVENT ENEMY USE

1-7.1 (A) Destruction of the generator set to prevent enemy use shall be in accordance with TM 750-244-3.

1-7.2 (F) (N) Air Force and Navy users shall refer to their service directives to obtain procedures for destruction of

Section II. EQUIPMENT DESCRIPTION AND DATA

1-8 GENERAL

The generator sets, models MEP-804A/MEP-804B and MEP-814A/MEP-814B (Figure 1-1), are fully enclosed, self-contained, skid-mounted, portable units. They are equipped with controls, instruments, and accessories necessary for operation as single units or in parallel with another unit of the same class and mode. The generator sets consist of a diesel engine, brushless generator, excitation system, speed-governing system, fuel system, 24 VDC starting system, control system, and fault system. An Isuzu diesel engine powers generator sets MEP-804A/MEP-814A. A Yanmar turbocharged diesel engine powers generator sets MEP-804B/MEP-814B.

1-9 TABULATED/ILLUSTRATED DATA

For generator set tabulated data, refer to Table 1-1.

Table 1-1. Tabulated Data.

Item	Characteristic
1. Generator Set:	
Model Number:	
15 kW 50/60 Hz with Isuzu Engine	MEP-804A
15 kW 50/60 Hz with Yanmar Engine	MEP-804B
15 kW 400 Hz with Isuzu Engine	MEP-814A
15 kW 400 Hz with Yanmar Engine	MEP-814B
National Stock Number (NSN):	
MEP-804A	6115-01-274-7388
MEP-804B	6115-01-530-1458
MEP-814A	6115-01-274-7393
MEP-814B	6115-01-529-9494
Overall Length:	
MEP-804A/MEP-804B	69.7 in. (177.2 cm)
MEP-814A/MEP-814B	69.7 in. (177.2 cm)
Overall Width:	
MEP-804A/MEP-804B	35.7 in. (90.8 cm)
MEP-814A/MEP-814B	35.7 in. (90.8 cm)
Overall Height:	
MEP-804A/MEP-804B	55 in. (139.7 cm)
MEP-814A/MEP-814B	55 in. (139.7 cm)
Dry Weight (Less Basic Issue Items):	
MEP-804A	1,885 lb. (855.0 kg.)
MEP-804B	1,785 lb. (809.6 kg.)
MEP-814A	2,015 lb. (914.0 kg.)
MEP-814B	1,915 lb. (868.6 kg.)
Wet Weight:	
MEP-804A	2,140 lb. (970.7 kg.)
MEP-804B	2,040 lb. (925.3 kg.)
MEP-814A	2,250 lb. (1022.6 kg.)
MEP-814B	2,150 lb. (975.2 kg.)

Table 1-1. Tabulated Data – Continued.

Item	Characteristic
2. Engine (MEP-804A/MEP-814A):	
Manufacturer	Isuzu
Model	C-240
Type	Four cylinder, four cycle, naturally aspirated diesel
Displacement	145 cu. in. (2.4 liters)
Altitude Degradation, 4,000 ft (1,220m) to 80,00 ft (2,440m)	3.5% per 1,000 ft (305m)
Firing Order	1, 3, 4, 2
Cold Weather Starting Aid System Use	40□F (4□C) or below
Valve Tappet Clearance Adjustment:	
Hot or Cold (Intake)	0.018 in. (0.45 mm)
Hot or Cold (Exhaust)	0.018 in. (0.45 mm)
3. Engine (MEP-804B/MEP-814B):	
Manufacturer	Yanmar
Model	4TNV84T
Part Number	4TNV84T-DFM
Type	Four cylinder, four cycle, turbocharged diesel
Displacement	121.7 cu. in. (1.995 liters)
Altitude Degradation, 4,000 ft (1,220m) to 8,000 ft (2,440m)	3.5% per 1,000 ft (305m)
Firing Order	1, 3, 4, 2
Cold Weather Starting Aid System Use	40□F (4□C) or below
Valve Tappet Clearance Adjustment:	
Hot or Cold (Intake)	0.040-0.055 in. (1.0-1.4 mm)
Hot or Cold (Exhaust)	0.045-0.070 in. (1.1-1.8 mm)
4. Cooling System:	
Type	Pressurized radiator and pump
Capacity:	
MEP-804A/MEP-814A	13.5 qts (12.8 liters)
MEP-804B/MEP-814B	11.2 qts (10.6 liters)
Normal Operating Temperature	170-200□F (77-93□C)
Temperature Indicating System Voltage Rating	24 VDC
5. Lubricating System:	
Type:	
MEP-804A/MEP-814A	Full flow, circulating pressure
MEP-804B/MEP-814B	Forced lubrication
Oil Pump Type:	
MEP-804A/MEP-814A	Positive displacement gear
MEP-804B/MEP-814B	Trochoid

Table 1-1. Tabulated Data – Continued.

Item	Characteristic
5. Lubricating System – Continued:	
Normal Operating Pressure	25-60 psi (172-414 kPa)
Oil Filter Type	Full flow, spin-on, replaceable element
Capacity	6 qts (5.7 liters)
Pressure Indicating System Voltage Rating	24 VDC
6. Fuel System:	
Type of Fuel	DF-1, DF-2, DF-A, JP4, JP5, JP8
Fuel Tank Capacity	14 gal. (53 liters)
Fuel Consumption Rate (50/60 Hz):	
MEP-804A	1.50 gal (5.7 liters) per hour
MEP-804B	1.2 gal (4.5 liters) per hour
Fuel Consumption Rate (400 Hz):	
MEP-814A	1.75 gal (6.6 liters) per hour
MEP-814B	1.4 gal (5.3 liters) per hour
Auxiliary Fuel Pump:	
Voltage Rating	24 VDC
Delivery Pressure	5.0-6.5 psi (34.5-65.5 kPa) (max)
Fuel Level Switch:	
Type	Float
Current	3.0 amps at 6-32 VDC
7. Engine Starting System:	
Batteries	Two 12 volt, connected in series
Starter (MEP-804A/MEP-814A):	
Manufacturer	Hitachi
Model	S25-121
Voltage Rating	24 VDC
Drive Type	Gear reduction
Starter (MEP-804B/MEP-814B):	
Manufacturer	Yanmar
Model	129612-77011
Voltage Rating	24 VDC
Drive Type	Direct drive
Battery Charging Alternator (MEP-804A/MEP-814A):	
Manufacturer	Hitachi
Model	LR220-24
Amperage Rating	20 amps at 24 VDC
Protective Fuse	30 amps
Battery Charging Alternator (MEP-804B/MEP-814B):	
Manufacturer	Yanmar
Part Number	129900-77240
Amperage Rating	35 amps at 24 VDC
Protective Fuse	50 amps

Table 1-1. Tabulated Data – Continued.

Item	Characteristic
8. AC Generator:	
Manufacturer Marathon Electric	
Type Rotating field, synchronous	
Load Capacity 15 kW	
Current Ratings: 50 Hz 60 Hz	400 Hz
120/208 (240/416) Volt Connection 43 (21) amps 52 (26) amps	52 (26) amps
Power Factor	0.8
Cooling Fan cooled	
Drive Type Direct coupling	
Duty Classification Continuous	
9. Governing System:	
Load Measuring Unit:	
Manufacturer Technology Research	
Model 19310	
Governor Control Unit: (MEP-804A)	
Manufacturer	Woodward Governor Co.
Model	8270-1002
Governor Control Unit: (MEP-804B)	
Manufacturer	Woodward Governor Co.
Model	8270-1096
Governor Control Unit: (MEP-814A/MEP-814B)	
Manufacturer	Woodward Governor Co.
Model	8270-1003
10. Protection Devices:	
Low Oil Pressure Switch:	
Trip Pressure 15□3 psi (103.4 □20.7 kPa)	
Voltage Rating	12/24 VDC
Current Rating 7 amps resistive; 4 amps inductive	
Coolant High Temperature Switch:	
Trip Temperature 220□3.5□F (104 □2□C)	
Voltage Rating	20-32 VDC
Current Rating 7 amps resistive; 4 amps inductive	
Overspeed Switch:	
Element Trip and Reset 2200□40 rpm	
Voltage Rating	28 VDC

Table 1-1. Tabulated Data – Continued.

Item	Characteristic
Overvoltage Relay:	
Trip Point Conditions	155□ 1 VAC for no less than 200 milliseconds (120 VAC coil winding)
Trip Point	No more than 1.25 seconds after trip conditions exist

1-10 DIFFERENCES BETWEEN MODELS

1-10.1 The differences between models of the generator sets covered in this manual are as follows:

Model MEP-804A is equipped with a 50/60 Hz generator and uses an Isuzu diesel engine.
Model MEP-804B is equipped with a 50/60 Hz generator and uses a Yanmar turbocharged diesel engine.
Model MEP-814A is equipped wit h a 400 Hz generator and uses an Isuzu diesel engine.
Model MEP-814B is equipped with a 400 Hz generator and uses a Yanmar turbocharged diesel engine.

1-10.2 For generator set performance characteristics, refer to Table 1-2.

Item	MEP-804A/MEP-804B	MEP-814A/MEP-814B

1. Voltage: a. Voltage Waveform Deviation Factor: 5% (max) 5% (max)

Single Voltage Harmonics: 2% (max) 2% (max)

b. Voltage Unbalance 5% of rated voltage (max) 5% of rated voltage (max) c. Phase Balance Voltage 1% of rated voltage (max) 1% of rated voltage (max) d. Voltage Modulation 2% (max) 2% (max) e. Voltage Regulation 1% (max) 1% (max) f. Short-term Stability (30 seconds) 1% of rated voltage 1% of rated voltage g. Long-term Stability (4 hours) 2% of rated voltage 2% of rated voltage h. Voltage Drift (60°F (16°C)) in 8 hour ±1% (max) ±1% (max)
period)

i. Dip and Rise for Rated Load 15% of rate d voltage (max) 12% of rated voltage (max)

Recovery Time 0.5 second 0.5 second

j. Dip for Low Power Factor Load 30% of no-load voltage (max) 25% of no-load voltage (max)

Recovery Time 0.7 second 0.7 second

95% of no-load voltage 95% of no-load voltage

Table 1-2. Performance Characteristics – Contin-

Item	MEP-804A/MEP-804B	MEP-814A/MEP-814B
k. Adjustment Range VAC		
	50 Hz	400 Hz
120/208V Connection	190-213V	197-229V
240/416V Connection	380-426V	395-458V
	60 Hz	
120/208V Connection	197-240V	
240/416V Connection	395-480V	
2. Frequency:		
a. Regulation	0.25% of rated frequency	0.25% of rated frequency
b. Short-term Steady-state Stability (30 seconds)	0.5% of rated frequency	0.5% of rated frequency
c. Long-term Steady-state Stability (4 hours)	1% of rated frequency	1% of rated frequency
d. Frequency Drift (60□F (16□C) in 8 hour period)	0.5% (max)	0.5% (max)
e. Undershoot with Application of Load	4% of ra ted frequency (max)	1.5% of rated frequency (max)
Recovery Time	2 seconds	1 second
f. Overshoot with Application of Load	4% of rat ed frequency (max)	1.5% of rated frequency (max)
Recovery Time	2 seconds	1 second
g. Adjustment Range	48-52 Hz, not below 45 Hz for 50 Hz operation 58-62 Hz, not above 65 Hz for 60 Hz operation	390-420 Hz, not below 370 Hz or above 430 Hz

Section III. PRINCIPLES OF OPERATION

1-11 INTRODUCTION

This section contains functional descriptions of the generator set, explains how the controls and indicators interact with the systems, and provides the location and description of major components.

1-12 PRINCIPLES OF OPERATION

1-12.1 Fault System.

1-12.1.1 The fault system (Figure 1-2) protects the generator set and any connected load against the potential faults described below and provides an indication of any incurred fault. The following summary of the fault system will assist in understanding the operation of the other generator set systems. Additional details relating to specific protection devices are also provided in the descriptions of the respective systems.

1-12.1.2 The fault system consists of the malfunction indicator, low fuel level float switch, fuel float switch module, fuel level relay, low oil pressure switch, coolant high temperature switch, overvoltage relay, overload/short circuit relay, overspeed switch, overspeed relay, OVERSPEED RESET switch, undervoltage relay, reverse power relay, engine fault relay, electrical fault relay, and BATTLE SHORT switch. In addition to the fault indicator lamps, the malfunction indicator includes the PUSH TEST & RESET LAMPS switch which, when depressed, illuminates all the lamps and resets any fault indication.

1-12.1.3 Activation of any one of the following protection devices will cause three events to occur. The AC circuit interrupter relay will open; the generator set engine will be shut down; and a fault indicator lamp will be illuminated to show which malfunction occurred.

1-12.1.3.1 Coolant High Temperature Switch. This device will activate when the engine coolant leaving the engine exceeds 220 ±3.5°F (104 ±2°C).

1-12.1.3.2 Low Oil Pressure Switch. This device activates when the engine lubrication oil pressure falls below 15 ±3 psi (103.4 ±20.7 kPa).

1-12.1.3.3 Low Fuel Level Float Switch. This device will activate when the fuel level falls to a point at which the operating time of the set at rated load is 4 minutes.

1-12.1.3.4 Overvoltage Relay. This device will activate when the 120 volt generator coil winding has risen to and remained at any value greater than 155 ± 1 volts.

1-12.1.3.5 Over-speed Relay. This device will activate when the engine speed exceeds 2200 ± 40 rpm.

1-12.1.4 The fuel float switch module is a device that prevents inadvertent engine shutdown by providing a 1 second delay after actuation of the low fuel level float switch.

1-12.1.5 Electrical protection devices will cause two events to occur. The AC circuit interrupter relay will open, and a fault indicator lamp will illuminate to indicate which fault occurred.

1-12.1.5.1 Short Circuit Relay. This device will activate when the set output current in any phase exceeds 425 percent of the rated value.

1-12.1.5.2 Overload Relay. This device will operate when the load current in any phase exceeds 110 percent of the rated value.

1-12.1.5.3 Undervoltage Relay. This device will activate instantaneously when the 120 volt generator coil winding has dropped to 48 volts and will trip after time delay when the coil voltage drops below 99 volts.

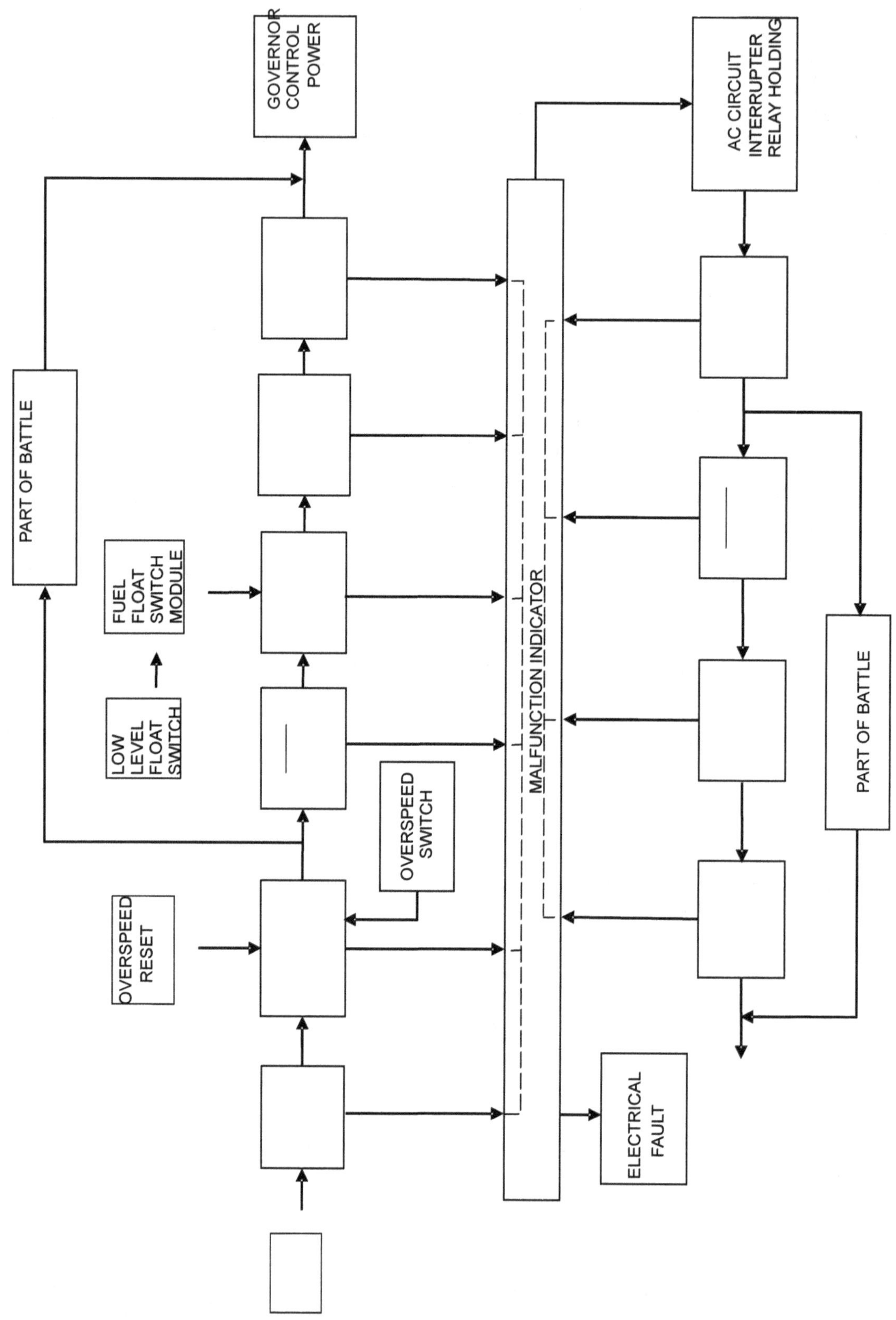

Figure 1-2. Fault System.

1-12.1.5.4 Reverse Power Relay. This device will operate if power flow into the generator set exceeds 20 percent of the rated value.

1-12.1.6 Although it is possible for more than one fault to occur at one time during operation, only the first fault to occur will be displayed by the malfunction indicator. The activated indicator lamp circuit remains illuminated until the malfunction indicator is reset. The lamp will be off with the MASTER SWITCH in the OFF position and will be reilluminated when the MASTER SWITCH is turned to one of the RUN positions. Resetting a fault indication is done in two steps. First, push the PUSH TEST & RESET LAMPS switch located on the malfunction indicator. Second, move the MASTER SWITCH to the OFF position. After a shutdown due to engine overspeed, the OVER-SPEED RESET switch must be actuated to reset the engine electrical control circuit before the engine can be restarted.

1-12.1.7 After the generator set engine has been started, the BATTLE SHORT switch can be used to override all of the potential faults except engine overspeed and short circuit.

1-12.2 Fuel System.

1-12.2.1 The fuel system (Figure 1-3) includes a primary subsystem and an auxiliary subsystem.

1-12.2.2 The primary subsystem consists of fuel lines, fittings, fuel tank, low fuel level float switch, fuel float switch module, fuel level sender, FUEL LEVEL indicator, transfer pump, fuel filter/water separator, injection pump, and injectors. The MEP-804B/MEP-814B generator sets have two additional fuel filters. One filter is mounted on the inlet of the fuel transfer pump to filter fuel pumped from the tank. The other filter is mounted on the engine to filter fuel prior to entering the fuel injection pump.

1-12.2.3 The injection pump output is controlled by the electronic governor control and governor actuator. When the electronic governor control is deenergized, electrical power is removed from the governor actuator which is spring-loaded to the fuel shutoff position. The electronic governor control is energized by turning the MASTER SWITCH to the START position or either of the two RUN positions. With the engine cranking or running, fuel is drawn from the fuel tank by the transfer pump. After reaching the transfer pump, fuel passes through a fuel filter/ water separator where water and small impurities are removed. The fuel then goes to the injection pump. With the governor system energized, the fuel is metered, pressurized, and pushed through the injectors by the injection pump. Fuel is sprayed by the injectors into the diesel engine combustion chambers where it is mixed with air and ignited. The fuel that is not used by the injectors is returned to the fuel tank by an excess fuel return line, power is removed from the electronic governor control, and the fuel is shut off whenever the MASTER SWITCH is turned to the OFF position. The electronic governor control is also deenergized by the fault system (paragraph 1-12.1.3). The FUEL LEVEL indicator displays the fuel level of the fuel tank from E (empty) to F (full) in quarter tank increments.

1-12.2.4 The auxiliary subsystem consists of an auxiliary fuel supply, fuel lines, fittings, auxiliary fuel filter, auxiliary fuel pump, auxiliary fuel pump float switch located in the fuel tank, and a fuel float switch module.

1-12.2.5 When the MASTER switch is set on PRIME & RUN AUX FUEL, it actuates the auxiliary fuel pump and transfers fuel from the auxiliary fuel supply to the fuel tank. The auxiliary fuel pump float switch shuts off the auxiliary fuel pump when the fuel tank is full and reactivates the pump as the level drops. The fuel float switch module allows the current used by the auxiliary fuel pump to bypass the float switch.

1-12.2.6 The 24 VDC control circuits provide control and power for indicators, float switches, fault system, governor system, and auxiliary fuel pump.

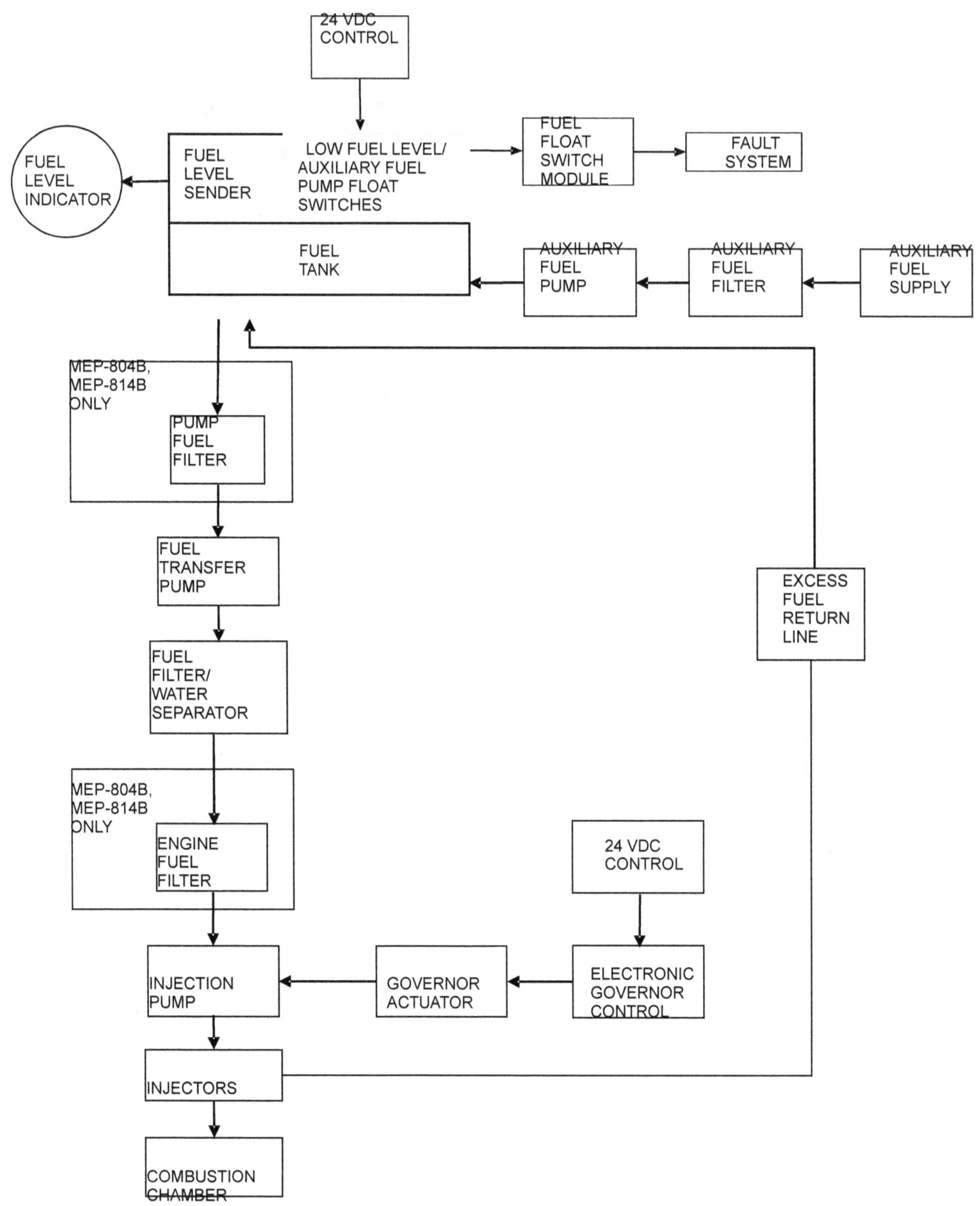

Figure 1-3. Generator Set Fuel System.

1-12.3 Generator Set Cooling System .

1-12.3.1 The generator set cooling system (Figure 1-4) includes air intake and exhaust grilles, baffles and ducting within the generator set housing, and engine-driven radiator cooling fan. The air intake grilles are located in panels on both sides of the generator set housing. The air exhaust grille is located in the housing top panel.

1-12.3.2 Air is drawn in through the air intake grilles and forced through the engine coolant radiator and out of the generator set through the exhaust grille by the radiator cooling fan. Most of the cooling air flows externally past the generator assembly and engine. Some cooling air is circulated internally through the generator assembly by a generator fan which is an integral part of the AC generator assembly. Baffles, ducting, and sound-absorbing material are used to control the airflow through the generator set and to reduce sound transmission through the grilles.

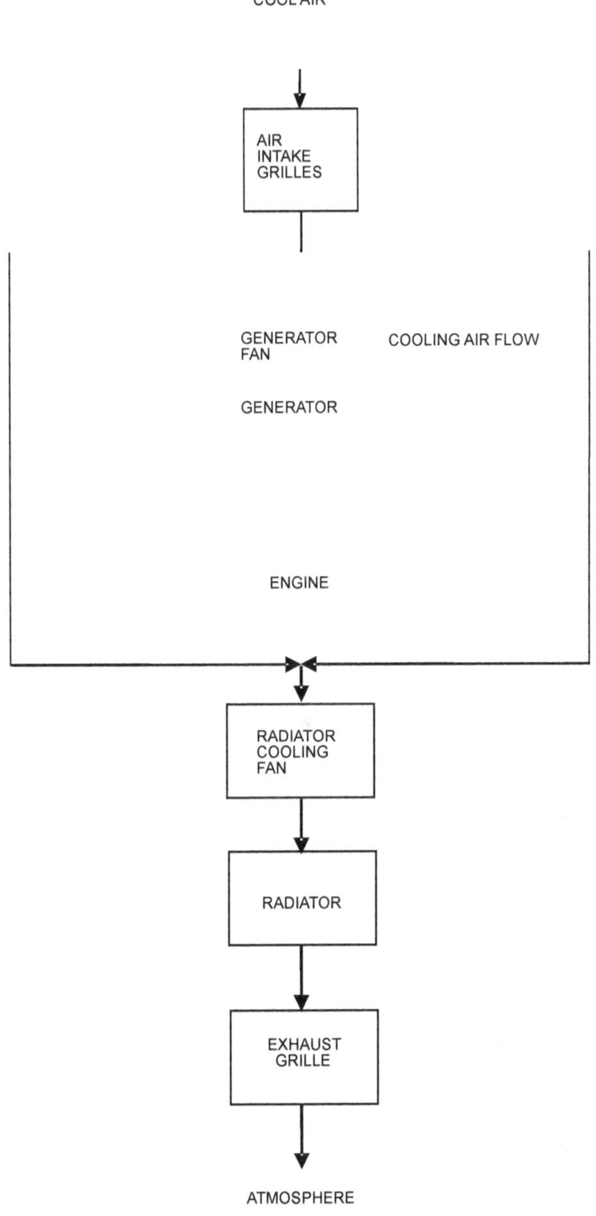

Figure 1-4. Generator Set Cooling System.

1-12.4 Engine Cooling System.

1-12.4.1 The engine cooling system (Figure 1-5) consists of a radiator, hoses, thermostat, temperature sender, coolant high temperature switch, COOLANT TEMP indicator, water pump, belt-driven fan, and cooling jackets (part of engine).

1-12.4.2 The water pump forces coolant through passages (cooling jackets) in the engine block and cylinder head where the coolant absorbs heat from the engine. When the engine reaches normal operating temperature, the thermostat opens and the heated coolant flows through the upper radiator hose assembly into the radiator. The cooling fan circulates air through the radiator where the coolant temperature is reduced.

1-12.4.3 A coolant high temperature switch in conjunction with the fault system provides automatic shutdown in the event that coolant temperature exceeds 220 ±3.5°F (104 ±2°C). The COOLANT TEMP indicator indicates the engine coolant temperature from 120°F to 240°F (48°C to 115°C).

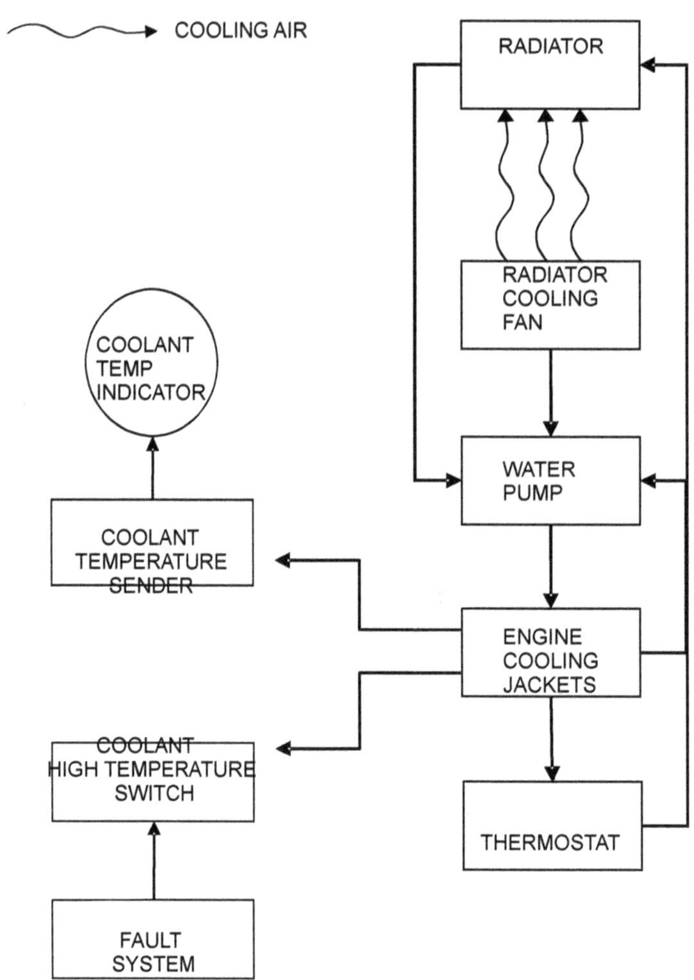

Figure 1-5. Engine Cooling System.

1-12.5 Engine Lubrication System.

1-12.5.1 The engine lubrication system (Figure 1-6) consists of an oil pan, dipstick, pump, oil sample valve, oil pressure sender, OIL PRESSURE indicator, low oil pressure switch, and filter.

1-12.5.2 The oil pan is a reservoir for engine lubricating oil. The dipstick indicates oil level in the pan. The oil level can be checked during engine operation. One side of the dipstick is used for checking oil level while the engine is running and the other side is used while the engine is shut down. The pump draws oil from the oil pan through a screen that removes large impurities. The oil then passes through a spin-on type filter where small impurities are removed. From the filter, oil is distributed to the engine's internal moving parts and then returns to the oil pan. The oil pressure sender located in the engine block senses oil pressure. The oil pressure is displayed on the OIL PRESSURE indicator. An Army Oil Analysis Program (AOAP) sample valve located in the block allows oil samples to be taken while the engine is operating. The low oil pressure switch, also located in the engine block, functions with the generator set fault system. The engine is automatically shut off if the oil pressure drops below 15 ±3 psi

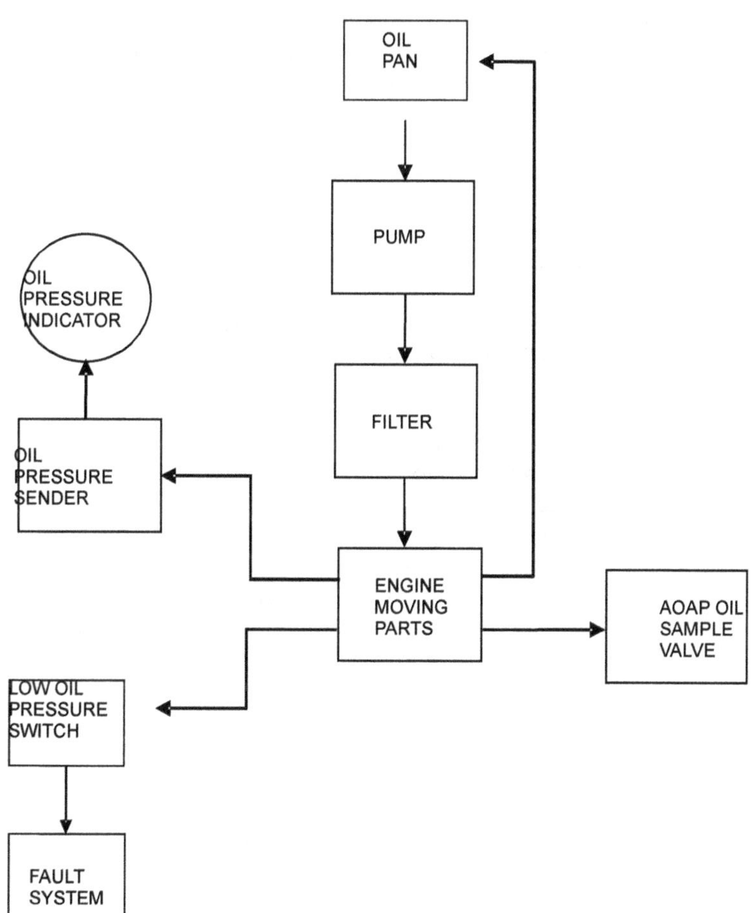

Figure 1-6. Engine Lubrication System.

1-12.6 Engine Air Intake and Exhaust System.

1-12.6.1 The engine air intake and exhaust system (Figure 1-7) consists of an air cleaner assembly, intake manifold, four glow plugs, contactor K22, positive crankcase ventilation (PCV) valve, exhaust manifold, and muffler. The air cleaner assembly includes a dust collector, filter element, dust evacuator valve, and restriction indicator. The Yanmar engine (MEP-804B/MEP-814B) also has a turbocharger.

1-12.6.2 Air is drawn into the dust collector and passes through the filter element. Airborne dirt is removed and trapped in the dust collector and filter element. Some dust can be removed from the dust collector by pinching the evacu-ator valve. The restriction indicator indicates when the filter should be serviced. On the Isuzu engine (MEP-804A/MEP-814A), filtered air is drawn out of the filter through air intake tubes to the intake manifold where it passes into the engine. On the Yanmar engine (MEP-804B/MEP-814B), filtered air is drawn out of the filter and pressurized by the turbocharger prior to being forced into the intake manifold where it passes into the engine.

1-12.6.3 The engine exhaust gases are expelled into the exhaust manifold. On the Isuzu engine (MEP-804A/ MEP-814A), the exhaust manifold channels the gases into the muffler to deaden the sound of the exhaust gases. On the Yanmar engine (MEP-804B/MEP-814B), exhaust gases from the exhaust manifold drive an impeller in the turbocharger. The impeller is mechanically connected to a fan on the intake which pressurizes the intake air prior to entering the engine. The exhaust from the turbocharger is routed to the muffler to deaden the sound of the exhaust.

1-12.6.4 The gases pass from the muffler through the muffler outlet and are vented upward from the generator set housing. A cover, which is held open by the pressure of the exhaust gases during operation, closes over the exhaust port to prevent rain, water, or other foreign matter from entering the exhaust port when the generator set is not in use. The cover is easily removed for connection of an exhaust pipe for indoor operation.

1-12.6.5 On the Isuzu engine (MEP-804A/MEP-814A), the PCV valve and associated tubing allow gases which build up in the crankcase to be recycled through the intake manifold. The PCV valve closes to retain vapors within the engine after the engine is shut down.

1-12.6.6 On the Yanmar engine (MEP-804B/MEP-814B), the closed crankcase ventilation (CCV) filter and associated hoses allow gases which build up in the crankcase to be recycled through the intake manifold instead of being vented to the atmosphere, thereby reducing pollution. The CCV filter also removes oil from the built-up gases and returns the oil to the crankcase, thereby reducing oil consumption.

1-12.6.7 To improve engine starting when ambient temperature is below 40°F (4°C), a preheat system is used. On the Isuzu engine (MEP-804A/MEP-814A), a glow plug is located in the engine head for each cylinder. On the Yanmar engine (MEP-804B/MEP-814B), two air intake preheaters are used. The glow plugs/air intake preheaters

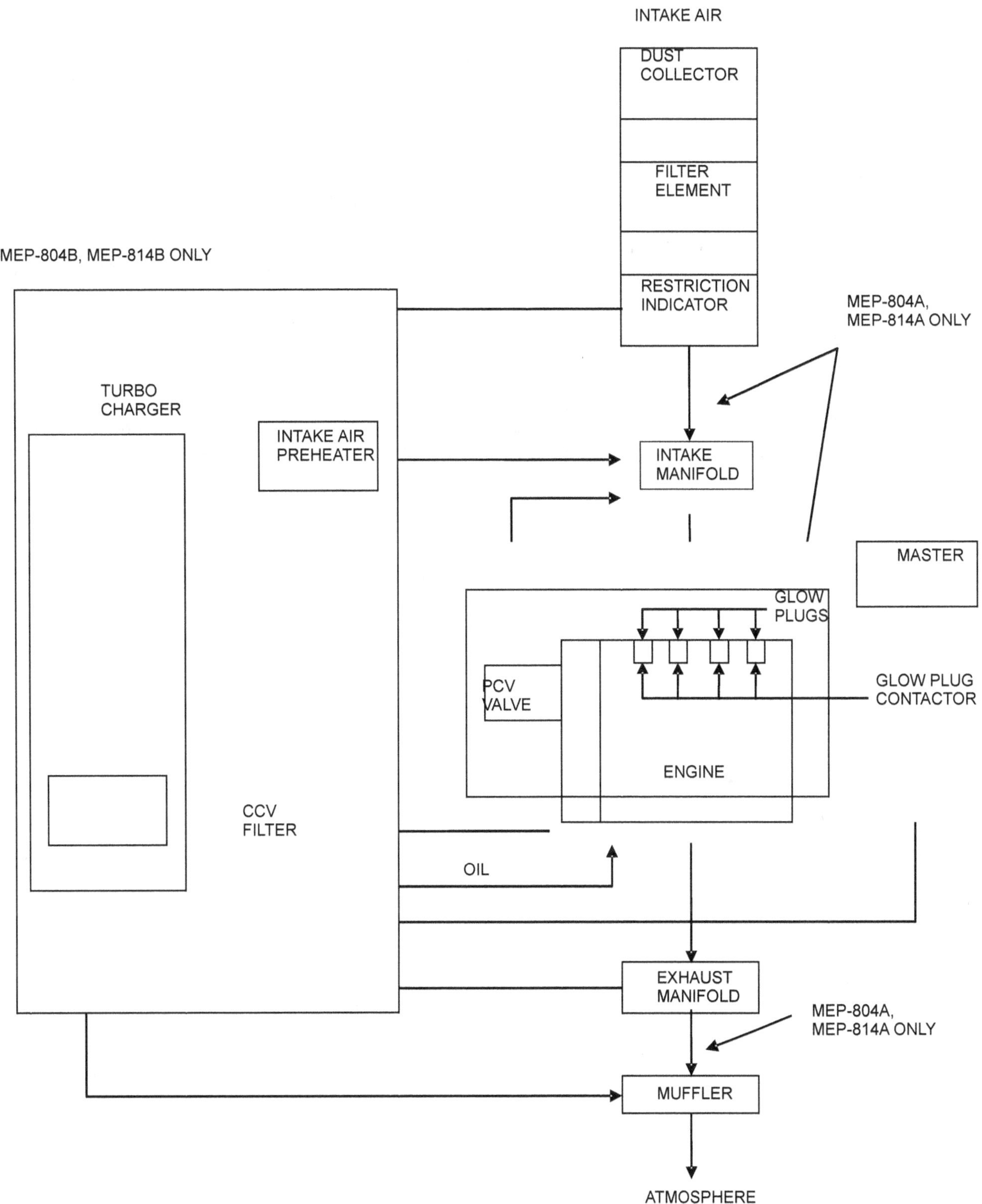

Figure 1-7. Engine Air Intake and Exhaust System.

1-12.7 Output Supply System.

1-12.7.1 The output supply system (Figure 1-8) consists of the AC generator, GROUND FAULT CIRCUIT INTERRUPTER, CONVENIENCE RECEPTACLE, current transformer, AC circuit interrupter relay, output terminals, AC CIRCUIT INTERRUPTER switch, kilowatt transducer, kilowattmeter (PERCENT POWER), VM-AM transfer switch,
AC voltmeter (VOLTS AC), and ammeter (PERCENT RATED POWER).

1-12.7.2 Power created by the generator is supplied through the current transformer, AC voltage reconnection board, and AC circuit interrupter relay to the output terminals. The AC voltage reconnection board allows configuration of the generator set for 120/208 volt connections or 240/416 volt connections. The AC CIRCUIT INTERRUPTER switch closes and opens the AC circuit interrupter relay. This enables or interrupts the power flow between the voltage reconnection board and the output terminals. The voltage regulation system (paragraph 1-12.10) senses generator output voltage and provides a control signal to the generator exciter to maintain the desired generator output voltage. Generator output frequency is controlled by the governor control system (paragraph 1-12.9) and is read on the FREQUENCY meter (HERTZ). The current transformer provides a reduced current signal to the kilowatt transducer and ammeter (PERCENT RATED CURRENT). The kilowatt transducer and kilowattmeter (PERCENT POWER) provide an indication of the power being used by the load. The ammeter (PERCENT RATED CURRENT) indicates the percent of rated current being supplied to the load. The position of the VM-AM transfer switch selects the output terminals from which current and voltage is measured. The AC circuit interrupter relay will open and disconnect the load whenever any of the following faults occur: reverse power, undervoltage, overload, or short circuit.

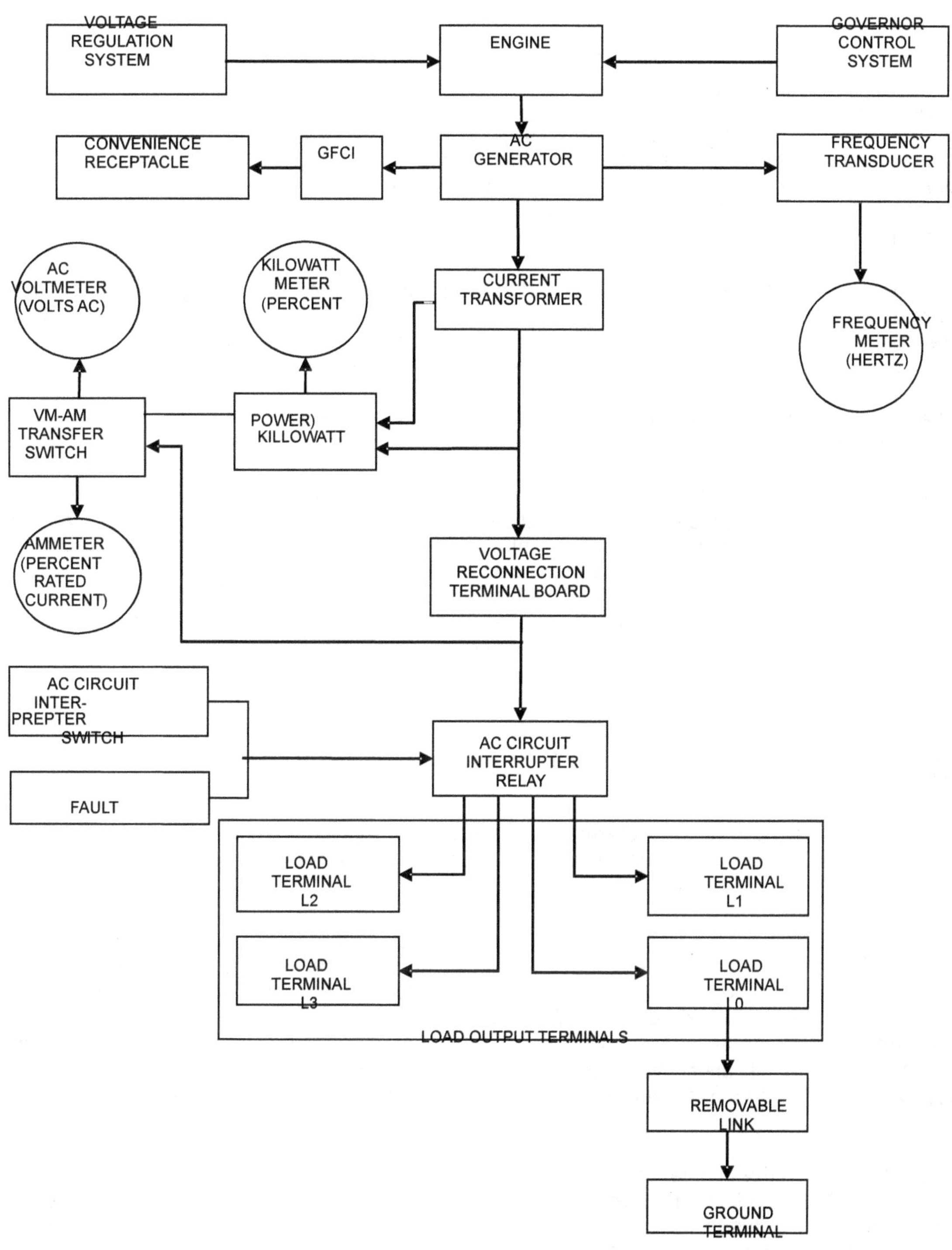

Figure 1-8. Output Supply System.

1-12.8 Generator Set Controls.

1-12.8.1 Engine Starting System.

1-12.8.1.1 Engine starting is accomplished primarily with two 12 volt batteries, connected in series to provide 24 VDC power, and a starter (Figure1-9). The starter includes a cranking motor and a solenoid. To permit engine starting, the DC CONTROL POWER circuit breaker must be pushed in, the DEAD CRANK switch must be in the NORMAL position, and the BATTLE SHORT switch must be in the OFF position. In addition, any ENGINE SHUT-DOWN fault previously registered on the malfunction indicator panel must have been corrected and the malfunction indicator panel must have been reset. When the MASTER SWITCH is then placed in the START position, the starting circuits supply 24 VDC power to the starter. As the engine accelerates to approximately 900 rpm, the starting circuits disconnect power from the starter.

1-12.8.1.2 When the MASTER SWITCH is first moved to the START position, the various instrument and control cir-cuits are energized. The engine starting system includes two control circuits. One starting control circuit energizes the start relay through closed switch contacts of the engine fault relay and the BATTLE SHORT switch. The other starting control circuit energizes the cranking relay coil through closed contacts of the crank disconnect switch and the start relay.
(The crank disconnect switch is an integral part of the electronic governor control.)
With the cranking relay energized, power passes from the batteries through closed contacts of the cranking relay to
energize the starter solenoid. With the starter solenoid energized, power passes from the starter solenoid to the cranking motor. The cranking motor then cranks the engine. Engine speed is sensed by the magnetic pick-up which sends a signal to the electronic governor control. As the engine accelerates to approximately 900 rpm, the signal from the magnetic pickup causes the crank disconnect switch to open one set of contacts and close another set of contacts. The open contacts break the circuit to the cranking relay and stop engine cranking. The closed contacts cause the field flash relay to be energized. When the MASTER SWITCH is moved to one of the two RUN positions, both starting control circuits are deenergized. The other generator set control and instrument circuits remain energized.

1-12.8.1.3 The engine may be cranked without starting by use of the DEAD CRANK switch. With the DEAD CRANK switch in the CRANK position, the cranking relay coil is energized to initiate engine cranking without energizing any other starting or control functions.

1-12.8.1.4 The generator set can be started without batteries by connecting an external 24 VDC power source to the NATO slave receptacle. The generator set can also supply starting power to another generator set through the NATO slave receptacle.

1-12.8.1.5 The batteries are charged by the battery charging alternator that is belt-driven by the engine. The BATTERY CHARGE ammeter indicates the charge/discharge rate of the batteries from -10 amps to +20 amps. A shunt provides a DC voltage signal, which is directly proportional to the actual battery current flow, to the BATTERY CHARGE ammeter. Normal operating indication on the BATTERY CHARGE ammeter depends on the state of the charge in the batteries. A low charge, which may exist immediately after engine starting, will cause a high reading (needle moves toward CHARGE area). When the charge in the batteries has been restored, the indicator moves near zero (0). The battery charging system is protected from reverse polarity in the battery connections by a fuse and diode.

1-12.8.2 Field Flash. When the engine reaches sufficient speed (900 rpm), the magnetic pickup causes a set of con-tacts in the crank disconnect switch to close and energize the field flash relay. This circuit provides current to the exciter field windings which sets up an electromagnetic field. The field current is necessary for the set to generate sufficient voltage for the voltage regulator to begin controlling the output voltage of the generator set. The field flash circuit is maintained until the MASTER SWITCH is released from the START position.

1-12.8.3 Operation. Placing the MASTER SWITCH in the PRIME & RUN or PRIME & RUN AUX FUEL position keeps the electronic governor control energized, and fuel will be supplied to the fuel injection pump as long as

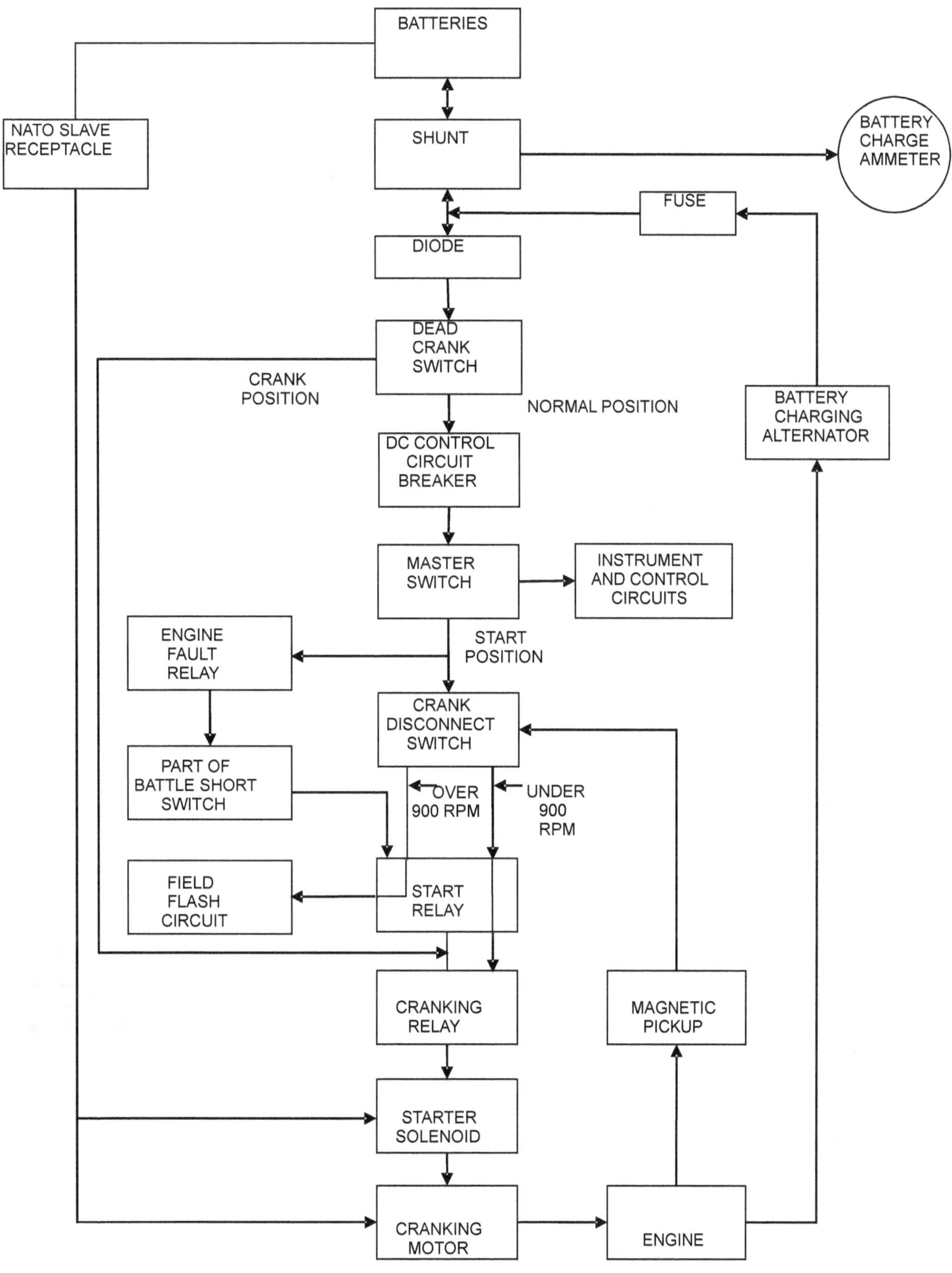

Figure 1-9. Engine Starting System.

1-12.8.4 Applying the Load. The load is applied by placing the AC CIRCUIT INTERRUPTER switch in the CLOSED position. This is a momentary contact switch that returns to the neutral or center position. The AC circuit interrupter relay is energized by this momentary contact and a holding circuit keeps it closed, bringing the load on line.

1-12.8.5 Shutdown.

1-12.8.5.1 The AC circuit interrupter relay is disengaged by placing the AC CIRCUIT INTERRUPTER switch in the OPEN position. This is a momentary contact switch which will break the AC circuit interrupter relay holding circuit and then return to the neutral or center position disconnecting the load from the line.

1-12.8.5.2 When the MASTER SWITCH is placed in the OFF position, all power is removed from the control circuit and the engine will stop.

1-12.8.5.3 The EMERGENCY STOP switch assembly will remove power from the control circuit by energizing the engine fault relay. This will cause the engine to shut down. The EMERGENCY STOP switch assembly is not to be used as an alternative for routine shutdown procedures. When the generator set is stopped using the EMERGENCY STOP switch assembly, some circuits remain energized causing a drain on the batteries until the MASTER SWITCH is placed in the OFF position.

1-12.8.6 Paralleling.

1-12.8.6.1 The generator set is capable of being operated in parallel with one other set of the same model number. This capability is provided by the PARALLELING RECEPTACLE, paralleling cable, LOAD SHARING ADJUST and REACTIVE CURRENT ADJUST rheostats, two SYNCHRONIZATION LIGHTS, UNIT-PARALLEL switch, reverse power relay, voltage sensing relay, droop current transformer, and permissive paralleling relay.

1-12.8.6.2 The paralleling cable is used to interconnect the governor and AC voltage regulator paralleling circuits of two generator sets. The UNIT-PARALLEL switch is used to select parallel operation. Voltage and frequency of the two generator sets are synchronized by adjusting each set's VOLTAGE and FREQUENCY controls. Phase synchronization is indicated by the SYNCHRONIZATION LIGHTS. The reverse power relay serves as a safety device by detecting any excessive out-of-phase condition and interrupting power to the AC circuit interrupter relay holding coil when that condition occurs. The permissive paralleling relay monitors the voltage phase relationship and prevents the AC circuit interrupter relay from closing when the units are not properly synchronized.

1-12.9 Governor Control System.

1-12.9.1 The governor control system (Figure 1-10) includes the electronic governor control, governor actuator, magnetic pickup, load measuring unit, frequency transducer, FREQUENCY meter (HERTZ), kilowatt transducer, fuel injection pump, FREQUENCY SELECT switch, and FREQUENCY adjust potentiometer.

1-12.9.2 The governor actuator is a linear electromechanical actuator which controls the output of the fuel injection pump in response to the electrical input from the electronic governor control. The FREQUENCY adjust potentiometer, located on the control panel and adjusted by the operator, provides a signal representing the desired engine speed/ generator frequency to the electronic governor control. A signal representative of the actual engine speed/ generator frequency is sent to the electronic governor control by the magnetic pickup. Any change in engine speed from that selected by the operator, as sensed by the magnetic pickup, causes the electronic governor control to increase or decrease the fuel pump output to maintain the desired speed. The load measuring unit senses changes in external load demand and provides a change signal to the electronic governor control allowing the control to start its response prior to any actual change in engine speed. The generator set frequency and power output are indicated by the FREQUENCY meter (HERTZ) and the kilowattmeter (PERCENT POWER) on the control panel. The FREQUENCY SELECT switch is used to set the generator for 50 or 60 Hz operating frequencies (50/60 Hz sets only).

1-12.9.3 The electronic governor control also contains the engine overspeed switch (fault system) and the crank disconnect switch (engine starting system). These switches function as a result of input from the magnetic pickup.

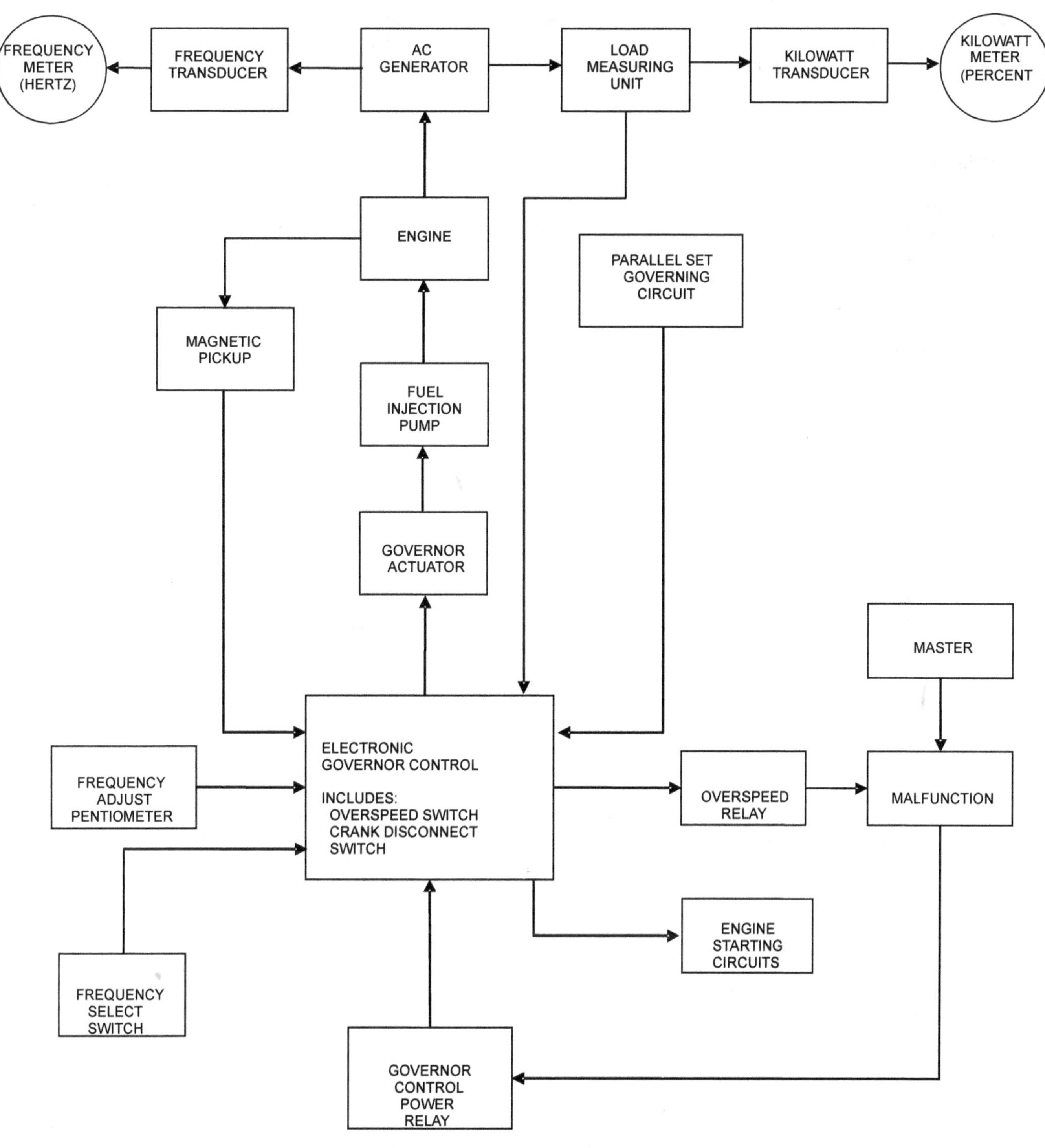

Figure 1-10. Governor Control System.

1-12.9.4 The 24 VDC power is supplied to the electronic governor control through the governor control power relay. The governor control relay is controlled by the fault system. The electronic governor controls of two generator sets operating in parallel are interconnected by the paralleling cable.

1-12.10 Voltage Regulation System.

The voltage regulation system (Figure 1-11) consists of the AC voltage regulator, VOLTAGE adjust potentiometer, and power transformer. The AC voltage regulator senses and controls the generator output voltage which is operator adjustable within the design limits by use of the VOLTAGE adjust potentiometer. The power transformer provides operating power to the AC voltage regulator. The output voltage is indicated by the AC voltmeter (VOLTS AC) on the control panel.

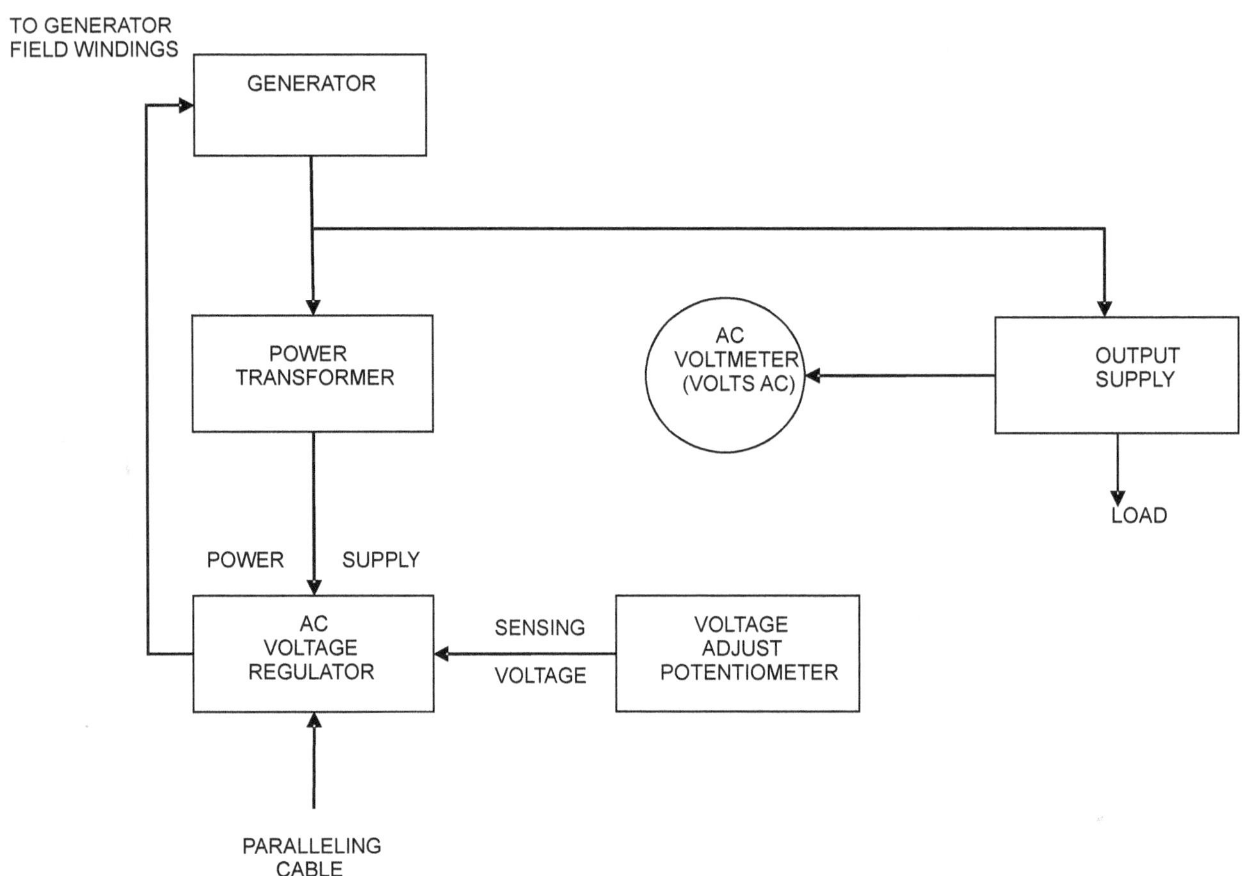

Figure 1-11. Voltage Regulation System.

1-13 LOCATION AND DESCRIPTION OF MAJOR COMPONENTS

NOTE

All locations (index numbers) referenced are given facing the control box side (rear) of the generator set. These index numbers refer to both Figure 1-12 (MEP-804A/MEP-814A) and Figure 1-13 (MEP-804B/MEP-814B), except for the turbocharger, CCV filter, fuel transfer pump filter, and engine fuel filter (Figure 1-13).

1-13.1 Malfunction Indicator Panel (1, Figure 1-12/1, Figure 1-13). The malfunction indicator panel is located to the left of the control panel. It indicates malfunctions of the generator set components.

1-13.2 Control Panel Assembly (2). The generator set control panel is located at the rear of the generator set and contains controls and instruments for operating the engine and the generator.

1-13.3 Muffler (3). On the Isuzu engine, the muffler and exhaust tubing are connected to the exhaust manifold on the engine. On the Yanmar engine, the muffler and exhaust tubing are connected to the turbocharger output. The exhaust exits from the top of the generator set housing. Gases are exhausted upward.

1-13.4 Skid Base (4). The skid base supports the generator set. It has fork lift access openings and cross members for short distance movement. The skid base has provisions in the bottom for installation of the generator set on a trailer.

1-13.5 Fuel Filter/Water Separator (5). The fuel filter/water separator is located in the engine compartment on the right side. The element removes impurities and water from the diesel fuel.

1-13.6 DEAD CRANK Switch (6). The DEAD CRANK switch is located in the engine compartment on the right side. The switch allows the engine to be turned over without starting for maintenance purposes.

1-13.7 Oil Filter (7). The oil filter is located in the engine compartment on the right side. The filter removes impurities from the engine lube oil.

1-13.8 Voltage Reconnection Terminal Board (8). The voltage reconnection terminal board is located on the right side (rear) of the generator set. The board allows reconfiguration from 120/208 to 240/416 VAC output.

1-13.9 Load Output Terminal Board (9). The load output terminal board is located on the right side (rear) of the generator set. Four AC output terminals are located on the board. They are marked L1, L2, L3, and L0. A fifth terminal, marked GND, is located next to the output terminals and serves as equipment ground for the generator set. A removable, solid copper bar is connected between the L0 and GND terminals.

1-13.10 CONVENIENCE RECEPTACLE (10). The CONVENIENCE RECEPTACLE is a 120 VAC receptacle used to operate small plug-in type equipment.

1-13.11 PARALLELING RECEPTACLE (11). The PARALLELING RECEPTACLE is used to connect the paralleling cable between two generator sets of the same size and mode to operate in parallel.

1-13.12 Diagnostic Connector (12). The diagnostic connector is a multi-pin plug that is wired to specific points in the generator set electrical system to enable monitoring and troubleshooting of the generator set operation at a single location.

1-13.13 Air Cleaner Assembly (13). The air cleaner assembly is located on the left side behind the air cleaner access door. It consists of a dry type, disposable paper filter and canister. The air cleaner assembly features a dust collector which traps large dust particles. The air cleaner assembly has a restriction indicator which will pop up during operation when the air cleaner requires servicing.

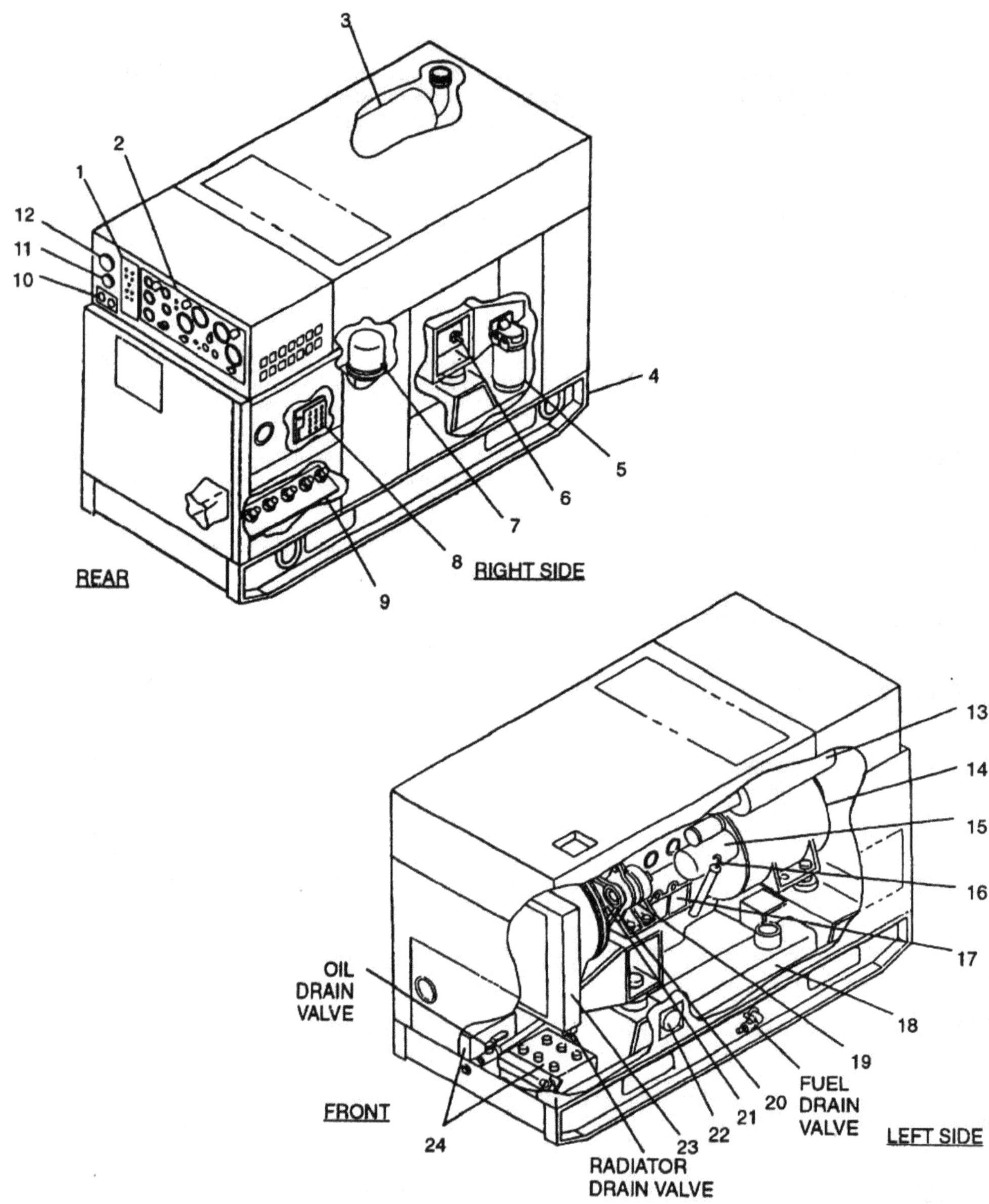

Figure 1-12. Generator Set Components – MEP-804A/MEP-814A.

1-13.14 AC Generator (14). The AC generator is a single bearing, drip-proof, synchronous, brushless, three phase, air-cooled generator. The generator is coupled directly to the rear of the diesel engine.

1-13.15 Starter (15). The starter is located on the left side of the engine. The electric cranking motor mechanically engages the engine flywheel in order to start the diesel engine.

1-13.16 Dipstick (16). On the Isuzu engine (MEP-804A/MEP-814A), the dipstick is located on the left side of the engine compartment. On the Yanmar engine (MEP-804B/MEP-814B), the dipstick is located on the right side of the engine compartment. The dipstick shows the lube oil level in the engine crankcase.

1-13.17 Engine (17). The generator is powered by one of two possible engines: an Isuzu engine (MEP-804A/ MEP-814B) or a Yanmar engine (MEP-804B/MEP-814B). The engine occupies the front half of the generator set. The Isuzu engine is a four cylinder, four cycle, fuel injected, naturally aspirated, liquid-cooled diesel engine. The Yanmar engine is similar, but turbocharged. The engine is also equipped with a fuel filter/water separator, oil filter, and an air cleaner assembly. Protection devices automatically stop the engine during conditions of high coolant temperature, low oil pressure, no fuel, overspeed, or overvoltage.

1-13.18 Fuel Tank (18). The 14 gallon (53 liters) fuel tank is located in the front of the generator set below the engine ~~and between th~~e skid base side members. The fuel tank is a fuel reservoir and has sufficient capacity to enable the generator set to operate for at least 8 hours without refueling.

1-13.19 Battery Charging Alternator (19). The battery charging alternator is located on the left side of the engine. It is capable of maintaining the batteries in a state of full charge in addition to providing the required 24 VDC control power.

1-13.20 Fan Belt (20). The fan belt is located in the engine compartment on the front of the engine. The belt drives the fan, water pump, and battery charging alternator.

1-13.21 Water Pump (21). The water pump is located in the engine compartment on the front of the engine. The pump circulates the engine coolant through the engine block and the radiator.

1-13.22 NATO Slave Receptacle (22). The NATO slave receptacle is located on the left side of the generator set under the engine compartment access door. It is a NATO receptacle used for remote battery connection.

1-13.23 Radiator (23). The radiator is located at the front of the generator set. It acts as a heat exchanger for the engine ~~coolant.~~

1-13.24 Batteries (24). Two batteries are located at the front of the generator set. The batteries are electrolyte servicea~~ble, lead acid,~~ 12 volt type. After starting, the generator set is capable of operating with batteries removed. A diode, located behind the control panel, protects the generator set if the batteries are incorrectly connected.

1-13.25 Turbocharger (25, Figure 1-13). The Yanmar engine (MEP-804B and MEP-814B) is turbocharged. The turbocharger increases the horsepower of the diesel engine in order to deliver the generator set maximum power.

1-13.26 CCV Filter (26). The Yanmar engine (MEP-804B/MEP-814B) contains a CCV filter, mounted on the engine, that filters oil out of the gases which build up in the crankcase and routes the gases to the intake manifold. A check valve in the CCV filter holds the oil in the filter until the engine stops. The oil then drains into the crank-case.

1-13.27 Fuel Transfer Pump Filter (27). The Yanmar engine (MEP-804B/MEP-814B) contains a fuel transfer pump filter, located on the inlet to the fuel transfer pump, filters contaminants and protects the pump.

1-13.28 Engine Fuel Filter (28). The Yanmar engine (MEP-804B/MEP-814B) contains an engine fuel filter,

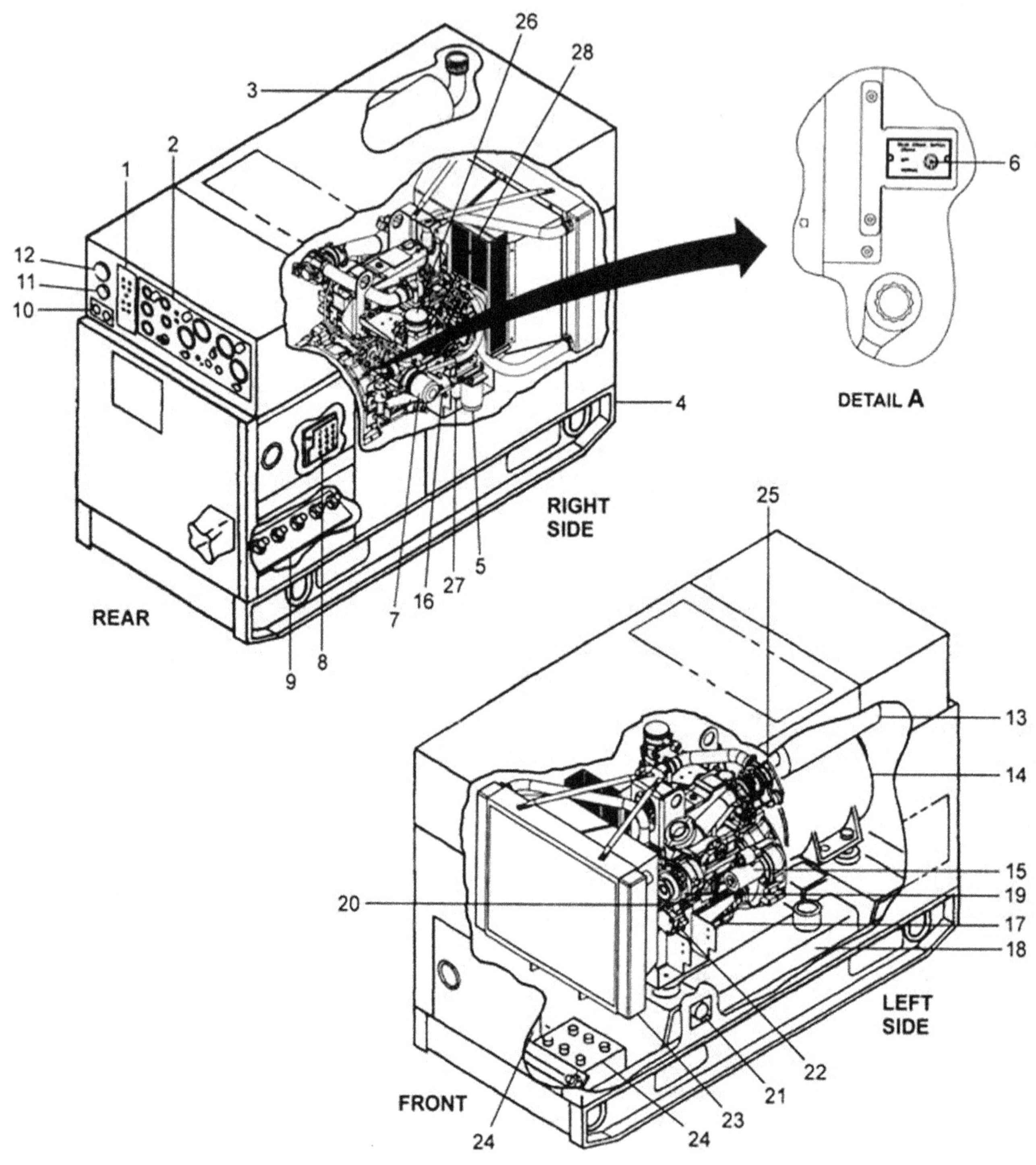

Figure 1-13. Generator Set Components – MEP-804B/MEP-814B.

CHAPTER 2

UNIT MAINTENANCE INSTRUCTIONS

Subject Index Page

Subject Index Page

Subject Index Page

Subject Index Page

Section I. SERVICE UPON RECEIPT OF EQUIPMENT

2-1 INSPECTING AND SERVICING THE EQUIPMENT

This section provides information and guidance for inspecting, servicing, and installing the generator set under normal conditions.

2-1.1 Inspection.

a. a. Unpack and inventory all end item components for serviceability.

b. b. Check that all packing materials have been removed.

c. c. Check generator set identificat ion plate for proper identification.

d. d. Inspect generator set exterior for shipping damage.

e. e. Open battery compartment access door and inspect batteries for damage.

f. f. Check battery cables for proper polarity connection, damage, and loose connections.

g. g. Open control panel access door and check panel for damage.

h. h. Lower control panel and check electrical components for damage or loose connections.

i. i. Raise control panel and secure fasteners.

j. j. Check air cleaner assembly for external damage and exhaust opening for obstruction.

k. k. Check fan belt for looseness and ensure it is not frayed or cracked.

l. l. Inspect generator set for loose or missing mounting hardware or damaged or missing parts.

m. Drain preservative from engine and fill with proper lubricating oil (paragraph 2-1.2.4), as required.

n. Unpack grounding rod from inside left engine access door, parallel cable, and auxiliary fuel hose from storage box. Inspect each item for damage and accountability.

2-1.2 Service.

2-1.2.1 Batteries. When servicing batteries, refer to TM 9-6140-200-14.

WARNING

Each battery weighs more than 70 pounds (32 kg) and requires a two-person lift. Lifting batteries can cause back strain. Ensure proper lifting techniques are used when lifting batteries. Failure to comply with this warning can cause injury to personnel.

WARNING

Batteries give off a flammable gas. Do not smoke or use open flame when performing maintenance. Failure to comply with this warning can cause injury or death to personnel and damage to the generator set.

WARNING

Battery acid can cause burns to unprotected skin. Wear safety goggles and chemical gloves and avoid acid splash while working on batteries. Failure to comply with this warning can cause injury to personnel.

WARNING

The connection of any electrical equipment and the disconnection of any electrical equipment may cause an explosion hazard. Do not connect any electrical equipment or disconnect any electrical equipment in an explosive atmosphere. Failure to comply with this warning can cause injury or death to personnel.

WARNING

Many components require a two-person lift. Lifting heavy components can cause back strain. Ensure proper lifting techniques are used when lifting heavy components. Failure to comply with this warning can cause injury to personnel.

2-1.2.2 Radiator.

WARNING

Do not operate generator set while servicing radiator. Failure to comply with this warning can cause injury to personnel and damage to the equipment.

WARNING

All metal jewelry can conduct electricity and become entangled in generator set components. Remove all jewelry when working on generator set. Failure to comply with this warning can cause injury or death to personnel.

WARNING

DO NOT wear loose clothing when performing checks, services and maintenance. Failure to comply with this warning can cause injury or death to personnel.

WARNING

Cooling system operates at high temperature and pressure. Contact with high pressure steam and/or liquids can result in burns and scalding. Shut down generator set, and allow system to cool before performing checks, services, and maintenance, or wear gloves and additional protective clothing and goggles as required. Failure to comply with this warning can cause injury or death to personnel.

a. Remove radiator cap.

b. b. Check that radiator drain valve is closed (Figure 1-12).

c. c. Fill radiator with proper coolant/antifreeze (Table 2-1). Fill radiator to a level 2 inches below fill opening.

d. d. Remove overflow bottle cap.

e. e. Fill overflow bottle to COLD level.

Table 2-1. Coolant.

Ambient Temperature	Radiator Coolant	Ratio
+40°F to +120°F (4°C to 49°C)	Water: MIL-A-53009A(1) Inhibitor, Corrosion	35:1
-25°F to +120°F (-32°C to 49°C)	Water: A-A-52624A Antifreeze	1:1
-25°F to +120°F (-32°C to 49°C)	A-A-52624A Antifreeze	N/A

f. Install overflow bottle and radiator caps.

WARNING

Cooling system operates at high temperature and pressure. Contact with high pressure steam and/or liquids can result in burns and scalding. Shut down generator set, and allow system to cool before performing checks, services, and maintenance, or wear gloves and additional protective clothing and goggles as required. Failure to comply with this warning can cause injury or death to personnel.

WARNING

Fan has sharp blades. Use caution and wear gloves when removing or installing belts. Failure to comply with this warning can cause injury to personnel.

g. After 30 minutes of operation check coolant/antifreeze level at overflow bottle. Add coolant/antifreeze to overflow bottle, as required.

WARNING

High voltage is produced when the generator set is in operation. Never attempt to start the generator set unless it is properly grounded. Failure to comply with this warning can cause injury or death to personnel.

2-1.2.3 Fuel Tank.

WARNING

Diesel fuel is flammable and toxic to eyes, skin, and respiratory tract. Skin and eye protection are required when working in contact with diesel fuel. Avoid repeated or prolonged contact. Provide adequate ventilation. Operators are to wash exposed skin and change chemical soaked clothing promptly if exposed to fuel. Failure to comply with this warning can cause injury or death to personnel.

a. Check that fuel drain valve is closed (Figure 1-12).

b. Remove fuel tank filler cap.

c. Fill fuel tank with fuel type (Table 2-2). Fuel tank capacity is 14 gallons (53 liters).

WARNING

Fuels used in the generator set are flammable. When filling the fuel tank, maintain metal-to-metal contact between filler nozzle and fuel tank opening to eliminate static electrical discharge. Failure to comply with this warning can cause injury or death to personnel, and damage to the generator set.

WARNING

Fuels used in the generator set are flammable. Do not smoke or use open flames when performing maintenance. Failure to comply with this warning can cause injury or death to personnel, and damage to the generator set.

d. Install fuel tank filler cap.

Table 2-2. Fuel.

Ambient Temperature	Diesel Fuel
+20°F to +120°F (-7°C to 49°C)	A-A-52557A, Grade 2-D MIL-DTL-83133E, JP-8
-25°F to +20°F (-32°C to -7°C)	A-A-52557A, Grade 1-D MIL-DTL-5624T, JP-5

2-1.2.4 Lubricating Oil.

a. Place suitable container under oil drain plug and remove plug.

WARNING

Oil filter base and housing springs are under tension and can act as projectiles when being removed. Use eye protection when removing springs. Failure to comply with this warning can cause injury to personnel.

b. b. Open battery access door, open oil drain valve (Figure 1-12), and drain oil.

c. c. Close oil drain valve and remove oil fill cap.

NOTE

Dipstick is marked indicating that oil level can be checked and oil added when engine is running or stopped. Make sure the correct side of dipstick is checked.

d. Fill engine with proper engine lubricating oil (Table 2-3) to FULL mark on dipstick. Lubrication system capacity is 6 quarts (5.7 Liters).

WARNING

The high pressure oil system operates at high temperature and pressure. Contact with hot oil can result in burns and scalding. Shut down generator set, and allow system to cool before performing checks, services, and maintenance. Wear heat resistant gloves and avoid contacting hot surfaces. Do not allow hot oil or components to contact skin or hands. Failure to comply with this warning can cause injury or death to personnel.

WARNING

Wear heat resistant gloves and avoid contacting hot metal surfaces with your hands after components have been heated. Wear additional protective clothing as required. Failure to comply with this warning can cause injury to personnel.

e. Install oil fill cap. Close battery access door.

Table 2-3. Lubricating Oil.

Ambient Temperature	Lubricating Oil
+20°F to +120°F (-7°C to 49°C)	MIL-PRF-2104H OE HDO-30 or OE HDO-15/40
0°F to +20°F (-17°C to 6°C)	MIL-PRF-2104H OE HDO-10
-25°F to 0°F (-32°C to -17°C)	MIL-PRF-46167C

2-2 GENERATOR SET INSTALLATION

WARNING

Exhaust discharge contains deadly gases including carbon monoxide. DO NOT operate generator set in enclosed areas unless exhaust discharge is properly vented outside. Failure to comply with this warning can cause injury or death to personnel.

WARNING

Hot exhaust gases can ignite flammable materials. Allow room for safe discharge of hot gases and sparks. Failure to comply with this warning can cause injury or death to personnel.

a. a. Ensure that installation site is as level as possible.

b. b. Provide adequate ventilation to prevent recirc ulation of hot air exhausted from generator set.

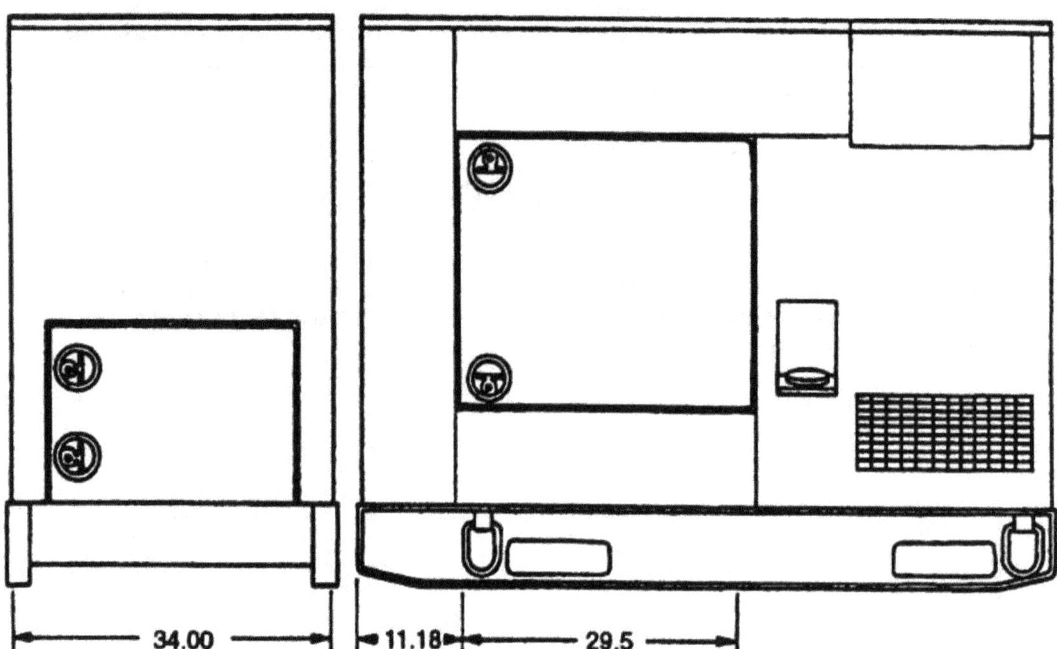

c. c. Refer to Figure 2-1 for base mounting measurements.

Figure 2-1. Base Mounting Measurements.

2-2.2 Outdoor Installation.

a. a. Make use of natural protective barriers.

b. b. Allow space on all sides for service and maintenance. Refer to Figure 2-2 for minimum clearance measurements.

c. c. Ensure that site soil is firm and well drained.

2-2.3 Indoor Installation.

WARNING

Hot exhaust gases can ignite flammable materials. Allow room for safe discharge of hot gases and sparks. Failure to comply with this warning can cause injury or death to personnel.

WARNING

Exhaust discharge contains deadly gases including carbon monoxide. DO NOT operate generator set in enclosed areas unless exhaust discharge is properly vented outside. Failure to comply with this warning can cause injury or death to personnel.

CAUTION

Never position generator set with the air inlets near a wall or other object that interferes with cooling air circulation. Damage to equipment could occur.

a. a. Provide ducts and vents to outside of building if good supply of cooling air is not available.

b. b. Make air intake and outlet openings in building same size or larger as those on the generator set.

c. Install a gas-tight metal pipe from exhaust pipe of generator set to outside of building.

NOTE

Make exhaust pipe extension as short and straight as possible with only one 90 degree bend, if needed.

d. d. Ensure that inside diameter of exhaust pipe extension is as large as or larger than generator set exhaust pipe.

WARNING

Hot exhaust gases can ignite flammable materials. Allow room for safe discharge of hot gases and sparks. Failure to comply with this warning can cause injury or death to personnel.

e. e. Provide for harmless discharge of hot gases and sparks. Do not direct exhaust into area containing flammable materials.

WARNING

If not shielded, hot exhaust pipe can ignite flammable wall materials. Failure to comply with this warning can cause injury or death to personnel.

f. Shield exhaust pipe with fireproof material at point where it passes through a flammable wall.

WARNING

An unwrapped exhaust pipe can cause injury if touched. Failure to comply with this warning can cause injury to personnel.

g. g. Wrap exhaust pipe in protective material.

h. h. Allow space on all sides for service and maintenance. Refer to Figure 2-2 for minimum clearance measurements.

WARNING

Exhaust discharge contains deadly gases including carbon monoxide. DO NOT operate generator set in enclosed areas unless exhaust discharge is properly vented outside. Failure to comply with this warning can cause injury or death to personnel.

WARNING

Engine exhaust fumes contain deadly poisonous gases.

Severe exposure can cause death or permanent brain damage.

Exhaust gases are most dangerous in places with poor airflow. Best defense against exhaust gas poisoning is very good airflow.

To protect yourself and your partners, always obey the following rules:

- **DO NOT run engine indoors unless you have VERY GOOD AIRFLOW.**
- **DO NOT idle engine for a long time unless there is VERY GOOD AIRFLOW.**
- **Be alert at all times. Check for smell of exhaust fumes.**
- **REMEMBER: Best defense against exhaust gas poisoning is VERY GOOD AIRFLOW.**
- **Exhaust gas poisoning causes dizziness, headache, loss of muscle control, sleepiness, coma, and death. If anyone shows signs of exhaust gas poisoning, get ALL PERSONNEL clear of exhaust area. Make sure they have lots of fresh air. KEEP THEM WARM, CALM, AND INACTIVE. GET MEDICAL HELP. If anyone stops breathing, give artificial respiration. See FM 4-25.11 for first aid.**

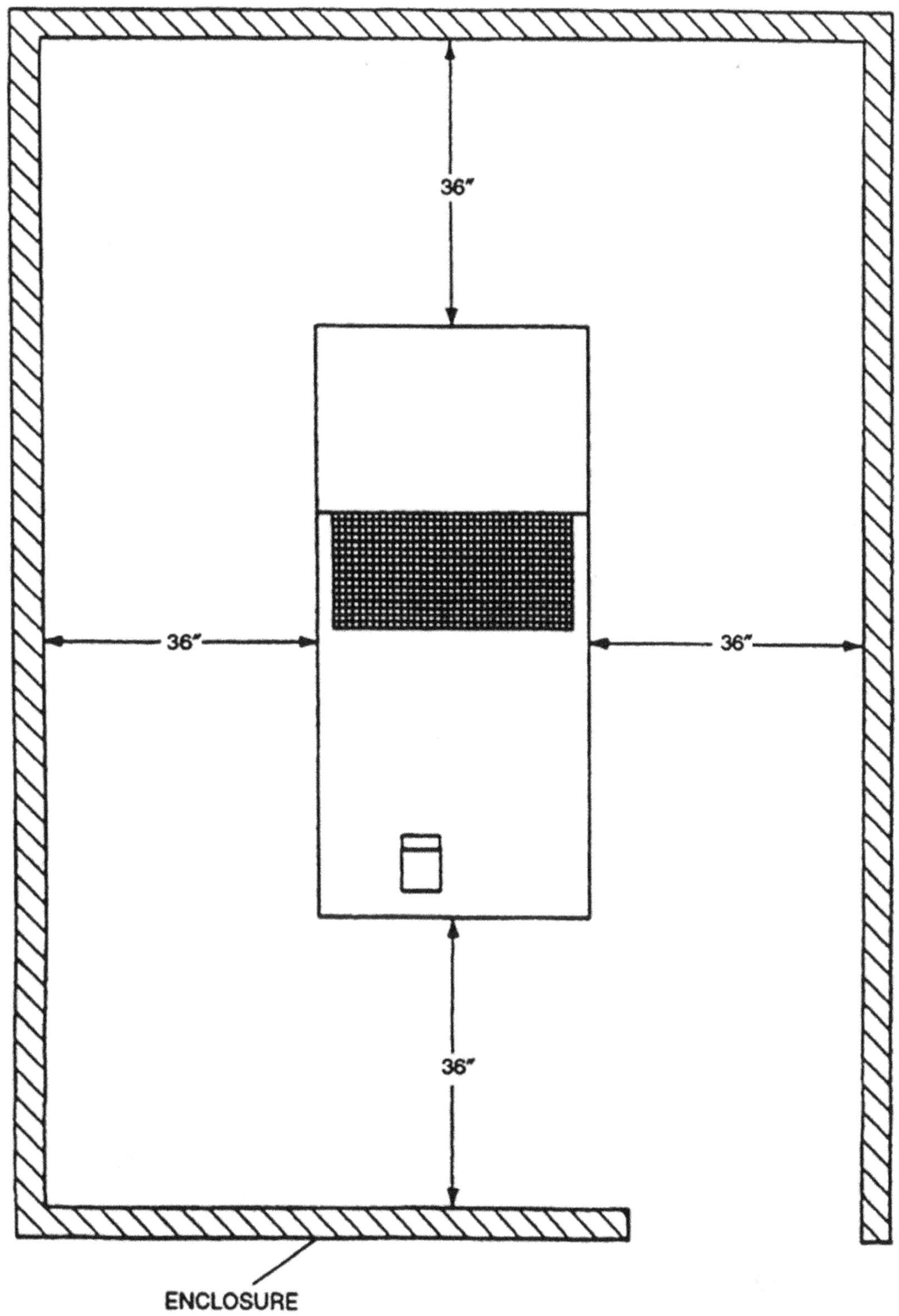

Figure 2-2. Minimum Enclosure Clearance Measurements.

2-3 FABRICATION/ASSEMBLY OF PARTS

Table 2-4 provides a list of generator set parts that require fabrication or assembly when replacing. Refer to Appendix D for fabrication and assembly instructions.

Table 2-4. Fabricated/Assembled Parts.

NAME	PART NUMBER
AC Power Cable Assembly	88-22126-1
AC Power Cable Assembly	88-22126-2
AC Power Cable Assembly	88-22126-3
AC Power Cable Assembly	88-22126-4
AC Power Cable Assembly	88-22126-5
AC Power Cable Assembly	88-22126-6
AC Power Cable Assembly	88-22126-7
Battery Cable Assembly	88-22123
Battery Cable Assembly	88-22179
Battery Cable Assembly	88-22309
Battery Cable Assembly	88-22310
Battery Cable Assembly	88-22311
Electromagnetic Interference (EMI) Capacitor Assembly	88-22758
Load Wrench Cord	88-22469
Diode Assembly	88-22418-2
Control Panel Holder	88-22120
Baffle Insulation	88-22594
Front Housing Insulation	88-22595
Resistor Assembly	01-21506-1
Resistor Assembly	01-21506-2
Volt Resistor Assembly	88-22631
Resistor-Diode Assembly	88-22106
Fuel Level Switch Assembly	88-22792
Transducer Assembly	88-22550
Varistor Wire L1	88-20305-1
Varistor Wire L2	88-20305-2
Varistor Wire L3	88-20305-3
Varistor Wire L0	88-20305-5

Section II. REPAIR PARTS; SPECIAL TOOLS; TEST, MEASUREMENT, AND DIAGNOSTIC EQUIPMENT (TMDE); AND SPECIAL SUPPORT EQUIPMENT

2-4 REPAIR PARTS AND SPECIAL TOOLS LIST (RPSTL)

2-4.1 Tools and Equipment. There are no special tools or support equipment required to perform unit level maintenance on the generator set. A list of recommended tools and support equipment required to maintain the generator set is contained in Appendix B, Section III.

2-4.2 Maintenance Repair Parts. Repair parts and equipment are listed and illustrated in the Repair Parts and Special Tools List (RPSTL) manual, TM 9-6115-643-24P.

Section III. SPECIAL LUBRICATION INSTRUCTIONS

NOTE

There are no special lubrication instructions. Refer to LO 9-6115-643-12 for generator set lubrication requirements.

Section IV. PREVENTIVE MAINTENANCE CHECKS AND SERVICES (PMCS)

2-5 PMCS PROCEDURES

2-5.1 General. To ensure that the generator set is ready for operation at all times, it must be inspected so that defects can be discovered and corrected before they result in serious damage or failure.

2-5.2 Purpose of PMCS Table. Your Preventive Maintenance Checks and Services table lists the inspections and care of your equipment required to keep it in good operating condition.

2-5.3 Purpose of Service Intervals. The interval column of your PMCS table tells you when to do a certain check or service.

2-5.4 "Procedures" Column. The procedures column of your PMCS table tells you how to do the required checks and services. Carefully follow these instructions.

2-5.5 The "Equipment Is Not Ready/Available If" Column. This column tells you when and why the generator set cannot be used.

NOTE

The terms ready/available and mission capable refer to the same status: generator set is on hand and is able to perform its combat missions (see DA Pam 738-750).

2-5.6 Reporting and Correcting Deficiencies. If your generator set does not perform as required, refer to Troubleshooting section for possible problems. Report any malfunctions or failures on DA Form 2404, or refer to DA Pam 738-750.

Table 2-5. Unit Preventive Maintenance Checks and Services.

M-Monthly Q-Quarterly S-Semi-annually A-Annually B-Bi-annually H-Hours

Item No.	Interval M	Q	S	A	B	H	Item to be Inspected	Procedures Check for and have repaired or adjusted as necessary	Equipment Is Not Ready/Available If:
								NOTE Oil filter should be changed with lube oil change (TM 9-2815-254-24/ TM 9-2815-538-24&P).	
1						300	Engine Lube Oil	Drain engine lube oil. Add proper lube oil (LO 9-6115-643-12).	
2						300	Fuel Filter/ Water Separator	Change fuel filter/water separator (paragraph 2-105).	
3						300	Engine Fuel Filter (MEP-804B/MEP-814B)	Check fuel filter for leaks and replace filter element (TM 9-2815-538-24&P).	Fuel filter leaks that would prevent op-era-tion; fuel filter not replaced, as required.
4						300	Fuel Transfer Pump Filter (MEP-804B/ MEP-814B)	Replace fuel filter (paragraph 2-106.3).	
5						1500	Cooling System	Drain coolant and flush cooling system or add proper coolant (paragraph 2-79.2).	
6							Radiator Cap	Inspect radiator cap for corrosion, torn or deteriorated seal, and obvious damage.	Radiator cap or seal is damaged.
7						300	Batteries	Remove batteries (paragraph 2-12.2). Clean batteries, cable terminals, and battery posts; test batteries for state of charge (paragraph 2-12.1).	Batteries will not hold charge.
8						300	Air Cleaner Assembly	Inspect air cleaner assembly and mounting bracket for cracks, dents, and other damage. Inspect element for clogs and damage. Clean or replace, as necessary. Clean housing with cleaning cloth.	

Table 2-5. Unit Preventive Maintenance Checks and Services – Continued.

M-Monthly Q-Quarterly S-Semi-annually A-Annually B-Bi-annually H-Hours

Item No.	Interval						Item to be Inspected	Procedures Check for and have repaired or adjusted as necessary	Equipment Is Not Ready/Available If:
	M	Q	S	A	B	H			
9				□		750	Air Cleaner Tubing and Breather	Remove, clean, and inspect tubing and breather (para-graph 2-77 (MEP-804A/ MEP-814A)/paragraph 2-78 (MEP-804B/MEP-814B))	
10				□		300	Hardware and Sound Insulation	Inspect for loose, damaged, or missing hardware and sound in-sulation. Tighten loose hardware. Repair or replace damaged or missing hardware and insulation.	Hardware or insula-tion loose, missing, or damaged.
11				□		1500	Radiator and	Clean radiator exterior surfaces (paragraph 2-80).	
12							Interior of Generator Set	Clean engine compartment.	
13				□		1500	Magnetic Pickup	Remove, inspect, and clean magnetic pickup (paragraph 2-121).	Magnetic pickup is damaged.
14				□			Wiring Harnesses	Inspect wiring harnesses for breaks and loose connec-tions. Repair and tighten wir-ing harnesses, as necessary.	Wiring harnesses are damaged or connec-tions are loose.
15				□		3000	Muffler	Check muffler for leaks, restriction, and accumulation of carbon. Replace or clean, as re-quired.	Muffler leaks, is restricted, or has excessive carbon accumulation.
16				□	□	300	Auxiliary Fuel Filter	**NOTE** **If the auxiliary fuel system is used as the primary fuel source, then the auxiliary fuel filter must be replaced semi-annually.** Check for proper operation using auxiliary fuel system as primary source.	

Section V. TROUBLESHOOTING

2-6 UNIT TROUBLESHOOTING PROCEDURES

2-6.1 Purpose of Troubleshooting Table. This section contains troubleshooting information for locating and correcting operating troubles which may develop in the generator set. Each malfunction for an individual component, unit, or system is followed by a list of tests or inspections which will help you to determine probable causes and corrective actions to take. You should perform the tests/inspections and corrective actions in the order listed.

This table cannot list all malfunctions that can occur, nor all tests or inspections and corrective actions. If a malfunction is not listed or cannot be corrected by listed corrective actions, notify your supervisor.

NOTE

Before you use this table, be sure you have performed your PMCS.

Before using this table, ensure that operator level troubleshooting steps have been performed.

Refer to the Diagnostic Connector Connection Points Table 2-9, Electrical Schematic FO-1 (MEP-804A/MEP-814A)/FO-3 (MEP-804B/MEP-814B), and Wiring Diagram FO-2 (MEP-804A/MEP-814A)/FO-4 (MEP-804B/MEP-814B) as troubleshooting aids.

SYMPTOM INDEX
GENERATOR SET

Table 2-6. Unit Troubleshooting.

MALFUNCTION
TEST OR INSPECTION
CORRECTIVE ACTION

1. ENGINE FAILS TO CRANK.

Step 1. Test for defective DEAD CRANK switch (paragraph 2-122.1).

a. If DEAD CRANK switch is not defective, do step 5.

b. b. If defective, replace DEAD CRANK switch (paragraph 2-122).

Step 2. Check for loose or corroded battery cable terminals or battery posts.

a. a. If terminals are tight and posts are clean, do step 3.

b. b. If not clean and tight, clean and tighten battery cable terminals and posts (paragraph 2-12.4).

Step 3. Check that batteries are in stalled correctly (paragraph 2-12.5).

a. If batteries are installed correctly, do step 4.

b. b. If not incorrectly installed, inst all batteries correctly (paragraph 2-12.5).

Step 4. Test for low or no battery charge (paragraph 2-12.1).

a. a. If fully charged, do step 5.

b. b. If not fully charged, replace batteries (paragraph 2-12).

Step 5. Test for defective DC CONTROL POWER circuit breaker (paragraph 2-50.2).

a. If DC CONTROL POWER circuit breaker is not defective, do step 6.

b. b. If defective, replace DC CONTROL POWER circuit breaker (paragraph 2-50).

Step 6. Test for battery voltage at input of MASTER SWITCH.

a. a. If battery voltage is present, do step 7.

b. b. If battery voltage is not present, do step 8.

Step 7. Test MASTER SWITCH output voltage in START position.

a. a. If battery voltage is present at MASTER SWITCH output terminal (7) and GND, do step
11.

b. b. If battery voltage is not present at output terminal, test MASTER SWITCH (paragraph 2-32.3).

Step 8. Test for defective battery charging ammeter shunt (paragraph 2-54.2).

a. If battery charging ammeter shunt is not defective, do step 9.

b. b. If defective, replace battery charging ammeter shunt (paragraph 2-54).

Step 9. Test for defective reverse battery diode (CR1) (paragraph 2-66.2).

a. a. If reverse battery diode is not defective, do step 10.

b. b. If defective, replace reverse battery diode (paragraph 2-66).

Step 10. Test for defective EMERGENCY STOP switch assembly (paragraph 2-40.2).

a. a. If EMERGENCY STOP switch assembly is not defective, do step 11.

b. b. If defective, replace EMERGENCY ST OP switch assembly (paragraph 2-40).

Table 2-6. Unit Troubleshooting – Continued.

MALFUNCTION TEST OR INSPECTION CORRECTIVE ACTION
Step 11. Test for defective start relay (K15) (paragraph 2-60.2). a. If start relay is not defective, do step 12. b. b. If defective, replace start relay (paragraph 2-60). Step 12. Test for defective cranking relay (K2) (paragraph 2-113.2). a. a. If cranking relay is not defective, do step 13. b. b. If defective, replace cranking relay (paragraph 2-113). Step 13. Test for defective crank disconnect relay (K16) (paragraph 2-60.2). a. a. If crank disconnect relay is not defective, do step 14. b. b. If defective, replace crank disconnect relay (paragraph 2-60). Step 14. Check starting circuit for breaks or loose connections. a. a. If starting circuit has no brea ks and connections are tight, do step 5. b. b. If defective, repair or replace defective wires or connections. Step 15. Test for defective starter solenoid (TM 9-2815-254-24 (MEP-804A/MEP-814A)) or test for defective starter (TM 9-2815-538-24&P (MEP-804B/MEP-814B)). a. a. If starter solenoid is not defective, do step 16. b. b. If defective, replace starter solenoid (TM 9-2815-254-24 (MEP-804A/MEP-814A)) or replace starter (TM 9-2815-538-24&P (MEP-804B/MEP-814B)). Step 16. Test for defective starting motor (TM 9-2815-254-24 (MEP-804A/MEP-814A)/TM 9-2815-538-24&P (MEP-804B/MEP-814B)). a. a. If defective, replace starting motor (TM 9-2815-254-24 (MEP-804A/MEP-814A)/TM 9-2815- 538-24&P (MEP-804B/MEP-814B)). b. b. If starting motor is not defective, notify next higher level of maintenance.
2. ENGINE CRANKS BUT FAILS TO START. Step 1. Test for low battery output (paragraph 2-12.1). a. If batteries are fully charged, do step 2. b. b. If not fully charged, replace batteries (paragraph 2-12). Step 2. Check for clogged or defective fuel filter/water separator. a. a. If fuel filter/water separator is not clogged or defective, do step 3. b. b. If clogged or defective, service fuel filter/water separator (paragraph 2-105.2). Step 3. Check for blocked fuel line(s) and/or components. a. a. If fuel lines are not blocked, do step 4. b. b. If blocked, unblock or replace fuel line(s) and/or components (paragraph 2-97/para- graph 2-

Table 2-6. Unit Troubleshooting – Continued.

MALFUNCTION TEST OR INSPECTION CORRECTIVE ACTION
Step 4. Check for air in fuel lines. a. If air in fuel lines, bleed fuel lines (TM 9-2815-254-24 (MEP-804A/MEP-814A)/paragraph 2-96 (MEP-804B/MEP-814B)). b. If no air in fuel lines, do step 5. Step 5. Check for contaminated or incorrect grade of fuel. a. If fuel is not contaminated and is correct grade, do step 6. b. If contaminated or incorrect grade, drain engine fuel system and service fuel filter/water separator (paragraph 2-105.2). Drain generator set tank. Service generator set fuel tank with clean fuel of a proper grade (Table 2-2). Prime engine fuel system (TM 9-2815-254-24 (MEP-804A/MEP-814A)) or move MASTER SWITCH to PRIME & RUN position for 1 minute prior to starting engine (MEP-804B/MEP-814B). Step 6. Check for defective fuel feed pump (TM 9-2815-254-24 (MEP-804A/MEP-814A)) or fuel transfer pump (paragraph 2-106 (MEP-804B/MEP-814B)). a. If defective, replace fuel feed pump (TM 9-2815-254-24 (MEP-804A/MEP-814A)/paragraph 2-106 (MEP-804B/MEP-814B)). b. If fuel feed pump is not defective, notify next higher level of maintenance.
3. ENGINE STOPS SUDDENLY. Step 1. Check for tripped protective devices. a. If no malfunction indicator lights are illuminated, do step 2. b. If illuminated, correct fault indicated. Step 2. Check for air lock in fuel supply line. a. If no air lock in fuel line, do step 3. b. If air in line, bleed fuel lines (TM 9-2815-254-24 (MEP-804A/MEP-814A)/paragraph 2-96 (MEP-804B/MEP-814B)). Step 3. Check for blocked fuel line(s) and/or components, starting at injection pump inlet line. a. If blocked, unblock or replace fuel line(s) and/or components (paragraph 2-97/paragraph 2-98). b. If fuel lines and/or components are not blocked, notify next higher level of maintenance.

Step 1. Check for obstruction in fuel line(s).

a. If fuel line(s) are not obstructed, do step 2.

b. b. If obstructed, unblock or replace fuel line(s) (paragraph 2-97).

Step 2. Check for blocked exhaust pipe or muffler.

a. a. If exhaust pipe and muffler are not blocked, do step 3.

b. b. If blocked, unblock exhaust pipe/muffler or replace exhaust pipe/muffler (paragraph

2-71

Table 2-6. Unit Troubleshooting – Continued.

MALFUNCTION
TEST OR INSPECTION
CORRECTIVE ACTION

Step 3. Check for contaminated or incorrect grade of fuel.

a. If contaminated or incorrect grade, drain engine fuel system and service fuel filter/water separator (paragraph 2-105.2). Drain generator set tank. Service generator set fuel tank with clean fuel of a proper grade (Table 2-2). Prime engine fuel system (TM 9-2815-254-24 (MEP-804A/MEP-814A)) or move MASTER SWITCH to PRIME & RUN position for 1 minute prior to starting engine (MEP-804B/MEP-814B).

b. If fuel is not contaminated or incorrect grade, notify next higher level of maintenance.

5. ENGINE MISFIRING.

Step 1. Check for contaminated or incorrect grade of fuel.

a. If fuel is not contaminated and is correct grade, do step 2. b. If contaminated or incorrect grade, drain engine fuel system and service fuel filter/water separator (paragraph 2-105.2). Drain generator set tank. Service generator set tank with clean fuel of a proper grade (Table 2-2). Prime engine fuel system (TM 9-2815-254-24 (MEP-804A/MEP-814A)) or move MASTER SWITCH to PRIME & RUN position for 1 minute prior to starting engine (MEP-804B/MEP-814B).

Step 2. Check for air in fuel lines.

a. If air in lines, bleed fuel lines (TM 9-2815-254-24 (MEP-804A/MEP-814A)/paragraph 2-96 (MEP-804B/MEP-814B)).

b. If no air in fuel lines, notify next higher level of maintenance.

6. ENGINE DOES NOT DEVELOP FULL POWER.

Step 1. Check for restricted fuel filter/water separator.

a. If fuel filter/water separator is not restricted, do step 2.

b. b. If restricted, service fuel filter/water separator (paragraph 2-105.2).

Step 2. Check for contaminated or incorrect grade of fuel.

a. a. If fuel is not contaminated and is correct grade, do step 3.

b. b. If contaminated or incorrect grade, drain engine fuel system and service fuel filter/water separator (paragraph 2-105.2). Drain generator set tank. Service generator set tank with clean fuel of a proper grade (Table 2-2). Prime engine fuel system (TM 9-2815-254-24 (MEP-804A/MEP-814A)) or move MASTER SWITCH to PRIME & RUN position for 1 minute prior to starting engine (MEP-804B/MEP-814B).

Step 3. Check for blocked air intake system.

a. If air intake system is not blocked, do step 4.

b. b. If blocked, unblock or replace air intake system components as required.

Step 4. Check blocked exhaust pipe or muffler.

a. a. If blocked, unblock or replace exhaust pipe/muffler (paragraph 2-71 (MEP-804A/MEP-814A)/paragraph 2-72 (MEP-804B/MEP-814B)).

b. b. If exhaust pipe and muffler are not blocked, notify next higher level of maintenance.

Table 2-6. Unit Troubleshooting – Continued.

MALFUNCTION
TEST OR INSPECTION
CORRECTIVE ACTION

7. ABNORMAL ENGINE NOISE.

Check for contaminated or incorrect grade of fuel.

a. If contaminated or incorrect grade, drain engine fuel system and service fuel filter/water separator (paragraph 2-105.2). Drain generator set tank. Service generator set fuel tank with clean fuel of a proper grade (Table 2-2). Prime engine fuel system (TM 9-2815-254-24 (MEP-804A/MEP-814A)) or move MASTER SWITCH to PRIME & RUN position for 1 minute prior to starting engine (MEP-804B/MEP-814B).

b. If fuel is not contaminated and is correct grade, notify next higher level of maintenance.

8. BLACK OR GREY SMOKE IN EXHAUST.

Check for improper grade of fuel.

a. If improper grade of fuel, drain engine fuel system. Drain generator set tank. Service generator set fuel tank with clean fuel of a proper grade (Table 2-2). Prime engine fuel system (TM 9-2815-254-24 (MEP-804A/MEP-814A)) or move MASTER SWITCH to PRIME & RUN position for 1 minute prior to starting engine (MEP-804B/MEP-814B).

b. If proper grade of fuel, notify next higher level of maintenance.

9. BLUE OR WHITE EXHAUST SMOKE.

Step 1. Check for excessive engine oil level.

a. a. If proper oil level, do step 2.

b. b. If excessive oil level, drain to proper level (paragraph 2-1.2.4).

Step 2. Check for improper grade of fuel.

a. a. If proper grade of fuel, do step 3.

b. b. If improper grade, drain engine fuel system and generator set tank. Service generator set fuel tank with clean fuel of a proper grade (Table 2-2). Prime engine fuel system (TM 9-2815-254-24 (MEP-804A/MEP-814A)) or move MASTER SWITCH to PRIME & RUN position for 1 minute prior to starting engine (MEP-804B/MEP-814B).

Step 3. Check for defective thermostat (T M 9-2815-254-24 (MEP-804A/MEP-814A)/TM 9-2815-538-
24&P

(MEP-804B/MEP-814B)).

a. a. If defective, replace thermostat (TM 9-2815-254-24 (MEP-804A/MEP-814A)/TM 9-2815-
538-

24&P (MEP-804B/MEP-814B)).

b. b. If thermostat is not defective, notify next higher level of maintenance.

2. HIGH OIL CONSUMPTION.

Step 1. Check for blocked air intake system.

a. a. If air intake system is not blocked, do step 2.

Table 2-6. Unit Troubleshooting – Continued.

MALFUNCTION TEST OR INSPECTION CORRECTIVE ACTION
Step 2. Check for improper lube oil type. a. If improper, drain oil and refill with proper lube oil type (Table 2-3 and paragraph 2-1.2.4). Replace oil filter (TM 9-2815-254-24 (MEP-804A/MEP-814A)/TM 9-2815-538-24&P (MEP-804B/MEP-814B)). b. If proper lube oil type, notify next higher level of maintenance.
11. LOW OIL PRESSURE. Step 1. Check for improper lube oil type. a. If proper lube oil type, do step 2. b. If improper, drain oil and refill with proper lube oil type (Table 2-3 and paragraph 2-1.2.4). Replace oil filter (TM 9-2815-254-24 (MEP-804A/MEP-814A)/TM 9-2815-538-24&P (MEP-804B/MEP-814B)). Step 2. Test for defective OIL PRESSURE indicator (paragraph 2-25.2). a. If OIL PRESSURE indicator is not defective, do step 3. b. If defective, replace OIL PRESSURE indicator (paragraph 2-25). Step 3. Test for defective oil pressure sender (paragraph 2-118.1). a. If defective, replace oil pressure sender (paragraph 2-118). b. If oil pressure sender is not defec tive, notify next higher level of maintenance.
12. HIGH OIL PRESSURE. Step 1. Check for improper lube oil type. a. If proper lube oil type, do step 2. b. If improper, drain oil and refill with proper lube oil type (Table 2-3 and paragraph 2-1.2.4). Replace oil filter (TM 9-2815-254-24 (MEP-804A/MEP-814A)/TM 9-2815-538-24&P (MEP-804B/MEP-814B)). Step 2. Test for defective OIL PRESSURE indicator (paragraph 2-25.2). a. If OIL PRESSURE indicator is not defective, do step 3. b. If defective, replace OIL PRESSURE indicator (paragraph 2-25). Step 3. Test for defective oil pressure sender (paragraph 2-118.1). a. If defective, replace oil pressure sender (paragraph 2-118). b. If oil pressure sender is not defec tive, notify next higher level of maintenance.

(MEP-804B/MEP-814B)).

a. a. If fan belt is not broken and tension is correct, do step 2.

b. b. If worn or broken, replace fan belt (paragraph 2-93 (MEP-804A/MEP-814A)/paragraph 2-94 (MEP-804B/MEP-814B)). If loose, adjust fan belt (paragraph 2-93.2 (MEP-804A/MEP-814A)/paragraph 2-94.2 (MEP-804B/MEP-814B)).

Table 2-6. Unit Troubleshooting – Continued.

MALFUNCTION
TEST OR INSPECTION
CORRECTIVE ACTION

Step 2. Check for defective radiator cap (paragraph 2-79.1).

a. If radiator cap is not defective, do step 3.

b. b. If defective, replace radiator cap (paragraph 2-79).

Step 3. Check for defective coolant hose(s).

a. a. If coolant hoses are not leaking or collapsed, do step 4.

b. b. If defective, replace coolant hose(s) (paragraphs 2-80 thru 2-88).

Step 4. Check for defective thermostat (TM 9-2815-254-24 (MEP-804A/MEP-814A/TM 9-2815-538-24&P (MEP-804B/MEP-814B)).

a. a. If thermostat is not defective, do step 5.

b. b. If defective, replace thermostat (TM 9-2815-254-24 (MEP-804A/MEP-814A)/TM 9-2815-538-24&P (MEP-804B/MEP-814B)).

Step 5. Check for clogged radiator (paragraph 2-79.2).

a. a. If radiator is not clogged, do step 6.

b. b. If clogged, remove obstruction or replace radiator (paragraph 2-89 (MEP-804A/MEP-
814A)/
paragraph 2-90 (MEP-804B/MEP-814B)).

Step 6. Check for defective water pump (TM 9-2815-254-24 (MEP-804A/MEP-814A)/TM 9-2815-538-24&P (MEP-804B/MEP-814B)).

a. a. If defective, replace water pump (TM 9-2815-254-24 (MEP-804A/MEP-814A)/TM
9-2815-
538-24&P (MEP-804B/MEP-814B)).

b. b. If water pump is not defective, notify next higher level of maintenance.

14. ENGINE COOLANT TEMPERATURE TOO LOW.

Step 1. Check for defective thermostat (TM 9-2815-254-24 (MEP-804A/MEP-814A)/TM 9-2815-538-24&P (MEP-804B/MEP-814B)).

a. a. If thermostat is operating correctly, do step 2.

b. b. If defective, replace thermostat (TM 9-2815-254-24 (MEP-804A/MEP-814A)/TM 9-2815-538-24&P (MEP-804B/MEP-814B)).

Step 2. Test for defective temper ature sender (paragraph 2-119.1).

a. If temperature sender is not defective, do step 3.

b. b. If defective, replace tem perature sender (paragraph 2-119).

Step 3. Test for defective TEMPERATURE indicator (paragraph 2-24.2).

a. a. If defective, replace TEMPERATURE indicator (paragraph 2-24).

Table 2-6. Unit Troubleshooting – Continued.

MALFUNCTION
TEST OR INSPECTION
CORRECTIVE ACTION

15. EXCESSIVE FUEL CONSUMPTION.

Step 1. Check for blocked air intake system.

a. If air intake system is not blocked, do step 2. b. If blocked, unblock or replace air intake system components as required (paragraph 2-74 (MEP-804A/MEP-814A)/paragraph 2-75 (MEP-804B/MEP-814B)).

Step 2. Check for leaks in fuel system.

a. If no leaks in fuel system, do step 3. b. If leaks are found, repair fuel system as required.

Step 3. Check for contaminated oil.

a. If oil is contaminated, change oil (paragraph 2-1.2.4). b. If oil is not contaminated or trouble persists, notify next higher level of maintenance.

16. COOLANT IN CRANKCASE OR OIL IN COOLANT.

Check for coolant or oil contamination.

If contaminated, notify next higher level of maintenance.

17. ENGINE VIBRATING.

Step 1. Check for bent or broken cooling fan blades.

a. If fan blades are not damaged, do step 2. b. If damaged, replace cooling fan (paragraph 2-91 (MEP-804A/MEP-814A)/paragraph 2-92 (MEP-804B/MEP-814B)).

Step 2. Check for loose or defective engine mounts.

a. Tighten loose mounting bolts. b. If bolts are tight or problem persists, notify next higher level of maintenance.

18. ENGINE FAILS TO START IN COLD WEATHER.

Step 1. Test for low or no battery charge (paragraph 2-12.1).

a. If batteries are fully charged, do step 2.

b. b. If not fully charged, replace batteries (paragraph 2-12).

Step 2. Check for improper lube oil type.

a. a. If proper lube oil type, do step 3.

b. b. If improper, drain oil and refill with proper lube oil type (Table 2-3). Replace oil filter

(TM 9-

Table 2-6. Unit Troubleshooting – Continued.

MALFUNCTION
TEST OR INSPECTION
CORRECTIVE ACTION

Step 3. Check for improper grade of fuel.

a. a. If proper grade of fuel, do step 4.

b. b. If improper, drain engine fuel system. Drain generator set tank. Service generator set fuel tank with clean fuel of a proper grade (Table 2-2). Prime engine fuel system (TM 9-2815-254-24 (MEP-804A/MEP-814A)) or move MASTER SWITCH to PRIME & RUN position for 1 minute prior to starting engine (MEP-804B/MEP-814B).

Step 4. MEP-804A/MEP-814A: Check for defective glow plug DC contactor (paragraph 2-123.2).

a. a. If glow plug DC contactor is not defective, do step 5.

b. b. If defective, replace glow plug DC contactor (paragraph 2-123).

Step 5. MEP-804A/MEP-814A: Check for defective glow plug(s) (TM 9-2815-254-24).

a. a. If glow plug(s) is not defective, do step 8.

b. b. If defective, replace glow plug(s) (TM 9-2815-254-24).

Step 6. MEP-80BA/MEP-814B: Check for defective contactor K22 (paragraph 2-124.2).

a. a. If contactor K22 is not defective, do step 7.

b. b. If defective, replace contactor K22 (paragraph 2-124).

Step 7. MEP-804B/MEP-814B: Check for defective air inlet preheater (TM 9-2815-538 -24&P).

a. If heater is not defective, do step 8.

b. b. If defective, replace engine inlet preheater element(s) (TM 9-2815-538-24&P).

Step 8. Test for defective MASTER SWITCH (paragraph 2-32.3).

a. a. If defective, replace MASTER SWITCH (paragraph 2-32).

b. b. If MASTER SWITCH is not defective and problem persists, notify next higher level of maintenance.

~~19. BATTERY CHARGE AMMETER SHOWS NO CHARGE WHEN BATTERIES ARE LOW OR DISCHARGED.~~

~~Step 1. Check for broken or loose fan belt (paragraph 2-93.1 (MEP-804A/MEP-814A)/paragraph 2-94.1~~ (MEP-804B/MEP-814B)).

a. a. If fan belt is not broken and tension is correct, do step 2.

b. b. If worn or broken, replace fan belt (paragraph 2-93 (MEP-804A/MEP-814A)/paragraph

2-94

(MEP-804B/MEP-814B)). If loose, adjust fan belt (paragraph 2-93.2 (MEP-804A/MEP-814A)/paragraph 2-94.2 (MEP-804B/MEP-814B)).

Step 2. Check for defective BATTERY CHARGER FUSE (paragraph 2-48.1 (MEP-804A/MEP-814A))/ BATTERY CHARGER CIRCUIT BREAKER (paragraph 2-49.1 (MEP-804B/MEP-814B)).

a. a. If BATTERY CHARGER FUSE/BATTERY CHARGER CIRCUIT BREAKER is not blown/defective, do step 3.

b. b. If defective, replace BATTERY CHARGER FUSE (paragraph 2-48 (MEP-804A/MEP-814A))/BATTERY CHARGER CIRCUIT BREAKER (paragraph 2-49 (MEP-804B/MEP-814B)).

Table 2-6. Unit Troubleshooting – Continued.

MALFUNCTION TEST OR INSPECTION CORRECTIVE ACTION
Step 3. Test for defective battery charging ammeter shunt (MT4) (paragraph 2-54.2). a. If battery charging ammeter shunt is not defective, do step 4. b. If defective, replace battery charging ammeter shunt (paragraph 2-54). Step 4. Test for defective battery charging alternator (TM 9-2815-254-24 (MEP-804A/MEP-814A)/ TM 9-2815-538-24&P (MEP-804B/MEP-814B)). a. If battery charging alternator is not defective, do step 5. b. If defective, replace battery charging alternator (TM 9-2815-254-24 (MEP-804A/ MEP-814A)/TM 9-2815-538-24&P (MEP-804B/MEP-814B)). Step 5. Test for defective BATTERY CHARGE AMMETER (paragraph 2-27.2). a. If BATTERY CHARGE AMMETER is not defective, do step 6. b. If defective, replace BATTERY CHARGE AMMETER (paragraph 2-27). Step 6. Check for breaks or loose connections in charging circuit. If breaks or loose connections are found, repair charging circuit. Refer to Electrical Schematic FO-1 (MEP-804A/MEP-814A)/FO-3 (MEP-804B/MEP-814B).
20. BATTERY CHARGE AMMETER SHOWS EXCESSIVE CHARGING AFTER PROLONGED OPERATION. Step 1. Test for defective batteries (paragraph 2-12.1). a. If batteries are not defective, do step 2. b. If defective, replace batteries (paragraph 2-12). Step 2. Test for defective BATTERY CHARGE AMMETER (paragraph 2-27.2). a. If BATTERY CHARGE AMMETER is not defective, do step 3. b. If defective, replace BATTERY CHARGE AMMETER (paragraph 2-27). Step 3. Test for defective battery charging alternator (TM 9-2815-254-24 (MEP-804A/MEP-814A)/ TM 9-2815-538-24&P (MEP-804B/MEP-814B)). a. If battery charging alternator is not defective, do step 4. b. If defective, replace battery charging alternator (TM 9-2815-254-24 (MEP-804A/ MEP-814A)/TM 9-2815-538-24&P (MEP-804B/MEP-814B)). Step 4. Check for short in charging circuit. If shorted, repair charging circuit. Refer to Electrical Schematic FO-1 (MEP-804A/MEP-814A/FO-3 (MEP-804B/MEP-814B).

21. AC VOLTMETER (VOLTS AC) DOES NOT INDICATE VOLTAGE.

Step 1. Test for defective AC voltmeter (VOLTS AC) (paragraph 2-31.2).

a. If AC voltmeter (VOLTS AC) is not defective, do step 2.

b. b. If defective, replace AC vo ltmeter (VOLTS AC) (paragraph 2-31).

Step 2. Test for defective VOLTAGE adjust potentiometer (paragraph 2-33.3).

a. a. If VOLTAGE adjust potentiometer is not defective, do step 3.

b. b. If defective, replace VOLTAGE adjust potentiometer (paragraph

Table 2-6. Unit Troubleshooting – Continued.

MALFUNCTION
TEST OR INSPECTION
CORRECTIVE ACTION

Step 3. Test for defective VM-AM transfer switch (paragraph 2-38.3).

a. a. If defective, replace VM-AM transfer switch (paragraph 2-38).

b. b. If VM-AM transfer switch is not defective, notify next higher level of maintenance.

1. AC VOLTMETER (VOLTS AC) INDICATES VOLTAGE, BUT FREQUENCY METER (HERTZ) IS OFF SCALE.

Step 1. Test for defective frequency transd ucer (paragraphs 2-52.2 and 2-52.3).

a. a. If frequency transducer is not defective, do step 2.

b. b. If defective, replace freque ncy transducer (paragraph 2-52).

Step 2. Test for defective FREQUENCY meter (HERTZ) (paragraphs 2-28.2 and 2-28.3).

a. a. If defective, replace FREQUENCY meter (HERTZ) (paragraph 2-28).

b. b. If FREQUENCY meter (HERTZ) is not defective, notify next higher level of maintenance.

1. AC VOLTMETER (VOLTS AC) VOLTAGE FLUCTUATES.

Step 1. Check for loose electrical connections. Refer to Electrical Schematic FO-1 MEP-804A/MEP-814A)/FO-3 (MEP-804B/MEP-814B).

a. a. If no loose connections, do step 2.

b. b. If loose, tighten electrical connections.

Step 2. Test for defective AC voltmeter (VOLTS AC) (paragraph 2-31.2).

a. If AC voltmeter (VOLTS AC) is not defective, do step 3.

b. b. If defective, replace AC voltmeter (VOLTS AC) (paragraph 2-31).

Step 3. Test for defective VOLTAGE adjust potentiometer (paragraph 2-33.3).

a. a. If VOLTAGE adjust potentiometer is not defective, do step 4.

b. b. If defective, replace VOLTAGE adjust potentiometer (paragraph 2-33).

Step 4. Test for defective VM-AM transfer switch (paragraph 2-38.3).

a. a. If VM-AM transfer switch is not defective, notify next higher level of maintenance.

b. b. If defective, replace VM-AM transfer switch (paragraph 2-38).

Step 5. Check for defective load measuring unit (paragraph 2-65.1).

a. a. If load measuring unit is not defective and trouble persists, notify next higher level of maintenance.

b. b. If defective, replace load measuring unit (paragraph 2-65).

---

1. FREQUENCY METER (HERTZ) FREQUENCY FLUCTUATES.

Step 1. Check for erratic engine operation. Refer to Malfunction 4, Engine Runs Erratically or Stalls Frequently. If engine is operating properly, do step 2.

Table 2-6. Unit Troubleshooting – Continued.

MALFUNCTION TEST OR INSPECTION CORRECTIVE ACTION
Step 2. Test for defective frequency transducer (paragraphs 2-52.2 and 2-52.3). a. If frequency transducer is not defective, do step 3. b. If defective, replace frequency transducer (paragraph 2-52). Step 3. Test for defective FREQUENCY meter (HERTZ) (paragraphs 2-28.2 and 2-28.3). If defective, replace FREQUENCY meter (HERTZ) (paragraph 2-28).
25. NO VOLTAGE AT CONVENIENCE RECEPTACLE. Step 1. Open control panel and inspect circuit interrupter on side of GROUND FAULT CIRCUIT INTERRUPTER. a. a. If device is tripped, reset device for generator sets, contract number DAAK01-94-D-0036 and contract number DAAK02-92-D-0034. b. b. For generator sets contract number DAAK01-88-D-0082, check in-line fuse on black lead of GROUND FAULT CIRCUIT INTERRUPTER. c. c. If device is not tripped, do step 2. Step 2. Check GROUND FAULT CIRCUIT INTERRUPTER. a. a. If indicator is tripped, reset by pressing RESET button. b. b. If Indicator is not tripped, do step 3. Step 3. Check voltage across CONVENIENCE RECEPTACLE. a. a. If voltage is present, replace CONVENIENCE RECEPTACLE. b. b. If voltage is not present, do step 4. Step 4. Check voltage across terminals 4 and 6 on TB-5. a. a. If voltage is present, replace GROUND FAULT CIRCUIT INTERRUPTER. b. b. If voltage is not present, search for loose or broken wires or loose pin in connectors.

Section VI. RADIO INTERFERENCE SUPPRESSION

2-7 GENERAL METHODS USED TO ATTAIN PROPER SUPPRESSION

Suppression is attained by providing a low resistance path to ground for stray currents. The methods used include shielding the ignition and high-frequency wires, grounding the frame with bonding straps, and using filtering systems.

2-8 INTERFERENCE SUPPRESSION COMPONENTS

2-8.1 Primary Suppression Components. Primary suppression components are those whose primary function is to suppress electromagnetic interference.

The primary suppression components on this generator set are the output box access door EMI seal (outer seal) (paragraph 2-18); the load output terminal board EMI filters (paragraph 2-114); and the voltage re-connection terminal board capacitors (paragraph 2-107). On MEP-804B/MEP-814B only, there are also grounding straps for access doors (paragraph 2-14), a control box panel opening EMI seal paragraph 2-68.2), and additional harness shielding.

2-8.2 Secondary Suppression Components. Secondary suppression components have electromagnetic interference suppression functions which are incidental or secondary to their primary function. The only secondary suppression component for the generator set is the housing. Refer to Section IX, Maintenance of Housing, for removal and installation procedures.

Section VII. SPECIAL INSTRUCTIONS

2-9 NUCLEAR, BIOLOGICAL, CHEMICAL (NBC) CONTAMINATION

The generator set is capable of being operated by personnel wearing nuclear, biological, or chemical (NBC) protective clothing without special tools or support equipment. Refer to FM 3-5, NBC decontamination, for information on decontamination procedures. The following are specific procedures for the generator set:

2-9.1 Control panel indicators, sealing gaskets, rubber sleeves, rope draw cords at output terminal access ports, control panel door gaskets, access door gaskets, rubber tubing and belts within the engine compartment, coverings for electrical conduits, external water drain tubing, and retaining cords for slave receptacle covers will absorb and retain chemical agents. Replacement of these items is the recommended method of decontamination.

2-9.2 Lubricants, fuel, coolant, or battery fluid may be present on the external surfaces of the generator set or components due to leaks or normal operation. These fluids will absorb NBC agents. The preferred method of decontamination is removal of these fluids using conventional decontamination methods in accordance with FM 3-5.

2-9.3 Continued decontamination of external generator set surfaces with super tropical bleach (STB)/ decontaminating solution number 2 (DS2) will degrade clear plastic indicator coverings to a point where reading indicators will become impossible. This problem will become more evident for soldiers wearing protective masks. Therefore, the use of STB or DS2 decontaminants in these areas should be minimized. Indicators should be decontaminated with warm soapy water.

2-9.4 External surfaces of the control panel assembly that are marked with painted or stamped lettering will not withstand repeated decontamination with STB or DS2 without degradation of this lettering. The recommended method of decontamination for these areas is warm soapy water.

2-9.5 Areas that will entrap contaminants, making efficient decontamination extremely difficult, include the following: space behind knobs and switches on the control panel, exposed heads of screws, areas adjacent to and behind exposed wiring conduits, hinged areas of access doors, spaces behind externally mounted equipment specification data plates, areas around external oil drain valve, retaining chains for external receptacle covers, areas behind external receptacle covers, access door locking mechanisms, recessed wells for access door handles, fuel caps, load output terminal board access door, slave receptacles, frequency adjustment controls, areas around tiedown/lifting rings, crevices around access doors, external screens covering ventilation areas, and areas adjacent to the external fuel drain valve. Replacement of these items, if available, is the preferred method of decontamination. Conventional decontamination methods should be used on these areas, while stressing the importance of thoroughness and the probability of some degree of continuing contact and vapor hazard.

2-9.6 In an NBC contaminated environment, the generator set should be operated with all access doors closed to reduce the effects of contamination.

2-9.7 The use of overhead shelters or chemical protective covers is recommended as an additional means of protection against contamination in accordance with FM 3-5. When using covers, care should be taken to provide adequate space for air flow and exhaust.

2-9.8 For additional NBC information, refer to FM 3-3 and 3-4. Other services use applicable publications for NBC.

Section VIII. MAINTENANCE OF DC ELECTRICAL SYSTEM

2-10 INTRODUCTION

This section contains unit maintenance procedures for the DC Electrical System. Deficiencies noted during inspection/repair which are beyond the scope of unit maintenance shall be reported to the next higher level of maintenance.

NOTE

Refer to TM 9-6115-643-10 for all operator procedures.

2-11 BATTERY AND SLAVE RECEPTACLE CABLES

WARNING

Slave receptacle (NATO connector) is electrically live at all times and is unfused. The Battery Disconnect Switch does not remove power from the slave receptacle. NATO slave receptacle has 24 VDC even when Battery Disconnect Switch is set to OFF. This circuit is only dead when the batteries are fully disconnected. Disconnect the batteries before performing maintenance on the slave receptacle. Failure to comply with this warning can cause injury or death to personnel.

WARNING

All metal jewelry can conduct electricity and become entangled in generator set components. Remove all jewelry when working on generator set. Failure to comply with this warning can cause injury or death to personnel.

WARNING

When disconnecting or removing batteries, disconnect the negative lead that connects directly to the grounding stud first; disconnect the negative end of the interconnection cable next. When installing batteries, reverse the connection sequence. Failure to comply with this warning can cause injury to personnel.

WARNING

Dangerous voltage exists on live circuits. Always observe precautions and never work alone. Failure to comply with this warning can cause injury or death to personnel.

NOTE

This procedure is typical for the positive, negative, and interconnect battery cables, and the positive and negative NATO slave receptacle cables.

2-11.1 Inspection.

a. Shut down generator set.

b. b. Open battery access door and left side engine access door.

c. c. Inspect battery/slave receptacle cables for security; cracked insulation; broken, burned, or corroded terminals; missing parts; or other damage. Close battery access door.

2-11.2 Removal.

a. a. Shut down generator set.

b. b. Open battery access door and left side engine access door.

NOTE

Tag all cables before removal.

c. c. Disconnect negative battery cable terminal lug (1, Figure 2-3) from battery.

d. d. Disconnect and remove applicable cable assembly (Figure 2-3/Figure 2-4).

2-11.3 Cleaning.

a. a. Remove terminal cover(s) (2, Figure 2-3) from battery post(s), if applicable.

b. b. Clean battery post(s) and cable terminals with battery terminal cleaner.

c. c. Install terminal cover(s) (2), if removed.

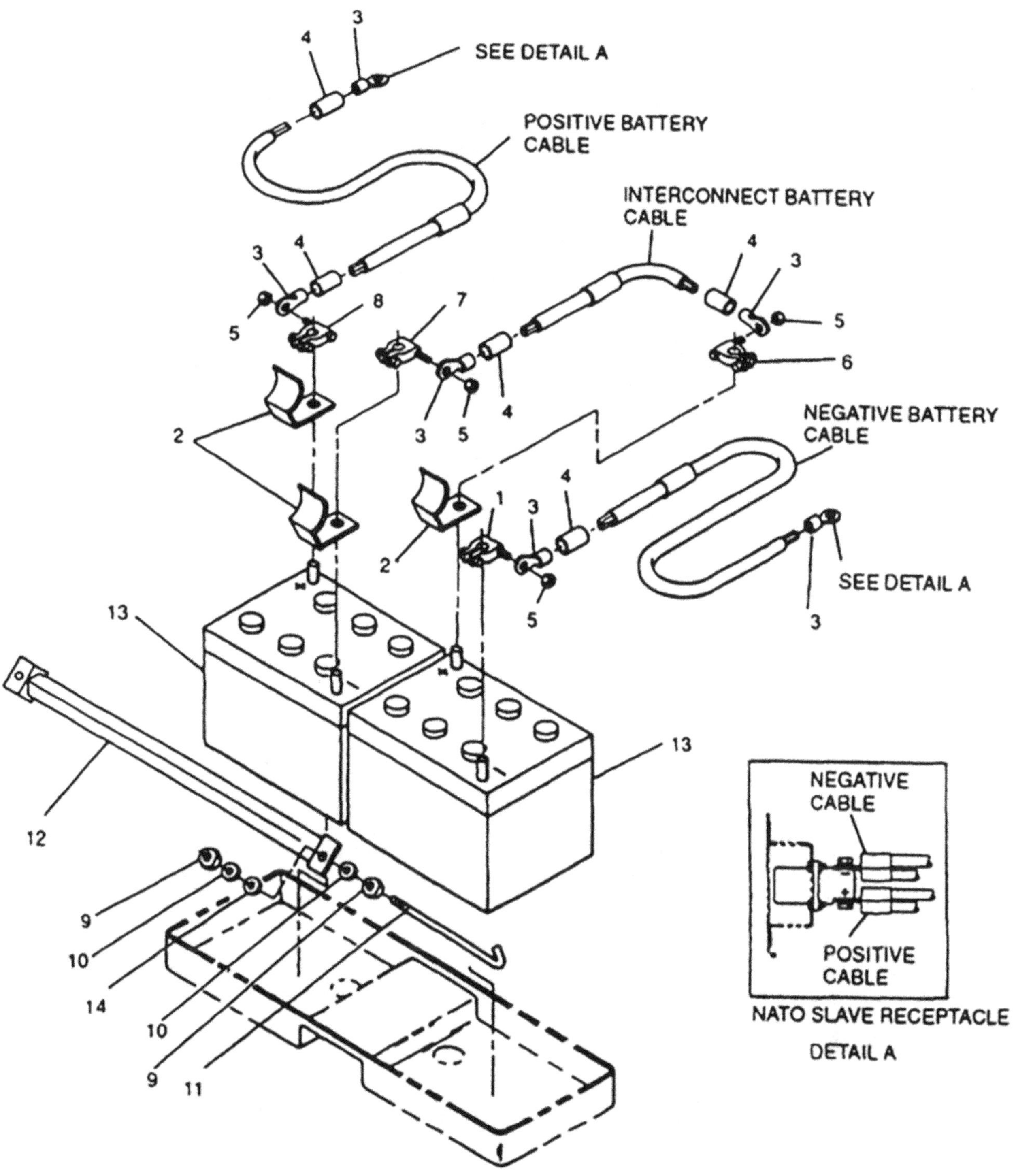

Figure 2-3. Batteries and Cables.

2-11.4 Repair.

WARNING

The connection of any electrical equipment and the disconnection of any electrical equipment may cause an explosion hazard. Do not connect any electrical equipment or disconnect any electrical equipment in an explosive atmosphere. Failure to comply with this warning can cause injury or death to personnel.

NOTE

If cable cannot be repaired, refer to Appendix D.

a. a. Remove nut(s) (5, Figure 2-3) and battery cable terminal lug(s), as necessary.

b. b. Remove broken or damaged terminal (3, Figur e 2-3/3, Figure 2-4) from cable assembly.

c. c. Slide new shrinkable tubing (4, Figure 2-3/4, Figure 2-4) over cable end.

d. d. Install terminal (3) to cable end (Appendix D).

e. e. Heat shrinkable tubing (4) with heat gun until secure.

f. f. Install battery cable terminal lug(s) with nuts(s) (5, Figure 2-3), as necessary.

g. g. Connect negative battery cable terminal lug (1, Figure 2-3) to battery. Close battery ac- cess door.

2-12 BATTERIES

WARNING

Batteries give off a flammable gas. Do not smoke or use open flame when performing maintenance. Failure to comply with this warning can cause inju- ry or death to personnel, and damage to the generator set.

a. a. Shut down generator set.

b. b. Open battery access door and disconnect negative battery cable.

c. c. Remove battery vent caps.

d. d. Test specific gravity of electrolyte in each battery cell with a hydrometer. Refer to Table 2-7 for state of charge with specific gravity corrected to 80□F (27□C). Refer to Table 2-8 for specific gravity temperature corrections.

e. e. Recharge or replace batteries, as nece ssary. Retest batteries per above instructions.

f. f. Install battery vent caps.

g. g. Connect negative battery c able and close battery access door.

Table 2-7. State of Charge with Specific Gravity Correct to 80°F (27°C).

Specific Gravity	Percent Charge
1.280	100
1.250	75
1.220	50
1.190	25
1.160	Little useful capacity
1.130	Discharged

Table 2-8. Specific Gravity Temperature Corrections.

Temperature □F	Correction Factor
+120□F (49□C)	+0.016
+115□F (46□C)	+0.014
+110□F (43□C)	+0.012
+105□F (41□C)	+0.010
+100□F (38□C)	+0.008
+95□F (35□C)	+0.006
+90□F (32□C)	+0.004
+85□F (29□C)	+0.002
+80□F (27□C)	0
+75□F (24□C)	-0.002
+70□F (21□C)	-0.004
+65□F (19□C)	-0.006
+60□F (16□C)	-0.008
+55□F (13□C)	-0.010
+50□F (10□C)	-0.012
+45□F (7□C)	-0.014
+40□F (5□C)	-0.016
+35□F (2□C)	-0.018
+30□F (-1□C)	-0.020
+25□F (-4□C)	-0.022
+20□F (-7□C)	-0.024
+15□F (-9□C)	-0.026
+10□F (-12□C)	-0.028
+5□F (-15□C)	-0.030
0□F (-18□C)	-0.032
-5□F (-20□C)	-0.034
-10□F (-23□C)	-0.036
-15□F (-26□C)	-0.038
-20□F (-29□C)	-0.040

2-12.2 Removal.

WARNING

Batteries give off a flammable gas. Do not smoke or use open flame when performing maintenance. Failure to comply with this warning can cause injury or death to personnel, and damage to the generator set.

WARNING

When disconnecting or removing batteries, disconnect the negative lead that connects directly to the grounding stud first; disconnect the negative end of the interconnection cable next. When installing batteries, reverse the connection sequence. Failure to comply with this warning can cause injury to personnel.

WARNING

Each battery weighs more than 70 pounds (32 kg) and requires a two-person lift. Lifting batteries can cause back strain. Ensure proper lifting techniques are used when lifting batteries. Failure to comply with this warning can cause injury to personnel.

WARNING

Many components require a two-person lift. Lifting heavy components can cause back strain. Ensure proper lifting techniques are used when lifting heavy components. Failure to comply with this warning can cause injury to personnel.

a. a. Shut down generator set.

b. b. Open battery compartment access door and disconnect negative battery cable terminal lug (1, Figure 2-3).

c. c. Disconnect interconnect battery cable terminals (6 and 7) and remove interconnect battery

cable

assembly. Disconnect positive battery cable terminal lug (8).

d. d. Remove nuts (9), washers (10), lock washers (14), hook bolts (11), and battery hold-down (12). Discard lockwashers (14).

e. e. Remove batteries (13).

2-12.3 Inspection.

f. a. Remove batteries.

b. b. Inspect batteries for cracked cases; broken, burned, or corroded posts; missing parts; and

other

damage.

c. c. Install batteries.

2-12.4 Service.

b. b. Clean cable terminal lugs and battery posts.

c. c. Install terminal covers (2) on batteries (13).

2-12.5 Installation.

WARNING

High voltage is produced when the generator set is in operation. Never attempt to start the generator set unless it is properly grounded. Failure to comply with this warning can cause injury or death to personnel.

a. a. Position batteries (13, Figure 2-3) in generator set. Ensure that batteries are serviced and fully charged.

b. b. Apply general purpose grease (Item 9, Appen dix C) to battery posts and cable terminal lugs (1, 6, 7, and 8).

c. c. Install hook bolts (11) and battery hold-down (12) with new lock washers (14), washers (10), and nuts (9).

WARNING

When disconnecting or removing batteries, disconnect the negative lead that connects directly to the grounding stud first; disconnect the negative end of the interconnection cable next. When installing batteries, reverse the connection sequence. Failure to comply with this warning can cause injury to personnel.

d. d. Connect positive cable terminal lug (8).

e. e. Position interconnect battery cable connect terminals (6 and 7).

f. f. Connect negative battery cable termina l lug (1). Close battery access door.

2-13 NATO SLAVE RECEPTACLE

WARNING

Slave receptacle (NATO connector) is electrically live at all times and is unfused. The Battery Disconnect Switch does not remove power from the slave receptacle. NATO slave receptacle has 24 VDC even when Battery Disconnect Switch is set to OFF. This circuit is only dead when the batteries are fully disconnected. Disconnect the batteries before performing maintenance on the slave receptacle. Failure to comply with this warning can cause injury or death to personnel.

2-13.1 Inspection.

a. a. Shut down generator set.

b. b. Open left side engine access door.

c. c. Inspect NATO slave receptacle for security, corrosion, missing hardware, and other damage. Close engine access door.

2-13.2 Removal.

a. a. Shut down generator set.

b. b. Open battery access door and disconnect negative battery cable.

c. c. Open left side engine access door and tag and disconnect battery and slave receptacle cables from slave receptacle (7, Figure 2-4) by removing bolts (1) and lock washers (2). Discard lock washers (2).

d. d. Remove bolts (5), nuts (6), and NATO slave receptacle (7) from generator set housing.

e. e. Remove nut (8), bolt (9), and cover (10) from generator set.

2-13.3 Installation.

a. a. Insert NATO slave receptacle (7, Figure 2-4) into left side lower door sill.

b. b. Secure receptacle (7) with bolts (5) and nuts (6).

c. c. Connect slave receptacle and battery cables to slave receptacle (7) with bolts (1) and new lock washers (2). Remove tags.

d. d. Install cover (10) on generator set with bolt (9) and nut (8).

e. e. Connect negative battery cable. Close battery access door and engine access door.

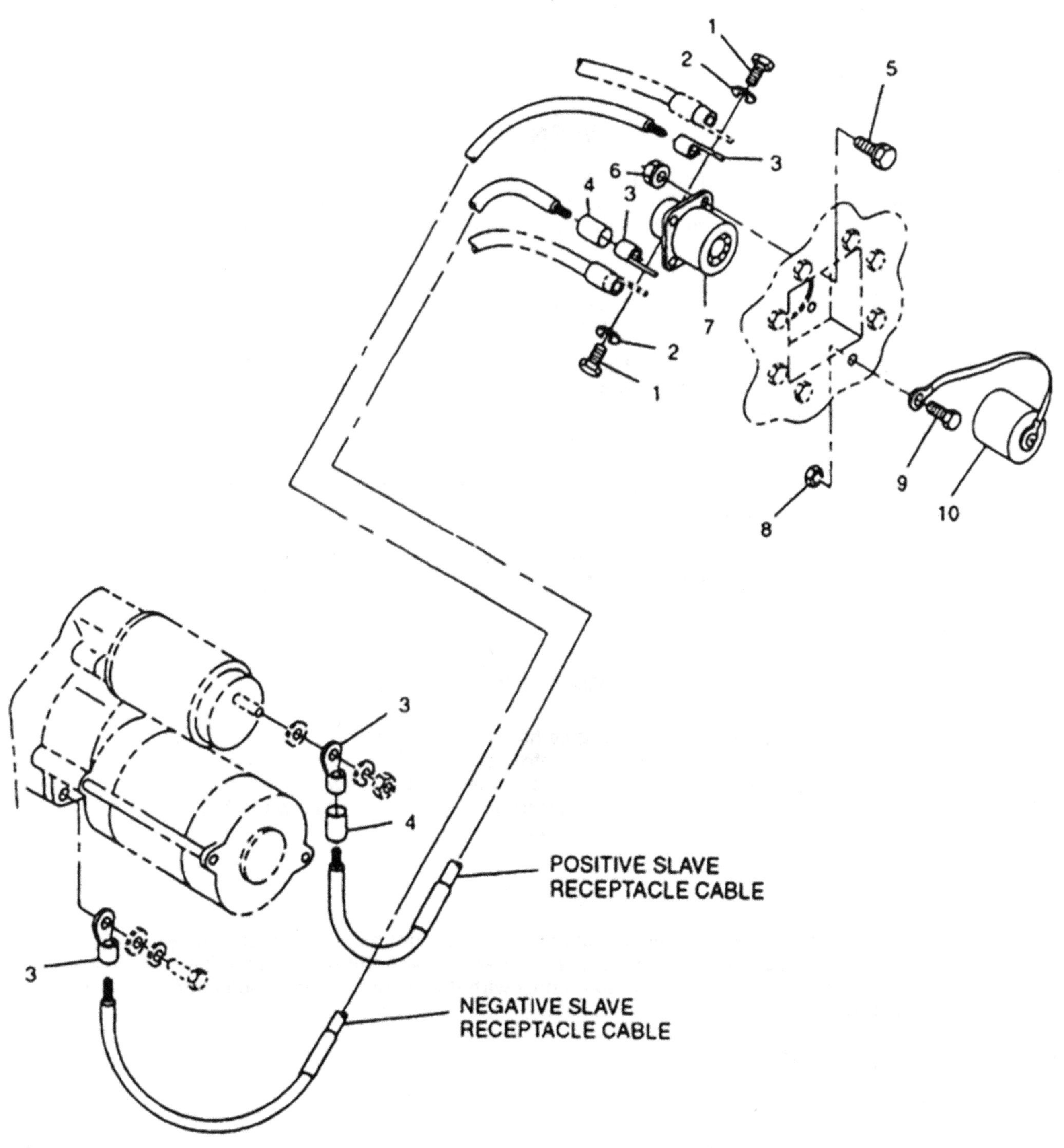

Figure 2-4. NATO Slave Receptacle and Cables.

Section IX. MAINTENANCE OF HOUSING

2-14 ACCESS DOORS

WARNING

All metal jewelry can conduct electricity and become entangled in generator set components. Remove all jewelry when working on generator set. Failure to comply with this warning can cause injury or death to personnel.

WARNING

DO NOT wear loose clothing when performing checks, services and maintenance. Failure to comply with this warning can cause injury or death to personnel.

NOTE

This procedure is written for the left engine access door, but is typical for all access doors, hinges, latches, and data plates.

When removing and installing battery access door, note position of spacers for door hold open mechanism.

2-14.1 Removal.

WARNING

When running, generator set engine has hot metal surfaces that will burn flesh on contact. Shut down generator set, and allow engine to cool before performing checks, services, and maintenance. Wear gloves and additional protective clothing as required. Failure to comply with this warning can cause injury or death to personnel.

WARNING

Wear heat resistant gloves and avoid contacting hot metal surfaces with your hands after components have been heated. Wear additional protective clothing as required. Failure to comply with this warning can cause injury to personnel.

a. a. Shut down generator set.

b. b. Open left side engine access door.

c. c. MEP-804A/MEP-814A: Remove nuts (1, Figure 2-5), lock washers (2), bolts (3), washers (4), and hinges (5). Discard lock washers (2).

MEP-804B/MEP-814B: Remove nuts (1, Figure 2-5), lock washers (2), ground straps (2.1), bolts (3), washers (4), and hinges (5). Discard lock washers (2).

d. d. Remove left side engine access door (6) from generator set.

e. e. Remove assembled nuts (7), bolts (8), and document box (9).

f. f. Remove assembled nuts (10), bolts (11), upper air baffle (12), and lower air baffle (13) from engine access door (6).

g. g. Remove assembled nuts (14) and bolts (15) to separate air baffles (12 and 13).

h. h. Remove assembled nuts (16), bolts (17), bracke t (18), and holding rod (19) from engine access door (6).

i. i. Remove assembled nuts (20), bolts (21), and bracket (22) from engine access door (6).

j. j. Remove assembled nuts (23), screws (24), and latches (25) from engine access door (6).

k. k. Drill out rivets (26) and remove fuel system diagram plate (27) from document box (9).

2-14.2 Inspection.

a. a. Shut down generator set.

b. b. Inspect access doors, hinges, latches, and baffles for loose and missing hardware, cracks, dents, loose paint, and corrosion.

c. Inspect data plates for readability and loose or missing rivets.

2-14.3 Repair.

WARNING

CARC paint is a health hazard, and is irritating to eyes, skin, and respiratory system. Wear protective eyewear, mask, and gloves when applying or removing CARC paint. Failure to comply with this warning can cause injury to personnel.

WARNING

Solvent used to clean parts is potentially dangerous to personnel and property. Clean parts in a well-ventilated area. Avoid inhalation of solvent fumes. Wear goggles and rubber gloves to protect eyes and skin. Wash exposed skin thoroughly. Do not smoke or use near open flame or excessive heat. Failure to comply with this warning can cause injury to personnel, and damage to the equipment.

a. a. Repair all dents and cracks and remove loose paint.

b. b. Remove light corrosion with fine grit abrasive paper (Item 15, Appendix C).

c. c. Repaint surfaces in accordance with TM 43-0139. (F) Use applicable directives.

d. d. Replace unreadable data plates.

e. e. Replace loose or missing rivets.

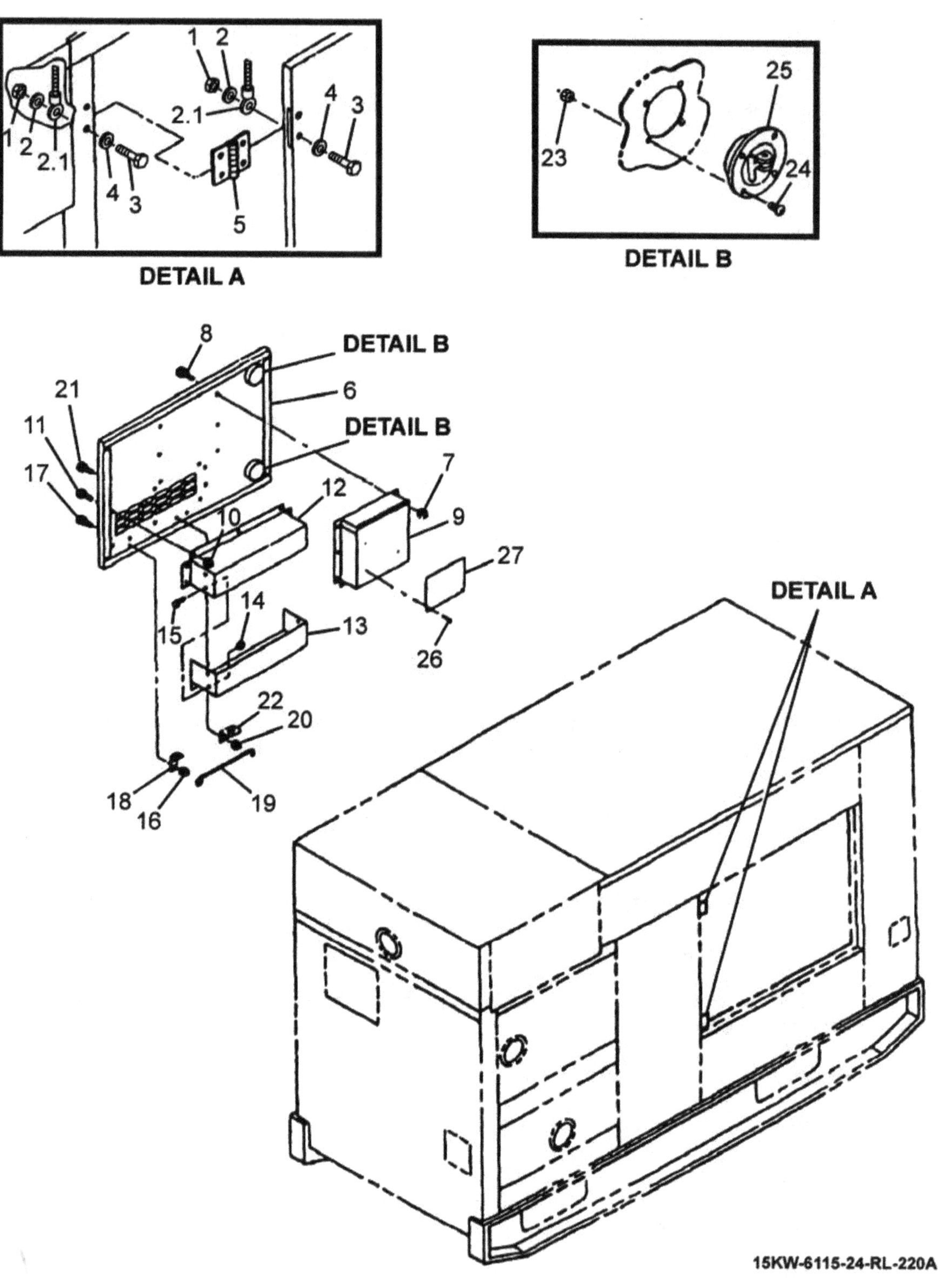

Figure 2-5. Generator Set Access Doors.

2-14.4 **Installation**.

a. Install latches (25, Figure 2-5) in left side engine access door (6) with screws (24) and assembled nuts (23).

b. Install bracket (22) on engine access door (6) with bolts (21) and assembled nuts (20).

c. Install bracket (18) and holding rod (19) on engine access door (6) with bolts (17) and assembled nuts (16).

d. d. Assemble top air baffle (12) to bottom air baffle (13) with assembled nuts (14) and bolts (15).

e. e. Install air baffles (12 and 13) on engine access door (6) with bolts (11) and assembled nuts (10).

f. f. Install fuel system diagram plate (27) on document box (9) with rivets (26).

g. g. Install document box (9) with bolts (8) and assembled nuts (7).

h. h. MEP-804A/MEP-814A: Install right side engine access door (6) and hinges (5) on generator set with bolts (3), washers (4), new lock washers (2), and nuts (1). Close engine access door.

MEP-804B/MEP-814B: Install right side engine access door (6) and hinges (5) on generator set with bolts (3), washers (4), ground straps (2.1), new lock washers (2), and nuts (1). Close engine access door.

2-15 CONTROL BOX TOP PANEL 2-15.1

Removal.

WARNING

All metal jewelry can conduct electricity and become entangled in generator set components. Remove all jewelry when working on generator set. Failure to comply with this warning can cause injury or death to personnel.

WARNING

DO NOT wear loose clothing when performing checks, services and maintenance. Failure to comply with this warning can cause injury or death to personnel.

WARNING

When running, generator set engine has hot metal surfaces that will burn flesh on contact. Shut down generator set and allow engine to cool before performing checks, services, and maintenance. Wear gloves and additional protective clothing as required. Failure to comply with this warning can cause injury or death to personnel.

WARNING

Top housing panels and exhaust system can get very hot. Shut down gen-

services, and maintenance. Failure to comply with this warning can cause severe burns and injury to personnel.

a. a. Shut down generator set.

b. b. Open battery access door and disconnect negative battery cable.

c. c. Remove control panel access door (paragraph 2-14.1).

d. d. Lower control panel and remove bolts (1, Figure 2-6), flat washers (2), lock washers (3), and assembled nuts (4) from top panel (5). Discard lock washers (3).

CAUTION

The control box top panel is attached to the generator set with a silicone sealant to prevent water from entering the control box. Care must be taken not to bend or scratch the control box top panel when separating.

e. e. Separate and remove control box top panel (5).

f. f. Remove gasket (15), if damaged.

g. g. Remove bolts (6), washers (7), lock washers (8), nuts (9), and stiffener (10) from control box assembly. Discard lock washers (8).

h. Remove bolts (11), assembled nuts (12), bracke t (13), and ring (14) from control box top panel (5).

2-15.2 Inspection.

a. a. Shut down generator set.

b. b. Open battery access door and disconnect negative battery cable.

c. c. Inspect control box top panel (5, Figure 2-6) for dents, cracks, loose paint, and corrosion.

2-15.3 Repair.

WARNING

CARC paint is a health hazard, and is irritating to eyes, skin, and respiratory system. Wear protective eyewear, mask, and gloves when applying or removing CARC paint. Failure to comply with this warning can cause injury to personnel.

WARNING

Solvent used to clean parts is potentially dangerous to personnel and property. Clean parts in a well-ventilated area. Avoid inhalation of solvent fumes. Wear goggles and rubber gloves to protect eyes and skin. Wash exposed skin thoroughly. Do not smoke or use near open flame or excessive heat. Failure to comply with this warning can cause injury to personnel, and damage to the equipment.

a. Repair all dents and cracks, and remove all loose paint.

b. b. Remove light corrosion with fine grit abrasive paper (Item 15, Appendix C).

c. c. Repaint surfaces in accordance with TM 43-0139. (F) Refer to applicable directives.

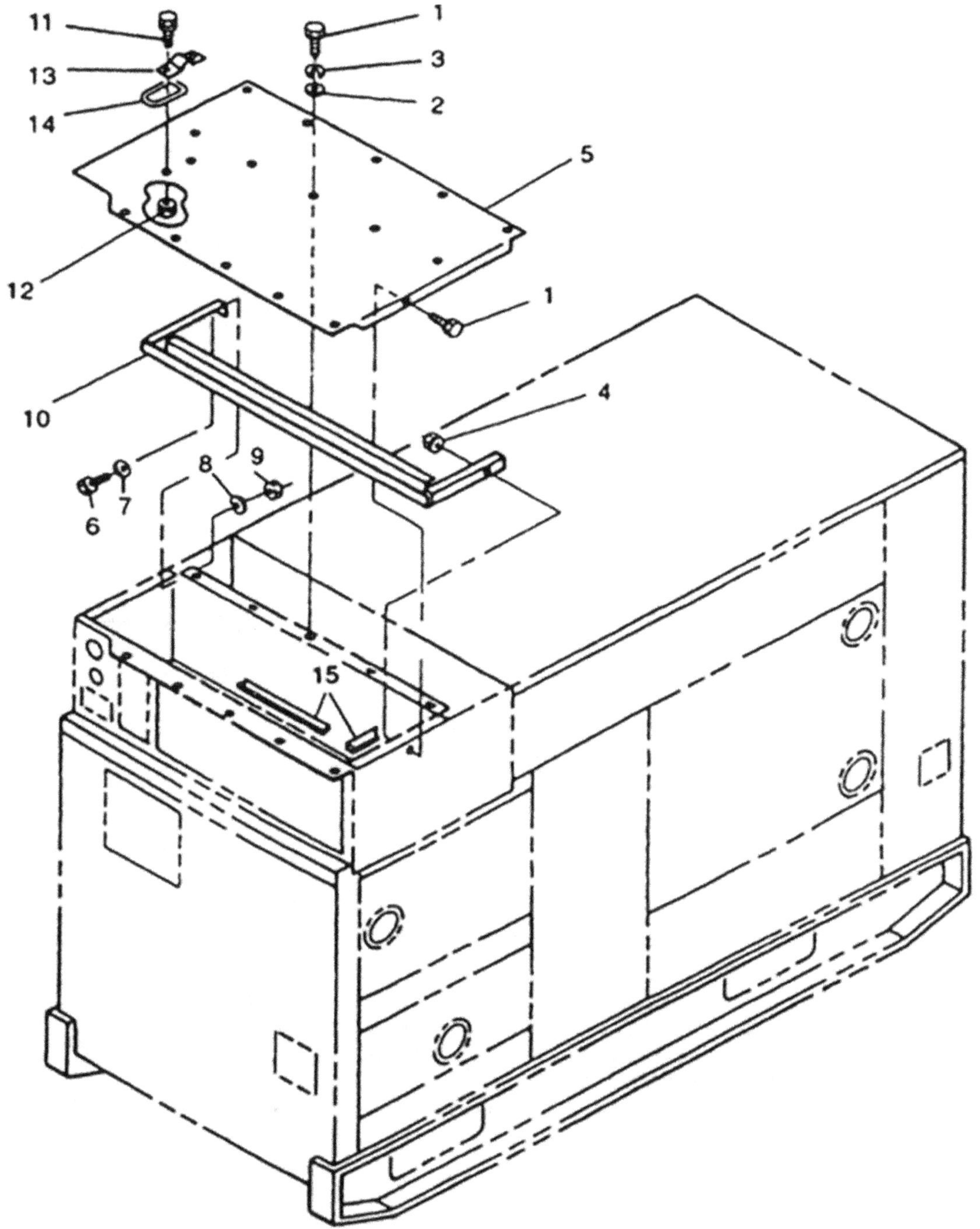

Figure 2-6. Control Box Top Panel.

2-15.4 Installation.

a. a. Install ring (14, Figure 2-6) and bracket (13) on control box top panel (5) with bolts (11) and assembled nuts (12).

b. b. Install stiffener (10) in control box assembly with bolts (6), washers (7), new lock washers (8), and nuts (9).

NOTE

When replacing old sealant with new gasket, ensure all old sealant residue is completely removed.

c. c. Install new gasket (15), if removed.

d. d. Immediately install top panel (5) with bolts (1), flat washers (2), new lock washers (3), and assembled nuts (4).

e. e. Install control panel access door (paragraph 2-14.4).

f. f. Connect negative battery cable. Close battery access door

2-16 TOP HOUSING SECTION 2-16.1

Removal.

WARNING

All metal jewelry can conduct electricity and become entangled in generator set components. Remove all jewelry when working on generator set. Failure to comply with this warning can cause injury or death to personnel.

WARNING

DO NOT wear loose clothing when performing checks, services and maintenance. Failure to comply with this warning can cause injury or death to personnel.

WARNING

When running, generator set engine has hot metal surfaces that will burn flesh on contact. Shut down generator set and allow engine to cool before performing checks, services, and maintenance. Wear gloves and additional protective clothing as required. Failure to comply with this warning can cause injury or death to personnel.

WARNING

Top housing panels and exhaust system can get very hot. Shut down generator set, and allow system to cool before performing checks, services, and maintenance. Failure to comply with this warning can cause severe burns and injury to personnel.

a. Shut down generator set.

b. b. Remove control box assembly (paragraph 2-19.2).

c. c. Remove bolts (1, Figure 2-7), flat washers (3), lock washers (2), mount (4), and exhaust cover (5) from top housing panel (6). Discard lock washers (2).

d. d. Remove bolts (7 and 10), flat washers (9 and 12), lock washers (8 and 11), and top housing panel (6) from generator set. Discard lock washers (8 and 11).

e. e. Disconnect radiator fill hose and overflow hose from radiator fill panel (13) and remove radiator fill panel (13) from generator set.

f. f. Remove bolts (14), flat washers (16), lock washer s (15), and frame (17) from generator set. Discard lock washers (15)

g. g. Remove muffler (paragraph 2-71.2 (MEP-804A/MEP-814A)/paragraph 2-72.2 (MEP-804B/MEP-814B)).

h. h. Remove bolts (18), assembled nuts (19), and air duct channels (20 and 21).

i. i. Remove bolts (22 and 24), assembled nuts (23 and 25), and panel (26).

j. j. Remove bolts (27 and 31), flat washers (30), lock washers (29), nuts (28), assembled nuts (32), and top side panels (33 and 34) with top section floor panel (35). Discard lock washers (29).

k. k. Remove bolts (36), assembled nuts (37) , and panel (38) from floor panel (35).

l. l. Remove bolts (39), assembled nuts (4 0), and support (41) from panel (38).

m. m. Remove bolts (42), assembled nuts (43), a nd side panels (34 and 33) from floor panel (35).

n. n. Remove bolt (44), assembled nut (4 5), and bracket (46) from angle (49).

NOTE

Open output box access door to reach assembled nuts.

o. Remove bolts (47), assembled nuts (4 8), and angle (49) from generator set.

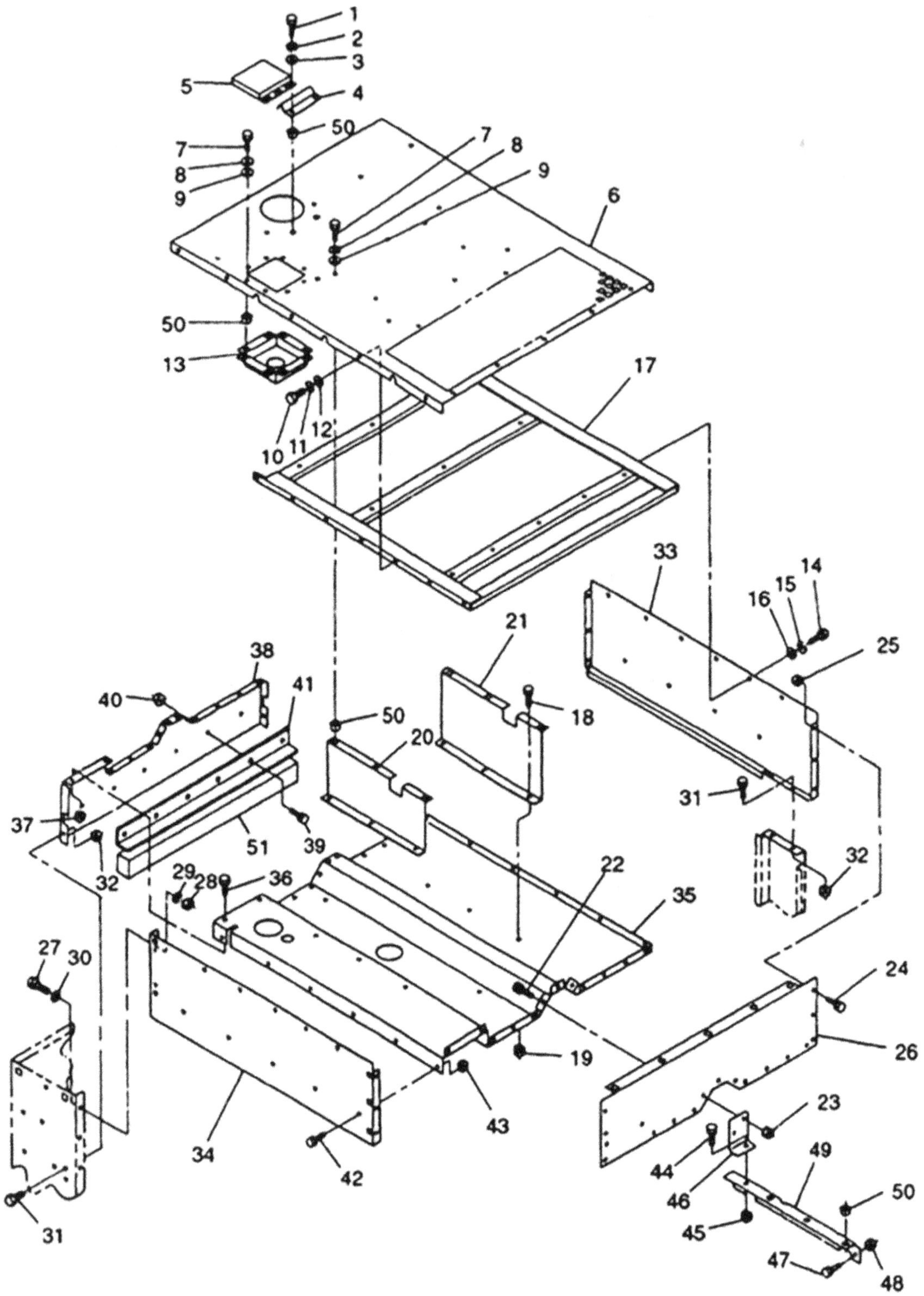

Figure 2-7. Generator Set Top Housing Section.

2-16.2 Inspection.

a. a. Shut down generator set.

b. b. Inspect all top housing section panels for dents, cracks, loose paint, and corrosion.

c. c. Inspect all cage nuts (50, Figure 2-7) for cracking or stripped threads.

d. d. Inspect seal (51) for tears, looseness, and deterioration.

WARNING

CARC paint is a health hazard, and is irritating to eyes, skin, and respiratory system. Wear protective eyewear, mask, and gloves when applying or removing CARC paint. Failure to comply with this warning can cause injury to personnel.

WARNING

Solvent used to clean parts is potentially dangerous to personnel and property. Clean parts in a well-ventilated area. Avoid inhalation of solvent fumes. Wear goggles and rubber gloves to protect eyes and skin. Wash exposed skin thoroughly. Do not smoke or use near open flame or excessive heat. Failure to comply with this warning can cause injury to personnel, and damage to the equipment.

a. a. Repair all dents and cracks, and remove all loose paint.

b. b. Remove light corrosion with fine grit abrasive paper (Item 15, Appendix C).

c. c. Repaint surfaces in accordance with TM 43-0139. (F) Refer to applicable directives.

d. d. Replace any cage nuts (50, Figure 2-7) that are stripped or cracked.

e. e. Replace loose or damaged seal (51).

2-16.4 Installation.

a. a. Install angle (49, Figure 2-7) in generator set with bolts (47) and assembled nuts (48).

b. b. Install bracket (46) to angle (49) with bolt (44) and assembled nut (45).

c. Install top side panels (34) and (33) on top section floor panel (35) with bolts (42) and assembled nuts (43).

d. d. Install support (41) on panel (38) wit h bolts (39) and assembled nuts (40).

e. e. Install panel (38) on top section floor panel (35) with bolts (36) and assembled nuts (37).

f. f. Position top section floor panel (35), with side and front panels attached, in generator set and secure with bolts (27 and 31), flat washers (30), new lock washers (29), nuts (28), and assembled nuts (32).

g. g. Install panel (26) with bolts (22 and 24) and assembled nuts (23 and 25).

h. h. Install air duct channels (20 and 21) with bolts (18) and assembled nuts (19).

i. i. Install muffler (paragraph 2-71.3 (MEP-804A/MEP-814A)/paragraph 2-72.3 (MEP-804B/MEP-814B)).

j. j. Install frame (17) with bolts (14), flat washers (16), and new lock washers (15).

k. k. Install radiator fill panel (13) in generator set and connect radiator fill hose and overflow hose to panel (13).

l. l. Install top housing panel (6) with bolts (7 and 10), flat washers (9 and 12), and new lock washers (8 and 11).

m. m. Install mount (4) and exhaust cover (5) with bolts (1), flat washers (3), and new lock washers
(2).

n. n. Install control box assembly (paragraph 2-19.4).

2-17 FRONT HOUSING SECTION

WARNING

All metal jewelry can conduct electricity and become entangled in generator set components. Remove all jewelry when working on generator set. Failure to comply with this warning can cause injury or death to personnel.

WARNING

DO NOT wear loose clothing when performing checks, services and maintenance. Failure to comply with this warning can cause injury or death to personnel.

WARNING

When running, generator set engine has hot metal surfaces that will burn flesh on contact. Shut down generator set, and allow engine to cool before performing checks, services, and maintenance. Wear gloves and additional protective clothing as required. Failure to comply with this warning can cause injury or death to personnel.

WARNING

Top housing panels and exhaust system can get very hot. Shut down generator set, and allow system to cool before performing checks, services, and maintenance. Failure to comply with this warning can cause severe burns and injury to personnel.

a. Shut down generator set.

b. b. Remove battery box access door (paragraph 2-14.1).

c. c. Remove engine access doors (paragraph 2-14.1).

d. d. Remove top housing section (paragraph 2-16.1).

e. e. Remove batteries (paragraph 2-12.2) and slave receptacle (paragraph 2-13.2).

f. f. Remove bolts (1, Figure 2-8) and grou nd rods (2) from brackets (54).

g. g. Remove bolts (3, 6, and 10), flat washers (5, 7, and 11), lock washers (4, 8, and 12), nuts (9 and 13), and front panel (14) from generator set. Discard lock washers (4, 8, and 12).

h. h. Remove bolts (15 and 18), flat washers (17 and 19), lock washers (16 and 20), nuts (21), and side panels (22 and 23) from generator set. Discard lock washers (16 and 20).

i. i. Remove clip halves (24) and insulation (25) from front panel (14).

j. j. Remove bolts (26 and 29), washers (28), lo ck washers (27), assembled nuts (30), and air deflector (31) from front panel (14). Discard lock washers (27).

k. k. Remove bolts (32), lock washers (33), washers (34), and panels (35 and 36) from front panel (14). Discard lock washers (33).

l. l. Remove bolts (37), assembled nuts (38), and s upports (39 and 40) from panels (35 and 36).

m. m. Remove bolts (41) and panel (42) from front panel (14).

n. n. Remove bolts (43), assembled nuts (44), and support (45) from panel (42).

o. o. Remove bolts (46), assembled nuts (47), and support channel (48) from front panel (14).

p. p. Remove bolts (49), assembled nuts (50), and slave receptacle box (51) from side panel (22).

q. q. Remove bolts (52), assembled nuts (53), an d brackets (54) from panels (14 and 22).

r. r. Drill out rivets (55, 56, and 57); remove identification plates (58, 59, and 60), if necessary.

2-17.2 Inspection.

a. a. Shut down generator set.

b. b. Inspect all front housing section panels for dents, cracks, loose paint, corrosion, and other damage.

c. c. Inspect all cage nuts (61, Figure 2-8) for cracking or stripped threads.

d. d. Inspect insulation for damage and missing clip halves (24).

e. e. Inspect seals (62 and 63) for tears, looseness, and deterioration.

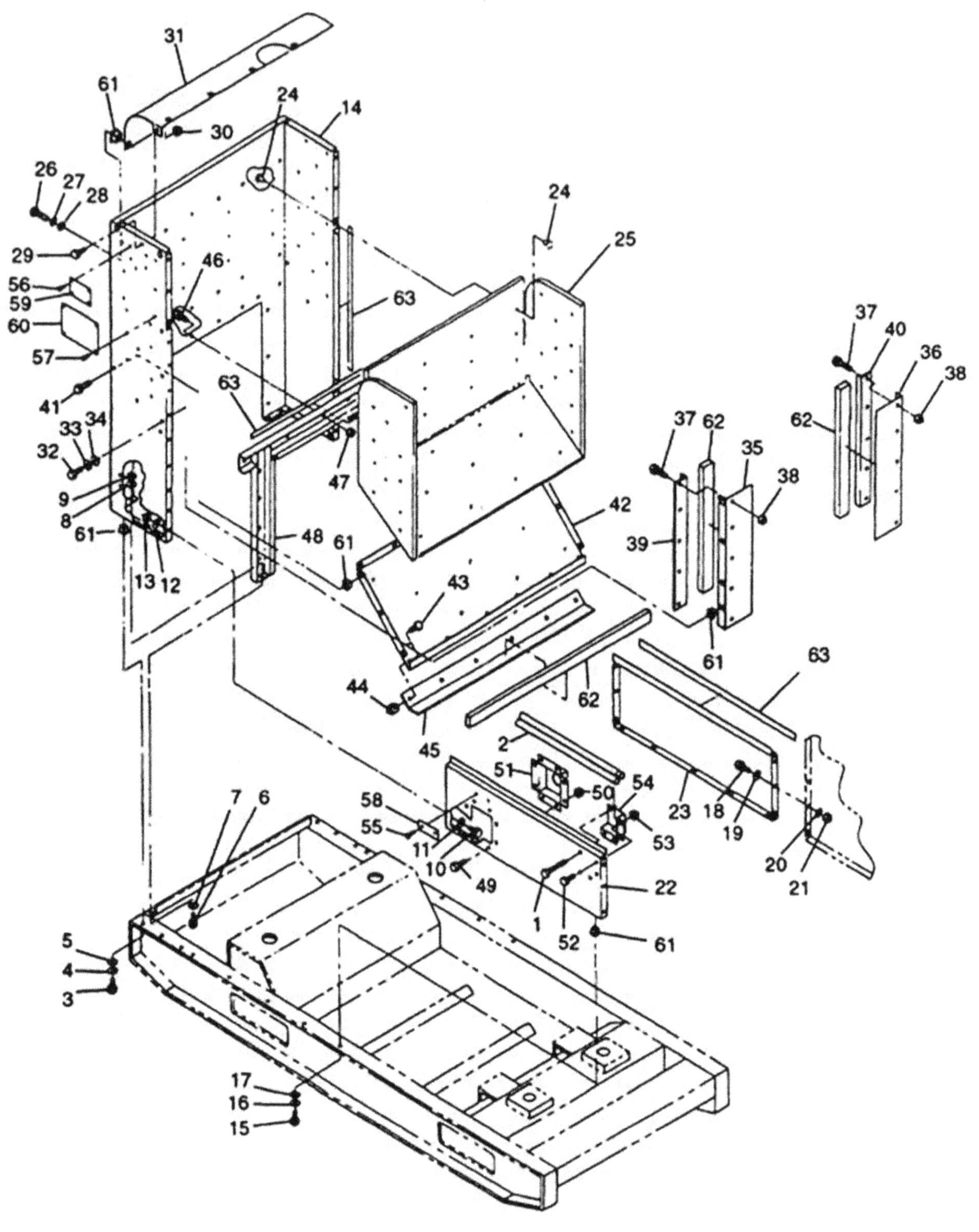

Figure 2-8. Generator Set Front Housing Section.

2-17.3 Repair.

WARNING

CARC paint is a health hazard, and is irritating to eyes, skin, and respiratory system. Wear protective eyewear, mask, and gloves when applying or removing CARC paint. Failure to comply with this warning can cause injury to personnel.

WARNING

Solvent used to clean parts is potentially dangerous to personnel and property. Clean parts in a well-ventilated area. Avoid inhalation of solvent fumes. Wear goggles and rubber gloves to protect eyes and skin. Wash exposed skin thoroughly. Do not smoke or use near open flame or excessive heat. Failure to comply with this warning can cause injury to personnel, and damage to the equipment.

a. a. Repair all dents and cracks and remove ail loose paint.

b. b. Remove light corrosion with fine grit abrasive paper (Item 15, Appendix C).

c. c. Repaint surfaces in accordance with TM 43-0139. (F) Refer to applicable directives.

d. d. Replace any cracked or stripped cage nuts (61, Figure 2-8).

e. e. Replace damaged insulation and missing clip halves (24). Refer to Appendix D for fabrication of insulation.

f. Replace loose or damaged seals (62 and 63).

2-17.4 Installation.

a. a. Install identification plates (58, 59, and 60, Figure 2-8) on front panel (14) and side panel (22) with rivets (55, 56, and 57), if removed.

b. b. Install brackets (54) on panels (14 and 22) with bolts (52) and assembled nuts (53).

c. c. Install slave receptacle box (51) on side panel (22) with bolts (49) and assembled nuts (50).

d. d. Install side panels (22 and 23) on generator set with bolts (15 and 18), flat washers (17 and 19), new lock washers (16 and 20), and nuts (21).

e. e. Install support channel (48) on front panel (14) with bolts (46) and assembled nuts (47).

f. f. Install support (45) on panel (42) wit h bolts (43) and assembled nuts (44).

g. g. Install panel (42) on front panel (14) with bolts (41).

h. h. Install supports (39 and 40) on panels (35 and 36) with bolts (37) and assembled nuts (38).

i. i. Install panels (35 and 36) on front panel (14) with bolts (32), new lock washers (33), and washers (34).

j. Install air deflector (31) on front panel (14) with bolts (26 and 29), washers (28), new lock washers (27), and assembled nuts (30).

k. k. Install insulation (25) on front panel (14) with clip halves (24).

l. l. Install front housing panel (14) on generator set with bolts (3, 6, and 10), flat washers (5, 7, and 11), new lock washers (4, 8, and 12), and nuts (9 and 13).

m. m. Install slave receptacle (paragraph 2-13.3) and batteries (paragraph 2-12.5).

n. n. Install battery box access door and engine access doors (paragraph 2-14.4).

o. o. Install top housing section (paragraph 2-16.4).

p. p. Install ground rods (2) in brackets (54) and secure with bolts (1). Close all access doors.

2-18 REAR HOUSING SECTION

2-18.1 Removal.

WARNING

All metal jewelry can conduct electricity and become entangled in generator set components. Remove all jewelry when working on generator set. Failure to comply with this warning can cause injury or death to personnel.

WARNING

DO NOT wear loose clothing when performing checks, services and maintenance. Failure to comply with this warning can cause injury or death to personnel.

WARNING

When running, generator set engine has hot metal surfaces that will burn flesh on contact. Shut down generator set, and allow engine to cool before performing checks, services, and maintenance. Wear gloves and additional protective clothing as required. Failure to comply with this warning can cause injury or death to personnel.

WARNING

Top housing panels and exhaust system can get very hot. Shut down generator set, and allow system to cool before performing checks, services, and maintenance. Failure to comply with this warning can cause severe burns and injury to personnel.

WARNING

Dangerous voltage exists on live circuits. Always observe precautions and never work alone. Failure to comply with this warning can cause injury or death to personnel.

WARNING

DC voltages are present at generator set electrical components even with generator set shut down. Avoid shorting any positive with ground/negative. Failure to comply with this warning can cause injury to personnel, and damage to equipment.

a. a. Shut down generator set.

b. b. Remove control box assembly (paragraph 2-19.2).

c. Remove air cleaner assembly (paragrap h 2-74.3 (MEP-804A/MEP-814A)/paragraph 2-75.3 (MEP-804B/ MEP-814B)).

d. d. Remove output box access door and load terminal board access door (paragraph 2-14.1).

e. e. Remove engine access doors and air cleaner access door (paragraph 2-14.1).

f. f. Remove fuel tank filler neck (paragraph 2-99.1).

g. g. Remove auxiliary fuel pump (paragraph 2-98.3).

h. h. Remove fuel float module (paragraph 2-104.3).

i. i. Remove top housing panel (paragraph 2-16.1, step d).

j. j. Remove bolts (1 and 4, Figure 2-9), flat washers (3 and 5), lock washers (2 and 6), nuts (7), and rear panel (8) from generator set. Discard lock washers (2 and 6).

k. k. Remove bolts (9), assembled nuts (10), and load cable entrance box (11) from rear panel (8).

l. l. Remove bolts (12 and 16), flat washers (13 and 18), lock washers (14 and 17), nuts (15), and left side rear panel (19) from generator set. Discard lock washers (14 and 17).

m. m. Remove bolts (20), assembled nuts (21), and baffle (22) from left side rear panel (19).

n. n. Remove clip halves (23) and insulation (24) from baffle (22).

o. o. Remove bolts (25), assembled nuts (26), and fuel tank filler neck panel (27) from left side rear panel (19).

NOTE

Ensure output box assembly is secured prior to removal of corner post and door sills.

p. p. Remove bolts (28, 32, and 34), flat washers (29), lock washers (30), nuts (31), assembled nuts (33 and 35), corner post (36), and door sills (37 and 38) from generator set. Discard lock washers (30).

q. q. Remove bolts (39, 43, 47, and 49), flat washers (40 and 44), lock washers (41 and 45), nuts (42 and 46), assembled nuts (48 and 50), and right side panel (51) from generator set. Discard lock washers (41 and 45).

r. r. Drill out rivets (52) and remove plate (53) from left side rear panel (19), if necessary.

s. s. Drill out rivets (54) and remove plate (55) from fuel filler neck panel (27), if necessary.

t. t. Remove output box EMI seals (56) from door sill (37), corner post (36), and right side panel (51), if necessary.

2-18.2 Inspection.

a. a. Shut down generator set.

b. b. Inspect rear housing section panels for dents, cracks, loose paint, corrosion, and other damage.

c. c. Inspect all cage nuts (57, Figure 2-9) for cracking or stripped threads.

d. d. Inspect all insulation for damage and missing clip halves.

e. e. Inspect EMI seals (56) for tears, looseness, and deterioration.

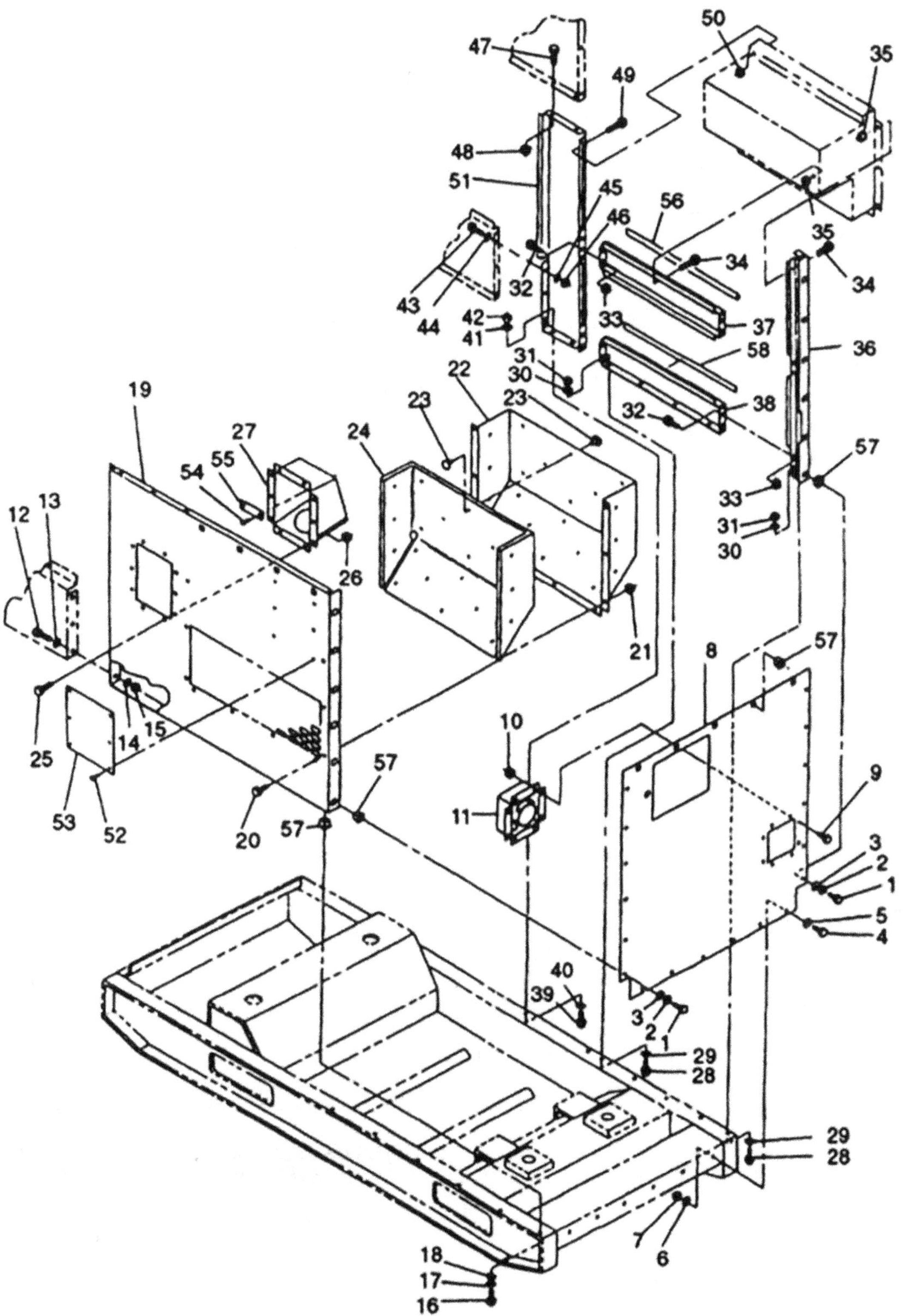

Figure 2-9. Generator Set Rear Housing Section.

2-18.3 Repair.

WARNING

CARC paint is a health hazard, and is irritating to eyes, skin, and respiratory system. Wear protective eyewear, mask, and gloves when applying or removing CARC paint. Failure to comply with this warning can cause injury to personnel.

WARNING

Solvent used to clean parts is potentially dangerous to personnel and property. Clean parts in a well-ventilated area. Avoid inhalation of solvent fumes. Wear goggles and rubber gloves to protect eyes and skin. Wash exposed skin thoroughly. Do not smoke or use near open flame or excessive heat. Failure to comply with this warning can cause injury to personnel, and damage to the equipment.

a. a. Repair all dents and cracks, and remove all loose paint.

b. b. Remove light corrosion with fine grit abrasive paper (Item 15, Appendix C).

c. c. Repaint surfaces in accordance with TM 43-0139. (F) Use applicable directives.

d. d. Replace all cracked or stripped cage nuts (57, Figure 2-9).

e. e. Replace damaged insulation and missing clip halves (23). Refer to Appendix D for fabrication
of
insulation.

f. f. Replace EMI seals (56) that are loose or show any evidence of damage.

g. g. Replace damaged door seals (58).

2-18.4 Installation.

a. a. Install plate (55, Figure 2-9) on fuel tank filler neck panel (27) with rivets (54), if removed.

b. b. Install plate (53) on left side rear panel (19) with rivets (52), if removed.

c. c. Install right side panel (51) on generator set with bolts (39, 43, 47, and 49), flat washers (40
and
44), new lock washers (41 and 45), nuts (42 and 46), and assembled nuts (48 and 50).

d. d. Install door sills (37 and 38) and corner post (36) on generator set with bolts (28, 32, and 34), flat washers (29), new lock washers (30), nuts (31), and assembled nuts (33 and 35).

e. e. Install fuel filler neck panel (27) on left side rear panel (19) with bolts (25) and assembled nuts (26).

f. f. Install insulation (24) on baffle (22) with clip halves (23).

g. g. Install baffle (22) on left side rear panel (19) with bolts (20) and assembled nuts (21).

h. h. Install left side rear panel (19) on generator set with bolts (12 and 16), flat washers (13 and

i. i. Install load entrance box (11) on rear panel (8) with bolts (9) and assembled nuts (10).

j. j. Install rear panel (8) on generator set with bolts (1 and 4), flat washers (3 and 5), new lock washers (2 and 6), and nuts (7).

k. k. Install output box EMI seals (56) on door sill (37), corner post (36), and right side panel (51) with adhesive (Item 1, Appendix C), if removed. Ensure closed side of seal faces outward.

l. l. Install top housing panel (paragraph 2-16.4, step l).

m. m. Install fuel float module (paragraph 2-104.4).

n. n. Install auxiliary fuel pump (paragraph 2-98.4).

o. o. Install fuel tank filler neck (paragraph 2-99.3).

p. p. Install air cleaner access door and engine access doors (paragraph 2-14.4).

q. q. Install load terminal board access door and output box access door (paragraph 2-14.4).

NOTE

Output box EMI seals are primary suppression components. Ensure that a complete seal is made between output box door and generator set.

r. r. Install air cleaner assembly (paragraph 2-74.4 (MEP-804A/MEP-814A)/paragraph 2-75.4 (MEP-804B/ MEP-814B)).

s. s. Install control box assembly (paragraph 2-19.4).

Section X. MAINTENANCE OF CONTROL BOX ASSEMBLY

2-19 CONTROL BOX ASSEMBLY 2-19.1

Inspection.

a. a. Shut down generator set.

b. b. Inspect control box assembly for cracks, breaks, corrosion, loose paint, and missing parts.

WARNING

All metal jewelry can conduct electricity and become entangled in generator set components. Remove all jewelry when working on generator set. Failure to comply with this warning can cause injury or death to personnel.

WARNING

DO NOT wear loose clothing when performing checks, services and maintenance. Failure to comply with this warning can cause injury or death to personnel.

WARNING

Dangerous voltage exists on live circuits. Always observe precautions and never work alone. Failure to comply with this warning can cause injury or death to personnel.

WARNING

High voltage is produced when this generator set is in operation. SHUT DOWN generator set and make sure it is free of any power source before attempting any repair or maintenance on the set, or when connecting or disconnecting load cables. Failure to comply with this warning can cause injury or death to personnel.

WARNING

DC voltages are present at generator set electrical components even with generator set shut down. Avoid shorting any positive with ground/negative. Failure to comply with this warning can cause injury to personnel, and damage to equipment.

WARNING

Ensure that the engine cannot be started while maintenance is being performed. (ENGINE CONTROL switch set to OFF/RESET; Battery Disconnect Switch is OFF; DEAD CRANK SWITCH is OFF). Failure to comply with this warning can cause injury or death to personnel.

WARNING

Top housing panels and exhaust system can get very hot. Shut down generator set, and allow system to cool before performing checks, services, and maintenance. Failure to comply with this warning can cause severe burns and injury to personnel.

a. a. Shut down generator set.

b. b. Open battery access door and disconnect negative battery cable.

c. c. Remove control box top panel (paragraph 2-15.1).

d. d. Open output box access door and disconnect harness connectors P5 and P6 from control box.

e. e. MEP-804B/MEP-814B: Open air cleaner access door (below control panel) and disconnect wire harness connector P61 from control box wire harness connector J61.

f. Remove bolts (1 and 5, Figure 2-10), lock washers (2), flat washers (3), nuts (4), and control box assembly (6) from generator set. Discard lock washers (2).

2-19.3 Repair.

Repair control box assembly by replacing damaged terminals, damaged or missing hardware, and damaged or defective components.

2-19.4 Installation.

a. a. Install control box assembly (6, Figure 2-10) on generator set with bolts (1 and 5), flat washers (3), new lock washers (2), and nuts (4).

b. b. MEP-804B/MEP-814B: Open air cleaner access door (below control panel) and connect control box wire harness connector J61 to wire harness connector P61.

c. c. Connect control box harness connectors P5 and P6.

d. d. Install control box top panel (paragraph 2-15.4).

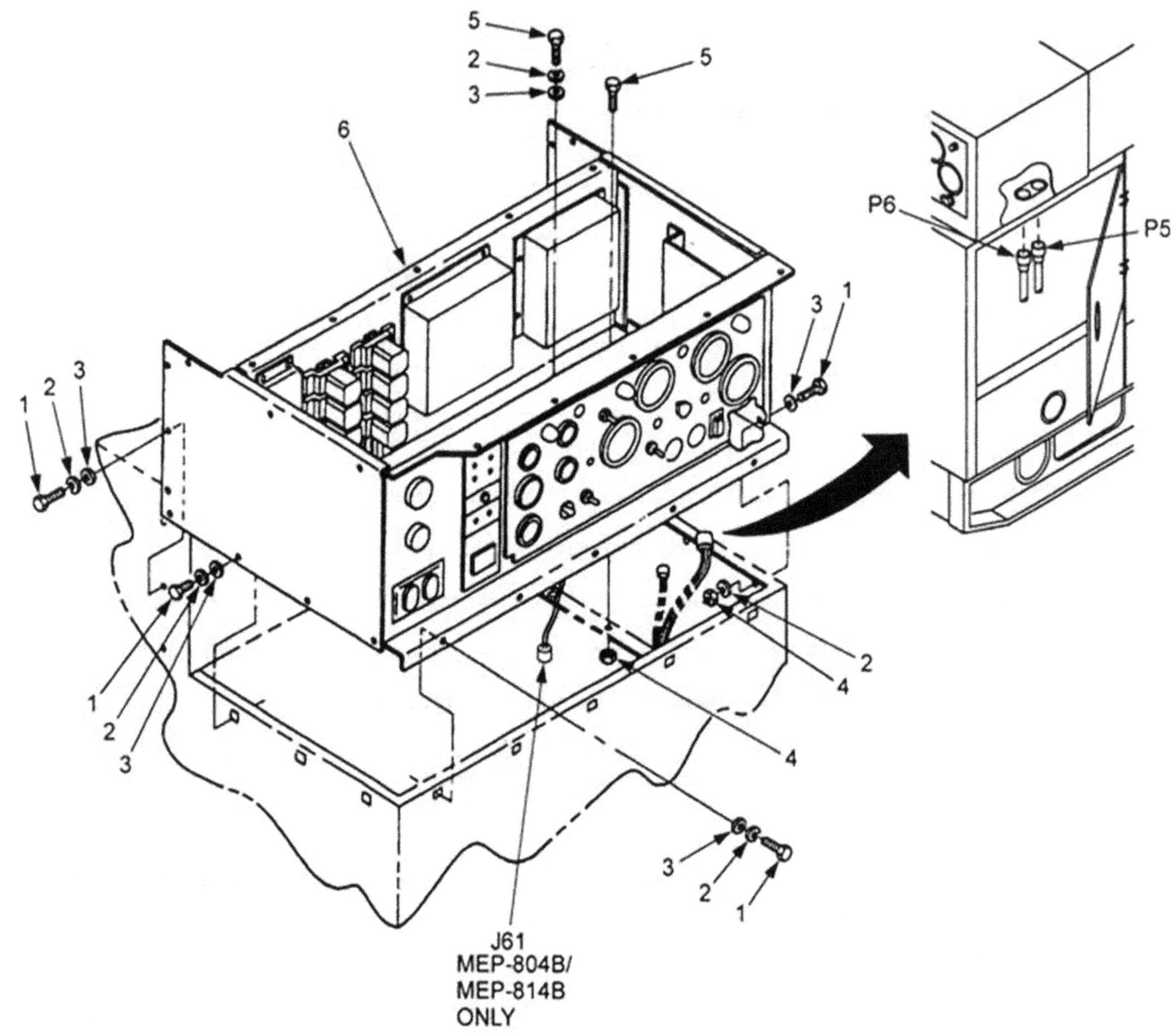

Figure 2-10. Control Box Assembly.

2-20 PANEL LIGHTS 2-20.1

Inspection.

Inspect panel light holder and directional cap for cracks, corrosion, stripped threads, and other damage.

2-20.2 Removal.

WARNING

All metal jewelry can conduct electricity and become entangled in generator set components. Remove all jewelry when working on generator set. Failure to comply with this warning can cause injury or death to personnel.

WARNING

DO NOT wear loose clothing when performing checks, services and maintenance. Failure to comply with this warning can cause injury or death to personnel.

WARNING

When running, generator set engine has hot metal surfaces that will burn flesh on contact. Shut down generator set and allow engine to cool before performing checks, services, and maintenance. Wear gloves and additional protective clothing as required. Failure to comply with this warning can cause injury or death to personnel.

WARNING

Dangerous voltage exists on live circuits. Always observe precautions and never work alone. Failure to comply with this warning can cause injury or death to personnel.

WARNING

High voltage is produced when this generator set is in operation. SHUT DOWN generator set and make sure it is free of any power source before attempting any repair or maintenance on the set, or when connecting or disconnecting load cables. Failure to comply with this warning can cause injury or death to personnel.

WARNING

DC voltages are present at generator set electrical components even with generator set shut down. Avoid shorting any positive with ground/negative. Failure to comply with this warning can cause injury to personnel, and damage to equipment.

WARNING

Ensure that the engine cannot be started while maintenance is being performed. (ENGINE CONTROL switch set to OFF/RESET; Battery Disconnect Switch is OFF; DEAD CRANK SWITCH is OFF). Failure to comply with this warning can cause injury or death to personnel.

WARNING

Top housing panels and exhaust system can get very hot. Shut down generator set, and allow system to cool before performing checks, services, and maintenance. Failure to comply with this warning can cause severe burns and injury to personnel.

a. a. Shut down generator set.

b. b. Open battery access door and disconnect negative battery cable.

c. c. Release control panel by turning two fasteners and lower control panel slowly.

d. d. Tag and disconnect panel light (1 , Figure 2-11) electrical leads.

e. e. Remove nut (2) and washer (3).

f. f. Remove panel light (1) from control panel.

WARNING

High voltage is produced when this generator set is in operation. SHUT DOWN generator set and make sure it is free of any power source before attempting any repair or maintenance on the set, or when connecting or disconnecting load cables. Failure to comply with this warning can cause injury or death to personnel.

a. a. Remove panel light directional cap (4, Fi gure 2-11) from panel light housing (1).

b. b. Remove panel light bulb (5).

2-20.4 Assembly.

a. a. Install panel light bulb (5, Figure 2-11) into panel light housing (1).

b. b. Install panel light directional cap (4).

2-20.5 Installation.

a. a. Insert panel light (1, F igure 2-11) into control panel.

b. b. Install washer (3) and nut (2).

c. c. Connect panel light electr ical leads. Remove tags.

d. d. Raise and secure control panel.

e. e. Connect negative battery c able and close battery access door.

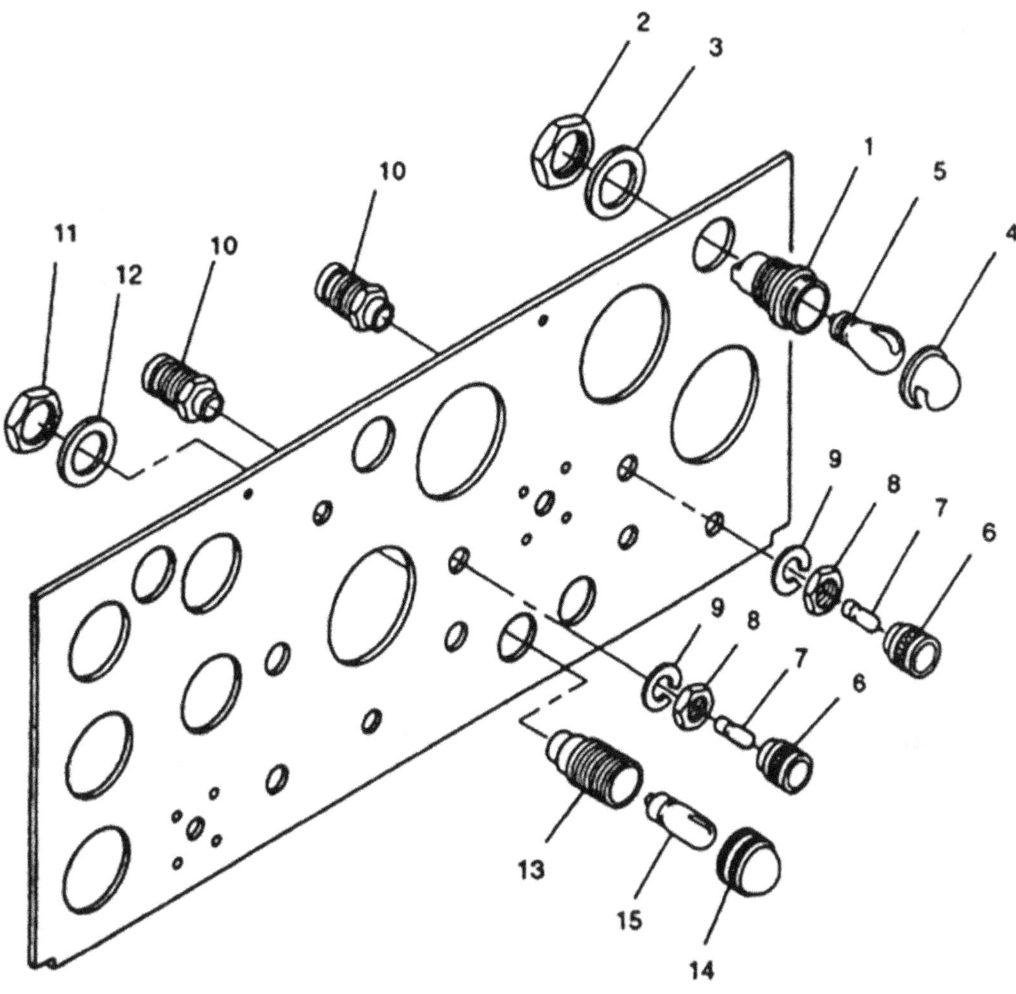

Figure 2-11. Control Panel Lights.

2-21 PRESS-TO-TEST LIGHTS 2-21.1

Inspection.

a. a. Inspect lights for cracks, corrosion, evidence of shorting, and other damage.

b. b. Replace or repair lights, as necessary.

2.21.2 Testing.

a. a. Place MASTER SWITCH in either PRIME & RUN position.

b. b. Press in lamp holders (6, Figure 2-11) and check that lamps are illuminated.

c. c. Perform steps d thru g below if lamp (7) fails to extinguish.

d. d. Release control panel by turning two fasteners and lower control panel slowly.

e. e. Set multimeter for DC volts and check for voltage at terminals 1 and 3 on receptacle

f. f. Replace press-to-test light assembly if voltage is present.

g. g. Raise and secure control panel.

WARNING

All metal jewelry can conduct electricity and become entangled in generator set components. Remove all jewelry when working on generator set. Failure to comply with this warning can cause injury or death to personnel.

WARNING

DO NOT wear loose clothing when performing checks, services and maintenance. Failure to comply with this warning can cause injury or death to personnel.

WARNING

When running, generator set engine has hot metal surfaces that will burn flesh on contact. Shut down generator set, and allow engine to cool before performing checks, services, and maintenance. Wear gloves and additional protective clothing as required. Failure to comply with this warning can cause injury or death to personnel.

WARNING

Dangerous voltage exists on live circuits. Always observe precautions and never work alone. Failure to comply with this warning can cause injury or death to personnel.

WARNING

High voltage is produced when this generator set is in operation. SHUT DOWN generator set and make sure it is free of any power source before attempting any repair or maintenance on the set, or when connecting or disconnecting load cables. Failure to comply with this warning can cause injury or death to personnel.

WARNING

DC voltages are present at generator set electrical components even with generator set shut down. Avoid shorting any positive with ground/negative. Failure to comply with this warning can cause injury to personnel, and damage to equipment.

WARNING

Ensure that the engine cannot be started while maintenance is being performed. (ENGINE CONTROL switch set to OFF/RESET; Battery Disconnect Switch is OFF; DEAD CRANK SWITCH is OFF). Failure to comply with this warning can cause injury or death to personnel.

WARNING

Top housing panels and exhaust system can get very hot. Shut down generator set, and allow system to cool before performing checks, services, and maintenance. Failure to comply with this warning can cause severe burns and injury to personnel.

a. a. Shut down generator set.

b. b. Open battery access door and disconnect negative battery cable.

c. c. Release control panel by turning two fasteners and lower control panel slowly.

WARNING

High voltage is produced when this generator set is in operation. SHUT DOWN generator set and make sure it is free of any power source before attempting any repair or maintenance on the set, or when connecting or disconnecting load cables. Failure to comply with this warning can cause injury or death to personnel.

d. d. Tag and disconnect electrical leads to light receptacles (10, Figure 2-11).

e. e. Remove lamp holders (6), nuts (8), lock washers (9), and receptacles (10) from control panel. Discard lock washers (9).

f. Remove lamps (7) from lamp holders (6), if necessary.

2-21.4 Installation.

a. a. Install press-to-test light receptacles (10, Figure 2-11) in control panel with new lock washers (9) and nuts (8).

b. b. Install lamps (7) in lamp holders (6), if removed.

c. c. Install lamp holders (6) on receptacles (10).

d. d. Connect electrical leads. Remove tags.

e. e. Raise and secure control panel.

f. f. Connect negative battery cabl e and close battery access door.

2-22 SYNCHRONIZING LIGHTS 2-22.1

Inspection.

Inspect synchronizing lights for cracks, corrosion, evidence of shorting, and other damage.

2-22.2 Removal.

WARNING

All metal jewelry can conduct electricity and become entangled in generator set components. Remove all jewelry when working on generator set. Failure to comply can cause injury or death to personnel by electrocution.

WARNING

DO NOT wear loose clothing when performing checks, services and maintenance. Failure to comply with this warning can cause injury or death to personnel.

WARNING

When running, generator set engine has hot metal surfaces that will burn flesh on contact. Shut down generator set and allow engine to cool before performing checks, services, and maintenance. Wear gloves and additional protective clothing as required. Failure to comply with this warning can cause injury or death to personnel.

WARNING

Dangerous voltage exists on live circuits. Always observe precautions and never work alone. Failure to comply with this warning can cause injury or death to personnel.

WARNING

High voltage is produced when this generator set is in operation. SHUT DOWN generator set and make sure it is free of any power source before attempting any repair or maintenance on the set, or when connecting or disconnecting load cables. Failure to comply with this warning can cause injury or death to personnel.

WARNING

DC voltages are present at generator set electrical components even with generator set shut down. Avoid shorting any positive with ground/negative. Failure to comply with this warning can cause injury to personnel, and damage to equipment.

WARNING

Ensure that the engine cannot be started while maintenance is being performed. (ENGINE CONTROL switch set to OFF/RESET; Battery Disconnect Switch is OFF; DEAD CRANK SWITCH is OFF). Failure to comply with this warning can cause injury or death to personnel.

WARNING

Top housing panels and exhaust system can get very hot. Shut down generator set, and allow system to cool before performing checks, services, and maintenance. Failure to comply with this warning can cause severe burns and injury to personnel.

a. a. Shut down generator set.

b. b. Open battery access door and disconnect negative battery cable.

c. c. Release control panel by turning two fasteners and lower control panel slowly.

WARNING

High voltage is produced when this generator set is in operation. SHUT DOWN generator set and make sure it is free of any power source before attempting any repair or maintenance on the set, or when connecting or disconnecting load cables. Failure to comply with this warning can cause injury or death to personnel.

d. d. Tag and disconnect synchronizing lights (13, Figure 2-11) electrical connections.

e. e. Remove nuts (11) and washers (12).

f. f. Remove synchronizing lights (13) from control panel.

g. g. Unscrew lenses (14) and remove lamps (15) from light receptacles (13).

2-22.3 Installation.

a. a. Insert synchronizing lights (13, Figure 2-11) into control panel.

b. b. Install washers (12) and nuts (11).

c. c. Connect electrical leads. Remove tags.

d. d. Raise and secure control panel.

e. e. Install lamps (15) and screw lenses (14) on light receptacles (13).

f. f. Connect negative battery cabl e and close battery access door.

2-23 FUEL LEVEL INDICATOR 2-23.1

Inspection.

a. a. Shut down generator set.

b. b. Inspect indicator for broken lens, cracked housing, and other damage.

2-23.2 Testing.

a. a. Shut down generator set.

b. b. Open battery access door and disconnect negative battery cable.

c. c. Release control panel by turning two fasteners and lower control panel slowly.

NOTE

For MEP-804A/MEP-804B, ensure frequency selector switch is in 50 Hz position.

d. d. Isolate generator set VOLTAGE adjust potentiometer by disconnecting wire 137A from AC voltage regulator terminal 5 and wire 107G from kilowatt transducer terminal V1.

e. e. Disconnect and isolate electrical lead from terminal S of FUEL LEVEL indicator.

f. f. Connect jumper wire between disconnected wire 137A and terminal S of FUEL LEVEL indicator.

g. Connect jumper wire between disconnected wire 107G and terminal G of FUEL LEVEL indicator.

h. h. Set multimeter for ohms and connect between wires 137A and 107G.

i. i. Adjust potentiometer until multimeter indicates between 216 and 264 ohms resistance.

j. j. Remove multimeter, but do not disturb potentiometer adjustment.

k. k. Connect negative battery cable and move generator set MASTER SWITCH to PRIME & RUN position.

l. l. FUEL LEVEL indicator should indicate EMPTY (±1/8 inch).

m. m. Move MASTER SWITCH to OFF position and disconnect negative battery cable.

n. n. Repeat steps h thru j above, setting potentiometer between 29.7 and 36.3 ohms.

o. o. Connect negative battery cable and mo ve MASTER SWITCH to PRIME & RUN position.

p. p. FUEL LEVEL indicator should indicate FULL (±1/8 inch).

q. q. Move MASTER SWITCH to OFF position and disconnect negative battery cable.

r. r. Replace FUEL LEVEL indicator if it fails to function properly.

s. s. Remove jumper wires and connect electrical le ads to FUEL LEVEL indicator, voltage regulator, and kilowatt transducer.

t. t. Raise and secure control panel.

u. u. Connect negative battery c able and close battery access door.

2-23.3 Removal.

WARNING

All metal jewelry can conduct electricity and become entangled in generator set components. Remove all jewelry when working on generator set. Failure to comply with this warning can cause injury or death to personnel.

WARNING

DO NOT wear loose clothing when performing checks, services and maintenance. Failure to comply with this warning can cause injury or death to personnel.

WARNING

When running, generator set engine has hot metal surfaces that will burn flesh on contact. Shut down generator set and allow engine to cool before performing checks, services, and maintenance. Wear gloves and additional protective clothing as required. Failure to comply with this warning can cause injury or death to personnel.

WARNING

Dangerous voltage exists on live circuits. Always observe precautions and never work alone. Failure to comply with this warning can cause injury or death to personnel.

WARNING

High voltage is produced when this generator set is in operation. SHUT DOWN generator set and make sure it is free of any power source before attempting any repair or maintenance on the set, or when connecting or disconnecting load cables. Failure to comply with this warning can cause injury or death to personnel.

WARNING

DC voltages are present at generator set electrical components even with generator set shut down. Avoid shorting any positive with ground/negative. Failure to comply with this warning can cause injury to personnel, and damage to equipment.

WARNING

Ensure that the engine cannot be started while maintenance is being performed. (ENGINE CONTROL switch set to OFF/RESET; Battery Disconnect Switch is OFF; DEAD CRANK SWITCH is OFF). Failure to comply can cause injury or death to personnel.

a. a. Shut down generator set.

b. b. Open battery access door and disconnect negative battery cable.

c. c. Release control panel by turning two fasteners and lower control panel slowly.

d. d. Tag and disconnect FUEL LEVEL indicator (4, Figure 2-12) electrical leads.

e. e. Remove nuts (1), washers (2), and clamp (3).

f. f. Remove FUEL LEVEL indicator (4) from control panel.

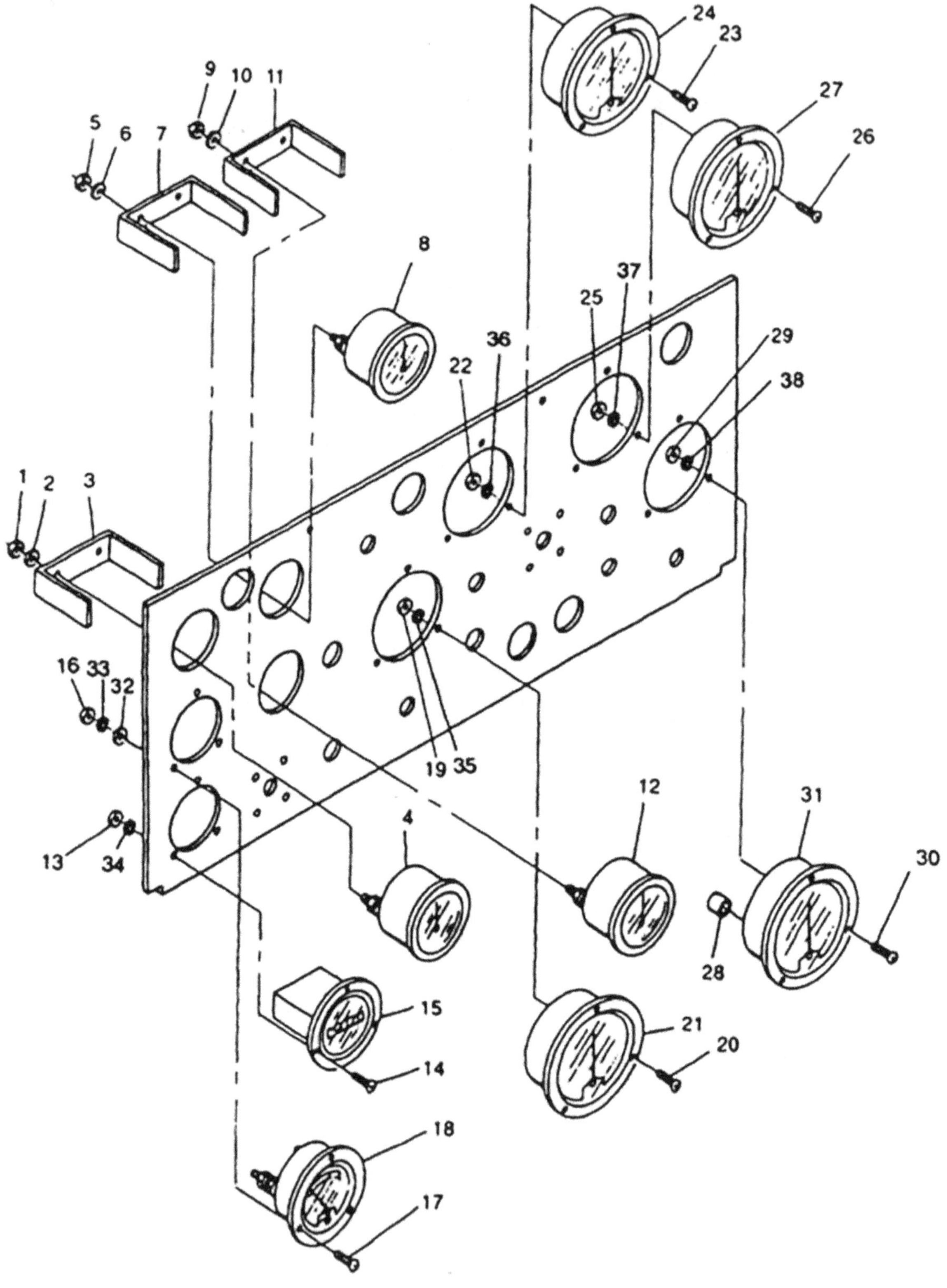

Figure 2-12. Control Panel Indicators.

2-23.4 Installation.

a. a. Insert FUEL LEVEL indicator (4, Figure 2-12) into control panel.

b. b. Install clamp (3), washers (2), and nuts (1).

c. c. Connect electrical leads. Remove tags.

d. d. Raise and secure control panel.

e. e. Connect negative battery c able and close battery access door.

2-24 TEMPERATURE INDICATOR 2-24.1

Inspection.

a. a. Shut down generator set.

b. b. Inspect indicator for broken lens, cracked housing, and other damage.

2-24.2 Testing.

a. a. Shut down generator set.

b. b. Open battery access door and disconnect negative battery cable.

c. c. Release control panel by turning two fasteners and lower control panel slowly.

NOTE

For MEP-804A/MEP-804B, ensure frequency selector switch is in 50 Hz position.

d. Isolate generator set VOLTAGE adjust potentiometer by disconnecting wire 137A from AC voltage regulator terminal 5 and wire 107G from kilowatt transducer terminal V1.

e. e. Disconnect and isolate electrical lead from terminal S of TEMPERATURE indicator.

f. f. Connect jumper wire between disconnected wire 137A and terminal S of TEMPERATURE indicator.

g. Connect jumper wire between disconnected wire 107G and terminal G of TEMPERATURE indicator.

h. h. Set multimeter for ohms and connect between wires 137A and 107G.

i. i. Adjust potentiometer until multimeter indicates between 117 and 143 ohms resistance.

j. j. Disconnect multimeter, but do not disturb potentiometer setting.

k. k. Connect negative battery cable and move generator set MASTER SWITCH to PRIME & RUN position.

l. l. TEMPERATURE indicator should indicate between 116°F and 164°F.

m. m. Move MASTER SWITCH to OFF position and disconnect negative battery cable.

n. n. Repeat steps h thru j above, setting potentiometer between 58.5 and 71.5 ohms.

o. o. Connect negative battery cable and mo ve MASTER SWITCH to PRIME & RUN position.

p. p. TEMPERATURE indicator should indicate between 156°F and 204°F.

q. q. Move MASTER SWITCH to OFF position and disconnect negative battery cable.

r. r. Replace TEMPERATURE indicator if indications are other than above.

s. s. Remove jumper wires and connect electrical leads to TEMPERATURE indicator, AC voltage regulator, and kilowatt transducer.

t. t. Raise and secure control panel.

u. u. Connect negative battery ca ble. Close battery access doors.

2-24.3 Removal.

WARNING

All metal jewelry can conduct electricity and become entangled in generator set components. Remove all jewelry when working on generator set. Failure to comply can cause injury or death to personnel by electrocution.

WARNING

DO NOT wear loose clothing when performing checks, services and maintenance. Failure to comply with this warning can cause injury or death to personnel.

WARNING

When running, generator set engine has hot metal surfaces that will burn flesh on contact. Shut down generator set and allow engine to cool before performing checks, services, and maintenance. Wear gloves and additional protective clothing as required. Failure to comply with this warning can cause injury or death to personnel.

WARNING

Dangerous voltage exists on live circuits. Always observe precautions and never work alone. Failure to comply with this warning can cause injury or death to personnel.

WARNING

High voltage is produced when this generator set is in operation. SHUT DOWN generator set and make sure it is free of any power source before attempting any repair or maintenance on the set, or when connecting or disconnecting load cables. Failure to comply with this warning can cause injury or death to personnel.

WARNING

DC voltages are present at generator set electrical components even with generator set shut down. Avoid shorting any positive with ground/negative. Failure to comply with this warning can cause injury to personnel, and damage to equipment.

WARNING

Ensure that the engine cannot be started while maintenance is being performed. (ENGINE CONTROL switch set to OFF/RESET; Battery Disconnect Switch is OFF; DEAD CRANK SWITCH is OFF). Failure to comply with this warning can cause injury or death to personnel.

WARNING

Top housing panels and exhaust system can get very hot. Shut down generator set, and allow system to cool before performing checks, services and maintenance. Failure to comply with this warning can cause severe burns and injury to personnel.

a. a. Shut down generator set.

b. b. Open battery access door and disconnect negative battery cable.

c. c. Release control panel by turning two fasteners and lower control panel slowly.

d. d. Tag and disconnect TEMPERATURE indicator (8, Figure 2-12) electrical leads.

e. e. Remove nuts (5), washers (6), and clamp (7).

f. f. Remove TEMPERATURE indicator (8) from control panel.

2-24.4 Installation.

WARNING

High voltage is produced when the generator set is in operation. Never attempt to start the generator set unless it is properly grounded. Failure to comply with this warning can cause injury or death to personnel.

a. a. Insert TEMPERATURE indicator (8, Figure 2-12) in control panel.

b. b. Install clamp (7), washers (6), and nuts (5).

c. c. Connect electrical leads. Remove tags.

d. d. Raise and secure control panel.

e. e. Connect negative battery c able and close battery access door.

2-25 OIL PRESSURE INDICATOR

2-25.1 Inspection.

a. a. Shut down generator set.

b. b. Inspect indicator for broken lens, cracked housing, and other damage.

2-25.2 Testing.

a. a. Shut down generator set.

b. b. Open battery access door and disconnect negative battery cable.

c. c. Release control panel by turning two fasteners and lower control panel slowly.

NOTE

For MEP-804A/MEP-804B, ensure frequency selector switch is in 50 Hz position.

d. d. Isolate generator set VOLTAGE adjust potentiometer by disconnecting wire 137A from AC voltage regulator terminal 5 and wire 107G from kilowatt transducer terminal V1.

e. e. Disconnect and isolate electrical lead from terminal S of OIL PRESSURE indicator.

f. f. Connect jumper wire between disconnected wire 137A and terminal S of OIL PRESSURE indicator.

g. g. Connect jumper wire between disconnected wire 107G and terminal G of OIL PRESSURE indicator.

h. h. Set multimeter for ohms and connect between wires 137A and 107G.

i. i. Adjust potentiometer until multimeter indicates between 92.7 and 113.3 ohms resistance.

j. j. Disconnect multimeter, but do not disturb potentiometer adjustment.

k. k. Connect negative battery cable and mo ve MASTER SWITCH to PRIME & RUN position.

l. l. OIL PRESSURE indicator should indicate between 32 and 48 PSI.

m. m. Move MASTER SWITCH to OFF position and disconnect negative battery cable.

n. n. Repeat steps h thru j above, setting potentiometer to between 30.15 and 36.85 ohms.

o. o. Connect negative battery cable and mo ve MASTER SWITCH to PRIME & RUN position.

p. p. OIL PRESSURE indicator should indicate between 72 and 80 PSI.

q. q. Move MASTER SWITCH to OFF position and disconnect negative battery cable.

r. r. Replace OIL PRESSURE indicator if it does not function properly.

s. s. Remove jumper wires and connect electrical leads to OIL PRESSURE indicator, AC voltage regulator, and kilowatt transducer.

t. Raise and secure control panel.

u. Connect negative battery cable and rais e control panel. Close battery access door.

2-25.3 Removal.

WARNING

All metal jewelry can conduct electricity and become entangled in generator set components. Remove all jewelry when working on generator set. Failure to comply with this warning can cause injury or death to personnel.

WARNING

DO NOT wear loose clothing when performing checks, services and maintenance. Failure to comply with this warning can cause injury or death to personnel.

WARNING

When running, generator set engine has hot metal surfaces that will burn flesh on contact. Shut down generator set and allow engine to cool before performing checks, services, and maintenance. Wear gloves and additional protective clothing as required. Failure to comply with this warning can cause injury or death to personnel.

WARNING

Dangerous voltage exists on live circuits. Always observe precautions and never work alone. Failure to comply with this warning can cause injury or death to personnel.

WARNING

High voltage is produced when this generator set is in operation. SHUT DOWN generator set and make sure it is free of any power source before attempting any repair or maintenance on the set, or when connecting or disconnecting load cables. Failure to comply with this warning can cause injury or death to personnel.

WARNING

DC voltages are present at generator set electrical components even with generator set shut down. Avoid shorting any positive with ground/negative. Failure to comply with this warning can cause injury to personnel, and damage to equipment.

WARNING

Ensure that the engine cannot be started while maintenance is being performed. (ENGINE CONTROL switch set to OFF/RESET; Battery Disconnect Switch is OFF; DEAD CRANK SWITCH is OFF). Failure to comply with this warning can cause injury or death to personnel.

WARNING

Top housing panels and exhaust system can get very hot. Shut down generator set, and allow system to cool before performing checks, services and maintenance. Failure to comply with this warning can cause severe burns and injury to personnel.

a. a. Shut down generator set.

b. b. Open battery access door and disconnect negative battery cable.

c. c. Release control panel by turning two fasteners and lower control panel slowly.

d. d. Tag and disconnect OIL PRESSURE indicator (12, Figure 2-12) electrical leads.

e. e. Remove nuts (9), washers (10), clamp (11), and ground wire.

f. f. Remove OIL PRESSURE indicator (12) from control panel.

2-25.4 Installation.

WARNING

High voltage is produced when the generator set is in operation. Never attempt to start the generator set unless it is properly grounded. Failure to comply with this warning can cause injury or death to personnel.

a. a. Insert OIL PRESSURE indicator (12, Figure 2-12) into control panel.

b. b. Install clamp (11), washers (10), and nuts (9).

c. c. Connect electrical leads. Remove tags.

d. d. Raise and secure control panel.

e. e. Connect negative battery cabl e and close battery access door.

2-26 TIME METER (TOTAL HOURS)

2-26.1 Inspection.

a. a. Shut down generator set.

b. b. Inspect meter for broken lens, cracked housing, and other damage.

2-26.2 Testing.

a. a. Release control panel by turning two fasteners and lower control panel slowly.

b. b. Turn MASTER SWITCH to PRIME & RUN position.

c. c. Crank engine momentarily to energize time meter relay.

d. d. Set multimeter for DC volts and connect across terminals 1 and 2 of time meter (TOTAL

HOURS).

e. e. If 24 VDC is present, wait approximately 6 minutes. Time meter (TOTAL HOURS) should move 1/10 of an hour.

f. f. If time meter (TOTAL HOURS) does not operate properly, meter is defective and must be replaced.

g. Raise and secure control panel.

2-26.3 Removal.

WARNING

All metal jewelry can conduct electricity and become entangled in generator set components. Remove all jewelry when working on generator set. Failure to comply with this warning can cause injury or death to personnel.

WARNING

DO NOT wear loose clothing when performing checks, services and maintenance. Failure to comply with this warning can cause injury or death to personnel.

WARNING

When running, generator set engine has hot metal surfaces that will burn flesh on contact. Shut down generator set and allow engine to cool before performing checks, services, and maintenance. Wear gloves and additional protective clothing as required. Failure to comply with this warning can cause injury or death to personnel..

WARNING

Dangerous voltage exists on live circuits. Always observe precautions and never work alone. Failure to comply with this warning can cause injury or death to personnel.

WARNING

High voltage is produced when this generator set is in operation. SHUT DOWN generator set and make sure it is free of any power source before attempting any repair or maintenance on the set, or when connecting or disconnecting load cables. Failure to comply with this warning can cause injury or death to personnel.

WARNING

DC voltages are present at generator set electrical components even with generator set shut down. Avoid shorting any positive with ground/negative. Failure to comply with this warning can cause injury to personnel, and damage to equipment.

WARNING

Ensure that the engine cannot be started while maintenance is being performed. (ENGINE CONTROL switch set to OFF/RESET; Battery Disconnect Switch is OFF; DEAD CRANK SWITCH is OFF). Failure to comply with this warning can cause injury or death to personnel.

WARNING

Top housing panels and exhaust system can get very hot. Shut down generator set, and allow system to cool before performing checks, services and maintenance. Failure to comply with this warning can cause severe burns and injury to personnel.

a. a. Shut down generator set.

b. b. Open battery access door and disconnect negative battery cable.

c. c. Release control panel by turning two fasteners and lower control panel slowly.

d. d. Tag and disconnect time meter (TOTAL HOURS) (15, Figure 2-12) electrical leads.

e. e. Remove screws (14), washers (34), and nuts (13).

f. f. Remove time meter (TOTAL HOURS) (15) from control panel.

2-26.4 Installation.

WARNING

High voltage is produced when the generator set is in operation. Never attempt to start the generator set unless it is properly grounded. Failure to comply with this warning can cause injury or death to personnel.

a. a. Insert time meter (TOTAL HOURS) (15, Figure 2-12) into control panel.

b. b. Install screws (14), washers (34), and nuts (13).

c. c. Connect electrical leads. Remove tags.

d. d. Raise and secure control panel.

e. e. Connect negative battery c able and close battery access door.

2-27 BATTERY CHARGE AMMETER 2-27.1

Inspection.

a. a. Shut down generator set.

b. b. Inspect ammeter for broken lens, cracked housing, and other damage.

2-27.2 Testing.

a. a. Start and operate generator set at rated voltage and frequency.

b. b. Release control panel by turning two fasteners and lower control panel slowly.

c. c. Set multimeter for DC volts and connect across BATTERY CHARGE ammeter terminals. Connect positive lead to positive terminal and negative lead to negative terminal if you observe or think battery is charging. Multimeter should indicate up to 50 mV (25 mV equals +10 amps on BATTERY CHARGE ammeter).

d. d. Reverse multimeter leads if you observe or think battery is discharging. Multimeter should indicate up to 25 mV (25 mV equals -10 amps on BATTERY CHARGE ammeter).

e. e. If multimeter indicates millivolt reading and battery charge is not within ±10 percent of equal ampere reading, or ammeter is not indicating, replace BATTERY CHARGE ammeter.

f. f. Raise and secure control panel.

2-27.3 Removal.

WARNING

All metal jewelry can conduct electricity and become entangled in generator set components. Remove all jewelry when working on generator set. Failure to comply with this warning can cause injury or death to personnel.

WARNING

DO NOT wear loose clothing when performing checks, services and maintenance. Failure to comply with this warning can cause injury or death to personnel.

WARNING

When running, generator set engine has hot metal surfaces that will burn flesh on contact. Shut down generator set and allow engine to cool before performing checks, services, and maintenance. Wear gloves and additional protective clothing as required. Failure to comply with this warning can cause injury or death to personnel.

WARNING

Dangerous voltage exists on live circuits. Always observe precautions and never work alone. Failure to comply with this warning can cause injury or death to personnel.

WARNING

High voltage is produced when this generator set is in operation. SHUT DOWN generator set and make sure it is free of any power source before attempting any repair or maintenance on the set, or when connecting or disconnecting load cables. Failure to comply with this warning can cause injury or death to personnel.

WARNING

DC voltages are present at generator set electrical components even with generator set shut down. Avoid shorting any positive with ground/negative. Failure to comply can cause injury to personnel and damage to equipment.

WARNING

Ensure that the engine cannot be started while maintenance is being performed. (ENGINE CONTROL switch set to OFF/RESET; Battery Disconnect Switch is OFF; DEAD CRANK SWITCH is OFF). Failure to comply with this warning can cause injury or death to personnel.

WARNING

Top housing panels and exhaust system can get very hot. Shut down generator set, and allow system to cool before performing checks, services and maintenance. Failure to comply with this warning can cause severe burns and injury to personnel.

a. a. Shut down generator set.

b. b. Open battery access door and disconnect negative battery cable.

c. c. Release control panel by turning two fasteners and lower control panel slowly.

d. d. Tag and disconnect BATTERY CHARGE ammete r (18, Figure 2-12) electrical leads.

e. e. Remove nuts (16), washers (32 and 33), and screws (17).

f. f. Remove BATTERY CHARGE ammeter (18).

2-27.4 Installation.

WARNING

High voltage is produced when the generator set is in operation. Never attempt to start the generator set unless it is properly grounded. Failure to comply with this warning can cause injury or death to personnel.

a. a. Insert BATTERY CHARGE ammeter (18, Figure 2-12) into control panel.

b. b. Install screws (17), washers (32, 33), and nuts (16).

c. c. Connect electrical leads. Remove tags.

d. d. Raise and secure control panel.

e. e. Connect negative battery c able and close battery access door.

2-28 FREQUENCY METER (HERTZ)

2-28.1 Inspection.

a. Shut down generator set.

b. Inspect meter for broken lens, cracked housing, and other damage.

2-28.2 Testing (MEP-804A/MEP-804B).

NOTE

Test frequency transducer in accordance with paragraph 2-52.2/paragraph 2-52.3 prior to testing FREQUENCY meter (HERTZ).

a. a. Shut down generator set.

b. b. Open battery access door and disconnect negative battery cable.

c. c. Release control panel by turning two fasteners and lower control panel slowly.

d. d. Disconnect wire 181A from positive (+) terminal of FREQUENCY meter (HERTZ).

e. Set multimeter for DC milliamps and connect negative lead to positive (+) terminal of FREQUENCY meter (HERTZ) and positive lead to wire 181A.

f. f. Position FREQUENCY SELECT switch to 60 Hz.

g. g. Connect negative battery cable, start and opera te generator set at rated voltage, and adjust frequency to 60 Hz.

h. Multimeter indication should be between 0.781 and 0.923 Ma.

I Position FREQUENCY SELECT switch to 50 Hz and adjust frequency to 50 Hz.

j. j. Multimeter indication should be between 0.071 and 0.213 Ma.

k. k. Replace FREQUENCY meter (HERTZ) if meter readings are not as stated above.

l. l. Raise and secure control panel.

2-28.3 Testing (MEP-814A/MEP-814B).

a. a. Shut down generator set.

b. b. Open battery access door and disconnect negative battery cable.

c. c. Release control panel by turning two fasteners and lower control panel slowly.

d. d. Disconnect wire 181A from positive (+) terminal of FREQUENCY meter (HERTZ).

e. e. Set multimeter for DC milliamps and connect negative lead to positive (+) terminal of FREQUENCY meter (HERTZ) and positive lead to wire 181A.

f. f. Connect negative battery cable, start and oper ate generator set at rated voltage, and adjust frequency to 400 Hz.

g. g. Multimeter indication should be between 0.240 and 0.260 DC milliamps.

h. h. Replace FREQUENCY meter (HERTZ) if multimeter readings are not as stated above.

i. i. Raise and secure control panel.

WARNING

All metal jewelry can conduct electricity and become entangled in generator set components. Remove all jewelry when working on generator set. Failure to comply with this warning can cause injury or death to personnel.

WARNING

DO NOT wear loose clothing when performing checks, services and maintenance. Failure to comply with this warning can cause injury or death to personnel.

WARNING

When running, generator set engine has hot metal surfaces that will burn flesh on contact. Shut down generator set and allow engine to cool before performing checks, services, and maintenance. Wear gloves and additional protective clothing as required. Failure to comply with this warning can cause injury or death to personnel.

WARNING

Dangerous voltage exists on live circuits. Always observe precautions and never work alone. Failure to comply with this warning can cause injury or death to personnel.

WARNING

High voltage is produced when this generator set is in operation. SHUT DOWN generator set and make sure it is free of any power source before attempting any repair or maintenance on the set, or when connecting or disconnecting load cables. Failure to comply with this warning can cause injury or death to personnel.

WARNING

DC voltages are present at generator set electrical components even with generator set shut down. Avoid shorting any positive with ground/negative. Failure to comply with this warning can cause injury to personnel, and damage to equipment.

WARNING

Ensure that the engine cannot be started while maintenance is being performed. (ENGINE CONTROL switch set to OFF/RESET; Battery Disconnect Switch is OFF; DEAD CRANK SWITCH is OFF). Failure to comply with this warning can cause injury or death to personnel.

WARNING

Top housing panels and exhaust system can get very hot. Shut down generator set, and allow system to cool before performing checks, services and maintenance. Failure to comply with this warning can cause severe burns and injury to personnel.

a. a. Shut down generator set.

b. b. Open battery access door and disconnect negative battery cable.

c. c. Release control panel by turning two fasteners and lower control panel slowly.

d. d. Tag and disconnect FREQUENCY meter (HERTZ) (21, Figure 2-12) electrical leads.

e. e. Remove nuts (19), washers (35), and screws (20).

f. f. Remove FREQUENCY meter (HERTZ) (21) from control panel.

2-28.5 Installation.

WARNING

High voltage is produced when the generator set is in operation. Never attempt to start the generator set unless it is properly grounded. Failure to comply with this warning can cause injury or death to personnel.

a. a. Insert FREQUENCY meter (HERTZ) (21, Figure 2-12) into control panel.

b. b. Install screws (20), washers (35), and nuts (19).

c. c. Connect electrical leads. Remove tags.

d. d. Raise and secure control panel.

e. e. Connect negative battery c able and close battery access door.

2-29 AMMETER (PERCENT RATED CURRENT

2-29.1 Inspection.

a. a. Shut down generator set.

b. b. Inspect ammeter for broken lens, cracked housing, and other damage.

2-29.2 Testing.

a. a. Shut down generator set.

b. b. Open battery access and disconnect negative battery cable.

c. c. Release control panel by turning two fasteners and lower control panel slowly.

d. d. Tag and disconnect wire 183A from terminal 1 of ammeter (PERCENT RATED CURRENT).

e. e. Set multimeter for AC amperes and connect between disconnected wire 183A and terminal 1 of ammeter (PERCENT RATED CURRENT).

f. f. Connect negative battery cable, start and operat e generator set at rated voltage and frequency, and apply some load to generator set.

g. g. Observe and note indications on multimeter and ammeter (PERCENT RATED CURRENT).

h. h. Shut down generator set.

i. i. Calculate percent of current from multimeter indication using the following formula:

Percent of current = 100 x Multimeter Indication
0.75 amperes

j. j. Compare calculated percent of current to ammeter (PERCENT RATED CURRENT) indication noted during operation. If difference is greater than 10 percent, replace ammeter.

k. k. Disconnect negative battery cable.

l. l. Remove multimeter and connect wire 183A to ammeter (PERCENT RATED CURRENT).

m. m. Raise and secure control panel.

n. n. Connect negative battery cabl e and close battery access door.

2-29.3 Removal.

WARNING

All metal jewelry can conduct electricity and become entangled in generator set components. Remove all jewelry when working on generator set. Failure to comply with this warning can cause injury or death to personnel.

WARNING

DO NOT wear loose clothing when performing checks, services and maintenance. Failure to comply with this warning can cause injury or death to personnel.

WARNING

When running, generator set engine has hot metal surfaces that will burn flesh on contact. Shut down generator set and allow engine to cool before performing checks, services, and maintenance. Wear gloves and additional protective clothing as required. Failure to comply with this warning can cause injury or death to personnel.

WARNING

Dangerous voltage exists on live circuits. Always observe precautions and never work alone. Failure to comply with this warning can cause injury or death to personnel.

WARNING

High voltage is produced when this generator set is in operation. SHUT DOWN generator set and make sure it is free of any power source before attempting any repair or maintenance on the set, or when connecting or disconnecting load cables. Failure to comply with this warning can cause injury or death to personnel.

WARNING

DC voltages are present at generator set electrical components even with generator set shut down. Avoid shorting any positive with ground/negative. Failure to comply with this warning can cause injury to personnel, and damage to equipment.

WARNING

Ensure that the engine cannot be started while maintenance is being performed. (ENGINE CONTROL switch set to OFF/RESET; Battery Disconnect Switch is OFF; DEAD CRANK SWITCH is OFF). Failure to comply with this warning can cause injury or death to personnel.

WARNING

Top housing panels and exhaust system can get very hot. Shut down generator set, and allow system to cool before performing checks, services and maintenance. Failure to comply with this warning can cause severe burns and injury to personnel.

a. a. Shut down generator set.

b. b. Open battery access door and disconnect negative battery cable.

c. c. Release control panel by turning two fasteners and lower control panel slowly.

d. d. Tag and disconnect ammeter (PERCENT RATED CURRENT) (24, Figure 2-12) electrical

e. e. Remove nuts (22), washers (36), and screws (23).

f. f. Remove ammeter (PERCENT RATED CURRENT) (24) from control panel.

WARNING

High voltage is produced when the generator set is in operation. Never attempt to start the generator set unless it is properly grounded. Failure to comply with this warning can cause injury or death to personnel.

a. a. Insert ammeter (PERCENT RATED CURRENT) (24, Figure 2-12) into control panel.

b. b. Install screws (23), washers (36), and nuts (22).

c. c. Connect electrical leads. Remove tags.

d. d. Raise and secure control panel.

e. e. Connect negative battery c able and close battery access door.

2-30 KILOWATTMETER (PERCENT POWER)

2-30.1 Inspection.

a. a. Shut down generator set.

b. b. Inspect meter for broken lens, cracked housing, and other damage.

2-30.2 Testing.

a. a. Shut down generator set.

b. b. Open battery access door and disconnect negative battery cable.

c. c. Release control panel by turning two fasteners and lower control panel slowly.

d. d. Tag and disconnect wire 120A from positive terminal of kilowattmeter (PERCENT POWER).

e. e. Set multimeter for milliamperes and connect between disconnected wire 120A and positive terminal of kilowattmeter.

f. f. Connect negative battery cable, start and operate generator set at rated voltage and frequency, and apply some load to generator set.

g. g. Observe and note indications on multimeter and kilowattmeter.

h. h. Shut down generator set.

i. i. Calculate percent of power from multimeter indication using the following formula:

$$\text{Percent of power} = \frac{133 \times \text{Multimeter Indication}}{1.2 \text{ Ma.}}$$

j. Compare calculated percent of power to kilowattmeter indication noted during operation. If difference is greater than 13 percent, replace kilowattmeter (PERCENT POWER).

k. k. Disconnect negative battery cable.

l. l. Remove multimeter and connect wire 120A to kilowattmeter (PERCENT POWER).

m. m. Raise and secure control panel.

n. n. Connect negative battery c able and close battery access door.

2-30.3 Removal.

WARNING

All metal jewelry can conduct electricity and become entangled in generator set components. Remove all jewelry when working on generator set. Failure to comply with this warning can cause injury or death to personnel.

WARNING

DO NOT wear loose clothing when performing checks, services and maintenance. Failure to comply with this warning can cause injury or death to personnel.

WARNING

When running, generator set engine has hot metal surfaces that will burn flesh on contact. Shut down generator set and allow engine to cool before performing checks, services, and maintenance. Wear gloves and additional protective clothing as required. Failure to comply with this warning can cause injury or death to personnel.

WARNING

Dangerous voltage exists on live circuits. Always observe precautions and never work alone. Failure to comply with this warning can cause injury or death to personnel.

WARNING

High voltage is produced when this generator set is in operation. SHUT DOWN generator set and make sure it is free of any power source before attempting any repair or maintenance on the set, or when connecting or disconnecting load cables. Failure to comply with this warning can cause injury or death to personnel.

WARNING

DC voltages are present at generator set electrical components even with generator set shut down. Avoid shorting any positive with ground/negative. Failure to comply with this warning can cause injury to personnel, and damage to equipment.

WARNING

Ensure that the engine cannot be started while maintenance is being performed. (ENGINE CONTROL switch set to OFF/RESET; Battery Disconnect Switch is OFF; DEAD CRANK SWITCH is OFF). Failure to comply with this warning can cause injury or death to personnel.

WARNING

Top housing panels and exhaust system can get very hot. Shut down generator set, and allow system to cool before performing checks, services and maintenance. Failure to comply with this warning can cause severe burns and injury to personnel.

a. a. Shut down generator set.

b. b. Open battery access door and disconnect negative battery cable.

c. c. Release control panel by turning two fasteners and lower control panel slowly.

d. d. Tag and disconnect kilowattmeter (PERCENT POWER) (27, Figure 2-12) electrical leads.

e. e. Remove nuts (25), washers (37), and screws (26).

f. f. Remove kilowattmeter (PERCENT POWER) (27) from control panel.

2-30.4 Installation.

WARNING

High voltage is produced when the generator set is in operation. Never attempt to start or maintain the generator set unless it is properly grounded. Failure to comply can cause injury or death to personnel.

a. a. Insert kilowattmeter (PERCENT POWER) (27, Figure 2-12) into control panel.

b. b. Install screws (26), washers (37), and nuts (25).

c. c. Connect electrical leads. Remove tags.

d. d. Raise and secure control panel.

e. e. Connect negative battery cabl e and close battery access door.

2-31 AC VOLTMETER (VOLTS AC) 2-31.1

Inspection.

a. a. Shut down generator set.

b. b. Inspect AC voltmeter (VOLTS AC) for br oken lens, cracked housing, and other damage.

a. a. Shut down generator set.

b. b. Release control panel by turning tw o fasteners and lower control panel slowly.

c. c. Set multimeter for AC volts and connect to AC voltmeter (VOLTS AC) terminals.

d. d. Move voltage reconnection board to 120/208 position.

e. e. Start and operate generator set at rated voltage and frequency.

f. f. Move VM-AM transfer switch to L3-L1 position. Note indications on multimeter and AC voltmeter (VOLTS AC).

g. g. Move VM-AM transfer switch to L3-L0 position. Note indications on multimeter and AC voltmeter (VOLTS AC).

h. h. Shut down generator set.

i. i. Move voltage reconnection board to 240/416 position.

j. j. Start and operate generator set at rated voltage and frequency.

k. k. Repeat steps f and g above.

l. l. Shut down generator set.

m. m. Compare AC voltmeter (VOLTS AC) readings to multimeter in each position.

n. n. Replace AC voltmeter (VOLTS AC) if readings differ more than ±5 VAC between 115 and 125 VAC or ±10 VAC between 200 and 250 VAC.

o. o. Remove multimeter.

p. p. Raise and secure control panel.

2-31.3 Removal.

WARNING

All metal jewelry can conduct electricity and become entangled in generator set components. Remove all jewelry when working on generator set. Failure to comply with this warning can cause injury or death to personnel.

WARNING

DO NOT wear loose clothing when performing checks, services and maintenance. Failure to comply with this warning can cause injury or death to personnel.

WARNING

When running, generator set engine has hot metal surfaces that will burn flesh on contact. Shut down generator set and allow engine to cool before performing checks, services, and maintenance. Wear gloves and additional protective clothing as required. Failure to comply with this warning can cause injury or death to personnel.

WARNING

Dangerous voltage exists on live circuits. Always observe precautions and never work alone. Failure to comply with this warning can cause injury or death to personnel.

WARNING

High voltage is produced when this generator set is in operation. SHUT DOWN generator set and make sure it is free of any power source before attempting any repair or maintenance on the set, or when connecting or disconnecting load cables. Failure to comply with this warning can cause injury or death to personnel.

WARNING

DC voltages are present at generator set electrical components even with generator set shut down. Avoid shorting any positive with ground/negative. Failure to comply with this warning can cause injury to personnel, and damage to equipment.

WARNING

Ensure that the engine cannot be started while maintenance is being performed. (ENGINE CONTROL switch set to OFF/RESET; Battery Disconnect Switch is OFF; DEAD CRANK SWITCH is OFF). Failure to comply with this warning can cause injury or death to personnel.

WARNING

Top housing panels and exhaust system can get very hot. Shut down generator set, and allow system to cool before performing checks, services and maintenance. Failure to comply with this warning can cause severe burns and injury to personnel.

a. a. Shut down generator set.

b. b. Open battery access door and disconnect negative battery cable.

c. c. Release control panel by turning two fasteners and lower control panel slowly.

d. d. Tag and disconnect AC voltmeter (VOLTS AC) (31, Figure 2-12) electrical leads and remove sleeves (28).

e. e. Remove nuts (29), washers (38), and screws (30).

f. f. Remove AC voltmeter (VOLTS AC) (31) from control panel.

WARNING

High voltage is produced when the generator set is in operation. Never attempt to start the generator set unless it is properly grounded. Failure to comply with this warning can cause injury or death to personnel.

a. a. Insert AC voltmeter (VOLTS AC) (31, Figure 2-12) into control panel.

b. b. Install screws (30), washers (38), and nuts (29).

c. c. Connect electrical leads. Remove tags.

d. d. Install sleeves (28) over terminals.

e. e. Raise and secure control panel.

f. f. Connect negative battery cabl e and close battery access door.

2-32 MASTER SWITCH 2-32.1

Inspection.

a. a. Shut down generator set.

b. b. Inspect switch for loose connec tions/mounting and other damage.

2-32.2 Removal.

WARNING

All metal jewelry can conduct electricity and become entangled in generator set components. Remove all jewelry when working on generator set. Failure to comply with this warning can cause injury or death to personnel.

WARNING

DO NOT wear loose clothing when performing checks, services and maintenance. Failure to comply with this warning can cause injury or death to personnel.

WARNING

When running, generator set engine has hot metal surfaces that will burn flesh on contact. Shut down generator set and allow engine to cool before performing checks, services, and maintenance. Wear gloves and additional protective clothing as required. Failure to comply with this warning can cause injury or death to personnel.

WARNING

Dangerous voltage exists on live circuits. Always observe precautions and never work alone. Failure to comply with this warning can cause injury or death to personnel.

WARNING

High voltage is produced when this generator set is in operation. SHUT DOWN generator set and make sure it is free of any power source before attempting any repair or maintenance on the set, or when connecting or disconnecting load cables. Failure to comply with this warning can cause injury or death to personnel.

WARNING

DC voltages are present at generator set electrical components even with generator set shut down. Avoid shorting any positive with ground/negative. Failure to comply with this warning can cause injury to personnel, and damage to equipment.

WARNING

Ensure that the engine cannot be started while maintenance is being performed. (ENGINE CONTROL switch set to OFF/RESET; Battery Disconnect Switch is OFF; DEAD CRANK SWITCH is OFF). Failure to comply with this warning can cause injury or death to personnel.

WARNING

Top housing panels and exhaust system can get very hot. Shut down generator set, and allow system to cool before performing checks, services and maintenance. Failure to comply with this warning can cause severe burns and injury to personnel.

a. a. Shut down generator set.

b. b. Open battery access door and disconnect negative battery cable.

c. c. Release control panel by turning two fasteners and lower control panel slowly.

d. d. Loosen setscrew (1, Figure 2-13) and remove knob (2) from MASTER SWITCH (5).

e. e. Remove nuts (3) and screws (4).

f. f. Remove MASTER SWITCH (5) from control panel.

NOTE

Ensure the diode is properly tagged so the polarity is not reversed during installation.

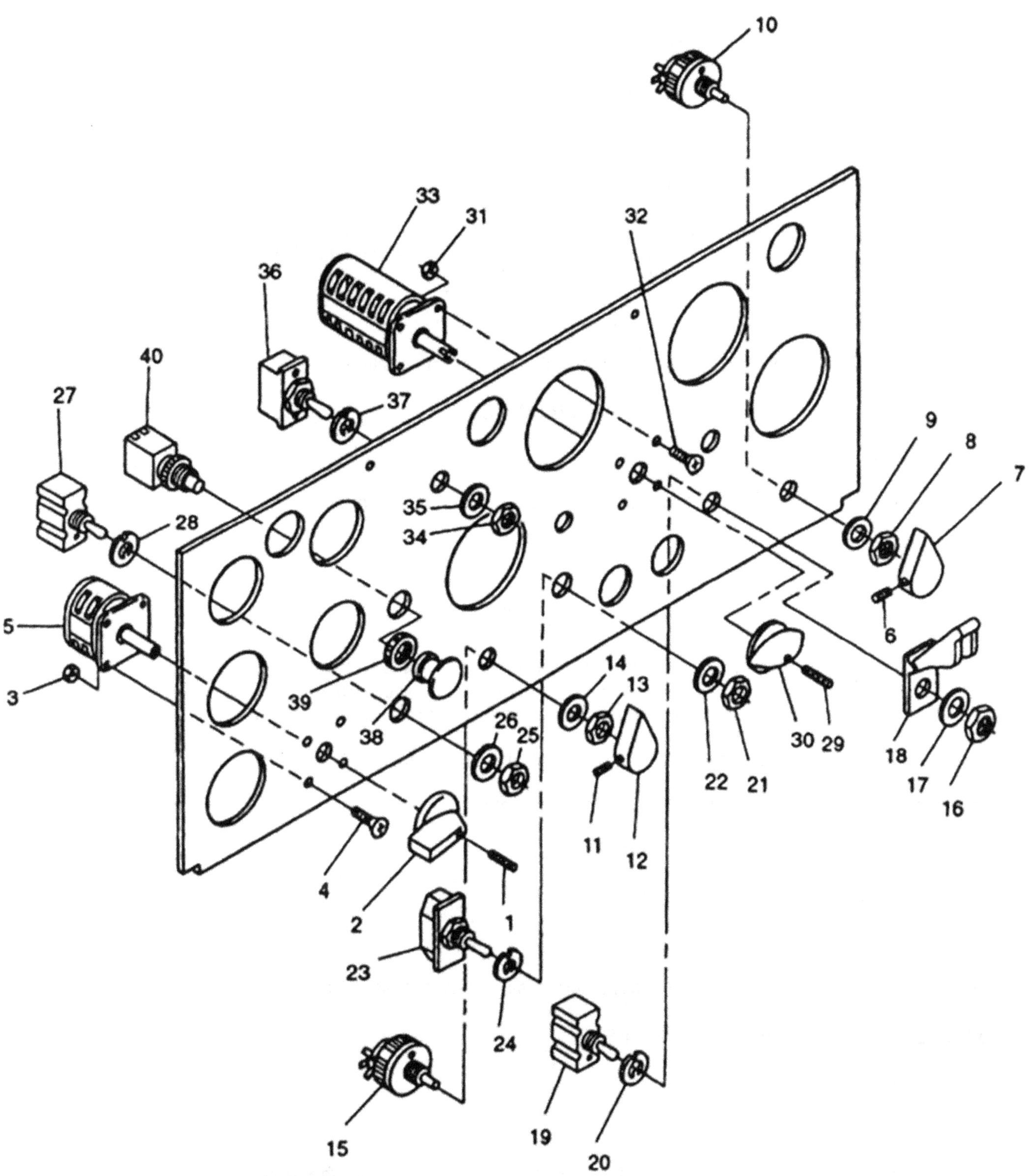

Figure 2-13. Control Panel Switches.

2-32.3 Testing.

a. a. Shut down generator set.

b. b. Open battery access door and disconnect negative battery cable.

c. c. Release control panel by turning two fasteners and lower control panel slowly.

d. d. Tag and disconnect MASTER SWITCH (5, Figure 2-13) electrical leads.

e. e. Set multimeter for ohms and check switch for continuity. Refer to Electrical Schematic FO-1 (5-1 Circuit Schedule) (MEP-804A/MEP-814A)/FO-3 (MEP-804B/MEP-814B) to determine circuits made to corresponding switch positions.

f. f. Check continuity until all five positions have been checked.

g. g. Replace switch if open circuit is noted in any switch position.

h. h. Connect electrical leads to MASTER SWITCH (5). Remove tags.

i. i. Raise and secure control panel.

j. j. Connect negative battery cabl e and close battery access door.

2-32.4 Installation.

WARNING

High voltage is produced when the generator set is in operation. Never attempt to start the generator set unless it is properly grounded. Failure to comply with this warning can cause injury or death to personnel.

a. a. Connect electrical leads to MASTER SWITCH (5, Figure 2-13). Remove tags.

b. b. Insert MASTER SWITCH (5) into control panel.

c. c. Install screws (4) and nuts (5).

d. d. Install knob (2) and tighten setscrew (1).

e. e. Raise and secure control panel.

f. f. Connect negative battery cabl e and close battery access door.

2-33 VOLTAGE ADJUST POTENTIOMETER

2-33.1 Inspection.

a. a. Shut down generator set.

b. b. Inspect potentiometer for loose connections/mounting and other damage.

2-33.2 Removal.

WARNING

All metal jewelry can conduct electricity and become entangled in generator set components. Remove all jewelry when working on generator set. Failure to comply with this warning can cause injury or death to personnel.

WARNING

DO NOT wear loose clothing when performing checks, services and maintenance. Failure to comply with this warning can cause injury or death to personnel.

WARNING

When running, generator set engine has hot metal surfaces that will burn flesh on contact. Shut down generator set and allow engine to cool before performing checks, services, and maintenance. Wear gloves and additional protective clothing as required. Failure to comply with this warning can cause injury or death to personnel.

WARNING

Dangerous voltage exists on live circuits. Always observe precautions and never work alone. Failure to comply with this warning can cause injury or death to personnel.

WARNING

High voltage is produced when this generator set is in operation. SHUT DOWN generator set and make sure it is free of any power source before attempting any repair or maintenance on the set, or when connecting or disconnecting load cables. Failure to comply with this warning can cause injury or death to personnel.

WARNING

DC voltages are present at generator set electrical components even with generator set shut down. Avoid shorting any positive with ground/negative. Failure to comply with this warning can cause injury to personnel, and damage to equipment.

WARNING

Ensure that the engine cannot be started while maintenance is being performed. (ENGINE CONTROL switch set to OFF/RESET; Battery Disconnect Switch is OFF; DEAD CRANK SWITCH is OFF). Failure to comply with this warning can cause injury or death to personnel.

WARNING

Top housing panels and exhaust system can get very hot. Shut down generator set, and allow system to cool before performing checks, services and maintenance. Failure to comply with this warning can cause severe burns and injury to personnel.

a. a. Shut down generator set.

b. b. Open battery access door and disconnect negative battery cable.

c. c. Release control panel by turning two fasteners and lower control panel slowly.

d. d. Tag and disconnect VOLTAGE adjust potentiomet er (10, Figure 2-13) rheostat electrical leads using soldering gun. Remove shrinkable tubing.

e. e. Remove setscrews (6) and knob (7).

f. f. Remove nut (8) and washer (9).

g. g. Remove VOLTAGE adjust potentiometer (10) from control panel.

2-33.3 Testing.

NOTE

Refer to TM 9-6115-643-24P to verify AC voltage regulator being tested.

a. a. Remove VOLTAGE adjust potentiometer (paragraph 2-33.2).

b. b. Set multimeter for ohms and connect across two outer terminals of potentiometer (10, Figure 2-13). Indication shall be as follows:

10,000 ohms (part number RV4NAYSD103A installed)
20,000 ohms (part number RV4NAYSD203A installed)

c. c. Rotate potentiometer shaft counterclockwise (CCW) as far as it will go.

d. d. Connect multimeter between center terminal and either outer terminal.

e. e. Slowly and at an even rate, rotate potentiometer shaft clockwise (CW) as far as it will go while observing multimeter.

f. f. Multimeter indication shall increase at an even rate as follows:

0 to 10,000 ohms (part number RV4NAYSD103A installed)
0 to 20,000 ohms (part number RV4NAYSD203A installed)

g. g. If multimeter indication changes erratically or is not at maximum ohms when rotation is complete, potentiometer is defective and must be replaced.

h. h. Install VOLTAGE adjust potentiometer (paragraph 2-33.4).

2-33.4 Installation.

a. a. Insert VOLTAGE adjust potentiometer (10, Figure 2-13) into control panel.

b. b. Install washer (9) and nut (8).

c. c. Install knob (7) and setscrews (6).

d. d. Install new shrinkable tubing and connect electrical leads using soldering gun. Remove tags.

e. e. Raise and secure control panel.

f. f. Connect negative battery cabl e and close battery access door.

2-34 FREQUENCY ADJUST POTENTIOMETER

2-34.1 Inspection.

a. a. Shut down generator set.

b. b. Inspect potentiometer for loose connections/mounting and other damage.

2-34.2 Removal.

WARNING

All metal jewelry can conduct electricity and become entangled in generator set components. Remove all jewelry when working on generator set. Failure to comply with this warning can cause injury or death to personnel.

WARNING

DO NOT wear loose clothing when performing checks, services and maintenance. Failure to comply with this warning can cause injury or death to personnel.

WARNING

When running, generator set engine has hot metal surfaces that will burn flesh on contact. Shut down generator set and allow engine to cool before performing checks, services, and maintenance. Wear gloves and additional protective clothing as required. Failure to comply with this warning can cause injury or death to personnel.

WARNING

Dangerous voltage exists on live circuits. Always observe precautions and never work alone. Failure to comply with this warning can cause injury or death to personnel.

WARNING

High voltage is produced when this generator set is in operation. SHUT DOWN generator set and make sure it is free of any power source before attempting any repair or maintenance on the set, or when connecting or disconnecting load cables. Failure to comply with this warning can cause injury or death to personnel.

WARNING

DC voltages are present at generator set electrical components even with generator set shut down. Avoid shorting any positive with ground/negative. Failure to comply with this warning can cause injury to personnel, and damage to equipment.

WARNING

Ensure that the engine cannot be started while maintenance is being performed. (ENGINE CONTROL switch set to OFF/RESET; Battery Disconnect Switch is OFF; DEAD CRANK SWITCH is OFF). Failure to comply with this warning can cause injury or death to personnel.

WARNING

Top housing panels and exhaust system can get very hot. Shut down generator set, and allow system to cool before performing checks, services and maintenance. Failure to comply with this warning can cause severe burns and injury to personnel.

a. a. Shut down generator set.

b. b. Open battery access door and disconnect negative battery cable.

c. c. Release control panel by turning two fasteners and lower control panel slowly.

d. d. Tag and disconnect FREQUENCY adjust potentiom eter (15, Figure 2-13) electrical leads using soldering gun. Remove shrinkable tubing.

e. e. Remove setscrews (11) and knob (12).

f. f. Remove nut (13) and washer (14).

g. g. Remove FREQUENCY adjust potentiometer (15) from control panel.

2-34.3 Testing.

a. a. Remove FREQUENCY adjust potentiometer (paragraph 2-34.2).

b. b. Set multimeter for ohms and connect across outer terminals of potentiometer (15, Figure 2-13). Multimeter should indicate between 4,500 and 5,500 ohms resistance.

c. Rotate potentiometer shaft CCW as far as it will go.

d. Connect multimeter between center terminal and either outer terminal. Multimeter should indicate 0 ohms resistance.

e. e. Slowly and at an even rate, rotate potentiometer shaft CW as far as it will go. Multimeter should increase at an even rate from 0 to 5,000 ohms.

f. f. If multimeter indication changes erratically or is not between 4,500 and 5,500 ohms when rotation is completed, potentiometer is defective and must be replaced.

g. Install FREQUENCY adjust potentiometer (paragraph 2-34.4).

2-34.4 Installation.

WARNING

High voltage is produced when the generator set is in operation. Never attempt to start the generator set unless it is properly grounded. Failure to comply with this warning can cause injury or death to personnel.

a. a. Insert FREQUENCY adjust potentiometer (15, Figure 2-13) into control panel.

b. b. Install washer (14) and nut (13).

c. c. Install knob (12) and setscrews (11).

d. d. Install new shrinkable tubing and solder leads. Remove tags.

e. e. Raise and secure control panel.

f. f. Connect negative battery cabl e and close battery access door.

2-35 BATTLE SHORT SWITCH 2-35.1

Inspection.

a. a. Shut down generator set.

b. b. Inspect switch for loose connec tions/mounting and other damage.

2-35.2 Removal.

WARNING

All metal jewelry can conduct electricity and become entangled in generator set components. Remove all jewelry when working on generator set. Failure to comply with this warning can cause injury or death to personnel.

WARNING

DO NOT wear loose clothing when performing checks, services and maintenance. Failure to comply with this warning can cause injury or death to personnel.

WARNING

When running, generator set engine has hot metal surfaces that will burn flesh on contact. Shut down generator set and allow engine to cool before performing checks, services, and maintenance. Wear gloves and additional protective clothing as required. Failure to comply with this warning can cause injury or death to personnel.

WARNING

Dangerous voltage exists on live circuits. Always observe precautions and never work alone. Failure to comply with this warning can cause injury or death to personnel.

WARNING

High voltage is produced when this generator set is in operation. SHUT DOWN generator set and make sure it is free of any power source before attempting any repair or maintenance on the set, or when connecting or disconnecting load cables. Failure to comply with this warning can cause injury or death to personnel.

WARNING

DC voltages are present at generator set electrical components even with generator set shut down. Avoid shorting any positive with ground/negative. Failure to comply with this warning can cause injury to personnel, and damage to equipment.

WARNING

Ensure that the engine cannot be started while maintenance is being performed. (ENGINE CONTROL switch set to OFF/RESET; Battery Disconnect Switch is OFF; DEAD CRANK SWITCH is OFF). Failure to comply with this warning can cause injury or death to personnel.

WARNING

Top housing panels and exhaust system can get very hot. Shut down generator set, and allow system to cool before performing checks, services and maintenance. Failure to comply with this warning can cause severe burns and injury to personnel.

a. a. Shut down generator set.

b. b. Open battery access door and disconnect negative battery cable.

c. c. Release control panel by turning two fasteners and lower control panel slowly.

d. d. Tag and disconnect BATTLE SHORT switch (19, Figure 2-13) electrical leads.

e. e. Remove nut (16), washer (17), and protective cover (18).

f. f. Remove BATTLE SHORT switch (19) from control panel and tab washer (20) from switch stem.

2-35.3 Testing.

a. a. Shut down generator set.

b. b. Open battery access door and disconnect negative battery cable.

c. c. Release control panel by turning two fasteners and lower control panel slowly.

d. d. Tag and disconnect BATTLE SHORT switch (19, Figure 2-13) electrical leads.

e. e. Place switch in ON position.

NOTE

Refer to Wiring Diagram FO-2 (MEP-804A/MEP-814A)/FO-4 (MEP-804B/MEP-814B) for terminal positions.

a. a. Set multimeter for ohms and check for continuity between terminals 2 and 3, 5 and 6, 8 and 9, and 11 and 12.

b. b. Place switch in OFF position.

c. c. Check for continuity between terminals 1 and 2, 4 and 5, 7 and 8, and 10 and 11.

d. d. Replace switch if any open circuit is indicated.

e. e. Connect electrical leads to switch (19). Remove tags.

f. f. Raise and secure control panel.

g. g. Connect negative battery cabl e and close battery access door.

2-35.4 Installation.

WARNING

High voltage is produced when the generator set is in operation. Never attempt to start the generator set unless it is properly grounded. Failure to comply with this warning can cause injury or death to personnel.

a. a. Install tab washer (20, Figure 2-13) and insert BATTLE SHORT switch (19) into control panel.

b. b. Install protective cover (18), washer (17), and nut (16).

c. c. Connect electrical leads. Remove tags.

d. d. Raise and secure control panel.

e. e. Connect negative battery c able and close battery access door.

2-36 AC CIRCUIT INTERRUPTER SWITCH

2-36.1 Inspection.

a. a. Shut down generator set.

b. b. Inspect switch for loose conne ctions/mounting and other damage.

WARNING

All metal jewelry can conduct electricity and become entangled in generator set components. Remove all jewelry when working on generator set. Failure to comply with this warning can cause injury or death to personnel.

WARNING

DO NOT wear loose clothing when performing checks, services and maintenance. Failure to comply with this warning can cause injury or death to personnel.

WARNING

When running, generator set engine has hot metal surfaces that will burn flesh on contact. Shut down generator set and allow engine to cool before performing checks, services, and maintenance. Wear gloves and additional protective clothing as required. Failure to comply with this warning can cause injury or death to personnel.

WARNING

Dangerous voltage exists on live circuits. Always observe precautions and never work alone. Failure to comply with this warning can cause injury or death to personnel.

WARNING

High voltage is produced when this generator set is in operation. SHUT DOWN generator set and make sure it is free of any power source before attempting any repair or maintenance on the set, or when connecting or disconnecting load cables. Failure to comply with this warning can cause injury or death to personnel.

WARNING

DC voltages are present at generator set electrical components even with generator set shut down. Avoid shorting any positive with ground/negative. Failure to comply with this warning can cause injury to personnel, and damage to equipment.

WARNING

Ensure that the engine cannot be started while maintenance is being performed. (ENGINE CONTROL switch set to OFF/RESET; Battery Disconnect Switch is OFF; DEAD CRANK SWITCH is OFF). Failure to comply with this warning can cause injury or death to personnel.

WARNING

Top housing panels and exhaust system can get very hot. Shut down generator set, and allow system to cool before performing checks, services and maintenance. Failure to comply with this warning can cause severe burns and injury to personnel.

a. a. Shut down generator set.

b. b. Open battery access door and disconnect negative battery cable.

c. c. Release control panel by turning two fasteners and lower control panel slowly.

d. d. Tag and disconnect AC CIRCUIT INTERRUPTER switch (23, Figure 2-13) electrical leads.

e. e. Remove nut (21) and washer (22).

f. f. Remove AC CIRCUIT INTERRUPTER switch (23) from control panel and tab washer (24) from switch stem.

2-36.3 Testing.

a. a. Shut down generator set.

b. b. Open battery access door and disconnect negative battery cable.

c. c. Release control panel by turning two fasteners and lower control panel slowly.

d. d. Tag and disconnect AC CIRCUIT INTERRUPTER switch (23, Figure 2-13) electrical leads.

NOTE

Refer to Wiring Diagram FO-2 (MEP-804A/MEP-814A)/FO-4 (MEP-804B/MEP-814B) for terminal positions.

e. e. Set multimeter for ohms and check for continuity between terminals 5 and 4 and terminals 2 and 3.

f. f. Check for open circuits between terminals 5 and 6 and terminals 1 and 2.

g. g. Place and hold AC CIRCUIT INTERRUPTER switch in CLOSED position.

h. h. Check for continuity between terminals 5 and 6 and terminals 2 and 3.

i. i. Check for open circuits between terminals 5 and 4 and terminals 2 and 1.

j. j. Place and hold AC CIRCUIT INTERRUPTER switch in OPEN position.

k. k. Check for continuity between terminals 5 and 4 and terminals 1 and 2.

l. l. Check for open circuits between terminals 5 and 6 and terminals 3 and 1.

m. m. Replace switch if any continuity check is other than indicated above.

n. n. Connect electrical leads to switch (23). Remove tags.

o. o. Raise and secure control panel.

p. p. Connect negative battery c able and close battery access door.

WARNING

High voltage is produced when the generator set is in operation. Never attempt to start the generator set unless it is properly grounded. Failure to comply with this warning can cause injury or death to personnel.

a. a. Install tab washer (24, Figure 2-13) and insert AC CIRCUIT INTERRUPTER switch (23) into control panel.

b. b. Install washer (22) and nut (21).

c. c. Connect electrical leads. Remove tags.

d. d. Raise and secure control panel.

e. e. Connect negative battery c able and close battery access door.

2-37 PARALLEL-UNIT SWITCH 2-37.1

Inspection.

a. a. Shut down generator set.

b. b. Inspect switch for loose conn ections/mounting and other damage.

2-37.2 Removal.

WARNING

All metal jewelry can conduct electricity and become entangled in generator set components. Remove all jewelry when working on generator set. Failure to comply with this warning can cause injury or death to personnel.

WARNING

DO NOT wear loose clothing when performing checks, services and maintenance. Failure to comply with this warning can cause injury or death to personnel.

WARNING

When running, generator set engine has hot metal surfaces that will burn flesh on contact. Shut down generator set and allow engine to cool before performing checks, services, and maintenance. Wear gloves and additional protective clothing as required. Failure to comply with this warning can cause injury or death to personnel.

WARNING

Dangerous voltage exists on live circuits. Always observe precautions and never work alone. Failure to comply with this warning can cause injury or death to personnel.

WARNING

High voltage is produced when this generator set is in operation. SHUT DOWN generator set and make sure it is free of any power source before attempting any repair or maintenance on the set, or when connecting or disconnecting load cables. Failure to comply with this warning can cause injury or death to personnel.

WARNING

DC voltages are present at generator set electrical components even with generator set shut down. Avoid shorting any positive with ground/negative. Failure to comply with this warning can cause injury to personnel, and damage to equipment.

WARNING

Ensure that the engine cannot be started while maintenance is being performed. (ENGINE CONTROL switch set to OFF/RESET; Battery Disconnect Switch is OFF; DEAD CRANK SWITCH is OFF). Failure to comply with this warning can cause injury or death to personnel.

WARNING

Top housing panels and exhaust system can get very hot. Shut down generator set, and allow system to cool before performing checks, services and maintenance. Failure to comply with this warning can cause severe burns and injury to personnel.

a. a. Shut down generator set.

b. b. Open battery access door and disconnect negative battery cable.

c. c. Release control panel by turning two fasteners and lower control panel slowly.

d. d. Tag and disconnect PARALLEL-UNIT switch (27, Figure 2-13) electrical leads.

e. e. Remove nut (25) and washer (26).

f. f. Remove PARALLEL-UNIT switch (27) from control panel and tab washer (28) from switch stem.

2-37.3 Testing.

a. a. Shut down generator set.

b. b. Open battery access door and disconnect negative battery cable.

c. c. Release control panel by turning two fasteners and lower control panel slowly.

d. d. Tag and disconnect PARALLEL-UNIT switch (27, Figure 2-13) electrical leads.

NOTE

Refer to Wiring Diagram FO-2 (MEP-804A/MEP-814A)/FO-4 (MEP-804B/MEP-814B) for terminal positions.

e. e. Place switch in PARALLEL position.

f. f. Set multimeter for ohms and check for continuity between terminals 1 and 2, 4 and 5, 7 and 8, and 10 and 11.

g. g. Place switch in UNIT position.

h. h. Check for continuity between terminals 2 and 3, 5 and 6, 8 and 9, and 11 and 12.

i. i. Replace switch if any open circuit is indicated.

j. j. Connect electrical leads to switch (27). Remove tags.

k. k. Raise and secure control panel.

l. l. Connect negative battery c able and close battery access door.

2-37.4 Installation.

WARNING

High voltage is produced when the generator set is in operation. Never attempt to start the generator set unless it is properly grounded. Failure to comply with this warning can cause injury or death to personnel.

a. a. Install tab washer (28, Figure 2-13) and insert PARALLEL-UNIT switch (27) into control panel.

b. b. Install washer (26) and nut (25).

c. c. Connect electrical leads. Remove tags.

d. d. Raise and secure control panel.

e. e. Connect negative battery c able and close battery access door.

2-38 VM-AM TRANSFER SWITCH 2.38.1

a. a. Shut down generator set.

b. b. Inspect switch for loose conn ections/mounting and other damage.

WARNING

All metal jewelry can conduct electricity and become entangled in generator set components. Remove all jewelry when working on generator set. Failure to comply with this warning can cause injury or death to personnel.

WARNING

DO NOT wear loose clothing when performing checks, services and maintenance. Failure to comply with this warning can cause injury or death to personnel.

WARNING

When running, generator set engine has hot metal surfaces that will burn flesh on contact. Shut down generator set and allow engine to cool before performing checks, services, and maintenance. Wear gloves and additional protective clothing as required. Failure to comply with this warning can cause injury or death to personnel.

WARNING

Dangerous voltage exists on live circuits. Always observe precautions and never work alone. Failure to comply with this warning can cause injury or death to personnel.

WARNING

High voltage is produced when this generator set is in operation. SHUT DOWN generator set and make sure it is free of any power source before attempting any repair or maintenance on the set, or when connecting or disconnecting load cables. Failure to comply with this warning can cause injury or death to personnel.

WARNING

DC voltages are present at generator set electrical components even with generator set shut down. Avoid shorting any positive with ground/negative. Failure to comply with this warning can cause injury to personnel, and damage to equipment.

WARNING

Ensure that the engine cannot be started while maintenance is being performed. (ENGINE CONTROL switch set to OFF/RESET; Battery Disconnect Switch is OFF; DEAD CRANK SWITCH is OFF). Failure to comply with this warning can cause injury or death to personnel.

WARNING

Top housing panels and exhaust system can get very hot. Shut down generator set, and allow system to cool before performing checks, services and maintenance. Failure to comply with this warning can cause severe burns and injury to personnel.

a. Shut down generator set.

b. Open battery access door and disconnect negative battery cable.

c. Release control panel by turning two fasteners and lower control panel slowly.

d. Remove setscrew (29, Figure 2-13) and knob (30).

e. Remove nuts (31) and screws (32).

f. Remove VM-AM transfer switch (33) from control panel.

g. Tag and disconnect VM-AM transfer switch (33) electrical leads.

2-38.3 Testing.

a. Shut down generator set.

b. Remove VM-AM transfer switch (paragraph 2-38.2).

c. Set multimeter for ohms and check VM-AM transfer switch for continuity. Refer to Electrical Schematic FO-1 (S-6 Circuit Schedule) (MEP-804A/MEP-814A)/FO-3 (MEP-804B/MEP-814B) to determine circuits made to corresponding switch positions.

d. Check continuity in all six switch positions.

e. Replace VM-AM transfer switch If open circuit is indicated.

f. Install VM-AM transfer switch (paragraph 2-38.4).

2-38.4 Installation.

WARNING

High voltage is produced when the generator set is in operation. Never attempt to start the generator set unless it is properly grounded. Failure to comply with this warning can cause injury or death to personnel.

a. Connect electrical leads to VM-AM transfer switch (33, Figure 2-13). Remove tags.

b. b. Insert VM-AM transfer switch (33) into control panel.

c. c. Install screws (32) and nuts (31).

d. d. Install knob (30) and setscrew (29).

e. e. Raise and secure control panel.

f. f. Connect negative battery cabl e and close battery access door.

2-39 PANEL LIGHTS SWITCH 2-39.1

Inspection.

a. a. Shut down generator set.

b. b. Inspect switch for loose connec tions/mounting and other damage.

2-39.2 Removal.

WARNING

All metal jewelry can conduct electricity and become entangled in generator set components. Remove all jewelry when working on generator set. Failure to comply with this warning can cause injury or death to personnel.

WARNING

DO NOT wear loose clothing when performing checks, services and maintenance. Failure to comply with this warning can cause injury or death to personnel.

WARNING

When running, generator set engine has hot metal surfaces that will burn flesh on contact. Shut down generator set and allow engine to cool before performing checks, services, and maintenance. Wear gloves and additional protective clothing as required. Failure to comply with this warning can cause injury or death to personnel.

WARNING

Dangerous voltage exists on live circuits. Always observe precautions and never work alone. Failure to comply with this warning can cause injury or death to personnel.

WARNING

High voltage is produced when this generator set is in operation. SHUT DOWN generator set and make sure it is free of any power source before attempting any repair or maintenance on the set, or when connecting or disconnecting load cables. Failure to comply with this warning can cause injury or death to personnel.

WARNING

DC voltages are present at generator set electrical components even with generator set shut down. Avoid shorting any positive with ground/negative. Failure to comply with this warning can cause injury to personnel, and damage to equipment.

WARNING

Ensure that the engine cannot be started while maintenance is being performed. (ENGINE CONTROL switch set to OFF/RESET; Battery Disconnect Switch is OFF; DEAD CRANK SWITCH is OFF). Failure to comply with this warning can cause injury or death to personnel.

WARNING

Top housing panels and exhaust system can get very hot. Shut down generator set, and allow system to cool before performing checks, services and maintenance. Failure to comply with this warning can cause severe burns and injury to personnel.

a. a. Shut down generator set.

b. b. Open battery access door and disconnect negative battery cable.

c. c. Release control panel by turning two fasteners and lower control panel slowly.

d. d. Tag and disconnect PANEL LIGHTS switch (36, Figure 2-13) electrical leads.

e. e. Remove nut (34) and washer (35).

f. f. Remove PANEL LIGHTS switch (36) from cont rol panel and tab washer (37) from switch stem.

2-39.3 Testing.

a. a. Shut down generator set.

b. b. Open battery access door and disconnect negative battery cable.

c. c. Release control panel by turning two fasteners and lower control panel slowly.

d. d. Tag and disconnect PANEL LIGHTS switch (36, Figure 2-13) electrical leads.

e. e. Set multimeter for ohms and connect across switch terminals.

f. f. Place switch in ON position. Multimeter should indicate continuity.

g. g. Place switch in OFF position. Multimeter should indicate open circuit.

h. h. Replace PANEL LIGHTS switch if readings are not as above.

i. i. Connect electrical leads to switch (36). Remove tags.

j. j. Raise and secure control panel.

k. k. Connect negative battery c able and close battery access
door.

WARNING

High voltage is produced when the generator set is in operation. Never attempt to start the generator set unless it is properly grounded. Failure to comply with this warning can cause injury or death to personnel.

a. a. Install tab washer (37, Figure 2-13) and insert PANEL LIGHTS switch (36) into control panel.

b. b. Install washer (35) and nut (34).

c. c. Install electrical leads. Remove tags.

d. d. Raise and secure control panel.

e. e. Connect negative battery c able and close battery access door.

2-40 EMERGENCY STOP SWITCH ASSEMBLY 2-40.1

Inspection.

a. a. Shut down generator set.

b. b. Inspect switch assembly for loose connections/mounting and other damage.

2-40.2 Testing.

a. a. Shut down generator set.

b. b. Open battery access door and disconnect negative battery cable.

c. c. Release control panel by turning two fasteners and lower control panel slowly.

d. d. Tag and disconnect EMERGENCY STOP switch assembly (40, Figure 2-13) electrical leads.

e. e. With switch in normal (out) position, set multimeter for ohms and check circuit between switch terminals. If no continuity is indicated, EMERGENCY STOP switch assembly is defective.

f. Push EMERGENCY STOP switch assembly to in position.

g. Check for continuity between EMERGENCY STOP switch assembly terminals. Replace EMERGENCY STOP switch assembly if continuity is indicated.

h. h. Connect electrical leads to EMERGENCY STOP switch assembly (40). Remove tags.

i. i. Raise and secure control panel.

j. j. Connect negative battery c able and close battery access door.

2-40.3 Removal.

WARNING

All metal jewelry can conduct electricity and become entangled in generator set components. Remove all jewelry when working on generator set. Failure to comply with this warning can cause injury or death to personnel.

WARNING

DO NOT wear loose clothing when performing checks, services and maintenance. Failure to comply with this warning can cause injury or death to personnel.

WARNING

When running, generator set engine has hot metal surfaces that will burn flesh on contact. Shut down generator set and allow engine to cool before performing checks, services, and maintenance. Wear gloves and additional protective clothing as required. Failure to comply with this warning can cause injury or death to personnel.

WARNING

Dangerous voltage exists on live circuits. Always observe precautions and never work alone. Failure to comply with this warning can cause injury or death to personnel.

WARNING

High voltage is produced when this generator set is in operation. SHUT DOWN generator set and make sure it is free of any power source before attempting any repair or maintenance on the set, or when connecting or disconnecting load cables. Failure to comply with this warning can cause injury or death to personnel.

WARNING

DC voltages are present at generator set electrical components even with generator set shut down. Avoid shorting any positive with ground/negative. Failure to comply with this warning can cause injury to personnel, and damage to equipment.

WARNING

Ensure that the engine cannot be started while maintenance is being performed. (ENGINE CONTROL switch set to OFF/RESET; Battery Disconnect Switch is OFF; DEAD CRANK SWITCH is OFF). Failure to comply with this warning can cause injury or death to personnel.

WARNING

Top housing panels and exhaust system can get very hot. Shut down generator set, and allow system to cool before performing checks, services and maintenance. Failure to comply with this warning can cause severe burns and injury to personnel.

a. a. Shut down generator set.

b. b. Open battery access door and disconnect negative battery cable.

c. c. Release control panel by turning two fasteners and lower control panel slowly.

d. d. Tag and disconnect EMERGENCY STOP switch assembly (40, Figure 2-13) electrical leads.

e. e. Remove knob (38) and nut (39).

f. f. Remove EMERGENCY STOP switch assembly (40) from control panel.

2-40.4 Installation.

WARNING

High voltage is produced when the generator set is in operation. Never attempt to start the generator set unless it is properly grounded. Failure to comply with this warning can cause injury or death to personnel.

a. a. Insert EMERGENCY STOP switch assemb ly (40, Figure 2-13) into control panel.

b. b. Install nut (39) and knob (38).

c. c. Connect electrical leads. Remove tags.

d. e. Raise and secure control panel.

e. f. Connect negative battery cabl e and close battery access door.

2-41 REACTIVE CURRENT ADJUST RHEOSTAT 2-41.1

Inspection.

a. a. Shut down generator set.

b. b. Inspect rheostat for loose connections/mounting and other damage.

2-41.2 Testing.

a. a. Shut down generator set.

b. b. Open battery access door and disconnect negative battery cable.

c. c. Remove control box top panel (paragraph 2-15.1).

d. d. Disconnect wires 143C from TB4 terminal 21, 142A from TB6 terminal 12, and 135B from TB5 terminal 4.

e. e. Mark reading of REACTIVE CURRENT ADJUST rheo stat to reposition at conclusion of testing steps.

f. f. Set multimeter for ohms and connect to wires 135B and 142A. Multimeter reading should be between 4.5 and 5.5 ohms.

g. g. Connect multimeter to wires 135B and 143C and turn REACTIVE CURRENT ADJUST rheostat to full CW position. Multimeter reading should be approximately 0 ohms. Turn REACTIVE CURRENT ADJUST rheostat slowly to full CCW position and observe multimeter. Multimeter reading should evenly increase to between 4.5 and 5.5 ohms.

h. h. Connect multimeter to wires 142A and 143C and turn REACTIVE CURRENT ADJUST rheostat to full CW position. Multimeter reading should be between 4.5 and 5.5 ohms. Turn REACTIVE CURRENT ADJUST rheostat slowly to full CCW position and observe multimeter. Multimeter reading should evenly decrease to approximately 0 ohms.

i. i. Replace REACTIVE CURRENT ADJUST rheostat if multimeter readings are other than above. j

Reposition REACTIVE CURRENT ADJUST rheostat as marked in step e above.

j. k. Connect electrical wires as tagged.

k. l. Install control box top panel (paragraph 2-15.4).

l. m. Connect negative battery c able and close battery access door.

2-41.3 Removal.

WARNING

All metal jewelry can conduct electricity and become entangled in generator set components. Remove all jewelry when working on generator set. Failure to comply with this warning can cause injury or death to personnel.

WARNING

DO NOT wear loose clothing when performing checks, services and maintenance. Failure to comply with this warning can cause injury or death to personnel.

WARNING

When running, generator set engine has hot metal surfaces that will burn flesh on contact. Shut down generator set and allow engine to cool before performing checks, services, and maintenance. Wear gloves and additional protective clothing as required. Failure to comply with this warning can cause injury or death to personnel.

WARNING

Dangerous voltage exists on live circuits. Always observe precautions and never work alone. Failure to comply with this warning can cause injury or death to personnel.

WARNING

High voltage is produced when this generator set is in operation. SHUT DOWN generator set and make sure it is free of any power source before attempting any repair or maintenance on the set, or when connecting or disconnecting load cables. Failure to comply with this warning can cause injury or death to personnel.

WARNING

DC voltages are present at generator set electrical components even with generator set shut down. Avoid shorting any positive with ground/negative. Failure to comply with this warning can cause injury to personnel, and damage to equipment.

WARNING

Ensure that the engine cannot be started while maintenance is being performed. (ENGINE CONTROL switch set to OFF/RESET; Battery Disconnect Switch is OFF; DEAD CRANK SWITCH is OFF). Failure to comply with this warning can cause injury or death to personnel.

WARNING

Top housing panels and exhaust system can get very hot. Shut down generator set, and allow system to cool before performing checks, services and maintenance. Failure to comply with this warning can cause severe burns and injury to personnel.

a. a. Shut down generator set.

b. b. Open battery access door and disconnect negative battery cable.

c. c. Remove control box top panel (paragraph 2-15.1).

d. d. Tag and disconnect REACTIVE CURRENT ADJUST rheostat (3, Figure 2-14) electrical leads by unsoldering and remove shrinkable tubing.

e. Remove nuts (1 and 2) and rheostat (3).

2-41.4 Installation.

WARNING

High voltage is produced when the generator set is in operation. Never attempt to start the generator set unless it is properly grounded. Failure to comply with this warning can cause injury or death to personnel.

a. a. Install rheostat (3, Figure 2-14) in mounting bracket.

b. b. Install nuts (1 and 2).

c. c. Install new shrinkable tubing and s older electrical leads. Remove tags.

d. d. Install control box top panel (paragraph 2-15.4).

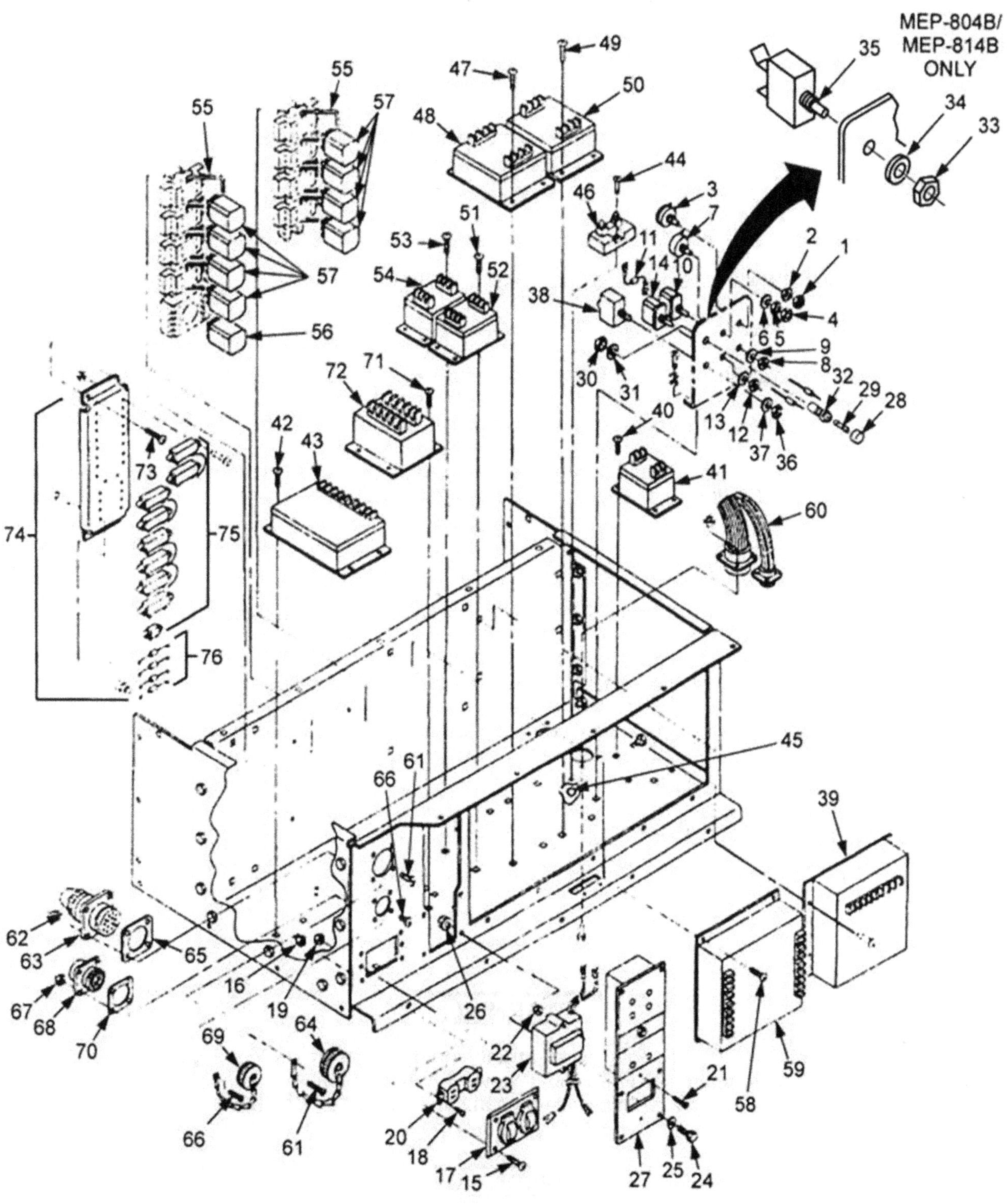

Figure 2-14. Control Box Components.

2-42 LOAD SHARING ADJUST RHEOSTAT 2-42.1

Inspection.

a. a. Shut down generator set.

b. b. Inspect rheostat for loose connections/mounting and other damage.

2.42.2 Testing.

a. a. Shut down generator set.

b. b. Open battery access door and disconnect negative battery cable.

c. c. Remove control box top panel (paragraph 2-15.1).

d. d. Tag and disconnect wire 161B from reverse po wer relay terminal 2. Insulate end of wire.

e. e. Disconnect wire 159A from governor control unit terminal 11.

f. f. Set multimeter for ohms and connect positive lead to wire 159A and negative lead of multimeter to terminal 12 of governor control unit. Record reading of rheostat.

g. g. Turn LOAD SHARING ADJUST rheostat to full CCW position. Multimeter reading should be between 4,500 and 5,500 ohms.

h. h. Turn LOAD SHARING ADJUST rheostat to f ull CW position. Multimeter reading should be approximately 0 ohms.

i. i. Replace LOAD SHARING ADJUST rheostat if multimeter readings are other than above.

j. j. If readings are within tolerance, return rheo stat to reading recorded in step f above.

k. k. Connect electrical wires as tagged.

l. l. Install control box top panel (paragraph 2-15.4).

m. m. Connect negative battery cabl e and close battery access door.

2-42.3 Removal.

WARNING

All metal jewelry can conduct electricity and become entangled in generator set components. Remove all jewelry when working on generator set. Failure to comply with this warning can cause injury or death to personnel.

WARNING

DO NOT wear loose clothing when performing checks, services and maintenance. Failure to comply with this warning can cause injury or death to personnel.

WARNING

When running, generator set engine has hot metal surfaces that will burn flesh on contact. Shut down generator set and allow engine to cool before performing checks, services, and maintenance. Wear gloves and additional protective clothing as required. Failure to comply with this warning can cause injury or death to personnel.

WARNING

Dangerous voltage exists on live circuits. Always observe precautions and never work alone. Failure to comply with this warning can cause injury or death to personnel.

WARNING

High voltage is produced when this generator set is in operation. SHUT DOWN generator set and make sure it is free of any power source before attempting any repair or maintenance on the set, or when connecting or disconnecting load cables. Failure to comply with this warning can cause injury or death to personnel.

WARNING

DC voltages are present at generator set electrical components even with generator set shut down. Avoid shorting any positive with ground/negative. Failure to comply with this warning can cause injury to personnel, and damage to equipment.

WARNING

Ensure that the engine cannot be started while maintenance is being performed. (ENGINE CONTROL switch set to OFF/RESET; Battery Disconnect Switch is OFF; DEAD CRANK SWITCH is OFF). Failure to comply with this warning can cause injury or death to personnel.

WARNING

Top housing panels and exhaust system can get very hot. Shut down generator set, and allow system to cool before performing checks, services and maintenance. Failure to comply with this warning can cause severe burns and injury to personnel.

a. a. Shut down generator set.

b. b. Open battery access door and disconnect negative battery cable.

c. c. Remove control box top panel (paragraph 2-15.1).

d. d. Tag and disconnect LOAD SHARING ADJUST r heostat (7, Figure 2-14) electrical

e. Remove nuts (4 and 5), lock washer (6), and rheostat (7). Discard lock washer (6).

2-42.4 Installation.

WARNING

High voltage is produced when the generator set is in operation. Never attempt to start the generator set unless it is properly grounded. Failure to comply with this warning can cause injury or death to personnel.

a. a. Install LOAD SHARING ADJUST rheostat (7, Figure 2-14) in mounting bracket.

b. b. Install new lock washer (6) and nuts (5 and 4).

c. c. Connect electrical leads. Remove tags.

d. d. Install control box top panel (paragraph 2-15.4).

e. e. Connect negative battery cabl e and close battery access door.

2-43 OVERSPEED RESET SWITCH 2-43.1

Inspection.

a. a. Shut down generator set.

b. b. Inspect switch for loose conn ections/mounting and other damage.

2-43.2 Testing.

a. a. Shut down generator set.

b. b. Open battery access door and disconnect negative battery cable.

c. c. Remove control box top panel (paragraph 2-15.1).

d. d. Tag and disconnect OVERSPEED RESET switch (10, Figure 2-14) electrical leads.

e. e. Set multimeter for ohms and connect across switch terminals. Multimeter should indicate continuity.

f. f. Position and hold switch in up position. Multimeter should indicate open circuit.

g. g. Replace OVERSPEED RESET switch if indications are other than above.

h. h. Connect electrical leads to switch (10). Remove tags.

i. i. Install control box top panel (paragraph 2-15.4).

j. j. Connect negative battery c able and close battery access door.

2-43.3 Removal.

WARNING

All metal jewelry can conduct electricity and become entangled in generator set components. Remove all jewelry when working on generator set. Failure to comply with this warning can cause injury or death to personnel.

WARNING

DO NOT wear loose clothing when performing checks, services and maintenance. Failure to comply with this warning can cause injury or death to personnel.

WARNING

When running, generator set engine has hot metal surfaces that will burn flesh on contact. Shut down generator set and allow engine to cool before performing checks, services, and maintenance. Wear gloves and additional protective clothing as required. Failure to comply with this warning can cause injury or death to personnel.

WARNING

Dangerous voltage exists on live circuits. Always observe precautions and never work alone. Failure to comply with this warning can cause injury or death to personnel.

WARNING

High voltage is produced when this generator set is in operation. SHUT DOWN generator set and make sure it is free of any power source before attempting any repair or maintenance on the set, or when connecting or disconnecting load cables. Failure to comply with this warning can cause injury or death to personnel.

WARNING

DC voltages are present at generator set electrical components even with generator set shut down. Avoid shorting any positive with ground/negative. Failure to comply with this warning can cause injury to personnel, and damage to equipment.

WARNING

Ensure that the engine cannot be started while maintenance is being performed. (ENGINE CONTROL switch set to OFF/RESET; Battery Disconnect Switch is OFF; DEAD CRANK SWITCH is OFF). Failure to comply with this warning can cause injury or death to personnel.

WARNING

Top housing panels and exhaust system can get very hot. Shut down generator set, and allow system to cool before performing checks, services and maintenance. Failure to comply with this warning can cause severe burns and injury to personnel.

a. a. Shut down generator set.

b. b. Open battery access door and disconnect negative battery cable.

c. c. Remove control box top panel (paragraph 2-15.1).

d. d. Tag and disconnect OVERSPEED RESET switch (10, Figure 2-14) electrical leads.

e. e. Remove nut (8), lock washer (9), and switch (10). Discard lock washer (9).

WARNING

High voltage is produced when the generator set is in operation. Never attempt to start the generator set unless it is properly grounded. Failure to comply with this warning can cause injury or death to personnel.

a. a. Install OVERSPEED RESET switch (10, Figure 2-14) in mounting bracket.

b. b. Install new lock washer (9) and nut (8).

c. c. Connect electrical leads. Remove tags.

d. d. Install control box top panel (paragraph 2-15.4).

e. e. Connect negative battery c able and close battery access door.

2-44 FREQUENCY SELECT SWITCH (MEP-804A/MEP-804B)

2-44.1 Inspection.

a. a. Shut down generator set.

b. b. Inspect switch for loose conn ections/mounting and other damage.

2-44.2 Testing.

a. a. Shut down generator set.

b. b. Open battery access door and disconnect negative battery cable.

c. c. Remove control box top panel (paragraph 2-15.1).

d. d. Tag and disconnect FREQUENCY SELECT switch (14, Figure 2-14) electrical leads.

e. e. Set multimeter for ohms and connect across switch terminals.

f. f. Place switch in up (60 Hz) position. Multimeter should indicate continuity.

g. g. Place switch in down (50 Hz) position. Multimeter should indicate open.

h. h. Replace FREQUENCY SELECT switch if indications are other than above.

i. i. Connect electrical leads to switch (14). Remove tags.

j. j. Install control box top panel (paragraph 2-15.4)

k. k. Connect negative battery cabl e and close battery access door.

WARNING

All metal jewelry can conduct electricity and become entangled in generator set components. Remove all jewelry when working on generator set. Failure to comply with this warning can cause injury or death to personnel.

WARNING

DO NOT wear loose clothing when performing checks, services and maintenance. Failure to comply with this warning can cause injury or death to personnel.

WARNING

When running, generator set engine has hot metal surfaces that will burn flesh on contact. Shut down generator set and allow engine to cool before performing checks, services, and maintenance. Wear gloves and additional protective clothing as required. Failure to comply with this warning can cause injury or death to personnel.

WARNING

Dangerous voltage exists on live circuits. Always observe precautions and never work alone. Failure to comply with this warning can cause injury or death to personnel.

WARNING

High voltage is produced when this generator set is in operation. SHUT DOWN generator set and make sure it is free of any power source before attempting any repair or maintenance on the set, or when connecting or disconnecting load cables. Failure to comply with this warning can cause injury or death to personnel.

WARNING

DC voltages are present at generator set electrical components even with generator set shut down. Avoid shorting any positive with ground/negative. Failure to comply with this warning can cause injury to personnel, and damage to equipment.

WARNING

Ensure that the engine cannot be started while maintenance is being performed. (ENGINE CONTROL switch set to OFF/RESET; Battery Disconnect Switch is OFF; DEAD CRANK SWITCH is OFF). Failure to comply with this warning can cause injury or death to personnel.

WARNING

Top housing panels and exhaust system can get very hot. Shut down generator set, and allow system to cool before performing checks, services and maintenance. Failure to comply with this warning can cause severe burns and injury to personnel.

a. a. Shut down generator set.

b. b. Open battery access door and disconnect negative battery cable.

c. c. Remove control box top panel (paragraph 2-15.1).

d. d. Tag and disconnect electrical leads from FR EQUENCY SELECT switch (14, Figure 2-14).

e. e. Tag and remove resistor (11) from FREQUENCY SELECT switch (14).

f. f. Remove nut (12), lock washer (13), and FREQUENCY SELECT switch (14). Discard lock washer (13).

2-44.4 <u>Installation</u>.

WARNING

High voltage is produced when the generator set is in operation. Never attempt to start the generator set unless it is properly grounded. Failure to comply with this warning can cause injury or death to personnel.

a. a. Position FREQUENCY SELECT switch (14, Figure 2-14) in mounting bracket.

b. b. Install new lock washer (13) and nut (12).

c. c. Install resistor (11) on FREQUENCY SELECT switch (14). Remove tags.

d. d. Connect electrical leads. Remove tags.

e. e. Install control box top panel (paragraph 2-15.4).

f. f. Connect negative battery cabl e and close battery access door.

2-45 CONVENIENCE RECEPTACLE 2-45.1

Inspection.

a. a. Shut down generator set.

b. b. Inspect CONVENIENCE RECEPTACLE for cracks, breaks, corrosion, bent terminals, and other damage.

c. c. Inspect cover for cracks, corrosion, or damaged springs.

d. d. Replace defective parts.

2-45.2 Testing.

a. a. Shut down generator set.

b. b. Remove control box top panel (paragraph 2-15.1).

c. c. Tag and disconnect CONVENIENCE RECEPTACLE (20, Figure 2-14) electrical leads.

d. d. Set multimeter for ohms and check for continuity between upper side terminals and lower side terminals of each plug outlet.

e. e. Replace CONVENIENCE RECEPTACLE if continuity is indicated between terminals.

f. f. Connect electrical leads to receptacle (20). Remove tags.

g. g. Install control box top panel (paragraph 2-15.4).

2-45.3 Removal.

WARNING

All metal jewelry can conduct electricity and become entangled in generator set components. Remove all jewelry when working on generator set. Failure to comply with this warning can cause injury or death to personnel.

WARNING

DO NOT wear loose clothing when performing checks, services and maintenance. Failure to comply with this warning can cause injury or death to personnel.

WARNING

When running, generator set engine has hot metal surfaces that will burn flesh on contact. Shut down generator set and allow engine to cool before performing checks, services, and maintenance. Wear gloves and additional protective clothing as required. Failure to comply with this warning can cause injury or death to personnel.

WARNING

Dangerous voltage exists on live circuits. Always observe precautions and never work alone. Failure to comply with this warning can cause injury or death to personnel.

WARNING

High voltage is produced when this generator set is in operation. SHUT DOWN generator set and make sure it is free of any power source before attempting any repair or maintenance on the set, or when connecting or disconnecting load cables. Failure to comply with this warning can cause injury or death to personnel.

WARNING

DC voltages are present at generator set electrical components even with generator set shut down. Avoid shorting any positive with ground/negative. Failure to comply with this warning can cause injury to personnel, and damage to equipment.

WARNING

Ensure that the engine cannot be started while maintenance is being performed. (ENGINE CONTROL switch set to OFF/RESET; Battery Disconnect Switch is OFF; DEAD CRANK SWITCH is OFF). Failure to comply with this warning can cause injury or death to personnel.

WARNING

Top housing panels and exhaust system can get very hot. Shut down generator set, and allow system to cool before performing checks, services and maintenance. Failure to comply with this warning can cause severe burns and injury to personnel.

a. a. Shut down generator set.

b. b. Open battery access door and disconnect negative battery cable.

c. c. Remove control box top panel (paragraph 2-15.1)

d. d. Remove screws (15, Figure 2-14) and nuts (16).

e. e. Remove CONVENIENCE RECEPTACLE cover (17).

f. f. Remove machine screws (18), nuts (19), and CONVENIENCE RECEPTACLE (20).

g. g. Tag and disconnect CONVENIENCE RECEPTACLE (20) electrical leads.

WARNING

High voltage is produced when the generator set is in operation. Never attempt to start the generator set unless it is properly grounded. Failure to comply with this warning can cause injury or death to personnel.

a. Connect electrical leads to CONVENIENCE RECEPTACLE (20, Figure 2-14). Remove tags.

b. Install CONVENIENCE RECEPTACLE (20) into panel cutout with machine screws (18) and nuts (19).

c. Install CONVENIENCE RECEPTACLE cover (17) with screws (15) and nuts (16).

d. Install control box top panel (paragraph 2-15.4).

e. Connect negative battery c able and close battery access door.

2-46 GROUND FAULT CIRCUIT INTERRUPTER

a. Shut down generator set.

b. Inspect GROUND FAULT CIRCUIT INTERRUPTER for cracks, corrosion, frayed wires, and other damage.

2-46.2 Testing.

a. Start and operate generator set at rated voltage and frequency.

b. Set multimeter for AC volts, press TEST bu tton, and check for zero voltage at CONVENIENCE RECEPTACLE.

c. Press RESET button on GROUND FAULT CIRCUIT INTERRUPTER and use multimeter to check for 120 VAC at CONVENIENCE RECEPTACLE.

d. Replace GROUND FAULT CIRCUIT INTERRUPTER if indications are other than above.

2-46.3 Removal.

WARNING

All metal jewelry can conduct electricity and become entangled in generator set components. Remove all jewelry when working on generator set. Failure to comply with this warning can cause injury or death to personnel.

WARNING

DO NOT wear loose clothing when performing checks, services and maintenance. Failure to comply with this warning can cause injury or death to personnel.

WARNING

When running, generator set engine has hot metal surfaces that will burn flesh on contact. Shut down generator set and allow engine to cool before performing checks, services, and maintenance. Wear gloves and additional protective clothing as required. Failure to comply with this warning can cause injury or death to personnel.

WARNING

Dangerous voltage exists on live circuits. Always observe precautions and never work alone. Failure to comply with this warning can cause injury or death to personnel.

WARNING

High voltage is produced when this generator set is in operation. SHUT DOWN generator set and make sure it is free of any power source before attempting any repair or maintenance on the set, or when connecting or disconnecting load cables. Failure to comply with this warning can cause injury or death to personnel.

WARNING

DC voltages are present at generator set electrical components even with generator set shut down. Avoid shorting any positive with ground/negative. Failure to comply with this warning can cause injury to personnel, and damage to equipment.

WARNING

Ensure that the engine cannot be started while maintenance is being performed. (ENGINE CONTROL switch set to OFF/RESET; Battery Disconnect Switch is OFF; DEAD CRANK SWITCH is OFF). Failure to comply with this warning can cause injury or death to personnel.

WARNING

Top housing panels and exhaust system can get very hot. Shut down generator set, and allow system to cool before performing checks, services and maintenance. Failure to comply with this warning can cause severe burns and injury to personnel.

a. a. Shut down generator set.

b. b. Open battery access door and disconnect negative battery cable.

c. c. Tag and disconnect GROUND FAULT CIRCUIT INTERRUPTER (23, Figure 2-14) electrical leads from TB4 and KS and CONVENIENCE RECEPTACLE (20).

d. Remove screws (21) and nuts (22).

e. Remove GROUND FAULT CIRCUIT INTERRUPTER (23) from malfunction indicator panel.

2-46.4 Installation.

WARNING

High voltage is produced when the generator set is in operation. Never attempt to start the generator set unless it is properly grounded. Failure to comply with this warning can cause injury or death to personnel.

a. a. Install GROUND FAULT CIRCUIT INTERRUPTER (23, Figure 2-14) in malfunction indicator panel with screws (21) and nuts (22).

b. b. Connect electrical leads to TB4 and KS and CONVENIENCE RECEPTACLE (20). Remove tags.

c. c. Connect negative battery c able and close battery access door.

2-46.4.1 In-Line Fuse Installation.

NOTE

The following procedure applies to generator sets under contract number DAAK01-88-D-0082.

When replacing GROUND FAULT CIRCUIT INTERRUPTER, use new GROUND FAULT CIRCUIT INTERRUPTER with integral circuit breaker. Refer to TM 9-6115-643-24P for new part number.

a. a. Shut down generator set.

b. b. Open left side engine access door and disconnect negative battery cable.

c. c. Remove malfunction indicator panel screws (24, Figure 2-14), washers (29), and nuts (26). Lay malfunction indicator panel to the side.

d. d. Cut black wire on load side of GROUND FAULT CIRCUIT INTERRUPTER (23).

e. e. Strip wires on in-line fuse holder (2, Figure 2-15) and install butt splices (3) at each end. Connect ends of black wire to in-line fuse holder butt splices (3).

f. Install fuse (1) in fuse holder (2).

Figure 2-15. In-Line Fuse Installation.

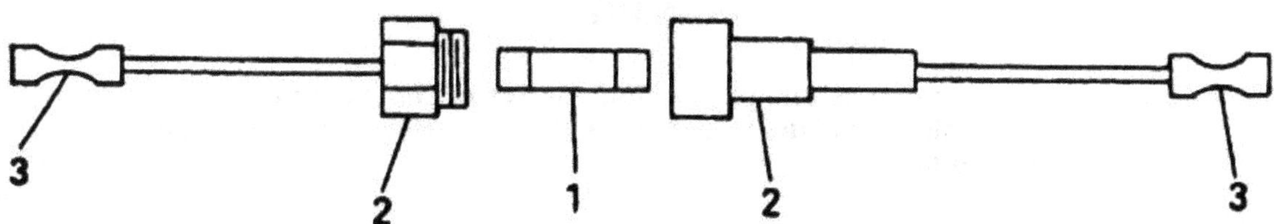

g. g. Secure excess wire to wiring harness using tie wrap.

h. h. Install malfunction indicator panel (27, Figure 2-14), screws (24), washers (25), and nuts (26).

i. Reconnect negative battery cable and close battery access door.

2-47 MALFUNCTION INDICATOR PANEL 2.47.1

Inspection.

a. a. Shut down generator set.

b. b. Inspect malfunction indicator panel for broken indicator lights, cracked housing, corrosion, and other damage.

2-47.2 Testing.

a. a. Depress TEST/RESET button and check that all indicators are illuminated.

b. b. Replace malfunction indicator panel if one or more indicators do not illuminate.

2-47.3 Removal.

WARNING

All metal jewelry can conduct electricity and become entangled in generator set components. Remove all jewelry when working on generator set. Failure to comply with this warning can cause injury or death to personnel.

WARNING

DO NOT wear loose clothing when performing checks, services and maintenance. Failure to comply with this warning can cause injury or death to personnel.

WARNING

When running, generator set engine has hot metal surfaces that will burn flesh on contact. Shut down generator set and allow engine to cool before performing checks, services, and maintenance. Wear gloves and additional protective clothing as required. Failure to comply with this warning can cause injury or death to personnel.

WARNING

Dangerous voltage exists on live circuits. Always observe precautions and never work alone. Failure to comply with this warning can cause injury or death to personnel.

WARNING

High voltage is produced when this generator set is in operation. SHUT DOWN generator set and make sure it is free of any power source before attempting any repair or maintenance on the set, or when connecting or disconnecting load cables. Failure to comply with this warning can

cause injury or death to personnel.

WARNING

DC voltages are present at generator set electrical components even with generator set shut down. Avoid shorting any positive with ground/negative. Failure to comply with this warning can cause injury to personnel, and damage to equipment.

WARNING

Ensure that the engine cannot be started while maintenance is being performed. (ENGINE CONTROL switch set to OFF/RESET; Battery Disconnect Switch is OFF; DEAD CRANK SWITCH is OFF). Failure to comply with this warning can cause injury or death to personnel.

WARNING

Top housing panels and exhaust system can get very hot. Shut down generator set, and allow system to cool before performing checks, services and maintenance. Failure to comply with this warning can cause severe burns and injury to personnel.

a. a. Shut off generator set.

b. b. Open battery access door and disconnect negative battery cable.

c. c. Remove control box top panel (paragraph 2-15.1).

d. d. Remove GROUND FAULT CIRCUIT INTERRUPTER (paragraph 2-46.3). (Do not disconnect electrical leads.)

e. e. Disconnect multi-pin connector at rear of m alfunction indicator panel (27, Figure 2-14).

f. f. Remove bolts (24), washers (25), and nuts (26).

g. g. Remove malfunction indicator panel (27) from control panel.

2-47.4 Installation.

WARNING

High voltage is produced when the generator set is in operation. Never attempt to start the generator set unless it is properly grounded. Failure to comply with this warning can cause injury or death to personnel.

a. a. Install malfunction indicator panel (27, Figure 2-14) in control panel with bolts (24), washers (25), and nuts (26).

b. b. Install multipin connector at rear of panel.

c. c. Install GROUND FAULT CIRCUIT INTERRUPTER (paragraph 2-46.4).

d. d. Install control box top panel (paragraph 2-15.4).

e. e. Connect negative battery c able and close battery access door.

2-48 BATTERY CHARGER FUSE (MEP-804A/MEP-814A) 2-48.1

Inspection.

a. a. Shut down generator set.

b. b. Inspect for blown fuse.

c. c. Inspect fuse, fuse holder, and cap for cracks, corrosion, and obvious damage.

d. d. Replace defective parts.

WARNING

All metal jewelry can conduct electricity and become entangled in generator set components. Remove all jewelry when working on generator set. Failure to comply with this warning can cause injury or death to personnel.

WARNING

DO NOT wear loose clothing when performing checks, services and maintenance. Failure to comply with this warning can cause injury or death to personnel.

WARNING

When running, generator set engine has hot metal surfaces that will burn flesh on contact. Shut down generator set and allow engine to cool before performing checks, services, and maintenance. Wear gloves and additional protective clothing as required. Failure to comply with this warning can cause injury or death to personnel.

WARNING

Dangerous voltage exists on live circuits. Always observe precautions and never work alone. Failure to comply with this warning can cause injury or death to personnel.

WARNING

High voltage is produced when this generator set is in operation. SHUT DOWN generator set and make sure it is free of any power source before attempting any repair or maintenance on the set, or when connecting or disconnecting load cables. Failure to comply with this warning can cause injury or death to personnel.

WARNING

DC voltages are present at generator set electrical components even with generator set shut down. Avoid shorting any positive with ground/negative. Failure to comply with this warning can cause injury to personnel, and damage to equipment.

WARNING

Ensure that the engine cannot be started while maintenance is being performed. (ENGINE CONTROL switch set to OFF/RESET; Battery Disconnect Switch is OFF; DEAD CRANK SWITCH is OFF). Failure to comply with this warning can cause injury or death to personnel.

WARNING

Top housing panels and exhaust system can get very hot. Shut down generator set, and allow system to cool before performing checks, services and maintenance. Failure to comply with this warning can cause severe burns and injury to personnel.

a. a. Shut down generator set.

b. b. Open battery access door and disconnect negative battery cable.

c. c. Remove control box top panel (paragraph 2-15.1).

d. d. Remove cap (28, Figure 2-14) and fuse (29) from fuse holder (32).

e. e. Tag and unsolder electrical leads from fuse holder (32).

f. f. Remove nut (30), lock washer (31), and fuse holder (32). Discard lock washer (31).

2-48.3 Installation.

WARNING

High voltage is produced when the generator set is in operation. Never attempt to start the generator set unless it is properly grounded. Failure to comply with this warning can cause injury or death to personnel.

a. a. Install fuse holder (32, Figure 2-14) and secure with new lock washer (31) and nut (30).

b. b. Solder electrical leads to fuse holder (32). Remove tags.

c. c. Install fuse (29) and cap (28).

d. d. Install control box top panel (paragraph 2-15.4).

2-49 BATTERY CHARGER CIRCUIT BREAKER (MEP-804B/MEP-814B)

2-49.1 Inspection.

a. a. Shut down generator set.

b. b. Inspect circuit breaker for obvious damage.

c. c. Replace circuit breaker, as required.

2-49.2 Removal.

WARNING

All metal jewelry can conduct electricity and become entangled in generator set components. Remove all jewelry when working on generator set. Failure to comply with this warning can cause injury or death to personnel.

WARNING

DO NOT wear loose clothing when performing checks, services and maintenance. Failure to comply with this warning can cause injury or death to personnel.

WARNING

When running, generator set engine has hot metal surfaces that will burn flesh on contact. Shut down generator set and allow engine to cool before performing checks, services, and maintenance. Wear gloves and additional protective clothing as required. Failure to comply with this warning can cause injury or death to personnel.

WARNING

Dangerous voltage exists on live circuits. Always observe precautions and never work alone. Failure to comply with this warning can cause injury or death to personnel.

WARNING

High voltage is produced when this generator set is in operation. SHUT DOWN generator set and make sure it is free of any power source before attempting any repair or maintenance on the set, or when connecting or disconnecting load cables. Failure to comply with this warning can cause injury or death to personnel.

WARNING

DC voltages are present at generator set electrical components even with generator set shut down. Avoid shorting any positive with ground/negative. Failure to comply with this warning can cause injury to personnel, and damage to equipment.

WARNING

Ensure that the engine cannot be started while maintenance is being performed. (ENGINE CONTROL switch set to OFF/RESET; Battery Disconnect Switch is OFF; DEAD CRANK SWITCH is OFF). Failure to comply with this warning can cause injury or death to personnel.

WARNING

Top housing panels and exhaust system can get very hot. Shut down generator set, and allow system to cool before performing checks, services and maintenance. Failure to comply with this warning can cause severe burns and injury to personnel.

a. a. Shut down generator set.

b. b. Open battery access door and disconnect negative battery cable.

c. c. Remove control box top panel (paragraph 2-15.1).

d. d. Tag and disconnect wiring from circuit breaker (35, Figure 2-14).

e. e. Remove nut (33), washer (34), and circuit breaker (35).

2-49.3 Installation.

WARNING

High voltage is produced when the generator set is in operation. Never attempt to start the generator set unless it is properly grounded. Failure to comply with this warning can cause injury or death to personnel.

a. a. Install circuit breaker (35, Figure 2-14) and secure with washer (34) and nut (33).

b. b. Remove tags and connect wiring to circuit breaker (35).

c. c. Install control box top panel (paragraph 2-15.4).

d. d. Connect negative battery cable and close battery access door.

2-50 DC CONTROL POWER CIRCUIT BREAKER

2-50.1 Inspection.

a. a. Shut down generator set.

b. b. Inspect circuit breaker for loose connections and mounting, cracked housing, and other damage.

2-50.2 Testing.

a. a. Shut down generator set.

b. b. Open battery access door and disconnect negative battery cable.

c. c. Remove control box top panel (paragraph 2-15.1).

d. d. Tag and disconnect DC CONTROL POWER circuit breaker (38, Figure 2-14) electrical leads.

e. e. Place circuit breaker in OPEN position.

f. f. Set multimeter for ohms and connect across circuit breaker terminals. Multimeter should indicate open circuit.

g. g. Place circuit breaker in CLOSED positi on. Multimeter should indicate continuity.

h. h. Replace circuit breaker if indications are not as above.

i Connect electrical leads to circ uit breaker (38). Remove tags.

j. j. Install control box top panel (paragraph 2-15.4).

k. k. Connect negative battery c able and close battery access door.

2-50.3 Removal.

WARNING

All metal jewelry can conduct electricity and become entangled in generator set components. Remove all jewelry when working on generator set. Failure to comply with this warning can cause injury or death to personnel.

WARNING

DO NOT wear loose clothing when performing checks, services and maintenance. Failure to comply with this warning can cause injury or death to personnel.

WARNING

When running, generator set engine has hot metal surfaces that will burn flesh on contact. Shut down generator set and allow engine to cool before performing checks, services, and maintenance. Wear gloves and additional protective clothing as required. Failure to comply with this warning can cause injury or death to personnel.

WARNING

Dangerous voltage exists on live circuits. Always observe precautions and never work alone. Failure to comply with this warning can cause injury or death to personnel.

WARNING

High voltage is produced when this generator set is in operation. SHUT DOWN generator set and make sure it is free of any power source before attempting any repair or maintenance on the set, or when connecting or disconnecting load cables. Failure to comply with this warning can cause injury or death to personnel.

WARNING

DC voltages are present at generator set electrical components even with generator set shut down. Avoid shorting any positive with ground/negative. Failure to comply with this warning can cause injury to personnel, and damage to equipment.

WARNING

Ensure that the engine cannot be started while maintenance is being performed. (ENGINE CONTROL switch set to OFF/RESET; Battery Disconnect Switch is OFF; DEAD CRANK SWITCH is OFF). Failure to comply with this warning can cause injury or death to personnel.

WARNING

Top housing panels and exhaust system can get very hot. Shut down generator set, and allow system to cool before performing checks, services and maintenance. Failure to comply with this warning can cause severe burns and injury to personnel.

a. a. Shut down generator set.

b. b. Open battery access door and disconnect negative battery cable.

c. c. Remove control box top panel (paragraph 2-15.1).

d. d. Remove nut (36, Figure 2-14) and flat washer (37).

e. e. Remove DC CONTROL POWER circuit breaker (38) from mounting bracket.

f. f. Tag and disconnect DC CONTROL POWER circuit breaker (38) electrical leads.

2-50.4 Installation.

WARNING

High voltage is produced when the generator set is in operation. Never attempt to start the generator set unless it is properly grounded. Failure to comply with this warning can cause injury or death to personnel.

a. a. Connect electrical leads. Remove tags.

b. b. Insert DC CONTROL POWER circuit breake r (38, Figure 2-14) into mounting brack-

c. c. Install flat washer (37) and nut (36).

d. d. Install control box top panel (paragraph 2-15.4).

e. e. Connect negative battery c able and close battery access door.

2-51 AC VOLTAGE REGULATOR 2-51.1

Inspection.

Inspect AC voltage regulator (39, Figure 2-14) for cracked case, broken wires, security, and other damage.

2-52 FREQUENCY TRANSDUCER 2-52.1

Inspection.

a. a. Shut down generator set.

b. b. Inspect transducer for cracked casing, burned or broken terminals, and other damage.

2.52.2 Testing (MEP-804A/MEP-804B).

a. a. Shut down generator set.

b. b. Open battery access door and disconnect negative battery cable.

c. c. Remove control box top panel (paragraph 2-15.1).

d. d. Release control panel by turning two fasteners and lower control panel slowly.

e. e. Disconnect wire 181A from positive (+) terminal of FREQUENCY meter (HERTZ).

f. f. Set multimeter for DC milliamperes (0 to 2 Ma range) and connect positive lead to disconnected wire 181A and negative lead to vacant terminal of FREQUENCY meter (HERTZ).

g. g. Move FREQUENCY SELECT switch to 60 Hz position.

h. h. Connect negative battery cable.

i. i. Start generator set and adjust frequency to 60 Hz.

j. j. Multimeter indication should be between 0.781 and 0.923 Ma.

k. k. Adjust frequency to 62 Hz and multimeter indication should be between 0.923 and 1.071 Ma.

l. l. Move FREQUENCY SELECT switch to 50 Hz position.

m. m. Adjust frequency to 50 Hz and multimeter indication should be between 0.071 and 0.213 Ma.

n. n. Adjust frequency to 52 Hz and multimeter indication should be between 0.213 and 0.355 Ma.

o. o. Shut down generator set.

p. p. Replace frequency transducer if readings are other than above.

q. q. If no repair is needed, remove multimeter and connect wire 181A to positive (+) terminal of FREQUENCY meter (HERTZ).

r. r. Raise and secure control panel.

s. s. Install control box top panel (paragraph 2-15.4).

2-52.3 Testing (MEP-814A/MEP-814B).

a. a. Shut down generator set.

b. b. Open battery access door and disconnect negative battery cable.

c. c. Release control panel by turning two fasteners and lower control panel slowly.

d. d. Disconnect wire 181A from positive terminal (+) of FREQUENCY meter (HERTZ).

e. e. Set multimeter for DC milliamperes (0 to 2 Ma range) and connect positive lead to free end of wire 181A and connect negative lead to positive terminal (+) of FREQUENCY meter (HERTZ).

f. f. Connect negative battery cable.

g. g. Start and operate generator set at rated voltage and adjust frequency to 400 Hz.

h. h. Multimeter indication should be between 0.229 and 0.271 Ma.

i. i. Adjust frequency to 412 Hz. Multimeter indication should be between 0.479 and 0.521 Ma.

j. j. Shut down generator set.

k. k. Replace frequency transducer if readings are other than above.

l. l. If no repair is needed, remove multimeter and connect wire 181A to positive terminal (+) of FREQUENCY meter (HERTZ).

m. m. Raise and secure control panel.

2-52.4 Removal.

WARNING

All metal jewelry can conduct electricity and become entangled in generator set components. Remove all jewelry when working on generator set. Failure to comply with this warning can cause injury or death to personnel.

WARNING

DO NOT wear loose clothing when performing checks, services and maintenance. Failure to comply with this warning can cause injury or death to personnel.

WARNING

When running, generator set engine has hot metal surfaces that will burn flesh on contact. Shut down generator set and allow engine to cool before performing checks, services, and maintenance. Wear gloves and additional protective clothing as required. Failure to comply with this warning can cause injury or death to personnel.

WARNING

Dangerous voltage exists on live circuits. Always observe precautions and never work alone. Failure to comply with this warning can cause injury or death to personnel.

WARNING

High voltage is produced when this generator set is in operation. SHUT DOWN generator set and make sure it is free of any power source before attempting any repair or maintenance on the set, or when connecting or disconnecting load cables. Failure to comply with this warning can cause injury or death to personnel.

WARNING

DC voltages are present at generator set electrical components even with generator set shut down. Avoid shorting any positive with ground/negative. Failure to comply with this warning can cause injury to personnel, and damage to equipment.

WARNING

Ensure that the engine cannot be started while maintenance is being performed. (ENGINE CONTROL switch set to OFF/RESET; Battery Disconnect Switch is OFF; DEAD CRANK SWITCH is OFF). Failure to comply with this warning can cause injury or death to personnel.

WARNING

Top housing panels and exhaust system can get very hot. Shut down generator set, and allow system to cool before performing checks, services and maintenance. Failure to comply with this warning can cause severe burns and injury to personnel.

a. a. Shut down generator set.

b. b. Open battery access door and disconnect negative battery cable.

c. c. Release control panel by turning two fasteners and lower control panel slowly.

d. d. Tag and disconnect frequency transducer (41, Figure 2-14) electrical leads.

e. Remove screws (40) and frequency transducer (41).

2-52.5 Installation.

WARNING

High voltage is produced when the generator set is in operation. Never attempt to start the generator set unless it is properly grounded. Failure to comply with this warning can cause injury or death to personnel.

a. a. Install frequency transducer (41, Figure 2-14) with screws (40).

b. b. Connect all electrical leads. Remove tags.

c. c. Raise and secure control panel.

d. d. Connect negative battery cable and close battery access door.

2-53 KILOWATT TRANSDUCER 2-53.1

Inspection.

a. a. Shut down generator set.

b. b. Inspect transducer for cracked casing, burned or broken terminals, and other damage.

2-53.2 Testing.

a. a. Start and operate generator set at rated voltage and frequency.

b. b. Apply some load to generator set.

c. c. Release control panel by turning two fasteners and lower control panel slowly.

d. d. Set multimeter for AC volts and take readings between terminals V1 and N1, V2 and N2, and
V3
and N3. Multimeter indication should be 120 VAC between each set of terminals.

e. e. Take readings between terminals S1 and –, S2 and –, S3 and –, L1 and –, L2 and –, and L3
and
–. Multimeter indication should be 0.1 to 3 VAC. (Reading will vary depending on amount of load applied to generator set.)

f. f. Change multimeter setting to DC millivolts and take reading between terminals + and –. Multimeter indication should be 0.1 to 50 mV (dependent on amount of load applied to generator set).

g. g. Shut down generator set.

h. h. Replace kilowatt transducer if multimeter indications are within ranges stated in steps d and e above, but not within range stated in step f above.

i. i. If no repair is needed, raise and secure control panel.

2-53.3 Removal.

WARNING

All metal jewelry can conduct electricity and become entangled in generator set components. Remove all jewelry when working on generator set. Failure to comply with this warning can cause injury or death to personnel.

WARNING

DO NOT wear loose clothing when performing checks, services and maintenance. Failure to comply with this warning can cause injury or death to personnel.

WARNING

When running, generator set engine has hot metal surfaces that will burn flesh on contact. Shut down generator set and allow engine to cool before performing checks, services, and maintenance. Wear gloves and additional protective clothing as required. Failure to comply with this warning can cause injury or death to personnel.

WARNING

Dangerous voltage exists on live circuits. Always observe precautions and never work alone. Failure to comply with this warning can cause injury or death to personnel.

WARNING

High voltage is produced when this generator set is in operation. SHUT DOWN generator set and make sure it is free of any power source before attempting any repair or maintenance on the set, or when connecting or disconnecting load cables. Failure to comply with this warning can cause injury or death to personnel.

WARNING

DC voltages are present at generator set electrical components even with generator set shut down. Avoid shorting any positive with ground/negative. Failure to comply with this warning can cause injury to personnel, and damage to equipment.

WARNING

Ensure that the engine cannot be started while maintenance is being performed. (ENGINE CONTROL switch set to OFF/RESET; Battery Disconnect Switch is OFF; DEAD CRANK SWITCH is OFF). Failure to comply with this warning can cause injury or death to personnel.

WARNING

Top housing panels and exhaust system can get very hot. Shut down generator set, and allow system to cool before performing checks, services and maintenance. Failure to comply with this warning can cause severe burns and injury to personnel.

a. a. Shut down generator set.

b. b. Open battery access door and disconnect negative battery cable.

c. c. Remove control box top panel (paragraph 2-15.1).

d. d. Tag and disconnect kilowatt transducer (43, Figure 2-14) electrical leads.

e. e. Remove screws (42) and kilowatt transducer (43).

2-53.4 Installation.

a. a. Install kilowatt transducer (43, Figure 2-14) with screws (42).

b. b. Connect electrical leads. Remove tags.

c. c. Install control box top panel (paragraph 2-15.4).

d. d. Connect negative battery c able and close battery access door.

2-54 SHUNT

2-54.1 Inspection.

a. a. Shut down generator set.

b. b. Inspect shunt for cracked casing, burned or broken terminals, and other damage.

2-54.2 Testing.

a. a. Shut down generator set.

b. b. Remove control box top panel (paragraph 2-15.1).

c. c. Tag and disconnect shunt (46, Figure 2-14) electrical leads.

d. d. Set multimeter for ohms and connect to shunt terminals 1 and 4. Multimeter should indicate less than 0.5 ohms.

e. e. Replace shunt if multimeter indication is greater than above.

f. f. If no repair is needed, connect electrical leads to shunt (46). Remove tags.

g. g. Install control box top panel (paragraph 2-15.4).

2-54.3 Removal.

WARNING

All metal jewelry can conduct electricity and become entangled in generator set components. Remove all jewelry when working on generator set. Failure to comply with this warning can cause injury or death to personnel.

WARNING

DO NOT wear loose clothing when performing checks, services and maintenance. Failure to comply with this warning can cause injury or death to personnel.

WARNING

When running, generator set engine has hot metal surfaces that will burn flesh on contact. Shut down generator set and allow engine to cool before performing checks, services, and maintenance. Wear gloves and additional protective clothing as required. Failure to comply with this warning can cause injury or death to personnel.

WARNING

Dangerous voltage exists on live circuits. Always observe precautions and never work alone. Failure to comply with this warning can cause injury or death to personnel.

WARNING

High voltage is produced when this generator set is in operation. SHUT DOWN generator set and make sure it is free of any power source before attempting any repair or maintenance on the set, or when connecting or disconnecting load cables. Failure to comply with this warning can cause injury or death to personnel.

WARNING

DC voltages are present at generator set electrical components even with generator set shut down. Avoid shorting any positive with ground/negative. Failure to comply with this warning can cause injury to personnel, and damage to equipment.

WARNING

Ensure that the engine cannot be started while maintenance is being performed. (ENGINE CONTROL switch set to OFF/RESET; Battery Disconnect Switch is OFF; DEAD CRANK SWITCH is OFF). Failure to comply with this warning can cause injury or death to personnel.

WARNING

Top housing panels and exhaust system can get very hot. Shut down generator set, and allow system to cool before performing checks, services and maintenance. Failure to comply with this warning can cause severe burns and injury to personnel.

a. a. Shut down generator set.

b. b. Open battery access door and disconnect negative battery cable.

c. c. Remove control box top panel (paragraph 2-15.1).

d. d. Remove control box assembly (paragraph 2-19.2).

e. e. Tag and disconnect shunt (46, Figure 2-14) electrical leads.

f. f. Open output box access door and remove screws (44), nuts (45), and shunt (46).

2-54.4 Installation.

WARNING

High voltage is produced when the generator set is in operation. Never attempt to start the generator set unless it is properly grounded. Failure to comply with this warning can cause injury or death to personnel.

a. a. Install shunt (46, Figure 2-14) and secure with screws (44) and nuts (45).

b. b. Close output box access door.

c. c. Connect all electrical leads. Remove tags.

d. d. Install control box assembly (paragraph 2-19.4).

e. e. Install control box top panel (paragraph 2-15.4).

f. f. Connect negative battery cabl e and close battery access door.

2-55 OVERVOLTAGE/UNDERVOLTAGE RELAY 2-55.1

Inspection.

a. a. Shut down generator set.

b. b. Inspect relay for cracked casing, burned or broken terminals, and other damage.

2-55.2 Testing.

a. a. Shut down generator set.

b. b. Open battery access door and disconnect negative battery cable.

c. c. Release control panel by turning two fasteners and lower control panel slowly.

CAUTION

The following procedure disables AC voltage regulator and allows generator to reach an overvoltage condition. Do not allow generator set to operate for an extended period of time in an extreme overvoltage condition.

d. d. Disconnect wire 137A from AC voltage regulator terminal 5 and insulate wire end.

e. e. Connect negative battery cable.

f. f. Start generator set. As generator accelerates to rated speed, it should instantly shut down and OVERVOLTAGE lamp on malfunction indicator panel should illuminate. If this does not occur, immediately shut down generator set.

g. g. Reconnect wire 137A and disconnect wire 141A from AC voltage regulator terminal 1. Insulate wire end.

h. h. Start generator set. As generator accelerates to rated speed, UNDERVOLTAGE lamp on malfunction indicator should illuminate. Move AC CIRCUIT INTERRUPTER switch to CLOSED position. AC circuit interrupter relay should not close.

i. i. Shut down generator set.

j. j. Replace overvoltage/undervoltage relay if generator set does not operate as above.

k. k. If no repair is needed, reconnect wire 141A at AC voltage regulator.

l. l. Raise and secure control panel.

2-55.3 Removal.

WARNING

All metal jewelry can conduct electricity and become entangled in generator set components. Remove all jewelry when working on generator set. Failure to comply with this warning can cause injury or death to personnel.

WARNING

DO NOT wear loose clothing when performing checks, services and maintenance. Failure to comply with this warning can cause injury or death to personnel.

WARNING

When running, generator set engine has hot metal surfaces that will burn flesh on contact. Shut down generator set and allow engine to cool before performing checks, services, and maintenance. Wear gloves and additional protective clothing as required. Failure to comply with this warning can cause injury or death to personnel.

WARNING

Dangerous voltage exists on live circuits. Always observe precautions and never work alone. Failure to comply with this warning can cause injury or death to personnel.

WARNING

High voltage is produced when this generator set is in operation. SHUT DOWN generator set and make sure it is free of any power source before attempting any repair or maintenance on the set, or when connecting or disconnecting load cables. Failure to comply with this warning can cause injury or death to personnel.

WARNING

DC voltages are present at generator set electrical components even with generator set shut down. Avoid shorting any positive with ground/negative. Failure to comply with this warning can cause injury to personnel, and damage to equipment.

WARNING

Ensure that the engine cannot be started while maintenance is being performed. (ENGINE CONTROL switch set to OFF/RESET; Battery Disconnect Switch is OFF; DEAD CRANK SWITCH is OFF). Failure to comply with this warning can cause injury or death to personnel.

WARNING

Top housing panels and exhaust system can get very hot. Shut down generator set, and allow system to cool before performing checks, services and maintenance. Failure to comply with this warning can cause severe burns and injury to personnel.

a. a. Shut down generator set.

b. b. Open battery access door and disconnect negative battery cable.

c. c. Remove control box top panel (paragraph 2-15.1).

d. d. Tag and disconnect overvoltage/undervoltage relay (48, Figure 2-14) electrical leads.

e. e. Remove screws (47) and overvoltage/undervoltage relay (48).

2-55.4 Installation.

WARNING

High voltage is produced when the generator set is in operation. Never attempt to start the generator set unless it is properly grounded. Failure to comply with this warning can cause injury or death to personnel.

a. a. Install overvoltage/undervoltage relay (48, Figure 2-14) with screws (47).

b. b. Connect electrical leads. Remove tags.

c. c. Install control box top panel (paragraph 2-15.4).

d. d. Connect negative battery c able and close battery access door.

2-56 SHORT CIRCUIT/OVERLOAD RELAY 2-56.1

Inspection.

a. a. Shut down generator set.

b. b. Inspect relay for cracked casing, burne d or broken terminals, and other damage.

2-56.2 Testing.

a. a. Shut down generator set.

b. b. Open battery access door and disconnect negative battery cable.

c. c. Release control panel by turning two fasteners and lower control panel slowly.

d. d. Tag, disconnect, and insulate the following wires from short circuit/overload relay:

Wires 111B and 111C from terminal 1
Wires 113B and 113C from terminal 2
Wires 115B and 115C from terminal 3
Wires 184B and 184C from terminal 4

e. e. Tag and disconnect wire 141A from terminal 1 on AC voltage regulator.

NOTE

Disconnecting wire 141A at AC voltage regulator terminal 1 disables AC voltage regulator and allows generator to develop very low AC output voltage.

f. f. Tag, disconnect, and isolate wire 217A from terminal 8 of overvoltage/undervoltage relay.

g. g. Connect a jumper wire from terminal 4 of sh ort circuit/overload relay to terminal L0 of load output terminal board.

h. h. Place voltage reconnection terminal board in 120/208 connection.

i. i. Connect negative battery cable.

j. j. Start generator set and operate at rated frequency.

k. k. Close AC CIRCUIT INTERRUPTER switch. After approximately 1 minute, AC circuit interrupter relay should open and OVERLOAD lamp should illuminate on malfunction indicator panel.

l. l. Shut down generator set.

m. m. Disconnect jumper wire from terminal 1 of short circuit/overload relay; connect it to terminal 2.

n. n. Repeat steps j and k above.

o. o. Shut down generator set.

p. p. Disconnect jumper wire from terminal 2 of short circuit/overload relay; connect it to terminal 3.

q. q. Repeat steps j and k above.

r. r. Shut down generator set.

s. s. Disconnect jumper wire from terminal L0 of load output terminal board; connect it to L2.

t. t. Start generator set and operate at rated frequency.

u. Close AC CIRCUIT INTERRUPTER switch. AC circuit interrupter relay should open immediately and SHORT CIRCUIT lamp should illuminate on malfunction indicator panel.

v. v. Shut down generator set.

w. w. Disconnect jumper wire from terminal 3 of short circuit/overload relay; connect it to terminal 2.

x. x. Repeat steps t and u above.

y. y. Shut down generator set.

z. z. Disconnect jumper wire from terminal 2 of short circuit/overload relay; connect it to terminal 1.

aa. Repeat steps t and u above. ab. Shut down generator set.

ac. Replace short circuit/overload relay if any test is not as above.

ad. If no repair is needed, remove jumper wires an d connect wires to short circuit/overload relay, overvoltage/ undervoltage relay, and AC voltage regulator as tagged. Remove tags.

ae. Raise and secure control panel.

2-56.3 Removal.

WARNING

All metal jewelry can conduct electricity and become entangled in generator set components. Remove all jewelry when working on generator set. Failure to comply with this warning can cause injury or death to personnel.

WARNING

DO NOT wear loose clothing when performing checks, services and maintenance. Failure to comply with this warning can cause injury or death to personnel.

WARNING

When running, generator set engine has hot metal surfaces that will burn flesh on contact. Shut down generator set and allow engine to cool before performing checks, services, and maintenance. Wear gloves and additional protective clothing as required. Failure to comply with this warning can cause injury or death to personnel.

WARNING

Dangerous voltage exists on live circuits. Always observe precautions and never work alone. Failure to comply with this warning can cause injury or death to personnel.

WARNING

High voltage is produced when this generator set is in operation. SHUT DOWN generator set and make sure it is free of any power source before attempting any repair or maintenance on the set, or when connecting or disconnecting load cables. Failure to comply with this warning can cause injury or death to personnel.

WARNING

DC voltages are present at generator set electrical components even with generator set shut down. Avoid shorting any positive with ground/negative. Failure to comply with this warning can cause injury to personnel, and damage to equipment.

WARNING

Ensure that the engine cannot be started while maintenance is being performed. (ENGINE CONTROL switch set to OFF/RESET; Battery Disconnect Switch is OFF; DEAD CRANK SWITCH is OFF). Failure to comply with this warning can cause injury or death to personnel.

WARNING

Top housing panels and exhaust system can get very hot. Shut down generator set, and allow system to cool before performing checks, services and maintenance. Failure to comply with this warning can cause severe burns and injury to personnel.

a. a. Shut down generator set.

b. b. Open battery access door and disconnect negative battery cable.

c. c. Remove control box top panel (paragraph 2-15.1).

d. d. Tag and disconnect short circuit/overload relay (50, Figure 2-14) electrical leads.

e. Remove screws (49) and short circuit/overload relay (50).

2-56.4 Installation.

WARNING

High voltage is produced when the generator set is in operation. Never attempt to start the generator set unless it is properly grounded. Failure to comply with this warning can cause injury or death to personnel.

a. a. Install short circuit/overload relay (50, Figure 2-14) with screws (49).

b. b. Connect electrical leads. Remove tags.

c. c. Install control box top panel (paragraph 2-15.4).

d. d. Connect negative battery cable and close battery access door.

2-57 PERMISSIVE PARALLELING RELAY

2-57.1 Inspection.

a. a. Shut down generator set.

b. b. Inspect relay for cracked casing, burned or broken terminals, and other damage.

2-57.2 Testing.

a. a. Shut down generator set.

b. b. Open battery access door and disconnect negative battery cable.

c. c. Release control panel by turning t wo fasteners and slowly lower control panel.

d. d. Disconnect and insulate wires 102D from terminal 1 and 196A and 196B from terminal 2 of permissive paralleling relay.

e. Mark a 10,000 ohm potentiometer as follows:

Center terminal = C
Two outside terminals = L and R

f. f. Set up a test circuit (Figure 2-16). Connect 120 VAC source (can be obtained from CONVENIENCE RECEPTACLE) to terminals L and R of potentiometer. Connect a wire from terminal C of potentiometer to terminal 1 of permissive paralleling relay. Connect a second wire from terminal R of potentiometer to terminal 2 of permissive paralleling relay. Set multimeter for AC volts and connect to terminals 1 and 2 of permissive paralleling relay.

g. g. Adjust 10,000 ohm potentiometer to full CCW position.

h. h. Connect negative battery cable.

i. i. Start and operate generator set at rated frequency and voltage. Multimeter indication should be 0 volts.

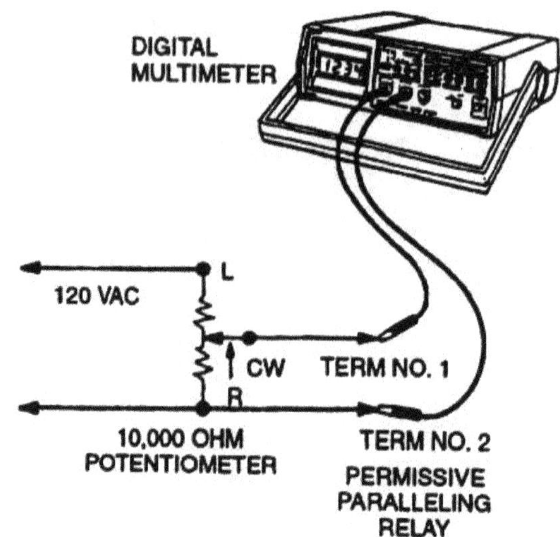

Figure 2-16. Permissive Paralleling Relay Test Setup.

j. j. Move AC CIRCUIT INTERRUPTER switch to CLOSED position; AC circuit interrupter relay should close. Move AC CIRCUIT INTERRUPTER switch to OPEN position; AC circuit interrupter relay should open. Observe AC CIRCUIT INTERRUPTER light for actuation of relay.

k. k. Adjust 10,000 ohm potentiometer CW until multimeter indicates 10 VAC.

l. l. Move AC CIRCUIT INTERRUPTER switch to CLOSED position. AC circuit interrupter relay should not close (AC CIRCUIT INTERRUPTER light should remain dark).

m. m. Shut down generator set.

n. n. Disconnect negative battery cable.

o. o. Replace permissive paralleling relay if operation is not as above.

p. p. If no repair is needed, remove multimeter a nd test circuit wires. Reconnect wires 102D, 196A and 196B to permissive paralleling relay.

q. q. Raise and secure control panel.

r. r. Connect negative battery c able and close battery access door.

2-57.3 Removal.

WARNING

All metal jewelry can conduct electricity and become entangled in generator set components. Remove all jewelry when working on generator set. Failure to comply with this warning can cause injury or death to personnel.

WARNING

DO NOT wear loose clothing when performing checks, services and maintenance. Failure to comply with this warning can cause injury or death to personnel.

WARNING

When running, generator set engine has hot metal surfaces that will burn flesh on contact. Shut down generator set and allow engine to cool before performing checks, services, and maintenance. Wear gloves and additional protective clothing as required. Failure to comply with this warning can cause injury or death to personnel.

WARNING

Dangerous voltage exists on live circuits. Always observe precautions and never work alone. Failure to comply with this warning can cause injury or death to personnel.

WARNING

High voltage is produced when this generator set is in operation. SHUT DOWN generator set and make sure it is free of any power source before attempting any repair or maintenance on the set, or when connecting or disconnecting load cables. Failure to comply with this warning can cause injury or death to personnel.

WARNING

DC voltages are present at generator set electrical components even with generator set shut down. Avoid shorting any positive with ground/negative. Failure to comply with this warning can cause injury to personnel, and damage to equipment.

WARNING

Ensure that the engine cannot be started while maintenance is being performed. (ENGINE CONTROL switch set to OFF/RESET; Battery Disconnect Switch is OFF; DEAD CRANK SWITCH is OFF). Failure to comply with this warning can cause injury or death to personnel.

WARNING

Top housing panels and exhaust system can get very hot. Shut down generator set, and allow system to cool before performing checks, services and maintenance. Failure to comply with this warning can cause severe burns and injury to personnel.

a. Shut down generator set.

b. b. Open battery access door and disconnect negative battery cable.

c. c. Release control panel by turning two fasteners and lower control panel slowly.

d. d. Tag and disconnect permissive paralleling relay (52, Figure 2-14) electrical leads.

e. e. Remove screws (51) and permissive paralleling relay (52).

WARNING

High voltage is produced when the generator set is in operation. Never attempt to start the generator set unless it is properly grounded. Failure to comply with this warning can cause injury or death to personnel.

a. a. Install permissive paralleling relay (52, Figure 2-14) with screws (51).

b. b. Connect all electrical leads. Remove tags.

c. c. Raise and secure control panel.

d. d. Connect negative battery cabl e and close battery access door.

2-58 REVERSE POWER RELAY 2-58.1

Inspection.

a. a. Shut down generator set.

b. b. Inspect relay for cracked casing, burne d or broken terminals, and other damage.

2-58.2 Testing.

a. a. Shut down generator set.

b. b. Open battery access door and disconnect negative battery cable.

c. c. Release control panel by turning two fasteners and lower control panel slowly.

d. d. Disconnect and insulate wires 158B and 158C from terminal 1 and wires 161A and 161B from terminal 2 of reverse power relay.

e. e. Mark a 5,000 ohm potentiometer as follows:

Center terminal = C
Two outside terminals = L and R

CAUTION

Voltage polarity is very important to prevent damage to generator set.

f. f. Set up a test circuit (Figure 2-17). Co nnect 24 VDC source to terminals L and R of potentiometer. Connect a wire between terminal C of potentiometer and terminal 1 of reverse power relay. Connect a second wire between terminal R of potentiometer and terminal 2 of reverse power relay. Set multimeter for DC volts and connect positive lead of multimeter to terminal 1 and negative lead to terminal 2 of reverse power relay.

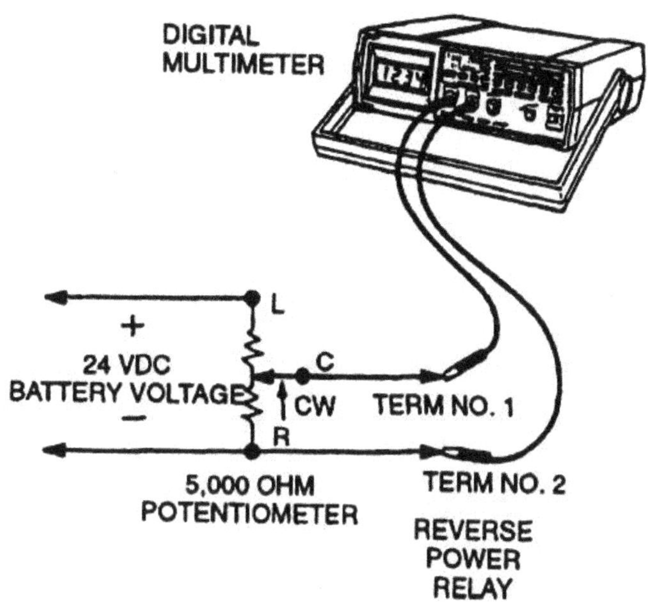

Figure 2-17. Reverse Power Relay Test Setup.

g. g. Adjust 5,000 ohm potentiometer to full CCW position. Multimeter should indicate 0 volts.

h. h. Connect negative battery cable.

i. i. Start and operate generator set at rated frequency and voltage.

j. j. Move AC CIRCUIT INTERRUPTER switch to CLOSED position. AC CIRCUIT INTERRUPTER light should illuminate.

k. k. Adjust 5,000 ohm potentiometer CW and at same time observe multimeter and AC CIRCUIT INTERRUPTER light. AC CIRCUIT INTERRUPTER light should go out at between 1.7 and 2.3 VDC indicating that AC circuit interrupter relay is open.

l. l. Shut down generator set.

m. m. Disconnect negative battery cable.

n. n. Replace reverse power relay if operation is not as above.

o. o. Remove multimeter and test circuit wires. Reconnect wires 158B, 158C, 161A, and 161B at reverse power relay.

p. p. Raise and secure control panel.

q. q. Connect negative battery c able and close battery access door.

2-58.3 Removal.

WARNING

All metal jewelry can conduct electricity and become entangled in generator set components. Remove all jewelry when working on generator set. Failure to comply with this warning can cause injury or death to personnel.

WARNING

DO NOT wear loose clothing when performing checks, services and maintenance. Failure to comply with this warning can cause injury or death to personnel.

WARNING

When running, generator set engine has hot metal surfaces that will burn flesh on contact. Shut down generator set and allow engine to cool before performing checks, services, and maintenance. Wear gloves and additional protective clothing as required. Failure to comply with this warning can cause injury or death to personnel.

WARNING

Dangerous voltage exists on live circuits. Always observe precautions and never work alone. Failure to comply with this warning can cause injury or death to personnel.

WARNING

High voltage is produced when this generator set is in operation. SHUT DOWN generator set and make sure it is free of any power source before attempting any repair or maintenance on the set, or when connecting or disconnecting load cables. Failure to comply with this warning can cause injury or death to personnel.

WARNING

DC voltages are present at generator set electrical components even with generator set shut down. Avoid shorting any positive with ground/negative. Failure to comply with this warning can cause injury to personnel, and damage to equipment.

WARNING

Ensure that the engine cannot be started while maintenance is being performed. (ENGINE CONTROL switch set to OFF/RESET; Battery Disconnect Switch is OFF; DEAD CRANK SWITCH is OFF). Failure to comply with this warning can cause injury or death to personnel.

WARNING

Top housing panels and exhaust system can get very hot. Shut down generator set, and allow system to cool before performing checks, services and maintenance. Failure to comply with this warning can cause severe burns and injury to personnel.

a. Shut down generator set.

b. b. Open battery access door and disconnect negative battery cable.

c. c. Release control panel by turning two fasteners and lower control panel slowly.

d. d. Tag and disconnect reverse power rela y (54, Figure 2-14) electrical leads.

e. e. Remove screws (53) and reverse power relay (54).

WARNING

High voltage is produced when the generator set is in operation. Never attempt to start the generator set unless it is properly grounded. Failure to comply with this warning can cause injury or death to personnel.

a. a. Install reverse power relay (54, Figure 2-14) with screws (53).

b. b. Connect electrical leads. Remove tags.

c. c. Raise and secure control panel.

d. d. Connect negative battery cable and close battery access door.

2-59 VOLTAGE SENSING RELAY 2-59.1

Inspection.

a. a. Shut down generator set.

b. b. Inspect relay for cracks, loose mounting, and other damage.

2-59.2 Testing.

a. a. Shut down generator set.

b. b. Remove voltage sensing relay (paragraph 2-59.3).

c. c. Set multimeter for ohms and check for open circuits between terminals 1 and 3 and terminals 8 and 6 of voltage sensing relay.

d. d. Check for continuity between terminals 1 and 4 and terminals 8 and 5.

e. e. Connect multimeter between terminals 2 and 7 of relay and check for between 1,260 and 1,890 ohms.

f. f. Depress reset button; check for open circuits between terminals 1 and 4 and terminals 8 and 5.

g. g. Continue to depress reset button and check for continuity between terminals 1 and 3 and terminals 8 and 6.

h. h. Replace voltage sensing relay if indications are not as above.

i. i. If no repair is needed, install voltage sensing relay (paragraph 2-59.4).

2-59.3 Removal.

WARNING

All metal jewelry can conduct electricity and become entangled in generator set components. Remove all jewelry when working on generator set. Failure to comply with this warning can cause injury or death to personnel.

WARNING

DO NOT wear loose clothing when performing checks, services and maintenance. Failure to comply with this warning can cause injury or death to personnel.

WARNING

When running, generator set engine has hot metal surfaces that will burn flesh on contact. Shut down generator set and allow engine to cool before performing checks, services, and maintenance. Wear gloves and additional protective clothing as required. Failure to comply with this warning can cause injury or death to personnel.

WARNING

Dangerous voltage exists on live circuits. Always observe precautions and never work alone. Failure to comply with this warning can cause injury or death to personnel.

WARNING

High voltage is produced when this generator set is in operation. SHUT DOWN generator set and make sure it is free of any power source before attempting any repair or maintenance on the set, or when connecting or disconnecting load cables. Failure to comply with this warning can cause injury or death to personnel.

WARNING

DC voltages are present at generator set electrical components even with generator set shut down. Avoid shorting any positive with ground/negative. Failure to comply with this warning can cause injury to personnel, and damage to equipment.

WARNING

Ensure that the engine cannot be started while maintenance is being performed. (ENGINE CONTROL switch set to OFF/RESET; Battery Disconnect Switch is OFF; DEAD CRANK SWITCH is OFF). Failure to comply with this warning can cause injury or death to personnel.

WARNING

Top housing panels and exhaust system can get very hot. Shut down generator set, and allow system to cool before performing checks, services and maintenance. Failure to comply with this warning can cause severe burns and injury to personnel.

a. a. Shut down generator set.

b. b. Open battery access door and disconnect negative battery cable.

c. c. Release control panel by turning two fasteners and lower control panel slowly.

d. d. Release wire clip (55, Figure 2-14) and remove voltage sensing relay (56) by gently pulling from socket.

WARNING

High voltage is produced when the generator set is in operation. Never attempt to start the generator set unless it is properly grounded. Failure to comply with this warning can cause injury or death to personnel.

a. a. Install voltage sensing relay (56, Figure 2-14) in socket and secure by snapping wire clip (55) over relay.

b. b. Raise and secure control panel.

c. c. Connect negative battery c able and close battery access door.

2-60 RELAYS

2-60.1 Inspection.

a. a. Shut down generator set.

b. b. Inspect relays for cracks, loose mounting, and other damage.

2-60.2 Testing.

a. a. Shut down generator set.

b. b. Remove applicable relay (paragraph 2-60.3).

c. c. Set multimeter for ohms and check for open circuits between terminals 7 and 4, 8 and 5, and 9 and 6. Check for closed circuits between terminals 7 and 1, 8 and 2, and 9 and 3.

d. d. Connect multimeter between terminals A and B and check for between 427.5 and 522.5 ohms. Using multimeter, check for closed circuits between terminals 7 and 1, 8 and 2, and 9 and 3. Check for open circuits between terminals 7 and 4, 8 and 5, and 9 and 6.

e. e. Replace relay if indications are other than above.

f. If no repair is needed, install relay (paragraph 2-60.4).

2-60.3 Removal.

WARNING

All metal jewelry can conduct electricity and become entangled in generator set components. Remove all jewelry when working on generator set. Failure to comply with this warning can cause injury or death to personnel.

WARNING

DO NOT wear loose clothing when performing checks, services and maintenance. Failure to comply with this warning can cause injury or death to personnel.

WARNING

When running, generator set engine has hot metal surfaces that will burn flesh on contact. Shut down generator set and allow engine to cool before performing checks, services, and maintenance. Wear gloves and additional protective clothing as required. Failure to comply with this warning can cause injury or death to personnel.

WARNING

Dangerous voltage exists on live circuits. Always observe precautions and never work alone. Failure to comply with this warning can cause injury or death to personnel.

WARNING

High voltage is produced when this generator set is in operation. SHUT DOWN generator set and make sure it is free of any power source before attempting any repair or maintenance on the set, or when connecting or disconnecting load cables. Failure to comply with this warning can cause injury or death to personnel.

WARNING

DC voltages are present at generator set electrical components even with generator set shut down. Avoid shorting any positive with ground/negative. Failure to comply with this warning can cause injury to personnel, and damage to equipment.

WARNING

Ensure that the engine cannot be started while maintenance is being performed. (ENGINE CONTROL switch set to OFF/RESET; Battery Disconnect Switch is OFF; DEAD CRANK SWITCH is OFF). Failure to comply with this warning can cause injury or death to personnel.

WARNING

Top housing panels and exhaust system can get very hot. Shut down generator set, and allow system to cool before performing checks, services and maintenance. Failure to comply with this warning can cause severe burns and injury to personnel.

a. Shut down generator set.

b. Open battery access door and disconnect negative battery cable.

c. Release control panel by turning two fasteners and lower control panel slowly.

d. Locate suspected defective relay (57, Figure 2-14), release wire clip (55), and remove relay by gently pulling from socket.

2-60.4 Installation.

WARNING

High voltage is produced when the generator set is in operation. Never attempt to start the generator set unless it is properly grounded. Failure to comply with this warning can cause injury or death to personnel.

a. Install relay (57, Figure 2-14) in socket and secure by snapping wire clip (55) over relay.

b. Raise and secure control panel.

c. Connect negative battery c able and close battery access door.

2-61 GOVERNOR CONTROL UNIT

2-61.1 Inspection.

b. Inspect governor control unit for loose connections and mounting, and other damage.

2-61.2 Removal.

WARNING

All metal jewelry can conduct electricity and become entangled in generator set components. Remove all jewelry when working on generator set. Failure to comply with this warning can cause injury or death to personnel.

WARNING

DO NOT wear loose clothing when performing checks, services and maintenance. Failure to comply with this warning can cause injury or death to personnel.

WARNING

When running, generator set engine has hot metal surfaces that will burn flesh on contact. Shut down generator set and allow engine to cool before performing checks, services, and maintenance. Wear gloves and additional protective clothing as required. Failure to comply with this warning can cause injury or death to personnel.

WARNING

Dangerous voltage exists on live circuits. Always observe precautions and never work alone. Failure to comply with this warning can cause injury or death to personnel.

WARNING

High voltage is produced when this generator set is in operation. SHUT DOWN generator set and make sure it is free of any power source before attempting any repair or maintenance on the set, or when connecting or disconnecting load cables. Failure to comply with this warning can cause injury or death to personnel.

WARNING

DC voltages are present at generator set electrical components even with generator set shut down. Avoid shorting any positive with ground/negative. Failure to comply with this warning can cause injury to personnel, and damage to equipment.

WARNING

Ensure that the engine cannot be started while maintenance is being performed. (ENGINE CONTROL switch set to OFF/RESET; Battery Disconnect Switch is OFF; DEAD CRANK SWITCH is OFF). Failure to comply with this warning can cause injury or death to personnel.

WARNING

Top housing panels and exhaust system can get very hot. Shut down generator set, and allow system to cool before performing checks, services and maintenance. Failure to comply with this warning can cause severe burns and injury to personnel.

a. a. Shut down generator set.

b. b. Open battery access door and disconnect negative battery cable.

c. c. Remove control box top panel (paragraph 2-15.1).

d. d. Tag and disconnect governor control unit (59, Figure 2-14) electrical leads.

e. e. Remove screws (58) and governor control unit (59).

2-61.3 Installation.

WARNING

High voltage is produced when the generator set is in operation. Never attempt to start the generator set unless it is properly grounded. Failure to comply with this warning can cause injury or death to personnel.

a. Install governor control unit (59, Figure 2-14) with screws (58).

b. Connect electrical leads. Remove tags.

c. Install control box top panel (paragraph 2-15.4).

d. Connect negative battery c able and close battery access door.

2-62 CONTROL BOX WIRING HARNESS

a. Shut down generator set.

b. Inspect control box harness (60, Figure 2-14) wiring for breaks, damaged insulation, and loose or damaged terminals.

2-62.2 Testing.

a. Shut down generator set.

b. Open battery access door and disconnect negative battery cable.

c. Remove control box top panel (paragraph 2-15.1).

d. Set multimeter for ohms and check wires for continuity using Wiring Diagram FO-2 (MEP-804A/MEP-814A)/FO-4 (MEP-804B/MEP-814B) as a guide.

e. If no repair is needed, install control box top panel (paragraph 2-15.4) and Connect negative battery cable and close battery access door.

2-62.3 Repair.

WARNING

All metal jewelry can conduct electricity and become entangled in generator set components. Remove all jewelry when working on generator set. Failure to comply with this warning can cause injury or death to personnel.

WARNING

DO NOT wear loose clothing when performing checks, services and maintenance. Failure to comply with this warning can cause injury or death to personnel.

WARNING

When running, generator set engine has hot metal surfaces that will burn flesh on contact. Shut down generator set and allow engine to cool before performing checks, services, and maintenance. Wear gloves and additional protective clothing as required. Failure to comply with this warning can cause injury or death to personnel.

WARNING

Dangerous voltage exists on live circuits. Always observe precautions and never work alone. Failure to comply with this warning can cause injury or death to personnel.

WARNING

High voltage is produced when this generator set is in operation. SHUT DOWN generator set and make sure it is free of any power source before attempting any repair or maintenance on the set, or when connecting or disconnecting load cables. Failure to comply with this warning can cause injury or death to personnel.

WARNING

DC voltages are present at generator set electrical components even with generator set shut down. Avoid shorting any positive with ground/negative. Failure to comply with this warning can cause injury to personnel, and damage to equipment.

WARNING

Ensure that the engine cannot be started while maintenance is being performed. (ENGINE CONTROL switch set to OFF/RESET; Battery Disconnect Switch is OFF; DEAD CRANK SWITCH is OFF). Failure to comply with this warning can cause injury or death to personnel.

WARNING

Top housing panels and exhaust system can get very hot. Shut down generator set, and allow system to cool before performing checks, services and maintenance. Failure to comply with this warning can cause severe burns and injury to personnel.

a. a. Replace individual wires, damaged terminal ends, clamps, and tie wraps.

b. b. Ensure proper connection of wires not indicating continuity.

2-63 DIAGNOSTIC CONNECTOR

NOTE

The diagnostic connector can be used as an aid in troubleshooting. Refer to Figure 2-18 and Table 2-9.

Diagnostic connector is a component of the control box harness assembly, but can be removed and installed separately.

2-63.1 Inspection.

a. a. Shut down generator set.

b. b. Open battery access door and disconnect negative battery cable.

c. c. Inspect diagnostic connector for cracks, breaks, corrosion, bent terminals, burns, and other damage.

d. d. Inspect cap for cracks, corrosion, or broken chain.

e. e. Inspect gasket for tears and deterioration.

f. f. Replace any defective part.

g. g. If no repair is needed, connect negative battery cable and close battery access door.

2-63.2 Removal.

WARNING

All metal jewelry can conduct electricity and become entangled in generator set components. Remove all jewelry when working on generator set. Failure to comply with this warning can cause injury or death to personnel.

WARNING

DO NOT wear loose clothing when performing checks, services and maintenance. Failure to comply with this warning can cause injury or death to personnel.

WARNING

When running, generator set engine has hot metal surfaces that will burn flesh on contact. Shut down generator set and allow engine to cool before performing checks, services, and maintenance. Wear gloves and additional protective clothing as required. Failure to comply with this warning can cause injury or death to personnel.

WARNING

Dangerous voltage exists on live circuits. Always observe precautions and never work alone. Failure to comply with this warning can cause injury or death to personnel.

WARNING

High voltage is produced when this generator set is in operation. SHUT DOWN generator set and make sure it is free of any power source before attempting any repair or maintenance on the set, or when connecting or disconnecting load cables. Failure to comply with this warning can cause injury or death to personnel.

WARNING

DC voltages are present at generator set electrical components even with generator set shut down. Avoid shorting any positive with ground/negative. Failure to comply with this warning can cause injury to personnel, and damage to equipment.

WARNING

Ensure that the engine cannot be started while maintenance is being performed. (ENGINE CONTROL switch set to OFF/RESET; Battery Disconnect Switch is OFF; DEAD CRANK SWITCH is OFF). Failure to comply with this warning can cause injury or death to personnel.

WARNING

Top housing panels and exhaust system can get very hot. Shut down generator set, and allow system to cool before performing checks, services and maintenance. Failure to comply with this warning can cause severe burns and injury to personnel.

a. a. Shut down generator set.

b. b. Open battery access door and disconnect negative battery cable.

c. c. Remove control box top panel (paragraph 2-15.1).

d. d. Loosen rear outer ring and plastic insert from diagnostic connector (63, Figure 2-14).

e. e. Tag and disconnect electrical leads to diagnostic connector (63) by inserting removal tool into pins of connector.

f. f. Remove screws (61) and nuts (62).

g. g. Remove diagnostic connector cap (64), diagnostic connector (63), and gasket (65).

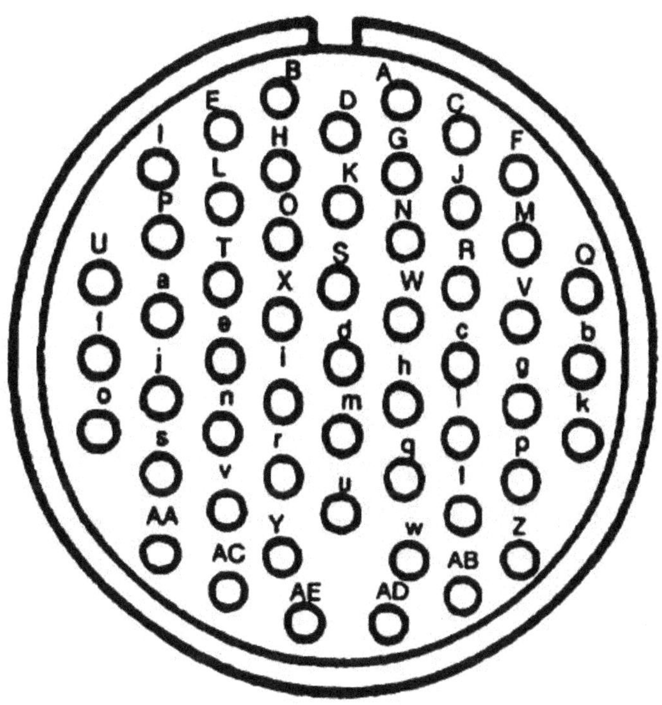

Figure 2-18. Diagnostic Connector Pin Positions.

Table 2-9. Diagnostic Connector Connection Points.

NOTE

The diagnostic connector can be used as a troubleshooting tool. Refer to Figure 2-18 for pin positions.

Pin	Description	Expected Output*
B	Chassis ground (GND)	Continuity 0 volts
C	DC paralleling voltage for governor synchronization	0-20 VDC (*,5)
D	Chassis ground (GND)	Continuity 0 volts
E	Paralleling voltage for AC voltage regulator	0-20 VAC set freq. (*,5)
F	DC exciter field voltage (positive)	0-60 VDC (*,6)
G	DC paralleling voltage for governor synchronization	0-20 VDC (*,5)
H	Paralleling voltage for AC voltage regulator	0-20 VAC set freq.
J	DC exciter field voltage (negative)	0-60 VDC (*,7)
M	DC voltage input to governor	24 VDC (*)
N	DC starter motor solenoid coil	24 VDC (2 or 3)
O	DC voltage across fuel pump (Aux)	24 VDC (1)
P	DC voltage S1 terminal 7	24 VDC (2)
S	DC voltage across engine fuel solenoid coil	24 VDC (*)
T	DC voltage across starter crank relay coil	24 VDC (2 or 3)
U	DC voltage (output of DC circuit breaker)	24 VDC
V	DC voltage across output circuit interrupter coil	24 VDC (*)
X	DC starter motor (motor side of solenoid contacts)	24 VDC (2 or 3)
Z	DC voltage across fuel level contacts	24 VDC (4)
a	DC voltage across low oil pressure switch	0 volts (*)
b	DC voltage across low oil pressure switch	24 VDC (4)
d	DC battery charging voltage	24-27.6 VDC
X	Input frequency sensing voltage to governor	2-6 volts, 0-4,000 Hz (3, 8)
Y	Input frequency sensing voltage to governor	2-6 volts, 0-4,000 Hz (3, 8)
(1)	With generator set operating	
(2)	Auxiliary fuel pump in operation (note fuel level); MASTER SWITCH in PRIME & RUN AUX FUEL position	
(3)	During engine starting	
(4)	Use DEAD CRANK switch	
(5)	MASTER SWITCH in PRIME & RUN position	
(6)	AC circuit interrupter closed	
(7)	Read between pins F and J	
(8)	Read between pins J and F	
	Read between pins X and Y	

2-63.3 Installation.

WARNING

High voltage is produced when the generator set is in operation. Never attempt to start the generator set unless it is properly grounded. Failure to comply with this warning can cause injury or death to personnel.

a. Install diagnostic connector (63, Figure 2-14), gasket (65), and cap (64) with screws (61) and nuts (62).

b. Connect electrical leads to diagnostic connector (63) by using insert tool. Remove tags.

c. Install plastic insert and tighten rear outer ring on connector.

d. Install control box top panel (paragraph 2-15.4).

e. Connect negative battery c able and close battery access door.

2-64 PARALLEL CONNECTOR

NOTE

The parallel connector is a component of the control box harness assem-

2-64.1 Inspection.

a. Shut down generator set.

b. Inspect parallel connector for cracks, corrosion, stripped or damaged threads, evidence of shorting, or other damage.

c. Inspect cap for cracks, corrosion, and broken chain.

d. Inspect gasket for tears and deterioration.

e. Replace defective parts.

2-64.2 Removal.

WARNING

All metal jewelry can conduct electricity and become entangled in generator set components. Remove all jewelry when working on generator set. Failure to comply with this warning can cause injury or death to personnel.

WARNING

DO NOT wear loose clothing when performing checks, services and maintenance. Failure to comply with this warning can cause injury or death to personnel.

WARNING

When running, generator set engine has hot metal surfaces that will burn flesh on contact. Shut down generator set and allow engine to cool before performing checks, services, and maintenance. Wear gloves and additional protective clothing as required. Failure to comply with this warning can cause injury or death to personnel.

WARNING

Dangerous voltage exists on live circuits. Always observe precautions and never work alone. Failure to comply with this warning can cause injury or death to personnel.

WARNING

High voltage is produced when this generator set is in operation. SHUT DOWN generator set and make sure it is free of any power source before attempting any repair or maintenance on the set, or when connecting or disconnecting load cables. Failure to comply with this warning can cause injury or death to personnel.

WARNING

DC voltages are present at generator set electrical components even with generator set shut down. Avoid shorting any positive with ground/negative. Failure to comply with this warning can cause injury to personnel, and damage to equipment.

WARNING

Ensure that the engine cannot be started while maintenance is being performed. (ENGINE CONTROL switch set to OFF/RESET; Battery Disconnect Switch is OFF; DEAD CRANK SWITCH is OFF). Failure to comply with this warning can cause injury or death to personnel.

WARNING

Top housing panels and exhaust system can get very hot. Shut down generator set, and allow system to cool before performing checks, services and maintenance. Failure to comply with this warning can cause severe burns and injury to personnel.

a. a. Shut down generator set.

b. b. Open battery access door and disconnect negative battery cable.

c. c. Remove control box top panel (paragraph 2-15.1).

d. d. Tag and disconnect parallel connector (68, Figure 2-14) electrical leads by inserting removal tool

e. e. Remove screws (66), nuts (67), and cap (69).

f. f. Remove parallel connector (68) and gasket
(70).

WARNING

High voltage is produced when the generator set is in operation. Never attempt to start the generator set unless it is properly grounded. Failure to comply with this warning can cause injury or death to personnel.

a. a. Install parallel connector (68, Figure 2-14), gasket (70), and cap (69) with screws (66) and nuts (67).

b. b. Connect electrical leads using insert tool. Remove tags.

c. c. Install control box top panel (paragraph 2-15.4).

d. d. Connect negative battery cable and close battery access door.

2-65 LOAD MEASURING UNIT

2-65.1 Inspection.

a. a. Shut down generator set.

b. b. Inspect load measuring unit for damaged case, cracked or broken terminal lugs, and loose or missing hardware.

2-65.2 Removal.

WARNING

All metal jewelry can conduct electricity and become entangled in generator set components. Remove all jewelry when working on generator set. Failure to comply with this warning can cause injury or death to personnel.

WARNING

DO NOT wear loose clothing when performing checks, services and maintenance. Failure to comply with this warning can cause injury or death to personnel.

WARNING

When running, generator set engine has hot metal surfaces that will burn flesh on contact. Shut down generator set and allow engine to cool before performing checks, services, and maintenance. Wear gloves and additional protective clothing as required. Failure to comply with this warning can cause injury or death to personnel.

WARNING

Dangerous voltage exists on live circuits. Always observe precautions and never work alone. Failure to comply with this warning can cause injury or death to personnel.

WARNING

High voltage is produced when this generator set is in operation. SHUT DOWN generator set and make sure it is free of any power source before attempting any repair or maintenance on the set, or when connecting or disconnecting load cables. Failure to comply with this warning can cause injury or death to personnel.

WARNING

DC voltages are present at generator set electrical components even with generator set shut down. Avoid shorting any positive with ground/negative. Failure to comply with this warning can cause injury to personnel, and damage to equipment.

WARNING

Ensure that the engine cannot be started while maintenance is being performed. (ENGINE CONTROL switch set to OFF/RESET; Battery Disconnect Switch is OFF; DEAD CRANK SWITCH is OFF). Failure to comply with this warning can cause injury or death to personnel.

WARNING

Top housing panels and exhaust system can get very hot. Shut down generator set, and allow system to cool before performing checks, services and maintenance. Failure to comply with this warning can cause severe burns and injury to personnel.

a. a. Shut down generator set.

b. b. Open battery access door and disconnect negative battery cable.

c. c. Remove control box top panel (paragraph 2-15.1).

d. d. Release control panel by turning two fasteners and lower control panel slowly.

e. e. Tag and disconnect load measuring unit (72, Figure 2-14) electrical leads.

f. f. Remove screws (71) and load measuring unit (72).

2-65.3 Installation.

WARNING

High voltage is produced when the generator set is in operation. Never attempt to start the generator set unless it is properly grounded. Failure to comply with this warning can cause injury or death to personnel.

a. a. Install load measuring unit (72, Figure 2-14) with screws (71).

b. b. Connect all electrical leads. Remove tags.

c. c. Install control box top panel (paragraph 2-15.4)

d. d. Raise and secure control panel.

e. e. Connect negative battery cable and close battery access door.

2-66 RESISTOR-DIODE ASSEMBLY 2-66.1

Inspection.

a. Shut down generator set.

b. Inspect resistor-diode assembly for cracks, breaks, corrosion, bent terminals, and other damage.

2-66.2 Testing.

a. a. Shut down generator set.

b. b. Open battery access door and disconnect negative battery cable.

c. c. Remove control box top panel (paragraph 2-15.1).

d. d. Release control panel by turning two fasteners and lower control panel slowly.

NOTE

Isolate component before testing.

e. e. Set multimeter for ohms and measure resistance across resistors R10, R11, and R12. Multimeter indication should be between 7.125 and 7.875 ohms for each resistor.

f. f. Using multimeter, measure resistance across resistor R14. Multimeter indication should be between 61.75 and 68.25 ohms (MEP-804A)/38 and 42 ohms (MEP-814A).

g. g. Using multimeter, measure resistance across resistor R15. Multimeter indication should be between 1,235 and 1,365 ohms.

h. h. Using multimeter, measure resistance across resistors R6 and R8. Multimeter indication should be between 4,750 and 5,250 ohms.

i. i. Using multimeter, measure resistance across resistors R7 and R9. Multimeter indication should be between 2,850 and 3,150 ohms.

j. j. Connect positive lead of multimeter to cathode side and negative lead to anode side of each diode (CR1, CR2, CR3, and CR4). Refer to Figure 2-19. Note ohms indication on multimeter for each diode.

k. k. Reverse multimeter leads so positive lead is connected to anode side and negative lead is connected to cathode side of each diode (CR1, CR2, CR3, and CR4). Note ohms indication on multimeter for each diode.

l. l. Multimeter indications should be 1:10 ratio or greater.

m. m. Replace defective component, if any indications are other than above.

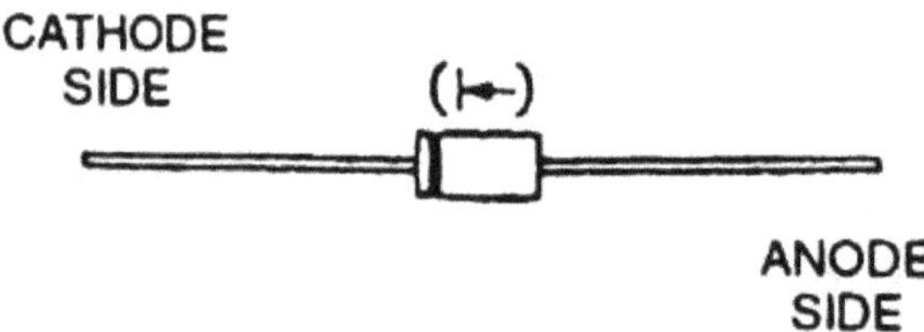

Figure 2-19. Diode Identification.

n. n. If no repair is needed, install control box top panel (paragraph 2-15.4).

o. o. Raise and secure control panel.

p. p. Connect negative battery c able and close battery access door.

WARNING

All metal jewelry can conduct electricity and become entangled in generator set components. Remove all jewelry when working on generator set. Failure to comply with this warning can cause injury or death to personnel.

WARNING

DO NOT wear loose clothing when performing checks, services and maintenance. Failure to comply with this warning can cause injury or death to personnel.

WARNING

When running, generator set engine has hot metal surfaces that will burn flesh on contact. Shut down generator set and allow engine to cool before performing checks, services, and maintenance. Wear gloves and additional protective clothing as required. Failure to comply with this warning can cause injury or death to personnel.

WARNING

Dangerous voltage exists on live circuits. Always observe precautions and never work alone. Failure to comply with this warning can cause injury or death to personnel.

WARNING

High voltage is produced when this generator set is in operation. SHUT DOWN generator set and make sure it is free of any power source before attempting any repair or maintenance on the set, or when connecting or disconnecting load cables. Failure to comply with this warning can cause injury or death to personnel.

WARNING

DC voltages are present at generator set electrical components even with generator set shut down. Avoid shorting any positive with ground/negative. Failure to comply with this warning can cause injury to personnel, and damage to equipment.

WARNING

Ensure that the engine cannot be started while maintenance is being performed. (ENGINE CONTROL switch set to OFF/RESET; Battery Disconnect Switch is OFF; DEAD CRANK SWITCH is OFF). Failure to comply with this warning can cause injury or death to personnel.

WARNING

Top housing panels and exhaust system can get very hot. Shut down generator set, and allow system to cool before performing checks, services and maintenance. Failure to comply with this warning can cause severe burns and injury to personnel.

a. a. Shut down generator set.

b. b. Open battery access door and disconnect negative battery cable.

c. c. Remove control box top panel (paragraph 2-15.1).

d. d. Tag and disconnect resistor-diode assembly (74, Figure 2-14) electrical leads.

e. e. Remove screws (73) and resistor-diode assembly (74).

2-66.4 Repair.

Repair resistor-diode assembly (74, Figure 2-14) by replacing resistors (75) and diodes (76).

2-66.5 Installation.

WARNING

High voltage is produced when the generator set is in operation. Never attempt to start the generator set unless it is properly grounded. Failure to comply with this warning can cause injury or death to personnel.

a. Install resistor-diode assembly (74, Figure 2-14) and secure with screws (73).

b. b. Connect electrical leads. Remove tags.

c. c. Install control box top panel (paragraph 2-15.4).

d. d. Connect negative battery c able and close battery access door.

2-67 CONTROL PANEL 2-67.1

Inspection.

Inspect control panel (10, Figure 2-20) for dents, cracks, loose paint, and corrosion.

WARNING

All metal jewelry can conduct electricity and become entangled in generator set components. Remove all jewelry when working on generator set. Failure to comply with this warning can cause injury or death to personnel.

WARNING

DO NOT wear loose clothing when performing checks, services and maintenance. Failure to comply with this warning can cause injury or death to personnel.

WARNING

When running, generator set engine has hot metal surfaces that will burn flesh on contact. Shut down generator set and allow engine to cool before performing checks, services, and maintenance. Wear gloves and additional protective clothing as required. Failure to comply with this warning can cause injury or death to personnel.

WARNING

Dangerous voltage exists on live circuits. Always observe precautions and never work alone. Failure to comply with this warning can cause injury or death to personnel.

WARNING

High voltage is produced when this generator set is in operation. SHUT DOWN generator set and make sure it is free of any power source before attempting any repair or maintenance on the set, or when connecting or disconnecting load cables. Failure to comply with this warning can cause injury or death to personnel.

WARNING

DC voltages are present at generator set electrical components even with generator set shut down. Avoid shorting any positive with ground/negative. Failure to comply with this warning can cause injury to personnel, and damage to equipment.

WARNING

Ensure that the engine cannot be started while maintenance is being performed. (ENGINE CONTROL switch set to OFF/RESET; Battery Disconnect Switch is OFF; DEAD CRANK SWITCH is OFF). Failure to comply with this warning can cause injury or death to personnel.

WARNING

Top housing panels and exhaust system can get very hot. Shut down generator set, and allow system to cool before performing checks, services and maintenance. Failure to comply with this warning can cause severe burns and injury to personnel.

a. a. Shut down generator set.

b. b. Open battery access door and disconnect negative battery cable.

c. c. Remove press-to-test lights from control panel (paragraph 2-21.3). (Do not unsolder wires.)

d. d. Remove VOLTAGE adjust potentiometer from control panel (paragraph 2-33.2). (Do not remove wires.)

e. e. Remove FREQUENCY adjust potentiometer from control panel (paragraph 2-34.2). (Do not remove wires.)

f. f. Tag and disconnect all electrical leads to remaining indicators and switches on control panel (10, Figure 2-20).

g. g. Remove screw (1), nuts (2), and strap (3).

h. h. Remove bolts (4 and 6), nuts (5 and 7), bracket (9), clamp (8), and control panel (10) from control box assembly.

i. MEP-804A/MEP-814A: Remove bolts (11), nuts (12), and hinge (13) from control box assembly.

MEP-804B/MEP-814B: Remove bolts (11), nuts (12), seal retainer (12.1), seal (12.2), and hinge (13) from control box assembly.

2-67.3 Repair.

WARNING

CARC paint is a health hazard, and is irritating to eyes, skin, and respira-

applying or removing CARC paint. Failure to comply with this warning can cause injury to personnel.

a. a. Repair all dents and cracks and remove all loose paint.

b. b. Remove light corrosion with fine grit abrasive paper (Item 15, Appendix C).

c. c. Repaint surface in accordance with TM 43-0139. (F) Refer to applicable directives.

WARNING

High voltage is produced when the generator set is in operation. Never attempt to start the generator set unless it is properly grounded. Failure to comply with this warning can cause injury or death to personnel.

a. a. MEP-804A/MEP-814A: Install hinge (13, Figure 2-20) on control box assembly with bolts (11) and nuts (12).

MEP-804B/MEP-814B: Install hinge (13, Figure 2-20), seal (12.2), and seal retainer (12.1) on control box assembly with bolts (11) and nuts (12).

b. b. Install control panel (10), clamp (8), and bracket (9) on control box assembly with bolts (4 and 6) and nuts (5 and 7).

c. c. Install strap (3) on control panel (10) with screw (1) and nuts (2).

d. d. Connect all electrical wires to indicators and switches as tagged. Remove tags.

e. e. Install press-to-test lights (paragraph 2-21.4).

f. f. Install VOLTAGE adjust potentiometer (paragraph 2-33.4).

g. g. Install FREQUENCY adjust potent iometer (paragraph 2-34.4).

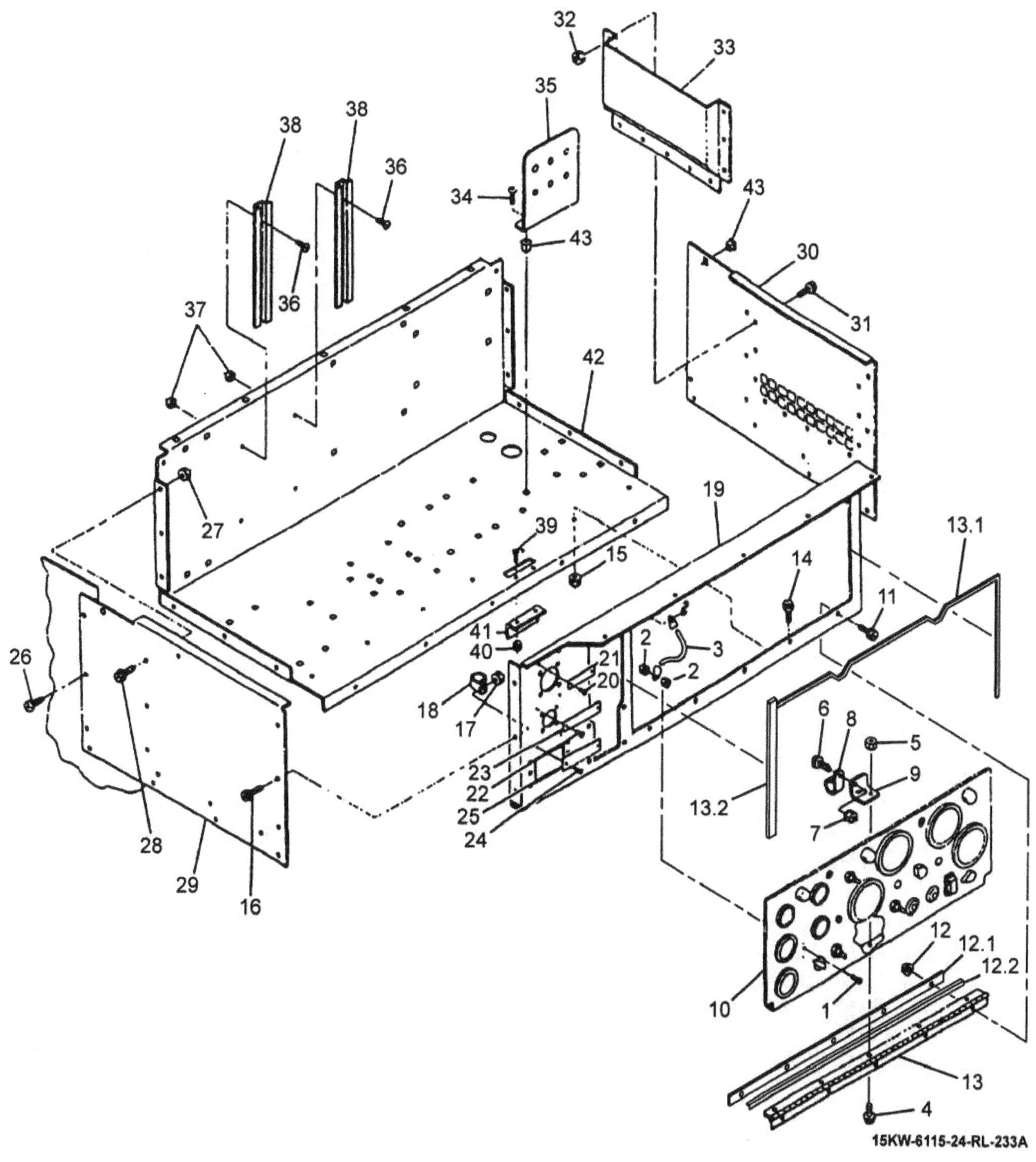

Figure 2-20. Control Box Panels.

2-68 CONTROL PANEL FRAME

2-68.1 Inspection.

Inspect control panel frame (19, Figure 2-20) for dents, cracks, loose paint, and corrosion.

WARNING

All metal jewelry can conduct electricity and become entangled in generator set components. Remove all jewelry when working on generator set. Failure to comply with this warning can cause injury or death to personnel.

WARNING

DO NOT wear loose clothing when performing checks, services and maintenance. Failure to comply with this warning can cause injury or death to personnel.

WARNING

When running, generator set engine has hot metal surfaces that will burn flesh on contact. Shut down generator set and allow engine to cool before performing checks, services, and maintenance. Wear gloves and additional protective clothing as required. Failure to comply with this warning can cause injury or death to personnel.

WARNING

Dangerous voltage exists on live circuits. Always observe precautions and never work alone. Failure to comply with this warning can cause injury or death to personnel.

WARNING

High voltage is produced when this generator set is in operation. SHUT DOWN generator set and make sure it is free of any power source before attempting any repair or maintenance on the set, or when connecting or disconnecting load cables. Failure to comply with this warning can cause injury or death to personnel.

WARNING

DC voltages are present at generator set electrical components even with generator set shut down. Avoid shorting any positive with ground/negative. Failure to comply with this warning can cause injury to personnel, and damage to equipment.

WARNING

Ensure that the engine cannot be started while maintenance is being performed. (ENGINE CONTROL switch set to OFF/RESET; Battery Disconnect Switch is OFF; DEAD CRANK SWITCH is OFF). Failure to comply with this warning can cause injury or death to personnel.

WARNING

Top housing panels and exhaust system can get very hot. Shut down generator set, and allow system to cool before performing checks, services and maintenance. Failure to comply with this warning can cause severe burns and injury to personnel.

a. a. Shut down generator set.

b. b. Open battery access door and disconnect negative battery cable.

c. c. Remove control box assembly (paragraph 2-19.2).

d. d. Remove control panel (paragraph 2-67.1).

MEP-804B/MEP-814B: Inspect and remove EMI gaskets (13.1 and 13.2, Figure 2-20), if necessary.

e. e. Remove diagnostic connector (paragraph 2-63.2). (Do not remove wires.)

f. f. Remove parallel connector (paragraph 2-64.2). (Do not remove wires.)

g. Remove CONVENIENCE RECEPTACLE (paragraph 2-45.3); GROUND FAULT CIRCUIT INTERRUPTER (paragraph 2-46.3); and malfunction indicator panel (paragraph 2-47.3).

h. h. Remove bolts (14 and 16), nuts (15 and 17), clamp (18), and control panel frame (19) from control box assembly.

i. i. Drill out rivets (20, 22, and 24) and remove identification plates (21, 23, and 25), if necessary.

2-68.3 Repair.

WARNING

CARC paint is a health hazard, and is irritating to eyes, skin, and respiratory system. Wear protective eyewear, mask, and gloves when applying or removing CARC paint. Failure to comply with this warning can cause injury to personnel.

a. a. Repair all dents and cracks and remove all loose paint.

b. b. Remove light corrosion with fine grit abrasive paper (Item 15, Appendix C).

c. c. Repaint surface in accordance with TM 43-0139. (F) Refer to applicable directives.

a. a. Install control panel frame (19, Figure 2-20) and clamp (18) to control box assembly with bolts (14 and 16) and nuts (15 and 17).

b. b. If removed, install identification plates (21, 23, and 25) on rear panel (19) with rivets (20, 22, and 24).

c. Install malfunction indicator panel (paragraph 2-47.4); GROUND FAULT CIRCUIT INTERRUPTER (paragraph 2-46.4); CONVENIENCE RECEPTACLE (paragraph 2-45.4); parallel connector (paragraph 2-64.3); and diagnostic connector (paragraph 2-63.3).

MEP-804B/MEP-814B: Install EMI gaskets (13.1 and 13.2), if necessary.

d. d. Install control panel (paragraph 2-67.4).

e. e. Install control box assembly (paragraph 2-19.4).

f. f. Connect negative battery cabl e and close battery access door.

2-69 CONTROL BOX SIDE PANELS 2-69.1

Inspection.

a. a. Inspect side panels (29 and 30, Figure 2-20) and air deflector (33) for dents, cracks, loose paint, and corrosion.

b. b. Inspect for missing or damaged cage nuts (43).

WARNING

All metal jewelry can conduct electricity and become entangled in generator set components. Remove all jewelry when working on generator set. Failure to comply with this warning can cause injury or death to personnel.

WARNING

DO NOT wear loose clothing when performing checks, services and maintenance. Failure to comply with this warning can cause injury or death to personnel.

WARNING

When running, generator set engine has hot metal surfaces that will burn flesh on contact. Shut down generator set and allow engine to cool before performing checks, services, and maintenance. Wear gloves and additional protective clothing as required. Failure to comply with this warning can cause injury or death to personnel.

WARNING

Dangerous voltage exists on live circuits. Always observe precautions

and never work alone. Failure to comply with this warning can cause injury

WARNING

High voltage is produced when this generator set is in operation. SHUT DOWN generator set and make sure it is free of any power source before attempting any repair or maintenance on the set, or when connecting or disconnecting load cables. Failure to comply with this warning can cause injury or death to personnel.

WARNING

DC voltages are present at generator set electrical components even with generator set shut down. Avoid shorting any positive with ground/negative. Failure to comply with this warning can cause injury to personnel, and damage to equipment.

WARNING

Ensure that the engine cannot be started while maintenance is being performed. (ENGINE CONTROL switch set to OFF/RESET; Battery Disconnect Switch is OFF; DEAD CRANK SWITCH is OFF). Failure to comply with this warning can cause injury or death to personnel.

WARNING

Top housing panels and exhaust system can get very hot. Shut down generator set, and allow system to cool before performing checks, services and maintenance. Failure to comply with this warning can cause severe burns and injury to personnel.

a. a. Shut down generator set.

b. b. Open battery access door and disconnect negative battery cable.

c. c. Remove control box top panel (paragraph 2-15.1).

d. d. Remove bolts (16, 26, and 28, Figure 2-20), nuts (17 and 27), and control box side panels (29 and 30) from generator set.

e. Remove bolts (31), nuts (32), and air deflector (33) from side panel (30).

2-69.3 Repair.

WARNING

CARC paint is a health hazard, and is irritating to eyes, skin, and respiratory system. Wear protective eyewear, mask, and gloves when applying or removing CARC paint. Failure to comply with this warning can cause injury to personnel.

a. a. Repair all dents and cracks and remove all loose paint.

b. b. Remove light corrosion with fine grit abrasive paper (Item 15, Appendix C).

c. c. Repaint surface in accordance with TM 43-0139. (F) Refer to applicable directives.

d. d. Replace missing or damaged cage nuts.

2-69.4 Installation.

a. a. Apply a light coat of sealant (Item 16, Appendix C) to flanges of air deflector (33, Figure 2-20).

b. b. Install air deflector (33) on control box side panel (30) with bolts (31) and nuts (32).

c. c. Install control box side panels (29 and 30, Figure 2-20) on generator set with bolts (16, 26, and 28) and nuts (17 and 27). Ensure center bolt (16) and nut (17) on left side panel secures clamp (18).

d. d. Install control box top panel (paragraph 2-15.4).

e. e. Connect negative battery c able and close battery access door.

2-70 CONTROL BOX TRAY (BOTTOM)

2-70.1 Inspection.

a. a. Inspect control box bottom (42, Figure 2-20) for dents, cracks, loose paint, and corrosion.

b. b. Inspect for missing or damaged cage nuts (43).

2-70.2 Removal.

WARNING

All metal jewelry can conduct electricity and become entangled in generator set components. Remove all jewelry when working on generator set. Failure to comply with this warning can cause injury or death to personnel.

WARNING

DO NOT wear loose clothing when performing checks, services and maintenance. Failure to comply with this warning can cause injury or death to personnel.

WARNING

When running, generator set engine has hot metal surfaces that will burn flesh on contact. Shut down generator set and allow engine to cool before performing checks, services, and maintenance. Wear gloves and additional protective clothing as required. Failure to comply with this warning can cause injury or death to personnel.

WARNING

Dangerous voltage exists on live circuits. Always observe precautions and never work alone. Failure to comply with this warning can cause injury or death to personnel.

WARNING

High voltage is produced when this generator set is in operation. SHUT DOWN generator set and make sure it is free of any power source before attempting any repair or maintenance on the set, or when connecting or disconnecting load cables. Failure to comply with this warning can cause injury or death to personnel.

WARNING

DC voltages are present at generator set electrical components even with generator set shut down. Avoid shorting any positive with ground/negative. Failure to comply with this warning can cause injury to personnel, and damage to equipment.

WARNING

Ensure that the engine cannot be started while maintenance is being performed. (ENGINE CONTROL switch set to OFF/RESET; Battery Disconnect Switch is OFF; DEAD CRANK SWITCH is OFF). Failure to comply with this warning can cause injury or death to personnel.

WARNING

Top housing panels and exhaust system can get very hot. Shut down generator set, and allow system to cool before performing checks, services and maintenance. Failure to comply with this warning can cause severe burns and injury to personnel.

a. a. Shut down generator set.

b. b. Open battery access door and disconnect negative battery cable.

c. c. Remove control box assembly (paragraph 2-19.2).

d. d. Remove control panel frame (paragraph 2-68.1).

e. e. Remove control box side panels (paragraph 2-69.1).

f. f. Remove control box components (paragraphs 2-41 thru 2-50, 2-52 thru 2-61, 2-65, and 2-66).

g. g. Contact Direct Support Maintenance to remove AC voltage regulator and control box harness.

h. h. Remove screws (34, Figure 2-20) and bracket (35) from control box bottom (42).

j. Remove screws (39), nuts (40), and latch plate (41) from control box bottom (42).

2-70.3 Repair.

WARNING

CARC paint is a health hazard, and is irritating to eyes, skin, and respiratory system. Wear protective eyewear, mask, and gloves when applying or removing CARC paint. Failure to comply with this warning can cause injury to personnel.

a. a. Repair all dents and cracks and remove all loose paint.

b. b. Remove light corrosion with fine grit abrasive paper (Item 15, Appendix C).

c. c. Repaint surface in accordance with TM 43-0139. (F) Refer to applicable directives.

d. d. Replace missing or damaged cage nuts.

2-70.4 Installation.

a. a. Install latch plate (41, Figure 2-20) on control box bottom (42) with screws (39) and nuts (40).

b. b. Install relay tracks (38) on control box bottom (42) with screws (36) and nuts (37).

c. c. Install bracket (35) on control box bottom (42) with screws (34).

d. d. Contact Direct Support Maintenance to install control box harness and AC voltage regulator.

e. e. Install control box components (paragraphs 2-41 thru 2-50, 2-52 thru 2-61, 2-65, and 2-66).

f. f. Install control box side panels (paragraph 2-69.4).

g. g. Install control panel frame (paragraph 2-68.4).

h. h. Install control box assembly (paragraph 2-19.4).

Section XI. MAINTENANCE OF AIR INTAKE AND EXHAUST SYSTEM

2-71 MUFFLER AND EXHAUST PIPE (MEP-804A/MEP-814A)

2-71.1 Inspection.

a. a. Shut down generator set.

b. b. Remove top housing panel and top housing frame (paragraph 2-16.1) and open engine access doors.

c. c. Inspect muffler and pipe for cracks, excessive corrosion, clogging, and other damage.

d. d. Replace damaged parts.

e. e. Install top housing frame and top housing panel (paragraph 2-16.4).

2-71.2 Removal.

WARNING

All metal jewelry can conduct electricity and become entangled in generator set components. Remove all jewelry when working on generator set. Failure to comply with this warning can cause injury or death to personnel.

WARNING

DO NOT wear loose clothing when performing checks, services and maintenance. Failure to comply with this warning can cause injury or death to personnel.

WARNING

When running, generator set engine has hot metal surfaces that will burn flesh on contact. Shut down generator set and allow engine to cool before performing checks, services, and maintenance. Wear gloves and additional protective clothing as required. Failure to comply with this warning can cause injury or death to personnel.

WARNING

Top housing panels and exhaust system can get very hot. Shut down generator set, and allow system to cool before performing checks, services and maintenance. Failure to comply with this warning can cause severe burns and injury to personnel.

WARNING

Wear heat resistant gloves and avoid contacting hot metal surfaces with your hands after components have been heated. Wear additional protective clothing as required. Failure to comply with this warning can cause injury to personnel.

a. a. Shut down generator set.

b. b. Remove top housing panel and top housing frame (paragraph 2-16.1) and open engine access doors.

c. c. Loosen nuts and remove clamp (1, Figure 2-21).

d. d. Open bands (2), separate muffler (3) and exhaust and adapter (12), and remove muffler (3) from generator set.

e. e. Remove nuts (4), lock washers (5), bolts (6), washers (7), and muffler supports (8). Discard lock washers (5).

f. f. Remove nuts (9), lock washers (10), washers (11), exhaust adapter (12), and gasket (13). Discard lock washers (10).

2-71.3 Installation.

a. a. Install muffler supports (8, Figure 2-21) with bolts (6), nuts (4), new lock washers (5), and washers (7).

b. b. Install exhaust adapter (12) and gasket (13) and secure with nuts (9), new lock washers (10), and washers (11).

c. c. Couple muffler (3) to exhaust adapter (12) with clamp (1). Do not tighten clamp.

d. d. Secure muffler (3) to supports (8) with bands (2).

e. e. Tighten clamp (1).

f. f. Install top housing frame and top housing panel (paragraph 2-16.4). Close engine access doors.

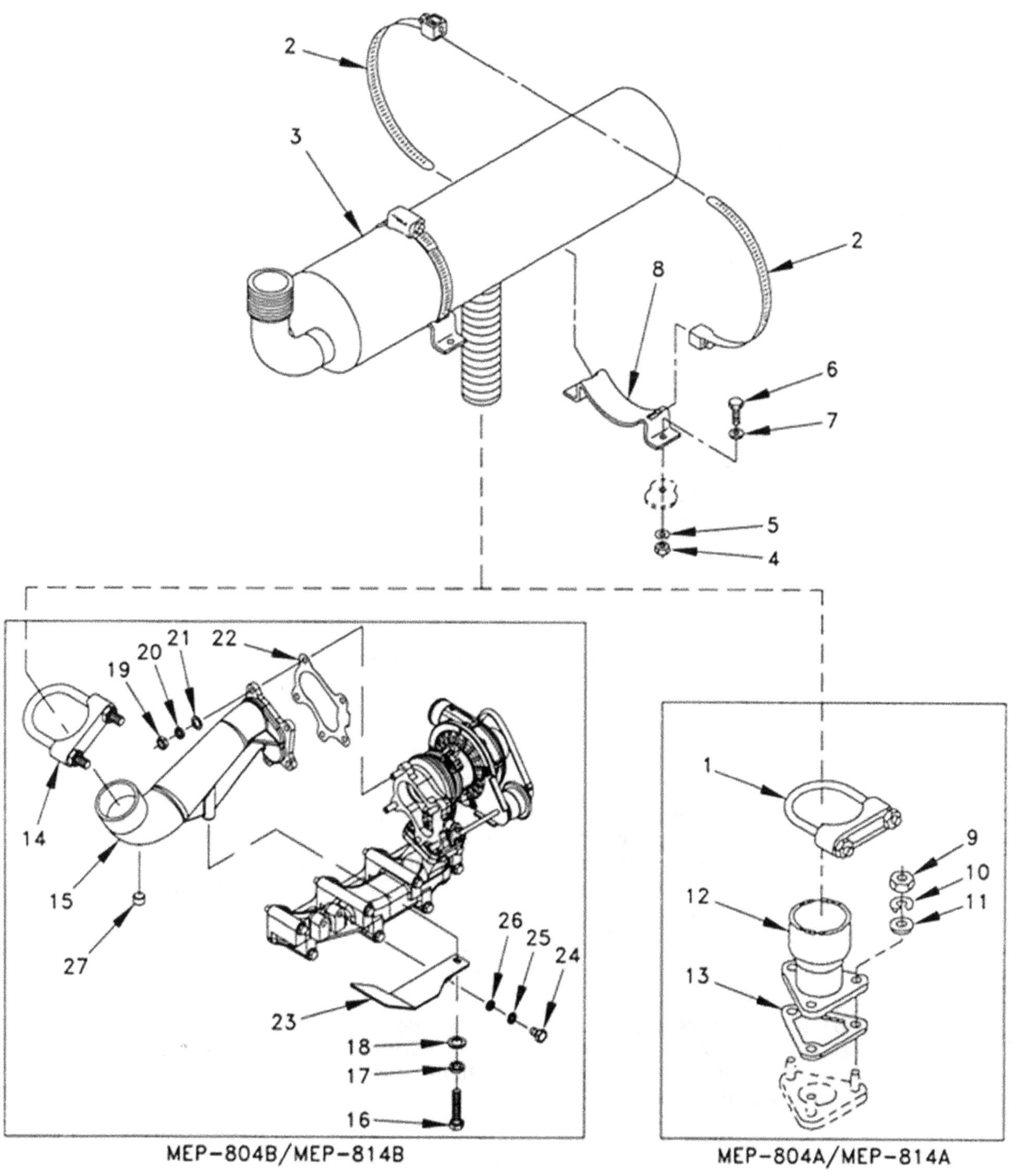

Figure 2-21. Muffler and Exhaust Pipe.

2-72 MUFFLER AND EXHAUST PIPE (MEP-804B/MEP-814B)

2-72.1 Inspection.

a. a. Shut down generator set.

b. b. Remove top housing panel and top housing frame (paragraph 2-16.1) and open engine access doors.

c. c. Inspect muffler and pipe for cracks, excessive corrosion, clogging, and other damage.

d. d. Replace damaged parts.

e. d. Install top housing frame and top housing panel (paragraph 2-16.4).

2-72.2 Removal.

WARNING

All metal jewelry can conduct electricity and become entangled in generator set components. Remove all jewelry when working on generator set. Failure to comply with this warning can cause injury or death to personnel.

WARNING

DO NOT wear loose clothing when performing checks, services and maintenance. Failure to comply with this warning can cause injury or death to personnel.

WARNING

When running, generator set engine has hot metal surfaces that will burn flesh on contact. Shut down generator set and allow engine to cool before performing checks, services, and maintenance. Wear gloves and additional protective clothing as required. Failure to comply with this warning can cause injury or death to personnel.

WARNING

Top housing panels and exhaust system can get very hot. Shut down generator set, and allow system to cool before performing checks, services and maintenance. Failure to comply with this warning can cause severe burns and injury to personnel.

WARNING

Wear heat resistant gloves and avoid contacting hot metal surfaces with your hands after components have been heated. Wear additional protective clothing as required. Failure to comply with this warning can cause injury to personnel.

a. Shut down generator set.

b. b. Remove top housing panel and top housing frame, paragraph 2-16.1 and open engine access doors.

c. c. Loosen bolts and remove clamp (14, Figure 2-21).

d. d. Open bands (2), separate muffler (3) and exhaust manifold (15), and remove muffler (3) from generator set.

e. e. Remove nuts (4), lock washers (5), bolts (6), washers (7), and muffler supports (8). Discard lock washers (5).

f. f. Remove screw (16), lock washer (17), and washer (18). Discard lock washer (17).

g. g. Remove nuts (19), lock washers (20), washers (21), exhaust manifold (15), and gasket (22). Discard lock washers (20).

h. h. Remove screws (24), lock washers (25), washers (26), and manifold bracket (23). Discard lock washers (25).

i. i. Remove plug (27) from exhaust manifold, if necessary.

2-72.3 Installation.

a. a. Install muffler supports (8, Figure 2-21) with bolts (6), nuts (4), new lock washers (5), and washers

(7).

b. b. Install manifold bracket (23) with screws (24), new lock washers (25), and washers (26).

c. c. Install plug (27) into exhaust manifold, if necessary.

d. d. Install exhaust manifold (15) and gasket (22) and secure loosely with nuts (19), new lock washers (20), and washers (21).

e. e. Loosely install screw (16), new lock washer (17), and washer (18).

f. f. Tighten nuts (19) and screw (16).

g. g. Couple muffler (3) to exhaust manifold (15) with clamp (14). Do not tighten clamp.

h. h. Secure muffler (3) to supports (8) with bands (2).

i. i. Tighten clamp (14).

j. j. Install top housing panel and top housing frame (paragraph 2-16.4). Close engine access doors.

2-73 AIR RESTRICTION INDICATOR 2-73.1

Inspection.

a. a. Shut down generator set.

b. b. Open left side engine access door.

c. c. Inspect air restriction indicator (1, Figure 2-22) for cracks, stripped threads, or other obvious damage. Close engine access door.

2-73.2 Removal.

WARNING

All metal jewelry can conduct electricity and become entangled in generator set components. Remove all jewelry when working on generator set. Failure to comply with this warning can cause injury or death to personnel.

WARNING

DO NOT wear loose clothing when performing checks, services and maintenance. Failure to comply with this warning can cause injury or death to personnel.

WARNING

When running, generator set engine has hot metal surfaces that will burn flesh on contact. Shut down generator set and allow engine to cool before performing checks, services, and maintenance. Wear gloves and additional protective clothing as required. Failure to comply with this warning can cause injury or death to personnel.

WARNING

Top housing panels and exhaust system can get very hot. Shut down generator set, and allow system to cool before performing checks, services and maintenance. Failure to comply with this warning can cause severe burns and injury to personnel.

WARNING

Wear heat resistant gloves and avoid contacting hot metal surfaces with your hands after components have been heated. Wear additional protective clothing as required. Failure to comply with this warning can cause injury to personnel.

a. a. Shut down generator set.

b. b. Open left side engine access door.

c. c. Unscrew air restriction indicator (1, Figure 2-22) from air cleaner housing (7).

2-73.3 Installation.

Install air restriction indicator (1, Figure 2-22) on air cleaner housing (7). Hand tighten only. Close engine access door.

2-74 AIR CLEANER ASSEMBLY (MEP-804A/MEP-814A) 2-74.1

Inspection.

a. Shut down generator set.

b. b. Open battery access door and disconnect negative battery cable.

c. c. Remove control box assembly (paragraph 2-19.2).

d. d. Inspect air cleaner assembly and mounting bracket for cracks, dents, and other damage.

e. e. Install control box assembly (paragraph 2-19.4).

f. f. Connect negative battery cabl e and close battery access door.

2-74.2 Service.

a. a. Remove air cleaner element (paragraph 2-76.1).

b. b. Wipe inside of air cleaner housing with cleaning cloth (Item 7, Appendix C).

c. c. Install new air cleaner element (paragraph 2-76.3).

2-74.3 Removal.

WARNING

All metal jewelry can conduct electricity and become entangled in generator set components. Remove all jewelry when working on generator set. Failure to comply with this warning can cause injury or death to personnel.

WARNING

DO NOT wear loose clothing when performing checks, services and maintenance. Failure to comply with this warning can cause injury or death to personnel.

WARNING

When running, generator set engine has hot metal surfaces that will burn flesh on contact. Shut down generator set and allow engine to cool before performing checks, services, and maintenance. Wear gloves and additional protective clothing as required. Failure to comply with this warning can cause injury or death to personnel.

WARNING

Top housing panels and exhaust system can get very hot. Shut down generator set, and allow system to cool before performing checks, services and maintenance. Failure to comply with this warning can cause severe burns and injury to personnel.

WARNING

Wear heat resistant gloves and avoid contacting hot metal surfaces with your hands after components have been heated. Wear additional protective clothing as required. Failure to comply with this warning can cause injury to personnel.

a. a. Shut down generator set.

b. b. Open battery access door and disconnect negative battery cable.

c. c. Remove control box assembly (paragraph 2-19.2).

d. d. Loosen clamp (2, Figure 2-22) and remove hose (18) from air cleaner housing (7).

e. e. Remove bolts (3), washers (4), lock washers (5), nuts (6), mounting bracket (12), and air cleaner assembly from generator set. Discard lock washers (5).

f. f. Remove bolts (8), washers (9), nuts (10), lock wa shers (11), and mounting bracket (12) from air cleaner assembly. Discard lock washers (11).

g. g. Remove and retain clamps (13) if replacing air cleaner.

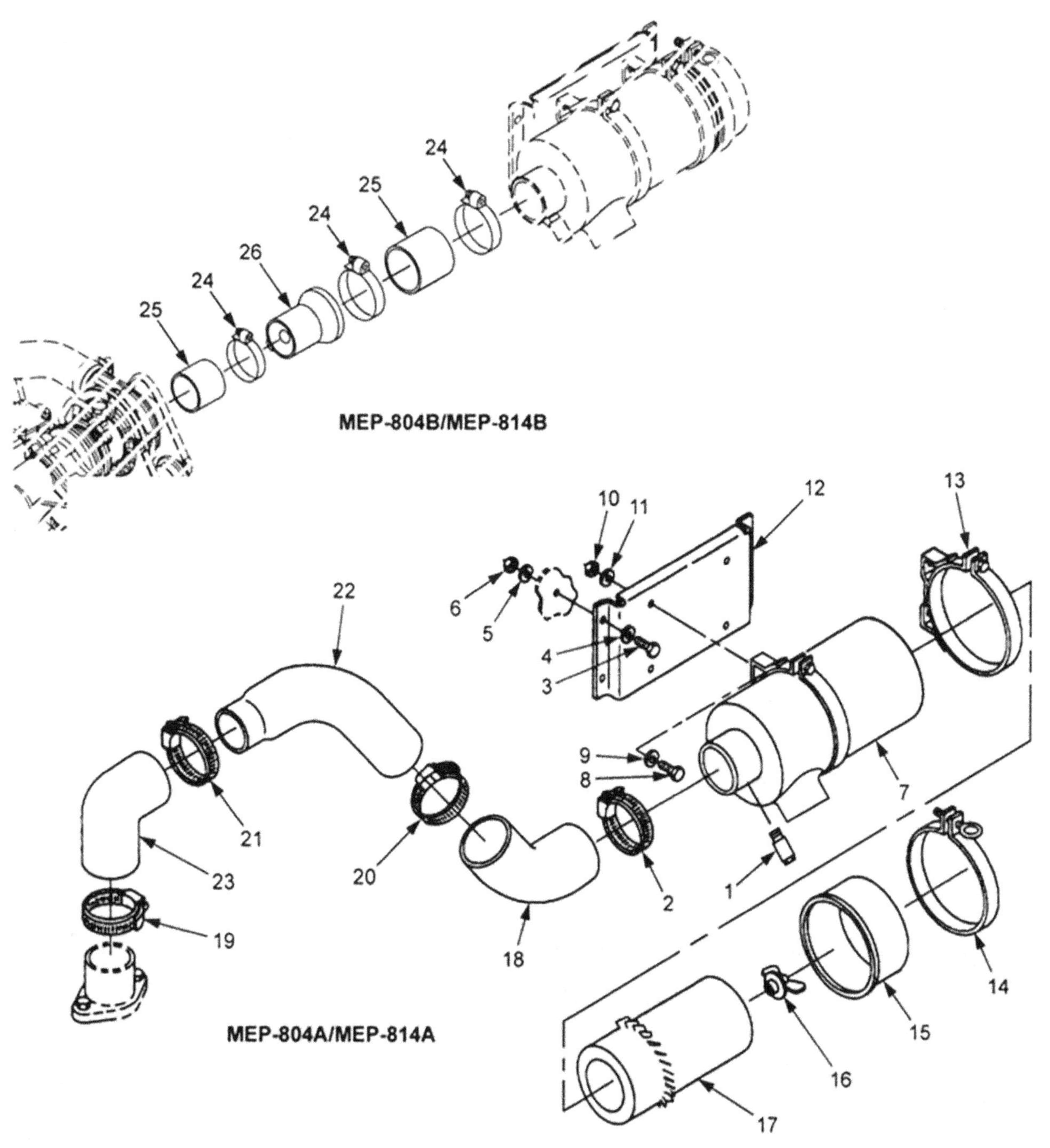

Figure 2-22. Air Cleaner Assembly.

2-74.4 Installation.

a. a. Install clamps (13, Figure 2-22) on air cleaner housing (7), if removed.

b. b. Install mounting bracket (12) on air cleaner assembly with bolts (8), washers (9), new lock washers (11), and nuts (10).

c. c. Install air cleaner assembly and mounting bracket (12) on generator set with bolts (3), washers (4), new lock washers (5), and nuts (6).

d. d. Install hose (18) and clamp (2) on air cleaner housing (7).

e. e. Tighten clamp (2).

f. f. Install control box assembly (paragraph 2-19.4).

g. g. Connect negative battery c able and close battery access door.

2-75 AIR CLEANER ASSEMBLY (MEP-804B/MEP-814B) 2-75.1

Inspection.

a. a. Shut down generator set.

b. b. Open battery access door and disconnect negative battery cable.

c. c. Remove control box assembly (paragraph 2-19.2).

d. d. Inspect air cleaner assembly and mounting bracket for cracks, dents, and other damage.

e. e. Install control box assembly (paragraph 2-19.4).

f. f. Connect negative battery cabl e and close battery access door.

2-75.2 Service.

a. a. Remove air cleaner element (paragraph 2-76.1).

b. b. Wipe inside of air cleaner housing with cleaning cloth (Item 7, Appendix C).

c. c. Install new air cleaner element (paragraph 2-76.3).

2-75.3 Removal.

WARNING

All metal jewelry can conduct electricity and become entangled in generator set components. Remove all jewelry when working on generator set. Failure to comply with this warning can cause injury or death to personnel.

WARNING

DO NOT wear loose clothing when performing checks, services and maintenance. Failure to comply with this warning can cause injury or death to personnel.

WARNING

When running, generator set engine has hot metal surfaces that will burn flesh on contact. Shut down generator set and allow engine to cool before performing checks, services, and maintenance. Wear gloves and additional protective clothing as required. Failure to comply with this warning can cause injury or death to personnel.

WARNING

Top housing panels and exhaust system can get very hot. Shut down generator set, and allow system to cool before performing checks, services and maintenance. Failure to comply with this warning can cause severe burns and injury to personnel.

WARNING

Wear heat resistant gloves and avoid contacting hot metal surfaces with your hands after components have been heated. Wear additional protective clothing as required. Failure to comply with this warning can cause injury to personnel.

a. a. Shut down generator set.

b. b. Open battery access door and disconnect negative battery cable.

c. c. Remove control box assembly (paragraph 2-19.2).

d. d. Loosen clamp (24, Figure 2-22).

e. e. Remove bolts (3), washers (4), lock washers (5), nuts (6), mounting bracket (12), and air
cleaner
assembly from generator set. Discard lock washers (5).

f. Remove bolts (8), washers (9), nuts (10), lock washers (11), and mounting bracket (12) from air cleaner assembly. Discard lock washers (11).

g. g. Remove and retain clamps (13) if replacing air cleaner.

2-75.4 Installation.

a. a. Installed clamps (13, Figure 2-22) on air cleaner housing, if removed.

b. b. Install mounting bracket on air cleaner assembly with bolts (8), washers (9), new lock washers (11), and nuts (10).

c. c. Install air cleaner assembly and mounting bracket (12) on generator set with bolts (3), washers
(4),
new lock washers (5), and nuts (6).

d. d. Tighten clamp (24).

e. e. Install control box assembly (paragraph 2-19.4).

2-76 AIR CLEANER ELEMENT

2-76.1 Removal.

WARNING

All metal jewelry can conduct electricity and become entangled in generator set components. Remove all jewelry when working on generator set. Failure to comply with this warning can cause injury or death to personnel.

WARNING

DO NOT wear loose clothing when performing checks, services and maintenance. Failure to comply with this warning can cause injury or death to personnel.

WARNING

When running, generator set engine has hot metal surfaces that will burn flesh on contact. Shut down generator set and allow engine to cool before performing checks, services, and maintenance. Wear gloves and additional protective clothing as required. Failure to comply with this warning can cause injury or death to personnel.

WARNING

Top housing panels and exhaust system can get very hot. Shut down generator set, and allow system to cool before performing checks, services and maintenance. Failure to comply with this warning can cause severe burns and injury to personnel.

WARNING

Wear heat resistant gloves and avoid contacting hot metal surfaces with your hands after components have been heated. Wear additional protective clothing as required. Failure to comply with this warning can cause injury to personnel.

a. a. Shut down generator set.

b. b. Open air cleaner access door.

c. c. Loosen clamp (14, Figure 2-22) and remove cover (15) from air cleaner housing (7).

d. d. Remove wing nut (16) and air cleaner element (17) from air cleaner housing (7).

2-76.2 Inspection.

a. a. Shut down generator set.

b. b. Remove air cleaner element (paragraph 2-76.1).

d. d. Wipe inside of air cleaner housing (7) with cleaning cloth (Item 7, Appendix C).

e. e. Install air cleaner element (paragraph 2-76.3).

2-76.3 Installation.

a. a. Install air cleaner element (17, Figure 2-22) in air cleaner housing (7).

b. b. Tighten wing nut (16).

c. c. Install cover (15) on air cleaner housing (7).

d. d. Tighten clamp (14). Close air cleaner access door.

2-77 AIR CLEANER TUBING (MEP-804A/MEP-814A)

2-77.1 Inspection.

a. a. Shut down generator set.

b. b. Open engine access doors.

c. c. Inspect all tubing for cracks, tears, and holes.

d. d. Inspect clamps for cracks.

e. e. Replace parts, as necessary . Close engine access doors.

2-77.2 Removal.

WARNING

All metal jewelry can conduct ~~electricity and~~ become entangled in generator set components. Remove all jewelry when working on generator set. Failure to comply with this warning can cause injury or death to personnel.

WARNING

DO NOT wear loose clothing when performing checks, services and maintenance. Failure to comply with this warning can cause injury or death to personnel.

WARNING

When running, generator set engine has hot metal surfaces that will burn flesh on contact. Shut down generator set and allow engine to cool before performing checks, services, and maintenance. Wear gloves and additional protective clothing as required. Failure to comply with this warning can cause injury or death to personnel.

WARNING

Top housing panels and exhaust system can get very hot. Shut down generator set, and allow system to cool before performing checks, services and maintenance. Failure to comply with this warning can cause severe burns and injury to personnel.

WARNING

Wear heat resistant gloves and avoid contacting hot metal surfaces with your hands after components have been heated. Wear additional protective clothing as required. Failure to comply with this warning can cause injury to personnel.

a. a. Shut down generator set.

b. b. Open engine access doors.

NOTE

Because of space restrictions, air cleaner tubing must be removed as an assembly.

c. c. Loosen clamps (2 and 19, Figure 2-22).

d. d. Remove hoses (18, 22, and 23).

e. e. Loosen clamps (20 and 21) and separate hoses (18, 22, and 23).

2-77.3 Installation.

a. a. Connect hoses (18 and 22, Figure 2- 22) with clamp (20). Do not tighten clamp.

b. b. Connect hose (22) to hose (23) with clamp (21). Do not tighten clamp.

c. c. Install hoses (18, 22, and 23), and secure with clamps (19 and 2).

d. d. Tighten clamps (2, 19, 20, and 21). Close engine access doors.

2-78 AIR CLEANER TUBING (MEP-804B/MEP-814B) 2-78.1 Inspection.

a. a. Shut down generator set.

b. b. Open engine access doors.

c. c. Inspect all tubing for cracks, tears, and holes.

d. d. Inspect clamps for cracks.

e. e. Replace parts, as necessary.

f. f. Close engine access doors.

2-78.2 Removal.

WARNING

All metal jewelry can conduct electricity and become entangled in generator set components. Remove all jewelry when working on generator set. Failure to comply with this warning can cause injury or death to personnel.

WARNING

DO NOT wear loose clothing when performing checks, services and maintenance. Failure to comply with this warning can cause injury or death to personnel.

WARNING

When running, generator set engine has hot metal surfaces that will burn flesh on contact. Shut down generator set and allow engine to cool before performing checks, services, and maintenance. Wear gloves and additional protective clothing as required. Failure to comply with this warning can cause injury or death to personnel.

WARNING

Top housing panels and exhaust system can get very hot. Shut down generator set, and allow system to cool before performing checks, services and maintenance. Failure to comply with this warning can cause severe burns and injury to personnel.

WARNING

Wear heat resistant gloves and avoid contacting hot metal surfaces with your hands after components have been heated. Wear additional protective clothing as required. Failure to comply with this warning can cause injury to personnel.

a. a. Shut down generator set.

b. b. Open engine access doors.

NOTE

Due to space restrictions, air cleaner tubing must be removed as an assembly.

c. Loosen clamps (24, Figure 2-22).

d. d. Remove hoses (25) and adapter (26).

2-78.3 <u>Installation</u>.

a. a. Connect hoses (24, Figure 2-22), adapter (26), and clamps (24).

b. b. Tighten clamps (24). Close engine access doors.

Section XII. MAINTENANCE OF COOLANT SYSTEM

2-79 COOLANT SYSTEM (MEP-804A/MEP-814A)

2-79.1 Testing.

a. a. Shut down generator set.

b. b. Slowly remove radiator cap (1, Figure 2-23).

c. c. Install coolant system pressure tester (ST255) in radiator neck and open engine access doors.

d. d. Pump pressure tester until 8 psi is indicated and check coolant system for leaks.

e. e. Pump pressure tester until 7 psi (± 1) is indicated and ensure radiator cap releases.

f. f. Release pressure from pressure tester and remove from radiator neck.

g. g. Install radiator cap (1). Close engine access doors.

WARNING

All metal jewelry can conduct electricity and become entangled in generator set components. Remove all jewelry when working on generator set. Failure to comply with this warning can cause injury or death to personnel.

WARNING

DO NOT wear loose clothing when performing checks, services and maintenance. Failure to comply with this warning can cause injury or death to personnel.

WARNING

Cooling system operates at high temperature and pressure. Contact with high pressure steam and/or liquids can result in burns and scalding. Shut down generator set, and allow system to cool before performing checks, services and maintenance, or wear gloves and additional protective clothing and goggles as required. Failure to comply with this warning can cause injury or death to personnel.

WARNING

Always remove radiator cap slowly to permit any pressure to escape. Failure to comply with this warning can cause injury to personnel.

a. a. Shut down generator set.

b. b. Open left side engine access door.

c. c. Flush or drain coolant system in accordance with TM 750-254.

d. d. Close left side engine access door.

2-80 RADIATOR FILLER HOSE AND PANEL (MEP-804A/MEP-814A)

2-80.1 Removal.

WARNING

All metal jewelry can conduct electricity and become entangled in generator set components. Remove all jewelry when working on generator set. Failure to comply with this warning can cause injury or death to personnel.

WARNING

DO NOT wear loose clothing when performing checks, services and maintenance. Failure to comply with this warning can cause injury or death to personnel.

WARNING

Cooling system operates at high temperature and pressure. Contact with high pressure steam and/or liquids can result in burns and scalding. Shut down generator set, and allow system to cool before performing checks, services and maintenance, or wear gloves and additional protective clothing and goggles as required. Failure to comply with this warning can cause injury or death to personnel.

WARNING

Always remove radiator cap slowly to permit any pressure to escape. Failure to comply with this warning can cause injury to personnel.

a. a. Shut down generator set.

b. b. Open engine access doors.

c. c. Remove generator housing top panel (paragraph 2-16.1).

d. d. Slowly remove radiator cap (1, Figure 2-23).

e. e. Open radiator drain valve (34) and drain coolant/antifreeze into suitable container to a level below radiator filler hose (3) connection at radiator.

f. f. Loosen clamps (2) and remove radiator filler hose (3) and clamps (2).

g. g. Loosen clamp (4) and disconnect overflow hose (5) from radiator filler neck (9).

h. h. Remove radiator fill panel (6) and filler neck (9) from generator set.

i. i. Remove bolts (7), nuts (8), cap (1), and radiator filler neck (9) from radiator fill panel (6).

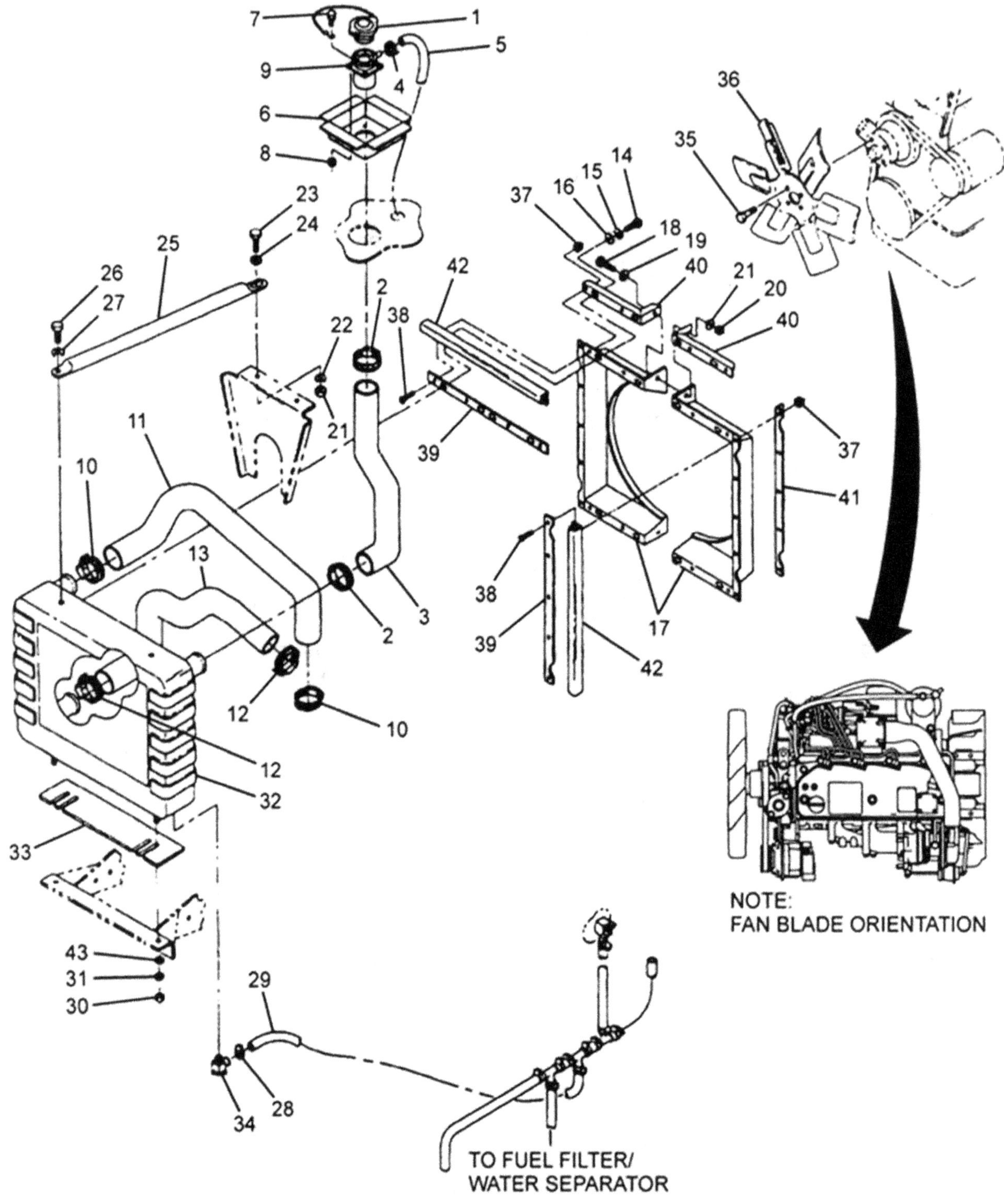

Figure 2-23. Coolant System – MEP-804A/MEP-814A.

2-80.2 Inspection and Cleaning.

a. a. Shut down generator set.

b. b. Remove radiator filler hose and panel, (paragraph 2-80.1).

c. c. Inspect radiator filler hose (3, Figure 2-23) for cracks, holes, and dry rot.

d. d. Inspect filler panel (6), filler neck (9), and cap (1) for cracks, excessive corrosion, and other damage.

e. e. Clean light corrosion from filler hose attaching points with fine grit abrasive paper (Item 15, Appendix C).

f. f. Replace damaged parts.

g. g. Install radiator filler hose and panel (paragraph 2-80.3).

2-80.3 Installation.

WARNING

Avoid breathing fumes generated by soldering. Eye protection is required. Good general ventilation is normally adequate. Failure to comply with this warning can cause injury to personnel.

a. a. Install radiator filler neck (9, Figure 2-23) and cap (1) on radiator fill panel (6) with bolts (7) and nuts (8).

b. b. Position radiator fill panel (6) and filler neck (9) in generator set and attach overflow hose (5) with clamp (4).

c. c. Install radiator filler hose (3) on filler neck (9) and radiator (32) with clamps (2).

d. d. Install generator housing top panel (paragraph 2-16.4).

e. e. Close radiator drain valve (34) and add coolant/antifreeze to proper level (paragraph 2-1.2.2).

f. f. Solder tab (Figure 2-24) and hook chain to tab if replacing radiator cap (1).

g. g. Install radiator cap (1, Figure 2-23).

h. h. Start generator set and allow unit to reach operating temperature and check for leaks.

i. i. Add coolant/antifreeze to overflow bottle, as required. Close engine access doors.

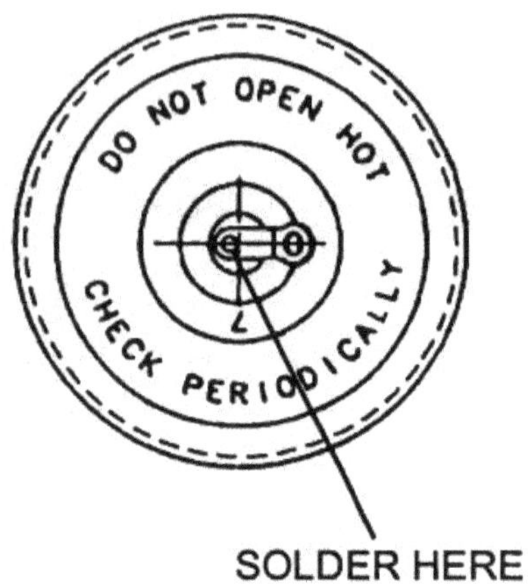

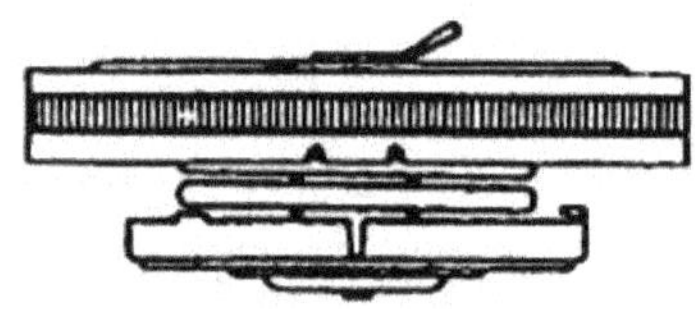

Figure 2-24. Radiator Cap.

2-81 FAN GUARDS (MEP-804A/MEP-814A) 2-81.1

Inspection.

a. a. Shut down generator set.

b. b. Open engine access doors.

c. c. Inspect fan guards, brackets, and attaching hardware for damage, corrosion, and loose or missing hardware.

d. Replace all damaged and missing components and tighten all loose attaching hardware. Close engine access doors.

2-81.2 Removal.

WARNING

All metal jewelry can conduct electricity and become entangled in generator set components. Remove all jewelry when working on generator set. Failure to comply with this warning can cause injury or death to personnel.

WARNING

DO NOT wear loose clothing when performing checks, services and maintenance. Failure to comply with this warning can cause injury or death to personnel.

WARNING

Fan has sharp blades. Use caution and wear gloves when removing or installing belts. Failure to comply with this warning can cause injury to personnel.

a. a. Shut down generator set.

b. b. Open battery access door, disconnect negative battery cable, and open both side engine access

doors.

c. c. Remove bolts (1 and 5, Figure 2-25), washers (2 and 6), nuts (3 and 7), and lock washers (4 and 8) securing fans guards (9 and 10). Discard lock washers (4 and 8).

d. d. Remove belts (11 and 15), lock washers (12 and 16), and washers (17) securing brackets (13, 14, and 18), if necessary. Discard lock washers (12 and 16).

2-81.3 Installation.

a. a. If removed, install bolts (11 and 15, Figure 2-25), new lock washers (12 and 16), and washers (17) securing brackets (13, 14, and 18).

NOTE

If damaged or if replacing fan guards, install protective edging. Cut to fit.

b. b. Install bolts (1 and 5), washers (2 and 6), nuts (3 and 7), and new lock washers (4 and 8) securing fan guards (9 and 10).

c. c. Close both side engine access doors, connect negative battery cable, and close battery access door.

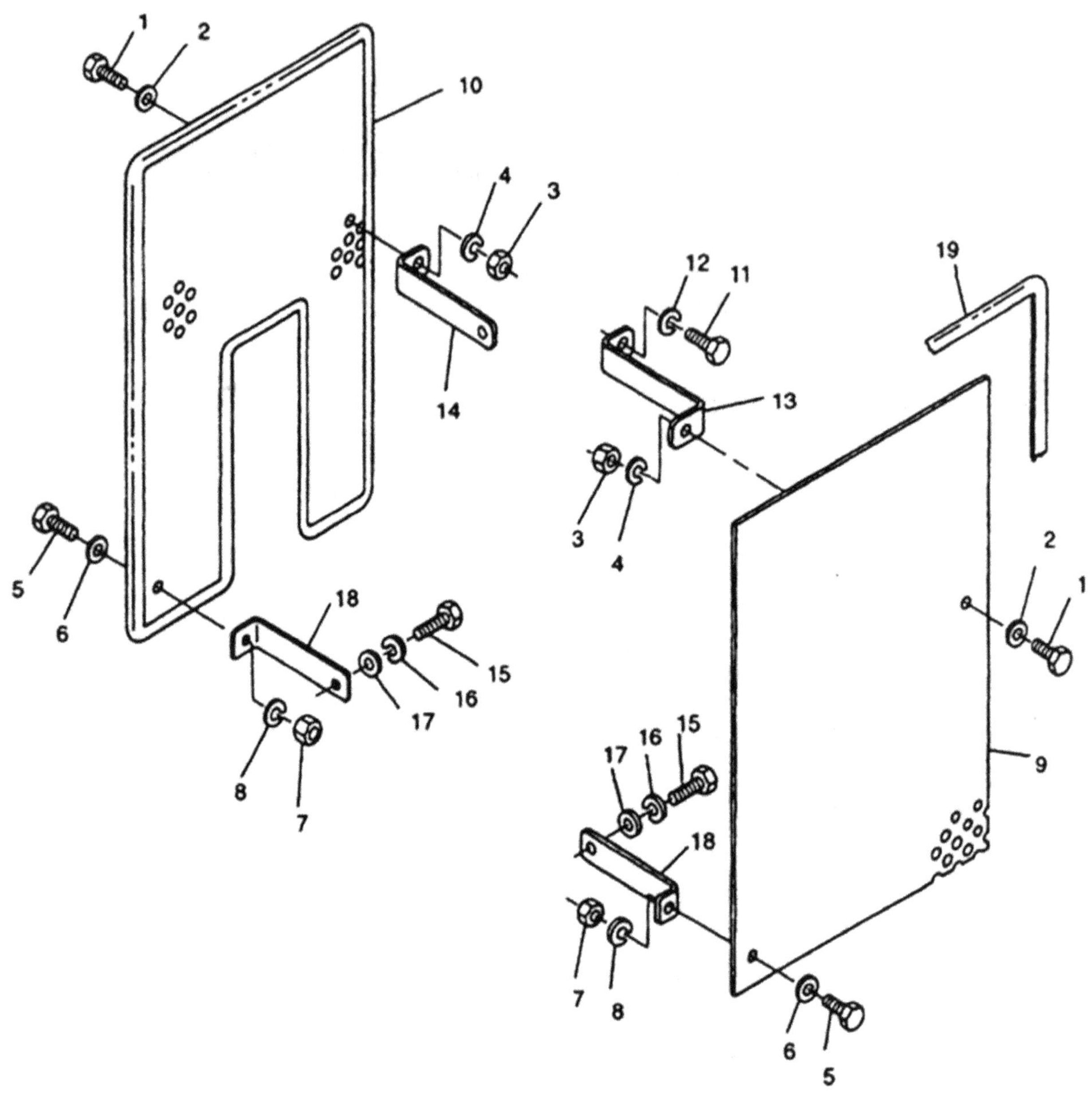

Figure 2-25. Fan Guards – MEP-804A/MEP-814A.

2-82 FAN GUARDS (MEP-804B/MEP-814B) 2-82.1

Inspection.

a. a. Shut down generator set.

b. b. Open both side engine access doors.

c. c. Inspect fan guards, brackets, and attaching hardware for damage, corrosion, and loose or missing hardware.

d. Replace all damaged and missing components and tighten all loose attaching hardware. Close both side engine access doors.

2-82.2 Removal.

WARNING

All metal jewelry can conduct electricity and become entangled in generator set components. Remove all jewelry when working on generator set. Failure to comply with this warning can cause injury or death to personnel.

WARNING

DO NOT wear loose clothing ~~when performin~~g checks, services and maintenance. Failure to comply with this warning can cause injury or death to personnel.

WARNING

Fan has sharp blades. Use caution and wear gloves when removing or installing belts. Failure to comply can cause injury to personnel.

a. a. Shut down generator set.

b. b. Open battery access door, disconnect negative battery cable, and open both side engine access doors.

c. c. Remove bolts (1, Figure 2-26), lock washers (2), washers (3), chain (4) for oil fill cap, and right fan guard (5). Discard lock washers (2).

d. d. Remove bolts (6), lock washers (7), washers (8), clamp (9), screw (10), lock washer (11), washer (12), nut (13), and left fan guard (14). Discard lock washers (7 and 11).

2-82.3 Installation.

a. a. Install bolts (6, Figure 2-26), new lock washers (7), washers (8), clamp (9), screw (10), new lock washer (11), washer (12), nut (13), and left fan guard (14).

b. b. Install bolts (1), new lock washers (2), washers (3), chain (4) for oil fill cap, and right fan guard (5).

c. c. Close both side engine access doors, connect negative battery cable, and close battery ac-

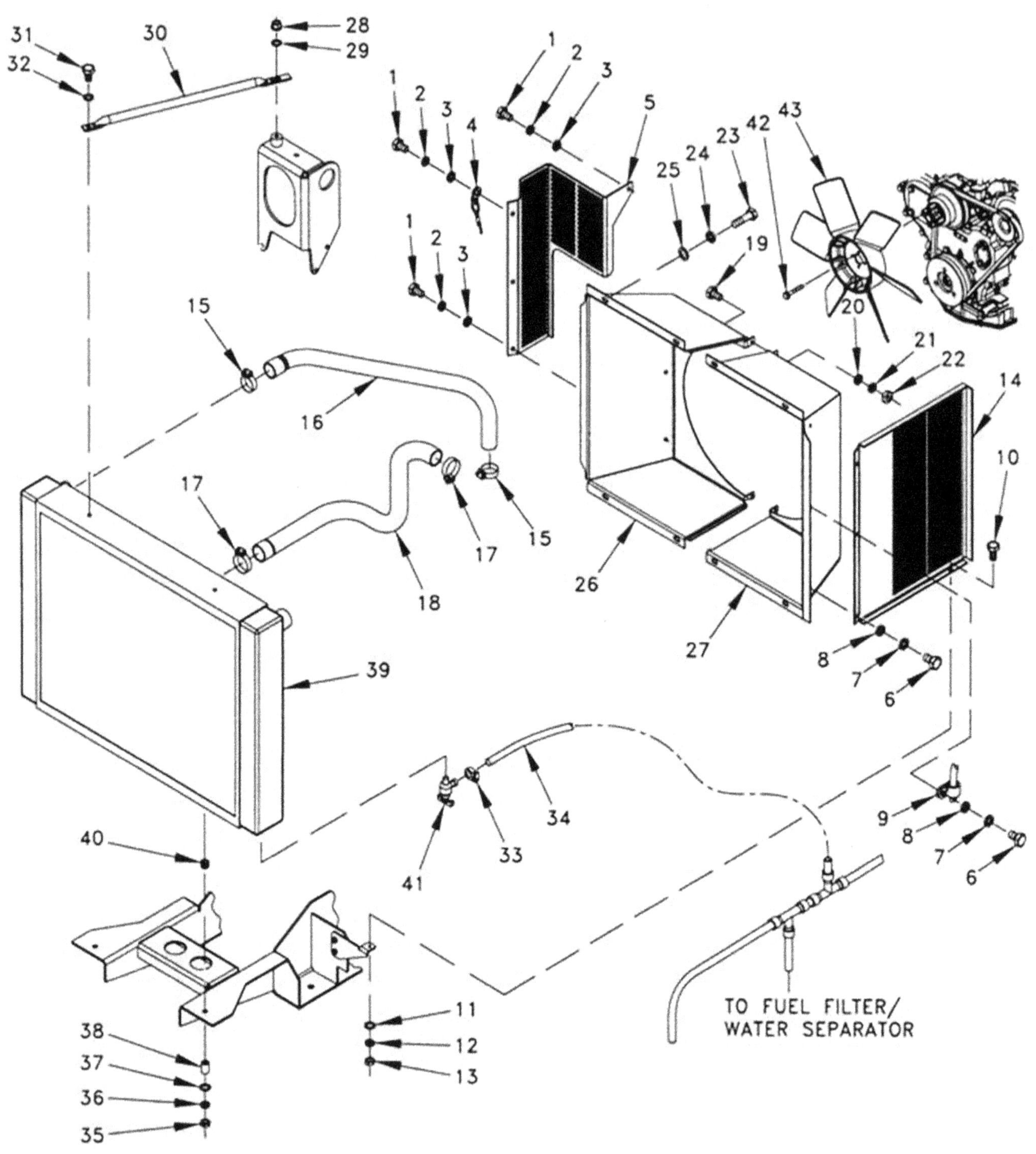

Figure 2-26. Coolant System – MEP-804B/MEP-814B.

2-83 UPPER COOLANT HOSE (MEP-804A/MEP-814A)

WARNING

All metal jewelry can conduct electricity and become entangled in generator set components. Remove all jewelry when working on generator set. Failure to comply with this warning can cause injury or death to personnel.

WARNING

DO NOT wear loose clothing when performing checks, services and maintenance. Failure to comply with this warning can cause injury or death to personnel.

WARNING

Cooling system operates at high temperature and pressure. Contact with high pressure steam and/or liquids can result in burns and scalding. Shut down generator set, and allow system to cool before performing checks, services and maintenance, or wear gloves and additional protective clothing and goggles as required. Failure to comply with this warning can cause injury or death to personnel.

WARNING

Always remove radiator cap slowly to permit any pressure to escape. Failure to comply with this warning can cause injury to personnel.

a. a. Shut down generator set.

b. b. Open battery access door, disconnect negative battery cable, and open engine access doors.

c. c. Slowly remove radiator cap (1, Figure 2-23).

d. d. Remove fan guard (paragraph 2-81.2).

e. e. Open radiator drain valve (34) and drain coolant/antifreeze into suitable container.

f. f. Loosen clamps (10) and remove upper coolant hose (11) and clamps (10).

2-83.2 Inspection and Cleaning.

a. a. Shut down generator set.

b. b. Remove upper coolant hose (paragraph 2-83.1).

c. c. Inspect upper coolant hose (11, Figu re 2-23) for cracks, holes, and dry rot.

d. d. Clean light corrosion from upper coolant hose attaching points with fine grit abrasive paper (Item 15, Appendix C).

e. e. Install upper coolant hose (paragraph 2-83.3).

2-83.3 Installation.

a. a. Install upper coolant hose (11, Figure 2-23) on thermostat housing opening and radiator (32) with clamps (10).

b. b. Close radiator drain valve (34) and add coolant/antifreeze to proper level (paragraph 2-1.2.2).

c. c. Install fan guards (paragraph 2-81.3).

d. d. Install radiator cap (1).

e. e. Connect negative battery c able and close battery access door.

f. f. Start generator set and allow unit to reach operating temperature and check for leaks.

g. g. Add coolant/antifreeze to overflow bottle, as required. Close engine access doors.

2-84 UPPER COOLANT HOSE (MEP-804B/MEP-814B)

2-84.1 Removal.

WARNING

All metal jewelry can conduct electricity and become entangled in generator set components. Remove all jewelry when working on generator set. Failure to comply with this warning can cause injury or death to personnel.

WARNING

DO NOT wear loose clothing when performing checks, services and maintenance. Failure to comply with this warning can cause injury or death to personnel.

WARNING

Cooling system operates at high temperature and pressure. Contact with high pressure steam and/or liquids can result in burns and scalding. Shut down generator set, and allow system to cool before performing checks, services and maintenance, or wear gloves and additional protective clothing and goggles as required. Failure to comply with this warning can cause injury or death to personnel.

WARNING

Always remove radiator cap slowly to permit any pressure to escape. Failure to comply with this warning can cause injury to personnel.

a. a. Shut down generator set.

b. b. Open battery access door, disconnect negative battery cable, and open engine access doors.

d. d. Remove fan guard (paragraph 2-82.2).

e. e. Open radiator drain valve (41, Figure 2-26) an d drain coolant/antifreeze into suitable container.

f. f. Loosen clamps (15).

g. g. Remove upper coolant hose (16) and clamps (15).

2-84.2 Inspection and Cleaning.

a. a. Shut down generator set.

b. b. Remove upper coolant hose (paragraph 2-84.1).

c. c. Inspect upper coolant hose (16, Figu re 2-26) for cracks, holes, and dry rot.

d. d. Clean light corrosion from upper coolant hose attaching points with fine grit abrasive paper (Item 15, Appendix C).

e. e. Install upper coolant hose (paragraph 2-84.3).

2-84.3 Installation.

a. a. Install upper coolant hose (16, Figure 2-26) on thermostat housing opening and radiator (39) with clamps (15).

b. b. Close radiator drain valve (41) and add coolant/antifreeze to proper level (paragraph 2-1.2.2).

c. c. Install fan guards (paragraph 2-82.3).

d. d. Install radiator cap (1, Figure 2-23).

e. e. Connect negative battery c able and close battery access door.

f. f. Start generator set and allow unit to reach operating temperature and check for leaks.

g. g. Add coolant/antifreeze to overflow bottle, as required. Close engine access doors.

2-85 LOWER COOLANT HOSE (MEP-804A/MEP-814A)

2-85.1 Removal.

WARNING

All metal jewelry can conduct electricity and become entangled in generator set components. Remove all jewelry when working on generator set. Failure to comply with this warning can cause injury or death to personnel.

WARNING

DO NOT wear loose clothing when performing checks, services and maintenance. Failure to comply with this warning can cause injury or death to personnel.

WARNING

Cooling system operates at high temperature and pressure. Contact with high pressure steam and/or liquids can result in burns and scalding. Shut down generator set, and allow system to cool before performing checks, services and maintenance, or wear gloves and additional protective clothing and goggles as required. Failure to comply with this warning can cause injury or death to personnel.

WARNING

Always remove radiator cap slowly to permit any pressure to escape. Failure to comply with this warning can cause injury to personnel.

a. a. Shut down generator set.

b. b. Open battery access door, disconnect negative battery cable, and open engine access doors.

c. c. Slowly remove radiator cap (1, Figure 2-23).

d. d. Remove fan guards (paragraph 2-81.2).

e. e. Open radiator drain valve (34) and drain coolant into suitable container.

f. f. Loosen clamps (12) and remove lower coolant hose (13).

2-85.2 Inspection and Cleaning.

a. a. Shut down generator set.

b. b. Remove lower coolant hose (paragraph 2-85.1).

c. c. Inspect lower coolant hose (13, Figure 2-23) for cracks, holes, and dry rot.

d. d. Clean lower coolant hose attaching points with fine grit abrasive paper (Item 15, Appendix C).

e. e. Install lower coolant hose (paragraph 2-85.3).

2-85.3 Installation.

a. a. Install lower coolant hose (13, Figure 2-23) on radiator outlet opening and water pump opening with clamps (12).

b. b. Close radiator drain valve (34) and add coolant/antifreeze to proper level (paragraph 2-1.2.2).

c. c. Install fan guards (paragraph 2-81.3).

d. d. Install radiator cap (1).

e. e. Connect negative battery c able and close battery access door.

f. f. Start generator set and allow unit to reach operating temperature and check for leaks.

g. g. Add coolant/antifreeze to overflow bottle, as required. Close engine access doors.

2-86 LOWER COOLANT HOSE (MEP-804B/MEP-814B)

WARNING

All metal jewelry can conduct electricity and become entangled in generator set components. Remove all jewelry when working on generator set. Failure to comply with this warning can cause injury or death to personnel.

WARNING

DO NOT wear loose clothing when performing checks, services and maintenance. Failure to comply with this warning can cause injury or death to personnel.

WARNING

Cooling system operates at high temperature and pressure. Contact with high pressure steam and/or liquids can result in burns and scalding. Shut down generator set, and allow system to cool before performing checks, services and maintenance, or wear gloves and additional protective clothing and goggles as required. Failure to comply with this warning can cause injury or death to personnel.

WARNING

Always remove radiator cap slowly to permit any pressure to escape. Failure to comply with this warning can cause injury to personnel.

a. a. Shut down generator set.

b. b. Open battery access door, disconnect negative battery cable, and open engine access doors.

c. c. Slowly remove radiator cap (1, Figure 2-23).

d. d. Remove fan guards (paragraph 2-82.2).

e. e. Open radiator drain valve (41, Figure 2-26) and drain coolant into suitable container.

f. f. Loosen clamps (17) and remove lower coolant hose (18).

2-86.2 Inspection and Cleaning.

a. a. Shut down generator set.

b. b. Remove lower coolant ho se (paragraph 2-86.1).

c. c. Inspect lower coolant hose (18, Figure 2-26) for cracks, holes, and dry rot.

d. d. Clean lower coolant hose attaching points with f ine grit abrasive paper (Item 15, Appendix C).

2-86.3 Installation.

a. a. Install lower coolant hose (18, Figure 2-26) on radiator outlet opening and water pump opening with clamps (16).

b. b. Close radiator drain valve (41) and add coolant/antifreeze to proper level (paragraph 2-1.2.2).

c. c. Install fan guards (paragraph 2-82.3).

d. d. Install radiator cap (1, Figure 2-23).

e. e. Connect negative battery c able and close battery access door.

f. f. Start generator set and allow unit to reach operating temperature and check for leaks.

2-87 COOLANT OVERFLOW AND DRAIN HOSES (MEP-804A/MEP-814A)

2-87.1 Inspection.

a. a. Shut down generator set.

b. b. Open engine access doors.

c. c. Inspect hoses for cracks, holes, and dry rot. Close engine access doors.

2-87.2 Removal.

WARNING

All metal jewelry can conduct electricity and become entangled in generator set components. Remove all jewelry when working on generator set. Failure to comply with this warning can cause injury or death to personnel.

WARNING

DO NOT wear loose clothing when performing checks, services and maintenance. Failure to comply with this warning can cause injury or death to personnel.

WARNING

Cooling system operates at high temperature and pressure. Contact with high pressure steam and/or liquids can result in burns and scalding. Shut down generator set, and allow system to cool before performing checks, services and maintenance, or wear gloves and additional protective clothing and goggles as required. Failure to comply with this warning can cause injury or death to personnel.

WARNING

Always remove radiator cap slowly to permit any pressure to escape. Failure to comply with this warning can cause injury to personnel.

a. Shut down generator set.

b. b. Open battery access door, disconnect negative battery cable, and open engine access doors.

c. c. Locate overflow or drain hose to be removed (Figure 2-23).

d. d. Disconnect hose at both ends and remove from generator set.

2-87.3 Installation.

a. a. Install overflow or drain hos e in generator set as removed.

b. b. Close engine access doors, connect negative battery cable, and close battery access door.

2-88 COOLANT OVERFLOW AND DRAIN HOSES (MEP-804B/MEP-814B)

2-88.1 Inspection.

a. a. Shut down generator set.

b. b. Open engine access doors.

c. c. Inspect hoses for cracks, holes, and dry rot. Close engine access doors.

2-88.2 Removal.

WARNING

All metal jewelry can conduct electricity and become entangled in generator set components. Remove all jewelry when working on generator set. Failure to comply with this warning can cause injury or death to personnel.

WARNING

DO NOT wear loose clothing when performing checks, services and maintenance. Failure to comply with this warning can cause injury or death to personnel.

WARNING

Cooling system operates at high temperature and pressure. Contact with high pressure steam and/or liquids can result in burns and scalding. Shut down generator set, and allow system to cool before performing checks, services and maintenance, or wear gloves and additional protective clothing and goggles as required. Failure to comply with this warning can cause injury or death to personnel.

WARNING

Always remove radiator cap slowly to permit any pressure to escape. Failure to comply with this warning can cause injury to personnel.

a. a. Shut down generator set.

b. b. Open battery access door, disconnect negative battery cable, and open engine access doors.

d. d. Disconnect hose at both ends and remove from generator set.

2-88.3 Installation.

a. a. Install overflow or drain hos e in generator set as removed.

b. b. Close engine access doors, connect negative battery cable, and close battery access door.

2-89 RADIATOR (MEP-804A/MEP-814A) 2-89.1

WARNING

All metal jewelry can conduct electricity and become entangled in generator set components. Remove all jewelry when working on generator set. Failure to comply with this warning can cause injury or death to personnel.

WARNING

DO NOT wear loose clothing when performing checks, services and maintenance. Failure to comply with this warning can cause injury or death to personnel.

WARNING

Cooling system operates at high temperature and pressure. Contact with high pressure steam and/or liquids can result in burns and scalding. Shut down generator set, and allow system to cool before performing checks, services and maintenance, or wear gloves and additional protective clothing and goggles as required. Failure to comply with this warning can cause injury or death to personnel.

WARNING

Always remove radiator cap slowly to permit any pressure to escape. Failure to comply with this warning can cause injury to personnel.

a. a. Shut down generator set.

b. b. Open engine access doors.

c. c. Remove generator set top housing section (paragraph 2-16.1).

d. d. Slowly remove radiator cap (1, Figure 2-23).

e. e. Remove fan guards (paragraph 2-81.2).

f. f. Open radiator drain valve (34) and drain coolant/antifreeze into suitable container.

g. g. Remove radiator filler hose assembly (paragraph 2-80.1).

h. h. Remove upper coolant hose (paragraph 2-83.1).

i. i. Remove lower coolant ho se (paragraph 2-85.1).

j. j. Remove bolts (14), lock washers (15), and flat washers (16); allow shroud (17) to rest on fan. Discard lock washers (15).

k. k. Remove bolts (18), washers (19), nuts (20), and lock washers (21) securing shroud halves (17), and shroud halves (17). Discard lock washers (21).

l. l. Remove nuts (21), lock washers (22), bolts (23), and washers (24) securing support rods (25) to bracket on engine. Discard lock washers (22).

m. m. Remove bolts (26), lock washers (27), and support rods (25) from radiator (32). Discard lock washers (27).

n. n. Loosen clamp (28) and disconnect radiator drain hose (29) from radiator drain valve (34).

o. o. Remove nuts (30) and washers (31 and 43) from radiator mounting points.

p. p. Lift radiator (32) up and out of generator set housing and remove shim(s) (33).

q. q. Remove radiator drain valve (34) from radiator (32).

2-89.2 Inspection and Cleaning.

a. a. Shut down generator set.

b. b. Remove radiator (paragraph 2-89.1).

c. c. Inspect radiator for excessive corrosion, cracks, or bent cooling fins.

d. d. Check inside of radiator for corrosion and scale.

WARNING

Eye protection is required when working with compressed air. Compressed air can propel particles at high velocity and injure eyes. Do not exceed 15 psi pressure when using compressed air. Failure to comply with this warning can cause injury to personnel.

WARNING

High pressure steam can blow particles or chemicals into eyes, can cause severe burns, and creates hazardous noise levels. Wear protective eye, skin, and hearing protection when using high pressure steam. Failure to comply with this warning can cause injury to personnel.

WARNING

Solvent used to clean parts is potentially dangerous to personnel and property. Clean parts in a well-ventilated area. Avoid inhalation of solvent fumes. Wear goggles and rubber gloves to protect eyes and skin. Wash exposed skin thoroughly. Do not smoke or use near open flame or excessive heat. Failure to comply with this warning can cause injury to personnel, and damage to the equipment.

e. e. Clean dirt particles from radiator core air passages using filtered, compressed air.

f. f. Clean exterior surface of radiator with dry cleaning solvent (Item 20, Appendix C).

g. g. Install radiator (paragraph 2-89.4).

WARNING

Avoid breathing fumes generated by soldering. Eye protection is required. Good general ventilation is normally adequate. Failure to comply with this warning can cause injury to personnel.

Repair radiator by straightening bent ra diator fins and soldering minor leaks.

2-89.4 Installation.

a. a. Install drain valve (34, Figure 2-23) in radiator (32).

b. b. Assemble shroud halves (17) with nuts (20), new lock washers (21), bolts (18), and washers (19).

c. c. Position radiator (32) on radiator mount and install shroud (17) on radiator with bolts (14), new lock washers (15), and flat washers (16).

d. d. Insert shim(s) (33) under radiator (32), as necessary, to obtain equal clearance between fan (36) and top and bottom of shroud (17). Secure shim(s) and radiator (32) to radiator mounting bracket with nuts (30) and washers (31 and 43).

e. e. Connect radiator drain hose (29) on radiator drain valve (34) with clamp (28).

f. f. Connect support rods (25) to radiator (32) with new lock washers (27) and bolts (26).

g. g. Attach support rods (25) to bracket on engine with bolts (23), washers (24), new lock washers (22), and nuts (21).

h. h. Install lower coolant hose (paragraph 2-85.3).

i. i. Install upper coolant hose (paragraph 2-83.3).

j. j. Install radiator filler hose and panel assembly (paragraph 2-80.3).

k. k. Install fan guards (paragraph 2-81.3).

l. l. Install generator set top housing section (paragraph 2-16.4).

m. m. Ensure radiator drain valve (34) is closed and add coolant/antifreeze to proper level (TM 9-6115-643-10).

n. n. Install radiator cap (1).

o. o. Start generator set and allow unit to reach operating temperature and check for leakage.

p. p. Add coolant to overflow bottle, as required. Close engine access doors.

2-90 RADIATOR (MEP-804B/MEP-814B) 2-90.1

Removal.

WARNING

All metal jewelry can conduct electricity and become entangled in generator set components. Remove all jewelry when working on generator set. Failure to comply with this warning can cause injury or death to personnel.

WARNING

DO NOT wear loose clothing when performing checks, services and maintenance. Failure to comply with this warning can cause injury or death to personnel.

WARNING

Cooling system operates at high temperature and pressure. Contact with high pressure steam and/or liquids can result in burns and scalding. Shut down generator set, and allow system to cool before performing checks, services and maintenance, or wear gloves and additional protective clothing and goggles as required. Failure to comply with this warning can cause injury or death to personnel.

WARNING

Always remove radiator cap slowly to permit any pressure to escape. Failure to comply with this warning can cause injury to personnel.

a. a. Shut down generator set.

b. b. Open engine access doors.

c. c. Remove generator set top housing section (paragraph 2-16.1).

d. d. Slowly remove radiator cap (1, Figure 2-23).

e. e. Remove fan guards (paragraph 2-82.2).

f. f. Open radiator drain valve (41, Figure 2-26) and drain coolant/antifreeze into suitable container.

g. g. Remove radiator filler hose assembly (paragraph 2-80.1).

h. h. Remove upper coolant hose (paragraph 2-84.1).

i. i. Remove lower coolant hose (paragraph 2-86.1).

j. j. Remove bolts (19), flat washers (20), lock washers (21), and nuts (22). Discard lock washers (21).

k. Remove bolts (23), lock washers (24), flat washers (25), and shroud halves (26 and 27). Discard lock washers (24).

l. l. Remove nuts (28) and flat washers (29) securing support rods (30) to bracket on engine.

m. m. Remove bolts (31), lock washers (32), and support rods (30) from radiator (39). Discard lock washers (32).

n. n. Loosen clamp (33) and disconnect radiator drain hose (34) from radiator drain valve (41).

o. o. Remove nuts (35), lock washers (36), flat washers (37), and bushing (38) from radiator mounting studs. Discard lock washers (36).

p. Lift radiator (39) up and out of generator set housing and remove grommets (40).

q. Remove radiator drain valve (41) from radiator (39).

WARNING

Eye protection is required when working with compressed air. Compressed air can propel particles at high velocity and injure eyes. Do not exceed 15 psi pressure when using compressed air. Failure to comply with this warning can cause injury to personnel.

WARNING

High pressure steam can blow particles or chemicals into eyes, can cause severe burns, and creates hazardous noise levels. Wear protective eye, skin, and hearing protection when using high pressure steam. Failure to comply with this warning can cause injury to personnel.

WARNING

Solvent used to clean parts is potentially dangerous to personnel and property. Clean parts in a well-ventilated area. Avoid inhalation of solvent fumes. Wear goggles and rubber gloves to protect eyes and skin. Wash exposed skin thoroughly. Do not smoke or use near open flame or excessive heat. Failure to comply with this warning can cause injury to personnel, and damage to the equipment.

a. a. Shut down generator set.

b. b. Remove radiator (paragraph 2-90.1).

c. c. Inspect radiator for excessive corrosion, cracks, or bent cooling fins.

d. d. Check inside of radiator for corrosion and scale.

e. e. Clean dirt particles from radiator core air passages using filtered, compressed air.

f. f. Clean exterior surface of radiator with dry cleaning solvent (Item 20, Appendix C).

g. g. Install radiator (paragraph 2-90.4).

WARNING

Avoid breathing fumes generated by soldering. Eye protection is required. Good general ventilation is normally adequate. Failure to comply with this warning can cause injury to personnel.

Repair radiator by straightening bent radiator fins and soldering minor leaks.

2-90.4 Installation.

a. a. Install drain valve (41, Figure 2-26) in radiator (39).

NOTE

Install radiator grommets beneath radiator to obtain proper clearance between fan blade and shroud.

b. b. Position radiator (39) and grommets (40) on radiator mount.

c. c. Install bushings (38), threaded end first, and secure with flat washers (37), new lock washers (36), and nuts (35).

d. d. Install shroud halves (26 and 27) and secure with bolts (23), new lock washers (24), and
flat
washers (25).

e. e. Install bolts (19), flat washers (20), new lock washers (21), and nuts (22).

f. f. Connect radiator drain hose (34) on radiator drain valve (41) with clamp (33).

g. g. Connect support rods (30) to radiator (39) with new lock washers (32) and bolts (31).

h. h. Attach support rods (30) to bracket on engine with nuts (28) and flat washers (29).

i. i. Install lower coolant hose (paragraph 2-86.3).

j. j. Install upper coolant hose (paragraph 2-84.3).

k. k. Install radiator filler hose and panel assembly (paragraph 2-80.3).

l. l. Install fan guards (paragraph 2-82.3).

m. m. Install generator set top housing section (paragraph 2-16.4).

n. n. Ensure radiator drain valve (41) is closed and add coolant/antifreeze to proper level (TM
9-6115-
643-10).

o. o. Install radiator cap (1, Figure 2-23).

p. p. Start generator set and allow unit to reach operating temperature and check for leakage.

q. q. Add coolant to overflow bottle, as required. Close engine access doors.

2-91 GENERATOR SET COOLING FAN (MEP-804A/MEP-814A) 2-91.1

WARNING

All metal jewelry can conduct electricity and become entangled in generator set components. Remove all jewelry when working on generator set. Failure to comply with this warning can cause injury or death to personnel.

WARNING

DO NOT wear loose clothing when performing checks, services and maintenance. Failure to comply with this warning can cause injury or death to personnel.

WARNING

Cooling system operates at high temperature and pressure. Contact with high pressure steam and/or liquids can result in burns and scalding. Shut down generator set, and allow system to cool before performing checks, services and maintenance, or wear gloves and additional protective clothing and goggles as required. Failure to comply with this warning can cause injury or death to personnel.

WARNING

Always remove radiator cap slowly to permit any pressure to escape. Failure to comply with this warning can cause injury to personnel.

a. a. Shut down generator set.

b. b. Open battery access door, disconnect negative battery cable, and open engine access doors.

c. c. Remove fan guards (paragraph 2-81.2).

d. d. Remove bolts (14, Figure 2-23), washers (16) , and lock washers (15); allow shroud (17) to rest on fan (36). Discard lock washers (15).

e. e. Remove bolts (18), washers (19), nuts (20), and lock washers (21). Separate shroud halves (17) and remove at least one half. Discard lock washers (21).

f. f. Remove bolts (35) and fan (36).

g. g. Remove nuts (37), screws (38), stiffeners (39) and (40), supports (41), and seals (42) from each side of shroud (17).

2-91.2 Inspection.

WARNING

Eye protection is required when working with compressed air. Compressed air can propel particles at high velocity and injure eyes. Do not exceed 15 psi pressure when using compressed air. Failure to comply with this warning can cause injury to personnel.

WARNING

High pressure steam can blow particles or chemicals into eyes, can cause severe burns, and creates hazardous noise levels. Wear protective eye, skin, and hearing protection when using high pressure steam. Failure to comply with this warning can cause injury to personnel.

WARNING

Solvent used to clean parts is potentially dangerous to personnel and property. Clean parts in a well-ventilated area. Avoid inhalation of solvent fumes. Wear goggles and rubber gloves to protect eyes and skin. Wash exposed skin thoroughly. Do not smoke or use near open flame or excessive heat. Failure to comply with this warning can cause injury to personnel, and damage to the equipment.

a. Shut down generator set.

b. Remove cooling fan, paragraph 2-91.1.

c. Inspect fan (36, Figure 2-23) and blades for cracks, bends, loose rivets, or other damage.

d. Inspect seals (42), supports (41), and stiffeners (39 and 40) for damage.

e. Replace damaged parts.

f. Install cooling fan (paragraph 2-91.3).

2-91.3 Installation.

a. Install seals (42, Figure 2-23), supports (41), and stiffeners (39 and 40) on each side of shroud (17) with screws (38) and nuts (37).

b. Carefully note orientation of fan blades, position fan (36), and secure with bolts (35). Torque bolts to 72 in.-lbs (8.0 Nm).

c. Install fan shroud halves (17) with nuts (20), new lock washers (21), bolts (18), and washers (19). Attach to radiator with bolts (14), new lock washers (15), and washers (16).

d. Install fan guards (paragraph 2-81.3).

e. Close engine access doors, connect negative battery cable, and close battery access door.

2-92 GENERATOR SET COOLING FAN (MEP-804B/MEP-814B)

WARNING

All metal jewelry can conduct electricity and become entangled in generator set components. Remove all jewelry when working on generator set. Failure to comply with this warning can cause injury or death to personnel.

WARNING

DO NOT wear loose clothing when performing checks, services and maintenance. Failure to comply with this warning can cause injury or death to personnel.

WARNING

Cooling system operates at high temperature and pressure. Contact with high pressure steam and/or liquids can result in burns and scalding. Shut down generator set, and allow system to cool before performing checks, services and maintenance, or wear gloves and additional protective clothing and goggles as required. Failure to comply with this warning can cause injury or death to personnel.

WARNING

Always remove radiator cap slowly to permit any pressure to escape. Failure to comply with this warning can cause injury to personnel.

a. a. Shut down generator set.

b. b. Open battery access door, disconnect negative battery cable, and open engine access doors.

c. c. Remove fan guards (paragraph 2-82.2).

d. d. Remove bolts (19, Figure 2-26), flat washers (20), lock washers (21), and nuts (22). Discard lock washers (21).

e. e. Remove bolts (23), lock washers (24), and flat washers (25); remove at least one shroud half (26 or 27). Discard lock washers (24).

f. f. Remove bolts (42) and fan (43).

2-92.2 Inspection.

WARNING

Eye protection is required when working with compressed air. Compressed air can propel particles at high velocity and injure eyes. Do not exceed 15 psi pressure when using compressed air. Failure to comply with this warning can cause injury to personnel.

WARNING

High pressure steam can blow particles or chemicals into eyes, can cause severe burns, and creates hazardous noise levels. Wear protective eye, skin, and hearing protection when using high pressure steam. Failure to comply with this warning can cause injury to personnel.

WARNING

Solvent used to clean parts is potentially dangerous to personnel and property. Clean parts in a well-ventilated area. Avoid inhalation of solvent fumes. Wear goggles and rubber gloves to protect eyes and skin. Wash exposed skin thoroughly. Do not smoke or use near open flame or excessive heat. Failure to comply with this warning can cause injury to personnel, and damage to the equipment.

a. a. Shut down generator set.

b. b. Remove cooling fan (paragraph 2-92.1).

c. c. Inspect fan (43, Figure 2-26) and blades for cracks, bends, loose rivets, and other damage.

d. d. Install cooling fan (paragraph 2-92.3).

2-92.3 Installation.

a. a. Position fan (43, Figure 2-26) and secure with bolts (42). Torque bolts to 7.2 to 8.1 ft-lbs (9.8 to 11.8 Nm).

b. b. Install fan shroud halves (26 or 27), as required, with flat washers (25), new lock washers (24), and bolts (23).

c. c. Install bolts (19), flat washers (20), new lock washers (21), and nuts (22).

d. d. Install fan guards (paragraph 2.82.3).

e. e. Close engine access doors, connect negative battery cable, and close battery access door.

2-93 FAN BELT (MEP-804A/MEP-814A) 2-93.1

Inspection.

WARNING

All metal jewelry can conduct ~~electricity and~~ become entangled in generator set components. Remove all jewelry when working on generator set. Failure to comply with this warning can cause injury or death to personnel.

WARNING

DO NOT wear loose clothing when performing checks, services and maintenance. Failure to comply with this warning can cause injury or death to personnel.

WARNING

When running, generator set engine has hot metal surfaces that will burn flesh on contact. Shut down generator set and allow engine to cool before performing checks, services, and maintenance. Wear gloves and additional protective clothing as required. Failure to comply with this warning can cause injury or death to personnel.

WARNING

Fan has sharp blades. Use caution and wear gloves when removing or installing belts. Failure to comply with this warning can cause injury to personnel.

a. a. Shut down generator set.

b. b. Open battery access door, disconnect negative battery cable, and open engine access doors.

c. c. Remove left side fan guard.

d. d. Inspect fan belt (6, Figure 2-27) for frays, cracks, oil soaking, and other damage.

e. e. Replace fan belt that shows any of above or cannot be adjusted for proper tension.

f. f. Install left side fan guard.

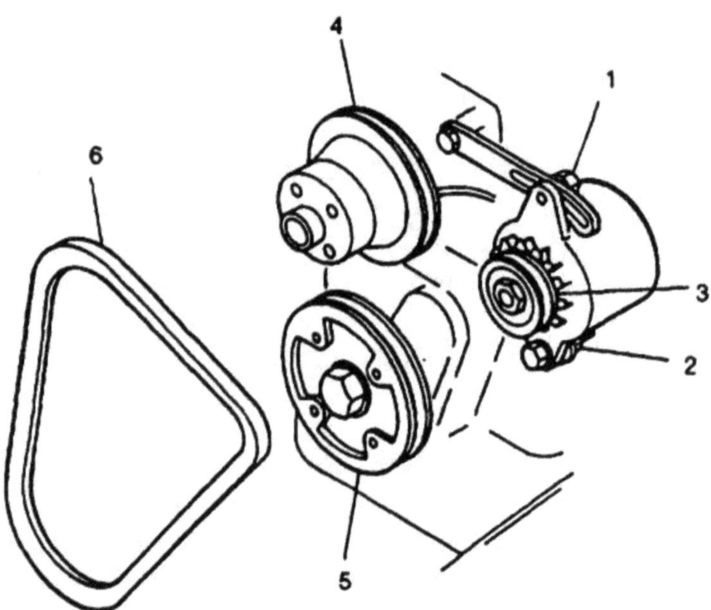

g. g. Close engine access doors, connect negative battery cable, and close battery access door.

Figure 2-27. Fan Belt.

2-93.2 Testing and Adjustment.

NOTE

Run engine for 5 minutes if belt is cold. If belt is hot, let cool for 10 to 15 minutes.

a. a. Shut down generator set.

b. b. Open battery access door, disconnect negative battery cable, and open engine access doors.

c. c. Check fan belt (6, Figure 2-27) for proper tension using a suitable belt tension gauge. Belt

d. d. If fan belt needs adjustment, loosen alternator mounting bolt (1) and nut (2).

CAUTION

Do not pry against alternator rear frame. Damage to alternator or mounting brackets could occur.

e. e. Apply outward pressure to alternator front frame until belt tension is correct.

f. f. Tighten alternator mounting bolt (1) and nut (2).

g. g. Close engine access doors, connect negative battery cable, and close battery access door.

2-93.3 Removal.

WARNING

Fan has sharp blades. Use caution and wear gloves when removing or installing belts. Failure to comply with this warning can cause injury to personnel.

a. a. Shut down generator set.

b. b. Open battery access door, disconnect negative battery cable, and open engine access doors.

c. c. Remove fan guards (paragraph 2-81.2).

d. d. Loosen alternator mounting bolt (1, Figure 2-27) and nut (2).

e. e. Pivot alternator to relieve tension on fan belt (6) and remove belt (6) from alternator pulley (3), fan pulley (4), and crankshaft pulley (5).

f. Slip belt (6) over fan (36, Figure 2-23) and remove belt from generator set.

2-93.4 Installation.

WARNING

Fan has sharp blades. Use caution and wear gloves when removing or installing belts. Failure to comply with this warning can cause injury to personnel.

a. a. Slip fan belt (6, Figure 2-27) over fan (36, Figure 2-23).

b. b. Install belt (6, Figure 2-27) onto alternator pulley (3), fan pulley (4), and crankshaft pulley (5).

c. c. Adjust tension on fan belt (6) (paragraph 2-93.2).

d. d. Install fan guards (paragraph 2-81.3).

e. e. Close engine access doors.

f. f. Connect negative battery cabl e and close battery access door.

2-94 FAN BELT (MEP-804B/MEP-814B) 2-94.1

WARNING

All metal jewelry can conduct electricity and become entangled in generator set components. Remove all jewelry when working on generator set. Failure to comply with this warning can cause injury or death to personnel.

WARNING

DO NOT wear loose clothing when performing checks, services and maintenance. Failure to comply with this warning can cause injury or death to personnel.

WARNING

When running, generator set engine has hot metal surfaces that will burn flesh on contact. Shut down generator set and allow engine to cool before performing checks, services, and maintenance. Wear gloves and additional protective clothing as required. Failure to comply with this warning can cause injury or death to personnel.

WARNING

Fan has sharp blades. Use caution and wear gloves when removing or installing belts. Failure to comply with this warning can cause injury to personnel.

a. a. Shut down generator set.

b. b. Open battery access door, disconnect negative battery cable, and open engine access doors.

c. c. Remove left side fan guard.

d. d. Inspect fan belt (6, Figure 2-27) for frays, cracks, oil soaking, and other damage.

e. e. Replace fan belt that shows any of above or cannot be adjusted for proper tension.

f. f. Replace fan guard.

g. g. Close engine access doors, connect negative battery cable, and close battery access door.

2-94.2 Testing and Adjustment.

NOTE

Run engine for 5 minutes if belt is cold. If belt is hot, let cool for 10 to 15 minutes.

a. a. Shut down generator set.

b. b. Open battery access door, disconnect negative battery cable, and open engine access doors.

c. c. Check fan belt (6, Figure 2-27) for proper tension by applying approximately 22 lbf (10 kgf) to middle of one belt span. Measure deflection. Compare deflection to values in Table 2-10.

d. If fan belt needs adjustment, loosen alternator mounting bolt (1) and nut (2).

CAUTION

Do not pry against alternator rear frame. Damage to alternator or mounting brackets could occur.

e. e. Apply outward pressure to alternator front frame until belt tension is correct.

f. f. Tighten alternator mounting bolt (1) and nut (2).

g. g. Close engine access doors, connect negative battery cable, and close battery access door.

Table 2-10. Alternator Belt Tension (Deflection).

Location	New Belt (Used less than 5 minutes)	Used Belt
Alternator to fan pulley	0.2 – 0.31 in. (5 – 8 mm)	0.28 – 0.39 in. (7 – 10 mm)
Alternator to crankshaft pulley	0.31 – 0.47 in. (8 – 12 mm)	0.39 – 0.56 in. (10 – 14 mm)
Fan pulley to crankshaft pulley	0.28 – 0.43 in. (7 – 11 mm)	0.35 – 0.51 in. (9 – 13 mm)

2-94.3 Removal.

WARNING

Fan has sharp blades. Use caution and wear gloves when removing or installing belts. Failure to comply with this warning can cause injury to personnel.

a. a. Shut down generator set.

b. b. Open battery access door, disconnect negative battery cable, and open engine access doors.

c. c. Remove fan guards (paragraph 2-82.2).

d. d. Loosen alternator mounting bolt (1, Figure 2-27) and nut (2).

e. e. Pivot alternator to relieve tension on fan belt (6) and remove belt (6) from alternator pulley (3), fan pulley (4), and crankshaft pulley (5).

f. Slip belt (6) over fan (43, Figure 2-26) and remove belt from generator set.

2-94.4 Installation.

WARNING

Fan has sharp blades. Use caution and wear gloves when removing or installing belts. Failure to comply with this warning can cause injury to personnel.

a. a. Slip fan belt (6, Figure 2-27) over fan (43, Figure 2-26).

b. b. Install belt (6, Figure 2-27) onto alternator pulley (3), fan pulley (4), and crankshaft pulley

c. c. Adjust tension on fan belt (6) (paragraph 2-94.2).

d. d. Install fan guards (paragraph 2-82.3).

e. e. Close engine access doors, connect negative battery cable, and close battery access door.

2-95 COOLANT RECOVERY SYSTEM 2-95.1

WARNING

Cooling system operates at high temperature and pressure. Contact with high pressure steam and/or liquids can result in burns and scalding. Shut down generator set, and allow system to cool before performing checks, services and maintenance, or wear gloves and additional protective clothing and goggles as required. Failure to comply with this warning can cause injury or death to personnel.

WARNING

In extreme cold weather, skin can stick to metal. Avoid contacting metal items with bare skin in extreme cold weather. Failure to comply with this warning can cause injury to personnel.

a. a. Shut down generator set.

b. b. Open left side engine access door.

c. c. Inspect coolant recovery system components for cracks, holes, or other damage.

d. d. Close left side engine access door.

2-95.2 Removal.

WARNING

All metal jewelry can conduct electricity and become entangled in generator set components. Remove all jewelry when working on generator set. Failure to comply with this warning can cause injury or death to personnel.

WARNING

DO NOT wear loose clothing when performing checks, services and maintenance. Failure to comply with this warning can cause injury or death to personnel.

WARNING

When running, generator set engine has hot metal surfaces that will burn flesh on contact. Shut down generator set and allow engine to cool before performing checks, services, and maintenance. Wear gloves and additional protective clothing as required. Failure to comply with this warning can cause injury or death to personnel.

WARNING

Cooling system operates at high temperature and pressure. Contact with high pressure steam and/or liquids can result in burns and scalding. Shut down generator set, and allow system to cool before performing checks, services and maintenance, or wear gloves and additional protective clothing and goggles as required. Failure to comply with this warning can cause injury or death to personnel.

WARNING

Always remove radiator cap slowly to permit any pressure to escape. Failure to comply with this warning can cause injury to personnel.

a. a. Shut down generator set.

b. b. Open left side engine access door.

c. c. Loosen clamp (1, Figure 2-28) and disconnect hose (2) from overflow bottle (5) and drain coolant into suitable container.

d. d. Loosen clamp (3) and disconnect hose (4) from overflow bottle (5).

e. e. Remove overflow bottle (5) from wire holder (10).

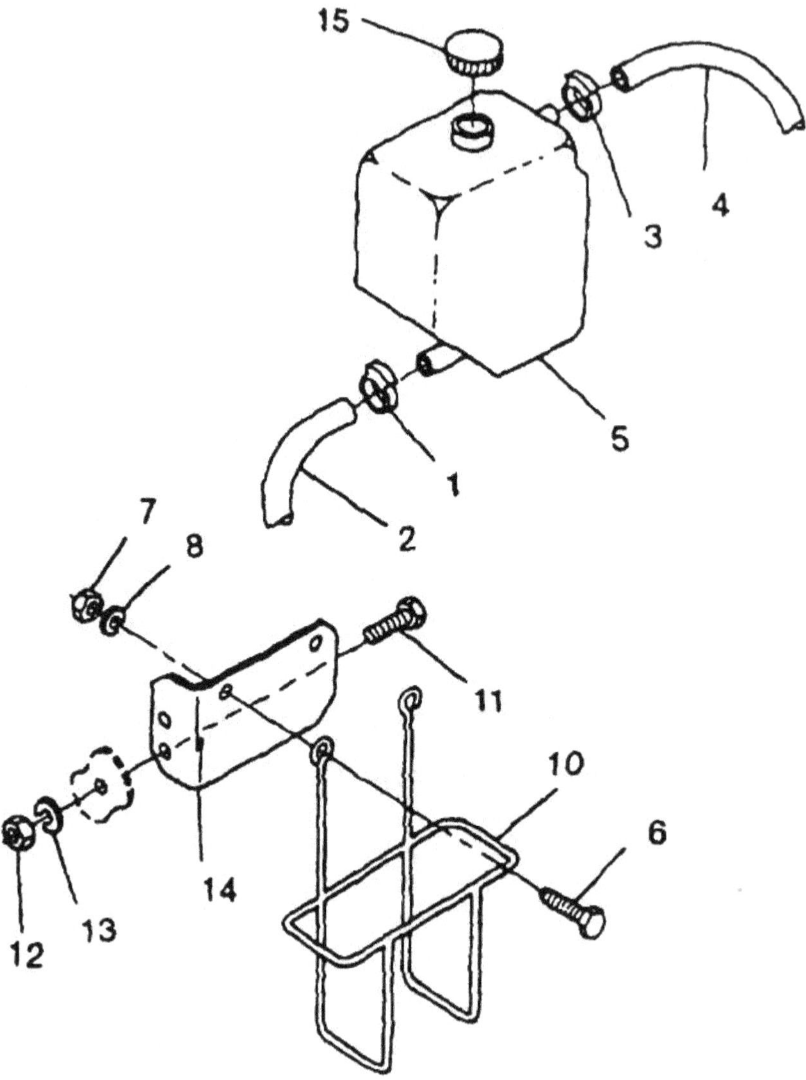

Figure 2-28. Coolant Recovery System.

a. a. Remove bolts (6), nuts (7), lock washers (8), and wire holder (10) from mount (14). Discard lock washers (8).

b. b. Remove bolts (11), nuts (12), lock washers (13), and mount (14) from engine. Discard lock washers (13).

WARNING

Always remove radiator cap slowly to permit any pressure to escape. Failure to comply with this warning can cause injury to personnel.

a. a. Install mount (14, Figure 2-28) on engine with bolts (11), nuts (12), and new lock washers (13).

c. c. Install coolant overflow bottle (5) in holder (10) and connect hoses (2 and 4) with clamps (1 and 3).

d. d. Remove cap (15).

e. e. Fill overflow bottle (5) with coolant to the COLD level. Refer to Table 2-1 for proper coolant.

f. f. Install cap (15).

g. g. Start generator set, check for leaks, and run until normal operating temperature is reached.

h. h. Remove cap (15).

i. i. Add coolant to HOT level of overflow bottle (5).

Section XIII. MAINTENANCE OF FUEL SYSTEM

2-96 BLEEDING AIR FROM FUEL SYSTEM (MEP-804B/MEP-814B)

WARNING

All metal jewelry can conduct electricity and become entangled in generator set components. Remove all jewelry when working on generator set. Failure to comply with this warning can cause injury or death to personnel.

WARNING

DO NOT wear loose clothing when performing checks, services and maintenance. Failure to comply with this warning can cause injury or death to personnel.

WARNING

When running, generator set engine has hot metal surfaces that will burn flesh on contact. Shut down generator set and allow engine to cool before performing checks, services, and maintenance. Wear gloves and additional protective clothing as required. Failure to comply with this warning can cause injury or death to personnel.

WARNING

Diesel fuel is flammable and toxic to eyes, skin, and respiratory tract. Skin and eye protection are required when working in contact with diesel fuel. Avoid repeated or prolonged contact. Provide adequate ventilation. Operators are to wash exposed skin and change chemical soaked clothing promptly if exposed to fuel. Failure to comply with this warning can cause injury or death to personnel.

WARNING

Fuels used in the generator set are flammable. Do not smoke or use open flames when performing maintenance. Failure to comply with this warning can cause injury or death to personnel, and damage to the generator set.

WARNING

Fuels used in the generator set are flammable. When filling the fuel tank, maintain metal-to-metal contact between filler nozzle and fuel tank opening to eliminate static electrical discharge. Failure to comply with this warning can cause injury or death to personnel, and damage to the generator set.

a. a. Shut down generator set.

b. b. Check fuel level in fuel tank. Refuel if necessary.

c. c. Open right side engine access door.

d. d. Loosen air bleeding bolt on fuel filter/water se parator (Figure 2-29); then, tighten finger tight.

e. e. Place MASTER SWITCH in PRIME & RUN position.

f. f. Loosen air bleeding bolt on fuel filter/water sepa rator until air or fuel comes out. Drain until fuel coming out is clear and not mixed with air bubbles.

g. Tighten air bleeding bolt.

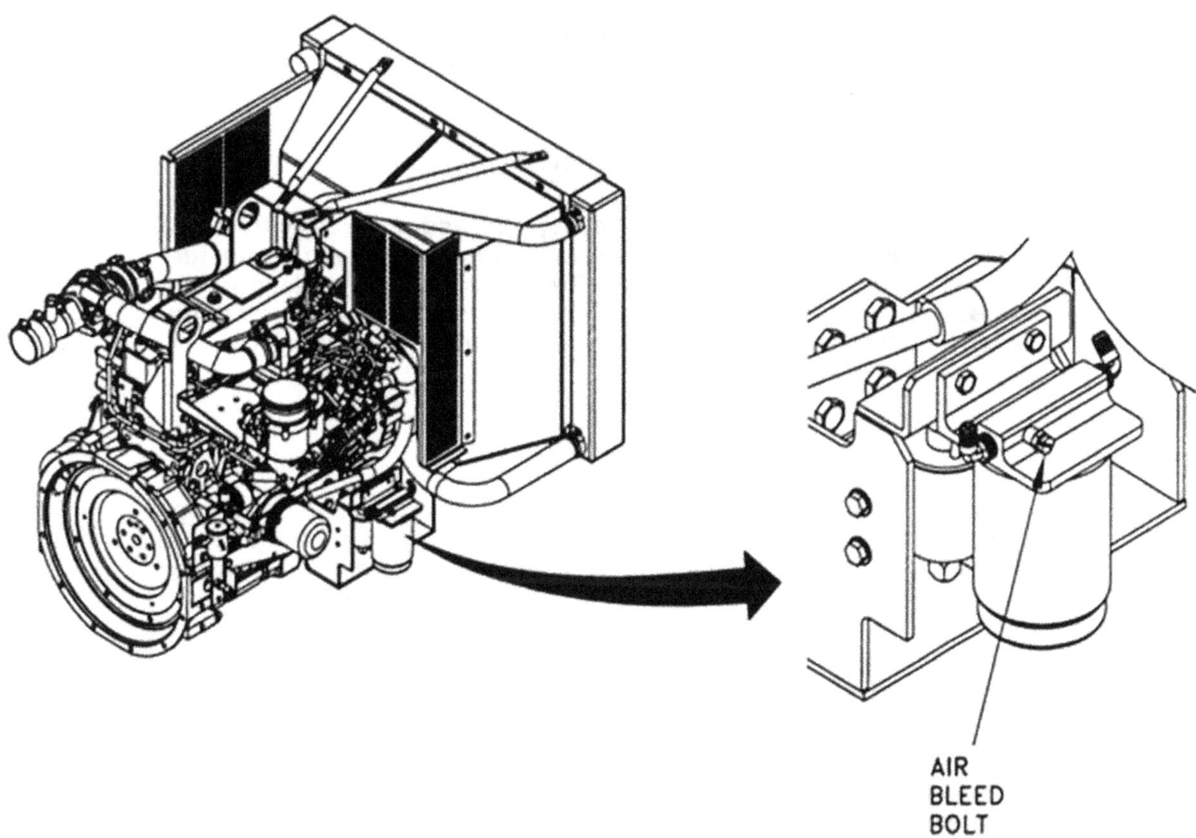

Figure 2-29. Bleeding Air From Fuel System – MEP-804B/MEP-814B. 2-97

LOW PRESSURE FUEL LINES AND FITTINGS 2-97.1 Removal.

WARNING

All metal jewelry can conduct ~~electricity and~~ become entangled in generator set components. Remove all jewelry when working on generator set. Failure to comply with this warning can cause injury or death to personnel.

WARNING

DO NOT wear loose clothing when performing checks, services and maintenance. Failure to comply with this warning can cause injury or death to personnel.

WARNING

When running, generator set engine has hot metal surfaces that will burn flesh on contact. Shut down generator set and allow engine to cool before performing checks, services, and maintenance. Wear gloves and additional protective clothing as required. Failure to comply with this warning can cause injury or death to personnel.

WARNING

Diesel fuel is flammable and toxic to eyes, skin, and respiratory tract. Skin and eye protection are required when working in contact with diesel fuel. Avoid repeated or prolonged contact. Provide adequate ventilation. Operators are to wash exposed skin and change chemical soaked clothing promptly if exposed to fuel. Failure to comply with this warning can cause injury or death to personnel.

WARNING

Fuels used in the generator set are flammable. Do not smoke or use open flames when performing maintenance. Failure to comply with this warning can cause injury or death to personnel, and damage to the generator set.

WARNING

Fuels used in the generator set are flammable. When filling the fuel tank, maintain metal-to-metal contact between filler nozzle and fuel tank opening to eliminate static electrical discharge. Failure to comply with this warning can cause injury or death to personnel, and damage to the generator set.

a. a. Shut down generator set.

b. b. Open battery access door and disconnect negative battery cable.

c. c. Identify fuel line or fitting that is damaged or leaking and must be removed (Figure 2-30).

d. d. Disconnect fuel line at both ends and remove any clamps.

e. e. Remove fuel line or fitting from generator set.

f. f. Cover or cap all openings.

2-97.2 Installation.

a. Remove any caps and position fuel line or fitting in generator set.

b. b. Install any clamps as removed and connect fuel line at both ends.

c. c. Connect negative battery c able and close battery access door.

d. d. Start generator set and check for fuel leaks.

e. e. Shut down generator set. Close all access doors.

2-98 AUXILIARY FUEL PUMP 2-98.1

Inspection.

a. Shut down generator set.

b. Inspect auxiliary fuel pump (10, Figure 2-30) for leaks, cracks, missing hardware, loose connections, and other damage.

2-98.2 Testing.

a. a. Shut down generator set.

b. b. Connect generator set to auxiliary fuel supply (ensure auxiliary fuel supply is no more than 6 feet (1.83m) below generator set).

c. c. Open left side engine access door.

d. d. Disconnect auxiliary fuel pump outlet line (15, Figure 2-30) at fuel tank fitting and place disconnected end in measuring container.

e. e. Move generator set MASTER SWITCH to PRIME & RUN AUX FUEL position for 1 minute and return MASTER SWITCH to OFF position.

f. f. Measuring container should have collected at least 36 ounces (1.06 liters) of fuel.

g. g. Replace auxiliary fuel pump if delivery amount is other than above.

h. h. Connect auxiliary fuel pump outlet line (15) at fuel tank fitting.

i. i. Disconnect generator set from auxiliary fuel supply. Close left side engine access door.

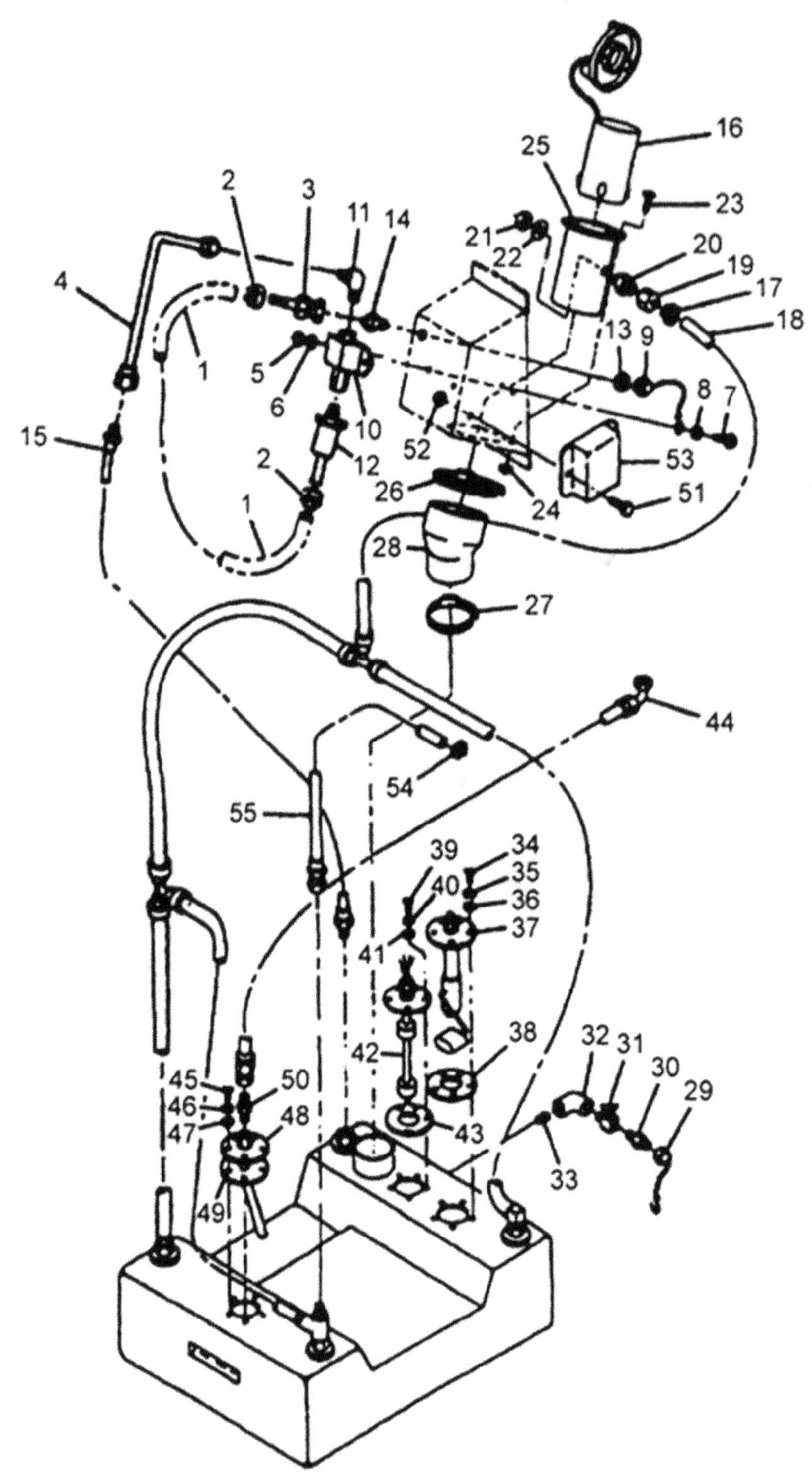

Figure 2-30. Fuel Tank Filler Neck and Low Pressure Fuel System.

2-98.3 Removal.

WARNING

All metal jewelry can conduct electricity and become entangled in generator set components. Remove all jewelry when working on generator set. Failure to comply with this warning can cause injury or death to personnel.

WARNING

DO NOT wear loose clothing when performing checks, services and maintenance. Failure to comply with this warning can cause injury or death to personnel.

WARNING

When running, generator set engine has hot metal surfaces that will burn flesh on contact. Shut down generator set and allow engine to cool before performing checks, services, and maintenance. Wear gloves and additional protective clothing as required. Failure to comply with this warning can cause injury or death to personnel.

WARNING

Diesel fuel is flammable and toxic to eyes, skin, and respiratory tract. Skin and eye protection are required when working in contact with diesel fuel. Avoid repeated or prolonged contact. Provide adequate ventilation. Operators are to wash exposed skin and change chemical soaked clothing promptly if exposed to fuel. Failure to comply with this warning can cause injury or death to personnel.

WARNING

Fuels used in the generator set are flammable. Do not smoke or use open flames when performing maintenance. Failure to comply with this warning can cause injury or death to personnel, and damage to the generator set.

WARNING

Fuels used in the generator set are flammable. When filling the fuel tank, maintain metal-to-metal contact between filler nozzle and fuel tank opening to eliminate static electrical discharge. Failure to comply with this warning can cause injury or death to personnel, and damage to the generator set.

a. a. Shut down generator set.

b. b. Open battery access door, disconnect negative battery cable, and open left side engine access door.

c. c. Tag and disconnect auxiliary fuel pump (10, Figure 2-30) electrical connector.

d. d. Loosen clamps (2). Disconnect filter (12) and remove adapter fitting (3) from filler neck panel fitting (14). Remove adapter fitting (3) from fuel inlet hose (1). Remove filter (12) from auxiliary fuel pump (10). Cap fuel inlet hose.

e. e. Disconnect auxiliary fuel outlet line (4) from auxiliary fuel pump (10).

f. f. Remove nuts (5), lock washers (6), bolts (7), washers (8), cap and chain assembly (9), and auxiliary fuel pump (10). Discard lock washers (6).

g. g. Remove fitting (11) from auxiliary fuel pump (10).

h. h. Remove nut (13) and fitting (14) from fuel filler panel, if necessary.

2-98.4 Installation.

a. a. Install fitting (14, Figure 2-30) and nut (13) in fuel filler panel.

b. b. Install fitting (11) on auxiliary fuel pump (10).

c. c. Install auxiliary fuel pump (10) and cap and chain assembly (9) in generator set with bolts (7), washers (8), new lock washers (6), and nuts (5).

d. d. Connect auxiliary fuel outlet line (4) to auxiliary fuel pump (10).

e. e. Remove caps from fuel inlet hose (1). Install filter (12) on auxiliary fuel pump (10). Install adapter fitting (3) on fuel inlet hose (1) and connect adapter fitting (3) to filler neck panel fitting (14). Connect other end of fuel inlet hose to filter (12).

f. f. Tighten clamps (2).

g. g. Connect auxiliary fuel pump electrical connector. Remove tag.

h. h. Connect negative battery cabl e and close battery access door.

i. i. Move generator set MASTER SWITCH to PRIME & RUN AUX FUEL position and check for fuel leaks.

j. Return MASTER SWITCH to OFF position. Close engine access door.

2-99 FUEL TANK FILLER NECK 2-99.1

Removal.

WARNING

All metal jewelry can conduct electricity and become entangled in generator set components. Remove all jewelry when working on generator set. Failure to comply with this warning can cause injury or death to personnel.

WARNING

DO NOT wear loose clothing when performing checks, services and maintenance. Failure to comply with this warning can cause injury or death to personnel.

WARNING

When running, generator set engine has hot metal surfaces that will burn flesh on contact. Shut down generator set and allow engine to cool before performing checks, services, and maintenance. Wear gloves and additional protective clothing as required. Failure to comply with this warning can cause injury or death to personnel.

WARNING

Diesel fuel is flammable and toxic to eyes, skin, and respiratory tract. Skin and eye protection are required when working in contact with diesel fuel. Avoid repeated or prolonged contact. Provide adequate ventilation. Operators are to wash exposed skin and change chemical soaked clothing promptly if exposed to fuel. Failure to comply with this warning can cause injury or death to personnel.

WARNING

Fuels used in the generator set are flammable. Do not smoke or use open flames when performing maintenance. Failure to comply with this warning can cause injury or death to personnel, and damage to the generator set.

WARNING

Fuels used in the generator set are flammable. When filling the fuel tank, maintain metal-to-metal contact between filler nozzle and fuel tank opening to eliminate static electrical discharge. Failure to comply with this warning can cause injury or death to personnel, and damage to the generator set.

a. a. Shut down generator set.

b. b. Open battery access door and disconnect negative battery cable.

c. c. Remove cap (29, Figure 2-30), open fuel drain valve (31), and drain fuel into suitable container.

d. d. Open left side engine access door.

e. e. Remove filler neck cap and tube assembly (16).

f. f. Loosen clamp (17) and disconnect hose (18) from adapter (19).

g. g. Remove adapter (19) from fitting (20).

h. h. Remove nut (21), washer (22), and fitting (20) from side of filler neck (25).

i. i. Remove bolts (23) and nuts (24) securing filler neck (25) to generator set housing.

j. j. Remove clamps (26 and 27), hose (28), and fille r neck (25) from fuel tank opening. Cover fuel tank opening.

2-99.2 Inspection.

a. a. Shut down generator set.

b. b. Remove fuel tank filler neck (paragraph 2-99.1).

c. c. Inspect hose (28, Figure 2-30) for cracking, wear, and other damage.

d. d. Inspect filler neck (25) for corrosion, cracking, and other damage.

e. e. Inspect filler neck cap and tube assembly (16) for damage.

f. f. Install fuel tank filler neck (paragraph 2-99.3).

2-99.3 Installation.

a. Position hose (28, Figure 2-30), clamps (26) and (27), and filler neck (25) on fuel tank.
b. Install bolts (23) and nuts (24) securing filler neck (25) to generator set housing.
c. Tighten clamps (26 and 27).
d. Install fitting (20), washer (22), and nut (21) inside of filler neck (25).
e. Install adapter (19) on fitting (20) and connect hose (18) to adapter (19) with clamp (17).
f. Install filler neck cap and tube assembly (16).
g. Close fuel drain valve (31), install cap (29), and service fuel tank. Refer to Table 2-2 for proper fuel.
h. Connect negative battery cable and close battery access door and engine access door.

2-100 FUEL DRAIN VALVE

2-100.1 Removal.

WARNING

All metal jewelry can conduct electricity and become entangled in generator set components. Remove all jewelry when working on generator set. Failure to comply with this warning can cause injury or death to personnel.

WARNING

DO NOT wear loose clothing when performing checks, services and maintenance. Failure to comply with this warning can cause injury or death to personnel.

WARNING

When running, generator set engine has hot metal surfaces that will burn flesh on contact. Shut down generator set and allow engine to cool before performing checks, services, and maintenance. Wear gloves and additional protective clothing as required. Failure to comply with this warning can cause injury or death to personnel.

WARNING

Diesel fuel is flammable and toxic to eyes, skin, and respiratory tract. Skin and eye protection are required when working in contact with diesel fuel. Avoid repeated or prolonged contact. Provide adequate ventilation. Operators are to wash exposed skin and change chemical soaked clothing promptly if exposed to fuel. Failure to comply with this warning can cause injury or death to personnel.

WARNING

Fuels used in the generator set are flammable. Do not smoke or use open flames when performing maintenance. Failure to comply with this warning can cause injury or death to personnel, and damage to the generator set.

WARNING

Fuels used in the generator set are flammable. When filling the fuel tank, maintain metal-to-metal contact between filler nozzle and fuel tank opening to eliminate static electrical discharge. Failure to comply with this warning can cause injury or death to personnel, and damage to the generator set.

a. a. Shut down generator set.

b. b. Open battery access door, disconnect negative battery cable, and open left side engine access door.

c. c. Remove cap and chain assembly (29, Figure 2-30) from adapter (30).

d. d. Open drain valve (31) and drain fuel into suitable container.

e. e. Remove drain valve (31) and adapter (30) from elbow (32).

f. f. Remove adapter (30) from drain valve (31).

g. g. Remove elbow (32) and adapter (33) from fuel tank fitting, if necessary.

2-100.2 Installation.

~~a. a. Install~~ adapter (33, Figure 2-30) and elbow (32) in fuel tank fitting, if removed.

b. b. Install fuel drain valve (31) into elbow (32).

c. c. Install adapter (30) into drain valve (31), and cap and chain assembly (29) on adapter (30).

d. d. Ensure fuel drain valve (31) is closed and service fuel tank. Refer to Table 2-2 for proper fuel.

e. e. Check fuel drain valve and fittings for leakage.

f. f. Close left side engine access door, connect negative battery cable, and close battery access door.

2-101 FUEL LEVEL SENDER

2-101.1 Inspection.

a. a. Shut down generator set.

b. b. Open left side engine access door.

c. c. Inspect fuel level sender (37, Figure 2-30) fo r loose connections/mounting and other damage.

d. d. Close left side engine access door.

WARNING

All metal jewelry can conduct electricity and become entangled in generator set components. Remove all jewelry when working on generator set. Failure to comply with this warning can cause injury or death to personnel.

WARNING

DO NOT wear loose clothing when performing checks, services and maintenance. Failure to comply with this warning can cause injury or death to personnel.

WARNING

When running, generator set engine has hot metal surfaces that will burn flesh on contact. Shut down generator set and allow engine to cool before performing checks, services, and maintenance. Wear gloves and additional protective clothing as required. Failure to comply with this warning can cause injury or death to personnel.

WARNING

Diesel fuel is flammable and toxic to eyes, skin, and respiratory tract. Skin and eye protection are required when working in contact with diesel fuel. Avoid repeated or prolonged contact. Provide adequate ventilation. Operators are to wash exposed skin and change chemical soaked clothing promptly if exposed to fuel. Failure to comply with this warning can cause injury or death to personnel.

WARNING

Fuels used in the generator set are flammable. Do not smoke or use open flames when performing maintenance. Failure to comply with this warning can cause injury or death to personnel, and damage to the generator set.

WARNING

Fuels used in the generator set are flammable. When filling the fuel tank, maintain metal-to-metal contact between filler nozzle and fuel tank opening to eliminate static electrical discharge. Failure to comply with this warning can cause injury or death to personnel, and damage to the generator set.

a. a. Shut down generator set.

b. b. Open battery access door, disconnect negative battery cable, and open left side engine access doors.

c. c. Disconnect fuel level sender (37, Figure 2-30) electrical lead.

d. d. Remove screws (34), lock washers (35), flat washers (36), fuel level sender (37), and gasket (38) from generator set fuel tank. Discard lock washers (35).

e. Cover opening in fuel tank.

2-101.3 Testing.

~~a. a. Sh~~ut down generator set.

b. b. Remove fuel level sender (paragraph 2-101.2).

c. c. Position fuel level sender in vertical position, similar to position as installed in fuel tank.

d. d. Set multimeter for ohms and connect positive lead to fuel level sender terminal and negative
lead to sender ground.

e. e. With fuel level sender arm resting freely in what would be an empty position, multimeter should indicate between 216 and 264 ohms.

f. f. Move fuel level sender arm up to what would be a full position and multimeter should indicate between 29.7 and 36.3 ohms.

g. g. Replace fuel level sender if indications are not as above.

h. h. Install fuel level sender (paragraph 2-101.4).

2-101.4 Installation.

~~a. a. Remo~~ve cover in fuel tank opening.

b. b. Clean, make flat and smooth mating surfaces to gasket (38, Figure 2-30), ensuring no foreign material enters fuel tank. Apply sealant (Item 17, Appendix C) to both sides of gasket (38).

c. c. Insert fuel level sender (37) and gasket (38) into fuel tank. Ensure float is in same position as removed.

d. d. Install screws (34), new lock washers (35), and flat washers (36).

e. e. Connect electrical lead. Remove tag.

f. Close left side engine access door, connect negative battery cable, and close battery access door.

2-102 LOW LEVEL/AUXILIARY FUEL PUMP FLOAT SWITCH

2-102.1 Inspection.

a. a. Shut down generator set.

b. b. Open left side engine access door.

c. Inspect low fuel level/auxiliary fuel pump float switch (42, Figure 2-30) for loose connections/mounting and other damage.

d. Close left side engine access door.

2-102.2 Removal.

WARNING

All metal jewelry can conduct electricity and become entangled in generator set components. Remove all jewelry when working on generator set. Failure to comply with this warning can cause injury or death to personnel.

WARNING

DO NOT wear loose clothing when performing checks, services and maintenance. Failure to comply with this warning can cause injury or death to personnel.

WARNING

When running, generator set engine has hot metal surfaces that will burn flesh on contact. Shut down generator set and allow engine to cool before performing checks, services, and maintenance. Wear gloves and additional protective clothing as required. Failure to comply with this warning can cause injury or death to personnel.

WARNING

Diesel fuel is flammable and toxic to eyes, skin, and respiratory tract. Skin and eye protection are required when working in contact with diesel fuel. Avoid repeated or prolonged contact. Provide adequate ventilation. Operators are to wash exposed skin and change chemical soaked clothing promptly if exposed to fuel. Failure to comply with this warning can cause injury or death to personnel.

WARNING

Fuels used in the generator set are flammable. Do not smoke or use open flames when performing maintenance. Failure to comply with this warning can cause injury or death to personnel, and damage to the generator set.

WARNING

Fuels used in the generator set are flammable. When filling the fuel tank, maintain metal-to-metal contact between filler nozzle and fuel tank opening to eliminate static electrical discharge. Failure to comply with this warning can cause injury or death to personnel, and damage to the generator set.

a. a. Shut down generator set.

b. b. Open battery access door, disconnect negative battery cable, and open left side engine access door.

c. c. Tag and disconnect low fuel level/auxiliary fuel pump float switch (42, Figure 2-30) electrical connector.

d. d. Remove screws (39), lock washers (40), flat washers (41), float switch (42), and gasket (43) from fuel tank. Discard lock washers (40).

e. e. Cover opening in fuel tank.

2-102.3 Testing.

~~a. a. Sh~~ut down generator set.

b. b. Remove low fuel level/auxiliary fuel pump float switch (paragraph 2-102.2).

c. c. Position float switch in vertical position, similar to position as installed in fuel tank.

d. d. For the top float, set multimeter for ohms and connect positive lead to pin 2 and negative lead to pin 1 of float switch electrical connector.

e. e. With upper or lower float moving toward the down position, multimeter should indicate continuity 1/4 inch before float reaches down position.

f. f. Move upper float to full up position. Multimeter should indicate open circuit.

g. g. Disconnect multimeter leads from pins 1 and 2. To check lower float, connect positive lead to pin 3 and negative lead to pin 4 of electrical connector.

h. h. Repeat steps e and f, except with lower float.

i. i. Replace low fuel level/auxiliary fuel pump float switch if indications are other than above.

j. j. Install low fuel level/auxiliary fuel pump float switch (paragraph 2-102.4).

2-102.4 Installation.

~~a. a. Loosen~~ float switch plate adjusting nut.

b. b. Remove cover in fuel tank opening.

c. c. Clean, make flat and smooth mating surfaces to gasket (43, Figure 2-30), ensuring no foreign material enters fuel tank. Apply sealant (Item 17, Appendix C) to both sides of gasket (43).

d. d. Position gasket (43) and float switch (42) in fuel tank.

e. e. Install screws (39), new lock washers (40), and flat washers (41).

f. f. Set float switch stem 1/16 inch from bottom of fuel tank and tighten float switch plate adjusting nut.

g. g. Connect electrical connector. Remove tag.

h. h. Close left side engine acce ss door, connect negative battery cable, and close battery access door.

2-103 FUEL PICKUP

2-103.1 Removal.

WARNING

All metal jewelry can conduct electricity and become entangled in generator set components. Remove all jewelry when working on generator set. Failure to comply with this warning can cause injury or death to personnel.

WARNING

DO NOT wear loose clothing when performing checks, services and maintenance. Failure to comply with this warning can cause injury or death to personnel.

WARNING

When running, generator set engine has hot metal surfaces that will burn flesh on contact. Shut down generator set and allow engine to cool before performing checks, services, and maintenance. Wear gloves and additional protective clothing as required. Failure to comply with this warning can cause injury or death to personnel.

WARNING

Diesel fuel is flammable and toxic to eyes, skin, and respiratory tract. Skin and eye protection are required when working in contact with diesel fuel. Avoid repeated or prolonged contact. Provide adequate ventilation. Operators are to wash exposed skin and change chemical soaked clothing promptly if exposed to fuel. Failure to comply with this warning can cause injury or death to personnel.

WARNING

Fuels used in the generator set are flammable. Do not smoke or use open flames when performing maintenance. Failure to comply with this warning can cause injury or death to personnel, and damage to the generator set.

WARNING

Fuels used in the generator set are flammable. When filling the fuel tank, maintain metal-to-metal contact between filler nozzle and fuel tank opening to eliminate static electrical discharge. Failure to comply with this warning can cause injury or death to personnel, and damage to the generator set.

a. a. Shut down generator set.

b. b. Open battery access door, disconnect negative battery cable, and open right side engine access
door.

c. c. Disconnect fuel line (44, Figure 2-30) from fitting (48).

NOTE

Mark position of fuel pickup before removing.

d. d. Remove screws (45), lock washers (46), flat washers (47), fuel pickup (48), and gasket (49) from fuel tank. Discard lock washers (46).

e. e. Remove fitting (50) from fuel pickup (48).

f. f. Cover opening in fuel tank.

2-103.2 Inspection.

a. a. Shut down generator set.

b. b. Remove fuel pickup (paragraph 2-103.1).

c. c. Inspect fuel pickup and fitting for clogs, stripped threads, and other damage.

d. d. Replace damaged parts.

e. e. Install fuel pickup (paragraph 2-103.3).

2-103.3 Installation.

NOTE

Ensure fuel pickup is in same position as marked on removal.

a. a. Remove cover in fuel tank opening.

b. b. Clean, make flat and smooth matting surfaces to gasket (49, Figure 2-30), ensuring no foreign material enters fuel tank. Apply sealant (Item 17, Appendix C) to both sides of gasket (49).

c. c. Install gasket (49) and fuel pickup (48) in fuel tank with screws (45), new lock washers (46),
and
flat washers (47).

d. d. Install fitting (50) in fuel pickup (48).

e. e. Connect fuel line (44) to fitting (48).

f. f. Close right side engine access door, connect negative battery cable, and close battery access door.

2-104 FUEL FLOAT MODULE

2-104.1 Inspection.

a. a. Shut down generator set.

b. b. Open left side engine access door.

c. c. Inspect fuel float module (53, Figure 2-30) for cracked housing, broken or damaged connectors and wiring, and other damage.

d. Close left side engine access door.

2-104.2 Testing.

a. a. Shut down generator set.

b. b. Open left side engine access door.

c. c. Disconnect fuel float module (5 3, Figure 2-30) electrical connector (J12) from fuel float switch connector (P12).

d. d. Connect pins 1 and 2 of fuel float module elec trical connector (J12) together with a jumper wire.

e. e. Move MASTER SWITCH to PRIME & RUN AUX FUEL position and auxiliary fuel pump should start operating. Remove jumper wire and auxiliary fuel pump should stop operating.

f. f. Start and operate generator set at rated voltage and frequency.

g. g. Using jumper wire, make connection between pins 3 and 4 of fuel float module electrical connector (J12). Generator set should shut down after approximately 2 seconds and NO FUEL lamp on malfunction indicator panel should light.

h. h. Replace fuel float module if operation is other than above.

i. i. Close left side engine access door.

2-104.3 Removal.

WARNING

All metal jewelry can conduct electricity and become entangled in generator set components. Remove all jewelry when working on generator set. Failure to comply with this warning can cause injury or death to personnel.

WARNING

DO NOT wear loose clothing when performing checks, services and maintenance. Failure to comply with this warning can cause injury or death to personnel.

WARNING

When running, generator set engine has hot metal surfaces that will burn flesh on contact. Shut down generator set and allow engine to cool before performing checks, services, and maintenance. Wear gloves and additional protective clothing as required. Failure to comply with this warning can cause injury or death to personnel.

WARNING

Diesel fuel is flammable and toxic to eyes, skin, and respiratory tract. Skin and eye protection are required when working in contact with diesel fuel. Avoid repeated or prolonged contact. Provide adequate ventilation. Operators are to wash exposed skin and change chemical soaked clothing promptly if exposed to fuel. Failure to comply with this warning can cause injury or death to personnel.

WARNING

Fuels used in the generator set are flammable. Do not smoke or use open flames when performing maintenance. Failure to comply with this warning can cause injury or death to personnel, and damage to the generator set.

WARNING

Fuels used in the generator set are flammable. When filling the fuel tank, maintain metal-to-metal contact between filler nozzle and fuel tank opening to eliminate static electrical discharge. Failure to comply with this warning can cause injury or death to personnel, and damage to the generator set.

a. a. Shut down generator set.

b. b. Open battery access door, disconnect negative battery cable, and open left side engine access door.

c. c. Tag and disconnect electrical connectors from fuel float module (53, Figure 2-30).

d. d. Remove bolts (51), nuts (52), and fuel float module (53) from generator set.

2-104.4 Installation.

a. a. Install fuel float module (53, Figure 2-30) in generator set with bolts (51) and nuts (52).

b. b. Connect electrical connectors. Remove tags.

c. c. Close left side engine access door, connect negative battery cable, and close battery access door.

2-105 FUEL FILTER/WATER SEPARATOR ELEMENT

2-105.1 Inspection.

a. a. Shut down generator set.

b. b. Open right side engine access door.

c. c. Inspect fuel filter/water separator, fuel lines , and fittings for loose connections, damage, and evidence of fuel leaks.

d. d. Close right side engine access door.

2-105.2 Service.

a. a. Shut down generator set.

b. b. Open battery access door and disconnect negative battery cable.

c. c. Open water drain valve (1, Figure 2-31) and drain all water from fuel filter/water separator.

d. d. Loosen clamp (3) and disconnect drain line (4).

NOTE

Catch fuel in suitable container.

e. e. Remove fuel filter/water separator element (2).

f. f. Wipe fuel filter/water separator head (11) with cleaning cloth (Item 7, Appendix C).

g. g. Check petcock valve operation for being only hand tight on fuel filter/water separator element (2) and apply a film of engine oil (Item 11, Appendix C) to sealing surface of fuel filter/water separator (2).

h. h. Fill new fuel filter/water separator element (2) with clean diesel fuel.

i. i. Install fuel filter/water separator element (2) hand tight. Do not overtighten.

j. j. Connect drain line (4) and install clamp (3).

k. k. Connect negative battery cable.

l. l. Bleed air from fuel system (TM 9-2815-254-24 (MEP-804A/MEP-814A)/paragraph 2-96 (MEP-804B/MEP-814B)).

m. Start engine and check for fuel leak. Close all access doors.

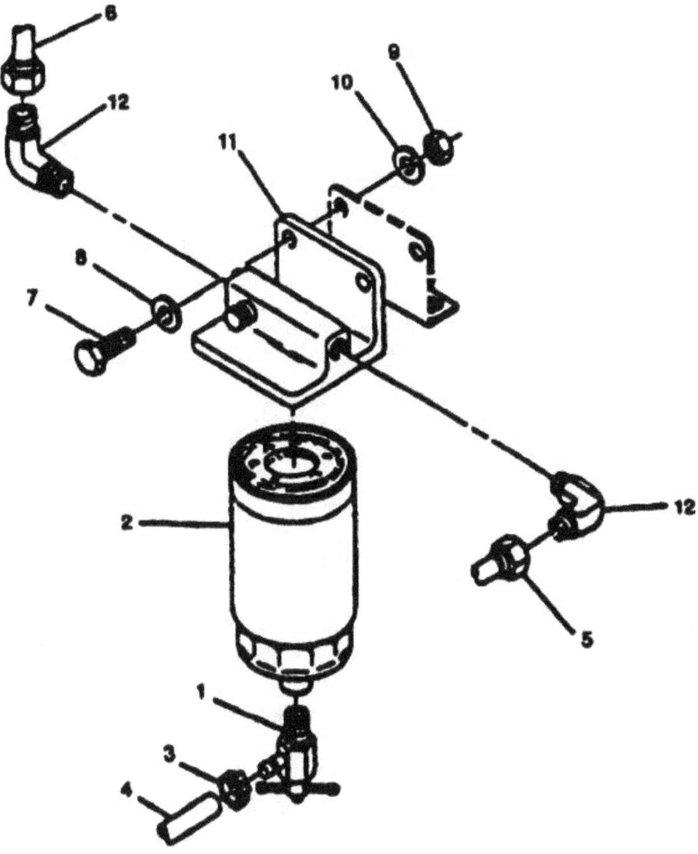

Figure 2-31. Fuel Filter/Water Separator.

2-105.3 Removal.

WARNING

All metal jewelry can conduct electricity and become entangled in generator set components. Remove all jewelry when working on generator set. Failure to comply with this warning can cause injury or death to personnel.

WARNING

DO NOT wear loose clothing when performing checks, services and maintenance. Failure to comply with this warning can cause injury or death to personnel.

WARNING

When running, generator set engine has hot metal surfaces that will burn flesh on contact. Shut down generator set and allow engine to cool before performing checks, services, and maintenance. Wear gloves and additional protective clothing as required. Failure to comply with this warning can cause injury or death to personnel.

WARNING

Diesel fuel is flammable and toxic to eyes, skin, and respiratory tract. Skin and eye protection are required when working in contact with diesel fuel. Avoid repeated or prolonged contact. Provide adequate ventilation. Operators are to wash exposed skin and change chemical soaked clothing promptly if exposed to fuel. Failure to comply with this warning can cause injury or death to personnel.

WARNING

Fuels used in the generator set are flammable. Do not smoke or use open flames when performing maintenance. Failure to comply with this warning can cause injury or death to personnel, and damage to the generator set.

WARNING

Fuels used in the generator set are flammable. When filling the fuel tank, maintain metal-to-metal contact between filler nozzle and fuel tank opening to eliminate static electrical discharge. Failure to comply with this warning can cause injury or death to personnel, and damage to the generator set.

a. a. Shut down generator set.

b. b. Open battery access door, disconnect negative battery cable, and open right side engine access
door.

c. Loosen clamp (3, Figure 2-31) and disconnect water drain line (4) from fuel filter/water separator.

d. d. Disconnect fuel lines (5 and 6).

e. e. Remove bolts (7), flat washers (8), nuts (9), lock washers (10), and fuel filter/water separator head (11) with element (2). Discard lock washers (10).

f. Cap all openings.

2-105.4 Repair.

a. a. Shut down generator set.

b. b. Open battery access door and disconnect negative battery cable.

c. c. Remove fuel filter/water separator (paragraph 2-105.3).

d. d. If damaged or if replacing fuel filter/water sepa rator, remove and retain elbows (12, Figure
2-31)
from head (11).

e. e. Install elbows (12).

f. f. Install fuel filter/water separator (paragraph 2-105.5).

2-105.5 Installation.

a. a. Remove caps.

b. b. Install fuel filter/water separator head (11, Figure 2-31) with nuts (9), new lock washers (10), bolts (7), and flat washers (8).

c. c. Install fuel lines (5 and 6).

d. d. Connect negative battery cable and close battery access door.

e. e. Bleed air from fuel system (TM 9-2815-254-24).

f. f. Start generator set and check for fuel leaks. Close right side engine access door.

2-106 FUEL TRANSFER PUMP (MEP-804B/MEP-814B)

2-106.1 Inspection.

a. a. Shut down generator set.

b. b. Open right side engine access door.

c. c. Inspect fuel transfer pump, fuel filter, fuel lines, and fittings for loose connections, damage, and evidence of fuel leaks. Remove fuel/water separator to provide access (paragraph 2-105.3), if necessary.

d. d. Close right side engine access door.

2-106.2 Testing.

a. a. Shut down generator set.

b. b. Open right side engine access door.

c. c. Disconnect fuel transfer pump connection at fuel/water separator (1, Figure 2-32) and place disconnected end in measuring container.

d. d. Ensure there is at least a couple of gallons (8 liters) of fuel in generator set fuel tank for test.

e. e. Move generator set MASTER SWITCH to PRIME & RUN position for 1 minute and return MASTER SWITCH to OFF position.

f. f. Measuring container should have collected at least 36 ounces (1.06 liters) of fuel.

g. g. Replace fuel transfer pump if delivery amount is other than above.

h. h. Connect fuel transfer pump hose to inlet of fuel water separator. Close right side engine access door.

2-106.3 Removal.

WARNING

All metal jewelry can conduct electricity and become entangled in generator set components. Remove all jewelry when working on generator set. Failure to comply with this warning can cause injury or death to personnel.

WARNING

DO NOT wear loose clothing when performing checks, services and maintenance. Failure to comply with this warning can cause injury or death to personnel.

WARNING

When running, generator set engine has hot metal surfaces that will burn flesh on contact. Shut down generator set and allow engine to cool before performing checks, services, and maintenance. Wear gloves and additional protective clothing as required. Failure to comply with this warning can cause injury or death to personnel.

WARNING

Diesel fuel is flammable and toxic to eyes, skin, and respiratory tract. Skin and eye protection are required when working in contact with diesel fuel. Avoid repeated or prolonged contact. Provide adequate ventilation. Operators are to wash exposed skin and change chemical soaked clothing promptly if exposed to fuel. Failure to comply with this warning can cause injury or death to personnel.

WARNING

Fuels used in the generator set are flammable. Do not smoke or use open flames when performing maintenance. Failure to comply with this warning can cause injury or death to personnel, and damage to the generator set.

WARNING

Fuels used in the generator set are flammable. When filling the fuel tank, maintain metal-to-metal contact between filler nozzle and fuel tank opening to eliminate static electrical discharge. Failure to comply with this warning can cause injury or death to personnel, and damage to the generator set.

a. a. Shut down generator set.

b. b. Open battery access door, disconnect negative battery cable, and open right side engine access door.

c. c. Remove fuel/water separator to provide access (paragraph 2-105.3).

d. d. Loosen clamp (2, Figure 2-32) and disconnect hose (3) from fuel filter (4).

e. e. Remove fuel filter (4) from fuel transfer pump (11) and fitting (5) from fuel filter, as required.

f. f. Loosen clamp (6) and disconnect hose (7) from top of fuel transfer pump (11).

g. g. Tag and disconnect wiring from fuel transfer pump (11).

h. h. Remove fuel transfer pump (11) by removing nuts (8), lock washers (9), and washers (10). Discard lock washers (9).

i. i. Remove elbow (12) from fuel transfer pump (11), as required.

j. j. Remove fitting (13) and elbow (14) from fuel transfer pump (11), as required.

k. k. Remove bracket (18) by removing bolts (15), lock washers (16), and washers (17). Discard lock washers (16).

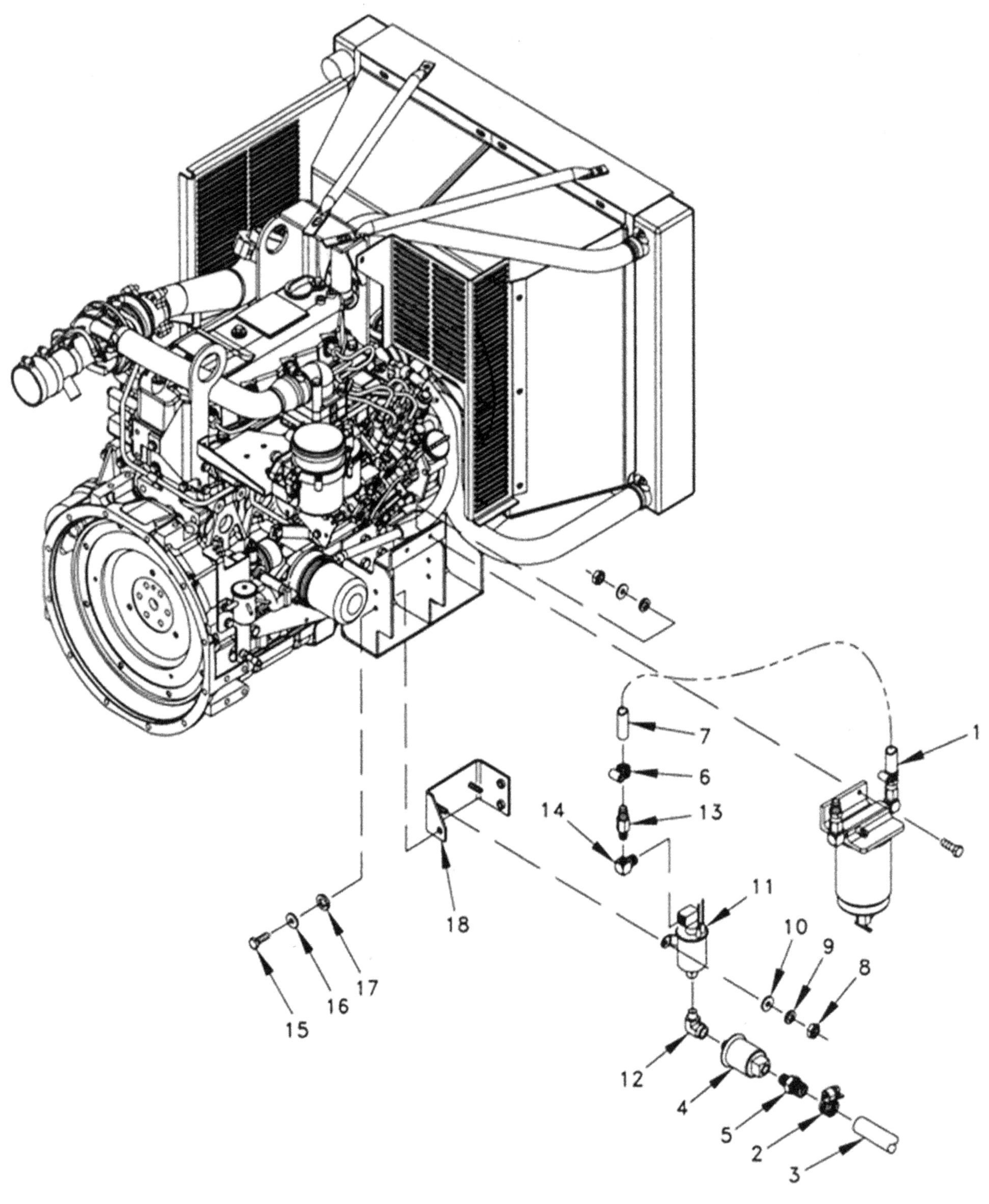

Figure 2-32. Fuel Transfer Pump – MEP-804B/MEP-814B.

2-106.4 Installation.

a. a. Install bracket (18, Figure 2-32) with bolts (15), new lock washers (16), and washers (17).

b. b. Install elbow (14) and fitting (13) to fuel transfer pump (11), as required.

c. c. Install elbow (12) to fuel transfer pump (11), as required.

d. d. Install fuel transfer pump (11) with nuts (8), new lock washers (9), and washers (10).

e. e. Install wiring to fuel transfer pump (11). Remove tags.

f. f. Connect hose (7) to top of fuel transfer pump (11).

g. g. Tighten clamp (6).

h. h. Install fuel filter (4) to fuel transfer pump (11) and install fitting (5), as required.

i. i. Connect hose (3) to fuel filter (4) and tighten clamp (2).

j. j. Install fuel/water separator (paragraph 2-105.5).

k. k. Close right side engine access door, connect negative battery cable, and close battery access door.

l. l. Bleed air from fuel system (paragraph 2-96).

m. m. Start generator set and check for leaks. Close right side engine access door.

Section XIV. MAINTENANCE OF OUTPUT BOX ASSEMBLY

2-107 VOLTAGE RECONNECTION TERMINAL BOARD

WARNING

All metal jewelry can conduct electricity and become entangled in generator set components. Remove all jewelry when working on generator set. Failure to comply with this warning can cause injury or death to personnel.

WARNING

DO NOT wear loose clothing when performing checks, services and maintenance. Failure to comply with this warning can cause injury or death to personnel.

WARNING

Dangerous voltage exists on live circuits. Always observe precautions and never work alone. Failure to comply with this warning can cause injury or death to personnel.

WARNING

DC voltages are present at generator set electrical components even with generator set shut down. Avoid shorting any positive with ground/negative. Failure to comply with this warning can cause injury to personnel, and damage to equipment.

WARNING

Ensure nuts on ground terminals are properly secured creating a good ground. Failure to comply with this warning can cause injury or death to personnel.

WARNING

Shut down generator set before performing inspection of wiring. Failure

WARNING

The connection of any electrical equipment and the disconnection of any electrical equipment may cause an explosion hazard. Do not connect any electrical equipment or disconnect any electrical equipment in an explosive atmosphere. Failure to comply with this warning can cause injury or death to personnel.

2-107.1 Inspection.

a. a. Shut down generator set.

b. b. Open output box access door.

c. c. Inspect protective cover (3, Figure 2-33) and moveable terminal board (5) for cracks, breaks, corrosion, and other damage.

d. d. Replace damaged parts.

e. e. Close output box access door.

2-107.2 Removal.

WARNING

Dangerous voltage exists on live circuits. Always observe precautions and never work alone. Failure to comply with this warning can cause injury or death to personnel.

WARNING

DC voltages are present at generator set electrical components even with generator set shut down. Avoid shorting any positive with ground/negative. Failure to comply with this warning can cause injury to personnel, and damage to equipment.

WARNING

Batteries give off a flammable gas. Do not smoke or use open flame when performing maintenance. Failure to comply with this warning can cause injury or death to personnel, and damage to the generator set.

WARNING

Battery acid can cause burns to unprotected skin. Wear safety goggles and chemical gloves and avoid acid splash while working on batteries. Failure to comply with this warning can cause injury to personnel.

WARNING

When disconnecting or removing batteries, disconnect the negative lead that connects directly to the grounding stud first; disconnect the negative end of the interconnection cable next. When installing batteries, reverse the connection sequence. Failure to comply with this warning can cause injury to personnel.

a. a. Shut down generator set.

b. b. Open battery access door, disconnect negative battery cable, and open output box access

c. c. Remove nuts (1, Figure 2-33), washers (2), and protective cover (3) from voltage reconnection terminal board assembly.

d. d. Remove nuts (4) and moveable terminal board (5) from voltage reconnection terminal board assembly.

2-107.3 Installation.

WARNING

When disconnecting or removing batteries, disconnect the negative lead that connects directly to the grounding stud first; disconnect the negative end of the interconnection cable next. When installing batteries, reverse the connection sequence. Failure to comply with this warning can cause injury to personnel.

a. a. Install removable terminal board (5, Figure 2-33) on voltage reconnection terminal board assembly with nuts (4).

b. b. Install protective cover (3) on voltage reconnection terminal board assembly with washers (2) and nuts (1).

c. c. Close output box access door.

d. d. Connect negative battery cable.

e. e. Close battery access door.

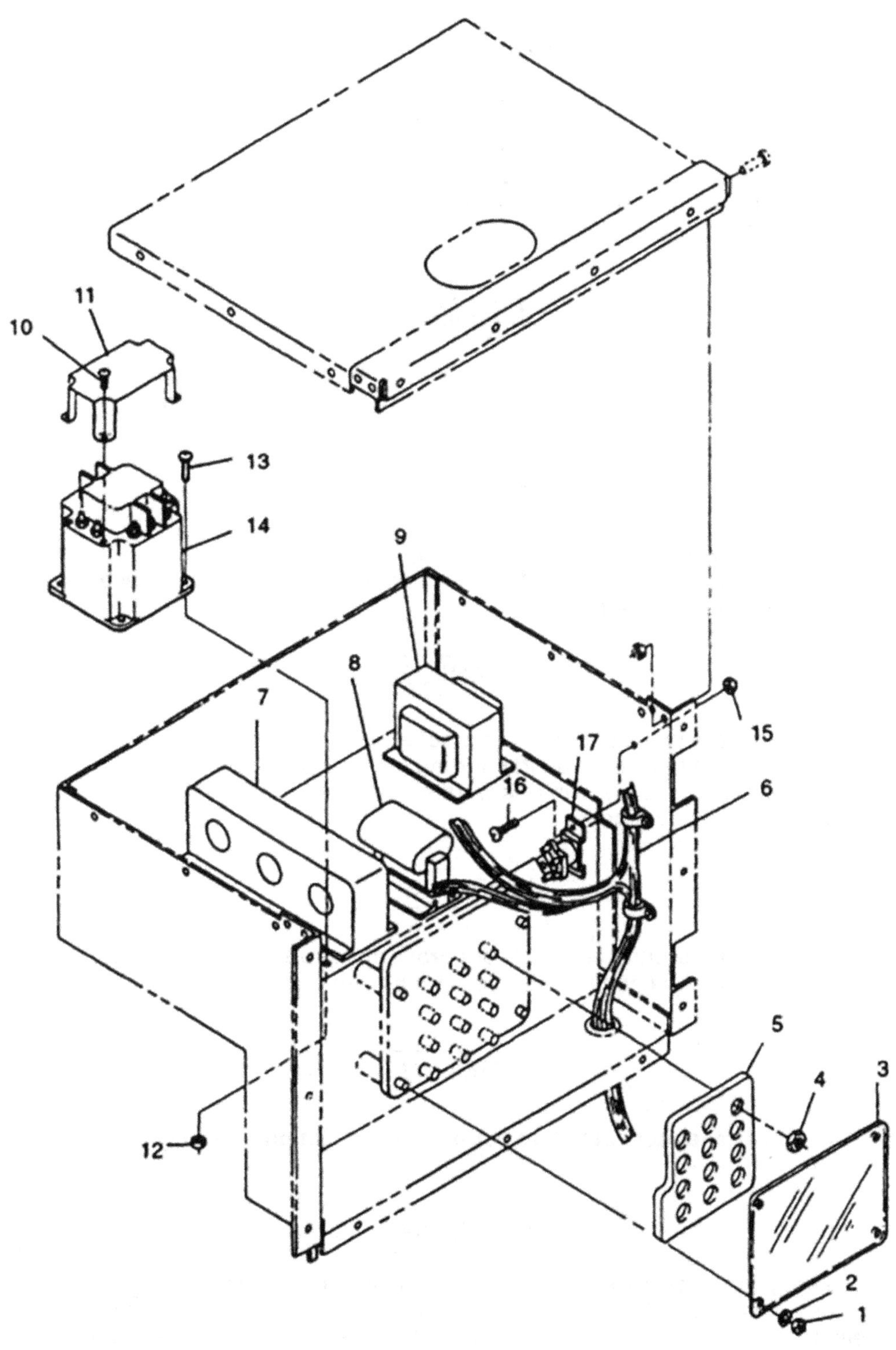

Figure 2-33. Output Box Assembly.

2-108 OUTPUT BOX WIRING HARNESS

2-108.1 Inspection.

WARNING

All metal jewelry can conduct electricity and become entangled in generator set components. Remove all jewelry when working on generator set. Failure to comply with this warning can cause injury or death to personnel.

WARNING

DO NOT wear loose clothing when performing checks, services and maintenance. Failure to comply with this warning can cause injury or death to personnel.

WARNING

Dangerous voltage exists on live circuits. Always observe precautions and never work alone. Failure to comply with this warning can cause injury or death to personnel.

WARNING

DC voltages are present at generator set electrical components even with generator set shut down. Avoid shorting any positive with ground/negative. Failure to comply with this warning can cause injury to personnel, and damage to equipment.

WARNING

Ensure nuts on ground terminals are properly secured creating a good ground. Failure to comply with this warning can cause injury or death to personnel.

WARNING

Shut down generator set before performing inspection of wiring. Failure

WARNING

The connection of any electrical equipment and the disconnection of any electrical equipment may cause an explosion hazard. Do not connect any electrical equipment or disconnect any electrical equipment in an explosive atmosphere. Failure to comply with this warning can cause injury or death to personnel.

a. a. Shut down generator set.

b. b. Inspect wiring harness (6, Figure 2-33) for burned, bent, corroded, and broken termi-

c. c. Inspect connectors for cracks, corrosion, stripped threads, bent or broken pins, and obvious damage.

d. d. Inspect wire insulation for burns, deterioration, and chafing.

2-108.2 Testing.

WARNING

When disconnecting or removing batteries, disconnect the negative lead that connects directly to the grounding stud first; disconnect the negative end of the interconnection cable next. When installing batteries, reverse the connection sequence. Failure to comply with this warning can cause injury to personnel.

a. a. Shut down generator set.

b. b. Open battery access door, disconnect negative battery cable, and open output box and engine access doors.

c. c. Set multimeter for ohms and test individual wires for continuity. Refer to Wiring Diagram FO-2 (MEP-804A/ MEP-814A)/FO-4 (MEP-804B/MEP-814B) for wire identification.

d. d. Close output box and engine access doors, connect negative battery cable, and close battery access door.

2-108.3 Repair.

WARNING

When disconnecting or removing batteries, disconnect the negative lead that connects directly to the grounding stud first; disconnect the negative end of the interconnection cable next. When installing batteries, reverse the connection sequence. Failure to comply with this warning can cause injury to personnel.

a. a. Shut down generator set.

b. b. Open battery access door, disconnect negative battery cable, and open applicable access doors.

c. c. Replace damaged terminals and secure hardware.

d. d. Connect negative battery cable and close all access doors.

2-109 CURRENT TRANSFORMER

2-109.1 Inspection.

WARNING

All metal jewelry can conduct electricity and become entangled in generator set components. Remove all jewelry when working on generator set. Failure to comply with this warning can cause injury or death to personnel.

WARNING

DO NOT wear loose clothing when performing checks, services and maintenance. Failure to comply with this warning can cause injury or death to personnel.

WARNING

Dangerous voltage exists on live circuits. Always observe precautions and never work alone. Failure to comply with this warning can cause injury or death to personnel.

WARNING

DC voltages are present at generator set electrical components even with generator set shut down. Avoid shorting any positive with ground/negative. Failure to comply with this warning can cause injury to personnel, and damage to equipment.

WARNING

Ensure nuts on ground terminals are properly secured creating a good ground. Failure to comply with this warning can cause injury or death to personnel.

WARNING

Shut down generator set before performing inspection of wiring. Failure

WARNING

The connection of any electrical equipment and the disconnection of any electrical equipment may cause an explosion hazard. Do not connect any electrical equipment or disconnect any electrical equipment in an explosive atmosphere. Failure to comply with this warning can cause injury or death to personnel.

a. a. Shut down generator set.

b. b. Open output box access door.

c. c. Inspect current transformer (7, Figure 2-33) for security, cracked housing, broken or stripped terminals, and loose or missing hardware.

d. Close output box access door.

2-109.2 Testing.

WARNING

When disconnecting or removing batteries, disconnect the negative lead that connects directly to the grounding stud first; disconnect the negative end of the interconnection cable next. When installing batteries, reverse the connection sequence. Failure to comply with this warning can cause injury to personnel.

a. a. Shut down generator set.

b. b. Open battery access door, disconnect negative battery cable, and open output box access door.

c. Tag and disconnect electrical leads from current transformer (7, Figure 2-33) secondary terminals.

d. d. Set multimeter for ohms and check for contin uity between secondary terminals A1 and A2, B1 and B2, and C1 and C2.

e. e. If continuity is not present, current transformer is defective. Notify next higher level of maintenance.

f. f. If continuity is present, connect electrical leads to secondary terminals. Remove tags.

g. g. Close output box access door, connect negativ e battery cable, and close battery access door.

2-110 DROOP CURRENT TRANSFORMER

Inspection.

WARNING

All metal jewelry can conduct electricity and become entangled in generator set components. Remove all jewelry when working on generator set. Failure to comply with this warning can cause injury or death to personnel.

WARNING

DO NOT wear loose clothing when performing checks, services and maintenance. Failure to comply with this warning can cause injury or death to personnel.

WARNING

Dangerous voltage exists on live circuits. Always observe precautions and never work alone. Failure to comply with this warning can cause injury or death to personnel.

WARNING

DC voltages are present at generator set electrical components even with generator set shut down. Avoid shorting any positive with ground/negative. Failure to comply with this warning can cause injury to personnel, and damage to equipment.

WARNING

Ensure nuts on ground terminals are properly secured creating a good ground. Failure to comply with this warning can cause injury or death to personnel.

WARNING

Shut down generator set before performing inspection of wiring. Failure

WARNING

The connection of any electrical equipment and the disconnection of any electrical equipment may cause an explosion hazard. Do not connect any electrical equipment or disconnect any electrical equipment in an explosive atmosphere. Failure to comply with this warning can cause injury or death to personnel.

a. a. Shut down generator set.

b. b. Open output box access door.

c. c. Inspect droop current transformer (8, Figure 2-33) for cracked housing, security, broken wire terminals, loose or missing hardware, and other damage.

d. Close output box access door.

2-111 POWER POTENTIAL TRANSFORMER

Inspection.

WARNING

All metal jewelry can conduct electricity and become entangled in generator set components. Remove all jewelry when working on generator set. Failure to comply with this warning can cause injury or death to personnel.

WARNING

DO NOT wear loose clothing when performing checks, services and maintenance. Failure to comply with this warning can cause injury or death to personnel.

WARNING

Dangerous voltage exists on live circuits. Always observe precautions and never work alone. Failure to comply with this warning can cause injury or death to personnel.

WARNING

DC voltages are present at generator set electrical components even with generator set shut down. Avoid shorting any positive with ground/negative. Failure to comply with this warning can cause injury to personnel, and damage to equipment.

WARNING

Ensure nuts on ground terminals are properly secured creating a good ground. Failure to comply with this warning can cause injury or death to personnel.

WARNING

Shut down generator set before performing inspection of wiring. Failure to comply with this warning can cause injury or death to personnel.

WARNING

The connection of any electrical equipment and the disconnection of any electrical equipment may cause an explosion hazard. Do not connect any electrical equipment or disconnect any electrical equipment in an explosive atmosphere. Failure to comply with this warning can cause injury or death to personnel.

a. a. Shut down generator set.

b. b. Open output box access door.

c. c. Inspect power potential transformer (9, Figure 2-33) for security, cracked housing, broken wire terminals, loose or missing hardware, and other damage.

d. Close output box access door.

2-112 AC CIRCUIT INTERRUPTER RELAY

2-112.1 Inspection.

WARNING

All metal jewelry can conduct electricity and become entangled in generator set components. Remove all jewelry when working on generator set. Failure to comply with this warning can cause injury or death to personnel.

WARNING

DO NOT wear loose clothing when performing checks, services and maintenance. Failure to comply with this warning can cause injury or death to personnel.

WARNING

Dangerous voltage exists on live circuits. Always observe precautions and never work alone. Failure to comply with this warning can cause injury or death to personnel.

WARNING

DC voltages are present at generator set electrical components even with generator set shut down. Avoid shorting any positive with ground/negative. Failure to comply with this warning can cause injury to personnel, and damage to equipment.

WARNING

Ensure nuts on ground terminals are properly secured creating a good ground. Failure to comply with this warning can cause injury or death to personnel.

WARNING

Shut down generator set before performing inspection of wiring. Failure

WARNING

The connection of any electrical equipment and the disconnection of any electrical equipment may cause an explosion hazard. Do not connect any electrical equipment or disconnect any electrical equipment in an explosive atmosphere. Failure to comply with this warning can cause injury or death to personnel.

a. a. Shut down generator set.

b. b. Open output box access door.

c. c. Inspect AC circuit interrupter relay (14, Figure 2-33) for security, cracked housing, broken wire terminals, and other damage.

d. Close output box access door.

2-112.2 Testing.

a. a. Shut down generator set.

b. b. Open battery access door, disconnect negative battery cable, and open output box access door.

c. c. Set multimeter for ohms and check for open circuits between terminals A1 and A2, B1 and B2, C1 and C2, and 11 and 12.

d. d. Connect jumper wire from cranking relay terminal A1 to AC circuit interrupter relay terminal X.

f. f. Check for closed circuits (continuity) between terminals A1 and A2, B1 and B2, C1 and C2, and 11 and 12.

g. g. Disconnect negative battery cable.

h. h. Replace AC circuit interrupter relay if indications are other than above.

i. i. Remove jumper wire if replacement is not needed.

j. j. Close output box access door , connect negative battery cable, and close battery access door.

2-112.3 Removal.

WARNING

When disconnecting or removing batteries, disconnect the negative lead that connects directly to the grounding stud first; disconnect the negative end of the interconnection cable next. When installing batteries, reverse the connection sequence. Failure to comply with this warning can cause injury to personnel.

a. a. Shut down generator set.

b. b. Open battery access door, disconnect negative battery cable, and open output box access door.

c. c. Remove screws (10, Figure 2-33) and cover (11) from AC circuit interrupter relay (14).

d. d. Tag and disconnect AC circuit interrupter relay (14) electrical leads.

e. e. Remove screws (13), nuts (12), and AC circuit interrupter relay (14) from output box.

2-112.4 Installation.

a. a. Install AC circuit interrupter relay (14, Figure 2-33) in output box with screws (13) and nuts (12).

b. b. Connect electrical leads. Remove tags.

c. c. Install cover (11) on AC circuit interrupter relay (14) with screws (10).

d. d. Close output box access door, connect negativ e battery cable, and close battery access door.

2-113 CRANKING RELAY 2-113.1

Inspection.

WARNING

All metal jewelry can conduct electricity and become entangled in generator set components. Remove all jewelry when working on generator set. Failure to comply with this warning can cause injury or death to personnel.

WARNING

DO NOT wear loose clothing when performing checks, services and maintenance. Failure to comply with this warning can cause injury or death to personnel.

WARNING

Dangerous voltage exists on live circuits. Always observe precautions and never work alone. Failure to comply with this warning can cause injury or death to personnel.

WARNING

DC voltages are present at generator set electrical components even with generator set shut down. Avoid shorting any positive with ground/negative. Failure to comply with this warning can cause injury to personnel, and damage to equipment.

WARNING

Ensure nuts on ground terminals are properly secured creating a good ground. Failure to comply with this warning can cause injury or death to personnel.

WARNING

Shut down generator set before performing inspection of wiring. Failure

WARNING

The connection of any electrical equipment and the disconnection of any electrical equipment may cause an explosion hazard. Do not connect any electrical equipment or disconnect any electrical equipment in an explosive atmosphere. Failure to comply with this warning can cause injury or death to personnel.

a. a. Shut down generator set.

b. b. Open output box and right side engine access doors.

c. c. Inspect cranking relay (17, Figure 2-33) fo r security, cracked housing, broken wire terminals, and other damage.

d. Close access doors.

2-113.2 Testing.

a. a. Shut down generator set.

b. b. Open battery access door, disconnect negative battery cable, and open output box access door.

c. c. Tag and disconnect wires from terminals X1, X2, and A2 of cranking relay (17, Figure 2-33).

d. d. Connect a jumper wire between terminals A1 and X1 of cranking relay.

e. e. Connect negative battery cable.

f. f. Connect X2 wire disconnected in step c above to cranking relay and listen for audible actuation.

g. g. Set multimeter for ohms and check for continuity between terminals A1 and A2 of cranking relay. If no continuity is indicated, cranking relay is defective and must be replaced.

h. h. Disconnect negative battery cable if replacement is not needed.

i. i. Remove jumper wire and connect remaining wires to cranking relay as tagged.

j. j. Close output box access do or, connect negative battery cable, and close battery access door.

2-113.3 Removal.

WARNING

When disconnecting or removing batteries, disconnect the negative lead that connects directly to the grounding stud first; disconnect the negative end of the interconnection cable next. When installing batteries, reverse the connection sequence. Failure to comply with this warning can cause injury to personnel.

a. a. Shut down generator set.

b. b. Open battery access door, disconnect negative battery cable, and open output box and right side engine access doors.

c. c. Tag and disconnect cranking relay (17, Figure 2-33) electrical leads.

d. d. Remove screws (16), nuts (15), and cranking relay (17) from output box.

2-113.4 Installation.

WARNING

When disconnecting or removing batteries, disconnect the negative lead that connects directly to the grounding stud first; disconnect the negative end of the interconnection cable next. When installing batteries, reverse the connection sequence. Failure to comply with this warning can cause injury to personnel.

a. a. Install cranking relay (17, Figure 2-33) with screws (16) and nuts (15).

b. b. Connect electrical leads. Remove tags.

c. c. Close output box and right side engine access doors, connect negative battery cable, and close battery access door.

2-114 LOAD OUTPUT TERMINAL BOARD

2-114.1 Removal.

WARNING

All metal jewelry can conduct electricity and become entangled in generator set components. Remove all jewelry when working on generator set. Failure to comply with this warning can cause injury or death to personnel.

WARNING

DO NOT wear loose clothing when performing checks, services and maintenance. Failure to comply with this warning can cause injury or death to personnel.

WARNING

Dangerous voltage exists on live circuits. Always observe precautions and never work alone. Failure to comply with this warning can cause injury or death to personnel.

WARNING

DC voltages are present at generator set electrical components even with generator set shut down. Avoid shorting any positive with ground/negative. Failure to comply with this warning can cause injury to personnel, and damage to equipment.

WARNING

Ensure nuts on ground terminals are properly secured creating a good ground. Failure to comply with this warning can cause injury or death to personnel.

WARNING

Shut down generator set before performing inspection of wiring. Failure

WARNING

The connection of any electrical equipment and the disconnection of any electrical equipment may cause an explosion hazard. Do not connect any electrical equipment or disconnect any electrical equipment in an explosive atmosphere. Failure to comply with this warning can cause injury or death to personnel.

a. a. Shut down generator set.

b. b. Open battery access door and disconnect negative battery cable, and open load terminal board

c. c. Disconnect load output cables L1, L2, L3, and L0 from load output terminal board.

d. d. Remove generator set housing rear panel (paragraph 2-18.1).

e. e. Remove bolt (1, Figure 2-34), nut (2), flat washer (3), and lock washer (4) securing ground strap (15) to skid base. Discard lock washer (4).

f. f. Remove bolts (5), lock washers (6), and washers (7) securing terminal board assembly (12) to supports. Discard lock washers (6)

g. g. Remove nuts (8) and washers (9). Tag and disconnect all main power leads (10) and varistor leads (11) from load terminals (23).

h. h. Remove terminal board assem bly (12) from generator set.

i. i. Remove load output termi nals (paragraph 2-115.1).

j. j. Remove EMI filter (24) positioned between L0 and GND terminals (23).

k. k. Remove varistor leads (11) from varistors (27).

l. l. Remove varistors (paragraph 2-116.1).

m. m. Remove EMI filters (28) positioned bet ween load terminals (23) and varistors (27).

n. n. Remove nuts (13), washers (14), ground strap (15), studs (16), bus bars (17 and 22), and ground plane bar (29) from terminal board (12).

o. o. Remove nuts (18) and washers (19) from studs (16).

p. p. Remove bolts (30), nuts (31), and load output terminal board supports (32 and 33) from generator set.

q. Remove nuts (34), bolts (35), cord (36), wren ch (37), and bracket (38) from support (33).

2-114.2 Inspection.

a. a. Shut down generator set.

b. b. Open load terminal board access door.

c. c. Inspect load output terminal board for cracks, corrosion, and obvious damage.

d. d. Inspect threaded components for stripped threads.

e. e. Inspect varistor electrical leads for damaged insulation and loose terminals.

f. f. Replace damaged and defective parts.

g. g. Close load terminal board access door.

2-114.3 Repair.

WARNING

When disconnecting or removing batteries, disconnect the negative lead that connects directly to the grounding stud first; disconnect the negative end of the interconnection cable next. When installing batteries, reverse the connection sequence. Failure to comply with this warning can cause injury to personnel.

Repair load output terminal board assembly by replacing damaged or defective wires, load terminals, EMI filters, and varistors.

2-114.4 Installation.

a. a. Install load output terminal board supports (32 and 33, Figure 2-34) in generator set with bolts (30) and nuts (31).

b. b. Install ground plane bar (29), bus bars (17 and 22), and ground strap (15) on terminal board (12) with studs (16), washers (14 and 19), and nuts (13 and 18).

c. c. Position EMI filters (28) between varistors (27) and L1, L2, and L3 load terminals (23) mounting holes.

d. d. Install varistors (paragraph 2-116.3).

e. e. Position EMI filter (24) between L0 and GND terminals (23) mounting holes.

f. f. Install load terminals (paragraph 2-115.3).

g. g. Connect leads (11) to varistors (27).

h. h. Position load output terminal board assembly in generator set and connect varistor leads (11) and main power leads (10) to load terminals (23) with washers (9) and nuts (8). Remove tags.

i. i. Secure terminal board assembly to supports with washers (7), new lock washers (6), and bolts (5).

j. j. Apply a thin coat of antiseize compound (Item 6, Appendix C) to skid at ground strap (15) attaching point.

k. k. Install bolt (1), flat washer (3), new lock washer (4), and nut (2) securing ground strap (15) to skid base.

l. l. Install bracket (38), wrench (37), and cord (36) on support (33) with bolts (35) and nuts (34).

m. m. Install generator set housing rear panel (paragraph 2-18.4).

n. n. Connect load output cables L1, L2, L3, and L0 at load output terminal board and close load terminal access door.

o. Connect negative battery c able and close battery access door.

2-115 LOAD OUTPUT TERMINALS

2-115.1 Removal.

WARNING

All metal jewelry can conduct electricity and become entangled in generator set components. Remove all jewelry when working on generator set. Failure to comply with this warning can cause injury or death to personnel.

WARNING

DO NOT wear loose clothing when performing checks, services and maintenance. Failure to comply with this warning can cause injury or death to personnel.

WARNING

Dangerous voltage exists on live circuits. Always observe precautions and never work alone. Failure to comply with this warning can cause injury or death to personnel.

WARNING

DC voltages are present at generator set electrical components even with generator set shut down. Avoid shorting any positive with ground/negative. Failure to comply with this warning can cause injury to personnel, and damage to equipment.

WARNING

Ensure nuts on ground terminals are properly secured creating a good ground. Failure to comply with this warning can cause injury or death to personnel.

WARNING

Shut down generator set before performing inspection of wiring. Failure

WARNING

The connection of any electrical equipment and the disconnection of any electrical equipment may cause an explosion hazard. Do not connect any electrical equipment or disconnect any electrical equipment in an explosive atmosphere. Failure to comply with this warning can cause injury or death to personnel.

a. a. Shut down generator set.

b. b. Open battery access door and disconnect negative battery cable.

c. c. Open load terminal board access door and disconnect load cables.

d. d. Remove load output terminal board assembly from generator set (paragraph 2-114.1).

e. e. Remove nuts (20, Figure 2-34), copper washers (21), and load terminals (23) from terminal board assembly (12).

---

a. a. Shut down generator set.

b. b. Open load terminal board access door.

c. c. Inspect load terminals for stripped threads and other obvious damage.

d. d. Replace damaged load terminals, as necessary.

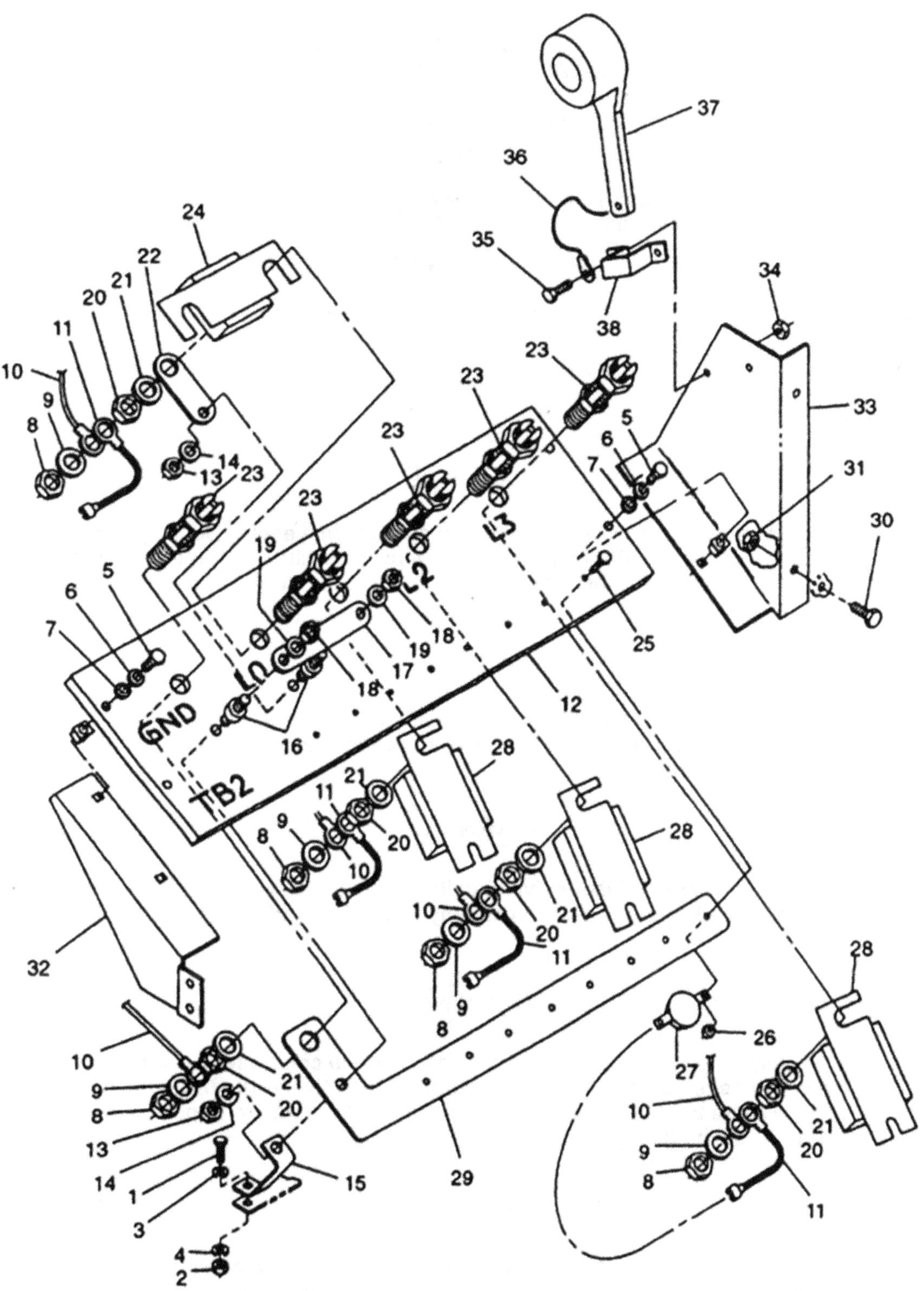

Figure 2-34. Load Terminal Board Assembly.

2-115.3 Installation.

a. a. Install load output terminals (23, Figure 2-34) on load output terminal board assembly (12) with copper washers (21) and nuts (20).

b. b. Install load output terminal board assembly in generator set (paragraph 2-114.4).

NOTE

Ensure GND load terminal passes through ground plane bracket (29) and L0 load terminal passes through bus bar (22).

c. c. Connect load cables and close load terminal board access door.

d. d. Connect negative battery cabl e and close battery access door.

2-116 VARISTORS

2-116.1 Removal.

WARNING

All metal jewelry can conduct electricity and become entangled in generator set components. Remove all jewelry when working on generator set. Failure to comply with this warning can cause injury or death to personnel.

WARNING

DO NOT wear loose clothing when performing checks, services and maintenance. Failure to comply with this warning can cause injury or death to personnel.

WARNING

Dangerous voltage exists on live circuits. Always observe precautions and never work alone. Failure to comply with this warning can cause injury or death to personnel.

WARNING

DC voltages are present at generator set electrical components even with generator set shut down. Avoid shorting any positive with ground/negative. Failure to comply with this warning can cause injury to personnel, and damage to equipment.

WARNING

Ensure nuts on ground terminals are properly secured creating a good ground. Failure to comply with this warning can cause injury or death to personnel.

WARNING

Shut down generator set before performing inspection of wiring. Failure

WARNING

The connection of any electrical equipment and the disconnection of any electrical equipment may cause an explosion hazard. Do not connect any electrical equipment or disconnect any electrical equipment in an explosive atmosphere. Failure to comply with this warning can cause injury or death to personnel.

a. a. Shut down generator set.

b. b. Open battery access door and disconnect negative battery cable.

c. c. Remove load output terminal board assemb ly from generator set (paragraph 2-114.1).

d. d. Tag and disconnect varistor leads (11, Figure 2-34) from varistors (27).

e. e. Remove nuts (26), bolts (25), and varistors (27) from load terminal board assembly (12).

2-116.2 Inspection and Testing.

a. a. Shut down generator set.

b. b. Remove varistors (paragraph 2-116.1).

c. c. Inspect varistors for obvious external damage.

d. d. Set multimeter for ohms and test each varistor by connecting multimeter to varistor terminals 1 and 2. Note multimeter indication.

e. e. Reverse multimeter leads and note multimeter indication.

f. f. Multimeter indications should be infinite ohms in both directions.

g. g. Varistors are defective and must be replaced if indications are other than above.

h. h. Install varistors (paragraph 2-116.3).

2-116.3 Installation.

a. a. Install varistors (27, Figure 2-34) on load output terminal board assembly (12) with bolts (25) and nuts (26).

b. b. Connect varistor leads (11) to varistors (27). Remove tags.

c. c. Install load output terminal board assembly in generator set (paragraph 2-114.4).

d. d. Connect negative battery c able and close battery access door.

Section XV. MAINTENANCE OF ENGINE ACCESSORIES

2-117 LOW OIL PRESSURE SWITCH

2-117.1 Testing.

WARNING

All metal jewelry can conduct electricity and become entangled in generator set components. Remove all jewelry when working on generator set. Failure to comply with this warning can cause injury or death to personnel.

WARNING

DO NOT wear loose clothing when performing checks, services and maintenance. Failure to comply with this warning can cause injury or death to personnel.

WARNING

Dangerous voltage exists on live circuits. Always observe precautions and never work alone. Failure to comply with this warning can cause injury or death to personnel.

WARNING

DC voltages are present at generator set electrical components even with generator set shut down. Avoid shorting any positive with ground/negative. Failure to comply with this warning can cause injury to personnel, and damage to equipment.

WARNING

Shut down generator set before performing inspection of wiring. Failure

WARNING

The connection of any electrical equipment and the disconnection of any electrical equipment may cause an explosion hazard. Do not connect any electrical equipment or disconnect any electrical equipment in an explosive atmosphere. Failure to comply with this warning can cause injury or death to personnel.

a. a. Shut down generator set.

b. b. Open battery access door, disconnect negative battery cable, and open right side engine access
door.

c. c. Tag and disconnect electrical leads from low oil pressure switch (1, Figure 2-35/Figure 2-36).

d. d. Set multimeter for ohms and connect across switch connector pins C and NO. Multimeter shall

e. Connect multimeter across switch connector pins C and NC. Multimeter shall indicate continuity.

f. f. Connect negative battery cable.

g. g. Start generator set. Place BATTLE SHORT switch in ON position before releasing MASTER SWITCH from START position.

h. Connect multimeter to switch connector pins C and NC. Multimeter shall indicate open circuit.

i. Connect multimeter across switch connector pins C and NO. Multimeter shall indicate continuity.

j. j. Shut down generator set. Return BATTLE SHORT switch to OFF position.

k. k. Disconnect negative battery cable.

l. l. If switch fails to meet continuity requirements, replace low oil pressure switch.

m. m. If replacement is not needed, connect electrical leads to low oil pressure switch. Remove tags.

n. n. Close right side engine access door.

o. o. Connect negative battery c able and close battery access door.

2-117.2 Removal.

WARNING

The high pressure oil system operates at high temperature and pressure. Contact with hot oil can result in burns and scalding. Shut down generator set, and allow system to cool before performing checks, services, and maintenance. Wear heat resistant gloves and avoid contacting hot surfaces. Do not allow hot oil or components to contact skin or hands. Failure to comply with this warning can cause injury or death to personnel.

a. a. Shut down generator set.

b. b. Open battery access door, disconnect negative battery cable, and open right side engine access

door.

c. Tag and disconnect low oil pressure switch (1, Figure 2-35/Figure 2-36) electrical leads.

d. d. Unscrew low oil pressure switch (1) from oil sample valve assembly.

2-117.3 Cleaning and Inspection.

a. a. Shut down generator set.

b. b. Remove low oil pressure switch (paragraph 2-117.2).

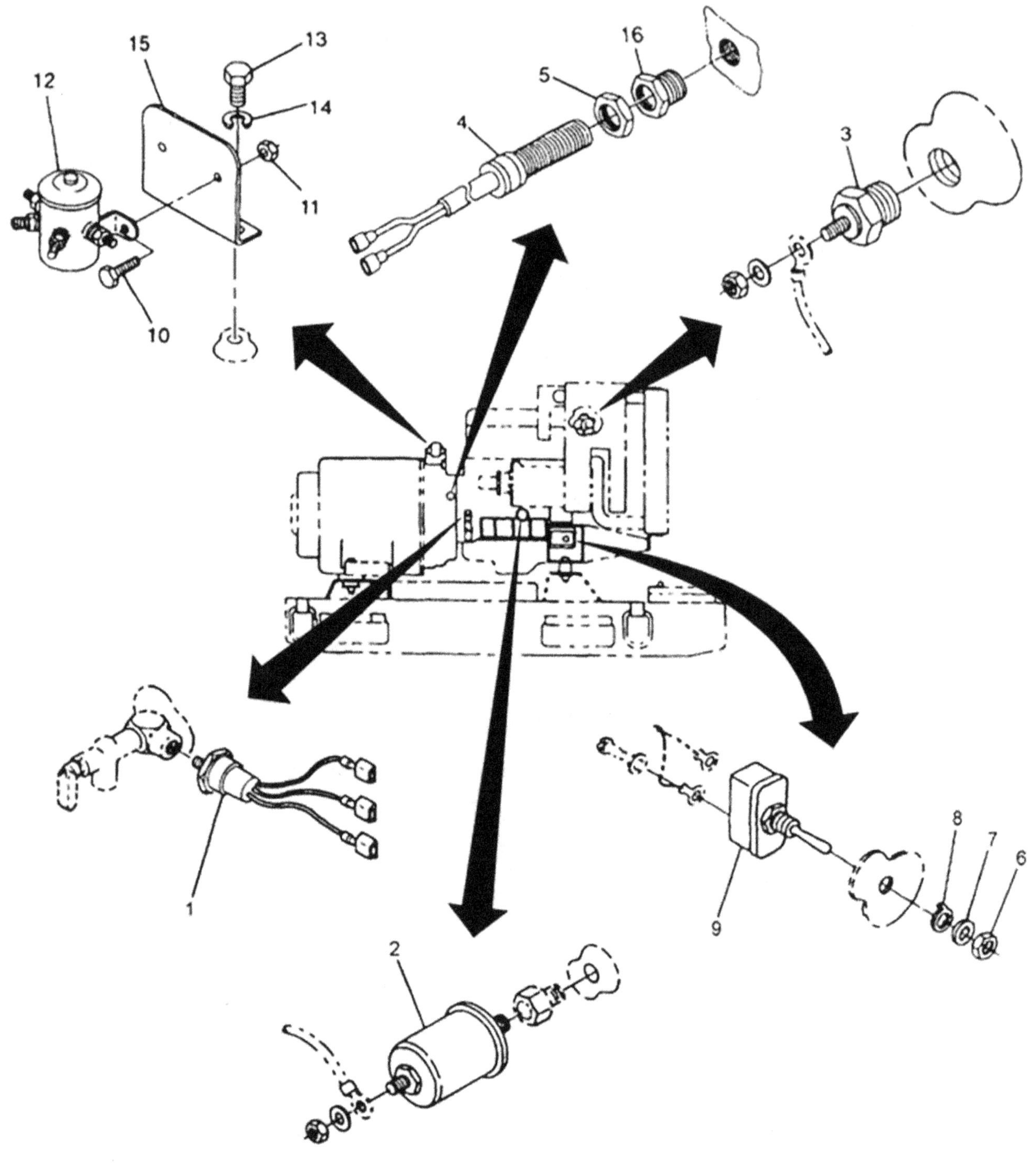

Figure 2-35. Right Side Engine Switches and Senders – MEP-804A/MEP-814A.

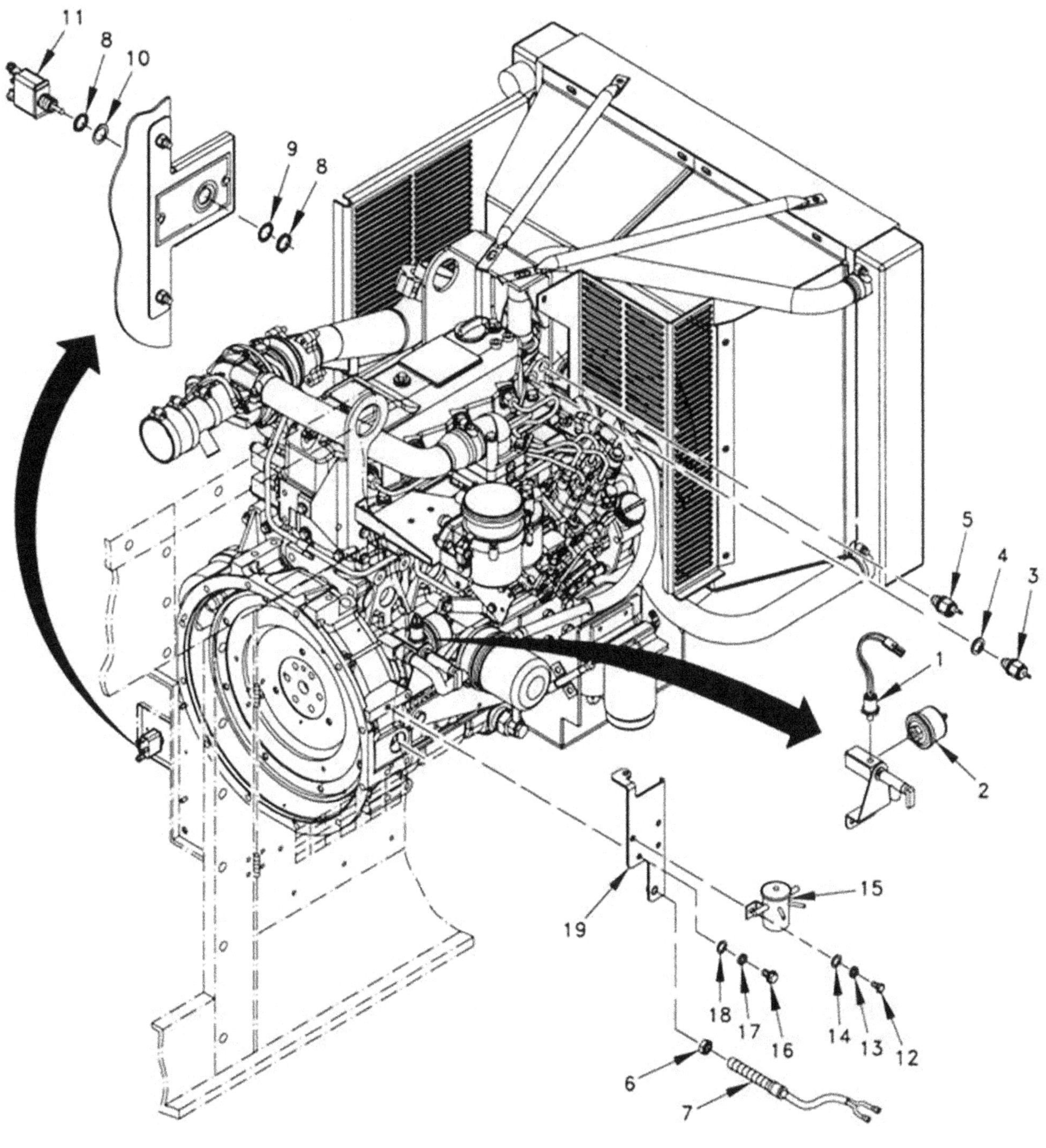

Figure 2-36. Right Side Engine Switches and Senders – MEP-804B/MEP-814B.

c. Clean low oil pressure switch with dry, filtered compressed air and wipe with a cleaning cloth (Item 7, Appendix C) lightly moistened with dry cleaning solvent (Item 20, Appendix C).

WARNING

Cleaning solvent is flammable and toxic to eyes, skin, and respiratory tract. Skin and eye protection are required when working in contact with cleaning solvent. Avoid repeated or prolonged contact. Work in ventilated area only. Failure to comply with this warning can cause injury or death to personnel.

WARNING

Solvent used to clean parts is potentially dangerous to personnel and property. Clean parts in a well-ventilated area. Avoid inhalation of solvent fumes. Wear goggles and rubber gloves to protect eyes and skin. Wash exposed skin thoroughly. Do not smoke or use near open flames or excessive heat. Failure to comply with this warning can cause injury to personnel, and damage to the equipment.

WARNING

Eye protection is required when working with compressed air. Compressed air can propel particles at high velocity and injure eyes. Do not exceed 15 psi pressure when using compressed air. Failure to comply with this warning can cause injury to personnel.

d. d. Inspect low oil pressure switch for cracked casing, stripped or damaged threads, corrosion, or other damage.

e. e. If no repair is needed, install low oil pressure switch (paragraph 2-117.4).

2-117.4 Installation.

a. a. Install low oil pressure switch (1, Figure 2-35/Figure 2-36) into oil sample valve assembly.

b. b. Connect electrical leads. Remove tags.

c. c. Close right side engine access door, connect negative battery cable, and close battery access door.

2-118 OIL PRESSURE SENDER

2-118.1 Testing.

WARNING

All metal jewelry can conduct electricity and become entangled in generator set components. Remove all jewelry when working on generator set. Failure to comply with this warning can cause injury or death to personnel.

WARNING

DO NOT wear loose clothing when performing checks, services and maintenance. Failure to comply with this warning can cause injury or death to personnel.

WARNING

Dangerous voltage exists on live circuits. Always observe precautions and never work alone. Failure to comply with this warning can cause injury or death to personnel.

WARNING

DC voltages are present at generator set electrical components even with generator set shut down. Avoid shorting any positive with ground/negative. Failure to comply with this warning can cause injury to personnel, and damage to equipment.

WARNING

Shut down generator set before performing inspection of wiring. Failure to comply with this warning can cause injury or death to personnel.

WARNING

The connection of any electrical equipment and the disconnection of any electrical equipment may cause an explosion hazard. Do not connect any electrical equipment or disconnect any electrical equipment in an explosive atmosphere. Failure to comply with this warning can cause injury or death to personnel.

a. a. Shut down generator set.

b. b. Open battery access door, disconnect negative battery cable, and open right side engine access door.

c. c. Disconnect electrical lead to oil pressure sender (2, Figure 2-35/Figure 2-36).

d. d. Set multimeter for ohms and connect between sender terminal and casing.

e. e. Multimeter indication shall be between 216 and 264 ohms.

f. f. Connect negative battery cable.

g. g. Start generator set.

h. h. As engine is cranking and accelerates to rated speed, observe multimeter. Indication shall decrease to between 100 and 33 ohms.

i. i. Shut down generator set and disconnect negative battery cable.

j. j. Replace oil pressure sender if indications are not as above.

k. k. If replacement is not needed, connect electrical lead to oil pressure sender.

l. l. Close right side engine access door, connect negative battery cable, and close battery access door.

2-118.2 Removal.

WARNING

The high pressure oil system operates at high temperature and pressure. Contact with hot oil can result in burns and scalding. Shut down generator set, and allow system to cool before performing checks, services, and maintenance. Wear heat resistant gloves and avoid contacting hot surfaces. Do not allow hot oil or components to contact skin or hands. Failure to comply with this warning can cause injury or death to personnel.

a. a. Shut down generator set.

b. b. Open battery access door, disconnect negative battery cable, and open right side engine access door.

c. Disconnect oil pressure sender (2, Figure 2-35/Figure 2-36) electrical lead.

d. d. Unscrew and remove oil pressure sender (2).

2-118.3 Cleaning and Inspection.

a. a. Shut down generator set.

b. b. Remove oil pressure sender (paragraph 2-118.2).

c. c. Clean oil pressure sender with dry, filtered compressed air and wipe with a cleaning cloth (Item 7, Appendix C) lightly moistened with dry cleaning solvent (Item 20, Appendix C).

WARNING

Cleaning solvent is flammable and toxic to eyes, skin, and respiratory tract. Skin and eye protection are required when working in contact with cleaning solvent. Avoid repeated or prolonged contact. Work in ventilated area only. Failure to comply with this warning can cause injury or death to personnel.

WARNING

Solvent used to clean parts is potentially dangerous to personnel and property. Clean parts in a well-ventilated area. Avoid inhalation of solvent fumes. Wear goggles and rubber gloves to protect eyes and skin. Wash exposed skin thoroughly. Do not smoke or use near open flames or excessive heat. Failure to comply with this warning can cause injury to personnel, and damage to the equipment.

WARNING

Eye protection is required when working with compressed air. Compressed air can propel particles at high velocity and injure eyes. Do not exceed 15 psi pressure when using compressed air. Failure to comply with this warning can cause injury to personnel.

d. d. Inspect oil pressure sender for cracked casing, stripped or damaged threads, corrosion, and other visible damage.

e. e. If no repair is needed, install oil pressure sender, paragraph 2-118.4.

2-118.4 Installation.

a. a. Install oil pressure sender (2, Figure 2-35/Figure 2-36) into engine block.

b. b. Connect electrical lead and close right side engine access door.

c. c. Connect negative battery c able and close battery access door.

2-119 COOLANT TEMPERATURE SENDER

2-119.1 Testing.

WARNING

All metal jewelry can conduct electricity and become entangled in generator set components. Remove all jewelry when working on generator set. Failure to comply with this warning can cause injury or death to personnel.

WARNING

DO NOT wear loose clothing when performing checks, services and maintenance. Failure to comply with this warning can cause injury or death to personnel.

WARNING

Dangerous voltage exists on live circuits. Always observe precautions and never work alone. Failure to comply with this warning can cause injury or death to personnel.

WARNING

DC voltages are present at generator set electrical components even with generator set shut down. Avoid shorting any positive with ground/negative. Failure to comply with this warning can cause injury to personnel, and damage to equipment.

WARNING

Shut down generator set before performing inspection of wiring. Failure to comply with this warning can cause injury or death to personnel.

WARNING

The connection of any electrical equipment and the disconnection of any electrical equipment may cause an explosion hazard. Do not connect any electrical equipment or disconnect any electrical equipment in an explosive atmosphere. Failure to comply with this warning can cause injury or death to personnel.

a. a. Shut down generator set.

b. b. Open battery access door, disconnect negative battery cable, and open right side engine access door.

c. c. Disconnect electrical lead from coolant temperature sender (3, Figure 2-35/Figure 2-36).

d. d. Set multimeter for ohms and connect positive lead to temperature sender terminal and negative lead to case. Multimeter indication shall be greater than 300 ohms.

e. e. Connect negative battery cable and start generator set.

f. f. Allow the engine to operate while observing multimeter.

g. g. Ohms indication should decrease as temperature rises.

h. h. Shut down generator set and disconnect negative battery cable.

i Replace coolant temperature sender if indications are not as above.

j. j. If replacement is not needed, connect electrical lead to sender.

k. k. Close right side engine access door.

l. l. Connect negative battery c able and close battery access door.

2-119.2 Removal.

a. a. Shut down generator set.

b. b. Open battery access door and disconnect negative battery cable.

WARNING

Cooling system operates at high temperature and pressure. Contact with high pressure steam and/or liquids can result in burns and scalding. Shut down generator set, and allow system to cool before performing checks, services and maintenance, or wear gloves and additional protective clothing and goggles as required. Failure to comply with this warning can cause injury or death to personnel.

WARNING

Always remove radiator cap slowly to permit any pressure to escape. Failure to comply with this warning can cause injury to personnel.

c. c. Slowly remove radiator cap (1, Figure 2-23).

d. d. For MEP-804A/MEP-814A, perform step (1) below; for MEP-804B/MEP-814B, perform step (2) below:

(1) Open left side engine access door, open engine block drain valve (1, Figure 2-37), and drain coolant into suitable container. Close drain valve.

(2) Open left side engine access door, open radiator drain valve (3), and drain coolant into suitable container. Close drain valve.

e. e. Open right side engine access door and disconnect coolant temperature sender (3, Figure 2-35/Figure 2-36) electrical lead.

f. f. Unscrew temperature sender (3) from engine head.

g. g. MEP-804B/MEP-814B: Remove copper washer (4, Figure 2-36).

2-119.3 Cleaning and Inspection.

a. a. Shut down generator set.

b. b. Remove coolant temperature sender (paragraph 2-119.2).

c. c. Clean temperature sender with dry, filtered compressed air and cleaning cloth (Item 7, Appendix C) lightly moistened with dry cleaning solvent (Item 20, Appendix C).

WARNING

Cleaning solvent is flammable and toxic to eyes, skin, and respiratory tract. Skin and eye protection are required when working in contact with cleaning solvent. Avoid repeated or prolonged contact. Work in ventilated area only. Failure to comply with this warning can cause injury or death to personnel.

WARNING

Solvent used to clean parts is potentially dangerous to personnel and property. Clean parts in a well-ventilated area. Avoid inhalation of solvent fumes. Wear goggles and rubber gloves to protect eyes and skin. Wash exposed skin thoroughly. Do not smoke or use near open flames or excessive heat. Failure to comply with this warning can cause injury to personnel, and damage to the equipment.

WARNING

Eye protection is required when working with compressed air. Compressed air can propel particles at high velocity and injure eyes. Do not exceed 15 psi pressure when using compressed air. Failure to comply with this warning can cause injury to personnel.

d. d. Inspect temperature sender for cracked casing, corrosion, and damaged threads and connector.

f. If no repair is needed, install coolant te mperature sender (paragraph 2-119.4).

2-119.4 Installation.

a. a. Install coolant temperature sender (3, Figure 2-35/Figure 2-36) and copper washer (4, Figure 2-36) (MEP-804B/MEP-814B only) in engine head.

b. b. Connect electrical lead and close right and left side engine access doors.

c. c. Add coolant to overflow bottle (5, Figure 2-28), as necessary, to replace drained coolant.

d. d. Connect negative battery c able and close battery access door.

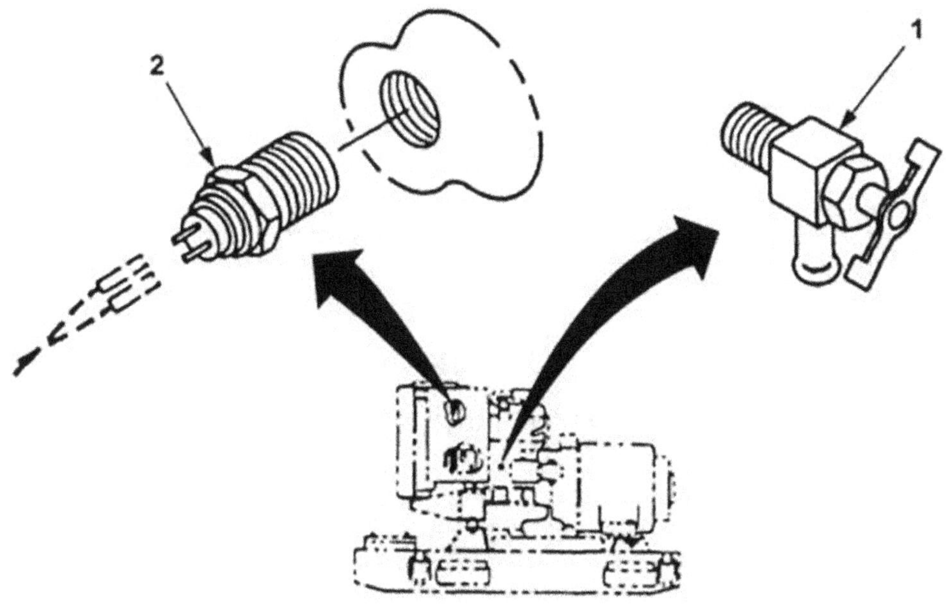

MEP-804A/MEP-814A

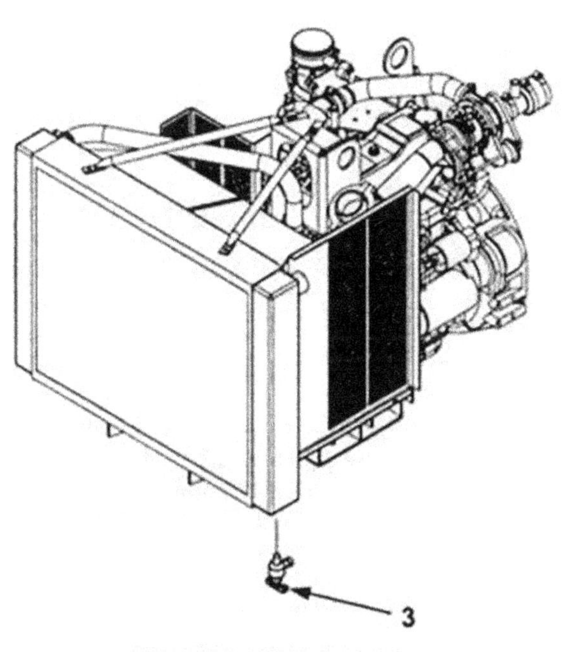

MEP-804B/MEP-814B

Figure 2-37. Left Side Engine Switches and Senders.

2-120 COOLANT HIGH TEMPERATURE SWITCH

2-120.1 Testing.

WARNING

All metal jewelry can conduct electricity and become entangled in generator set components. Remove all jewelry when working on generator set. Failure to comply with this warning can cause injury or death to personnel.

WARNING

DO NOT wear loose clothing when performing checks, services and maintenance. Failure to comply with this warning can cause injury or death to personnel.

WARNING

Dangerous voltage exists on live circuits. Always observe precautions and never work alone. Failure to comply with this warning can cause injury or death to personnel.

WARNING

DC voltages are present at generator set electrical components even with generator set shut down. Avoid shorting any positive with ground/negative. Failure to comply with this warning can cause injury to personnel, and damage to equipment.

WARNING

Shut down generator set before performing inspection of wiring. Failure

WARNING

The connection of any electrical equipment and the disconnection of any electrical equipment may cause an explosion hazard. Do not connect any electrical equipment or disconnect any electrical equipment in an explosive atmosphere. Failure to comply with this warning can cause injury or death to personnel.

a. a. Shut down generator set.

b. b. Remove coolant high temperature switch (paragraph 2-120.2).

c. c. Suspend high temperature switch in a container of 50/50 mixture of antifreeze and water so that sensing element is completely immersed but not touching sides or bottom of container.

d. d. Suspend a reliable thermometer in container. Do not allow end of thermometer to rest on bottom of container.

e. e. Set multimeter for ohms and check for continuity between switch terminals. Switch operates under open conditions.

f. Gradually heat antifreeze/water mixture, stirring so that heat will be evenly distributed and observe thermometer and multimeter.

g. g. At between 216.5°F and 223.5°F (102.5°C an d 106°C), multimeter should indicate continuity.

h. h. Replace high temperature switch if it fails to operate as above.

i. i. If replacement is not needed, install coolant high temperature switch (paragraph 2-120.4).

2-120.2 Removal.

a. a. Shut down generator set.

b. b. Open battery access door and disconnect negative battery cable.

WARNING

Cooling system operates at high temperature and pressure. Contact with high pressure steam and/or liquids can result in burns and scalding. Shut down generator set, and allow system to cool before performing checks, services and maintenance, or wear gloves and additional protective clothing and goggles as required. Failure to comply with this warning can cause injury or death to personnel.

WARNING

Always remove radiator cap slowly to permit any pressure to escape. Failure to comply with this warning can cause injury to personnel.

c. c. Slowly remove radiator cap (1, Figure 2-23).

d. d. For MEP-804A/MEP-814A, perform steps (1) and (2) below; for MEP-804B/MEP-814B), perform

step (3) below:

(1) Open left side engine access door, open engine block drain valve (1, Figure 2-37), and drain coolant into suitable container. Close drain valve.

(2) Remove left side fan guard (paragraph 2-81.2).

(3) Open left side engine access door, open radiator drain valve (3, Figure 2-37), and drain coolant into suitable container. Close drain valve.

e. Tag and disconnect coolant high temperature switch (2, Figure 2-37/5, Figure 2-36) electrical leads.

f. f. Unscrew high temperature switch (2, Figure 2-37/5, Figure 2-36) from thermostat housing.

2-120.3 Cleaning and Inspection.

a. a. Shut down generator set.

b. b. Remove coolant high temperature switch (paragraph 2-120.2).

WARNING

Cleaning solvent is flammable and toxic to eyes, skin, and respiratory tract. Skin and eye protection are required when working in contact with cleaning solvent. Avoid repeated or prolonged contact. Work in ventilated area only. Failure to comply with this warning can cause injury or death to personnel.

WARNING

Solvent used to clean parts is potentially dangerous to personnel and property. Clean parts in a well-ventilated area. Avoid inhalation of solvent fumes. Wear goggles and rubber gloves to protect eyes and skin. Wash exposed skin thoroughly. Do not smoke or use near open flames or excessive heat. Failure to comply with this warning can cause injury to personnel, and damage to the equipment.

WARNING

Eye protection is required when working with compressed air. Compressed air can propel particles at high velocity and injure eyes. Do not exceed 15 psi pressure when using compressed air. Failure to comply with this warning can cause injury to personnel.

c. c. Clean high temperature switch with dry, filtered compressed air and cleaning cloth (Item 7, Appendix C) lightly moistened with dry cleaning solvent (Item 20, Appendix C).

d. d. Inspect high temperature switch for cracked casing, corrosion, stripped or damaged threads, and bent or broken connector pins.

e. e. If no repair is needed, install coolant high temperature switch (paragraph 2-120.4).

2-120.4 Installation.

a. a. Install coolant high temperature switch (2, Figure 2-37/5, Figure 2-36) in thermostat housing.

b. b. Connect electrical leads. Remove tags.

c. c. Add coolant to overflow bottle (5, Figure 2- 28), as necessary, to replace drained coolant.

d. d. MEP-804A/MEP-814A: Install left side fan guard (paragraph 2-81.3).

e. e. Close left side engine acce ss door, connect negative battery cable, and close battery access door.

2-121 MAGNETIC PICKUP

2-121.1 Removal.

WARNING

All metal jewelry can conduct electricity and become entangled in generator set components. Remove all jewelry when working on generator set. Failure to comply with this warning can cause injury or death to personnel.

WARNING

DO NOT wear loose clothing when performing checks, services and maintenance. Failure to comply with this warning can cause injury or death to personnel.

WARNING

Dangerous voltage exists on live circuits. Always observe precautions and never work alone. Failure to comply with this warning can cause injury or death to personnel.

WARNING

DC voltages are present at generator set electrical components even with generator set shut down. Avoid shorting any positive with ground/negative. Failure to comply with this warning can cause injury to personnel, and damage to equipment.

WARNING

Shut down generator set before performing inspection of wiring. Failure to comply with this warning can cause injury or death to personnel.

WARNING

The connection of any electrical equipment and the disconnection of any electrical equipment may cause an explosion hazard. Do not connect any electrical equipment or disconnect any electrical equipment in an explosive atmosphere. Failure to comply with this warning can cause injury or death to personnel.

a. a. Shut down generator set.

b. b. Open battery access door, disconnect negative battery cable, and open right side engine access door.

c. c. Tag and disconnect magnetic pickup (4, Figure 2-35/7, Figure 2-36) electrical leads.

d. d. For MEP-804A/MEP-814A, perform step (1) below; for MEP-804B/MEP-814B, perform step (2) below:

(1) Loosen locknut (5, Figure 2-35) and remove magnetic pickup (4) from flywheel housing. Remove bushing (16).

(2) Loosen locknut (6, Figure 2-36) and remove magnetic pickup (7) from flywheel housing.

2-121.2 Cleaning and Inspection.

a. a. Shut down generator set.

b. b. Remove magnetic pickup (paragraph 2-121.1).

WARNING

Cleaning solvent is flammable and toxic to eyes, skin, and respiratory tract. Skin and eye protection are required when working in contact with cleaning solvent. Avoid repeated or prolonged contact. Work in ventilated area only. Failure to comply with this warning can cause injury or death to personnel.

WARNING

Solvent used to clean parts is potentially dangerous to personnel and property. Clean parts in a well-ventilated area. Avoid inhalation of solvent fumes. Wear goggles and rubber gloves to protect eyes and skin. Wash exposed skin thoroughly. Do not smoke or use near open flames or excessive heat. Failure to comply with this warning can cause injury to personnel, and damage to the equipment.

WARNING

Eye protection is required when working with compressed air. Compressed air can propel particles at high velocity and injure eyes. Do not exceed 15 psi pressure when using compressed air. Failure to comply with this warning can cause injury to personnel.

c. c. Clean magnetic pickup with dry, filtered compressed air and wipe with a cleaning cloth (Item 7, Appendix C) lightly moistened with dry cleaning solvent (Item 20, Appendix C).

d. d. Inspect magnetic pickup for cracked casing, stripped or damaged threads, corrosion, or other visible damage.

e. e. If no repair is needed, install magnetic pickup (paragraph 2-121.3).

2-121.3 Installation.

a. a. Install bushing (16, Figure 2-35) into flywheel housing.

b. b. Screw magnetic pickup (4, Figure 2-35/7, Figure 2-36) into flywheel housing until pickup bottoms out on top surface of gear tooth on flywheel. Back magnetic pickup out one turn and tighten locknut (5, Figure 2-35/6, Figure 2-36).

c. c. Connect electrical leads. Remove tags.

d. d. Connect negative battery cabl e and close battery access door.

e. e. Adjust magnetic pickup (paragraph 2-121.4).

f. f. Close right side access door.

2-121.4 Adjustment.

a. a. Release control panel by turning two fasteners and lower control panel slowly.

b. b. Disconnect wire 147C from terminal 16 and wire 148C from terminal 17 of governor control unit.

c. c. Set multimeter for ohms and connect to ends of disconnected wires 147C and 148C. Multimeter should indicate between 800 and 900 ohms.

d. d. Leave multimeter connected to wires 147C and 148C and set multimeter for AC volts.

e. e. Crank engine with DEAD CRANK switch and observe multimeter. Multimeter indication should be between 2.0 and 3.0 VAC.

CAUTION

Do not adjust magnetic pickup inward more than one eighth turn each time or damage to magnetic pickup may result.

f. f. To adjust output voltage in step e above, loosen jam nut and turn magnetic pickup in no more than one-eighth turn at a time to increase or decrease output voltage. Tighten jam nut.

g. g. Repeat steps e and f above until proper output voltage is achieved.

h. h. Remove multimeter and connect wires to governor control unit.

i. i. Raise and secure control panel.

2-122 DEAD CRANK SWITCH

2-122.1 Testing.

WARNING

All metal jewelry can conduct electricity and become entangled in generator set components. Remove all jewelry when working on generator set. Failure to comply with this warning can cause injury or death to personnel.

WARNING

DO NOT wear loose clothing when performing checks, services and maintenance. Failure to comply with this warning can cause injury or death to personnel.

WARNING

Dangerous voltage exists on live circuits. Always observe precautions and never work alone. Failure to comply with this warning can cause injury or death to personnel.

WARNING

DC voltages are present at generator set electrical components even with generator set shut down. Avoid shorting any positive with ground/negative. Failure to comply with this warning can cause injury to personnel, and damage to equipment.

WARNING

Shut down generator set before performing inspection of wiring. Failure to comply with this warning can cause injury or death to personnel.

WARNING

The connection of any electrical equipment and the disconnection of any electrical equipment may cause an explosion hazard. Do not connect any electrical equipment or disconnect any electrical equipment in an explosive atmosphere. Failure to comply with this warning can cause injury or death to personnel.

a. a. Shut down generator set.

b. b. Open battery access door, disconnect negative battery cable, and open right side engine access

door.

c. c. Tag and disconnect electrical leads from DEAD CRANK switch (9, Figure 2-35/11, Figure 2-36).

d. d. Set multimeter for ohms and, with switch in NORMAL position, check for continuity between contacts 2 and 3.

e. e. Move switch to CRANK position and check for continuity between contacts 1 and 2.

f. f. If DEAD CRANK switch fails continuity checks, replace switch.

g. g. If replacement is not needed, connect electrical leads to switch. Remove tags.

h. h. Close right side engine access door.

i. i. Connect negative battery c able and close battery access door.

2-122.2 Removal.

WARNING

When disconnecting or removing batteries, disconnect the negative lead that connects directly to the grounding stud first; disconnect the negative end of the interconnection cable next. When installing batteries, reverse the connection sequence. Failure to comply with this warning can cause injury to personnel.

a. a. Shut down generator set.

b. b. Open battery access door and disconnect negative battery cable.

c. c. For MEP-804A/MEP-814A, perform step (1) below; for MEP-804B/MEP-814B, perform step (2) below:

(1) Open right side engine access door and remove nut (6, Figure 2-35), lock washer (7), tab washer (8), and DEAD CRANK switch (9). Discard lock washer (7).

(2) Open right side engine access door and remove nut (8, Figure 2-36), lock washer (9), tab washer (10), and DEAD CRANK switch (11). Discard lock washer (9).

d. Tag and disconnect DEAD CRANK switch electrical leads.

2-122.3 Installation.

a. a. Connect electrical leads to DEAD CRANK switch (9, Figure 2-35/11, Figure 2-36). Remove tags.

b. b. For MEP-804A/MEP-814A, perform step (1) below; for MEP-804B/MEP-814B, perform step (2) below:

(1) Install DEAD CRANK switch (9, Figure 2-35), with tab washer (8), new lock washer (7), and nut (6).

(2) Install DEAD CRANK switch (11, Figure 2-36) with tab washer (10), new lock washer (9), and nut (8).

c. c. Connect negative battery cable and close battery access door and right side engine access door.

2-123 GLOW PLUGS DC CONTACTOR (MEP-804A/MEP-814A)

2-123.1 Removal.

WARNING

All metal jewelry can conduct electricity and become entangled in generator set components. Remove all jewelry when working on generator set. Failure to comply with this warning can cause injury or death to personnel.

WARNING

DO NOT wear loose clothing when performing checks, services and maintenance. Failure to comply with this warning can cause injury or death to personnel.

WARNING

Dangerous voltage exists on live circuits. Always observe precautions and never work alone. Failure to comply with this warning can cause injury or death to personnel.

WARNING

DC voltages are present at generator set electrical components even with generator set shut down. Avoid shorting any positive with ground/negative. Failure to comply with this warning can cause injury to personnel, and damage to equipment.

WARNING

Shut down generator set before performing inspection of wiring. Failure

WARNING

The connection of any electrical equipment and the disconnection of any electrical equipment may cause an explosion hazard. Do not connect any electrical equipment or disconnect any electrical equipment in an explosive atmosphere. Failure to comply with this warning can cause injury or death to personnel.

a. a. Shut down generator set.

b. b. Open battery access door, disconnect negative battery cable, and open right side engine access door.

c. c. Tag and disconnect glow plugs DC contactor (12, Figure 2-35) electrical leads.

d. d. Remove bolts (10), nuts (11), and contactor (12).

e. e. Remove bolts (13), lock washers (14), and mounting bracket (15) from flywheel housing, if necessary. Discard lock washers (14).

2-123.2 Testing.

CAUTION

Remove contactor before testing; otherwise, damage could occur to the multimeter.

a. a. Set multimeter for ohms and check for open circuit between 5/16-inch stud terminals (larger studs at front and rear of contractor).

b. b. Apply 24 VDC to 10-32 stud size terminals (smaller studs protruding at 450 angle from contractor).

c. c. Listen for audible actuation of contactor and check for closed circuit between 5/16-inch stud terminals.

d. d. Replace glow plugs DC contactor if indications are other than above.

2-123.3 Installation.

a. a. If removed, install mounting bracket (15, Figure 2-35) on flywheel housing with bolts (13) and new lock washers (14).

b. b. Install glow plugs DC contactor (12) on mount ing bracket (15) with bolts (10) and nuts (11).

c. c. Connect electrical leads. Remove tags.

d. d. Close right side engine access door, connect negative battery cable, and close battery access door.

2-124 CONTACTOR K22 (MEP-804B/MEP-814B)

2-124.1 Removal.

WARNING

All metal jewelry can conduct electricity and become entangled in generator set components. Remove all jewelry when working on generator set. Failure to comply with this warning can cause injury or death to personnel.

WARNING

DO NOT wear loose clothing when performing checks, services and maintenance. Failure to comply with this warning can cause injury or death to personnel.

WARNING

Dangerous voltage exists on live circuits. Always observe precautions and never work alone. Failure to comply with this warning can cause injury or death to personnel.

WARNING

DC voltages are present at generator set electrical components even with generator set shut down. Avoid shorting any positive with ground/negative. Failure to comply with this warning can cause injury to personnel, and damage to equipment.

WARNING

Shut down generator set before performing inspection of wiring. Failure

WARNING

The connection of any electrical equipment and the disconnection of any electrical equipment may cause an explosion hazard. Do not connect any electrical equipment or disconnect any electrical equipment in an explosive atmosphere. Failure to comply with this warning can cause injury or death to personnel.

a. a. Shut down generator set.

b. b. Open battery access door, disconnect negative battery cable, and open right side engine access

door.

c. c. Tag and disconnect contactor K22 (15, Figure 2-36) electrical leads.

d. d. Remove bolts (12), lock washers (13), washers (14), and contactor (15). Discard lock washers (13).

e. Remove bolts (16), lock washers (17), washers (18), and mounting bracket (19) from flywheel housing, if necessary. Discard lock washers (17).

CAUTION

Remove contactor before testing: otherwise, damage could occur to the multimeter.

a. a. Set multimeter for ohms and check for open circuit between 5/16-inch stud terminals (larger studs at front and rear of contactor).

b. b. Apply 24 VDC to 10-32 stud size terminals (smaller studs protruding at 450 angle from contractor).

c. c. Listen for audible actuation of contactor and check for closed circuit between 5/16-inch stud terminals.

d. d. Replace contactor K22 if indications are other than above.

2-124.3 Installation.

a. a. Install mounting bracket (19, Figure 2-36) on flywheel housing with bolts (16), new lock washers (17), and washers (18), if removed.

b. b. Install contactor K22 (15) on mounting bracke t (19) with bolts (12), new lock washers (13), and washers (14).

c. c. Connect electrical leads. Remove tags.

d. d. Close right side engine access door, connect negative battery cable, and close battery access door.

2-125 CLOSED CRANKCASE VENTILATION (CCV) FILTER (MEP-804B/MEP-814B)

2-125.1 Inspection.

a. a. Shut down generator set.

b. b. Open right side engine access door.

c. c. Inspect CCV filter, lines, and fittings for loose connections, damage, and evidence of leaks.

d. d. Loosen clamps on sides of filter assembly and remove top of case.

e. e. Remove and inspect replaceable filter and replace, as required.

f. f. Replace top of case and fasten clamps. g

Close right side engine access door.

2-125.2 Removal.

WARNING

All metal jewelry can conduct electricity and become entangled in generator set components. Remove all jewelry when working on generator set. Failure to comply with this warning can cause injury or death to personnel.

WARNING

DO NOT wear loose clothing when performing checks, services and maintenance. Failure to comply with this warning can cause injury or death to personnel.

WARNING

The high pressure oil system operates at high temperature and pressure. Contact with hot oil can result in burns and scalding. Shut down generator set, and allow system to cool before performing checks, services, and maintenance. Wear heat resistant gloves and avoid contacting hot surfaces. Do not allow hot oil or components to contact skin or hands. Failure to comply with this warning can cause injury or death to personnel.

WARNING

Wear heat resistant gloves and avoid contacting hot metal surfaces with your hands after components have been heated. Wear additional protective clothing as required. Failure to comply with this warning can cause injury to personnel.

a. a. Shut down generator set.

b. b. Open battery access door, disconnect negative battery cable, and open right side engine access door.

c. c. Loosen clamp (1, Figure 2-38) and disconnect outlet hose (2) also connected to turbocharger.

d. d. Loosen clamp (3) and disconnect inlet hose (4) also connected to top of engine.

e. e. Loosen clamp (5) and disconnect hose (6) also connected to fitting on other side of engine.

f. f. Remove CCV filter (10) by removing bolts (7), lock washers (8), and washers (9). Discard lock washers (8).

g. g. Remove fuel filter (11) (paragraph 2-126.2).

h. h. Remove filter bracket (16) by removing bolts (12), lock washers (13), washers (14), and nut (15). Discard lock washers (13).

2-125.3 Installation.

a. a. Install filter bracket (16, Figure 2-38) with bolts (12), new lock washers (13), washers (14), and nut (15).

b. b. Install fuel filter (11) (paragraph 2-126.3).

c. c. Install CCV filter (10) with bolts (7), new lock washers (8), and washers (9).

d. d. Connect hose (6) and tighten clamp (5).

e. e. Connect inlet hose (4) and tighten clamp (3).

f. f. Connect outlet hose (2) and tighten clamp (1).

g. g. Close right side engine access door.

h. h. Connect negative battery c able and close battery access door.

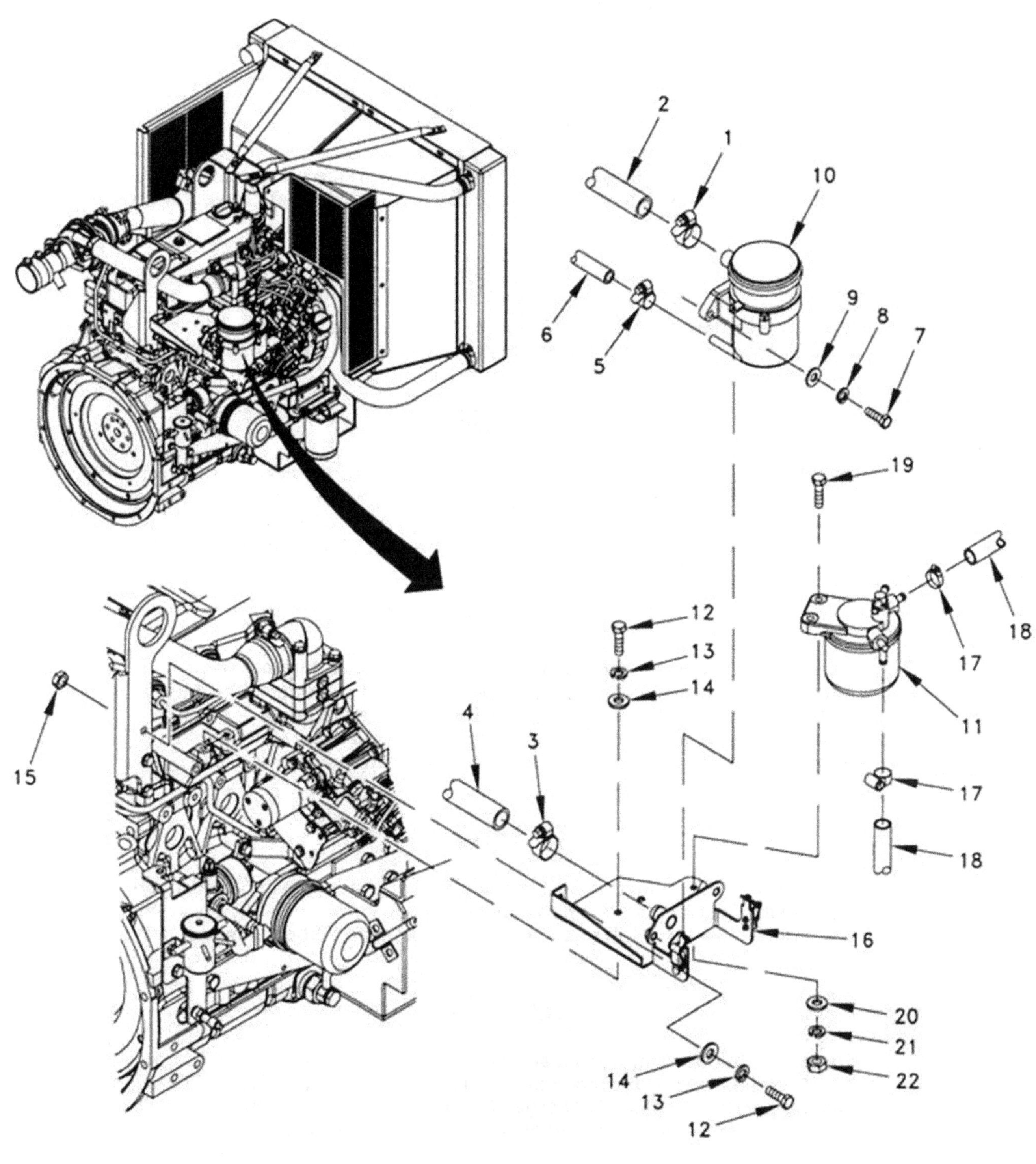

Figure 2-38. CCV Filter and Engine Fuel Filter – MEP-804B/MEP-814B.

2-126 ENGINE FUEL FILTER (MEP-804B/MEP-814B)

2-126.1 Inspection.

WARNING

All metal jewelry can conduct electricity and become entangled in generator set components. Remove all jewelry when working on generator set. Failure to comply with this warning can cause injury or death to personnel.

WARNING

DO NOT wear loose clothing when performing checks, services and maintenance. Failure to comply with this warning can cause injury or death to personnel.

WARNING

When running, generator set engine has hot metal surfaces that will burn flesh on contact. Shut down generator set and allow engine to cool before performing checks, services, and maintenance. Wear gloves and additional protective clothing as required. Failure to comply with this warning can cause injury or death to personnel.

WARNING

Diesel fuel is flammable and toxic to eyes, skin, and respiratory tract. Skin and eye protection are required when working in contact with diesel fuel. Avoid repeated or prolonged contact. Provide adequate ventilation. Operators are to wash exposed skin and change chemical soaked clothing promptly if exposed to fuel. Failure to comply with this warning can cause injury or death to personnel.

WARNING

Fuels used in the generator set are flammable. Do not smoke or use open flames when performing maintenance. Failure to comply with this warning can cause injury or death to personnel, and damage to the generator set.

WARNING

Fuels used in the generator set are flammable. When filling the fuel tank, maintain metal-to-metal contact between filler nozzle and fuel tank opening to eliminate static electrical discharge. Failure to comply with this warning can cause injury or death to personnel, and damage to the generator set.

a. a. Shut down generator set.

b. b. Open right side engine access door.

c. c. Inspect fuel filter (11, Figure 2-38), li nes, and fittings for loose connections, damage, and evidence of leaks.

d. Replace fuel filter (TM 9-6115-538-24&P).

e Close right side engine access door.

WARNING

Diesel fuel is flammable and toxic to eyes, skin, and respiratory tract. Skin and eye protection are required when working in contact with diesel fuel. Avoid repeated or prolonged contact. Provide adequate ventilation. Operators are to wash exposed skin and change chemical soaked clothing promptly if exposed to fuel. Failure to comply with this warning can cause injury or death to personnel.

a. a. Shut down generator set.

b. b. Open battery access door, disconnect negative battery cable, and open right side engine access

door.

c. c. Remove clamps (17, Figure 2-38) and hoses (18) from fuel filter (11).

d. d. Remove screws (19), washers (20), lock washers (21), nuts (22), and fuel filter (11). Discard lock washers (21).

WARNING

Diesel fuel is flammable and toxic to eyes, skin, and respiratory tract. Skin and eye protection are required when working in contact with diesel fuel. Avoid repeated or prolonged contact. Provide adequate ventilation. Operators are to wash exposed skin and change chemical soaked clothing promptly if exposed to fuel. Failure to comply with this warning can cause injury or death to personnel.

a. a. Install fuel filter (11, Figure 2-38) with screws (19), washers (20), new lock washers (21), and nuts (22).

b. b. Connect hoses (18) and clamps (17) and tighten clamps.

c. c. Close right side engine access door.

d. d. Connect negative battery cabl e and close battery access door.

e. e. Start generator set and check for leaks.

Section XVI. MAINTENANCE OF LUBRICATION SYSTEM

2-127 OIL DRAIN LINE

2-127.1 Inspection.

WARNING

The high pressure oil system operates at high temperature and pressure. Contact with hot oil can result in burns and scalding. Shut down generator set, and allow system to cool before performing checks, services, and maintenance. Wear heat resistant gloves and avoid contacting hot surfaces. Do not allow hot oil or components to contact skin or hands. Failure to comply can cause injury or death to personnel.

WARNING

Oil filter base and housing springs are under tension and can act as projectiles when being removed. Use eye protection when removing springs. Failure to comply with this warning can cause injury to personnel.

a. a. Shut down generator set.

b. b. Open battery and right side engine access doors.

c. c. Inspect oil drain line for cracks, holes, loose or missing hardware, and other damage.

d. d. Close battery and right side engine access doors.

2-127.2 Removal.

WARNING

All metal jewelry can conduct electricity and become entangled in generator set components. Remove all jewelry when working on generator set. Failure to comply with this warning can cause injury or death to personnel.

WARNING

DO NOT wear loose clothing when performing checks, services and maintenance. Failure to comply with this warning can cause injury or death to personnel.

WARNING

When running, generator set engine has hot metal surfaces that will burn flesh on contact. Shut down generator set and allow engine to cool before performing checks, services, and maintenance. Wear gloves and additional protective clothing as required. Failure to comply with this warning can cause injury or death to personnel.

a. a. Shut down generator set.

b. b. Open battery access door, disconnect negative battery cable, and open right side engine access door.

c. c. Remove plug (1, Figure 2-39), open drain valve (2), and drain engine oil into suitable container.

d. d. Remove clamps (3) and drain hose (4) from oil pan fittings and adapter (5).

e. e. Remove drain valve (2) from skid fitting.

f. f. Remove adapter (5) and pipe fitting (6) from oil drain valve (2).

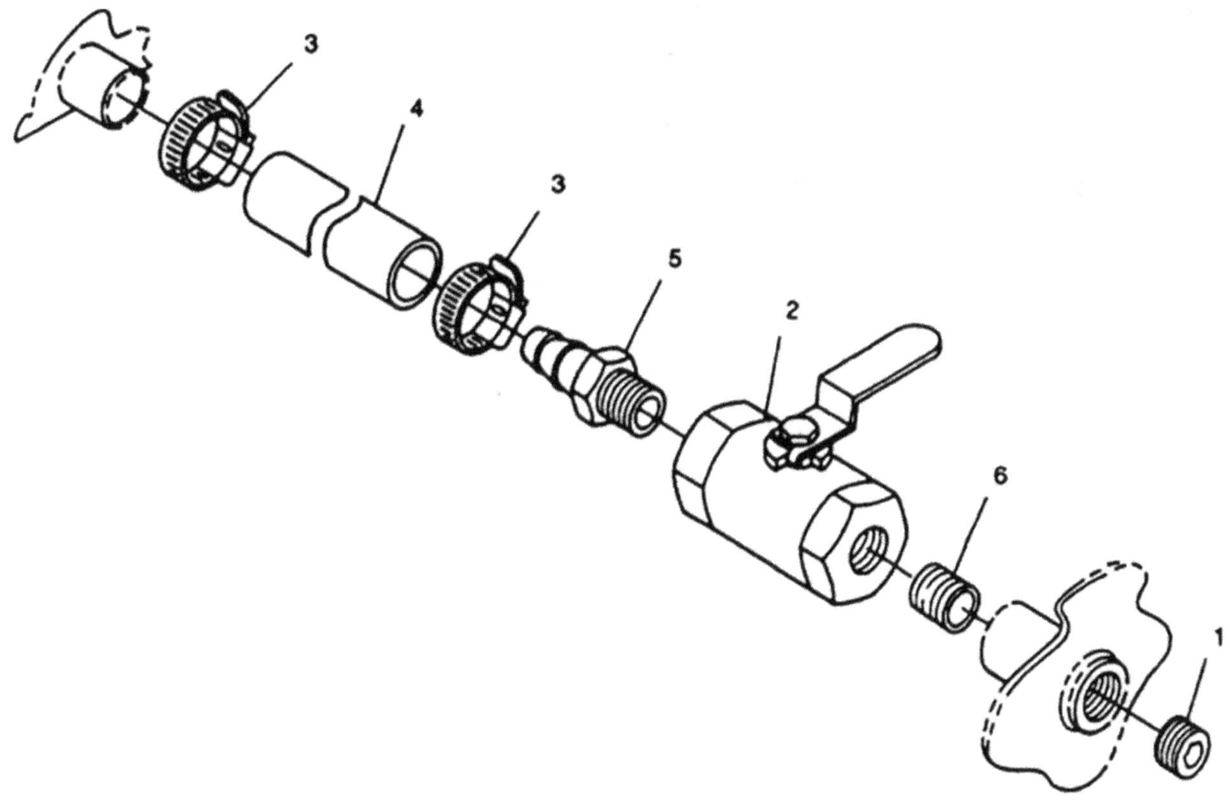

Figure 2-39. Oil Drain Line.

2-127.3 Installation.

a. a. Install adapter (5, Figure 2-39) and pipe fitting (6) in oil drain valve (2).

b. b. Install drain valve (2) on skid fitting.

c. c. Install drain hose (4) on oil pan fittings and adapter (5). Secure with clamps (3).

d. d. Ensure drain valve (2) is closed. Install plug (1) and service engine oil (paragraph 2-1.2.4).

e. e. Check engine oil drain line and valve for leakage.

Section XVII. PREPARATION FOR SHIPMENT AND STORAGE

2-128 PRESERVATION

Preserve generator sets in accordance with levels A/A, A/B, C/B or C/C of PPP-G-2919.

Preserve generator set cooling systems in accordance with method II of PPP-G-2919 or the antifreeze and water procedure of ATPD 2232.

2-129 PACKING

Pack generator sets in accordance with levels A/A, A/B, C/B or C/C of PPP-G-2919.

2-130 MARKING

Mark for shipment or storage in accordance with MIL-STD-129P(4).

2-131 USE OF CORROSION-PREVENTIVE COMPOUNDS, MOISTURE BARRIERS, AND DESICCANT MATERIALS

(A) Refer to Corrosion and Corrosion Prevention/Metal, MIL-HDBK-729 NOT 1.

2-132 STORAGE

(A) For storage information, refer to TB 740-97-2. (F) Refer to TO 38-1-5.

CHAPTER 3

GENERAL MAINTENANCE INSTRUCTIONS

Subject Index Page

Section I. REPAIR PARTS; SPECIAL TOOLS; TEST, MEASUREMENT, AND DIAGNOSTIC EQUIPMENT (TMDE); AND SPECIAL SUPPORT EQUIPMENT

3-1 MAINTENANCE REPAIR PARTS

Repair parts and equipment are listed and illustrated in the repair parts and special tools list manuals TM 9-6115-643-24P and TM 9-2815-254-24P, and in the new YANMAR engine manual with repair parts and special tools TM 9-2815-538-24&P.

3-2 TOOLS AND EQUIPMENT

There are no special tools or support equipment required to perform any level of maintenance on the generator set. A list of recommended tools and support equipment required to maintain the generator set is contained in Appendix B, Section III.

3-3 FABRICATION OF TOOLS AND EQUIPMENT

No requirement exists for fabrication of tools and equipment for maintenance of the generator set.

Section II. TROUBLESHOOTING

3-4 DIRECT SUPPORT TROUBLESHOOTING PROCEDURES

3-4.1 Purpose of Troubleshooting Table. This section contains troubleshooting information for locating and correcting operating troubles which may develop in the generator set. Each malfunction for an individual component, unit, or system is followed by a list of tests or inspections which will help you to determine probable causes and corrective actions to take. You should perform the tests/inspections and corrective actions in the order listed.

This table cannot list all malfunctions that can occur, nor all test or inspections and corrective actions. If a malfunction is not listed or cannot be corrected by listed corrective actions, notify your supervisor.

NOTE

Before you use this table, be sure you have performed your Preventive Maintenance Checks and Services (PMCS).

Before you use this table, be sure unit level troubleshooting steps have been performed.

Refer to the Electrical Schematic FO-1 (MEP-804A/MEP-814A)/FO-3 (MEP-804B/ MEP-814B) and Wiring Diagram FO-2 (MEP-804A/ MEP-814A)/FO-4 (MEP-804B/MEP-814B) as troubleshooting aids.

SYMPTOM INDEX
GENERATOR SET

Troubleshooting Procedure Page

Table 3-1. Direct Support Troubleshooting.

MALFUNCTION
TEST OR INSPECTION
CORRECTIVE ACTION

1. ENGINE CRANKS BUT FAILS TO START.

Step 1. Check for defective fuel feed pump (TM 9-2815-254-24 (MEP-804A/MEP-814A)) or defective fuel transfer pump (paragraph 2-106.2 (MEP-804B/MEP-814B)).

a. If pump is not defective, do step 2.

b. If defective, replace pump (TM 9-2815-254-24 (MEP-804A/MEP-814A)/paragraph 2-106 (MEP-804B/MEP-814B)).

Step 2. Test fuel injection pump ti ming (TM 9-2815-254-24 (MEP-804A/MEP-814A)).

a. a. If injection pump timing is correct, do step 3.

b. b. If not correctly timed, time fuel inject ion pump (TM 9-2815-254-24 (MEP-804A/MEP-814A).

Step 3. Test for defective governor actuator (paragraph 4-14.1 (MEP-804A/MEP-814A)/paragraph 4-15.1 (MEP-804B/MEP-814B)).

a. a. If governor actuator is not defective, do step 4.

b. b. If defective, replace governor actuator (paragraph 4-14.3 (MEP-804A/MEP-814A)/paragraph 4-15.3 (MEP-804B/MEP-814B)).

Step 4. Test for defective fuel injection pump (TM 9-2815-254-24 (MEP-804A/MEP-814A)/TM 9-2815-538-24&P (MEP-804B/MEP-814B)).

a. a. If fuel injection pump is not defective, do step 5.

b. b. If defective, repair or replace fuel in jection pump (TM 9-2815-254-24 (MEP-804A/MEP-814A)/ TM 9-2815-538-24&P (MEP-804B/MEP-814B)).

Step 5. Test for defective governor control unit (paragraph 4-2.1). If defective, replace governor control unit (paragraph 4-2.3).

2. ENGINE RUNS ERRATICALLY OR STALLS FREQUENTLY.

Step 1. Test fuel injection pump timing (TM 9-2815-254-24 (MEP-804A/MEP-814A)).

a. If fuel injection pump timing is correct, do step 2.

b. b. If not correctly timed, time fuel injectio n pump (TM 9-2815-254-24 (MEP-804A/MEP-814A)).

Step 2. Test for defective fuel injection pump (TM 9-2815-254-24 (MEP-804A/MEP-814A)/TM 9-2815-538-24&P (MEP-804B/MEP-814B)).

a. a. If fuel injection pump is not defective, do step 3.

b. b. If defective, repair or replace fuel in jector pump (TM 9-2815-254-24 (MEP-804A/MEP-814A/TM 9-2815-538-24&P (MEP-804B/MEP-814B)).

Table 3-1. Direct Support Troubleshooting – Contin-

MALFUNCTION
TEST OR INSPECTION
CORRECTIVE ACTION

Step 4. Check cylinder head gasket for leaks (T M 9-2815-254-24 (MEP-804A/MEP-814A)/TM 9-2815-538-24&P (MEP-804B/MEP-814B)).

a. a. If cylinder head gasket is not leaking, do step 5.

b. b. If leaking, replace cylinder head ga sket (TM 9-2815-254-24 (MEP-804A/MEP-814A)/ TM 9-2815-538-24&P (MEP-804B/MEP-814B)).

Step 5. Check for stuck or burnt va lves (TM 9-2815-254-24 (MEP-804A/MEP-814A)).

a. a. If valves are not stuck or burnt, do step 6.

b. b. If stuck or burnt, replace va lves (TM 9-2815-254-24 (MEP-804A/MEP-814A)).

c. c. If burnt, replace cylinder head (TM 9-2815-538-24&P (MEP-804B/MEP-814B)).

d. d. If stuck, replace engine (paragraph 3-9.3 (MEP-804B/MEP-814B)).

Step 6. Test for defective governor control unit, paragraphs 4-2.1 and 4-2.2.

a. a. If governor control unit is not defective, do step 7.

b. b. If necessary, adjust governor control unit (par agraph 4-2.3) or replace governor control unit (paragraph 4-2.4).

Step 7. Check for low engine compression (TM 9-2815-254-24 (MEP-804A/MEP-814A)/TM 9-2815-538-24&P (MEP-804B/MEP-814B)).

If compression is low, repair or repl ace engine (TM 9-2815-254-24 (MEP-804A/MEP-814A)/ paragraph 3-9.3 (MEP-804B/MEP-814B)).

3. ENGINE MISFIRING.

Step 1. Check for dirty fuel injector.

a. a. Remove and inspect fuel injector (TM 9-2815-254-24 (MEP-804A/MEP-814A)/TM 9-2815-538-24&P (MEP-804B/MEP-814B)).

b. b. If fuel injectors are not dirty, do step 2.

c. If dirty, clean fuel injector or if de fective, replace fuel injector (TM 9-2815-254-24 (MEP-804A/MEP-814A)/TM 9-2815-538-24&P (MEP-804B/MEP-814B)).

Step 2. Test for defectiv e governor (paragraph 4-14.1 (MEP-804A/MEP-814A)/paragraph 4-15.1 (MEP-804B/MEP-814B)).

a. a. If governor is not defective, do step 3.

b. b. If defective, replace governor (parag raph 4-14 (MEP-804A/MEP-814A)/paragraph 4-15 (MEP-804B/MEP-814B)).

Step 3. Test fuel injection pump timing (TM 9-2815-254-24 (MEP-804A/MEP-814A)).

a. If fuel injection pump timing is correct, do step 4.

Table 3-1. Direct Support Troubleshooting – Contin-

MALFUNCTION
TEST OR INSPECTION
CORRECTIVE ACTION

Step 4. Test for defective fuel injection pump (TM 9-2815-254-24 (MEP-804A/MEP-814A)).

a. a. If fuel injection pump is not defective, do step 5.

b. b. If defective, repair or replace fuel injector pump (TM 9-2815-254-24 (MEP-804A/MEP-814A)/
TM 9-2815-538-24&P (MEP-804B/MEP-814B)).

Step 5. Check valve adjustment (TM 9-2815-254-24 (MEP-804A/MEP-814A)/TM 9-2815-538-24&P (MEP-804B/MEP-814B)).

a. a. If valves are properly adjusted, do step 6.

b. b. If improperly adjusted, adjust valves (TM 9-2815-254-24 (MEP-804A/MEP-814A)/TM 9-2815-
538-24&P (MEP-804B/MEP-814B)).

Step 6. Check for weak valve springs (TM 9-2815-254-24 (MEP-804A/MEP-814A)).

a. a. If valve springs are not weak, do step 7.

b. b. If weak, replace valve springs (TM 9- 2815-254-24 (MEP-804A/MEP-814A)) or replace cylinder
head (TM 9-2815-538-24&P (MEP-804B/MEP-814B)).

Step 7. Check for stuck or burnt va lves (TM 9-2815-254-24 (MEP-804A/MEP-814A)).

a. a. If valves are not stuck or burnt, do step 8.

b. b. If stuck or burnt, replace valves (TM 9-2815-254-24 (MEP-804A/MEP-814A)).

c. c. If burnt, replace cylinder head (TM 9-2815-538-24&P (MEP-804B/MEP-814B)).

d. d. If stuck, replace engine (paragraph 3-9.3 (MEP-804B/MEP-814B)).

Step 8. Check for low engine compression (TM 9-2815-254-24 (MEP-804A/MEP-814A)/TM 9-2815-538-24&P (MEP-804B/MEP-814B)).

If compression is low, repair or repl ace engine (TM 9-2815-254-24 (MEP-804A/MEP-814A)/ TM 9-2815-538-24&P (MEP-804B/MEP-814B)).

4. <u>ENGINE DOES NOT DEVELOP FULL POWER.</u>

Step 1. Check for dirty fuel injectors.

a. a. Remove and inspect fuel injector (TM 9-2815-254-24 (MEP-804A/MEP-814A)/TM 9-2815-538-
24&P (MEP-804B/MEP-814B)).

b. b. If fuel injectors are not dirty, do step 2.

c. If dirty, clean fuel injector or if defective, replace fuel injector (TM 9-2815-254-24 (MEP-804A/MEP-814A)/TM 9-2815-538-24&P (MEP-804B/MEP-814B)).

Step 2. Test for defective governor (par agraph 4-14.1 (MEP-804A/MEP-814A)/paragraph 4-15.1 (MEP-804B/MEP-814B)).

a. a. If governor is not defective, do step 3.

Table 3-1. Direct Support Troubleshooting – Contin-

MALFUNCTION TEST OR INSPECTION CORRECTIVE ACTION
Step 3. Test for defective governor c ontrol unit (paragraphs 4-2.1 and 4-2.2). a. a. If governor control unit is not defective, do step 4. b. b. If necessary, adjust governor control unit (par agraph 4-2.3) or replace governor control unit (paragraph 4-2.4). Step 4. Test fuel injection pump timing (TM 9-2815-254-24 (MEP-804A/MEP-814A)). a. If fuel injection pump timing is correct, do step 5. b. If timing is not correct, time fuel inje ction pump (TM 9-2815-254-24 (MEP-804A/MEP-814A)/ TM 9-2815-538-24&P (MEP-804B/MEP-814B)). Step 5. Check valve adjustment (TM 9-2815-254-24 (MEP-804A/MEP-814A)/TM 9-2815-538-24&P (MEP-804B/MEP-814B)). a. a. If valves are properly adjusted, do step 6. b. b. If improperly adjusted, adjust valves (TM 9-2815-254-24 (MEP-804A/MEP-814A)/TM 9-2815-538-24&P (MEP-804B/MEP-814B)). Step 6. Check for weak valve springs (TM 9-2815-254-24 (MEP-804A/MEP-814A)). a. a. If valve springs are not weak, do step 7. b. b. If weak, replace valve springs (TM 9- 2815-254-24 (MEP-804A/MEP-814A)) or replace cylinder head (TM 9-2815-538-24&P (MEP-804B/MEP-814B)). Step 7. Check for stuck or burnt va lves (TM 9-2815-254-24 (MEP-804A/MEP-814A)). a. a. If valves are not stuck or burnt, do step 8. b. b. If stuck or burnt, replace valves (TM 9-2815-254-24 (MEP-804A/MEP-814A)). c. c. If burnt, replace cylinder head (TM 9-2815-538-24&P (MEP-804B/MEP-814B)). d. d. If stuck, replace engine (paragraph 3.9.3 (MEP-804B/MEP-814B)). Step 8. Check cylinder head gasket for leaks (T M 9-2815-254-24 (MEP-804A/MEP-814A)/TM 9-2815-538-24&P (MEP-804B/MEP-814B)). a. a. If cylinder head gasket is not leaking, do step 9. b. b. If leaking, replace cylinder head gasket (TM 9-2815-254-24 (MEP-804A/MEP-814A)/ TM 9-2815-538-24&P (MEP-804B/MEP-814B)). Step 9. Test for low engine compression (TM 9-2815-254-24 (MEP-804A/MEP-814A)/TM 9-2815-538-24&P (MEP-804B/MEP-814B)).

Table 3-1. Direct Support Troubleshooting – Contin-

MALFUNCTION TEST OR INSPECTION CORRECTIVE ACTION
5. BLACK OR GREY SMOKE IN EXHAUST. Step 1. Check for dirty fuel injectors. a. Remove and inspect fu el injectors (TM 9-2815-254-24 (MEP-804A/MEP-814A)/TM 9-2815-538- 24&P (MEP-804B/MEP-814B)). b. If fuel injectors are not dirty, do to step 2. c. If dirty, clean fuel injectors or if defect ive, re- place fuel injectors (TM 9-2815-254-24 (MEP-804A/MEP-814A)/TM 9-2815-538-24&P (MEP-804B/MEP-814B)). Step 2. Test fuel injection pump timi ng, TM 9-2815-254-24 (MEP-804A/MEP-814A). a. If timing is correct, do step 3. b. If timing is not correct, time fuel inje ction pump (TM 9-2815- 254-24 (MEP-804A/MEP-814A)/ TM 9-2815-538-24&P (MEP-804B/MEP-814B)). Step 3. MEP-804B/MEP-814B: Replace turbocharger (TM 9-2815-538-24&P).
6. BLUE OR WHITE EXHAUST SMOKE. Test fuel injection pump timing, TM 9-2815-254-24 (MEP-804A/MEP-814A) a. If timing is not correct, time fuel injection pump (M 9-2815-254-24 (MEP-804A/MEP-814A)/ TM 9-2815-538-24&P (MEP-804B/MEP-814B)). b. MEP-804B/MEP-814B: If timing is correc t, replace turbocharger (TM 9-2815-538-24&P).
7. LOW OIL PRESSURE. Check oil relief valve for proper o peration (TM 9-2815-254-24 (MEP-804A/MEP-814A)). a. If oil relief valve is operating proper ly, notify next higher level of maintenance. b. If not oper- ating properly, replace oil relief valve (TM 9-2815-254-24 (MEP-804A/MEP-814A)) or replace oil cooler (TM 9-2815-538-24&P (MEP-804B/MEP-814B)).
8. HIGH OIL PRESSURE. Test for defective oil relief valve (T M 9-2815-254-24 (MEP-804A/MEP-814A)). If defective, replace oil relief valve (TM 9-2815-254-24 (MEP-804A/MEP-814A)) or replace oil cooler (TM 9-2815-538-24&P (MEP-804B/MEP-814B)).

Step 1. Check for cracked cylinder head (TM 9-2815-254-24 (MEP-804A/MEP-814A)).

a. a. If cylinder head is not cracked, do step 2.

b. b. If cracked, replace cylinder head (T M 9-2815-254-24 (MEP-804A/MEP-814A)/TM

9-2815-538-

Table 3-1. Direct Support Troubleshooting – Contin-

MALFUNCTION TEST OR INSPECTION CORRECTIVE ACTION
Step 2. Check for defective cylinder head gasket (TM 9-2815-254-24 (MEP-804A/MEP-814A)/ TM 9-2815-538-24&P (MEP-804B/MEP-814B)). a. a. If defective, replace cylinder head gasket (TM 9-2815-254-24 (MEP-804A/MEP-814A)/ TM 9-2815-538-24&P (MEP-804B/MEP-814B)). b. b. If not defective and problem persists, replace engine (paragraph 3-8 (MEP-804A/MEP-814A)/ paragraph 3-9.3 (MEP-804B/MEP-814B)).
10. ABNORMAL ENGINE NOISE. Step 1. Check for defective engine mounts. a. a. If engine mounts are not defective, do step 2. b. b. If engine mounts are defective, replac e mounts (paragraph 3-6.2 (MEP-804A/MEP-814A)/ paragraph 3-7.2 (MEP-804B/MEP-814B)). Step 2. Check for dirty fuel injectors. a. a. Remove and inspect fuel injectors (TM 9-2815-254-24 (MEP-804A/MEP-814A)/TM 9-2815-538-24&P (MEP-804B/MEP-814B)). b. b. If fuel injectors are not dirty, do step 3. c. c. If dirty, clean fuel injectors, or if defe ctive, replace fuel injectors (TM 9-2815-254-24 (MEP-804A/MEP-814A)/TM 9-2815-538-24&P (MEP-804B/MEP-814B)). Step 3. Check for worn rocker arm shaft (TM 9-2815-254-24 (MEP-804A/MEP-814A)). a. a. If rocker arm shaft is not worn, do step 4. b. b. If worn, replace shaft and/or rocker arms (TM 9-2815-254-24 (MEP-804A/MEP-814A)) or replace cylinder head (TM 9-2815-538-24&P (MEP-804B/MEP-814B)). Step 4. Check for bent push rods (TM 9-2815-254-24 (MEP-804A/MEP-814A)). a. a. If push rods are not bent, do step 5. b. b. If bent, replace push rods (TM 9-2815-2 54-24 (MEP-804A/MEP-814A)) or replace engine (paragraph 3-9.3 (MEP-804B/MEP-814B)). Step 5. Check for worn idler gear s (TM 9-2815-254-24 (MEP-804A/MEP-814A)). a. a. If idler gears are not worn, do step 6. b. b. If worn, replace idler ge ars (TM 9-2815-254-24 (MEP-804A/MEP-814A)) or replace engine (paragraph 3-9.3 (MEP-804B/MEP-814B)). Step 6. Test fuel injection pump timing (TM 9-2815-254-24 (MEP-804A/MEP-814A)).

Table 3-1. Direct Support Troubleshooting – Contin-

MALFUNCTION TEST OR INSPECTION CORRECTIVE ACTION
Step 7. Check for foreign material in combustion chambers (TM 9-2815-254-24 (MEP-804A/MEP-814A)). a. a. If foreign material is found, clean combustion chamber(s) (TM 9-2815-254-24 (MEP-804A/ MEP-814A)). b. b. If no foreign matter is fo und and problem persists, refer to TM 9-2815-254-24 (MEP-804A/ MEP-814A) or replace engine (paragraph 3-9.3 (MEP-804B/MEP-814B)).
11. <u>GENERATOR SET FAILS TO GENERATE POWER.</u> Step 1. Test for defective governor c ontrol unit (paragraphs 4-2.1 and 4-2.2). a. a. If a governor control unit is not defective, do step 2. b. b. If necessary, adjust governor control unit (par agraph 4-2.3) or replace governor control unit (paragraph 4-2.4). Step 2. Test for defective AC voltage regulator (paragraph 4-1). a. a. If AC voltage regulator is not defective, do step 3. b. b. If defective, replace AC volt age regulator (paragraph 4-1.5). Step 3. Test for defective ge nerator exciter stator (paragraph 4-21.1 (MEP-804A/MEP-804B)/paragraph 4-29.2 (MEP-814A/MEP-814B)). a. a. If exciter stator is not defective, do step 4. b. b. If defective, replace exci ter stator (paragraph 4-21 (MEP-804A/MEP-804B)/paragraph 4-29 (MEP-814A/MEP-814B)). Step 4. Test for defective generator stator (paragraph 4-24-1 (MEP-804A/MEP-804B)/paragraph 4-32.2 (MEP-814A/MEP-814B)). a. a. If generator stator is not defective, do step 5. b. b. If defective, replace gener ator stator and housing assembly (paragraph 4-24 (MEP-804A/ MEP-804B)/paragraph 4-32 (MEP-814A/MEP-814B)). Step 5. Test for defective diode(s) in generato r rotating rectifier (paragraph 4-18.1 (MEP-804A/MEP-804B)/ paragraph 4-26.2 (MEP-814A/MEP-814B)). a. a. If diodes are not defective, do step 6. b. b. If defective, replace diode(s) (paragra ph 4-18 (MEP-804A/MEP-804B)/paragraph 4-26 (MEP-814A/MEP-814B)). Step 6. Test for defective generator rotor (paragraph 4-23.1 (MEP-804A/MEP-804B)/paragraph 4-31.2 (MEP-814A/MEP-814B)). a. a. If generator rotor is not defective, do step 7. b. b. If defective, replace ge nerator rotor assembly (paragraph 4-23 (MEP-804A/MEP-

Table 3-1. Direct Support Troubleshooting – Contin-

MALFUNCTION
TEST OR INSPECTION
CORRECTIVE ACTION

1. GENERATOR SET FAILS TO GENERATE SUFFICIENT VOLTAGE (MEP-804A/MEP-814A).

Step 1. Check for low engine speed (TM 9-2815-254-24).

a. If engine is operating correctly, do step 2.

b. b. If engine is not operating correctly, repair engine (TM 9-2815-254-

24). Step 2. Check for defective fuel injection nozzles (TM 9-2815-254-24).

a. a. If fuel injection nozzles ar e not defective, do step 3.

b. b. If defective, repair or replace fuel injection nozzles (TM 9-2815-254-

24). Step 3. Test for defective gene rator stator (paragraph 4-24.1).

If defective, replace generator stator and housing assembly (paragraph 4-24).

1. GENERATOR SET FAILS TO GENERATE SUFFICIENT VOLTAGE (MEP-804B/MEP-814B).

Step 1. Test for defective AC voltage regulator (paragraph 4-1).

a. If AC voltage regulator is defective, do step 2.

b. If defective, replace AC volt age regulator (paragraph 4-1.5). Step

2. MEP-804B: Test for defectiv e generator exciter stator, 4-21.1. If defective,

replace exciter stator (paragraph 4-21/paragraph 4-29).

1. GENERATOR SET OUTPUT FLUCTUATES.

Step 1. Check for irregular en gine speed (frequency fluctuation).

a. If engine is operating correctly, do step 2.

b. If engine is not operating correctly, re pair engine (TM 9-2815-254-24 (MEP-804A/MEP-
814A)/

TM 9-2815-538-24&P (MEP-804B/MEP-814B)).

Step 2. Test for defective governor (par agraph 4-14.1 (MEP-804A/MEP-814A)/paragraph 4-15.1 (MEP-804B/MEP-814B)).

a. a. If governor is not defective, do step 3.

b. b. If defective, replace governor (paragra ph 4-14.2 (MEP-804A/MEP-814A)/paragraph
4-15.2

(MEP-804B/MEP-814B)).

Step 3. MEP-804B/MEP-814B: Test for defectiv e governor control unit (paragraph 4-2.1).

a. a. If governor control unit is not defective, do step 4.

b. b. If defective, replace governor control unit (paragraph 4-2.3).

Step 4. Check for loose terminations. Refer to Electrical Schematic FO-1 (MEP-804A/MEP-814A)/ FO-3 (MEP-804B/MEP-814B).

a. a. If terminations are tight, do step 5.

Table 3-1. Direct Support Troubleshooting – Contin-

MALFUNCTION TEST OR INSPECTION CORRECTIVE ACTION
Step 5. Check AC voltage regulator for incorrect output (paragraph 4-1). a. If AC voltage regulator is operating properly, do step 6. b. If AC voltage regulator is inoperative, replace AC voltage regulator (paragraph 4-1.5). Step 6. Test for intermediate short in generator exciter field (paragraph 4-21.1 (MEP-804A/MEP-804B)/ paragraph 4-29.2 (MEP-814A/MEP-814B)). If shorted, replace exciter stator (par agraph 4-21 (MEP-804A/MEP-804B)/paragraph 4-29 (MEP-814A/MEP-814B)).
15. GENERATOR OVERHEATS. Step 1. Check for clog ged air intake screens. a. If air intake screens are not clogged, do step 2. b. If clogged, clean air intake screens. Step 2. Check for de fective generator fan. a. If fan is not defective, do step 3. b. If defective, replace fan (paragraph 4.2.2 (MEP-804A/ MEP-804B)/paragraph 4.31 (MEP-814A/MEP-814B)). Step 3. Check for dry generato r main bearing (paragraph 4-19 (MEP-804A/MEP-804B)/paragraph 4-27 (MEP-814A/MEP-814B)). If dry, replace main bearing (paragra ph 4-19 (MEP-804A/MEP-804B)/paragraph 4-27 (MEP-814A/MEP-814B)).

16. GENERATOR NOISY WHEN RUNNING.

Step 1. Check for defective generator main bearing (paragraph 4-19 (MEP-804A/MEP-804B)/paragraph 4-27 (MEP-814A/MEP-814B)).

a. a. If generator main bearing is not defective, do step 2.

b. b. If defective, replace main bearing (par agraph 4-19 (MEP-804A/MEP-804B)/paragraph 4-27 (MEP-814A/MEP-814B)).

Step 2. Check for loose engi ne/generator coupling (paragraph 3-8.2, step j (MEP-804A/MEP-814A)/ paragraph 3-9.2, step h (MEP-804B/MEP-814B)).

a. a. If engine/generator coupling is not loose, do step 3.

b. b. If loose, tighten engine/generator coup ling bolts (paragraph 3-8.2, step j (MEP-804A/ MEP-814A)/paragraph 3-9.2, step k (MEP-804B/MEP-814B)).

Step 3. Test for defective generator.

Section III. REMOVAL AND INSTALLATION OF MAJOR COMPONENTS

3-5 GENERAL

The engine and generator are bolted together at the engine flywheel and flywheel housing adapter. The engine and generator may be removed as an assembly or individually. The engine and generator assembly is mounted on the skid base at four points, two at the engine and two at the generator. There are also brackets installed on both sides of the engine which can be adjusted to support the rear of the engine when removing the generator separately. There are bolts and nuts which can be adjusted to support the front of the generator when removing the engine.

3-6 ENGINE AND GENERATOR ASSEMBLY (MEP-804A/MEP-814A) 3-6.1

Removal (MEP-804A/MEP-814A).

WARNING

All metal jewelry can conduct electricity and become entangled in generator set components. Remove all jewelry when working on generator set. Failure to comply with this warning can cause injury or death to personnel.

WARNING

DO NOT wear loose clothing when performing checks, services and maintenance. Failure to comply with this warning can cause injury or death to personnel.

WARNING

Dangerous voltage exists on live circuits. Always observe precautions and never work alone. Failure to comply with this warning can cause injury or death to personnel.

WARNING

DC voltages are present at generator set electrical components even with generator set shut down. Avoid shorting any positive with ground/negative. Failure to comply with this warning can cause injury to personnel, and damage to equipment.

WARNING

Slave receptacle (NATO connector) is electrically live at all times and is unfused. The Battery Disconnect Switch does not remove power from the slave receptacle. NATO slave receptacle has 24 VDC even when Battery Disconnect Switch is set to OFF. This circuit is only dead when the batteries are fully disconnected. Disconnect the batteries before performing maintenance on the slave receptacle. Failure to comply with this warning can cause injury or death to personnel.

WARNING

Ensure that the engine cannot be started while maintenance is being performed. (ENGINE CONTROL switch set to OFF/RESET; Battery Disconnect Switch is OFF; DEAD CRANK SWITCH is OFF). Failure to comply with this warning can cause injury or death to personnel.

WARNING

When disconnecting or removing batteries, disconnect the negative lead that connects directly to the grounding stud first; disconnect the negative end of the interconnection cable next. When installing batteries, reverse the connection sequence. Failure to comply with this warning can cause injury to personnel.

WARNING

Many components require a two-person lift. Lifting heavy components can cause back strain. Ensure proper lifting techniques are used when lifting heavy components. Failure to comply with this warning can cause injury to personnel.

WARNING

Each battery weighs more than 70 pounds (32 kg) and requires a two-person lift. Lifting batteries can cause back strain. Ensure proper lifting techniques are used when lifting batteries. Failure to comply with this warning can cause injury to personnel.

WARNING

Support components when removing attaching hardware or component may fall. Failure to comply with this warning can cause injury to personnel, and damage to equipment.

WARNING

The generator set, engine, and generator are extremely heavy and require an assistant and a lifting device (forklift, overhead lifting device) with sufficient capacity. Failure to comply with this warning can cause injury or death to personnel.

WARNING

The connection of any electrical equipment and the disconnection of any electrical equipment may cause an explosion hazard. Do not connect any electrical equipment or disconnect any electrical equipment in an explosive atmosphere. Failure to comply with this warning can cause injury or death to personnel.

a. Shut down generator set.

b. b. Open battery access door and remove batteries (paragraph 2-12.2).

c. c. Remove generator set housing (par agraphs 2-16.1, 2-17.1, and 2-18.1).

d. d. Remove radiator (paragraph 2-89.1).

e. e. Drain engine oil and disconnect oil drain line at oil pan (paragraph 2-127.2).

f. f. Disconnect fuel line at fuel transfer pump from fuel pickup (paragraph 2-97.1).

g. g. Disconnect engine excess fuel return line at tank fitting and fuel filter/water separator fuel lines at engine fittings (paragraph 2.97.1).

h. h. Disconnect coolant overflow hose fr om overflow bottle (paragraph 2-87.1).

i. i. Disconnect block coolant drain line at engine fitting (paragraph 2-87.1).

NOTE

On 400 Hz generator set (MEP-814A), attaching parts mount from underside of generator.

j. j. Remove nuts (1, Figure 3-1), washers (2), snubbing washers (3), and bolts (4) from engine support bracket (5).

k. k. Remove rear forklift guide from generator set (paragraph 4-34.1).

l. l. Remove nuts (6), washers (7), snubbing washers (8), bolts (9), and Belleville washers (10) from generator mounts.

m. m. Tie wrap generator power leads and secure out of the way.

CAUTION

Rated capacity of overhead hoist should be at least 2,000 pounds (907 Kg). Arrange lifting device so that it supports both engine and generator to avoid undue stress on the engine-generator coupling.

n. n. Attach lifting harness to engine and generator lifting points. Raise engine and generator assembly from skid base.

o. Move engine and generator assembly to maintenance work area. Support assembly on maintenance stand or fixture.

p. p. Remove engine shock mounts (14) from skid base by removing nuts (11), washers (12), and bolts (13).

q. q. Remove shock mounts (15) from generator mounting points on skid base.

r. r. Remove engine support bracket (5) by removing nuts (16), washers (17), and bolts (18).

s. s. Remove engine support brackets (22) by removing bolts (19), lock washers (20), and washers (21). Discard lock washers (20).

t. Remove generator mount angles (29), plates (28), nuts (23 and 27), washers (24 and 26), and bolts (25) from generator.

u. Remove bolts (30), lock washers (31), washers (32), nuts (33), and bracket (34) from engine. Discard lock washers (31).

v. Remove bolts (35) and bracket (36) from engine.

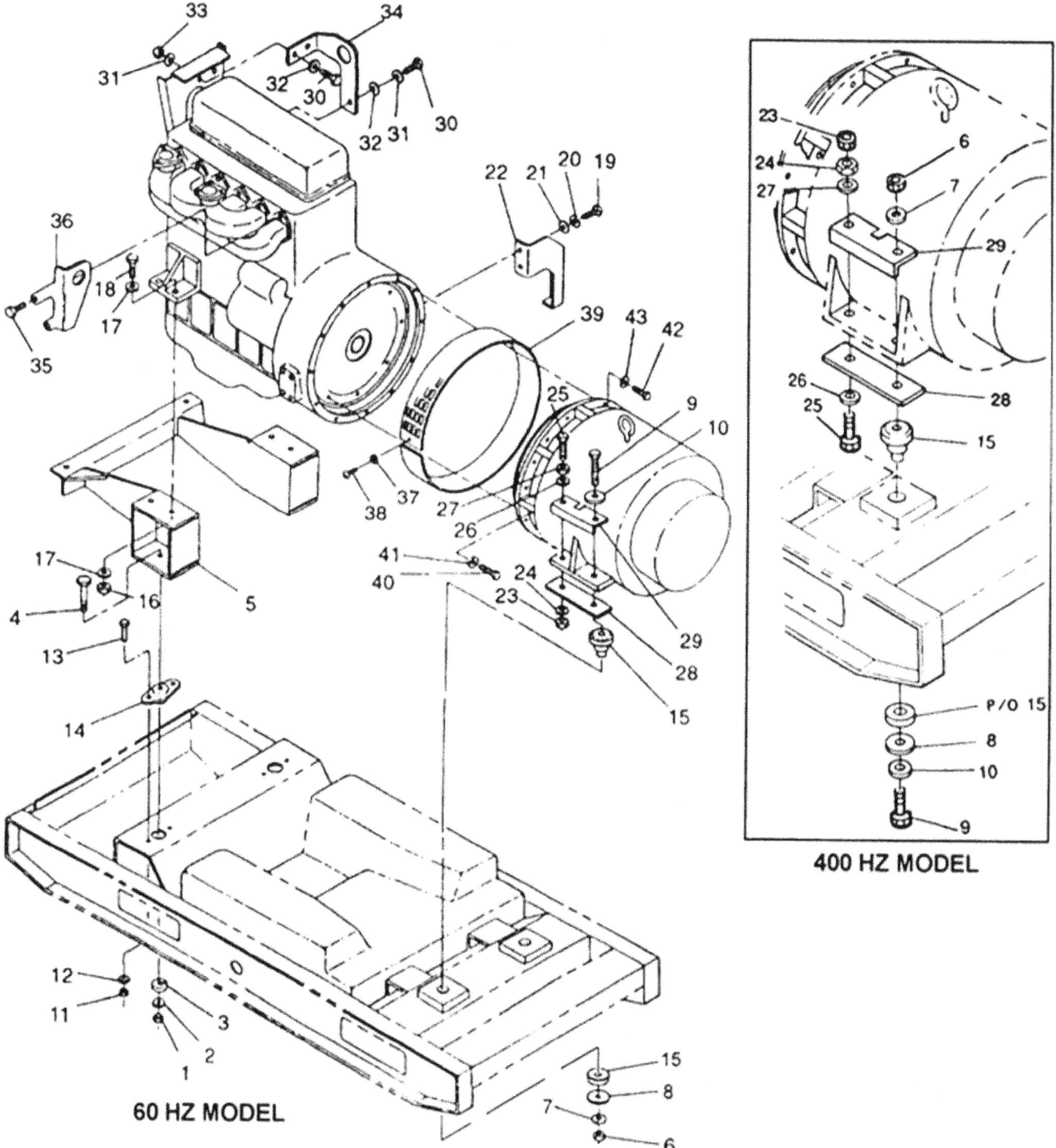

Figure 3-1. Engine and Generator Assembly – MEP-804A/MEP-814A.

3-6.2 Installation (MEP-804A/MEP-814A).

a. a. Install bracket (36, Figure 3-1) on engine with bolts (35).

b. b. Install bracket (34) on engine with bolts (30), new lock washers (31), washers (32), and nuts (33).

c. c. Install engine shock mounts (14) on skid base with bolts (13), washers (12), and nuts (11).

d. d. Position generator shock mounts (15) in skid base.

e. e. Install engine support bracket (5) on engine with bolts (18), washers (17), and nuts (16). Torque nuts to 31 ft-lbs (42 Nm).

f. f. Install engine support brackets (22) on engine with bolts (19), new lock washers (20), and washers (21).

g. g. Install generator mount angles (29) and plate (28) on generator with bolts (25), washers (24 and 26), and nuts (23 and 27).

CAUTION

Rated capacity of overhead hoist should be at least 2,000 pounds (907 Kg). Arrange lifting device so that it supports both engine and generator to avoid undue stress on the engine and generator coupling.

h. h. Attach lifting harness to engine and generator lifting points. Raise engine and generator assembly from maintenance stand or fixture.

i. i. Position engine and generator assembly on skid base, aligning mounting holes and brackets.

j. j. Install bolts (9), Belleville washers (10), snubbing washers (8), washers (7), and nuts (6) to secure generator to skid base. Torque bolts to 210 ft-lbs (285 Nm).

k. k. Adjust nuts (23 and 27) to obtain 0.5 inch (12.7 mm) minimum clearance between ends of bolts
(25)
and skid base.

l. l. Install bolts (4), snubbing washers (3), washers (2), nuts (1), and engine support bracket (5) to secure engine to skid base. Torque bolts to 75 ft-lbs (102 Nm).

m. m. Install rear forklift guide in skid base (paragraph 4-34.3).

n. n. Connect oil drain line to oil pan (paragraph 2-127.3).

o. o. Connect coolant overflow hose to overflow bottle (paragraph 2-87.3).

p. p. Connect block coolant drain line at engine fitting (paragraph 2-87.3).

q. q. Install radiator (paragraph 2-89.4).

r. r. Connect fuel line to fuel transfer pump from fuel pickup (paragraph 2-97.2).

s. s. Connect engine excess fuel return line at tank fitting and fuel filter/water separator fuel lines at engine fittings (paragraph 2-97.2).

t. Install generator set housing (paragraphs 2-16.4, 2-17.4, and 2-18.4).

u. u. Service engine lubrication system (paragraph 2-1.2.4).

v. v. Service coolant system (paragraph 2-1.2.2).

w. w. Install batteries (paragraph 2-12.5). Close battery access door.

x. x. Start generator set. Perform operational checks and check for leaks.

3-7 ENGINE AND GENERATOR ASSEMBLY (MEP-804B/MEP-814B) 3-7.1

WARNING

All metal jewelry can conduct electricity and become entangled in generator set components. Remove all jewelry when working on generator set. Failure to comply with this warning can cause injury or death to personnel.

WARNING

DO NOT wear loose clothing when performing checks, services and maintenance. Failure to comply with this warning can cause injury or death to personnel.

WARNING

Dangerous voltage exists on live circuits. Always observe precautions and never work alone. Failure to comply with this warning can cause injury or death to personnel.

WARNING

DC voltages are present at generator set electrical components even with generator set shut down. Avoid shorting any positive with ground/negative. Failure to comply with this warning can cause injury to personnel, and damage to equipment.

WARNING

Slave receptacle (NATO connector) is electrically live at all times and is unfused. The Battery Disconnect Switch does not remove power from the slave receptacle. NATO slave receptacle has 24 VDC even when Battery Disconnect Switch is set to OFF. This circuit is only dead when the batteries are fully disconnected. Disconnect the batteries before performing maintenance on the slave receptacle. Failure to comply with this warning can cause injury or death to personnel.

WARNING

Ensure that the engine cannot be started while maintenance is being performed. (ENGINE CONTROL switch set to OFF/RESET; Battery Disconnect Switch is OFF; DEAD CRANK SWITCH is OFF). Failure to comply with this warning can cause injury or death to personnel.

WARNING

When disconnecting or removing batteries, disconnect the negative lead that connects directly to the grounding stud first; disconnect the negative end of the interconnection cable next. When installing batteries, reverse the connection sequence. Failure to comply with this warning can cause injury to personnel.

WARNING

Many components require a two-person lift. Lifting heavy components can cause back strain. Ensure proper lifting techniques are used when lifting heavy components. Failure to comply with this warning can cause injury to personnel.

WARNING

Each battery weighs more than 70 pounds (32 kg) and requires a two-person lift. Lifting batteries can cause back strain. Ensure proper lifting techniques are used when lifting batteries. Failure to comply with this warning can cause injury to personnel.

WARNING

Support components when removing attaching hardware or component may fall. Failure to comply with this warning can cause injury to personnel, and damage to equipment.

WARNING

The generator set, engine, and generator are extremely heavy and require an assistant and a lifting device (forklift, overhead lifting device) with sufficient capacity. Failure to comply with this warning can cause injury or death to personnel.

WARNING

The connection of any electrical equipment and the disconnection of any electrical equipment may cause an explosion hazard. Do not connect any electrical equipment or disconnect any electrical equipment in an explosive atmosphere. Failure to comply with this warning can cause injury or death to personnel.

a. a. Shut down generator set.

b. b. Open battery access door and remove batteries (paragraph 2-12.2).

c. c. Drain oil and engine coolant (paragraph 2-79).

d. d. Remove generator set housing (par agraphs 2-16.1, 2-17.1, and 2-18.1).

e. e. Remove radiator filler hose and panel (paragraph 2-80.1).

f. f. Remove coolant recovery system overflow bottle (paragraph 2-95.2).

g. g. Remove radiator (paragraph 2-90.1).

h. h. Remove air cleaner tubing (paragraph 2-78.2).

i. i. Disconnect oil drain line at oil pan (paragraph 2-127.2).

j. j. Remove clamp (1, Figure 3-2) and disconnect hose (2) from bottom of fuel/water separator (3).

k. k. Remove clamp (5) and disconnect hose (4) from fuel filter (6).

l. l. Remove clamp (7) and disconnect hose (8) from fuel filter (9) on fuel pump.

m. m. Remove clamp (10) and disconnect hose (11) from fuel return (12) to tank.

n. n. Remove bolts (1, Figure 3-3), lock washers (2), washers (3), and rear engine support brackets (4). Discard lock washers (2).

o. o. Tag and disconnect wiring from engine components, as required.

CAUTION

Rated capacity of overhead hoist should be at least 2,000 pounds (907 Kg). Arrange lifting device so that it supports both engine and generator to avoid undue stress on the engine-generator coupling.

p. p. Attach lifting harness to engine and generator lifting points. Take up slack.

q. q. Remove nuts (5), washers (6), snubbing washers (7), and bolts (8) from engine mounts (19).

r. r. Remove nuts (11), washers (12), snubbing washers (13), bolts (14), and Belleville washers (15) from generator mounts.

s. Tie wrap generator power leads and secure out of the way.

WARNING

Keep hands and feet from underside of engine and generator while using lifting device to remove them from the skid base. Failure to comply with this warning can cause injury or death to personnel.

t. t. With aid of an assistant, slowly lift engine and generator assembly and move to maintenance work area. Support assembly on maintenance stand or fixture.

u. u. Remove engine shock mounts (19) from skid base by removing nuts (16), washers (17), and bolts (18).

v. v. Remove shock mounts (20) from generator mounting points on skid base.

w. w. Remove crossmember (25) by removing bolts (21) , lock washers (22), washers (23), and nuts (24).

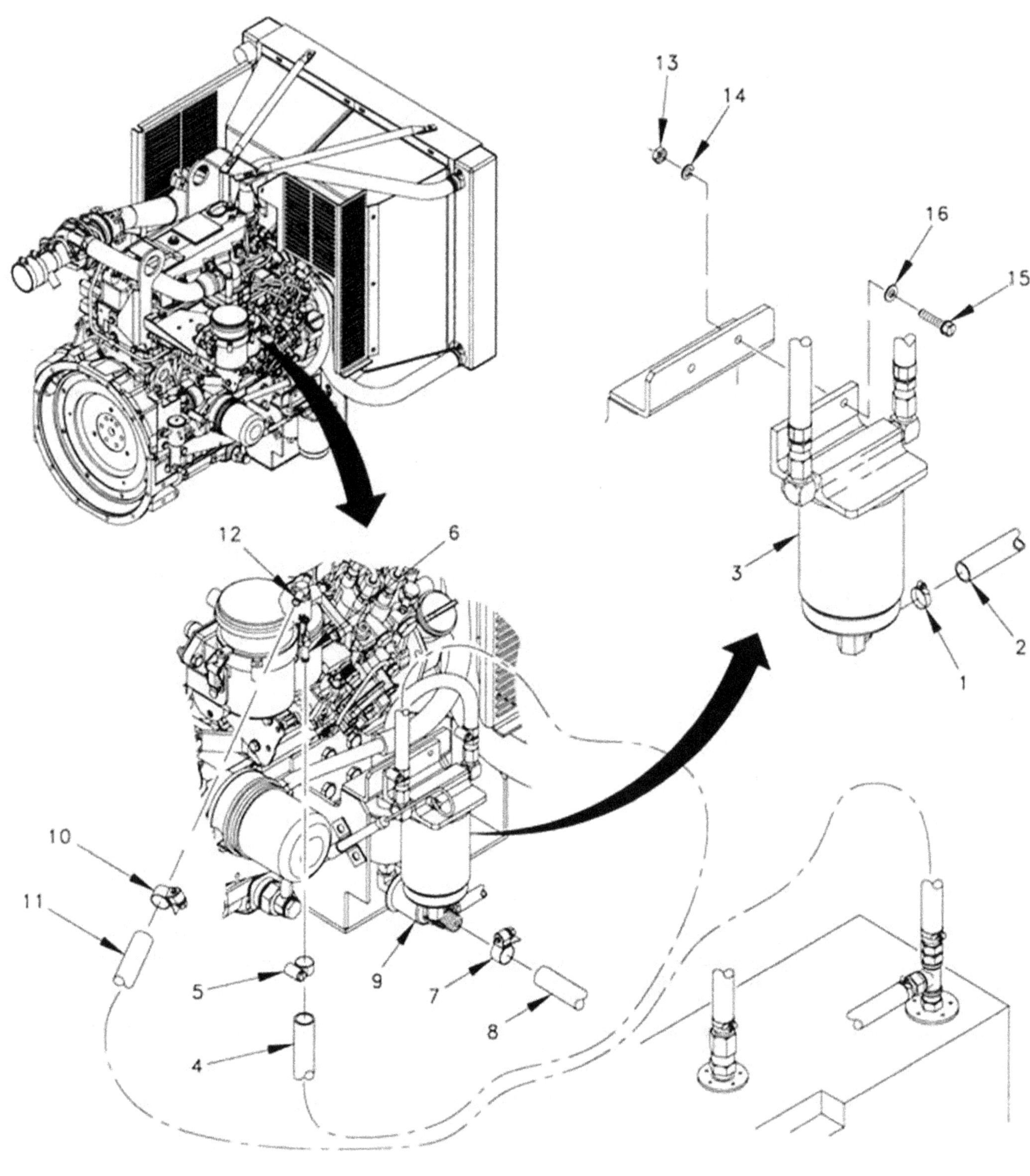

Figure 3-2. Engine Fuel Lines – MEP-804B/MEP-814B.

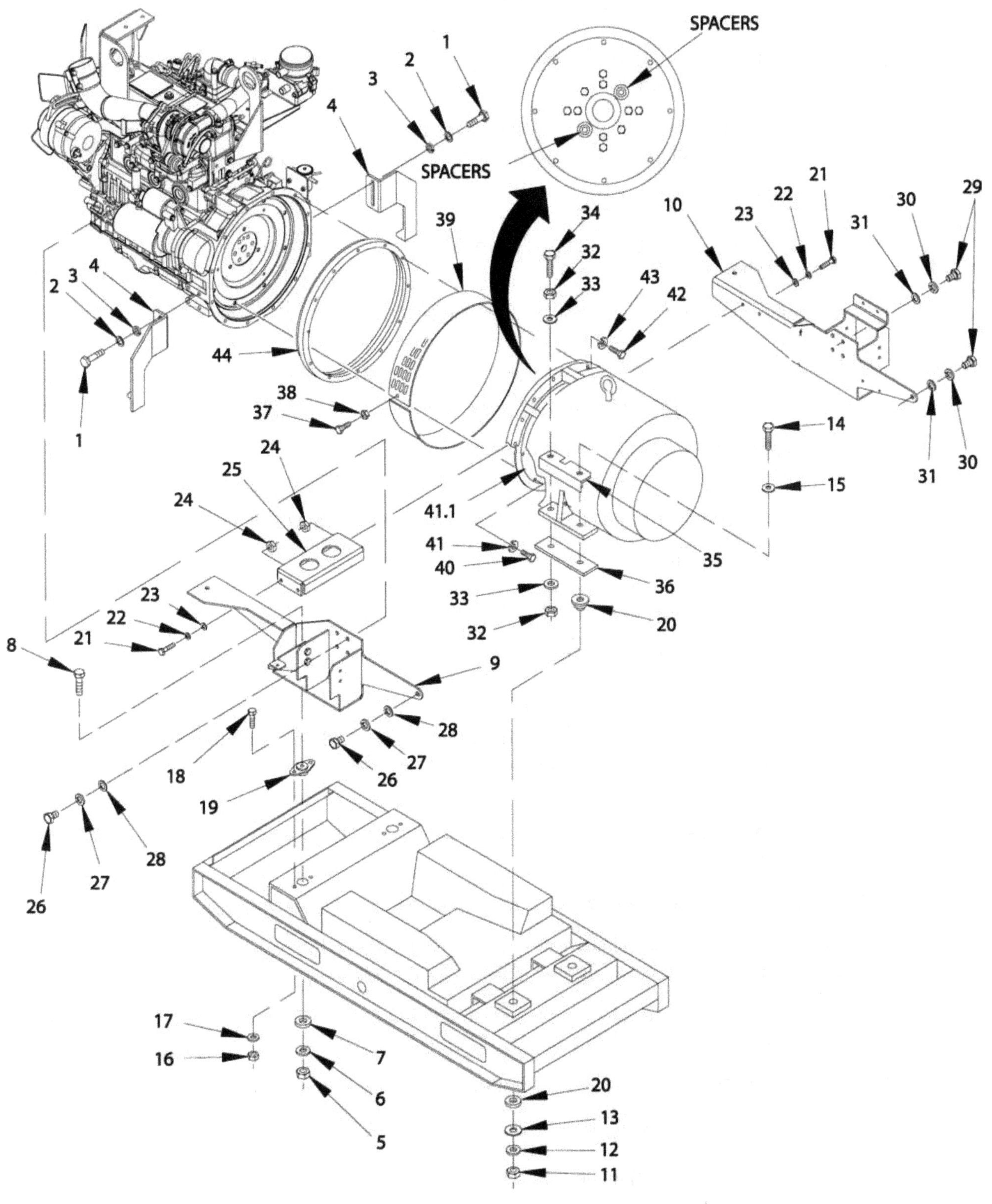

Figure 3-3. Engine and Generator Assembly – MEP-804B/MEP-814B.

x. x. Remove left engine support bracket (9) by removing bolts (26), lock washers (27), and washers (28). Discard lock washers (27).

y. y. Remove right engine support bracket (10) by removing bolts (29), lock washers (30), and washers (31). Discard lock washers (30).

z. z. Remove generator mount angles (35) and plates (36) by removing nuts (32), washers (33), and bolts (34) from generator.

3-7.2 Installation (MEP-804B/MEP-814B).

a Install right engine support bracket (10, Figure 3-3) on engine with bolts (29), new lock washers (30), and washers (31).

b. b. Install left engine support bracket (9) on engine wi th bolts (26), new lock washers (27), and wash-

ers

(28).

c. c. Install crossmember (25) with bolts (21), new lock washers (22), washers (23), and nuts (24).

d. d. Connect hose (4, Figure 3-2) from fuel/water separator (3) to fuel filter (6) and secure with clamp (5).

e. e. Install generator mount angles (35, Figure 3-3) and plates (36) with bolts (34), washers (33), and nuts (32). Adjust bolts to where they will not make contact when engine/generator is installed onto skid base. Tighten nuts.

f. f. Install engine shock mounts (19) to skid base with nuts (16), washers (17), and bolts (18).

g. g. Install shock mounts (20) to skid base.

CAUTION

Rated capacity of overhead hoist should be at least 2,000 pounds (907 Kg). Arrange lifting device so that it supports both engine and generator to avoid undue stress on the engine-generator coupling.

h. h. Attach lifting harness to engine and generator li fting points. Raise engine and generator assembly from maintenance stand or fixture.

i. i. Position engine and generator assembly on skid base, aligning mounting holes and brackets.

j. j. Install bolts (14), Belleville washers (15), snubbing washers (13), washers (12), and nuts (11) to secure generator to skid base. Torque bolts to 210 ft-lbs (285 Nm).

k. k. Install bolts (8), snubbing washers (7), washers (6), and nuts (5) to secure engine to skid base. Torque bolts to 75 ft-lbs (102 Nm).

l. l. Install rear engine support brackets (4) with bolts (1), new lock washers (2), and washers (3).

m. m. Connect oil drain line to oil pan (paragraph 2-127.3).

n. n. Install hose (8, Figure 3-2) to fuel filter (9) on fuel pump and secure with clamp (7).

o. o. Install hose (11) to fuel return (12) (to tank) and secure with clamp (10).

p. p. Install hose (2) to bottom of fuel/water separator (3) and secure with clamp (1).

q. q. Install air cleaner tubing (paragraph 2-78.4).

r. r. Connect tagged wiring to e ngine components. Remove tags.

s. s. Install radiator (paragraph 2-90.4).

t. t. Install coolant recovery system overflow bottle (paragraph 2-95.4).

u. u. Install radiator filler hose and panel (paragraph 2-80.3).

v. v. Install generator set housing (paragraphs 2-16.4, 2-17.4, and 2-18.4).

w. w. Service engine lubrication system (paragraph 2-1.2.4).

x. x. Service coolant system (paragraph 2-1.2.2).

y. y. Service fuel system (paragraph 2-1.2.3).

z. z. Install batteries (paragraph 2-12.5). Close battery access door.

aa. Start generator set. Perform operational checks and check for leaks.

WARNING

All metal jewelry can conduct electricity and become entangled in generator set components. Remove all jewelry when working on generator set. Failure to comply with this warning can cause injury or death to personnel.

WARNING

DO NOT wear loose clothing when performing checks, services and maintenance. Failure to comply with this warning can cause injury or death to personnel.

WARNING

Dangerous voltage exists on live circuits. Always observe precautions and never work alone. Failure to comply with this warning can cause injury or death to personnel.

WARNING

DC voltages are present at generator set electrical components even with generator set shut down. Avoid shorting any positive with ground/negative. Failure to comply with this warning can cause injury to personnel, and damage to equipment.

WARNING

Slave receptacle (NATO connector) is electrically live at all times and is unfused. The Battery Disconnect Switch does not remove power from the slave receptacle. NATO slave receptacle has 24 VDC even when Battery Disconnect Switch is set to OFF. This circuit is only dead when the batteries are fully disconnected. Disconnect the batteries before performing maintenance on the slave receptacle. Failure to comply with this warning can cause injury or death to personnel.

WARNING

Ensure that the engine cannot be started while maintenance is being performed. (ENGINE CONTROL switch set to OFF/RESET; Battery Disconnect Switch is OFF; DEAD CRANK SWITCH is OFF). Failure to comply with this warning can cause injury or death to personnel.

WARNING

When disconnecting or removing batteries, disconnect the negative lead that connects directly to the grounding stud first; disconnect the negative end of the interconnection cable next. When installing batteries, reverse the connection sequence. Failure to comply with this warning can cause injury to personnel.

WARNING

Many components require a two-person lift. Lifting heavy components can cause back strain. Ensure proper lifting techniques are used when lifting heavy components. Failure to comply with this warning can cause injury to personnel.

WARNING

Each battery weighs more than 70 pounds (32 kg) and requires a two-person lift. Lifting batteries can cause back strain. Ensure proper lifting techniques are used when lifting batteries. Failure to comply with this warning can cause injury to personnel.

WARNING

Support components when removing attaching hardware or component may fall. Failure to comply with this warning can cause injury to personnel, and damage to equipment.

WARNING

The generator set, engine, and generator are extremely heavy and require an assistant and a lifting device (forklift, overhead lifting device) with sufficient capacity. Failure to comply with this warning can cause injury or death to personnel.

WARNING

The connection of any electrical equipment and the disconnection of any electrical equipment may cause an explosion hazard. Do not connect any electrical equipment or disconnect any electrical equipment in an explosive atmosphere. Failure to comply with this warning can cause injury or death to personnel.

3-8.1 Removal (MEP-804A/MEP-814A).

a. a. Shut down generator set.

b. b. Open battery access door and remove batteries (paragraph 2-12.2).

c. c. Using a suitable container, drain engine oil (paragraph 2-1.2.4).

WARNING

Cooling system operates at high temperature and pressure. Contact with high pressure steam and/or liquids can result in burns and scalding. Shut down generator set, and allow system to cool before performing checks, services and maintenance, or wear gloves and additional protective clothing and goggles as required. Failure to comply with this warning can cause injury or death to personnel.

WARNING

Always remove radiator cap slowly to permit any pressure to escape. Failure to comply with this warning can cause injury to personnel.

d. d. Using a suitable container, drain engine coolant from radiator and engine block (paragraph 2-79.2).

WARNING

The high pressure oil system operates at high temperature and pressure. Contact with hot oil can result in burns and scalding. Shut down generator set, and allow system to cool before performing checks, services, and maintenance. Wear heat resistant gloves and avoid contacting hot surfaces. Do not allow hot oil or components to contact skin or hands. Failure to comply with this warning can cause injury or death to personnel.

e. e. Remove control box assembly (paragraph 2-19.2).

f. f. Remove bolts (1 and 4, Figure 3-4), lock washers (2 and 5), washers (3 and 6), and top housing

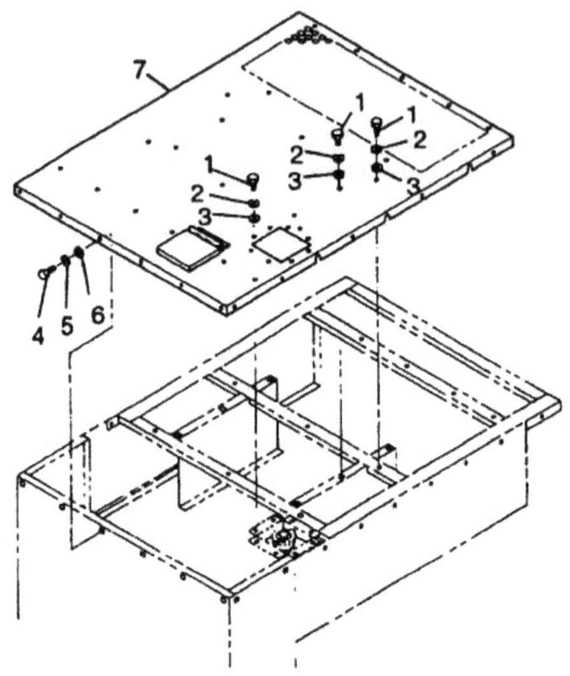

Figure 3-4. Generator Set Top Housing Panel.

g. g. Loosen clamp (1, Figure 3-5) at radiator (2) and disconnect filler hose and panel assembly (5) from radiator.

h. h. Loosen clamp (3) and disconnect overflow hose (4) from filler hose and panel assembly (5).

i. i. Remove filler hose and panel assembly (5) from generator set.

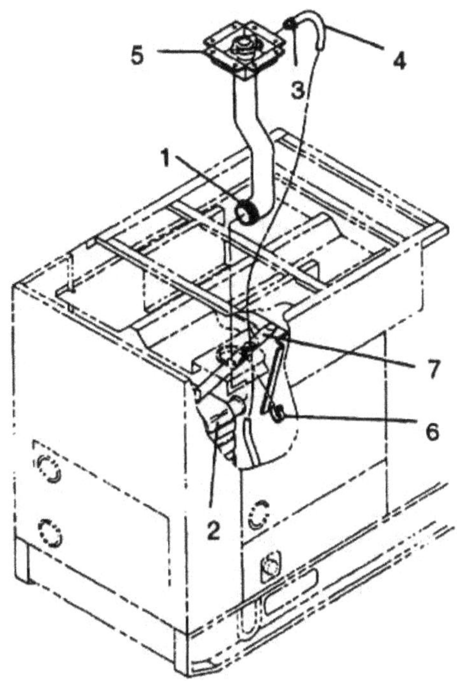

Figure 3-5. Filler Hose and Panel Assembly.

k. k. Remove clamp (1, Figure 3-6) and disconnect hose (2) from overflow bottle assembly (6).

l. l. Remove bolts (3), nuts (4), lock washers (5), and overflow bottle assembly (6) from left side of engine. Discard lock washers (5).

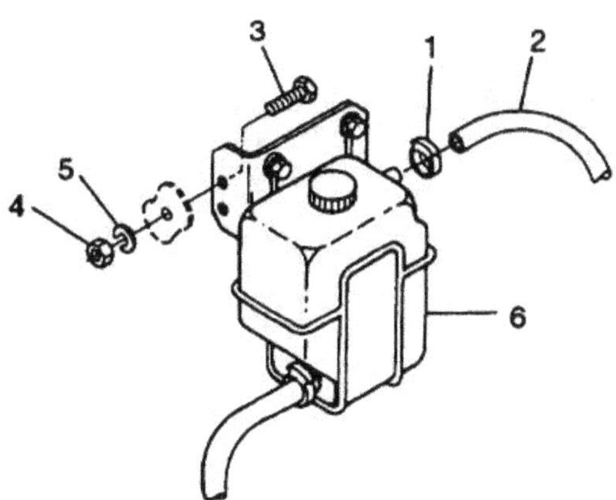

Figure 3-6. Overflow Bottle Assembly.

m. Remove exhaust pipe clamp (1, Figure 3-7) at exhaust manifold adapter (2).

n. n. Remove nuts (3), lock washers (4), washers (6), and bolts (5) securing top housing assembly (11) to front housing. Discard lock washers (4).

o. o. Remove assembled nuts (7) and bolts (8) securing top housing assembly (11) to rear side panels.

p. p. Remove assembled nut (9) and bolt (10) securing top housing assembly (11) to output box angle.

q. q. Using a lifting device, remove top housing assembly (11).

r. r. Remove fan guards and mounting brackets (paragraph 2-81.2 (MEP-804A/MEP-814A)/paragraph 2-82.2 (MEP-804B/MEP-814B)).

s. s. Remove nuts (1, Figure 3-8), lock washers (2 and 6), bolts (3 and 5), washers (4), and support

rods

(7) from engine lifting bracket and radiator assembly (27). Discard lock washers (2 and 6).

t. t. Remove bolts (10 and 12), washers (11 and 14), lock washers (9 and 13), nuts (8), and shroud halves (15) from radiator assembly (27). Discard lock washers (9 and 13).

u. u. Remove bolts (16) and fan (17) from water pump.

v. v. Loosen clamp (18) at thermostat housing and disconnect hose (19).

w. w. Loosen clamp (20) at water pump and disconnect hose (21).

x. x. Remove clamp (22) at radiator drain valve (24) and disconnect hose (23).

y. y. Remove nuts (25) and washers (26) securing radiator assembly (27). With aid of an assistant, remove radiator assembly (27) and shims (28).

z. Remove bolts (1 and 4, Figure 3-9), washers (3 and 5), lock washers (2 and 6), nuts (7), and rear housing panel (8) from generator set. Discard lock washers (2 and 6).

aa. Loosen clamp (1, Figure 3-10) securing fuel filler panel assembly (6) to fuel tank. Cap openings.

ab. Loosen clamp (2) and disconnect hose (3) from fuel filler panel assembly (6). Cap openings. ac.

Disconnect fuel hose (4) from fuel line (5). Cap openings.

ad. Tag and disconnect auxiliary fuel pump (7) and fuel float module (8) electrical connectors.

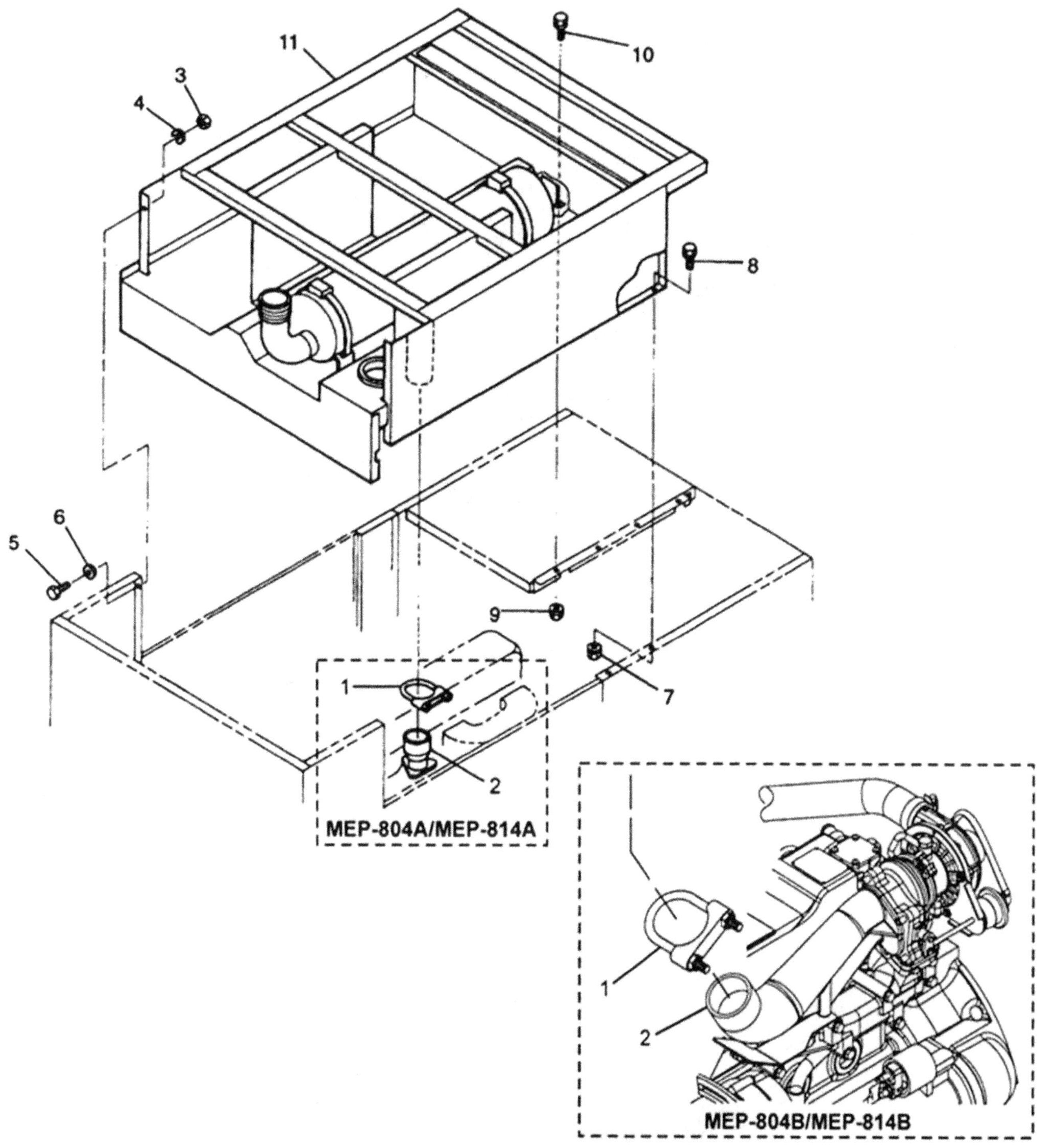

Figure 3-7. Top Housing Assembly.

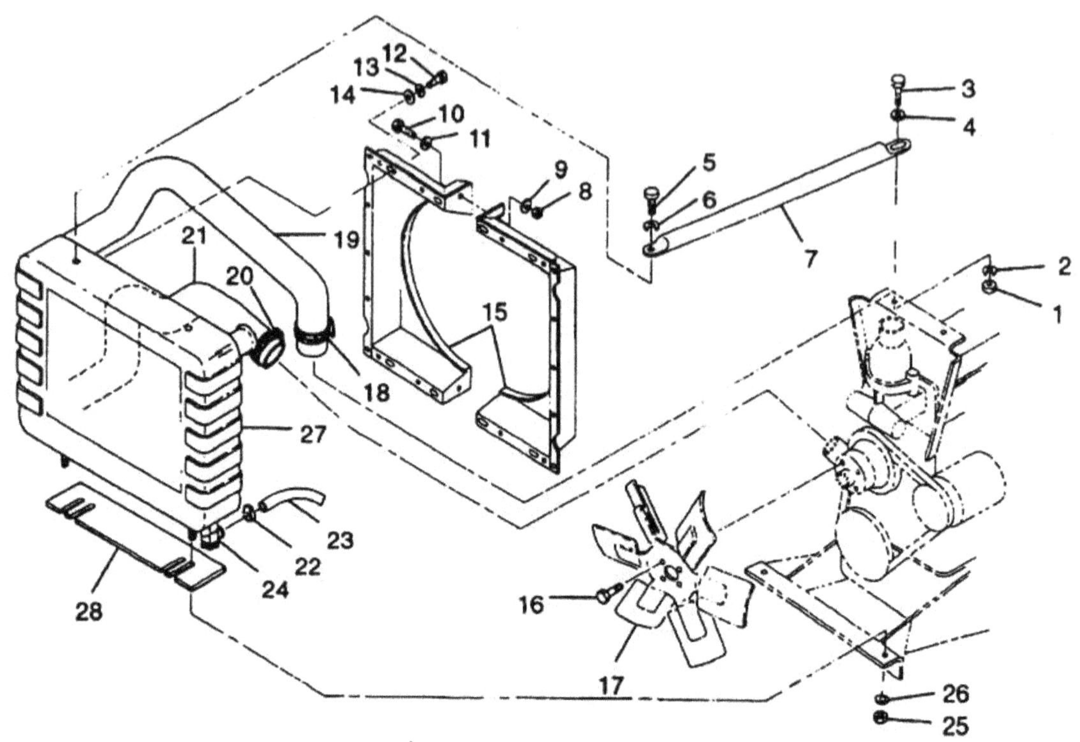

Figure 3-8. Radiator Assembly, Shroud and Fan – MEP-804A/MEP-814A.

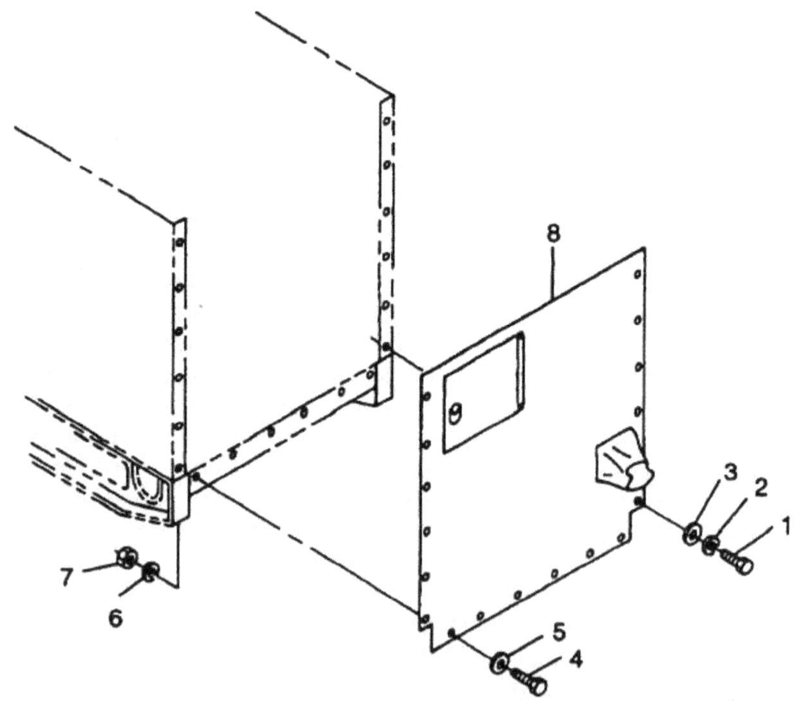

Figure 3-9. Rear Housing Panel.

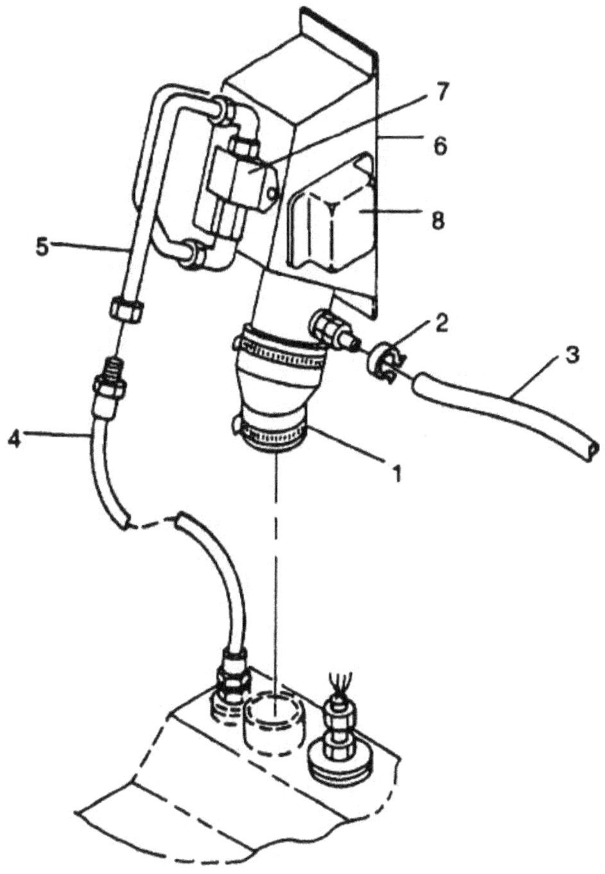

Figure 3-10. Fuel Filler Panel Assembly.

ae. Remove bolts (1, Figure 3-11), lock washers (2), battery cables (3), and slave cables (4) from NATO slave receptacle (5). Discard lock washers (2).

af. Remove bolts (1, Figure 3-12), lock washers (2), an d washers (3) securing left rear side panel (4) to skid base. Discard lock washers (2).

ag. Remove nuts (8), lock washers (6 and 9), washer s (7 and 11), and bolts (5 and 10) securing lower left side panel (12) to front housing and skid base. Discard lock washers (6 and 9).

ah. With aid of an assistant, remove left rear si de panel (4) and lower left side panel (12) as an assembly.

ai. Loosen clamps (1, Figure 3-13) and remove air intake hoses (2, 3, and 4) as an assembly. aj.

Loosen clamp (1, Figure 3-14) and disconnect oil drain hose (2) from engine oil pan.

ak. On right side of engine, tag and disconnect electrical leads from glow plug contactor (1, Figure 3-15), low oil pressure switch (2), oil pressure sender (3), coolant temperature sender (4), magnetic pickup (5), DEAD CRANK switch (6), and fuel injection pump governor actuator (7).

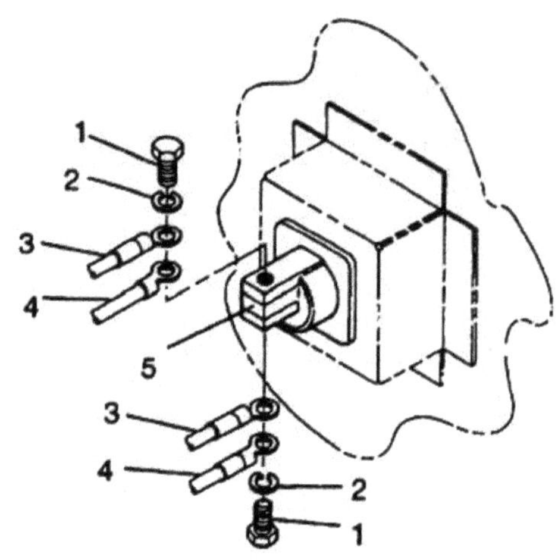

Figure 3-11. NATO Slave Receptacle.

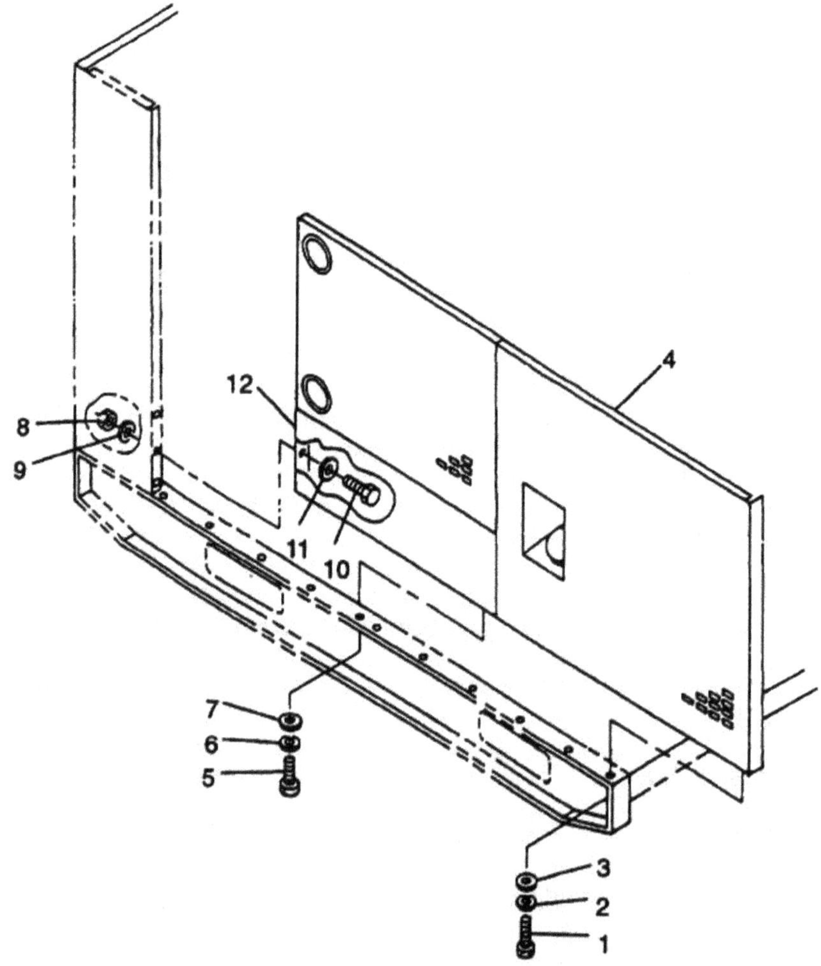

Figure 3-12. Left Side Housing Panels.

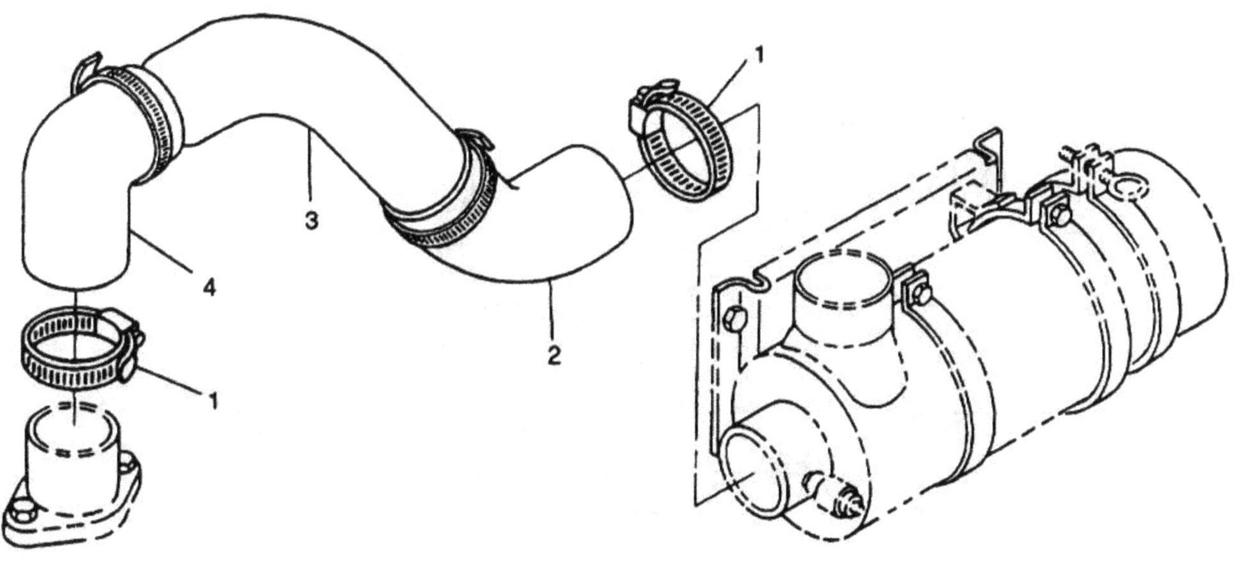

Figure 3-13. Air Cleaner Assembly – MEP-804A/MEP-814A.

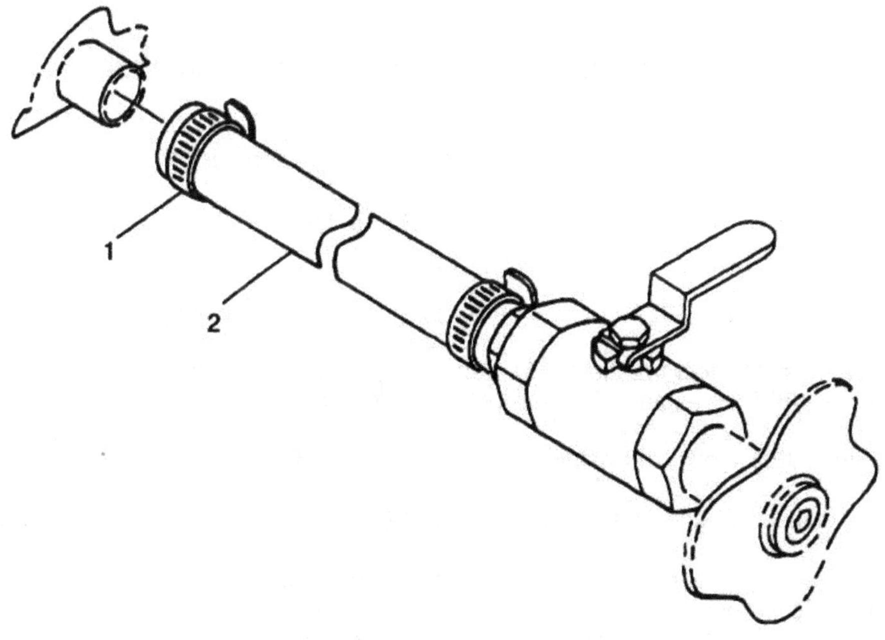

Figure 3-14. Oil Drain Line.

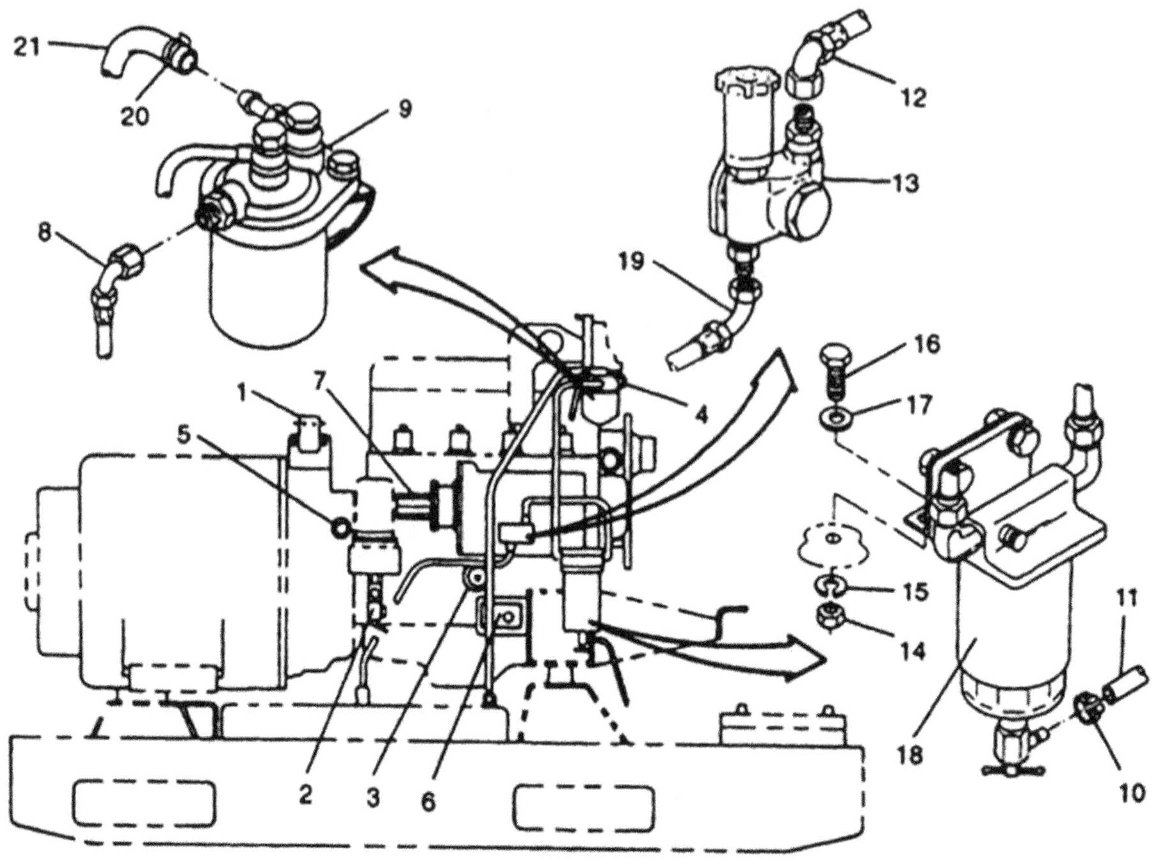

Figure 3-15. Right Side Engine Components – MEP-804A/MEP-814A.

al. Disconnect fuel line (8) from engine fuel filter (9). Cap openings.

am. Remove clamp (10) and disconnect hose (11) from fuel filter/water separator assembly (18). Cap openings.

an. Disconnect fuel line (12) from fuel transfer pump (13). Cap openings.

ao. Remove nuts (14), lock washers (15), bolts (16) , washers (17), and fuel filter/water separator assembly (18) from engine. Discard lock washers (15).

ap. Disconnect fuel line (19) from fuel transfer pump (13). Cap openings.

aq. Loosen clamp (20) and remove fuel return line (2 1) from engine fuel filter (9). Cap openings.

ar. On left side of engine, tag and disconnect electrical leads from battery charging alternator (1, Figure 3-16), starter solenoid (2), starter motor (3), and coolant high temperature switch (4). Move engine electrical harness to rear of generator set and clear of engine.

as. Loosen clamp (5) and disconnect coolant drain hose (6) from engine block coolant drain valve (7).

at. Remove nut (1, Figure 3-17), lock washers (2), and bolt (3) securing ground strap (4) to skid base. Discard lock washers (2).

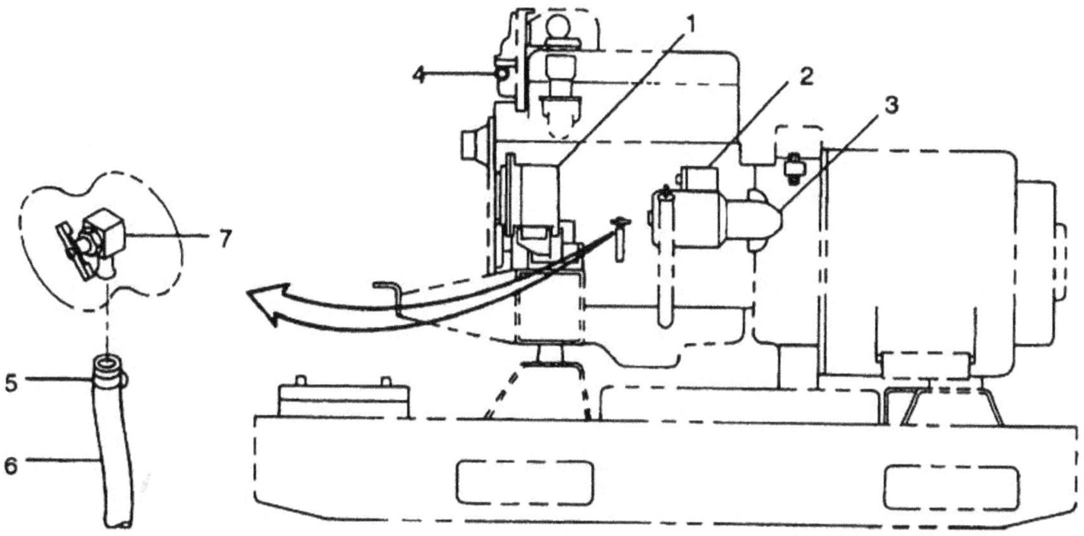

Figure 3-16. Left Side Engine Components – MEP-804A/MEP-814A.

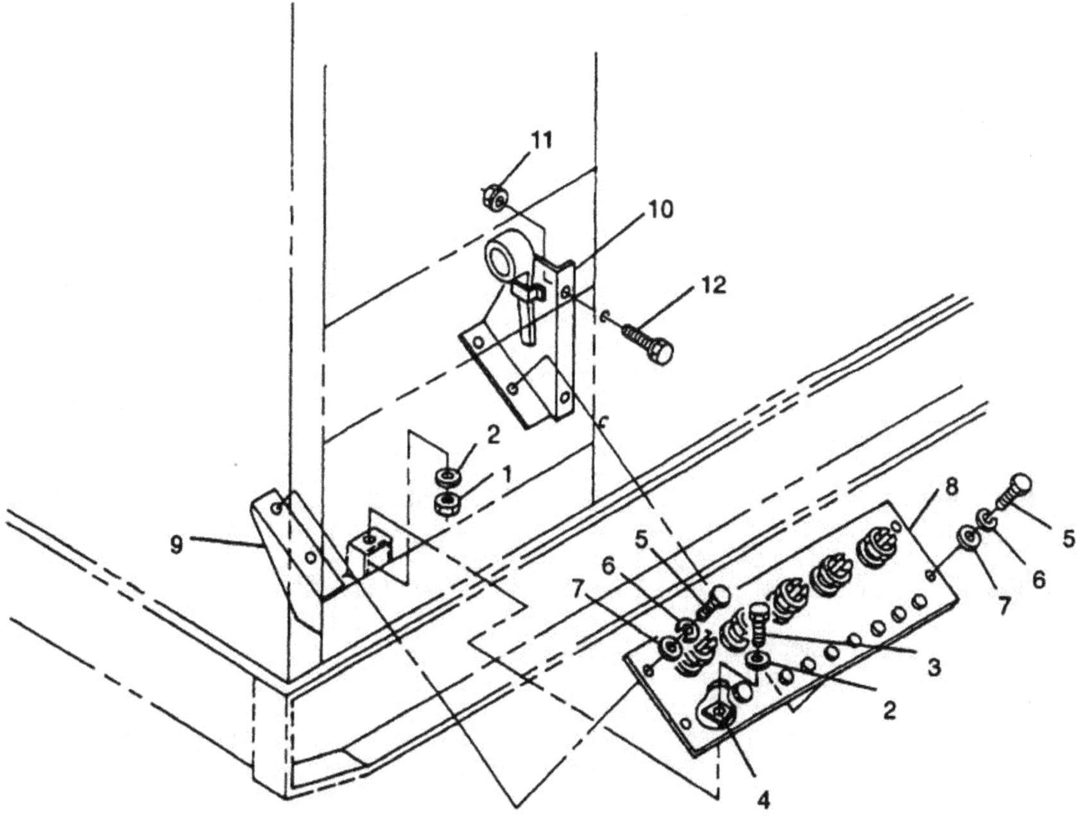

Figure 3-17. Load Output Terminal Board.

au. Remove bolts (5), washers (7), and lock washers (6) securing load output terminal board assembly (8) to supports (9 and 10). Pull load output terminal board assembly out through access door. Discard lock washers (6).

av. Remove assembled nuts (11), bolts (12), and support (10) from right side panel.

aw. Loosen nuts (23 and 27, Figure 3-1), turn bolts (25) to contact skid base, and tighten nuts (23 and 27).

ax. Remove screws (38), washers (37), and screen/cover (39) from generator case.

ay. Scribe mark on generator drive disc and engine flywheel for alignment of bolts during installation.

az. Remove bolts (40) and lock wash ers (41) securing generator drive disc to engine flywheel. Discard lock washers (41).

WARNING

Rated capacity of overhead hoist should be at least 1,500 pounds (680 kg). Do not use a hoist with less capacity. Failure to comply with this warning can cause injury or death to personnel, and damage to equipment.

ba. Attach lifting harness to engine and overhead hoist. Take up slack.

bb. Remove bolts (42) and lock washers (43) securi ng generator to flywheel housing. Discard lock washers (43).

bc. Remove nuts (1), washers (2), snubbing washers (3), and bolts (4) securing support bracket (5) to skid base.

WARNING

Keep hands and feet from underside of engine and generator while using lifting device to remove them from the skid base. Failure to comply with this warning can cause injury or death to personnel.

bd. With aid of an assistant, slowly lift engine assembly from skid base, ensuring that engine flywheel housing separates smoothly from generator without binding. Remove engine from generator set housing and place on engine stand.

be. Remove nuts (11), lock washers (12), bolts (13), and shock mounts (14) from skid base. Discard lock washers (12).

bf. Remove nuts (16), washers (17), bolts (18), and support frame (5) from engine.

bg. Remove bolts (19), lock washers (20), washers (21), and rear support brackets (22) from engine block. Discard lock washers (20).

bh. Remove bolts (30), lock washers (31), washers (32), nuts (33), and front engine lifting bracket (34) from engine assembly. Discard lock washers (31).

bi. Remove bolts (35) and rear engine lifting bracket (36) from engine assembly.

3-8.2 Installation (MEP-804A/MEP-814A).

a. a. Install rear engine lifting bracket (36, Fi gure 3-1) on engine assembly with bolts (35).

b. b. Install front engine lifting bracke t (34) on engine assembly with bolts (30), new lock washers (31), washers (32), and nuts (33).

c. c. Install rear support brackets (22) on engine block with bolts (19), new lock washers (20), and washers (21).

d. d. Install shock mounts (14) on skid base with bolts (13), new lock washers (12), and nuts (11).

e. e. Install support frame (5) on engine with bolts (18), washers (17), and nuts (16). Torque nuts to 31

WARNING

Rated capacity of overhead hoist should be at least 1,500 pounds (680 kg). Do not use a hoist with less capacity. Failure to comply with this warning can cause injury or death to personnel, and damage to equipment.

f. f. Attach lifting harness to engine and overhead hoist. Take up slack.

g. g. With aid of an assistant, lift engine from engine stand and position engine on skid base, aligning mounting holes, brackets, and generator to flywheel housing.

h. h. Install bolts (42) and new lock washers (43), tightening bolts (42) slowly to ensure even and proper seating of generator housing lip to flywheel housing. Torque bolts to 31 ft-lbs (42 Nm).

i. i. Secure engine support bracket (5) to skid base with bolts (4), snubbing washers (3), washers (2), and nuts (1). Torque bolts to 75 ft-lbs (102 Nm).

j. j. Align scribe mark on generator drive disc and engine flywheel, and install bolts (40) and new lock washers (41) securing generator drive disc to engine flywheel. Torque bolts to 35 ft-lbs (47 Nm).

k. k. Install screen/cover (39) on generator case with screw (38) and washer (37).

l. l. Loosen nuts (23 and 27). Adjust bolts (25) to obtain 0.5 inch (12.7 mm) minimum clearance with skid base. Torque nuts to 150 ft-lbs (204 Nm).

m. m. Install support (10, Figure 3-17) on right side panel with bolts (12) and assembled nuts (11).

n. n. Install load output terminal board assembly (8) on supports (9 and 10) with bolts (5), new lock washers (6), and washers (7).

o. o. Secure ground strap (4) to skid base with bolt (3), new lock washers (2), and nut (1).

p. p. Position engine electrical harness and connect electrical leads to battery charging alternator (1, Figure 3-16), starter solenoid (2), starter motor (3), and coolant high temperature switch (4) on left side of engine. Remove tags.

q. q. Connect coolant drain hose (6) at engine block drain valve (7) and tighten clamp (5).

r. r. Remove caps and connect fuel line (19, Fi gure 3-15) to fuel transfer pump (13).

s. s. Remove caps, connect fuel return line (21), a nd tighten clamp (20) at engine fuel filter (9).

t. t. Install fuel filter/water separator assembly (18) on engine with bolts (16), washers (17), new lock washers (15), and nuts (14).

u. u. Remove caps and connect fuel line (12) to fuel transfer pump (13).

v. v. Remove caps, connect hose (11), and install clamp (1 0) at fuel filter/water separator assembly (18).

w. w. Remove caps and connect fuel line (8) to engine fuel filter (9).

x. x. Connect electrical leads to fuel injection pump governor actuator (7), DEAD CRANK switch (6), magnetic pickup (5), coolant temperature sender (4), oil pressure sender (3), low oil pressure switch (2), and glow plug contactor (1). Remove tags.

y. y. Connect oil drain hose (2, Figure 3-14) to engine oil pan and tighten clamp (1).

z. z. Position air intake hoses (2, 3, and 4, Figure 3-13) in generator set and tighten clamps (1).

aa. With aid of an assistant, position left rear side housing panel (4, Figure 3-12) and lower side panel (12) on generator set.

ab. Secure left rear side panel (4) to skid base with bolts (1), new lock washers (2), and washers (3).

ac. Secure lower left side panel (12) to front housing and skid base with bolts (5 and 10), washers (7 and 11), new lock washers (6 and 9), and nuts (8).

ad. Connect battery cables (3, Figure 3-11) and slave cables (4) to NATO slave receptacle (5) with bolts (1) and new lock washers (2).

ae. Connect auxiliary fuel pump (7, Figure 3-10) and fuel float module (8) electrical connectors. Remove tags.

af. Remove caps and connect fuel hose (4) to fuel line (5).

ag. Remove caps and connect hose (3) to fuel filler panel assembly (6) and tighten clamp (2).

ah. Remove caps and connect fuel filler panel assembly (6) to fuel tank and tighten clamp (1).

ai. Install rear housing panel (8, Figure 3-9) on generator set with bolts (1 and 4), washers (3 and 5), new lock washers (2 and 6), and nuts (7).

aj. With aid of an assistant, position radiator assembly (27, Figure 3-8) and shims (28) in generator set and secure with nuts (25) and washers (26).

ak. Connect hose (21) to water pump and tighten clamp (20).

al. Connect hose (19) at thermostat housing and tighten clamp (18).

am. Connect hose (23) at radiator drain valve (24) and install clamp (22).

an. Install fan (17) on water pump with bolts (16). Torque bolts to 24 ft-lbs (33 Nm).

ao. Install shroud halves (15) on radiator assembly (27) with bolts (10 and 12), washers (11 and 14), new lockwashers (9 and 13), and nuts (8).

ap. Install support rods (7) to radiator assembly (27) and front engine lifting bracket with bolts (3 and 5), washers (4), new lock washers (2 and 6), and nuts (1).

aq. Install fan guards and mounting brackets (paragraph 2-81.3).

ar. Using lifting device, position top housing assembly (11, Figure 3-7) on generator set.

as. Secure top housing assembly (11) to output box angle with bolt (10) and assembled nut (9).

at. Secure top housing assembly (11) to rear side panels with bolts (8) and assembled nuts (7).

au. Secure top housing assembly (11) to front housing with bolts (5), washers (6), new lock washers (4), and nuts (3).

av. Connect exhaust pipe to exhaust manifold adapter (2) and install clamp (1).

aw. Install overflow bottle assembly (6, Figure 3-6) on left side of engine with bolts (3), new lock washers (5), and nuts (4).

ax. Connect hose (2) on overflow bottle assembly (6) and install clamp (1).

ay. Position filler hose and panel assembly (5, Figure 3-5) in generator set, connect to radiator (2), and tighten clamp (1).

az. Connect overflow hose (4) to filler hose and panel assembly (5) and tighten clamp (3).

ba. Install new ties (6) on radiator supports (7) to secure overflow hose (4).

bb. Install top housing panel (7, Figure 3-4) with bolts (1 and 4), new lock washers (2 and 5), and washers (3 and 6).

bc. Service coolant system (paragraph 2-1.2.2). bd.

Service lubrication system (paragraph 2-1.2.4). be.

Install control box assembly (paragraph 2-19.4).

bf. Install batteries (paragraph 2-12.5). Close battery access door.

bg. Start generator set and check for leaks and proper operation. bh. Shut down

generator set and service fluid levels, as necessary. **3-8.3** **Replacement**

(MEP-804A/MEP-814A).

a. a. Remove old engine assembly from generator set (paragraph 3-8.1).

b. b. Remove the following components that come with new engine and return with old engine:

(1) Fan belt.
(2) Mechanical governor on rear of fuel injection pump.
(3) Filter element for engine mounted fuel filter.

c. Remove the following components from old engine and install on new engine:

(1) Oil sample valve, low oil pressure switch, and fitting assembly (1, Figure 3-18).
(2) Loop clamp (2, Figure 3-18). Use hardware from old engine.
(3) Fuel filter fitting (3, Figure 3-18).
(4) Fuel transfer pump fittings (4, Figure 3-18).
(5) Oil pressure sender and fitting assembly (5, Figure 3-18).

(6) Radiator support bracket (6, Figure 3-18). Use hardware from old engine.

(7) Glow plug contactor (12, Figure 2-35) and mounting bracket (15). Use hardware from old engine.

(8) Exhaust adapter (12, Figure 2-21) and gasket (13). Use hardware from old engine.

(9) Governor actuator (3, Figure 4-6) and fuel Injection pump interface plate (10) (paragraph 4-14.2).

(10) Coolant drain valve (1, Figure 2-37).

(11) Coolant temperature sender (3, Figure 2-35).

(12) Coolant high temperature switch (2, Figure 2-37).

(13) Magnetic pickup (4, Figure 2-35). Screw into flywheel housing until magnetic pickup bottoms out, back out 1-1/2 turns, and tighten jam nut.

(14) Fan belt (6, Figure 2-27).

(15) Lower starter mounting bolt. Bolt replaces stud in new engine.

d. d. Install new engine in generator set (paragraph 3-8.2).

e. e. Check and adjust magnetic pickup, as necessary (paragraph 2-121.4).

f. f. Check and adjust fan belt, as necessary (paragraph 2-93.2).

g. g. Start generator set and check for leaks and proper operation.

3-9 ENGINE ASSEMBLY (MEP-804B/MEP-814B)

WARNING

All metal jewelry can conduct electricity and become entangled in generator set components. Remove all jewelry when working on generator set. Failure to comply with this warning can cause injury or death to personnel.

WARNING

DO NOT wear loose clothing when performing checks, services and maintenance. Failure to comply with this warning can cause injury or death to personnel.

WARNING

Dangerous voltage exists on live circuits. Always observe precautions and never work alone. Failure to comply with this warning can cause injury or death to personnel.

WARNING

DC voltages are present at generator set electrical components even with generator set shut down. Avoid shorting any positive with ground/negative. Failure to comply with this warning can cause injury to personnel, and damage to equipment.

WARNING

Slave receptacle (NATO connector) is electrically live at all times and is unfused. The Battery Disconnect Switch does not remove power from the slave receptacle. NATO slave receptacle has 24 VDC even when Battery Disconnect Switch is set to OFF. This circuit is only dead when the batteries are fully disconnected. Disconnect the batteries before performing maintenance on the slave receptacle. Failure to comply with this warning can cause injury or death to personnel.

WARNING

Ensure that the engine cannot be started while maintenance is being performed. (ENGINE CONTROL switch set to OFF/RESET; Battery Disconnect Switch is OFF; DEAD CRANK SWITCH is OFF). Failure to comply with this warning can cause injury or death to personnel.

WARNING

When disconnecting or removing batteries, disconnect the negative lead that connects directly to the grounding stud first; disconnect the negative end of the interconnection cable next. When installing batteries, reverse the connection sequence. Failure to comply with this warning can cause injury to personnel.

WARNING

Many components require a two-person lift. Lifting heavy components can cause back strain. Ensure proper lifting techniques are used when lifting heavy components. Failure to comply with this warning can cause injury to personnel.

WARNING

Each battery weighs more than 70 pounds (32 kg) and requires a two-person lift. Lifting batteries can cause back strain. Ensure proper lifting techniques are used when lifting batteries. Failure to comply with this warning can cause injury to personnel.

WARNING

Support components when removing attaching hardware or component may fall. Failure to comply with this warning can cause injury to personnel, and damage to equipment.

WARNING

The generator set, engine, and generator are extremely heavy and require an assistant and a lifting device (forklift, overhead lifting device) with sufficient capacity. Failure to comply with this warning can cause injury or death to personnel.

WARNING

The connection of any electrical equipment and the disconnection of any electrical equipment may cause an explosion hazard. Do not connect any electrical equipment or disconnect any electrical equipment in an explosive atmosphere. Failure to comply with this warning can cause injury or death to personnel.

3-9.1 Removal (MEP-804B/MEP-814B).

a. a. Shut down generator set.

b. b. Open battery access door and remove batteries (paragraph 2-12.2).

c. c. Using a suitable container, drain engine oil (paragraph 2-1.2.4).

WARNING

Cooling system operates at high temperature and pressure. Contact with high pressure steam and/or liquids can result in burns and scalding. Shut down generator set, and allow system to cool before performing checks, services and maintenance, or wear gloves and additional protective clothing and goggles as required. Failure to comply with this warning can cause injury or death to personnel.

WARNING

Always remove radiator cap slowly to permit any pressure to escape. Failure to comply with this warning can cause injury to personnel.

d. d. Using a suitable container, drain engine coolant from radiator and engine block (paragraph 2-79.2).

WARNING

The high pressure oil system operates at high temperature and pressure. Contact with hot oil can result in burns and scalding. Shut down generator set, and allow system to cool before performing checks, services, and maintenance. Wear heat resistant gloves and avoid contacting hot surfaces. Do not allow hot oil or components to contact skin or hands. Failure to comply with this warning can cause injury or death to personnel.

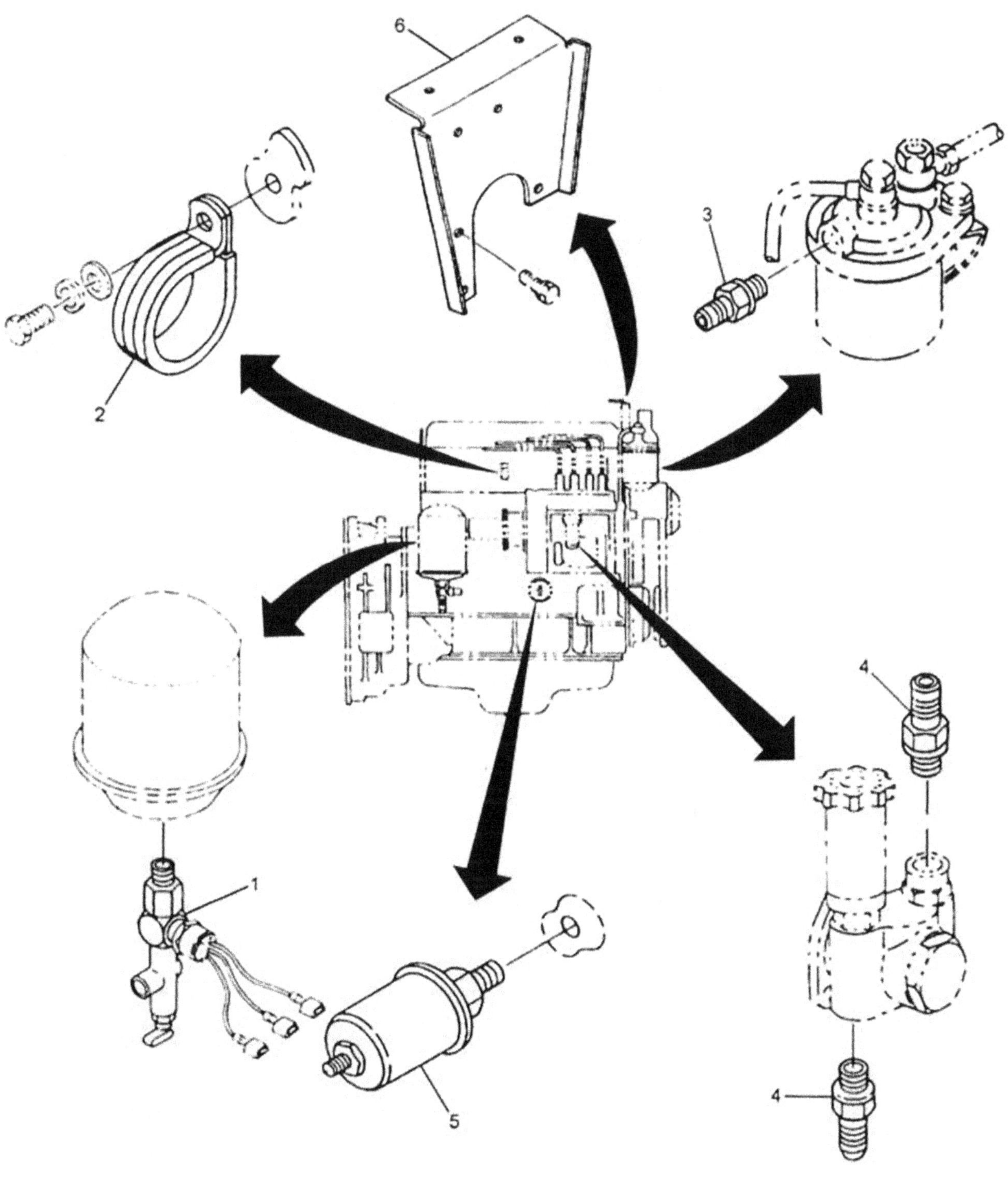

Figure 3-18. Engine Components – MEP-804A/MEP-814A.

f. f. Remove control box assembly (paragraph 2-19.2, steps d thru f).

g. g. Remove fan guards (paragraph 2-82.2, steps c and d).

h. h. Remove bolts (1 and 4, Figure 3-4), lock washers (2 and 5), washers (3 and 6), and top housing panel (7) from generator set. Discard lock washers (2 and 5).

i. i. Loosen clamp (1, Figure 3-5) at radiator (2) and disconnect filler hose and panel assembly (5) from radiator.

j. j. Loosen clamp (3) and disconnect overflow hose (4) from filler hose panel assembly (5) and remove filler hose and panel assembly.

k. k. Remove exhaust pipe clamp (1, Figure 3-7) at exhaust manifold adapter (2).

l. l. Remove nuts (3), lock washers (4), washers (6), and bolts (5) securing top housing assembly (11) to front housing. Discard lock washers (4).

m. m. Remove assembled nuts (7) and bolts (8) securing top housing assembly (11) to rear side panels.

n. n. Remove assembled nut (9) and bolt (10) securing t op housing assembly (11) to output box angle.

o. o. Remove bolts (1 and 4, Figure 3-9), washers (3 and 5), lock washers (2 and 6), nuts (7), and rear housing panel (8). Discard lock washers (2 and 6).

p. p. Loosen clamp (1, Figure 3-10) securing fuel filler panel assembly (6) to fuel tank.

q. q. Loosen clamp (2) and disconnect hose (3) from fuel filler panel assembly (6). Cap openings.

r. r. Disconnect fuel hose (4) from fuel line (5). Cap openings.

s. s. Tag and disconnect electrical connectors from auxiliary fuel pump (7) and fuel float module (8).

t. t. Tag and disconnect all connectors and wires from engine.

u. u. Remove bolts (1, Figure 3-12), lock washers (2), and washers (3) securing left rear side panel (4) to skid base. Discard lock washers (2).

v. v. Remove nuts (8), lock washers (6 and 9), washers (7 and 11), and bolts (5 and 10) securing lower left side panel (12) to front housing and skid base. Discard lock washers (6 and 9).

w. w. Remove left rear side panel (4) and lowe r left side panel (12) as an assembly.

x. x. Loosen clamp (1, Figure 3-6) and disconnect overflow hose (2) from overflow bottle (6).

y. y. Remove nuts (4), lock washers (5), bolts (3), and overflow bottle (6) with mount and wire holder. Discard lock washers (5).

z. Remove clamps (3, Figure 2-39) and drain hose (4) from adapter (5) and oil pan fitting.

aa. Remove nuts (28, Figure 2-26), washers (29), bolts (31), lock washers (32), and support rods (30) from engine lifting bracket and radiator assembly (39). Discard lock washers (32).

ab. Remove bolt (19), flat washer (20), lock washer (21), and nut (22). Discard lock washer (21).

ac. Remove bolts (23), lock washers (24), and flat washers (25) and remove shroud halves (26 and 27). Discard lock washers (24).

ad. Loosen clamps (15) and remove upper coolant hose (16) and clamps (15).

ae. Loosen clamps (17) and remove lower coolant hose (18) and clamps (17).

af. Loosen clamp (33) and disconnect radiator drain hose (34) from radiator drain valve (41).

ag. Remove nuts (35), lock washers (36), flat washers (37), and bushings (38) from radiator mounting studs. Discard lock washers (36).

ah. Lift radiator (39) up and out of generator set housing and remove grommets (40). ai.

Loosen clamps (13, Figure 2-22) on air cleaner housing (7).

aj. Loosen hose clamps (24) and slide air cleaner housing (7) back to separate hose (25) from adapter (26).

ak. Loosen clamp (3, Figure 2-38) and disconnect inlet hose (4) from filter bracket (16).

al. Remove clamp (1, Figure 3-2) and disconnect hos e (2) from bottom of fuel/ water separator (3).

am. Remove clamp (5) and disconnect hose (4) from fuel filter (6).

an Remove clamp (7) and disconnect hose (8) from fuel filter (9) on fuel pump.

ao. Remove nuts (13), lock washers (14), bolts (15), washers (16), and fuel/ water separator with bracket (3). Discard lock washers (14).

ap. Remove clamp (10) and disconnect hose (11) from fuel return (12) to tank.

aq. Remove four bolts (1, Figure 3-3), lock washers (2), washers (3), and left and right rear engine supports (4). Discard lock washers (2).

ar. Tag and disconnect wiring from engine components.

as. Loosen nuts (32) on each side of generator and screw bolts (34) down to contact skid base. Tighten bolt and nuts enough to support generator after engine is removed.

WARNING

Rated capacity of overhead hoist should be at least 1,500 pounds (680 kg). Do not use a hoist with less capacity. Failure to comply with this warning can cause injury or death to personnel, and damage to equipment.

at. Attach lifting harness to engine lifting points. Take up slack.

au. Remove nuts (5), washers (6), snubbing washers (7), and bolts (8) from engine shock mounts (19).

av. Remove screw (37), washer (38), and screen (39) from generator case.

aw. Remove screws (40) and lock washers (41) securing generator drive disc and inertia ring (41.1) to engine flywheel. Inertia ring will be captive inside generator behind generator drive disc and can remain there until installation. Discard lock washers (41).

ax. Remove bolts (42) and lock washers (43) securing generator to engine flywheel housing. Discard lock washers (43).

WARNING

Keep hands and feet from underside of engine and generator while using lifting device to remove them from the skid base. Failure to comply with this warning can cause injury or death to personnel.

ay. With aid of an assistant, slowly lift engine assembly from skid base, ensuring that engine flywheel housing separates smoothly from generator without binding. Remove engine from generator set housing and place on engine stand.

az. Remove engine shock mounts (19) from skid base by removing nuts (16), washers (17), and bolts (18).

ba. Remove crossmember (25) by removing bolts (21) , lock washers (22), washers (23), and nuts (24). Discard lock washers (22).

bb. Remove left engine support bracket (9) by removing bolts (26), lock washers (27), and washers (28). Discard lock washers (27).

bc. Remove right engine support bracket (10) by removing bolts (29), lock washers (30), and washers (31). Discard lock washers (30).

3-9.2 Installation (MEP-804B/MEP-814B).

a. a. Install right engine support bracket (10, Figure 3-3) on engine with bolts (29), new lock washers (30), and washers (31).

b. b. Install left engine support bracket (9) on engine with bolts (26), new lock washers (27), and washers (28).

c. c. Install crossmember (25) with bolts (21), new lock washers (22), washers (23), and nuts (24).

d. d. Install engine shock mounts (19) to skid base with nuts (16), washers (17), and bolts (18).

WARNING

Rated capacity of overhead hoist should be at least 1,500 pounds (680 kg). Do not use a hoist with less capacity. Failure to comply with this warning can cause injury or death to personnel, and damage to equipment.

e. e. Attach lifting harness to engine lifting points. Raise engine assembly from maintenance stand or fixture.

f. f. Position engine assembly and generator spacer (44, Figure 3-3) on skid base, aligning mounting holes and brackets.

g. g. Locate two spacers on generator coupling disk (Figure 3-3) 45 degrees off from bolts in engine flywheel.

CAUTION

Ensure generator drive disc is fully seated and flush with engine flywheel or drive disc may be damaged.

h. h. Install screws (40) and new lock washers (41) securing generator drive disc and inertia ring (41.1)

i. i. Install bolts (42) and new lock washers (43) securing generator to flywheel housing. Tighten bolts slowly to ensure proper seating of generator case lip in engine flywheel housing. Torque bolts to 31 ft-lbs (42 Nm).

j. j. Loosen nuts (32) and screw bolt (34) up approximately 1/2 inch (1.3 cm) so it does not contact frame and tighten nuts.

k. k. Install bolts (8), snubbing washers (7), washers (6), and nuts (5) to secure engine to skid base. Torque bolts to 75 ft-lbs (102 Nm).

l. l. Install screen (39) with washer (38) and screw (37).

m. m. Install left and right rear engine supports (4) with bolts (1), new lock washers (2), and washers (3).

n. n. Install hose (11, Figure 3-2) to fuel retu rn (12) (to tank) and secure with clamp (10).

o. o. Connect tagged wiring to e ngine components. Remove tags.

p. p. Install fuel/ water separator with bracket (3) with washers (16), bolts (15), new lock washers (14), and nuts (13).

q. q. Install hose (8) and clamp (7) onto fuel filter (9).

r. r. Install hose (4) and clamp (5) onto fuel filter (6).

s. s. Install hose (2) and clamp (1) onto bottom of fuel/ water separator (3).

t. t. Install inlet hose (4, Figure 2-38) clamp (3) onto filter bracket (16).

u. u. Slide air cleaner housing (7, Figure 2-22) into position and connect hose (25) onto adapter (26) with clamps (24)..

v. v. Rotate air cleaner housing (7) so inlet tube points down 45 degrees from horizontal and tighten clamps (13).

w. w. Install grommets (40, Figure 2-26), radiator (39), bushing (38), washers (37), new lock washers (36), and nuts (35).

x. x. Install shroud halves (26 and 27), washers (25), new lock washers (24), and bolts (23).

y. y. Install bolt (19), washer (20), new lock washer (21), and nut (22).

z. z. Install lower coolant hose (18) and clamps (17).

aa. Install upper coolant hose (16) and clamps (15).

ab. Install radiator drain hose (34) and clamp (33) onto radiator drain valve (41).

ac. Install support rods (30), new lock washers (32), bolts (31), washers (29), and nuts (28).

ad. Install drain hose (4, Figure 2-39) and clamps (3) onto adapter (5) and oil pan fitting.

ae. Install overflow bottle (6, Figure 3-6) with mount and wire holder and secure with bolts (3), new lock washers (5), and nuts (4).

af. Connect overflow hose (2) and clamp (1) to overflow bottle (6).

ag. Place lower left side panel (12, Figure 3-12) and left rear side panel (4) as an assembly into position and install bolts (1, 5 and 10), washers (3, 7 and 11), new lock washers (2, 6 and 9), and nuts (8).

ah Connect electrical connectors to auxiliary fuel pump (7, Figure 3-10) and fuel float module (8). Remove tags.

ai. Connect all connectors and wires to engine as tagged. Remove tags. aj.

Remove caps and connect fuel hose (4) to fuel line (5).

ak. Remove caps and connect hose (3) and clamp (2) to fuel filler panel assembly (6).

al. Remove caps and connect hose from fuel filler assembly (6) to fuel tank and tighten clamp (1).

am. Install rear housing panel (8, Figure 3-9) with bolts (1 and 4), washers (3 and 5), new lock washers (2 and 6), and nuts (7).

an. Ensure muffler pipe engages exhaust manifold adapter (2, Figure 3-7) when placing top housing (11) into position and install bolt (10) and nut (9).

ao. Secure top housing assembly (11) to rear side panels by installing bolts (8) and nuts (7).

ap. Secure top housing assembly (11) to front housing by installing bolts (5), washers (6), new lock washers (4), and nuts (3).

aq. Install pipe clamp (1) to secure muffler pipe and exhaust manifold adapter (2).

ar. Install filler hose of filler hose panel assembly (5, Figure 3-5) down through top housing assembly (11, Figure 3-7) and connect filler hose to radiator with clamp (1, Figure 3-5).

as. Route overflow hose (4) up through top housing assembly (11, Figure 3-7) and connect to filler hose panel assembly (5, Figure 3-5) with clamp (3).

at. Install top housing panel (7, Figure 3-4) with washers (3 and 6), new lock washers (2 and 5), and bolts (1 and 4).

au. Install fan guards (paragraph 2-82.3, steps a and b).

av. Install control box assembly (paragraph 2-19.4, steps a thru c).

aw. Install control box top panel (paragraph 2-15.4, steps a thru d). ax.

Service engine lubrication system (paragraph 2-1.2.4). ay. Service

coolant system (paragraph 2-1.2.2). az. Service fuel system (para-

graph 2-1.2.3).

ba. Install batteries (paragraph 2-12.5). Close battery access door. bb.

Start generator set. Perform operational checks and check for leaks. bc.

Adjust magnetic pickup, as required (paragraph 2-121.4). bd. Check and ad-

just fan belt, as necessary (paragraph 2-94.2). be. Adjust governor actuator,

as required (paragraph 4-15).

bf. Adjust governor control module, as required (paragraph 4-2.3).

3-9.3 Replacement (MEP-804B/MEP-814B).

a. a. Remove old engine assembly from generator set (paragraph 3-8.1).

b. b. Remove the following from the old engine and install on new engine (Discard all lock washers removed and install with new lock washers):

(1) Oil sample valve (1, Figure 3-19), low oil pressure switch (2), oil pressure sender (3), fitting assembly (4), bracket (5), and mounting hardware (6).

(2) Magnetic pickup (7, Figure 2-36), and locking nut (6). Screw into bracket until magnetic pickup contacts top surface of gear tooth of flywheel; back out one turn, and tighten locking nut.

(3) Intake air heater contactor (15), mounting hardware (12, 13, and 14), contactor bracket (19), and bracket mounting hardware (16, 17, and 18).

(4) Governor actuator (7, Figure 3-19) O-ring (8) and mounting hardware (9). Inspect and replace O-ring, as required.

NOTE

Crankcase ventilation filter bracket must be removed in order to remove rear lift ring; however, rear lift ring must be installed on new engine prior to installing crankcase ventilation filter bracket. The two brackets share a mounting hole and hardware. Mount rear lift ring using lower two bolts, etc., then mount filter bracket.

(5) Crankcase ventilation filter (10, Figure 2-38), hoses (2, 4, and 6), and mounting bracket (16) and mounting hardware. Remove fuel filter from bracket (16) (TM 9-6115-538-24&P). Fuel filter stays with old engine.

(6) Rear lift ring (10, Figure 3-19) and mounting hardware (11).

(7) Oil drain fitting (12), hose (13), clamps (14), barb fitting (15), clamp (16), and hose (17). (This is other end of hose (6, Figure 2-38) identified in step (5) above).

(8) Oil drain adapter (18, Figure 3-19).

(9) Upper radiator support (19) and mounting hardware (20).

NOTE

New engine is shipped with a box of parts. A new gasket for the turbocharger connection is in the box. Remove this gasket and discard remainder of parts.

(10) Exhaust manifold (15, Figure 2-21), mounting hardware (19, 20, and 21), gasket (22), mounting bracket (23), and bracket mounting hardware (24, 25, and 26). Replace gasket.

(11) Coolant temperature sender (3, Figure 2-36) and copper washer (4).

(12) Coolant high temperature switch (5).

c. c. Install new engine in generator set (paragraph 3-9.2).

d. d. Start generator set and check for leaks and proper operation.

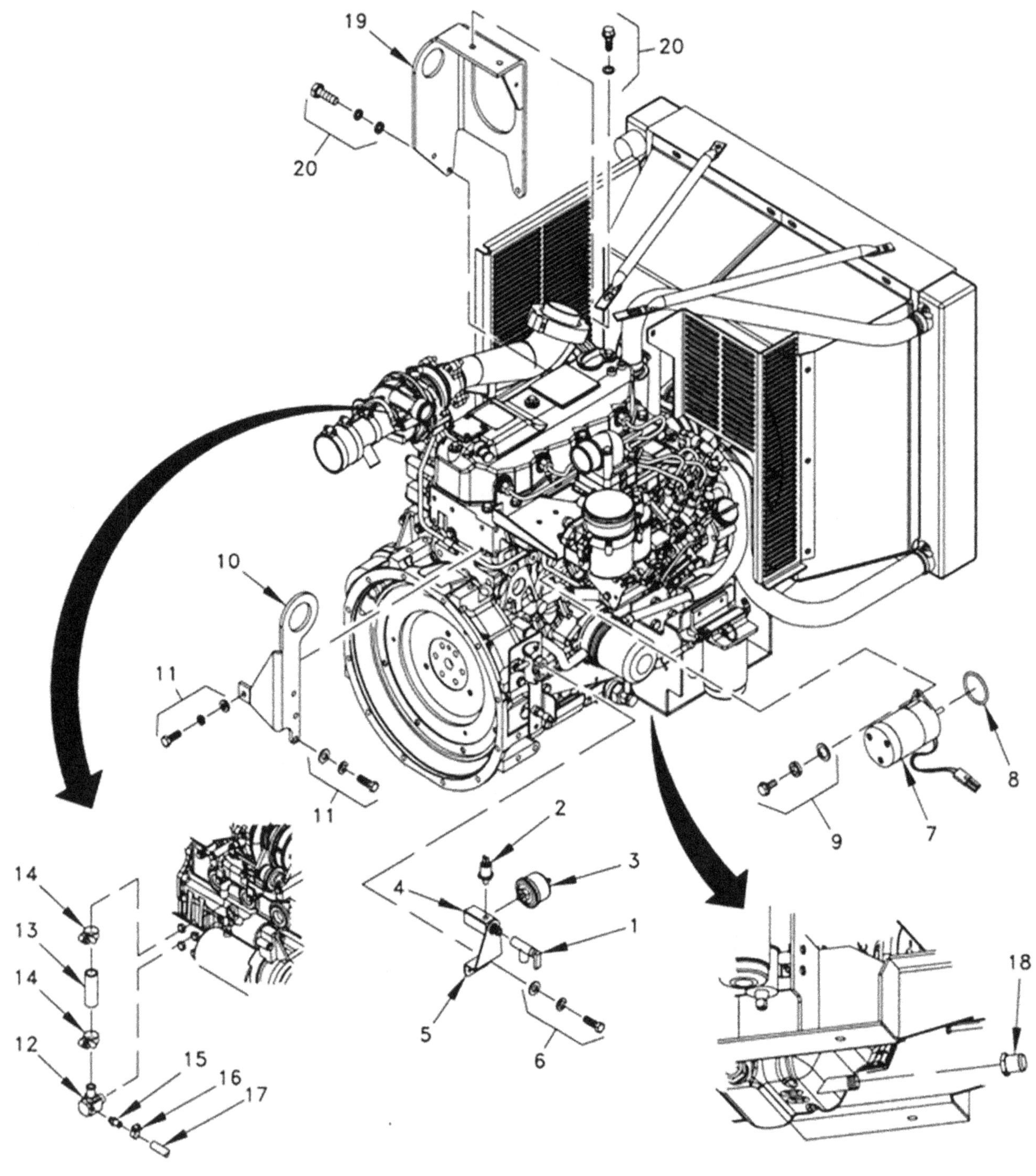

Figure 3-19. Engine Components – MEP-804B/MEP-814B.

CHAPTER 4

DIRECT SUPPORT MAINTENANCE INSTRUCTIONS

Subject Index Page

Subject Index Page

Section I. MAINTENANCE OF CONTROL BOX ASSEMBLY

4-1 AC VOLTAGE REGULATOR

WARNING

High voltage is produced when the generator set is in operation. Never attempt to start the generator set unless it is properly grounded. Failure to comply with this warning can cause injury or death to personnel.

WARNING

High voltage is produced when this generator set is in operation. Make sure generator set is completely shut down and free of any power source before attempting any repair or maintenance on the set, or when connecting or disconnecting load cables. Failure to comply with this warning can cause injury or death to personnel.

WARNING

Dangerous voltage exists on live circuits. Always observe precautions and never work alone. Failure to comply with this warning can cause injury or death to personnel.

WARNING

Ensure that the engine cannot be started while maintenance is being performed. (ENGINE CONTROL switch set to OFF/RESET; Battery Disconnect Switch is OFF; DEAD CRANK SWITCH is OFF). Failure to comply with this warning can cause injury or death to personnel.

WARNING

DC voltages are present at generator set electrical components even with generator set shut down. Avoid shorting any positive with ground/negative. Failure to comply with this warning can cause injury to personnel, and damage to equipment.

4-1.1 Testing (AC Voltage Regulator, P/N 01-21501-1, for MEP-804A/MEP-804B).

a. a. Shut down generator set.

b. b. Open output box access door.

c. c. Note position of voltage reconnection terminal board and set FREQUENCY SELECT switch to 60 Hz position.

d. Start generator set and turn VOLTAGE adjust potentiometer to ensure adjustment ranges on Table 4-1 are met, depending on position of voltage reconnection terminal board.

WARNING

High voltage is produced when the generator set is in operation. Never attempt to start the generator set unless it is properly grounded. Failure to comply with this warning can cause injury or death to personnel.

WARNING

DO NOT wear loose clothing when performing checks, services and maintenance. Failure to comply with this warning can cause injury or death to personnel.

WARNING

All metal jewelry can conduct electricity and become entangled in generator set components. Remove all jewelry when working on generator set. Failure to comply with this warning can cause injury or death to personnel.

WARNING

High voltage is produced when the generator set is in operation. DO NOT touch live voltage connections. Never attempt to connect or disconnect load cables or paralleling cables while the generator set is running. Failure to comply with this warning can cause injury or death to personnel.

WARNING

Wear heat resistant gloves and avoid contacting hot metal surfaces with your hands after components have been heated. Wear additional protective clothing as required. Failure to comply with this warning can cause injury to personnel.

e. e. Shut down generator set.

f. f. If no voltage or low voltage wa s indicated or voltage adjustment range could not be achieved, perform the following steps. Otherwise, AC voltage regulator (2, Figure 4-1) is serviceable.

g. g. Disconnect wire 141A from terminal 1 of AC voltage regulator.

h. h. Set multimeter for DC volts and connect positi ve lead to wire 141A. Connect negative lead of multimeter to terminal 3 of AC voltage regulator. Start generator set and operate at rated frequency. Move and hold MASTER SWITCH in START position. Multimeter should indicate between 4 and 12 VDC.

i. i. Shut down generator set. Isolate wire 141A.

j. j. Set multimeter for AC volts and connect to terminals 10 and 11 of AC voltage regulator. Start generator set and operate at rated frequency. Move and hold MASTER SWITCH in START position. Multimeter should indicate 210 to 280 VAC.

k. k. Shut down generator set.

l. l. Disconnect wire 137A from terminal 5 of AC voltage regulator. Set multimeter for ohms and connect positive lead to wire 137A and negative lead to terminal 4 of AC voltage regulator. Move VOLTAGE adjust potentiometer to full counterclockwise (CCW) position. Multimeter should indicate approximately 3,000 ohms with FREQUENCY SELECT switch in 60 Hz position, and approximately 0 ohms with FREQUENCY SELECT switch in 50 Hz position. Move VOLTAGE adjust potentiometer slowly clockwise (CW) while observing multimeter. Multimeter should increase smoothly to approximately 10,000 ohms.

m. m. If steps h, j, and l are as indicated above, AC voltage regulator is defective and must be replaced.

n. n. Connect all wires previously disconnected. Close output box access door.

Voltage Reconnection Terminal Board Position	Adjustment Range
120/208	197-240 volts
240/416	395-480 volts

4-1.2 Testing (AC Voltage Regulator, P/N 01-21501-2, for MEP-804A/MEP-804B).

a. a. Shut down generator set.

b. b. Open output box access door.

c. c. Note position of voltage reconnection terminal board and set FREQUENCY SELECT switch to 60 Hz position.

d. Start generator set and turn VOLTAGE adjust potentiometer to ensure adjustment ranges on Table 4-1 are met, depending on position of voltage reconnection terminal board.

e. e. Shut down generator set.

f. f. If no voltage or low voltage was indicated or voltage adjustment range could not be achieved, perform the following steps. Otherwise, AC voltage regulator (2, Figure 4-1) is serviceable.

g. g. Disconnect wire 141A from terminal 1 of AC voltage regulator.

h. h. Set multimeter for DC volts and connect positive lead to wire 141A. Connect negative lead of multimeter to terminal 3 of AC voltage regulator. Start generator set and operate at rated frequency. Move and hold MASTER SWITCH in START position. Multimeter should indicate between 4 and 12 VDC.

i. i. Shut down generator set. Isolate wire 141A.

j. j. Set multimeter for AC volts and connect to terminals 10 and 11 of AC voltage regulator. Start generator set and operate at rated frequency. Move and hold MASTER SWITCH in START position. Multimeter should indicate 110 to 160 VAC.

k. k. Shut down generator set.

l. l. Disconnect wire 137A from terminal 5 of AC voltage regulator. Set multimeter for ohms and connect positive lead to wire 137A and negative lead to terminal 4 of AC voltage regulator. Move

VOLTAGE adjust potentiometer to full CCW position. Multimeter should indicate no more than 2 ohms with FREQUENCY SELECT switch in 60 Hz position or 50 Hz position. Move VOLTAGE adjust potentiometer slowly CW while observing multimeter. Multimeter should increase smoothly to approximately 20,000 ohms.

m. If steps h, j, and l are as indicated above, AC voltage regulator is defective and must be replaced.

n. n. Connect all wires previously disconnected. Close output box access door.

4-1.3 Testing (AC Voltage Regulator, P/N 01-21507-1, for MEP-814A/MEP-814B).

a. a. Shut down generator set.

b. b. Open output box access door.

c. c. Note position of voltage reconnection terminal board.

d. d. Start generator set and turn VOLTAGE adjust pot entiometer to ensure adjustment ranges on Table 4-2 are met, depending on position of voltage reconnection terminal board.

e. e. Shut down generator set.

f. f. If no voltage or low voltage wa s indicated or voltage adjustment range could not be achieved, perform the following steps. Otherwise, AC voltage regulator (2, Figure 4-1) is serviceable.

g. g. Disconnect wire 141A from terminal 1 of AC voltage regulator.

h. h. Set multimeter for DC volts and connect positi ve lead to wire 141A. Connect negative lead of multimeter to terminal 3 of AC voltage regulator. Start generator set and operate at rated frequency. Move and hold MASTER SWITCH in START position. Multimeter should indicate between 4 and 12 VDC.

i. i. Shut down generator set. Isolate wire 141A.

j. j. Set multimeter for AC volts and connect to terminals 10 and 11 of AC voltage regulator. Start generator set and operate at rated frequency. Move and hold MASTER SWITCH in START position. Multimeter should indicate 110 to 160 VAC.

k. k. Shut down generator set.

l. l. Disconnect wire 137A from terminal 5 of AC voltage regulator. Set multimeter for ohms and connect positive lead to wire 137A and negative lead to terminal 4 of AC voltage regulator. Move VOLTAGE adjust potentiometer to full CCW position. Multimeter should indicate approximately 0 ohms. Move VOLTAGE adjust potentiometer slowly CW while observing multimeter. Multimeter should increase smoothly to approximately 20,000 ohms.

m. m. If steps h, j, and l are as indicated above, AC voltage regulator is defective and must be replaced.

n. n. Connect all wires previously disconnected. Close output box access door.

Table 4-2. Voltage Adjustment Range – MEP-814A/MEP-814B.

Voltage Reconnection Terminal Board Position	Adjustment Range
120/208	197-229 volts
240/416	395-458 volts

4-1.4 **Removal.**

a. a. Shut down generator set.

b. b. Open battery access door and disconnect negative battery cable.

c. c. Release control panel by turning two fasteners and lower control panel slowly.

d. d. Tag and disconnect AC voltage regulator (2, Figure 4-1) electrical leads.

e. e. Remove screws (1) and AC voltage regulator (2).

4-1.5 **Installation.**

CAUTION

The components of the AC voltage regulator kit are not interchangeable. Ensure the part number of the component to be installed is the same as the removed component. Failure to observe this caution will result in equipment damage.

a. a. Install AC voltage regulator (2, Figure 4-1) with screws (1).

b. b. Connect electrical leads. Remove tags.

c. c. Raise and secure control panel.

d. d. Connect negative battery c able. Close battery access door.

4-2 GOVERNOR CONTROL UNIT

WARNING

High voltage is produced when the generator set is in operation. Never attempt to start the generator set unless it is properly grounded. Failure to comply with this warning can cause injury or death to personnel.

WARNING

High voltage is produced when this generator set is in operation. Make sure generator set is completely shut down and free of any power source before attempting any repair or maintenance on the set, or when connecting or disconnecting load cables. Failure to comply with this warning can cause injury or death to personnel.

WARNING

Dangerous voltage exists on live circuits. Always observe precautions and never work alone. Failure to comply with this warning can cause injury or death to personnel.

WARNING

Ensure that the engine cannot be started while maintenance is being performed. (ENGINE CONTROL switch set to OFF/RESET; Battery Disconnect Switch is OFF; DEAD CRANK SWITCH is OFF). Failure to comply with this warning can cause injury or death to personnel.

WARNING

DC voltages are present at generator set electrical components even with generator set shut down. Avoid shorting any positive with ground/negative. Failure to comply with this warning can cause injury to personnel, and damage to equipment.

4-2.1 Internal Crank Relay Test.

a. a. While holding Master Switch (S-1) in START position, check for voltage from pin positions 1,3,5 on GCU (A-5) ground. Readings should be battery voltage. If there is no voltage between any connections, GCU (A-5) is not receiving proper voltage and wiring harness should be checked. Refer to Electrical Schematic FO-1 (MEP-804A/MEP-814A)/FO-3 (MEP-804B/MEP-814B).

NOTE

With the generator running, the voltage will read zero.

b. b. Place master switch in OFF position. Connect multimeter leads to terminals 19 and 20 on governor control unit (A-5). Then, move master switch to START position. Reading should be battery voltage when generator does not crank. If there is no voltage between terminal positions, GCU (A-5) is not functioning properly and should be replaced.

NOTE

With the generator running, there will be battery voltage.

c. c. Place master switch in OFF position. Connect multimeter leads to terminals 19 and 18 on governor control unit (A-5). Then move master switch to START position. Reading should be ZERO voltage when generator does not crank. If there is voltage between terminal positions, GCU (A-5) is not functioning properly and should be replaced.

4-2.2 Testing.

a. a. Shut down generator set.

b. b. Open output box access door. Remove protective cover and attach voltage and frequency recorder to terminals 9 and 12 of voltage reconnection terminal board.

c. c. Open load terminal board access door and attach load bank to generator set (4-wire connection).

NOTE

Ensure load bank and generator set voltage reconnection are set for same configuration (i.e., 120/208 or 240/416 VAC).

d. d. Start and operate generator set at rated voltage and frequency.

WARNING

High voltage is produced when the generator set is in operation. Never attempt to start the generator set unless it is properly grounded. Failure to comply with this warning can cause injury or death to personnel.

WARNING

DO NOT wear loose clothing when performing checks, services and maintenance. Failure to comply with this warning can cause injury or death to personnel.

WARNING

All metal jewelry can conduct electricity and become entangled in generator set components. Remove all jewelry when working on generator set. Failure to comply with this warning can cause injury or death to personnel.

WARNING

High voltage is produced when the generator set is in operation. DO NOT touch live voltage connections. Never attempt to connect or disconnect load cables or paralleling cables while the generator set is running. Failure to comply with this warning can cause injury or death to personnel.

WARNING

Wear heat resistant gloves and avoid contacting hot metal surfaces with your hands after components have been heated. Wear additional protective clothing as required. Failure to comply with this warning can cause injury to personnel.

WARNING

Operating the generator set exposes personnel to a high noise level. Hearing protection must be worn when operating or working near the generator set when the generator set is running. Failure to comply with this warning can cause hearing damage to personnel.

e. e. Turn on voltage and frequency recorder and operate at minimum chart speed of 5 mm/sec (chart resolution of 0.2 mm/sec). Adjust recorder voltage amplifier for a minimum chart resolution of 1.0 volt/mm, and frequency deviation amplifier for a minimum resolution of 0.2 Hz/mm.

WARNING

High voltage power is available when the main contactor is closed. Avoid accidental contact with live components. Ensure load cables are properly connected and the load cable door is shut before closing main contactor. Ensure load is turned off before closing main contactor. Ensure that soldiers working with/on loads connected to the generator set are aware that main contactor is about to be closed before closing main contactor. Failure to comply with this warning can cause injury or death to personnel.

f. f. Set load bank for a load equal to 75% of generator set rated load.

g. g. Apply and remove 75% load to generator set at 40-second intervals three times.

h. h. Repeat steps f and g above at 50 percent rated load.

i. i. Repeat steps f and g above at 25 percent rated load.

j. j. Repeat steps f and g above at 100 percent rated load.

k. k. Shut down generator set.

l. Turn off voltage and frequency recorder.

m. Examine voltage and frequency recorder char t. Generator set should meet the following performance criteria:

(1) Frequency regulation shall not exceed 1/4 of 1 percent of rated frequency.

(2) Frequency short-term stability (30 seconds), frequency will remain constant within a bandwidth equal to 1/2 of 1 percent rated frequency, without repetitive frequency variations, commonly called hunting.

(3) Generator set will reestablish stable engine operating conditions within 2 seconds of a sudden load change (within 1 second for 400 Hz unit). Maximum transient frequency change above or below (overshoot or undershoot) new steady state frequency shall not be more than 4 percent of rated frequency (not more than 1-1/2 percent for 400 Hz unit).

n. n. If above criteria is not met, adjust governor control unit (3, Figure 4-1) (paragraph 4-2.3).

o. o. If above criteria cannot be met by adjustment, governor control unit must be replaced.

p. p. Disconnect load ban k. Close load terminal board access door.

q. Disconnect voltage and frequency recorder and install voltage reconnection terminal board protective cover. Close output box access door.

4-2.3 Adjustment.

a. a. Shut down generator set.

b. b. Open output box access door, remove cover from voltage reconnection terminal board, and attach voltage and frequency recorder to terminals 9 and 12 of voltage reconnection terminal board.

NOTE

The following procedures require monitoring frequency, voltage, current, and power. All readings except for frequencies will utilize the generator set control panel gauges. However, since the designed overspeed trip frequency is greater than the range of the control panel FREQUENCY meter (HERTZ), a frequency counter will be required.

c. c. Attach frequency counter to voltage reconnection terminal board terminals 9 and 12.

d. d. Open load terminal board access door and attach l oad bank to generator set (4-wire connection).

NOTE

Ensure load bank and generator set voltage reconnection are set for same configuration (i.e., 120/208 or 240/416 VAC).

e. Check for proper adjustment of governor actuator (paragraph 4-14.3 (MEP-804A/MEP-814A)/paragraph 4-15.3 (MEP-804B/MEP-814B)). Adjust as necessary.

f. f. Check for proper adjustment of magnetic picku p (paragraph 2-121.4). Adjust as necessary.

g. g. Lower generator set control panel and turn INTEG, GOV GAIN, and LOAD PULSE potentiometer on governor control unit to their full CCW positions.

h. Start generator set and operate at rated voltage and frequency.

i. Observing frequency counter, slowly increase operating frequency of generator set by turning governor control unit FREQ range potentiometer CW until frequency counter indicates between 65.94 and 66.06 Hz (MEP-804A)/436 and 444 Hz (MEP-814A). At this point generator set has reached overspeed trip frequency and generator set should shut down.

NOTE

Perform steps j and k below if generator set does not shut down within limits noted in step i above. Otherwise, proceed to step l below.

j. If generator set has not shut down at upper limit of frequency noted in step i above, proceed as follows:

(1) Turn FREQ range potentiometer CCW until frequency counter indicates midrange of overspeed trip frequency. (Example: 66 Hz for MEP-804A)

(2) Turn OVERSPD control potentiometer on governor control unit CCW until generator set shuts down.

(3) Activate OVERSPD RESET switch.

(4) Repeat steps h and i above.

k. k. If generator set shuts down prior to reaching lower limit of frequency noted in step i above, proceed as follows:

(1) Turn OVERSPD control potentiometer CW one turn for each hertz generator set shut down prior to lower frequency limit.

(2) Activate OVERSPD RESET switch.

(3) Repeat steps h and i above.

l. l. Actuate OVERSPD RESET switch.

m. m. Turn FREQ range potentiometer on governor control unit two turns CCW.

n. n. Start and operate generator set at rated volt age and turn FREQUENCY adjust potentiometer on control panel to midrange.

o. o. Turn FREQ range potentiometer until rated frequency (50, 60, or 400 Hz) is indicated on control panel FREQUENCY meter (HERTZ).

p. p. Set load bank for generator set rated load and appl y load. Observe generator set instruments and adjust load as needed to en sure rated load is applied.

q. q. Set multimeter for DC volts and connect to terminals 11 and 12 of governor control unit (3, Figure 4-1).

r. Adjust LOAD SHARING ADJUST rheostat unt il multimeter indicates 6 VDC. Disconnect multimeter.

s. s. Remove load.

t. t. Turn on voltage and frequency recorder and operate at minimum chart speed of 5 mm/sec (chart resolution of 0.2 mm/sec). Adjust recorder voltage amplifier for minimum chart resolution of 1.0 volt/mm and frequency deviation of 0.2 Hz/mm.

u. u. Adjust GOV GAIN potentiometer on governor control unit as follows:

(1) Turn GOV GAIN potentiometer to its full CW position.

(2) Momentarily actuate and turn off LOAD switch on control panel.

(3) Observe strip chart on recorder for frequency oscillation (hunting). If required, slowly turn GOV GAIN CCW until frequency oscillation disappears.

v. v. Apply and remove rated load to generator set at 40-second intervals. Repeat this step two more times.

w. w. Shut down generator set and turn off recorder.

x. x. Examine voltage and frequency strip chart for the following performance criteria:

(1) Frequency regulation shall not exceed 1/4 of 1 percent of rated frequency.

(2) Frequency short-term stability (30 seconds), frequency will remain constant within a bandwidth equal to 1/2 of 1 percent rated frequency, without repetitive frequency variations (hunting).

(3) Generator set will reestablish stable engine operation within 2 seconds of a sudden load change (i.e., from a load to no-load condition) (within 1 second for 400 Hz units). Maximum transient frequency change above (overshoot) and below (undershoot) new steady state frequency shall not be more than 4 percent of rated frequency (not more than 1-1/2 percent for 400 Hz units).

NOTE

All required INTEG and LOAD PULSE potentiometer adjustments will be in 10 percent increments.

y. y. Adjust INTEG potentiometer on governor control unit CW to decrease recovery time of load transients.

z. z. Adjust LOAD PULSE potentiometer on governor control unit CW to decrease frequency overshoot and undershoot and to decrease recovery time of overshoot/undershoot transients.

aa. Start generator set.

NOTE

Steps u(1) and u(2) are not required when doing step ab.

ab. Repeat steps t thru v until generator set m eets performance requirements stated in step x.

ac. Apply and remove 75 percent rated load to generator set at 40-second intervals. Repeat this step two more times.

ad. Apply and remove 50 percent rated load to generator set at 40-second intervals. Repeat this step two more times.

ae. Apply and remove 25 percent rated load to generator set at 40-second intervals. Repeat this step two more times.

af. Shut down generator set and turn off strip chart recorder.

ag. Examine voltage and frequency strip chart for the following performance criteria:

(1) Frequency regulation shall not exceed 1/4 of 1 percent of rated frequency.

(2) Frequency short-term stability (30 seconds), frequency will remain constant within a bandwidth equal to 1/2 of 1 percent rated frequency, without repetitive frequency variations (hunting).

(3) Generator set will reestablish stable engine operation within 2 seconds of a sudden load change (i.e., from a load to no-load condition) (within 1 second for 400 Hz units). Maximum transient frequency change above (overshoot) and below (undershoot) new steady state frequency shall not be more than 4 percent of rated frequency (not more than 1-1/2 percent for 400 Hz units).

ah. Disconnect load ban k. Close load terminal board access door.

ai. Disconnect frequency counter and voltage and frequency recorder from voltage reconnection terminal board.

aj. Install voltage reconnection terminal board cover. Close output box access door.

4-2.4 Replacement.

a. a. Remove governor control unit (paragraph 2-61.2).

b. b. Install new governor control unit (paragraph 2-61.3).

c. c. Perform adjustment of governor control unit (paragraph 4-2.3).

4-3 CONTROL BOX WIRING HARNESS

WARNING

High voltage is produced when the generator set is in operation. Never attempt to start the generator set unless it is properly grounded. Failure to comply with this warning can cause injury or death to personnel.

WARNING

High voltage is produced when this generator set is in operation. Make sure generator set is completely shut down and free of any power source before attempting any repair or maintenance on the set, or when connecting or disconnecting load cables. Failure to comply with this warning can cause injury or death to personnel.

WARNING

Dangerous voltage exists on live circuits. Always observe precautions and never work alone. Failure to comply with this warning can cause injury or death to personnel.

WARNING

Ensure that the engine cannot be started while maintenance is being performed. (ENGINE CONTROL switch set to OFF/RESET; Battery Disconnect Switch is OFF; DEAD CRANK SWITCH is OFF). Failure to comply with this warning can cause injury or death to personnel.

WARNING

DC voltages are present at generator set electrical components even with generator set shut down. Avoid shorting any positive with ground/negative. Failure to comply with this warning can cause injury to personnel, and damage to equipment.

WARNING

The connection of any electrical equipment and the disconnection of any electrical equipment may cause an explosion hazard. Do not connect any electrical equipment or disconnect any electrical equipment in an explosive atmosphere. Failure to comply with this warning can cause injury or death to personnel.

WARNING

A qualified technician must make the power connections and perform all continuity checks. The power source may be a generator or commercial power. Failure to comply with this warning can cause injury or death to personnel.

4-3.1 Removal.

a. a. Shut down generator set.

b. b. Open battery access door and disconnect negative battery cable.

c. c. Remove control box assembly (paragraph 2-19.2).

d. d. Remove voltage sensing relay (paragraph 2-59.3).

e. e. Remove relays (paragraph 2-60.3).

f. f. Remove screws (4, Figure 4-1), washers (5), insulators (6), and clips (7).

g. g. Tag and remove sockets (8) from tracks (9).

h. h. Remove screws (10) and nuts (11) to free terminal boards (12).

i. i. Tag and disconnect electrical leads from AC voltage regulator (2) and governor control unit (3).

j. j. Tag and disconnect electrical leads from shunt (46, FIGURE 2-14), BATTERY CHARGER FUSE assembly (32) (MEP-804A/MEP-814A)/BATTERY CHARGER CIRCUIT BREAKER (35) (MEP-804B/MEP-814B), DC CONTROL POWER circuit breaker (38), REACTIVE CURRENT ADJUST rheostat (3), LOAD SHARING ADJUST rheostat (7), FREQUENCY SELECT switch (14), OVERSPEED RESET switch (10), overvoltage/undervoltage relay (48), frequency transducer (41), short circuit/overload relay (50), permissive paralleling relay (54), reverse power relay (52), load measuring unit (72), watt transducer (43), and resistor-diode assembly (74).

k. k. Tag and disconnect electrical leads to CONVENIENCE RECEPTACLE (20), GROUND FAULT CIRCUIT INTERRUPTER (23), and connector to malfunction indicator panel (27).

l. l. Tag and disconnect electrical leads to all in dicators, switches, and lights on control panel.

m. m. Remove screws (13 and 17, Figure 4-1), nuts (14 and 18), caps (16 and 20), diagnostic connector (15), parallel connector (19), and gaskets (21 and 22) from control box panel.

n. Remove screws (23) and nuts (24) from harness connectors and remove harness assembly (25) from control box.

4-3.2 Inspection.

a. Inspect control box harness wiring for breaks, damaged insulation, and loose or damaged terminals.

b. Inspect harness connectors, sockets, and terminal boards for cracks, corrosion, stripped threads, broken pins, and other visible damage.

4-3.3 Testing.

a. a. Set multimeter for ohms.

b. b. Check individual wires, connectors, and terminal boards for continuity. Refer to Wiring Diagram FO-2 (MEP-804A/MEP-814A)/FO-4 (MEP-804B/MEP-814B) for wire identification.

4-3.4 Repair.

a. a. Replace damaged cable assemblies, terminals, connectors, sockets, and terminal boards.

b. b. Replace or ensure proper connec tion of all wires not indicating continuity.

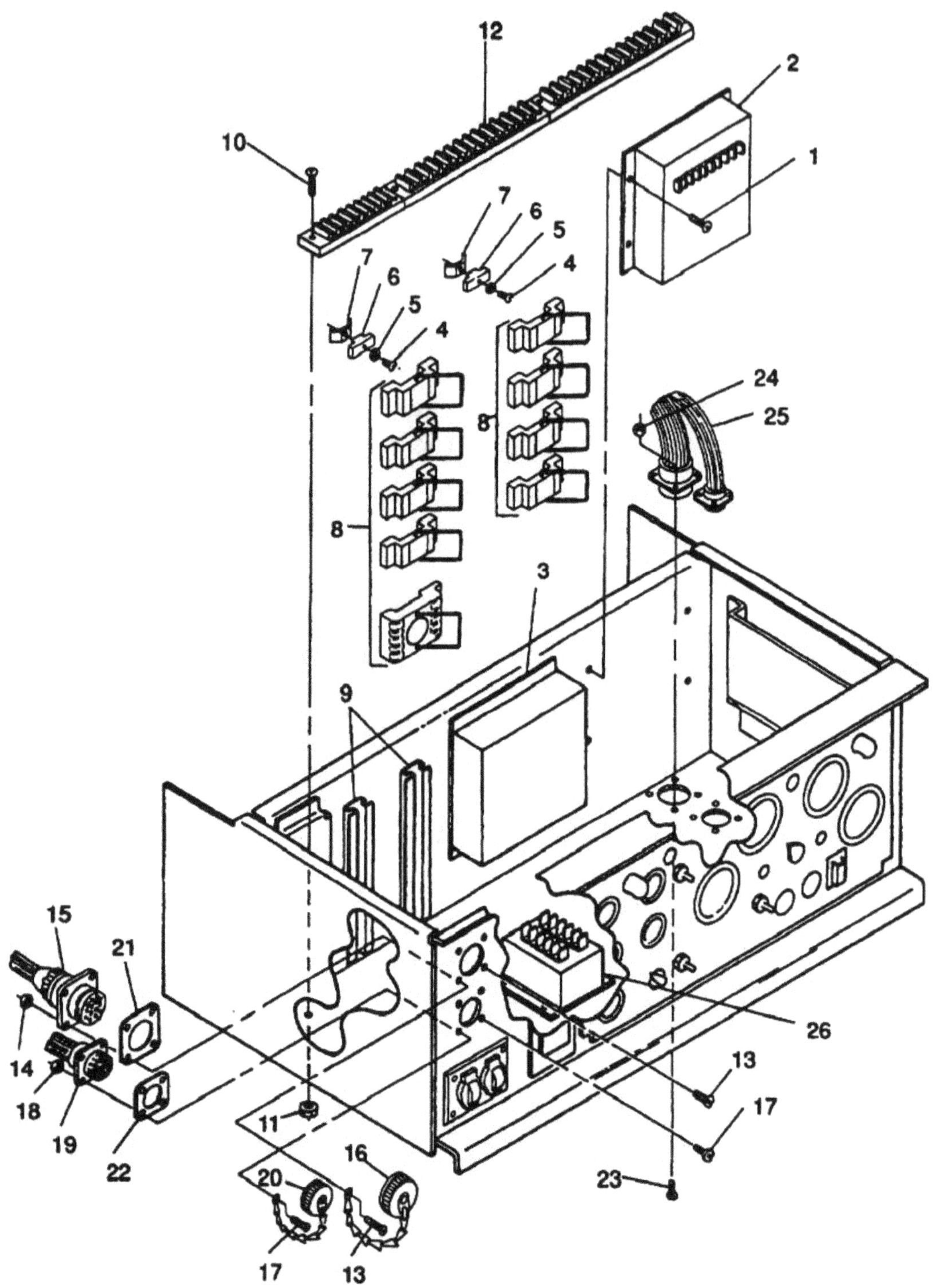

Figure 4-1. Control Box Components.

4-3.5 Installation.

WARNING

High voltage is produced when the generator set is in operation. Never attempt to start the generator set unless it is properly grounded. Failure to comply with this warning can cause injury or death to personnel.

WARNING

DO NOT wear loose clothing when performing checks, services and maintenance. Failure to comply with this warning can cause injury or death to personnel.

WARNING

All metal jewelry can conduct electricity and become entangled in generator set components. Remove all jewelry when working on generator set. Failure to comply with this warning can cause injury or death to personnel.

WARNING

High voltage is produced when the generator set is in operation. DO NOT touch live voltage connections. Never attempt to connect or disconnect load cables or paralleling cables while the generator set is running. Failure to comply with this warning can cause injury or death to personnel.

WARNING

Wear heat resistant gloves and avoid contacting hot metal surfaces with your hands after components have been heated. Wear additional protective clothing as required. Failure to comply with this warning can cause injury to personnel.

a. a. Position control box wiring harness assem bly (25, Figure 4-1) in control box.

b. b. Secure harness connectors to cont rol box with screws (23) and nuts (24).

c. c. Install gaskets (21 and 22), diagnostic connector (15), parallel connector (19), and caps (16 and 20) in control box panel with screws (13 and 17) and nuts (14 and 18).

d. d. Connect all electrical leads to control panel indicators, switches, and lights as tagged during removal. Remove tags.

e. e. Connect electrical connector to malfunction indicator panel (27, FIGURE 2-14) and electrical leads to CONVENIENCE RECEPTACLE (20) and GROUND FAULT CIRCUIT INTERRUPTER (23).

f. f. Connect electrical leads, as t agged during removal, to watt transducer (43), load measuring unit (72), reverse power relay (52), permissive paralleling relay (54), resistor-diode assembly (74), short circuit/overload relay (50), frequency transducer (41), overvoltage/undervoltage relay (48), DC CONTROL POWER circuit breaker (38), BATTERY CHARGER FUSE assembly (32) (MEP-804A/MEP-814A)/BATTERY CHARGER CIRCUIT BREAKER (35) (MEP-804B/MEP-814B), OVERSPEED RESET switch (10), FREQUENCY SELECT switch (14), REACTIVE CURRENT ADJUST rheostat (3), LOAD SHARING ADJUST rheostat (7), and shunt (46). Remove tags.

g. Connect electrical lead s to AC voltage regulator (2, Figure 4-1) and governor control unit (3). Remove tags.

h. h. Secure terminal boards (12) to control box with screws (10) and nuts (11).

i. i. Position sockets (8) on tracks (9) as tagged during removal.

j. j. Secure sockets (8) to tracks (9) with clips (7), insulators (6), washers (5), and screws (4). Remove tags.

k. k. Install relays (paragraph 2-60.4).

l. l. Install voltage sensing relay (paragraph 2-59.4).

m. m. Install control box assembly (paragraph 2-19.4).

n. n. Connect negative battery cable.

o. o. Start and operate generator set at rated voltage and frequency.

p. p. Close battery access door.

4-4 LOAD MEASURING UNIT

WARNING

High voltage is produced when the generator set is in operation. Never attempt to start the generator set unless it is properly grounded. Failure to comply with this warning can cause injury or death to personnel.

WARNING

High voltage is produced when this generator set is in operation. Make sure generator set is completely shut down and free of any power source before attempting any repair or maintenance on the set, or when connecting or disconnecting load cables. Failure to comply with this warning can cause injury or death to personnel.

WARNING

Dangerous voltage exists on live circuits. Always observe precautions and never work alone. Failure to comply with this warning can cause injury or death to personnel.

WARNING

Ensure that the engine cannot be started while maintenance is being performed. (ENGINE CONTROL switch set to OFF/RESET; Battery Disconnect Switch is OFF; DEAD CRANK SWITCH is OFF). Failure to comply with this warning can cause injury or death to personnel.

WARNING

DC voltages are present at generator set electrical components even with generator set shut down. Avoid shorting any positive with ground/negative. Failure to comply with this warning can cause injury to personnel, and damage to equipment.

WARNING

The connection of any electrical equipment and the disconnection of any electrical equipment may cause an explosion hazard. Do not connect any electrical equipment or disconnect any electrical equipment in an explosive atmosphere. Failure to comply with this warning can cause injury or death to personnel.

WARNING

A qualified technician must make the power connections and perform all continuity checks. The power source may be a generator or commercial power. Failure to comply with this warning can cause injury or death to personnel.

WARNING

High voltage is produced when the generator set is in operation. DO NOT touch live voltage connections. Never attempt to connect or disconnect load cables or paralleling cables while the generator set is running. Failure to comply with this warning can cause injury or death to personnel.

4-4.1 Testing.

a. a. Start and operate generator set at rated frequency and voltage.

b. b. Apply a load to generator set.

c. c. Note reading on kilowattmeter (PERCENT POWER).

d. d. Determine DC voltage (calculated value) from ki lowattmeter reading using the following formula:

$$\frac{10.8 \text{ X (kilowattmeter reading)}}{100}$$

e. e. Release control panel by turning two fasteners and lower control panel slowly.

f. f. Set multimeter for DC volts and connect positive lead to terminal 11 of load measuring unit (26, Figure 4-1) and negative lead to terminal 12.

g. g. Compare DC voltage (calculated value) to multimeter reading.

h. h. Load measuring unit must be replaced if difference is more than ±1.2 VDC.

i. i. Shut down generator set.

j. j. Remove multimeter from load measuring unit terminals.

k. k. Raise and secure control panel.

4-4.2 Replacement.

a. a. Remove load measuring unit (paragraph 2-65.2).

b. b. Install new load measuring unit (paragraph

Section II. MAINTENANCE OF COOLANT SYSTEM

4-5 RADIATOR REPAIR

Repair radiator in accordance with TM 750-254.

Section III. MAINTENANCE OF FUEL SYSTEM

4-6 FUEL TANK

WARNING

All metal jewelry can conduct electricity and become entangled in generator set components. Remove all jewelry when working on generator set. Failure to comply with this warning can cause injury or death to personnel.

WARNING

DO NOT wear loose clothing when performing checks, services and maintenance. Failure to comply with this warning can cause injury or death to personnel.

WARNING

High voltage is produced when the generator set is in operation. Never attempt to start the generator set unless it is properly grounded. Failure to comply with this warning can cause injury or death to personnel.

WARNING

Diesel fuel is flammable and toxic to eyes, skin, and respiratory tract. Skin and eye protection are required when working in contact with diesel fuel. Avoid repeated or prolonged contact. Provide adequate ventilation. Operators are to wash exposed skin and change chemical soaked clothing promptly if exposed to fuel. Failure to comply with this warning can cause injury or death to personnel.

WARNING

Fuels used in the generator set are flammable. Do not smoke or use open flames when performing maintenance. Failure to comply with this warning can cause injury or death to personnel, and damage to the generator set.

WARNING

Fuels used in the generator set are flammable. When filling the fuel tank, maintain metal-to-metal contact between filler nozzle and fuel tank opening to eliminate static electrical discharge. Failure to comply with this warning can cause injury or death to personnel, and damage to the generator set.

4-6.1 Inspection.

a. a. Shut down generator set.

b. b. Remove fuel tank (paragraph 4-6.2).

c. c. Inspect fuel tank (6, Figure 4-2) for lea ks, cracks, missing hardware, and other damage.

d. d. If no damage is found, install fuel tank (paragraph 4-6.3).

4-6.2 Removal.

a. a. Shut down generator set.

b. b. Open battery access door and disconnect negative battery cable.

c. c. Drain fuel tank.

d. d. Remove engine and generator assembly (paragraph 3-6.1 (MEP-804A/MEP-814A)/paragraph
3-7.1 (MEP-804B/MEP-814B)).

e. e. Remove fuel tank drain valve (paragraph 2-100.1).

f. f. Remove nuts (1, Figure 4-2), lock washers (2), washers (3), bolts (4), and plates (5) securing fuel tank to skid base. Discard lock washers (2).

g. g. Remove fuel tank (6) from skid base.

h. h. Remove low fuel level/auxiliary fuel pump float switch (paragraph 2-102.2).

i. i. Remove fuel level sender (paragraph 2-101.2).

j. j. Remove fuel pickup (paragraph 2-103.1).

k. k. Remove fittings (7) from excess fuel studs (9) on right side of tank.

l. l. Remove fitting (8) from excess fuel stud (9) on left side of tank.

m. m. Remove studs (9), washers (10), and bushings (11) from fuel tank.

n. n. Cover all openings.

4-6.3 Installation.

a. a. Remove covers placed over openings.

b. b. Install bushings (11, Figure 4-2), washers (10), and studs (9) in fuel tank (6).

c. c. Install fitting (8) in excess fuel stud (9) on left side of fuel tank.

d. d. Install fittings (7) in excess fuel studs (9) on right side of fuel tank.

e. e. Install low fuel level/auxiliary fuel pump float switch (paragraph 2-102.4).

f. f. Install fuel level sender (paragraph 2-101.4).

g. g. Install fuel pickup (paragraph 2-103.3).

h. h. Install fuel tank (6) in skid base and secure with plates (5), bolts (4), washers (3), new lock washers

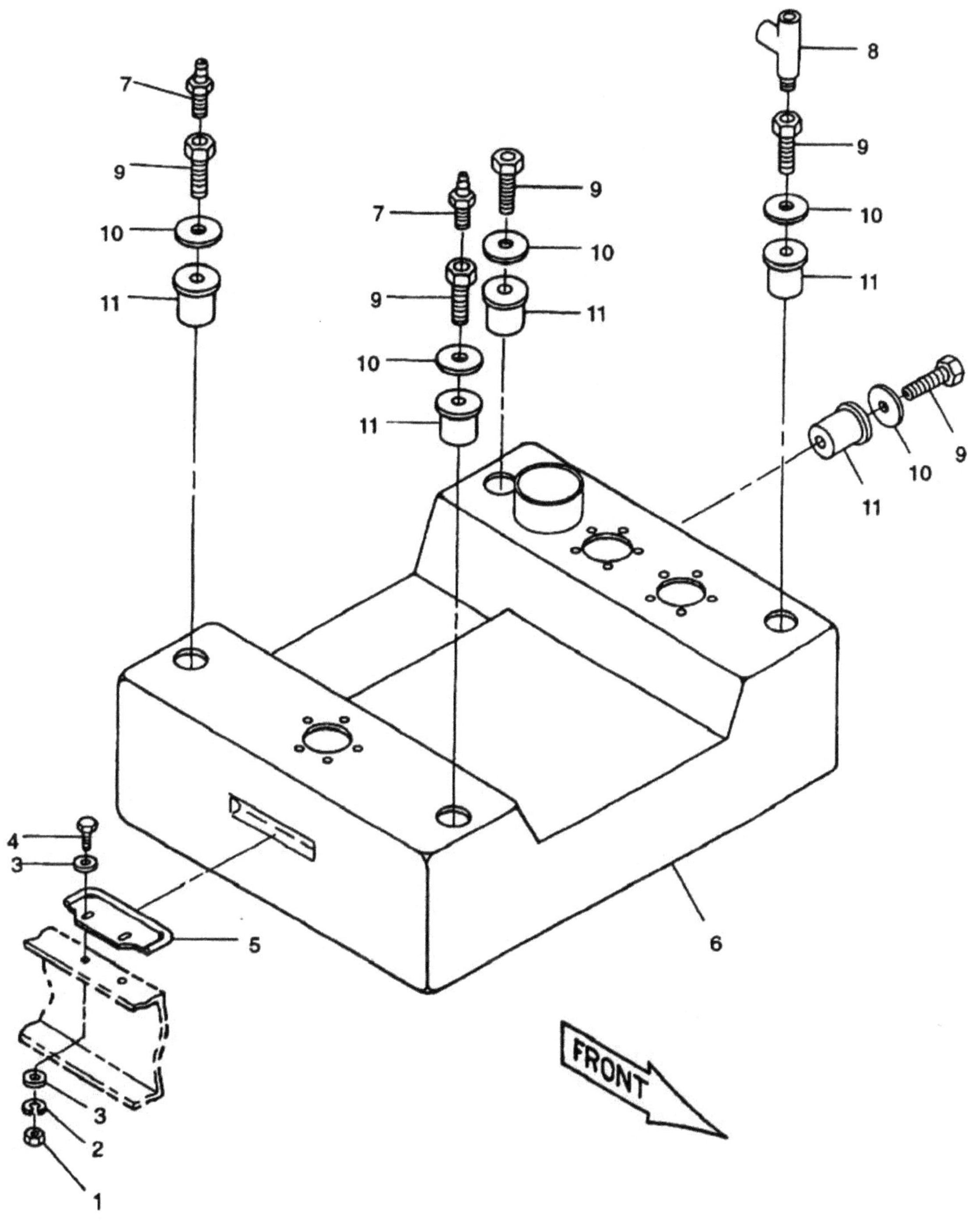

Figure 4-2. Fuel Tank.

i. Install engine and generator assembly (paragraph 3-6.2 (MEP-804A/MEP-814A)/paragraph 3-7.2 (MEP-804B/MEP-814B)).

j. j. Install fuel tank drain valve (paragraph 2-100.2).

k. k. Service fuel tank. Refer to Table 2-2 for proper fuel.

l. l. Connect negative battery c able. Close battery access door.

Section IV. MAINTENANCE OF OUTPUT BOX ASSEMBLY

4-7 OUTPUT BOX ASSEMBLY

WARNING

High voltage is produced when the generator set is in operation. Never attempt to start the generator set unless it is properly grounded. Failure to comply with this warning can cause injury or death to personnel.

WARNING

High voltage is produced when this generator set is in operation. Make sure generator set is completely shut down and free of any power source before attempting any repair or maintenance on the set, or when connecting or disconnecting load cables. Failure to comply with this warning can cause injury or death to personnel.

WARNING

Dangerous voltage exists on live circuits. Always observe precautions and never work alone. Failure to comply with this warning can cause injury or death to personnel.

WARNING

Ensure that the engine cannot be started while maintenance is being performed. (ENGINE CONTROL switch set to OFF/RESET; Battery Disconnect Switch is OFF; DEAD CRANK SWITCH is OFF). Failure to comply with this warning can cause injury or death to personnel.

WARNING

DC voltages are present at generator set electrical components even with generator set shut down. Avoid shorting any positive with ground/negative. Failure to comply with this warning can cause injury to personnel, and damage to equipment.

WARNING

The connection of any electrical equipment and the disconnection of any electrical equipment may cause an explosion hazard. Do not connect any electrical equipment or disconnect any electrical equipment in an explosive atmosphere. Failure to comply with this warning can cause injury or death to personnel.

WARNING

A qualified technician must make the power connections and perform all continuity checks. The power source may be a generator or commercial power. Failure to comply with this warning can cause injury or death to personnel.

WARNING

High voltage is produced when the generator set is in operation. DO NOT touch live voltage connections. Never attempt to connect or disconnect load cables or paralleling cables while the generator set is running. Failure to comply with this warning can cause injury or death to personnel.

4-7.1 Removal.

a. a. Shut down generator set.

b. b. Open battery access door and disconnect negative battery cable.

c. c. Remove control box assembly (paragraph 2-19.2).

d. d. Remove air cleaner assembly (paragraph 2-74.3 (MEP-804A/MEP-814A)/paragraph 2-75.3 (MEP-804B/ MEP-814B)).

e. e. Remove output box access door (paragraph 2-14.1).

f. f. Remove rear housing panel (paragraph 2-18.1).

g. g. Open left side engine access door. Tag and disconnect electrical leads from battery charging alternator, starter solenoid, starter motor, coolant high temperature switch, fuel level sender, fuel float module, and auxiliary fuel pump.

h. h. Open right side engine access door. Tag and disconnect electrical leads from contactor K22, DEAD CRANK switch, low oil pressure switch, oil pressure sender, governor actuator, fuel injection pump, coolant temperature sender, and magnetic pickup.

i. i. Note locations and remove loop clamps securing output box harness to engine.

j. j. Remove voltage reconnection terminal board (paragraph 4-8.1).

k. k. Remove nuts (1, Figure 4-3), bolts (2), and output box top panel (3) from output box assembly.

NOTE

Record number and direction of wraps when removing main generator cables from transformers to aid installation.

l. l. Unwrap main generator cables from droop current transformer (34) and current transformer (31).

m. m. Remove screws (4) and cover (5) from AC circuit interrupter relay (40).

n. n. Tag and disconnect output cables from terminals A2, B2, and C2 of AC circuit interrupter relay (40).

o. o. Tag and disconnect exciter leads F1 and F2 from terminals 1 and 2 of terminal board (26).

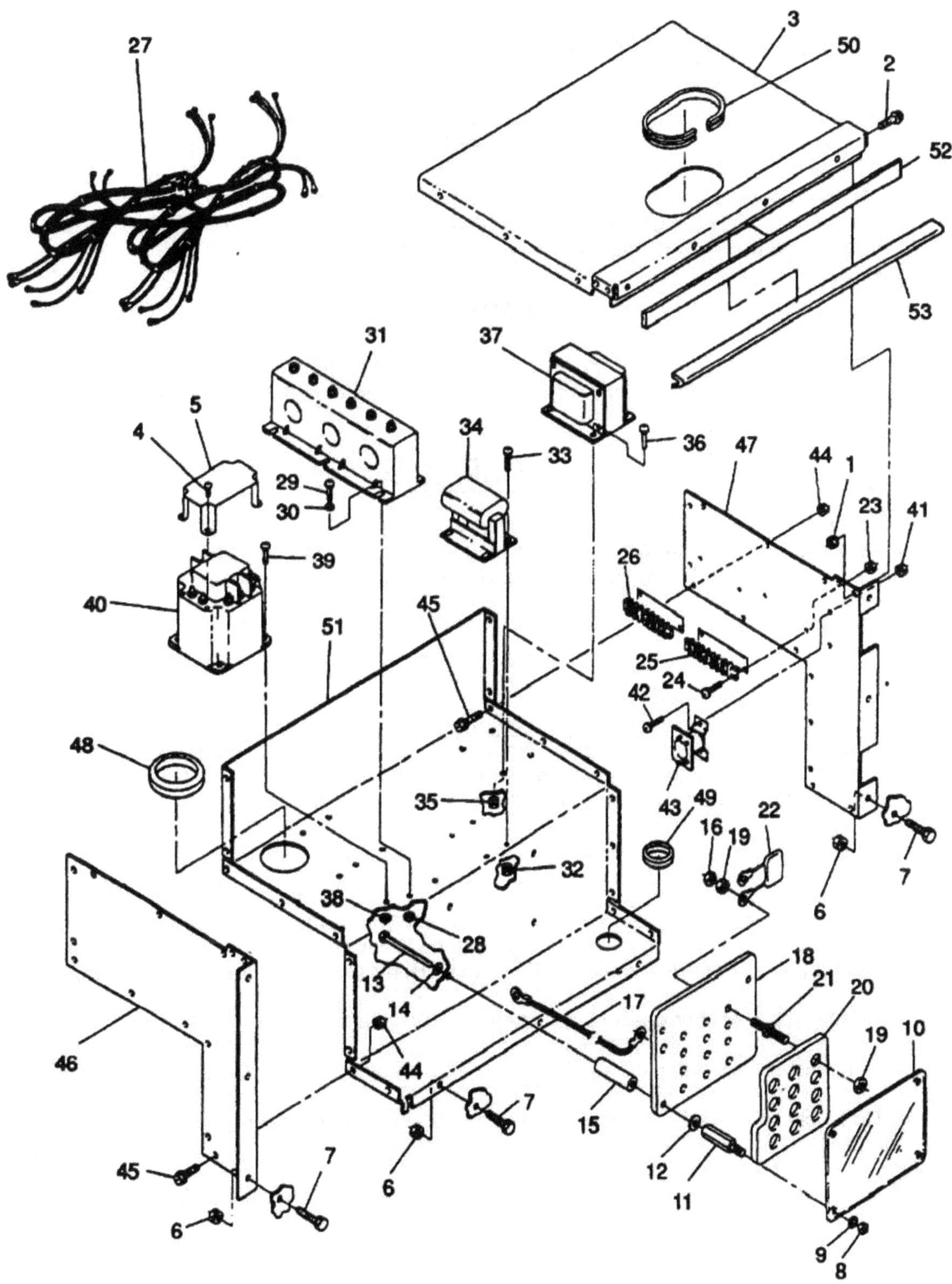

Figure 4-3. Output Box Assembly.

p. p. Remove bolts (44, Figure 2-7) and nuts (45) securing output box assembly to angle (46).

q. q. Remove nuts (6, Figure 4-3), bolts (7), and output box assembly from generator set.

WARNING

High voltage is produced when the generator set is in operation. Never attempt to start the generator set unless it is properly grounded. Failure to comply with this warning can cause injury or death to personnel.

WARNING

DO NOT wear loose clothing when performing checks, services and maintenance. Failure to comply with this warning can cause injury or death to personnel.

WARNING

All metal jewelry can conduct electricity and become entangled in generator set components. Remove all jewelry when working on generator set. Failure to comply with this warning can cause injury or death to personnel.

WARNING

High voltage is produced when the generator set is in operation. DO NOT touch live voltage connections. Never attempt to connect or disconnect load cables or paralleling cables while the generator set is running. Failure to comply with this warning can cause injury or death to personnel.

WARNING

Wear heat resistant gloves and avoid contacting hot metal surfaces with your hands after components have been heated. Wear additional protective clothing as required. Failure to comply with this warning can cause injury to personnel.

a. a. Install output box assembly in generator set with bolts (7, Figure 4-3) and nuts (6).

b. b. Install bolts (44, Figure 2-7) and nuts (45) securing rear of output box assembly to angle (46).

c. c. Connect exciter leads F1 and F2 to terminals 1 and 2 of terminal board (26, Figure 4-3). Remove tags.

d. d. Connect output cables to terminals A2, B2, and C2 of AC circuit interrupter relay (40). Remove tags.

e. e. Install cover (5) on AC circuit interrupter (40) with screws (4).

f. f. Install main generator cables onto current transformer (31) and droop current transformer (34) the same number of wraps recorded during removal.

g. g. Install voltage reconnection terminal board (paragraph 4-8.2).

h. h. On right side of engine, connect electrical leads to magnetic pickup, coolant temperature sender, governor actuator, fuel injection pump, oil pressure sender, low oil pressure switch, DEAD CRANK switch, and contactor K22. Remove tags.

i. i. On left side of engine, connect electrical leads to starter motor, starter solenoid, battery charging alternator, coolant high temperature switch, fuel level sender, fuel float module, and auxiliary fuel pump. Remove tags.

j. j. Install loop clamps securing output box harness to engine as noted during removal. Close engine access doors.

k. k. Install output box top panel (3) with bolts (2) and nuts (1).

l. l. Install air cleaner assembly (paragraph 2-74.4 (MEP-804A/MEP-814A)/paragraph 2-75.4 (MEP-804B/MEP-814B)).

m. m. Install rear housing panel (paragraph 2-18.4).

n. n. Install control box assembly (paragraph 2-19.4).

o. o. Install output box access door (paragraph 2-14.4).

p. p. Connect negative battery c able. Close battery access door.

4-8 VOLTAGE RECONNECTION TERMINAL BOARD

WARNING

High voltage is produced when the generator set is in operation. Never attempt to start the generator set unless it is properly grounded. Failure to comply with this warning can cause injury or death to personnel.

WARNING

High voltage is produced when this generator set is in operation. Make sure generator set is completely shut down and free of any power source before attempting any repair or maintenance on the set, or when connecting or disconnecting load cables. Failure to comply with this warning can cause injury or death to personnel.

WARNING

Dangerous voltage exists on live circuits. Always observe precautions and never work alone. Failure to comply with this warning can cause injury or death to personnel.

WARNING

Ensure that the engine cannot be started while maintenance is being performed. (ENGINE CONTROL switch set to OFF/RESET; Battery Disconnect Switch is OFF; DEAD CRANK SWITCH is OFF). Failure to comply with this warning can cause injury or death to personnel.

WARNING

DC voltages are present at generator set electrical components even with generator set shut down. Avoid shorting any positive with ground/negative. Failure to comply with this warning can cause injury to personnel, and damage to equipment.

WARNING

The connection of any electrical equipment and the disconnection of any electrical equipment may cause an explosion hazard. Do not connect any electrical equipment or disconnect any electrical equipment in an explosive atmosphere. Failure to comply with this warning can cause injury or death to personnel.

WARNING

A qualified technician must make the power connections and perform all continuity checks. The power source may be a generator or commercial power. Failure to comply with this warning can cause injury or death to personnel.

WARNING

High voltage is produced when the generator set is in operation. DO NOT touch live voltage connections. Never attempt to connect or disconnect load cables or paralleling cables while the generator set is running. Failure to comply with this warning can cause injury or death to personnel.

4-8.1 Removal.

a. a. Shut down generator set.

b. b. Open battery access door and disconnect negative battery cable.

c. c. Open output box access door and load terminal board access door.

d. d. Remove nuts (8, Figure 4-3), washers (9), and cover (10) from voltage reconnection board (18).

e. e. Unscrew standoffs (11) and remove washers (12), bolts (13), washers (14), and mounts (15).

f. f. Remove nuts (16). Tag and disconnect electrical cables (17) and main generator cables from voltage reconnection board (18).

g. g. Remove voltage reconnection board (18) and moveable terminal board (20) from generator set as an assembly.

h. h. Tag position of capacitors (22) on voltage reconnection board (18).

i. i. Remove nuts (19), movable terminal board (20), studs (21), and capacitors (22) from voltage reconnection board (18).

4-8.2 Installation.

WARNING

High voltage is produced when the generator set is in operation. Never attempt to start the generator set unless it is properly grounded. Failure to comply with this warning can cause injury or death to personnel.

WARNING

DO NOT wear loose clothing when performing checks, services and maintenance. Failure to comply with this warning can cause injury or death to personnel.

WARNING

All metal jewelry can conduct electricity and become entangled in generator set components. Remove all jewelry when working on generator set. Failure to comply with this warning can cause injury or death to personnel.

WARNING

High voltage is produced when the generator set is in operation. DO NOT touch live voltage connections. Never attempt to connect or disconnect load cables or paralleling cables while the generator set is running. Failure to comply with this warning can cause injury or death to personnel.

WARNING

Wear heat resistant gloves and avoid contacting hot metal surfaces with your hands after components have been heated. Wear additional protective clothing as required. Failure to comply with this warning can cause injury to personnel.

a. Insert studs (21, Figure 4-3) in voltage reconnection board (18). Position capacitors (22), as tagged, align moveable terminal board (20) with voltage reconnection board (18), and install nuts (19). Remove tags.

b. b. Position voltage reconnection board (18) and moveable terminal board (20) in generator set.

c. c. Connect electrical cables (17) and main generator cables to voltage reconnection board (18) with nuts (16). Remove tags.

d. d. Secure voltage reconnection board (18) to generator set with bolts (13), washers (14), mounts (15), washers (12), and standoffs (11).

e. e. Install cover (10) with washers (9) and nuts (8). Close output box and load terminal board access doors.

f. f. Connect negative battery cabl e. Close battery access door.

4-9 OUTPUT BOX WIRING HARNESS

WARNING

High voltage is produced when the generator set is in operation. Never attempt to start the generator set unless it is properly grounded. Failure to comply with this warning can cause injury or death to personnel.

WARNING

High voltage is produced when this generator set is in operation. Make sure generator set is completely shut down and free of any power source before attempting any repair or maintenance on the set, or when connecting or disconnecting load cables. Failure to comply with this warning can cause injury or death to personnel.

WARNING

Dangerous voltage exists on live circuits. Always observe precautions and never work alone. Failure to comply with this warning can cause injury or death to personnel.

WARNING

Ensure that the engine cannot be started while maintenance is being performed. (ENGINE CONTROL switch set to OFF/RESET; Battery Disconnect Switch is OFF; DEAD CRANK SWITCH is OFF). Failure to comply with this warning can cause injury or death to personnel.

WARNING

DC voltages are present at generator set electrical components even with generator set shut down. Avoid shorting any positive with ground/negative. Failure to comply with this warning can cause injury to personnel, and damage to equipment.

WARNING

The connection of any electrical equipment and the disconnection of any electrical equipment may cause an explosion hazard. Do not connect any electrical equipment or disconnect any electrical equipment in an explosive atmosphere. Failure to comply with this warning can cause injury or death to personnel.

WARNING

A qualified technician must make the power connections and perform all continuity checks. The power source may be a generator or commercial power. Failure to comply with this warning can cause injury or death to personnel.

WARNING

High voltage is produced when the generator set is in operation. DO NOT touch live voltage connections. Never attempt to connect or disconnect load cables or paralleling cables while the generator set is running. Failure to comply with this warning can cause injury or death to personnel.

4-9.1 Inspection.

a. a. Remove output box wiring harness (paragraph 4-9.2).

b. b. Inspect wiring harness for burned, bent, corroded, and broken terminals.

c. c. Inspect connectors for cracks, corrosion, stripped threads, bent or broken pins, and obvious damage.

d. d. Inspect wire insulation for burns, deterioration, and chafing.

e. e. Install output box wiring harness (paragraph 4-9.5).

4-9.2 Removal.

a. a. Shut down generator set.

b. b. Open battery access door and disconnect negative battery cable.

c. c. Remove control box assembly (paragraph 2-19.2).

d. d. Remove output box top panel (paragraph 4-7.1, step k).

e. e. Open left side engine access door. Tag and disconnect electrical leads from battery charging alternator, starter solenoid, starter motor, coolant high temperature switch, fuel level sender, fuel float module, and auxiliary fuel pump.

f. f. Open right side engine access door. Tag and disconnect electrical leads from contactor K22, DEAD CRANK switch, low oil pressure switch, oil pressure sender, governor actuator, fuel injection pump, coolant temperature sender, and magnetic pickup.

g. g. Open output box access door. Remove screws (4, Figure 4-3) and cover (5) from AC circuit interrupter relay (40). Tag and disconnect electrical leads from cranking relay (43), current transformer (31), AC circuit interrupter relay (40), power potential transformer (37), and voltage reconnection board (18).

h. h. Tag and disconnect electrical leads for droop cu rrent transformer (34) from terminal board (25).

i. i. Remove screws (24) and nuts (23) securing terminal boards (25 and 26) in output box.

j. j. Remove all clamps securing output box harness to generator set.

k. k. Remove output box harness (27) from output box and generator set.

4-9.3 Testing.

a. a. Set multimeter for ohms.

b. b. Check individual wires, connectors, and terminal boards for continuity. Refer to Wiring Diagram FO-2 (MEP-804A/MEP-814A)/FO-4 (MEP-804B/MEP-814B) for wire identification.

4-9.4 Repair.

a. a. Replace damaged cable assemblies, terminals, connectors, sockets, and terminal boards.

b. b. Replace wires with damaged insulation and those that do not indicate continuity.

4-9.5 Installation.

WARNING

High voltage is produced when the generator set is in operation. Never attempt to start the generator set unless it is properly grounded. Failure to comply with this warning can cause injury or death to personnel.

WARNING

DO NOT wear loose clothing when performing checks, services and maintenance. Failure to comply with this warning can cause injury or death to personnel.

WARNING

All metal jewelry can conduct electricity and become entangled in generator set components. Remove all jewelry when working on generator set. Failure to comply with this warning can cause injury or death to personnel.

WARNING

High voltage is produced when the generator set is in operation. DO NOT touch live voltage connections. Never attempt to connect or disconnect load cables or paralleling cables while the generator set is running. Failure to comply with this warning can cause injury or death to personnel.

WARNING

Wear heat resistant gloves and avoid contacting hot metal surfaces with your hands after components have been heated. Wear additional protective clothing as required. Failure to comply with this warning can cause injury to personnel.

a. a. Position output box wiring harness (27, F igure 4-3) in output box and generator set.

b. b. Install all clamps, as removed, securing output box wiring harness in generator set.

c. c. Secure terminal boards (25 and 26) in output box with screws (24) and nuts (23).

d. d. Connect electrical leads for droop current transformer (34) to terminal board (25). Remove tags.

e. e. Connect electrical lead s to AC circuit interrupter relay (40), current transformer (31), cranking relay (43), power potential transformer (37), and voltage reconnection board (18). Install cover (5) with screws (4) on AC circuit interrupter relay (40). Remove tags. Close output box access door.

f. f. On right side of engine, connect electrical leads to magnetic pickup, coolant temperature sender, governor actuator, fuel injection pump, oil pressure sender, low oil pressure switch, DEAD CRANK switch, and contactor K22. Remove tags. Close right side engine access door.

g. g. On left side of engine, connect electrical leads to starter motor, starter solenoid, battery charging alternator, coolant high temperature switch, fuel level sender, fuel float module, and auxiliary fuel pump. Remove tags. Close left side engine access door.

h. h. Install output box top panel (paragraph 4-7.2, step k).

i. i. Install control box assembly (paragraph 2-19.4).

j. j. Connect negative battery c able. Close battery access door.

4-10 CURRENT TRANSFORMER

WARNING

High voltage is produced when the generator set is in operation. Never attempt to start the generator set unless it is properly grounded. Failure to comply with this warning can cause injury or death to personnel.

WARNING

High voltage is produced when this generator set is in operation. Make sure generator set is completely shut down and free of any power source before attempting any repair or maintenance on the set, or when connecting or disconnecting load cables. Failure to comply with this warning can cause injury or death to personnel.

WARNING

Dangerous voltage exists on live circuits. Always observe precautions and never work alone. Failure to comply with this warning can cause injury or death to personnel.

WARNING

Ensure that the engine cannot be started while maintenance is being performed. (ENGINE CONTROL switch set to OFF/RESET; Battery Disconnect Switch is OFF; DEAD CRANK SWITCH is OFF). Failure to comply with this warning can cause injury or death to personnel.

WARNING

DC voltages are present at generator set electrical components even with generator set shut down. Avoid shorting any positive with ground/negative. Failure to comply with this warning can cause injury to personnel, and damage to equipment.

WARNING

The connection of any electrical equipment and the disconnection of any electrical equipment may cause an explosion hazard. Do not connect any electrical equipment or disconnect any electrical equipment in an explosive atmosphere. Failure to comply with this warning can cause injury or death to personnel.

WARNING

A qualified technician must make the power connections and perform all continuity checks. The power source may be a generator or commercial power. Failure to comply with this warning can cause injury or death to personnel.

WARNING

High voltage is produced when the generator set is in operation. DO NOT touch live voltage connections. Never attempt to connect or disconnect load cables or paralleling cables while the generator set is running. Failure to comply with this warning can cause injury or death to personnel.

4-10.1 Removal.

a. a. Shut down generator set.

b. b. Open battery access door and disconnect negative battery cable.

c. c. Remove control box assembly (paragraph 2-19.2).

d. d. Remove output box top panel (paragraph 4-7.1, step k).

e. e. Open output box and right side engine access doors.

f. f. Tag and disconnect current transformer (31, Figure 4-3) electrical leads.

g. g. Tag and disconnect main generator cables T2 and T8 from voltage reconnection board (18).

h. h. Unwrap main generator cables from droop current transformer (34) and current transformer (31). Note number and direction of wraps.

i. Remove screws (29), flat washers (30), nuts (2 8), and current transformer (31) from output box.

4-10.2 Testing.

a. a. Remove current transformer (paragraph 4-10.1).

b. b. Set multimeter for ohms and check for continuity between secondary terminals A1 and A2, B1 and B2, and C1 and C2.

c. c. If continuity is present, continue with test. If continuity is not present, current transformer is defective and must be replaced.

d. d. Set up a test circuit using 10 gauge wire (Figure 4-4). Make ten passes with wire through phase A window.

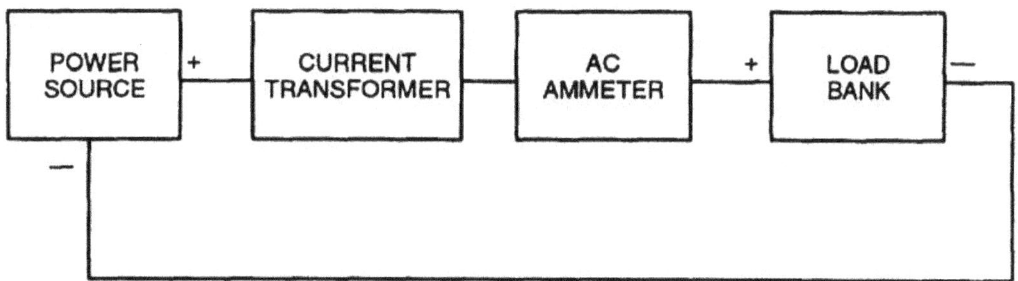

Figure 4-4. Testing Current Transformer.

e. e. Turn on power source and load bank. Adjust loa d bank until 27.7 amps is indicated on ammeter.

f. f. Set multimeter for amperes and connect to secondary terminals A1 and A2. Multimeter indication must be 0.9 to 1.1 amps.

g. g. Repeat steps d, e, and f above using phase B window and secondary terminals B1 and B2.

h. h. Repeat steps d, e, and f above using phas e C window and secondary terminals C1 and C2.

i. i. Replace current transformer if multimeter indication in any phase is other than stated in step f above.

j. j. Remove current transformer from test circuit .

k. k. Install current transformer (paragraph 4.10.3).

4-10.3 Installation.

WARNING

High voltage is produced when the generator set is in operation. Never attempt to start the generator set unless it is properly grounded. Failure to comply with this warning can cause injury or death to personnel.

WARNING

DO NOT wear loose clothing when performing checks, services and maintenance. Failure to comply with this warning can cause injury or death to personnel.

WARNING

All metal jewelry can conduct electricity and become entangled in generator set components. Remove all jewelry when working on generator set. Failure to comply with this warning can cause injury or death to personnel.

WARNING

High voltage is produced when the generator set is in operation. DO NOT touch live voltage connections. Never attempt to connect or disconnect load cables or paralleling cables while the generator set is running. Failure to comply with this warning can cause injury or death to personnel.

WARNING

Wear heat resistant gloves and avoid contacting hot metal surfaces with your hands after components have been heated. Wear additional protective clothing as required. Failure to comply with this warning can cause injury to personnel.

a. a. Install current transformer (31, Figure 4-3) with screws (29), flat washers (30), and nuts (28).

b. b. Wrap main generator cables around current transformer (31) and droop current transformer (34) using same number of wraps noted during removal.

c. c. Connect main generator c ables to voltage reconnection board (18). Remove tags.

d. d. Connect electrical leads to current transformer (31). Remove tags. Close output box and right side engine access doors.

e. e. Install output box top panel (paragraph 4-7.2, step k).

f. f. Install control box assembly (paragraph 2-19.4).

g. g. Connect negative battery c able. Close battery access door.

4-11 DROOP CURRENT TRANSFORMER

WARNING

High voltage is produced when the generator set is in operation. Never attempt to start the generator set unless it is properly grounded. Failure to comply with this warning can cause injury or death to personnel.

WARNING

High voltage is produced when this generator set is in operation. Make sure generator set is completely shut down and free of any power source before attempting any repair or maintenance on the set, or when connecting or disconnecting load cables. Failure to comply with this warning can cause injury or death to personnel.

WARNING

Dangerous voltage exists on live circuits. Always observe precautions and never work alone. Failure to comply with this warning can cause injury or death to personnel.

WARNING

Ensure that the engine cannot be started while maintenance is being performed. (ENGINE CONTROL switch set to OFF/RESET; Battery Disconnect Switch is OFF; DEAD CRANK SWITCH is OFF). Failure to comply with this warning can cause injury or death to personnel.

WARNING

DC voltages are present at generator set electrical components even with generator set shut down. Avoid shorting any positive with ground/negative. Failure to comply with this warning can cause injury to personnel, and damage to equipment.

WARNING

The connection of any electrical equipment and the disconnection of any electrical equipment may cause an explosion hazard. Do not connect any electrical equipment or disconnect any electrical equipment in an explosive atmosphere. Failure to comply with this warning can cause injury or death to personnel.

WARNING

A qualified technician must make the power connections and perform all continuity checks. The power source may be a generator or commercial power. Failure to comply with this warning can cause injury or death to personnel.

WARNING

High voltage is produced when the generator set is in operation. DO NOT touch live voltage connections. Never attempt to connect or disconnect load cables or paralleling cables while the generator set is running. Failure to comply with this warning can cause injury or death to personnel.

4-11.1 Removal.

a. a. Shut down generator set.

b. b. Open battery access door and disconnect negative battery cable.

c. c. Open output box and right side engine access doors.

d. d. Tag and disconnect main generator cables T2 a nd T8 from voltage reconnection board (18, Figure 4-3).

e. e. Unwrap main generator cables T2 and T8 from droop current transformer (34). Note number and direction of wraps.

f. f. Tag and disconnect droop current transformer (34) electrical leads from terminal board (25).

g. Remove screws (33), nuts (32), and droop current transformer (34) from output box.

4-11.2 Testing.

a. a. Remove droop current transformer (paragraph 4-11.1).

b. b. Set multimeter for ohms and check for continuity between secondary leads 1 and 2.

c. c. If continuity is present, continue with test. If contin uity is not present, droop current transformer is defective and must be replaced.

d. d. Set up a test circuit using 10 gauge wire (Figure 4-5). Make ten passes with wire through window of droop current transformer.

e. e. Turn on power supply and load bank. Adjust load bank until 20.8 amps is indicated on AC ammeter.

f. f. Set multimeter for AC amperes and connect to secondary leads 1 and 2. Multimeter indication must be between 0.9 and 1.1 amps.

g. g. Replace droop current transformer if multimeter indication is other than above.

h. h. Remove droop current transformer from test circuit.

i. i. Install droop current transformer (paragraph 4-11.3).

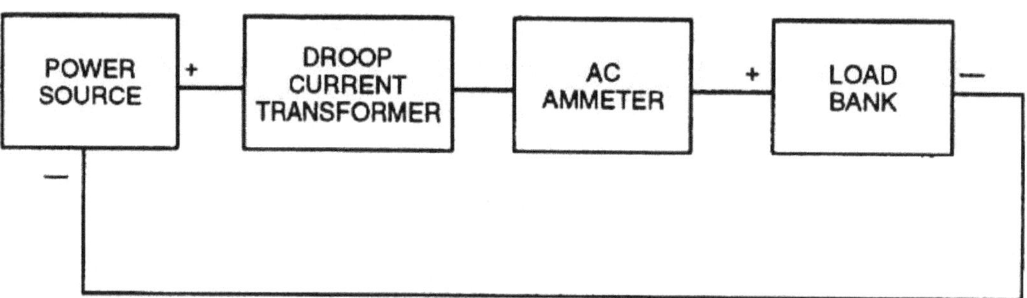

Figure 4-5. Testing Droop Current Transformer.

4-11.3 Installation.

WARNING

High voltage is produced when the generator set is in operation. Never attempt to start the generator set unless it is properly grounded. Failure to comply with this warning can cause injury or death to personnel.

WARNING

DO NOT wear loose clothing when performing checks, services and maintenance. Failure to comply with this warning can cause injury or death to personnel.

WARNING

All metal jewelry can conduct electricity and become entangled in generator set components. Remove all jewelry when working on generator set. Failure to comply with this warning can cause injury or death to personnel.

WARNING

High voltage is produced when the generator set is in operation. DO NOT touch live voltage connections. Never attempt to connect or disconnect load cables or paralleling cables while the generator set is running. Failure to comply with this warning can cause injury or death to personnel.

WARNING

Wear heat resistant gloves and avoid contacting hot metal surfaces with your hands after components have been heated. Wear additional protective clothing as required. Failure to comply with this warning can cause injury to personnel.

a. a. Install droop current transformer (34, Figure 4-3) in output box with screws (33) and nuts (32).

b. b. Wrap main generator cables around droop current transformer (34) using same number of wraps noted during removal.

c. c. Connect main generator c ables to voltage reconnection board (18). Remove tags.

d. d. Connect droop current transformer (34) electrical leads to terminal board (25). Remove tags. Close output box and engine access doors.

e. Connect negative battery c able. Close battery access door.

4-12 POWER POTENTIAL TRANSFORMER

WARNING

High voltage is produced when the generator set is in operation. Never attempt to start the generator set unless it is properly grounded. Failure to comply with this warning can cause injury or death to personnel.

WARNING

High voltage is produced when this generator set is in operation. Make sure generator set is completely shut down and free of any power source before attempting any repair or maintenance on the set, or when connecting or disconnecting load cables. Failure to comply with this warning can cause injury or death to personnel.

WARNING

Dangerous voltage exists on live circuits. Always observe precautions and never work alone. Failure to comply with this warning can cause injury or death to personnel.

WARNING

Ensure that the engine cannot be started while maintenance is being performed. (ENGINE CONTROL switch set to OFF/RESET; Battery Disconnect Switch is OFF; DEAD CRANK SWITCH is OFF). Failure to comply with this warning can cause injury or death to personnel.

WARNING

DC voltages are present at generator set electrical components even with generator set shut down. Avoid shorting any positive with ground/negative. Failure to comply with this warning can cause injury to personnel, and damage to equipment.

WARNING

The connection of any electrical equipment and the disconnection of any electrical equipment may cause an explosion hazard. Do not connect any electrical equipment or disconnect any electrical equipment in an explosive atmosphere. Failure to comply with this warning can cause injury or death to personnel.

WARNING

A qualified technician must make the power connections and perform all continuity checks. The power source may be a generator or commercial power. Failure to comply with this warning can cause injury or death to personnel.

WARNING

High voltage is produced when the generator set is in operation. DO NOT touch live voltage connections. Never attempt to connect or disconnect load cables or paralleling cables while the generator set is running. Failure to comply with this warning can cause injury or death to personnel.

4-12.1 Removal.

a. a. Shut down generator set.

b. b. Open battery access door and disconnect negative battery cable.

c. c. Remove control box assembly (paragraph 2-19.2).

d. d. Remove output box top panel (paragraph 4-7.1, step k).

e. e. Tag and disconnect power potential transformer (37, Figure 4-3) electrical leads.

f. f. Remove screws (36), nuts (35), and power potential transformer (37) from output

box. **4-12.2 Testing (Power Potential Transformer, P/N 17910).**

a. a. Remove power potential trans former (paragraph 4-12.1).

b. b. Connect terminals 1 and 2 of power potential transformer to an AC power source (polarity is not important).

c. c. Set power supply for 208 VAC, 60 Hz. Turn on power supply.

d. d. Set multimeter for AC volts and take readings between the following terminals and compare to voltages shown:

Terminals 3 and 7 = 144-176 VAC
Terminals 4 and 6 = 27-33 VAC

e. e. Replace power potential transformer if readings are not as voltages above.

f. f. Install power potential transformer (paragraph 4-12.4).

4-12.3 Testing (Power Potential Transformer, P/N A1497B, for MEP-804A).

a. a. Remove power potential trans former (paragraph 4-12.1).

b. b. Connect terminals 1 and 2 of power potential transf ormer to an AC power source. (Polarity is not important.)

c. c. Set power supply at 208 VAC, 60 Hz. Turn on power supply.

d. d. Set multimeter for AC volts and take readings between the following terminals and compare to voltages shown:

Terminals 5 and 6 = 54-66 VAC
Terminals 5 and 4 = 54-66 VAC
Terminals 5 and 3 = 126-154 VAC
Terminals 5 and 7 = 126-154 VAC

e. e. Replace power potential transformer if readings are not as above.

f. f. Install power potential transformer (paragraph 4-12.4).

4-12.4 Installation.

WARNING

High voltage is produced when the generator set is in operation. Never attempt to start the generator set unless it is properly grounded. Failure to comply with this warning can cause injury or death to personnel.

WARNING

DO NOT wear loose clothing when performing checks, services and maintenance. Failure to comply with this warning can cause injury or death to personnel.

WARNING

All metal jewelry can conduct electricity and become entangled in generator set components. Remove all jewelry when working on generator set. Failure to comply with this warning can cause injury or death to personnel.

WARNING

High voltage is produced when the generator set is in operation. DO NOT touch live voltage connections. Never attempt to connect or disconnect load cables or paralleling cables while the generator set is running. Failure to comply with this warning can cause injury or death to personnel.

WARNING

Wear heat resistant gloves and avoid contacting hot metal surfaces with your hands after components have been heated. Wear additional protective clothing as required. Failure to comply with this warning can cause injury to personnel.

a. a. Install power potential transformer (37, Figure 4-3) with screws (36) and nuts (35).

b. b. Connect electrical le ads and remove tags.

c. c. Install output box top panel (paragraph 4-7.2, step k).

d. d. Install control box assembly (paragraph 2-19.4).

e. e. Connect negative battery c able. Close battery access door.

4-13 OUTPUT BOX PANELS

WARNING

High voltage is produced when the generator set is in operation. Never attempt to start the generator set unless it is properly grounded. Failure to comply with this warning can cause injury or death to personnel.

WARNING

High voltage is produced when this generator set is in operation. Make sure generator set is completely shut down and free of any power source before attempting any repair or maintenance on the set, or when connecting or disconnecting load cables. Failure to comply with this warning can cause injury or death to personnel.

WARNING

Dangerous voltage exists on live circuits. Always observe precautions and never work alone. Failure to comply with this warning can cause injury or death to personnel.

WARNING

Ensure that the engine cannot be started while maintenance is being performed. (ENGINE CONTROL switch set to OFF/RESET; Battery Disconnect Switch is OFF; DEAD CRANK SWITCH is OFF). Failure to comply with this warning can cause injury or death to personnel.

WARNING

DC voltages are present at generator set electrical components even with generator set shut down. Avoid shorting any positive with ground/negative. Failure to comply with this warning can cause injury to personnel, and damage to equipment.

WARNING

The connection of any electrical equipment and the disconnection of any electrical equipment may cause an explosion hazard. Do not connect any electrical equipment or disconnect any electrical equipment in an explosive atmosphere. Failure to comply with this warning can cause injury or death to personnel.

WARNING

A qualified technician must make the power connections and perform all continuity checks. The power source may be a generator or commercial power. Failure to comply with this warning can cause injury or death to personnel.

WARNING

High voltage is produced when the generator set is in operation. DO NOT touch live voltage connections. Never attempt to connect or disconnect load cables or paralleling cables while the generator set is running. Failure to comply with this warning can cause injury or death to personnel.

4-13.1 Inspection.

a. a. Inspect output box panels (3, 46, 47, and 51, Figure 4-3) for cracks, dents, loose paint, corrosion, and other damage.

b. b. Inspect grommets (48, 49, and 50); door seal (52); and electromagnetic interference (EMI) seal (53) for looseness, tears, deterioration, and other damage.

4-13.2 Removal.

a. a. Shut down generator set.

b. b. Open battery access door and disconnect negative battery cable.

c. c. Remove control box assembly (paragraph 2-19.2).

d. d. Remove generator housing rear panel (paragraph 2-18.1).

e. e. Remove output box access door (paragraph 2-14.1).

f. f. Remove air cleaner assembly (paragraph 2-74.2 (MEP-804A/MEP-814A)/paragraph 2-75.2 (MEP-804B/ MEP-814B)).

g. g. Remove bolts (2, Figure 4-3), nuts (1), and output box top panel (3).

h. h. Remove voltage reconnection terminal board (paragraph 4-8.1).

i. i. Remove droop current transformer (paragraph 4-11.1).

j. j. Remove power potential trans former (paragraph 4-12.1).

k. k. Remove current transformer (paragraph 4-10.1).

l. l. Remove output box wiring harness (paragraph 4-9.2).

m. m. Remove nuts (38), screws (39), and AC circuit interrupter relay (40).

n. n. Remove nuts (41), screws (42), and cranking relay (43).

o. o. Remove nuts (6), bolts (7), and output box panels from generator set.

p. p. Remove bolts (45), nuts (44), and output box side panels (46 and 47) from output box bottom panel (51).

q. q. Remove grommets (48, 49, and 50) from output box panels (3 and 51).

r. r. Remove door seal (52) and EMI seal (53) from output box top panel (3), if necessary.

WARNING

CARC paint is a health hazard, and is irritating to eyes, skin, and respiratory system. Wear protective eyewear, mask, and gloves when applying or removing CARC paint. Failure to comply with this warning can cause injury to personnel.

a. a. Repair all dents and cracks and remove all loose paint.

b. b. Remove light corrosion with fine grit abrasive paper (Item 15, Appendix C).

c. c. Replace damaged seals and grommets.

d. d. Repaint surfaces in accordance with TM 43-0139. (F) Refer to applicable directives.

4-13.4 Installation.

WARNING

High voltage is produced when the generator set is in operation. Never attempt to start the generator set unless it is properly grounded. Failure to comply with this warning can cause injury or death to personnel.

WARNING

DO NOT wear loose clothing when performing checks, services and maintenance. Failure to comply with this warning can cause injury or death to personnel.

WARNING

All metal jewelry can conduct electricity and become entangled in generator set components. Remove all jewelry when working on generator set. Failure to comply with this warning can cause injury or death to personnel.

WARNING

High voltage is produced when the generator set is in operation. DO NOT touch live voltage connections. Never attempt to connect or disconnect load cables or paralleling cables while the generator set is running. Failure to comply with this warning can cause injury or death to personnel.

WARNING

Wear heat resistant gloves and avoid contacting hot metal surfaces with your hands after components have been heated. Wear additional protective clothing as required. Failure to comply with this warning can cause injury to personnel.

a. a. Install grommets (48, 49, and 50, Figure 4-3) in output box panels (3 and 51).

b. b. Install output box side panels (46 and 47) on output box bottom panel (51) with bolts (45) and nuts (44).

c. c. Install output box panels in generator set with bolts (7) and nuts (6).

d. d. Install cranking relay (43) with screws (42) and nuts (41).

e. e. Install AC circuit interrupter relay (40) with screws (39) and nuts (38).

f. f. Install current transformer (paragraph 4-10.3).

g. g. Install power potential transformer (paragraph 4-12.4).

h. h. Install droop current transformer (paragraph 4-11.3).

i. i. Install voltage reconnection terminal board (paragraph 4-8.2).

j. j. Install output box wiring harness (paragraph 4-9.5).

k. k. Install air cleaner assembly (paragraph 2-74.4 (MEP-804A/MEP-814A)/paragraph 2-75.4 (MEP-804B/ MEP-814B)).

l. l. Install output box top panel (3) with bolts (2) and nuts (1).

m. m. If removed, install self-adhesive door seal (52) and EMI seal (53) with adhesive (Item 1, Appendix C) on output box top panel (3).

n. n. Install output box access door (paragraph 2-14.4).

o. o. Install generator housing rear panel (paragraph 2-18.4).

p. p. Install control box assembly (paragraph 2-19.4).

q. q. Connect negative battery c able. Close battery access door.

Section V. MAINTENANCE OF ENGINE ACCESSORIES

4-14 GOVERNOR ACTUATOR (MEP-804A/MEP-814A)

WARNING

High voltage is produced when the generator set is in operation. Never attempt to start the generator set unless it is properly grounded. Failure to comply with this warning can cause injury or death to personnel.

WARNING

High voltage is produced when this generator set is in operation. Make sure generator set is completely shut down and free of any power source before attempting any repair or maintenance on the set, or when connecting or disconnecting load cables. Failure to comply with this warning can cause injury or death to personnel.

WARNING

Dangerous voltage exists on live circuits. Always observe precautions and never work alone. Failure to comply with this warning can cause injury or death to personnel.

WARNING

Ensure that the engine cannot be started while maintenance is being performed. (ENGINE CONTROL switch set to OFF/RESET; Battery Disconnect Switch is OFF; DEAD CRANK SWITCH is OFF). Failure to comply with this warning can cause injury or death to personnel.

WARNING

DC voltages are present at generator set electrical components even with generator set shut down. Avoid shorting any positive with ground/negative. Failure to comply with this warning can cause injury to personnel, and damage to equipment.

WARNING

The connection of any electrical equipment and the disconnection of any electrical equipment may cause an explosion hazard. Do not connect any electrical equipment or disconnect any electrical equipment in an explosive atmosphere. Failure to comply with this warning can cause injury or death to personnel.

WARNING

A qualified technician must make the power connections and perform all continuity checks. The power source may be a generator or commercial power. Failure to comply with this warning can cause injury or death to personnel.

WARNING

High voltage is produced when the generator set is in operation. DO NOT touch live voltage connections. Never attempt to connect or disconnect load cables or paralleling cables while the generator set is running. Failure to comply with this warning can cause injury or death to personnel.

WARNING

Batteries give off a flammable gas. Do not smoke or use open flame when performing maintenance. Failure to comply with this warning can cause injury or death to personnel, and damage to the generator set.

WARNING

Battery acid can cause burns to unprotected skin. Wear safety goggles and chemical gloves and avoid acid splash while working on batteries. Failure to comply with this warning can cause injury to personnel.

WARNING

Diesel fuel is flammable and toxic to eyes, skin, and respiratory tract. Skin and eye protection are required when working in contact with diesel fuel. Avoid repeated or prolonged contact. Provide adequate ventilation. Operators are to wash exposed skin and change chemical soaked clothing promptly if exposed to fuel. Failure to comply with this warning can cause injury or death to personnel.

WARNING

Fuels used in the generator set are flammable. Do not smoke or use open flames when performing maintenance. Failure to comply with this warning can cause injury or death to personnel, and damage to the generator set.

WARNING

Fuels used in the generator set are flammable. When filling the fuel tank, maintain metal-to-metal contact between filler nozzle and fuel tank opening to eliminate static electrical discharge. Failure to comply with this warning can cause injury or death to personnel, and damage to the generator set.

4-14.1 Testing.

a. a. Shut down generator set.

b. b. Open battery access door and disconnect negative battery cable.

c. c. Open right side engine access door and disconnect connector P10 (Figure 4-6) from engine electrical harness.

d. Remove rear fuel injection pump access plug located on top of injection pump housing.

e. e. Connect 5 ohm, 25 watt re sistor in series with pin 2 of connector P10 and ground.

f. f. Connect jumper wire from pin 1 of connector P10 to battery positive terminal (24 VDC).

g. g. While looking through rear fuel injection pump access plug hole, connect negative battery cable.

Governor actuator shaft should extend to full fuel position.

h. h. Disconnect negative battery cable. Governor ac tuator shaft should retract to no fuel position.

i. i. Replace governor actuator if it does not function as above.

j. j. Remove resistor and jumper wire from governor actuator.

k. k. Install rear fuel injection pump access plug.

l. l. Connect connector P10 to engine electrical harness. Close right side engine access door.

m. m. Connect negative battery c able. Close battery access door.

4-14.2 Removal.

a. a. Shut down generator set.

b. b. Open battery access door and disconnect negative battery cable.

c. c. Open right side engine access door.

d. d. Tag and disconnect connector P10 (Figure 4-6) from engine electrical harness.

e. e. Remove rear fuel injection pump access plug located on top of fuel injection pump housing.

f. f. Remove Allen head screws (1) and lock washers (2) securing governor actuator (3) to mounting plate (10). Discard lock washers (2).

g. g. While looking through rear fuel injection pump access plug hole, tilt governor actuator (3) down so that clevis (6) clears stud (19) of injection pump rack linkage. Remove governor actuator (3) and gasket (4) from fuel injection pump.

h. h. Loosen nut (5) and unscrew clevis (6) and retention spring (7) from governor actuator shaft.

NOTE

Hold suitable container under fuel injection pump to catch oil that will drain from pump housing when mounting plate (10) is removed.

i. i. Remove bolts (8), lock washers (9), mounting plate (10), and gasket (11) from fuel injection pump. Discard lock washers (9).

j. j. Remove nut (12), lock washer (13), bolt (14), bracket (15), spring retainer (16), and spring (17) from injection pump rack lever (inside pump). Discard lock washer (13).

k. k. Remove nuts (18 and 20) and stud (19) from bracket (15).

l. l. Remove bolt (21) from mounting plate (10).

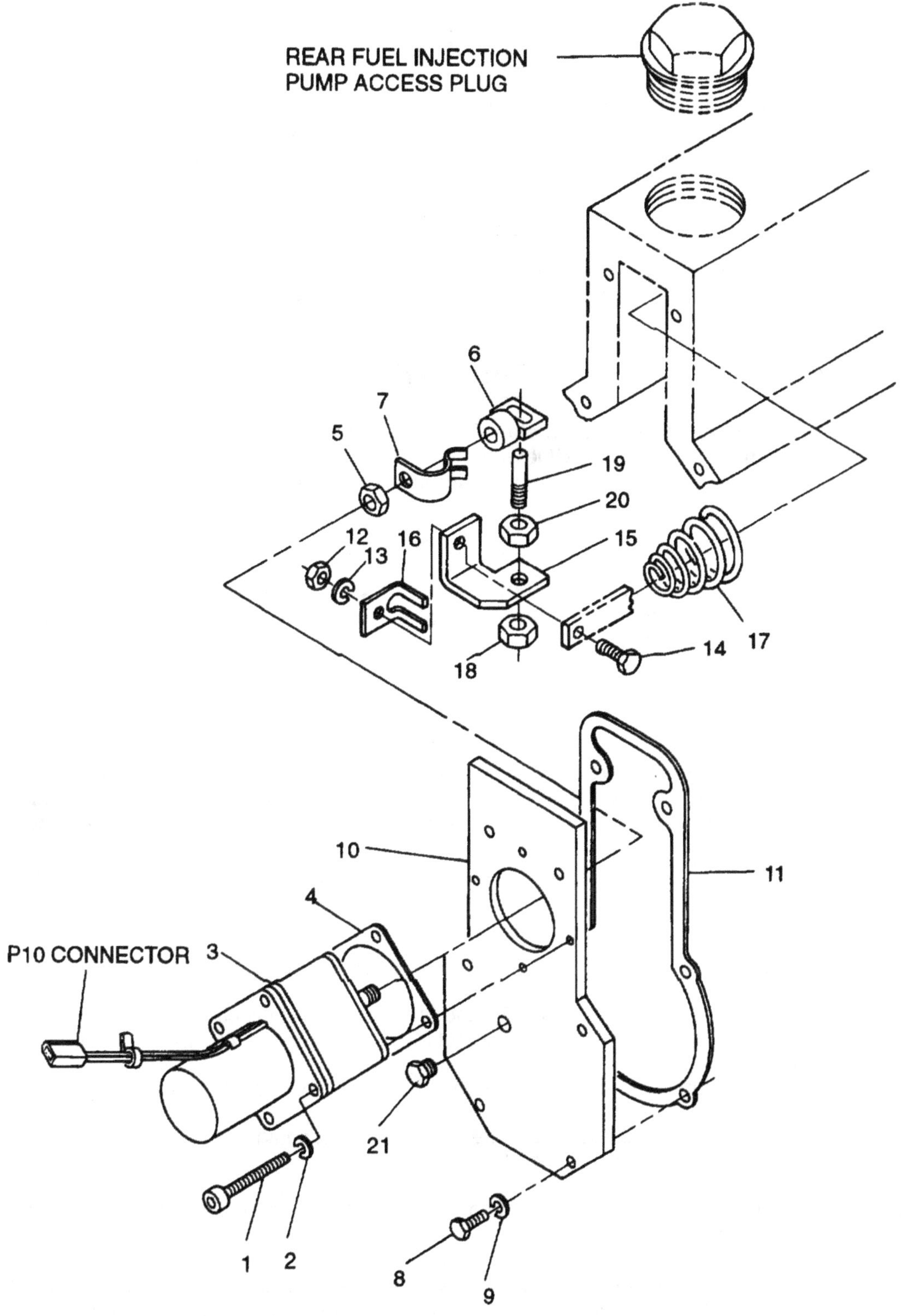

Figure 4-6. Governor Actuator Assembly – MEP-804A/MEP-814A.

4-14.3 Installation and Adjustment.

WARNING

High voltage is produced when the generator set is in operation. Never attempt to start the generator set unless it is properly grounded. Failure to comply with this warning can cause injury or death to personnel.

WARNING

DO NOT wear loose clothing when performing checks, services and maintenance. Failure to comply with this warning can cause injury or death to personnel.

WARNING

All metal jewelry can conduct electricity and become entangled in generator set components. Remove all jewelry when working on generator set. Failure to comply with this warning can cause injury or death to personnel.

WARNING

High voltage is produced when ~~the generator~~ set is in operation. DO NOT touch live voltage connections. Never attempt to connect or disconnect load cables or paralleling cables while the generator set is running. Failure to comply with this warning can cause injury or death to personnel.

WARNING

Wear heat resistant gloves and ~~avoid contact~~ing hot metal surfaces with your hands after components have been heated. Wear additional protective clothing as required. Failure to comply with this warning can cause injury to personnel.

a. a. Install nut (20, Figure 4-6) onto stud (19) until only two or three threads are left exposed on stud shaft.

b. b. Install stud (19) and nut (20) in bracket (15) and secure with nut (18).

c. c. Install spring (17), spring retainer (16), and brac ket (15) on injection pump rack lever (inside pump) with bolt (14), new lock washer (13), and nut (12). Torque nut to 48 in.-lbs (5.4 Nm).

d. d. Apply sealant (Item 18, Appendix C) to both sides of gasket (11). Install gasket and mounting plate (10) on injection pump with bolts (8) (threads coated with sealant (Item 18, Appendix C)) and new lock washers (9). Torque bolts to 100 in.-lbs (11.3 Nm).

e. e. Position clevis (6) within retention spring (7). Slotted flat end of clevis (6) is to be positioned between the two fingers of spring (7).

f. f. Turn clevis and retention spring assembly onto shaft of actuator (3) so that side of hole (slot) in clevis (6) (closest to actuator face) is between 0.96 to 1.00 inch (2.43 to 2.54 cm) from face of actuator (3). Tighten nut (5) against clevis.

g. g. Apply sealant (Item 18, Appendix C) to both side s of gasket (4) and position gasket on actuator base with bolt (1) and new lock washer (2) in upper hole of actuator.

NOTE

Polarity of power to the actuator is not important.

h. h. Connect 5 ohm, 25 watt resistor in series with one pin of connector P10 and ground. Connect a jumper wire from other pin of connector P10 to positive battery terminal.

i. i. Connect negative battery cable to energize and extend actuator shaft.

j. j. Insert medium size screwdriver through access hole at top of injection pump housing and hold injection pump rack fully toward rear of engine.

k. k. With actuator (3) energized, insert actuator shaft through hole in mounting plate (10) at a slight upward angle. Move shaft inward until lower finger of retention spring (7) compresses sufficiently allowing top of stud (19) to align itself under slot of clevis (6).

l. l. Use tip of screwdriver that held fuel pump rack to compress upper finger of retention spring (7) so that clevis (6) can drop down on stud (19).

m. m. Deenergize actuator (3).

n. n. Secure actuator (3) to mounting plate (10) with Allen head screws (1) (threads coated with sealant (Item 18, Appendix C)) and new lock washers (2). Torque screws to 100 in.-lbs (11.3 Nm).

o. o. While looking through rear injection pump acce ss plug hole, energize and reenergize actuator a few times to ensure there is no binding. Lubricate actuator shaft with oil (Item 11, Appendix C).

p. p. Using rear injection pump access hole, fill pump housing with engine oil (Table 2-3) to a level even with bottom of hole for bolt (21).

q. q. Install rear injection pump access plug in top of injection pump and bolt (21) in mounting plate (10).

r. r. Remove jumper wires and connect connector P10 to engine harness.

s. s. Connect battery cable. Close battery access doors.

4-15 GOVERNOR ACTUATOR (MEP-804B/MEP-814B)

WARNING

High voltage is produced when the generator set is in operation. Never attempt to start the generator set unless it is properly grounded. Failure to comply with this warning can cause injury or death to personnel.

WARNING

High voltage is produced when this generator set is in operation. Make sure generator set is completely shut down and free of any power source before attempting any repair or maintenance on the set, or when connecting or disconnecting load cables. Failure to comply with this warning can cause injury or death to personnel.

WARNING

Dangerous voltage exists on live circuits. Always observe precautions and never work alone. Failure to comply with this warning can cause injury or death to personnel.

WARNING

Ensure that the engine cannot be started while maintenance is being performed. (ENGINE CONTROL switch set to OFF/RESET; Battery Disconnect Switch is OFF; DEAD CRANK SWITCH is OFF). Failure to comply with this warning can cause injury or death to personnel.

WARNING

DC voltages are present at generator set electrical components even with generator set shut down. Avoid shorting any positive with ground/negative. Failure to comply with this warning can cause injury to personnel, and damage to equipment.

WARNING

The connection of any electrical equipment and the disconnection of any electrical equipment may cause an explosion hazard. Do not connect any electrical equipment or disconnect any electrical equipment in an explosive atmosphere. Failure to comply with this warning can cause injury or death to personnel.

WARNING

A qualified technician must make the power connections and perform all continuity checks. The power source may be a generator or commercial power. Failure to comply with this warning can cause injury or death to personnel.

WARNING

High voltage is produced when the generator set is in operation. DO NOT touch live voltage connections. Never attempt to connect or disconnect load cables or paralleling cables while the generator set is running. Failure to comply with this warning can cause injury or death to personnel.

4-15.1 Testing.

a. a. Shut down generator set.

b. b. Open battery access door and disconnect negative battery cable.

c. c. Open right side engine access door and disconnect connector P10 (Figure 4-7) from engine electrical harness.

d. Remove rear fuel injection pump access plug located on top of injection pump housing.

e. e. Connect 5 ohm, 25 watt re sistor in series with pin 2 of connector P10 and ground.

f. f. Connect jumper wire from pin 1 of connector P10 to battery positive terminal (24 VDC).

g. g. While looking through rear fuel injection pump access plug hole, connect negative battery cable.

Governor actuator shaft should extend to full fuel position.

h. h. Disconnect negative battery cable. Governor ac tuator shaft should retract to no fuel position.

i. i. Replace governor actuator if it does not function as above.

j. j. Remove resistor and jumper wire from governor actuator.

k. k. Install rear fuel injection pump access plug.

l. l. Connect connector P10 to engine electrical harness. Close right side engine access door.

m. m. Connect negative battery c able. Close battery access door.

4-15.2 Removal.

a. a. Shut down generator set.

b. b. Open battery access door and disconnect negative battery cable.

c. c. Open right side engine access door.

d. d. Tag and disconnect connector P10 from electrical harness.

e. e. Remove screws (1, Figure 4-7), lock washers (2), flat washers (3), governor actuator (4), and O-ring (5). Discard lock washers (2).

f. f. Inspect and replace O-ring, as required.

4-15.3 Installation and Adjustment.

WARNING

High voltage is produced when the generator set is in operation. Never attempt to start the generator set unless it is properly grounded. Failure to comply with this warning can cause injury or death to personnel.

WARNING

DO NOT wear loose clothing when performing checks, services and maintenance. Failure to comply with this warning can cause injury or death to personnel.

WARNING

All metal jewelry can conduct electricity and become entangled in generator set components. Remove all jewelry when working on generator set. Failure to comply with this warning can cause injury or death to personnel.

WARNING

High voltage is produced when the generator set is in operation. DO NOT touch live voltage connections. Never attempt to connect or disconnect load cables or paralleling cables while the generator set is running. Failure to comply with this warning can cause injury or death to personnel.

WARNING

Wear heat resistant gloves and avoid contacting hot metal surfaces with your hands after components have been heated. Wear additional protective clothing as required. Failure to comply with this warning can cause injury to personnel.

a. a. Install governor actuator (4, Figure 4-7) and O-ring (5) with screws (1), new lock washers (2), and flat washers (3).

b. b. Connect connector P 10 to electrical harness.

c. c. Connect battery cable. Close battery access doors.

NOTE

Governor actuator is calibrated at the factory.

d. Loosen lock nut (6) and adjust throttle minimum adjust screw (7) to where throttle (8) is within 3/16 inch (0.48 cm) of throttle maximum fuel stop (9).

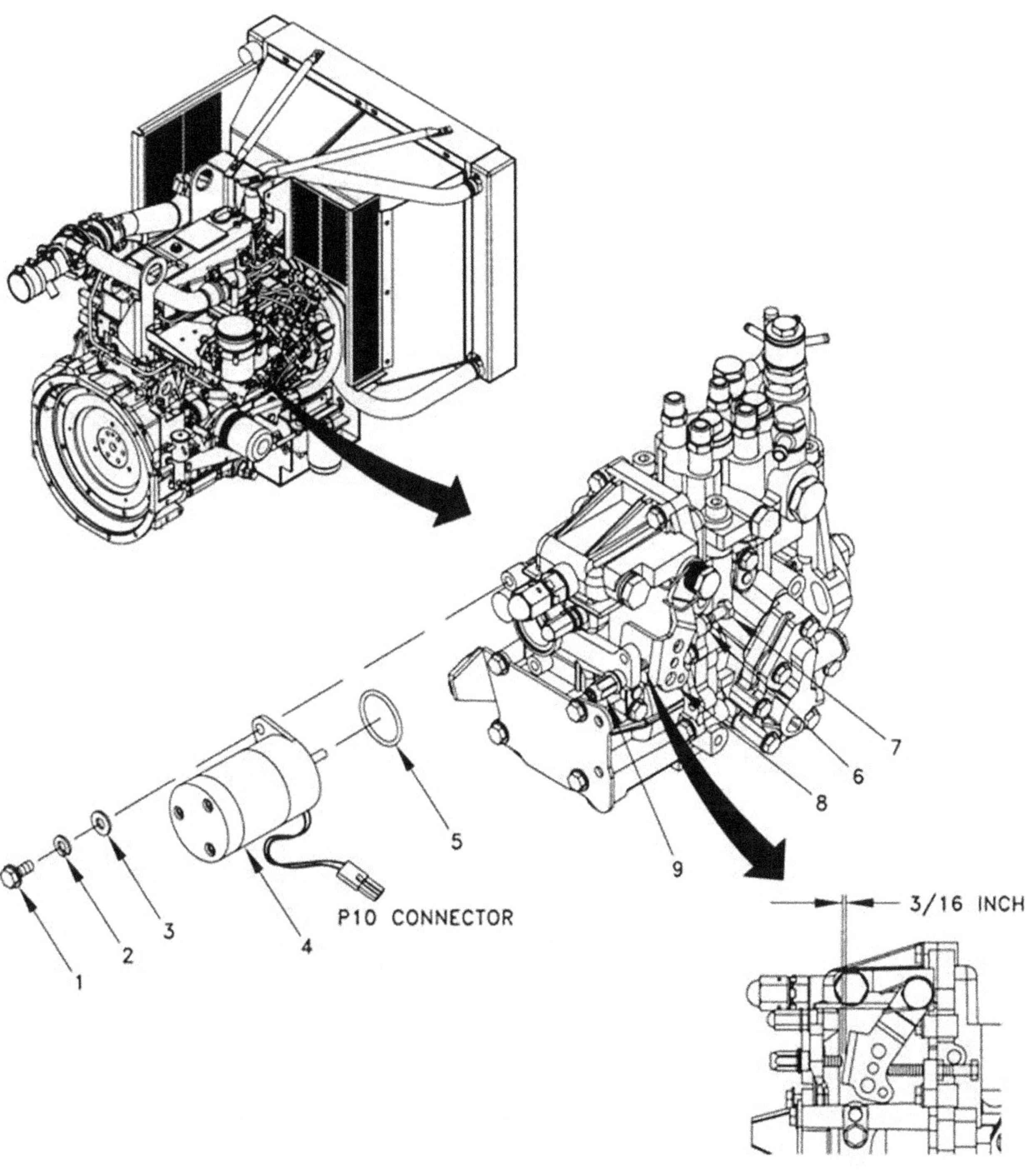

Figure 4-7. Governor Actuator Assembly – MEP-804B/MEP-814B.

Section VI. MAINTENANCE OF GENERATOR ASSEMBLY

4-16 TECHNICAL DESCRIPTION

4-16.1 General. Revolving field type generators have a DC field revolving within a stationary AC winding called the stator. AC power is distribution from the generator through leads connected to the stator windings. There are no sliding contacts between the AC winding and the load, therefore, great amounts of power may be drawn from this generator.

NOTE

Refer to Figure 4-8 as needed.

To energize the field, DC excitation must be applied to the generator field coils. The excitation current is supplied from a brushless exciter mounted on the generator shaft.

The brushless exciter is actually an AC generator with its output rectified through a full wave bridge circuit. This type of brushless exciter will provide the necessary excitation current. The generator set field flash circuit, activated during each engine start, applies voltage to the exciter stator to begin the voltage buildup process to energize the generator field.

The generator output voltage is controlled by controlling the alternating field current. This is accomplished by regulating the exciter field coil voltage. The exciter field coil voltage is regulated with a solid state-type AC voltage regulator.

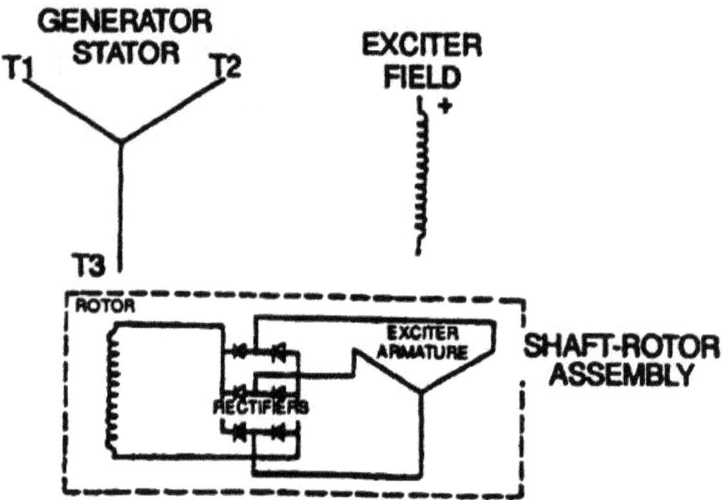

Figure 4-8. Brushless Generator Schematic.

4-16.2 Damper Bars. Damper bars are inserted through the field laminations and welded at the end to a solid copper plate. The damper windings provide stable parallel operation, reduce damping current losses, and limit the increase of third harmonic voltage with increase in load.

4-16.3 Brushless Exciter. The brushless exciter consists of an armature with a three-phase AC winding and rotating rectifier assembly within a stationary field.

The stationary exciter field assembly is mounted in the main generator frame. The exciter armature is press-fit and keyed onto the shaft assembly. The rotating rectifier assembly slides over the bearing end of the generator rotor shaft and is secured with bolts and washers to an adapter hub which is shrunk on the generator shaft.

4-16.4 Rotating Rectifier Bridge. The rotating rectifier bridge consists of rectifying diodes mounted on a brass heat sink which is in turn mounted on an insulating ring. The entire assembly bolts to the adapter on the generator shaft. Therefore, the rotating rectifier assembly will rotate with the exciter armature eliminating the need for any sliding contacts between the exciter output and the alternator field.

4-16.5 Exciter Field. The exciter field on the high frequency exciter consists of laminated segments of high carbon steel which are fitted together to make up the field poles. The field coils are placed into the slots of the field poles.

4-16.6 Exciter Field Coil Voltage Source. Field coil DC voltage is obtained by rectifying the voltage from a phase to neutral line of the generator output or other appropriate terminal to provide the needed voltage reference.

The rectifier bridge is an integral part of the static regulator. The static regulator senses a change in the generator output and automatically regulates current flow in the exciter field coil circuit to increase or decrease the exciter field strength. An external adjust rheostat sized to be compatible with the regulator is used to provide adjustment to the regulator sensing circuit.

4-16.7 Balance. The rotor assembly is precision balanced to a high degree of static and dynamic balance. Balance is achieved with the balance lugs on the field pole tips. The balance will remain dynamically stable at speed in excess of the design frequencies.

4-16.8 Bearing. The generator rotor assembly is suspended on shielded, factory-lubricated ball bearings. They are greased for life and do not require lubrication.

4-16.9 Stator Assembly. The stator assembly consists of laminations of steel mounted in a rolled steel frame. Random wound stator coils are fitted into the insulated slots.

4-17 GENERATOR ASSEMBLY

WARNING

High voltage is produced when the generator set is in operation. Never attempt to start the generator set unless it is properly grounded. Failure to comply with this warning can cause injury or death to personnel.

WARNING

High voltage is produced when this generator set is in operation. Make sure generator set is completely shut down and free of any power source before attempting any repair or maintenance on the set, or when connecting or disconnecting load cables. Failure to comply with this warning can cause injury or death to personnel.

WARNING

Dangerous voltage exists on live circuits. Always observe precautions and never work alone. Failure to comply with this warning can cause injury or death to personnel.

WARNING

Ensure that the engine cannot be started while maintenance is being performed. (ENGINE CONTROL switch set to OFF/RESET; Battery Disconnect Switch is OFF; DEAD CRANK SWITCH is OFF). Failure to comply with this warning can cause injury or death to personnel.

WARNING

DC voltages are present at generator set electrical components even with generator set shut down. Avoid shorting any positive with ground/negative. Failure to comply with this warning can cause injury to personnel, and damage to equipment.

WARNING

The connection of any electrical equipment and the disconnection of any electrical equipment may cause an explosion hazard. Do not connect any electrical equipment or disconnect any electrical equipment in an explosive atmosphere. Failure to comply with this warning can cause injury or death to personnel.

WARNING

A qualified technician must make the power connections and perform all continuity checks. The power source may be a generator or commercial power. Failure to comply with this warning can cause injury or death to personnel.

WARNING

High voltage is produced when the generator set is in operation. DO NOT touch live voltage connections. Never attempt to connect or disconnect load cables or paralleling cables while the generator set is running. Failure to comply with this warning can cause injury or death to personnel.

4-17.1 Removal.

WARNING

When disconnecting or removing batteries, disconnect the negative lead that connects directly to the grounding stud first; disconnect the negative end of the interconnection cable next. When installing batteries, reverse the connection sequence. Failure to comply with this warning can cause injury to personnel.

a. a. Shut down generator set.

b. b. Open battery access door and disconnect negative battery cable.

c. c. Remove generator set rear housing section (paragraph 2-18.1).

d. Loosen bolts (1, Figure 4-9) and lower engine support brackets (2) to rest on skid base. Tighten bolts. If necessary, place wooden shims under brackets to ensure contact with skid base.

WARNING

Many components require a two-person lift. Lifting heavy components can cause back strain. Ensure proper lifting techniques are used when lifting heavy components. Failure to comply with this warning can cause injury to personnel.

WARNING

The generator set, engine, and generator are extremely heavy and require an assistant and a lifting device (forklift, overhead lifting device) with sufficient capacity. Failure to comply with this warning can cause serious injury or death to personnel.

WARNING

Rated capacity of overhead hoist should be at least 1,500 pounds (680 kg). Do not use a hoist with less capacity. Failure to comply with this warning can cause injury or death to personnel, and damage to equipment.

WARNING

Keep hands and feet from underside of engine and generator while using lifting device to remove them from the skid base. Failure to comply with this warning can cause injury or death to personnel.

e. e. Attach lifting harness to overhead hoist and generator lifting eye and take up slack.

f. f. Remove screw (3), washer (4), and screen/cover (5) from generator case.

g. g. Remove bolts (6) and lock washers (7) securing generator drive disc and inertia ring (7.1) to engine flywheel. Discard lock washers (7).

h. h. For MEP-804A/MEP-814A, perform step (1) below; for MEP-804B/MEP-814B, perform step (2) below:

(1) Remove bolts (8) and lock washers (9) securing generator to engine flywheel housing. Discard lock washers (9).

(2) Remove bolts (8) and lock washers (9) securing generator and spacer (21) to engine flywheel housing. Discard lock washers (9).

i. i. Remove nuts (10), washers (11), snubbing washers (12), bolts (13), and Belleville washers (14) securing generator to skid base.

j. Lift generator slowly from skid base, ensuring that engine flywheel housing and generator separate smoothly to avoid any undue stress.

NOTE

Mark location of spacers, washers, and bolts to ensure correct positioning during installation.

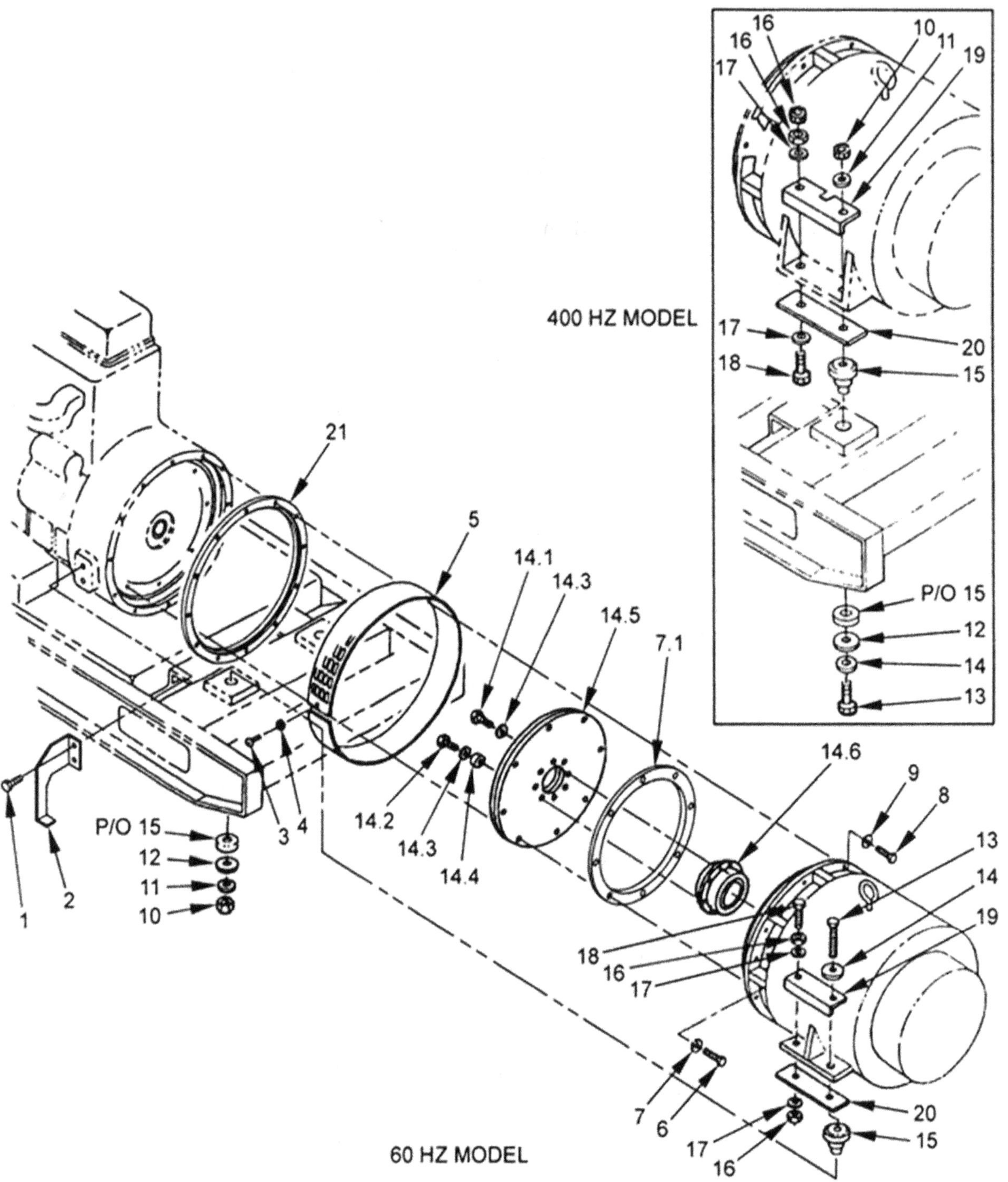

Figure 4-9. Generator Assembly.

(1) Remove bolts (14.1 and 14.2), washers (14.3), spacers (14.4), and drive discs (14.5) from drive hub (14.6). Remove inertia ring (7.1).

(2) Install bolts (14.1 and 14.2), washers (14.3), spacers (14.4), and drive discs (14.5) to drive hub (14.6). Torque bolts (14.1 and 14.2) to 35 ft-lbs (47 Nm).

k. k. Remove shock mounts (15) from skid base.

l. l. Remove nuts (16), washers (17), bolts (18), angles (19), and plates (20) from generator.

4-17.2 Installation.

a. a. Position generator shock mounts (15, Figure 4-9) in skid base.

b. b. Install angles (19) and plates (20) on generator with bolts (18), washers (17), and nuts (16).

(1) Remove bolts (14.1 and 14.2), washers (14.3), spacers (14.4), and drive discs (14.5) from drive hub (14.6).

(2) Place inertia ring (7.1) on shaft just inside drive hub (14.6).

NOTE

Ensure correct positioning of spacers, washers, and bolts as marked on removal.

(3) Install bolts (14.1 and 14.2), washers (14.3), spacers (14.4), and drive discs (14.5) to drive hub (14.6).

(4) Torque bolts (14.1 and 14.2) to 35 ft-lbs (47 Nm).

~~WARNING~~

Many components require a two-person lift. Lifting heavy components can cause back strain. Ensure proper lifting techniques are used when lifting heavy components. Failure to comply with this warning can cause injury to personnel.

WARNING

The generator set, engine, and generator are extremely heavy and require an assistant and a lifting device (forklift, overhead lifting device) with sufficient capacity. Failure to comply with this warning can cause serious injury or death to personnel.

WARNING

Rated capacity of overhead hoist should be at least 1,500 pounds (680 kg). Do not use a hoist with less capacity. Failure to comply with this warning can cause injury or death to personnel, and damage to equipment.

WARNING

Keep hands and feet from underside of engine and generator while using lifting device to remove them from the skid base. Failure to comply with this warning can cause injury or death to personnel.

c. c. Attach lifting harness to overhead hoist and generator lifting eye.

d. d. Position generator on skid base aligning mounting holes with mounts and engine flywheel housing.

e. e. Install bolts (6) and new lock washers (7) securing generator drive disc and inertia ring (7.1) to engine flywheel. Torque bolts to 35 ft-lbs (47 Nm).

f. f. For MEP-804A/MEP-814A, perform step (1) below; for MEP-804B/MEP-814B, perform step (2) below:

(1) Install bolts (8) and new lock washers (9) securing generator to flywheel housing. Tighten bolts slowly to ensure proper seating of generator case lip in engine flywheel housing. Torque bolts to 31 ft-lbs (42 Nm).

(2) Install bolts (8) and new lock washers (9) securing generator and spacer (21) to flywheel housing. Tighten bolts slowly to ensure proper seating of generator case lip in engine flywheel housing. Torque bolts to 31 ft-lbs (42 Nm).

g. g. Install bolts (13), Belleville washers (14), snub bing washers (12), washers (11), and nuts (10) securing generator assembly to skid base. Torque bolts to 210 ft-lbs (285 Nm).

h. h. Install screen/cover (5) on generator case with screw (3) and washer (4).

i. i. For MEP-804A/MEP-814A, perform step (1) below; for MEP-804B/MEP-814B, perform step (2) below:

(1) Adjust nuts (16) to obtain 0.5 inch (12.7 mm) minimum clearance between ends of bolts (18) and skid base.

(2) Loosen nuts (16) and torque bolts (18) to 120 ft-lbs (163 Nm). Tighten nuts.

j. j. Loosen bolts (1) and raise lower engine support brackets (2) to uppermost position. Tighten bolts. Remove wooden shims, if used.

k. k. Install generator set rear housing section (paragraph 2-18.4).

l. l. Connect negative battery cable. Close battery access door.

m. m. Start generator set and check for proper operation.

4-18 ROTATING RECTIFIER DIODES (MEP-804A/MEP-804B)

WARNING

High voltage is produced when the generator set is in operation. Never attempt to start the generator set unless it is properly grounded. Failure to comply with this warning can cause injury or death to personnel.

WARNING

Dangerous voltage exists on live circuits. Always observe precautions and never work alone. Failure to comply with this warning can cause injury or death to personnel.

WARNING

Ensure that the engine cannot be started while maintenance is being performed. (ENGINE CONTROL switch set to OFF/RESET; Battery Disconnect Switch is OFF; DEAD CRANK SWITCH is OFF). Failure to comply with this warning can cause injury or death to personnel.

WARNING

DC voltages are present at generator set electrical components even with generator set shut down. Avoid shorting any positive with ground/negative. Failure to comply with this warning can cause injury to personnel, and damage to equipment.

4-18.1 Testing.

a. a. Shut down generator set. Allow generator to cool to ambient temperature.

b. b. Open battery access door and disconnect negative battery cable.

c. c. Remove generator set housin g rear panel (paragraph 2-18.1).

d. d. Remove generator end bell (paragraph 4-19.1).

e. e. Tag main rotor leads (2, Figure 4-10). Remove screws (1) and main rotor leads (2) from diode mounting plate assembly (5).

f. f. Tag exciter rotor leads (4). Remove screws (3) and exciter rotor leads (4) from diode mounting plate assembly (5).

g. g. Set multimeter for ohms and connect positive lead to one side and negative lead to other side of each diode (7). Record multimeter reading for each diode.

h. h. Repeat step g above with multimeter leads reversed.

i. i. Resistance (ohms) readings should be front to back ratio of 1:10 or greater. If any reading is less than 1:10, diode is defective and must be replaced.

j. j. Install exciter rotor leads (4) and main rotor leads (2) to diode mounting plate assembly (5) with screws (1) and (3). Remove tags.

k. k. Install generator end bell (paragraph 4-19.2).

l. l. Install generator set housing rear panel (paragraph 2-18.4).

m. Connect negative battery c able. Close battery access door.

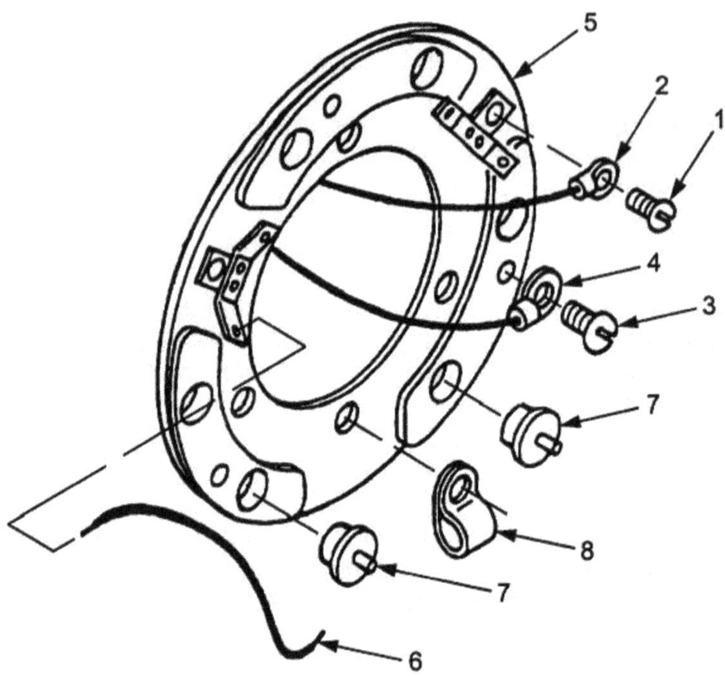

Figure 4-10. Rectifier Assembly – MEP-804A/MEP-804B.

4-18.2 Removal.

WARNING

Many components require a two-person lift. Lifting heavy components can cause back strain. Ensure proper lifting techniques are used when lifting heavy components. Failure to comply with this warning can cause injury to personnel.

WARNING

The generator set, engine, and generator are extremely heavy and require an assistant and a lifting device (forklift, overhead lifting device) with sufficient capacity. Failure to comply with this warning can cause serious injury or death to personnel.

WARNING

All metal jewelry can conduct electricity and become entangled in generator set components. Remove all jewelry when working on generator set. Failure to comply with this warning can cause injury or death to personnel.

WARNING

DO NOT wear loose clothing when performing checks, services and maintenance. Failure to comply with this warning can cause injury or death to personnel.

a. a. Shut down generator set.

b. b. Open battery access door and disconnect negative battery cable.

c. c. Remove generator set housin g rear panel (paragraph 2-18.1).

d. d. Remove generator end bell (paragraph 4-19.1).

e. e. Remove rotating rectifier (paragraph 4-20.1).

f. f. Unsolder electrical lead (6, Figure 4-10) from diode (7) being removed.

g. g. Press diode (7) from diode mounting plate assembly (5).

4-18.3 Installation.

a. a. Run bead of thermal-electric compound (Item 8, Appendix C) around base of diode (7, Figure 4-10) prior to installing.

b. b. Press diode (7) into diode mounting plate assembly (5).

c. c. Using solder (Item 19, Appendix C) and soldering iron, solder electrical lead (6) to diode (7).

d. d. Install rotating rectifier (paragraph 4-20.2).

e. e. Install generator end bell (paragraph 4-19.2).

f. f. Install generator set housing rear panel (paragraph 2-18.4).

g. g. Connect negative battery c able. Close battery access door.

4-19 END BELL AND MAIN BEARING (MEP-804A/MEP-804B)

WARNING

High voltage is produced when the generator set is in operation. Never attempt to start the generator set unless it is properly grounded. Failure to comply with this warning can cause injury or death to personnel.

WARNING

Dangerous voltage exists on live circuits. Always observe precautions and never work alone. Failure to comply with this warning can cause injury or death to personnel.

WARNING

Ensure that the engine cannot be started while maintenance is being performed. (ENGINE CONTROL switch set to OFF/RESET; Battery Disconnect Switch is OFF; DEAD CRANK SWITCH is OFF). Failure to comply with this warning can cause injury or death to personnel.

WARNING

DC voltages are present at generator set electrical components even with generator set shut down. Avoid shorting any positive with ground/negative. Failure to comply with this warning can cause injury to personnel, and damage to equipment.

4-19.1 Removal.

WARNING

Many components require a two-person lift. Lifting heavy components can cause back strain. Ensure proper lifting techniques are used when lifting heavy components. Failure to comply with this warning can cause injury to personnel.

WARNING

The generator set, engine, and generator are extremely heavy and require an assistant and a lifting device (forklift, overhead lifting device) with sufficient capacity. Failure to comply with this warning can cause serious injury or death to personnel.

WARNING

All metal jewelry can conduct electricity and become entangled in generator set components. Remove all jewelry when working on generator set. Failure to comply with this warning can cause injury or death to personnel.

WARNING

DO NOT wear loose clothing when performing checks, services and maintenance. Failure to comply with this warning can cause injury or death to personnel.

a. a. Shut down generator set.

b. b. Open battery access door and disconnect negative battery cable.

c. c. Remove generator set housin g rear panel (paragraph 2-18.1).

d. d. Remove bolts (1, Figure 4-11), lock washers (2), and end bell (3) from generator assembly. Discard lock washers (2).

e. e. Remove plug (4) and packing (5) from end bell (3), if necessary.

CAUTION

If bearing needs to be removed for any reason, always install new bearing. Main bearing is easily damaged when removed from rotor shaft. Damage to equipment could result.

f. f. Using bearing puller, remove bearing (6) from main rotor assembly (23) shaft and discard bearing.

g. g. Remove retaining ring (7) from main rotor assembly (23) shaft, if necessary.

4-19.2 Installation.

a. a. If removed, install retaining ring (7, Figure 4-11) on main rotor assembly (23) shaft.

CAUTION

If bearing needs to be removed for any reason, always install new bearing. Main bearing is easily damaged when removed from rotor shaft. Damage to equipment could result.

b. b. Install new bearing (6) on main rotor assembly (23) shaft. Ensure that bearing is seated squarely against retaining ring (7) by applying pressure to inner race only.

c. c. If removed, install packing (5) and plug (4) in end bell (3).

d. d. Position end bell (3) on bearing (6).

e. e. Install end bell (3) on generator assembly with bolts (1) and new lock washers (2).

f. f. Install generator set housin g rear panel (paragraph 2-18.4).

g. g. Connect negative battery c able. Close battery access door.

4-20 ROTATING RECTIFIER (MEP-804A/MEP-804B)

WARNING

High voltage is produced when the generator set is in operation. Never attempt to start the generator set unless it is properly grounded. Failure to comply with this warning can cause injury or death to personnel.

WARNING

Dangerous voltage exists on live circuits. Always observe precautions and never work alone. Failure to comply with this warning can cause injury or death to personnel.

WARNING

Ensure that the engine cannot be started while maintenance is being performed. (ENGINE CONTROL switch set to OFF/RESET; Battery Disconnect Switch is OFF; DEAD CRANK SWITCH is OFF). Failure to comply with this warning can cause injury or death to personnel.

WARNING

DC voltages are present at generator set electrical components even with generator set shut down. Avoid shorting any positive with ground/negative. Failure to comply with this warning can cause injury to personnel, and damage to equipment.

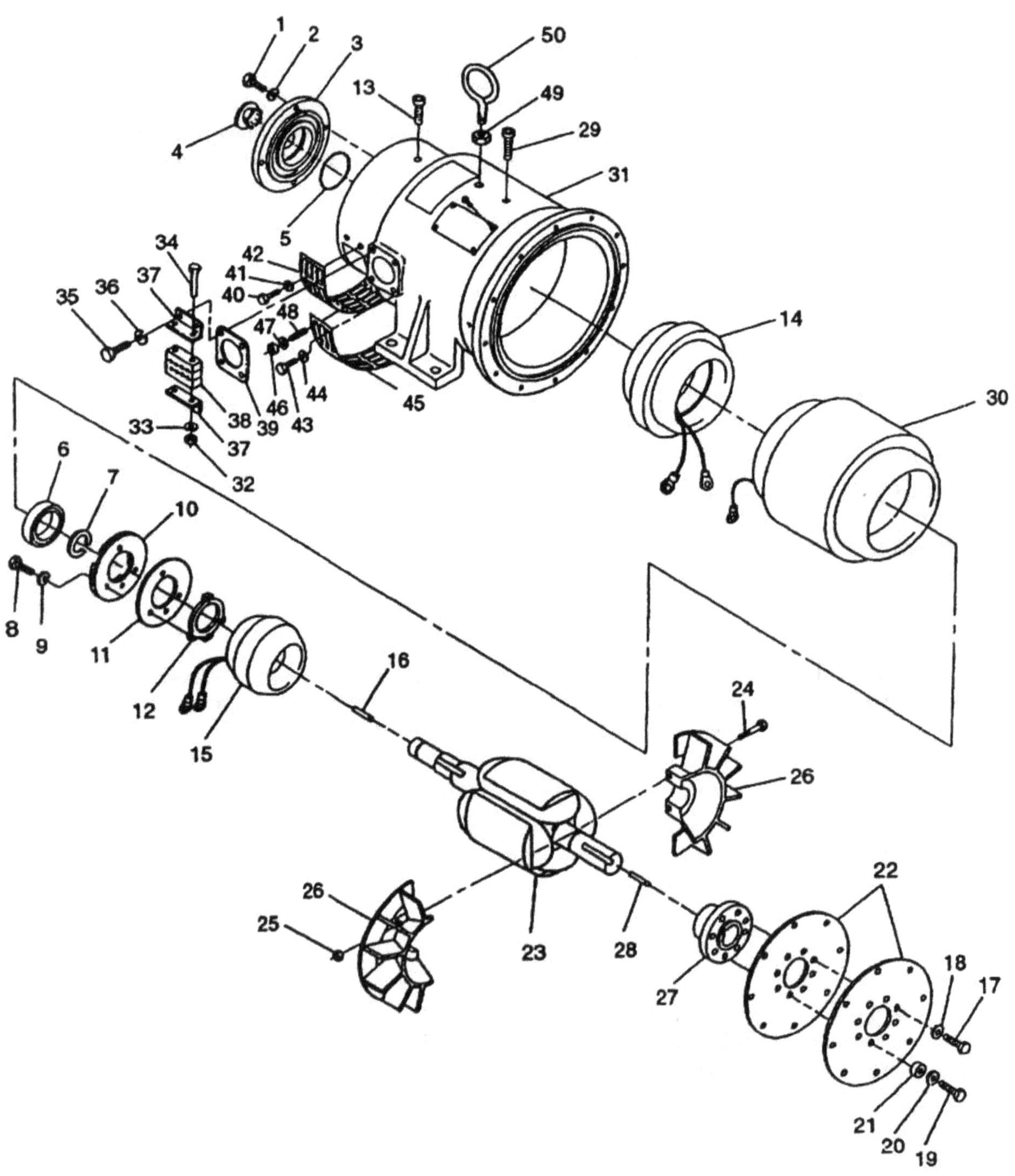

Figure 4-11. Generator Assembly – MEP-804A/MEP-804B.

4-20.1 Removal.

WARNING

Many components require a two-person lift. Lifting heavy components can cause back strain. Ensure proper lifting techniques are used when lifting heavy components. Failure to comply with this warning can cause injury to personnel.

WARNING

The generator set, engine, and generator are extremely heavy and require an assistant and a lifting device (forklift, overhead lifting device) with sufficient capacity. Failure to comply with this warning can cause serious injury or death to personnel.

WARNING

All metal jewelry can conduct electricity and become entangled in generator set components. Remove all jewelry when working on generator set. Failure to comply with this warning can cause injury or death to personnel.

WARNING

DO NOT wear loose clothing when performing checks, services and maintenance. Failure to comply with this warning can cause injury or death to personnel.

a. a. Shut down generator set.

b. b. Open battery access door and disconnect negative battery cable.

c. c. Remove generator set housin g rear panel (paragraph 2-18.1).

d. d. Remove generator end bell and main bearing (paragraph 4-19.1)

e. e. Remove screws (3, Figure 4-10) securing exciter rotor leads (4) to rotating rectifier (3 places). Tag and remove exciter rotor leads (4).

f. f. Remove screws (1). Tag and remove main rotor leads (2) from rotating rectifier (2 places).

g. g. Remove bolts (8, Figure 4-11), lock washers (9), rotating rectifier (10), and insulating plate (11) from rectifier hub (12). Discard lock washers (9).

h. h. Remove clamp (8, Figure 4-10) from diode mounting plate assembly (5), if necessary.

4-20.2 Installation.

a. a. If removed, install clamp (8, Figure 4-10) on diode mounting plate assembly (5).

b. b. Install rotating rectifier (10, Figure 4-11) and insulating plate (11) on rectifier hub (12) with bolts (8) and new lock washers (9).

c. c. Connect two main rotor leads (2, Figure 4-10) to rotating rectifier with screws (1). Remove tags.

d. d. Connect three exciter rotor leads (4) to rotating rectifier with screws (3). Remove tags.

e. e. Install main bearing and generat or end bell (paragraph 4-19.2).

f. f. Install generator set housing rear panel (paragraph 2-18.4).

g. g. Connect negative battery c able. Close battery access door.

WARNING

High voltage is produced when the generator set is in operation. Never attempt to start the generator set unless it is properly grounded. Failure to comply with this warning can cause injury or death to personnel.

WARNING

Dangerous voltage exists on live circuits. Always observe precautions and never work alone. Failure to comply with this warning can cause injury or death to personnel.

WARNING

Ensure that the engine cannot be started while maintenance is being performed. (ENGINE CONTROL switch set to OFF/RESET; Battery Disconnect Switch is OFF; DEAD CRANK SWITCH is OFF). Failure to comply with this warning can cause injury or death to personnel.

WARNING

DC voltages are present at generator set electrical components even with generator set shut down. Avoid shorting any positive with ground/negative. Failure to comply with this warning can cause injury to personnel, and damage to equipment.

4-21.1 Testing.

a. a. Shut down generator set. Allow generator to cool to ambient temperature.

b. b. Open battery access door and disconnect negative battery cable.

c. c. Open output box access door and disconnect exciter field leads F1 and F2 from terminals 1 and 2 of TB8.

d. d. Set multimeter for ohms and connect between disconnected exciter field leads. Multimeter reading should be as shown in Table 4-3.

e. e. Reading other than shown in Table 4-3 indicates that open or shorted windings and exciter stator must be replaced.

f. f. Connect multimeter between each exciter field lead and generator frame in turn.

g. Reading of less than infinity indicates defective ground insulation and exciter stator must be replaced.

h. h. Connect exciter field leads to terminals 1 and 2 of TB8. Close output box access door.

i. i. Connect negative battery c able. Close battery access door.

Table 4-3. Generator Resistance Values at 25°C (77°F)

Component	Resistance	
	MEP-804A/MEP-804B	MEP-814A/MEP-814B
Exciter Stator	Between 28.657 and 38.771 ohms	Between 25.0335 and 33.8715 ohms
Exciter Rotor	Between 0.709 and 0.959 ohms	Between 0.2025 and 0.2739 ohms
Generator Rotor	Between 1.247 and 2.073 ohms	Between 1.2688 and 1.7166 ohms
Generator Stator	Between 0.146 and 0.198 ohms	Between 0.0717 and 0.0971 ohms

4-21.2 Removal.

WARNING

Many components require a two-person lift. Lifting heavy components can cause back strain. Ensure proper lifting techniques are used when lifting heavy components. Failure to comply with this warning can cause injury to personnel.

WARNING

The generator set, engine, and generator are extremely heavy and require an assistant and a lifting device (forklift, overhead lifting device) with sufficient capacity. Failure to comply with this warning can cause serious injury or death to personnel.

WARNING

All metal jewelry can conduct electricity and become entangled in generator set components. Remove all jewelry when working on generator set. Failure to comply with this warning can cause injury or death to personnel.

WARNING

DO NOT wear loose clothing when performing checks, services and maintenance. Failure to comply with this warning can cause injury or death to personnel.

a. a. Shut down generator set.

b. b. Open battery access door and disconnect negative battery cable.

c. c. Remove generator set housin g rear panel (paragraph 2-18.1).

d. d. Remove generator assembly from generator set (paragraph 4-17.1).

e. e. Remove generator end bell (paragraph 4-19.1).

f. f. Remove main rotor assembly (paragraph 4-23.2).

g. g. Remove main stator (paragraph 4-24.2).

h. h. Remove setscrews (13, Figure 4-11) from generator housing (31).

WARNING

Hot metal surfaces can cause burns to skin. Wear protective gloves and eye protection when applying heat to generator housing. Failure to comply with this warning can cause injury to personnel.

CAUTION

Ensure wires are not touching generator housing and are away from open flame when heat is applied. Be sure to distribute heat evenly to prevent damage to the housing.

i. Apply heat evenly to ribs of generator housing (31) at end bell section using an acetylene torch with a rose bud tip. Refer to Figure 4-12 for location of ribs. Heat ribs until exciter stator (14) drops down and lift off housing (31).

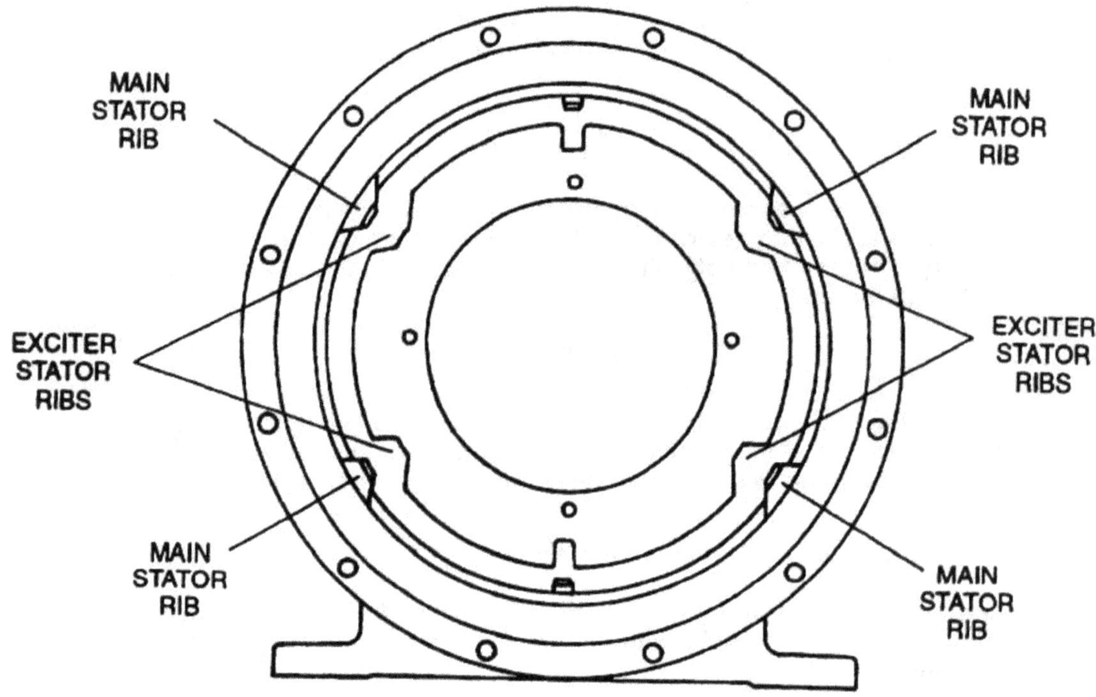

Figure 4-12. Generator Housing Rib Locations.

4-21.3 Installation.

a. a. Stand generator housing (31, Figure 4-11) on end with end bell opening in a downward position.

b. b. Position exciter stator (14) in generator housing (31). Ensure exciter stator wires are pointing upward and are in alignment with wiring port on side of generator housing (31).

c. c. Apply heat evenly to ribs of generator housing (31) at end bell section using an acetylene torch with a rose bud tip. Refer to Figure 4-12 for rib locations. Heat ribs until exciter stator (14) drops into position.

d. d. Install setscrews (13) in generator housing (31).

CAUTION

Special care must be taken when feeding wires to prevent damaging or breaking off wires.

e. e. Feed exciter stator (14) wires through wiring port on side of generator housing (31).

f. f. Retest exciter stator to ensure stator was no t damaged during installation (paragraph 4.21.1).

g. g. Install main stator (paragraph 4-24.3).

h. h. Install generator assembly (paragraph 4-17.2).

4-22 EXCITER ROTOR (MEP-804A/MEP-804B)

WARNING

High voltage is produced when the generator set is in operation. Never attempt to start the generator set unless it is properly grounded. Failure to comply with this warning can cause injury or death to personnel.

WARNING

Dangerous voltage exists on live circuits. Always observe precautions and never work alone. Failure to comply with this warning can cause injury or death to personnel.

WARNING

Ensure that the engine cannot be started while maintenance is being performed. (ENGINE CONTROL switch set to OFF/RESET; Battery Disconnect Switch is OFF; DEAD CRANK SWITCH is OFF). Failure to comply with this warning can cause injury or death to personnel.

WARNING

DC voltages are present at generator set electrical components even with generator set shut down. Avoid shorting any positive with ground/negative. Failure to comply with this warning can cause injury to personnel, and damage to equipment.

4-22.1 Testing.

a. a. Shut down generator set. Allow generator to cool to ambient temperature.

b. b. Open battery access door and disconnect negative battery cable.

c. c. Remove generator set housing rear panel (paragraph 2-18.1).

d. d. Remove generator end bell (paragraph 4-19.1).

e. e. Tag and disconnect exciter rotor leads (4, Figure 4-10) from rotating rectifier (3 places) by removing screws (3).

f. Connect resistance bridge between two exciter rotor leads (4) and note resistance reading. Continue this procedure until readings are noted for each combination of leads (i.e., 1 and 2, 1 and 3, and 2

g. g. Resistance readings should be as shown in Table 4-3 for each combination of leads. Reading other than shown in Table 4-3 indicates that open or shorted windings and exciter rotor must be replaced.

h. h. Set multimeter for ohms and connect between each exciter rotor lead (4) and generator housing in turn.

i. Reading of less than infinity indicates defective ground insulation and exciter rotor must be replaced.

j. j. Connect exciter rotor leads (4) to rotating rectifier with screws (3). Remove tags.

k. k. Install end bell (paragraph 4-19.2).

l. l. Install generator set housing rear panel (paragraph 2-18.4).

m. m. Connect negative battery c able. Close battery access door.

NOTE

Ambient temperature must be expressed in °C. To convert °F to °C, use °F = °C x 9/5 + 32.

(1) To determine the resistance values at current ambient temperature, use the following formula:

$$_1 = R_{25}[1 + 0.00385 (T-25)]$$

Where:

R_1 = Unknown resistance
$_{25}$ = Known resistance at 25°C (77°F)
T = Current ambient temperature

(2) Example for exciter stator leads at 5°C (41°F) (MEP-804A/MEP-804B):

$_1 = 33.714 [1 + 0.00385 (5-25)]$ R
$_1 = 33.714 [1 + 0.00385 (-20)]$ $_1$
$= 33.714 [1 \pm 0.077)]$
$_1 = 33.714 [0.923]$ R
$_1 = 31.118 \pm 15\%$ ohms

4-22.2 Removal.

WARNING

Many components require a two-person lift. Lifting heavy components can cause back strain. Ensure proper lifting techniques are used when lifting heavy components. Failure to comply with this warning can cause injury to personnel.

WARNING

The generator set, engine, and generator are extremely heavy and require an assistant and a lifting device (forklift, overhead lifting device) with sufficient capacity. Failure to comply with this warning can cause serious injury or death to personnel.

WARNING

All metal jewelry can conduct electricity and become entangled in generator set components. Remove all jewelry when working on generator set. Failure to comply with this warning can cause injury or death to personnel.

WARNING

DO NOT wear loose clothing when performing checks, services and maintenance. Failure to comply with this warning can cause injury or death to personnel.

a. a. Shut down generator set.

b. b. Open battery access door and disconnect negative battery cable.

c. c. Remove generator set housing rear panel (paragraph 2.18.1).

d. d. Remove generator end bell and main bearing (paragraph 4.19.1).

e. e. Remove rotating rectifier (paragraph 4.20.1).

WARNING

Use protective gloves when handling heated rectifier hub. Failure to comply with this warning can cause injury to personnel.

NOTE

Rectifier hub is heat shrunk onto rotor assembly shaft and will require the use of heat to remove.

a. a. Using bearing puller, remove rectifier hub (12, Figure 4-11) from rotor assembly (23) shaft.

b. b. Pull two main rotor leads out of holes in exciter rotor (15).

c. c. Attach hub puller to two lead holes and remove exciter rotor (15) from rotor assembly (23) shaft.

d. d. Remove key (16) from rotor assembly (23) shaft.

4-22.3 <u>Installation</u>.

a. a. Install key (16, Figure 4-11) on main rotor assembly (23) shaft.

b. b. Press exciter rotor (15) onto shaft over key (16) and against shoulder of rotor assembly (23) shaft.

c. c. Heat rectifier hub (12) until it slides into place on rotor assembly (23) shaft with slight pressure.

d. d. Once rectifier hub (12) has cooled, pull two ma in rotor leads (2, Figure 4-10)) through holes in exciter rotor (15, Figure 4-11) and position for attachment to rotating rectifier (10).

e. Install rotating rectifier (paragraph 4-20.2).

f. f. Install main bearing and generator end bell (paragraph 4-19.2).

g. g. Install generator set housin g rear panel (paragraph 2-18.4).

h. h. Connect negative battery c able. Close battery access door.

WARNING

High voltage is produced when the generator set is in operation. Never attempt to start the generator set unless it is properly grounded. Failure to comply with this warning can cause injury or death to personnel.

WARNING

Dangerous voltage exists on live circuits. Always observe precautions and never work alone. Failure to comply with this warning can cause injury or death to personnel.

WARNING

Ensure that the engine cannot be started while maintenance is being performed. (ENGINE CONTROL switch set to OFF/RESET; Battery Disconnect Switch is OFF; DEAD CRANK SWITCH is OFF). Failure to comply with this warning can cause injury or death to personnel.

WARNING

DC voltages are present at generator set electrical components even with generator set shut down. Avoid shorting any positive with ground/negative. Failure to comply with this warning can cause injury to personnel, and damage to equipment.

4-23.1 Testing.

a. a. Shut down generator set. Allow generator to cool to ambient temperature.

b. b. Open battery access door and disconnect negative battery cable.

c. c. Remove generator set housing rear panel (paragraph 2-18.1).

d. d. Remove generator end bell (paragraph 4-19.1).

e. e. Tag and disconnect main rotor leads (2, Figure 4-10) from rotating rectifier (2 places) by removing screws (1).

f. f. Set multimeter for ohms and connect between disconnected main rotor leads. Multimeter reading should be as shown in Table 4-3.

g. g. Reading other than shown in Table 4-3 indicates that shorted or open windings and main rotor must be replaced.

h. h. Connect multimeter between each main ro tor lead and generator housing in turn.

i. i. Reading of less than infinity indicates defective ground insulation and main rotor must be replaced.

j. j. Connect main rotor leads (2) to rotating rectifier (2 places) with screws (1).

k. k. Install generator end bell (paragraph 4-19.2).

l. l. Install generator set housin g rear panel (paragraph 2-18.4).

m. m. Connect negative battery c able. Close battery access door.

4-23.2 Removal.

WARNING

Many components require a two-person lift. Lifting heavy components can cause back strain. Ensure proper lifting techniques are used when lifting heavy components. Failure to comply with this warning can cause injury to personnel.

WARNING

The generator set, engine, and generator are extremely heavy and require an assistant and a lifting device (forklift, overhead lifting device) with sufficient capacity. Failure to comply with this warning can cause serious injury or death to personnel.

WARNING

All metal jewelry can conduct electricity and become entangled in generator set components. Remove all jewelry when working on generator set. Failure to comply with this warning can cause injury or death to personnel.

WARNING

DO NOT wear loose clothing when performing checks, services and maintenance. Failure to comply with this warning can cause injury or death to personnel.

a. a. Shut down generator set.

b. b. Remove generator assembly from generator set (paragraph 4-17.1).

NOTE

Mark location of spacers, washers, and bolts to ensure correct positioning during installation.

c. c. Remove bolts (17 and 19, Figure 4-11), washers (18 and 20), spacers (21), and drive discs (22) from drive hub (27).

d. d. Attach a suitable rotor lifting device to drive hub (27) and overhead hoist (Figure 4-13).

e. e. Remove generator end bell (paragraph 4-19.1).

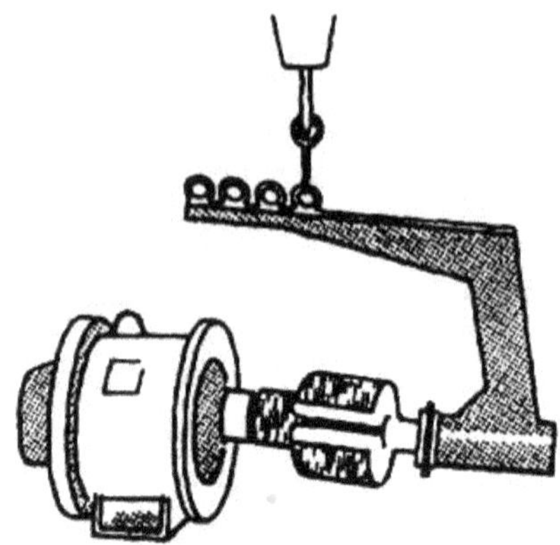

Figure 4-13. Rotor Assembly Lifting Device (Typical).

CAUTION

Special care should be taken when removing rotor assembly. Winding damage could result if rotor is allowed to hit main stator.

f. f. Carefully remove rotor assembly (23) and atta ched components from generator housing (31).

g. g. Remove main bearing (paragraph 4-19.1)

h. h. Remove rotating rectifier (paragraph 4-20.1).

i. i. Remove exciter rotor (paragraph 4-22.2).

j. j. Remove bolts (24), nuts (25), and fan halves (26) from rotor assembly (23).

k. k. Using bearing puller, remove drive hub (27) and key (28) from rotor assembly (23).

4-23.3 Installation.

a. a. Place key (28, Figure 4-11) on rotor assembly (2 3) and install drive hub (27) over key (28) and onto main rotor assembly (23).

NOTE

Ensure gap between fan halves are equal from side to side.

b. b. Install fan halves (26) on rotor assembly (23) with bolts (24) and nuts (25).

c. c. Install exciter rotor (paragraph 4-22.3).

d. d. Install rotating rectifier (paragraph 4-20.2).

e. e. Install main bearing (paragraph 4-19.2).

CAUTION

Special care must be taken when installing rotor assembly. Winding damage could result if rotor is allowed to hit main stator.

f. Attach a suitable rotor lifting device to drive hub (27) and overhead hoist (Figure 4-13).

g. g. Carefully install rotor assembly (23) and attached components into generator housing (31).

h. h. Install generator end bell (paragraph 4-19.2) and remove rotor lifting device.

NOTE

Ensure correct positioning of spacers, washers, and bolts as marked on removal.

i. i. Install drive discs (22) on drive hub (27) with spacers (21), washers (18 and 20), and bolts (17 and 19). Torque bolts (17) to 28 ft-lbs (38 Nm) and bolts (19) to 35 ft-lbs (47 Nm).

j. j. Install generator assembly in generator set, paragraph 4-17.2.

4-24 MAIN STATOR (MEP-804A/MEP-804B)

WARNING

High voltage is produced when the generator set is in operation. Never attempt to start the generator set unless it is properly grounded. Failure to comply with this warning can cause injury or death to personnel.

WARNING

Dangerous voltage exists on live circuits. Always observe precautions and never work alone. Failure to comply with this warning can cause injury or death to personnel.

WARNING

Ensure that the engine cannot be started while maintenance is being performed. (ENGINE CONTROL switch set to OFF/RESET; Battery Disconnect Switch is OFF; DEAD CRANK SWITCH is OFF). Failure to comply with this warning can cause injury or death to personnel.

WARNING

DC voltages are present at generator set electrical components even with generator set shut down. Avoid shorting any positive with ground/negative. Failure to comply with this warning can cause injury to personnel, and damage to equipment.

4-24.1 Testing.

a. a. Shut down generator set. Allow generator to cool to ambient temperature.

b. b. Open battery access door and disconnect negative battery cable.

c. c. Remove protective cover and moveable termina l board from voltage reconnection terminal board (paragraph 2-107.2).

d. d. Disconnect two electrical connectors from bottom of control box assembly.

e. e. Tag and disconnect wires 107C and 109J fr om terminals 1 and 2 of power potential transformer (37, Figure 4-3).

f. f. Connect resistance bridge and note readings between terminals T1 and T4, T2 and T5, T3 and T6, T7 and T0, T8 and T0, and T9 and T0 of voltage reconnection terminal board.

g. g. All resistance readings should be as shown in Table 4-3.

h. h. If resistance is low, there are shorted windings. If resistance is high, stator windings are open. In either case, stator must be replaced.

i. i. Disconnect removable grounding link (17, Figure 2-34) on load output terminal board.

j. j. Set multimeter for ohms and connect between each coil lead and ground in turn.

k. k. If multimeter indicates resistance on any connection, stator windings are grounded and stator must be replaced.

l. l. Connect removable grounding link (17) on load output terminal board.

m. m. Connect wires 107C and 109J to terminals 1 and 2 of power potential transformer (37, Figure 4-3).

Remove tags.

n. n. Connect two electrical connectors to bottom of control box assembly.

o. o. Install moveable terminal board and protective cover on voltage reconnection terminal board (paragraph 2-107.3).

p. Connect negative battery c able. Close battery access door.

4-24.2 Removal.

WARNING

Many components require a two-person lift. Lifting heavy components can cause back strain. Ensure proper lifting techniques are used when lifting heavy components. Failure to comply with this warning can cause injury to personnel.

WARNING

The generator set, engine, and generator are extremely heavy and require an assistant and a lifting device (forklift, overhead lifting device) with sufficient capacity. Failure to comply with this warning can cause serious injury or death to personnel.

WARNING

All metal jewelry can conduct electricity and become entangled in generator set components. Remove all jewelry when working on generator set. Failure to comply with this warning can cause injury or death to personnel.

WARNING

DO NOT wear loose clothing when performing checks, services and maintenance. Failure to comply with this warning can cause injury or death to personnel.

a. a. Shut down generator set.

b. b. Remove generator assembly from generator set (paragraph 4-17.1).

c. c. Remove generator rotor assembly (paragraph 4-23.2).

d. d. Remove nuts (32, Figure 4-11), lock washers (33), bolts (34), and lead clamp assembly (38) from brackets (37). Discard lock washers (33).

e. e. Remove main stator setscrews (29) from housing (31).

f. f. Set generator housing (31) on end with end bell opening in the upright position.

CAUTION

Special care must be taken when feeding wires to prevent damaging or breaking off wires.

g. Feed main stator (30) and exciter stator (14) wires through generator housing (31) and up through end bell opening.

WARNING

Hot metal surfaces can cause burns to skin. Wear protective gloves and eye protection when applying heat to generator housing. Failure to comply with this warning can cause injury to personnel.

CAUTION

Ensure wires are not touching generator housing and are away from open flame when heat is applied. Be sure to distribute heat evenly to prevent damage to housing.

h. Apply heat evenly to ribs of generator housing (3 1) at main stator (30) end using an acetylene torch with a rose bud tip. Refer to Figure 4-12 for rib locations. Heat ribs until main stator (30) drops down and lift off housing (31).

4-24.3 Installation.

a. a. Stand generator housing (31, Figure 4-11) on end with end bell opening in a downward position.

b. b. Position main stator (30) in generator housing (31). Ensure stator wires are pointing downward
and

WARNING

Hot metal surfaces can cause burns to skin. Wear protective gloves and eye protection when applying heat to generator housing. Failure to comply with this warning can cause injury to personnel.

CAUTION

Ensure wires are not touching generator housing and are away from open flame when heat is applied. Be sure to distribute heat evenly to prevent damage to housing.

c. Apply heat evenly to ribs of generator housing (3 1) at main stator (30) end using an acetylene torch with a rose bud tip. Refer to Figure 4-12 for rib locations. Heat generator housing (31) ribs until main stator (30) drops into position.

d. d. Install setscrews (29) in generator housing (31).

CAUTION

Special care must be taken when feeding wires to prevent damaging or breaking off wires.

e. e. Feed main stator (30) and exciter stator (14) wires through wiring port on side of generator housing (31).

f. f. Retest main stator to ensure stator was not damaged during installation (paragraph 4-24.1).

g. g. Position exciter stator (14) and main stator (30) leads in lead clamp assembly (38).

h. h. Install lead clamp assembly (38) in brackets (37) with bolts (34), new lock washers (33), and nuts (32).

i. i. Install rotor assembly (paragraph 4-23.3).

j. j. Install generator assembly (paragraph 4-17.2).

4-25 GENERATOR HOUSING (MEP-804A/MEP-804B)

WARNING

High voltage is produced when the generator set is in operation. Never attempt to start the generator set unless it is properly grounded. Failure to comply with

WARNING

Dangerous voltage exists on live circuits. Always observe precautions and never work alone. Failure to comply with this warning can cause injury or death to personnel.

WARNING

Ensure that the engine cannot be started while maintenance is being performed. (ENGINE CONTROL switch set to OFF/RESET; Battery Disconnect Switch is OFF; DEAD CRANK SWITCH is OFF). Failure to comply with this warning can cause injury or death to personnel.

WARNING

DC voltages are present at generator set electrical components even with generator set shut down. Avoid shorting any positive with ground/negative. Failure to comply with this warning can cause injury to personnel, and damage to equipment.

4-25.1 Removal.

WARNING

Many components require a two-person lift. Lifting heavy components can cause back strain. Ensure proper lifting techniques are used when lifting heavy components. Failure to comply with this warning can cause injury to personnel.

WARNING

The generator set, engine, and generator are extremely heavy and require an assistant and a lifting device (forklift, overhead lifting device) with sufficient capacity. Failure to comply with this warning can cause serious injury or death to personnel.

WARNING

All metal jewelry can conduct electricity and become entangled in generator set components. Remove all jewelry when working on generator set. Failure to comply with this warning can cause injury or death to personnel.

WARNING

DO NOT wear loose clothing when performing checks, services and maintenance. Failure to comply with this warning can cause injury or death to personnel.

a. a. Shut down generator set.

b. b. Remove generator assembly from generator set (paragraph 4-17.1).

c. c. Remove rotor assembly (paragraph 4-23.2).

d. d. Remove exciter stator (paragraph 4-21.2).

e. e. Remove main stator (paragraph 4-24.2).

f. f. Remove bolts (35, Figure 4-11), lock washers (36), brackets (37), and gasket (39) from housing assembly (31). Discard lock washers (36).

g. g. Remove bolts (40), washers (41), and screen (42) from housing (31).

h. h. Remove bolts (43), washers (44), and screen (45) from housing (31).

i. i. Remove nut (46), washer (47), and ground stud (48) from housing (31).

j. J. Loosen nut (49) and unscrew lifting eye (50) from housing (31).

4-25.2 <u>Installation</u>.

a. a. Install lifting eye (50, Figure 4-11) in generator housing (31) and tighten nut (49).

b. b. Install ground stud (48), washer (47), and nut (46) in housing (31).

c. c. Install screen (45) on housing (31) with bolts (43) and washers (44).

d. d. Install screen (42) on housing (31) with bolts (40) and washers (41).

e. e. Install gasket (39) and brackets (37) on housing assembly (31) with bolts (35) and new lock washers (36).

f. f. Install exciter stator (paragraph 4-21.3).

g. g. Install main stator (paragraph 4-24.3).

h. h. Install rotor assembly (paragraph 4-23.3).

i. i. Install generator assembly in generator set (paragraph 4-17.2).

WARNING

High voltage is produced when the generator set is in operation. Never attempt to start the generator set unless it is properly grounded. Failure to comply with this warning can cause injury or death to personnel.

WARNING

Dangerous voltage exists on live circuits. Always observe precautions and never work alone. Failure to comply with this warning can cause injury or death to personnel.

WARNING

Ensure that the engine cannot be started while maintenance is being performed. (ENGINE CONTROL switch set to OFF/RESET; Battery Disconnect Switch is OFF; DEAD CRANK SWITCH is OFF). Failure to comply with this warning can cause injury or death to personnel.

WARNING

DC voltages are present at generator set electrical components even with generator set shut down. Avoid shorting any positive with ground/negative. Failure to comply with this warning can cause injury to personnel, and damage to equipment.

4-26.1 Removal.

WARNING

Many components require a two-person lift. Lifting heavy components can cause back strain. Ensure proper lifting techniques are used when lifting heavy components. Failure to comply with this warning can cause injury to personnel.

WARNING

The generator set, engine, and generator are extremely heavy and require an assistant and a lifting device (forklift, overhead lifting device) with sufficient capacity. Failure to comply with this warning can cause serious injury or death to personnel.

WARNING

All metal jewelry can conduct electricity and become entangled in generator set components. Remove all jewelry when working on generator set. Failure to comply with this warning can cause injury or death to personnel.

WARNING

DO NOT wear loose clothing when performing checks, services and maintenance. Failure to comply with this warning can cause injury or death to personnel.

a. a. Shut down generator set.

b. b. Open battery access door and disconnect negative battery cable.

c. c. Remove generator housing rear panel (paragraph 2-18.1).

d. d. Remove end bell plates and screens (paragraph 4-27.1, step d).

NOTE

It will be necessary to bar (turn) engine in order to position a specific area of the rotating rectifier at one of the end bell access holes.

e. e. Unsolder electrical lead from diode (6, Figure 4-14) being removed.

NOTE

Mark polarity of diodes to ensure correct positioning during installation.

f. f. Remove nut (7), lock washer (8), and diode (6) from rotating rectifier plate (5) through access hole in end bell. Discard lock washer (8).

4-26.2 Testing.

a. a. Shut down generator set. Allow generator to cool to ambient temperature.

b. b. Open battery access door and disconnect negative battery cable.

c. c. Remove generator set housing rear panel (paragraph 2-18.1).

d. d. Remove generator end bell plates and screens (paragraph 4-27.1, step d).

NOTE

It will be necessary to bar (turn) engine in order to position a specific area of the rotating rectifier at one of the end bell access holes.

e. e. Remove nuts (1, Figure 4-14) and lock washers (2) from rotating rectifier terminals. Discard lock washers (2).

f. f. Tag and remove main rotor and diode leads from rotating rectifier terminals.

g. Tag exciter rotor leads and remove bolts (3), washers (4), and exciter rotor leads from rectifier mounting plate assemblies.

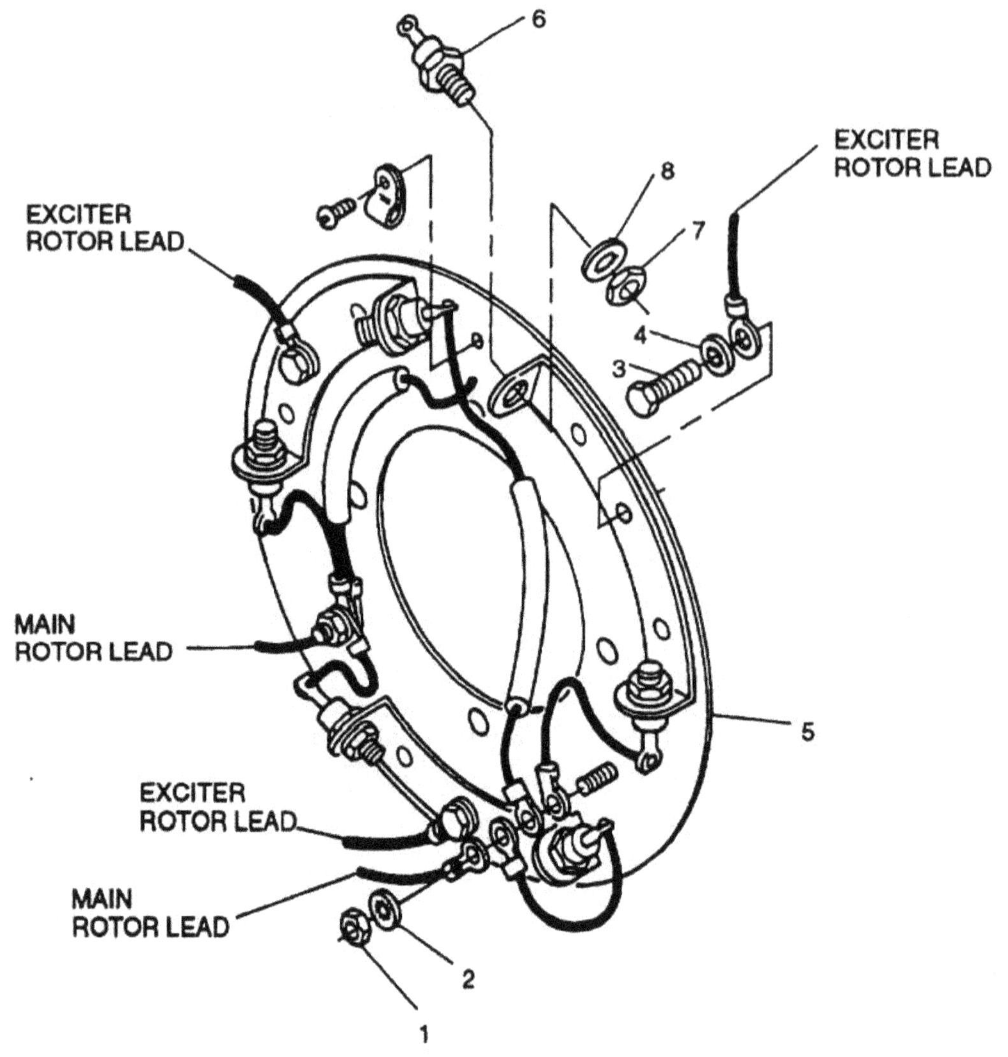

Figure 4-14. Rectifier Assembly – MEP-814A/MEP-814B.

h. h. Set multimeter for ohms and connect positive lead to one side and negative lead to other side of diode. Repeat procedure for each diode. Record multimeter reading for each diode.

i. i. Repeat step h above, except with multimeter leads reversed. Resistance (ohms) readings should be low in one direction and high in reversed direction. If any reading is high or low in both directions, diode is defective and must be replaced.

j. j. Install diode and main rotor leads to rotating rectifier terminals with new lock washers (2) and nuts (1). Remove tags.

k. k. Install exciter rotor leads to rectifier mounting plate assemblies with washers (4) and bolts (3). Remove tags.

l. l. Install generator end bell plates and screens (paragraph 4-27.2, step e).

m. m. Install generator set housing rear panel (paragraph 2-18.4).

n. n. Connect negative battery c able. Close battery access door.

4-26.3 Installation.

a. a. Run bead of thermo compound (Item 8, Appendix C) around base of diode (6, Figure 4-14) prior to installing. Do not coat threads.

b. b. Insert diode (6) through generator end bell access hole and install on rotating rectifier plate (5) with new lock washer (8) and nut (7) as marked during removal. Torque nut 28 to 30 in.-lbs (3.2 to 3.4 Nm).

c. c. Using solder (Item 19, Appendix C) and soldering iron, solder electrical lead to diode (6).

d. d. Install generator end bell plates and screens (paragraph 4-27.2, step e).

e. e. Install generator set housin g rear panel (paragraph 2-18.4).

f. f. Connect negative battery cabl e. Close battery access door.

4-27 END BELL AND MAIN BEARING (MEP-814A/MEP-814B)

WARNING

High voltage is produced when the generator set is in operation. Never attempt to start the generator set unless it is properly grounded. Failure to comply with this warning can cause injury or death to personnel.

WARNING

Dangerous voltage exists on live circuits. Always observe precautions and never work alone. Failure to comply with this warning can cause injury or death to personnel.

WARNING

Ensure that the engine cannot be started while maintenance is being performed. (ENGINE CONTROL switch set to OFF/RESET; Battery Disconnect Switch is OFF; DEAD CRANK SWITCH is OFF). Failure to comply with this warning can cause injury or death to personnel.

WARNING

DC voltages are present at generator set electrical components even with generator set shut down. Avoid shorting any positive with ground/negative. Failure to comply with this warning can cause injury to personnel, and damage to equipment.

4-27.1 Removal.

WARNING

Many components require a two-person lift. Lifting heavy components can cause back strain. Ensure proper lifting techniques are used when lifting heavy components. Failure to comply with this warning can cause injury to personnel.

WARNING

The generator set, engine, and generator are extremely heavy and require an assistant and a lifting device (forklift, overhead lifting device) with sufficient capacity. Failure to comply with this warning can cause serious injury or death to personnel.

WARNING

All metal jewelry can conduct electricity and become entangled in generator set components. Remove all jewelry when working on generator set. Failure to comply with this warning can cause injury or death to personnel.

WARNING

DO NOT wear loose clothing when performing checks, services and maintenance. Failure to comply with this warning can cause injury or death to personnel.

a. a. Shut down generator set.

b. b. Open battery access door and disconnect negative battery cable.

c. c. Remove generator set housin g rear panel (paragraph 2-18.1).

d. d. Remove bolts (1, Figure 4-15), lock washers (2), plates (3), and screens (4) from end bell (5). Discard lock washers (2).

CAUTION

The end bell supports the main rotor, thus the rotor will drop on the stator once the end bell is removed. Prior to proceeding, turn engine until two main rotor poles are vertical in generator stator. Having the rotor in this position will limit the amount of drop.

e. e. Remove bolts (6) and lock washers (7) from end bell (5). Discard lock washers (7).

f. f. Install two bolts (6) in backout holes in end bell flange (Figure 4-16).

g. g. Loosen lead clamp assembly (37, Figure 4-15) clamping generator leads at side of generator housing. Ensure that wires F1 and F2 are free to slide in and out of generator housing.

h. h. Remove end bell (5), with exciter stator (17) attached, by tightening bolts (6) evenly into backout holes.

CAUTION

If bearing needs to be removed for any reason, always install new bearing. Main beating is easily damaged when removed from rotor shaft. Damage to equipment could result.

i. i. Using bearing puller, remove bearing (8) from main rotor shaft (9). Discard bearing (8).

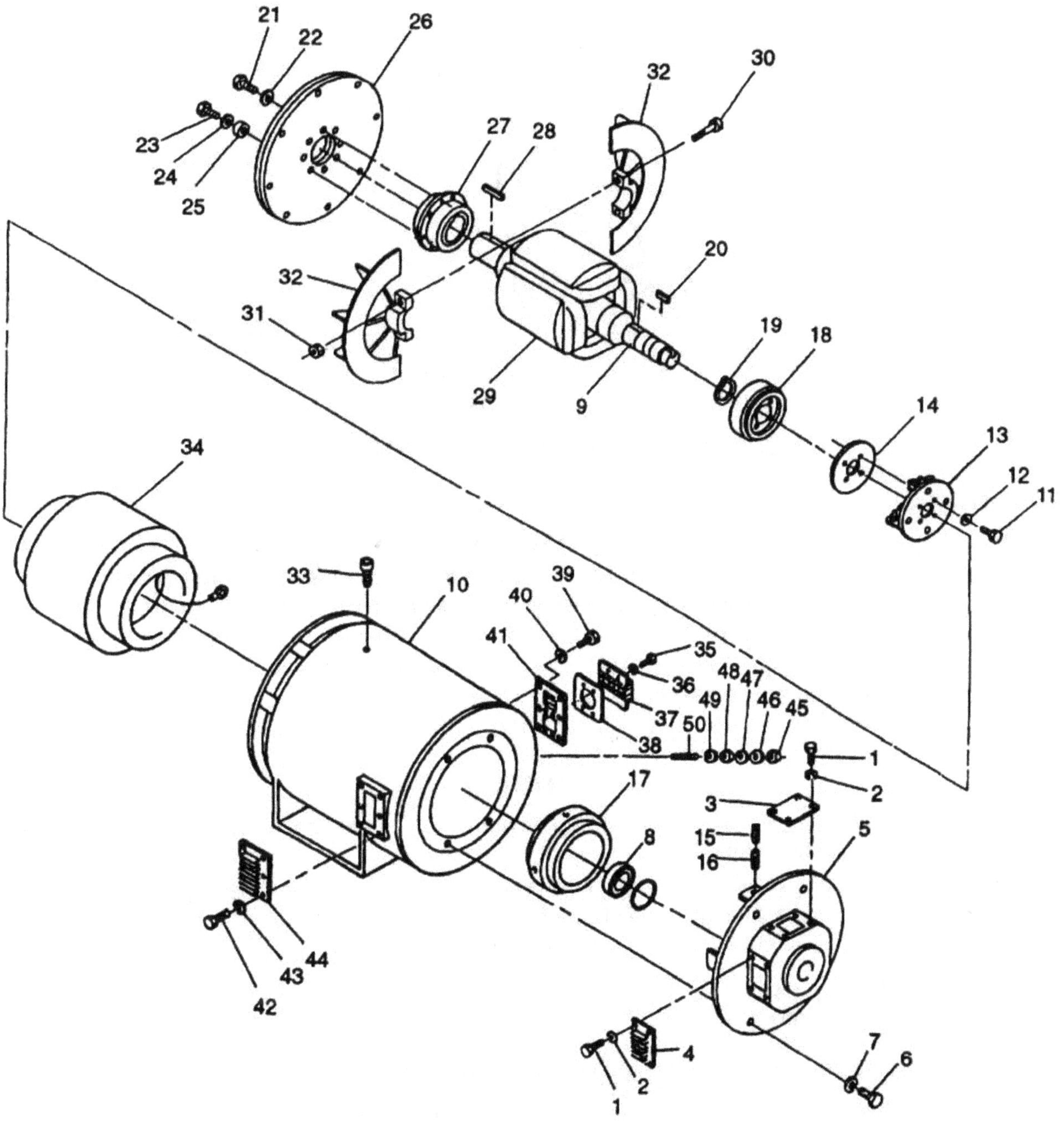

Figure 4-15. Generator Assembly – MEP-814A/MEP-814B.

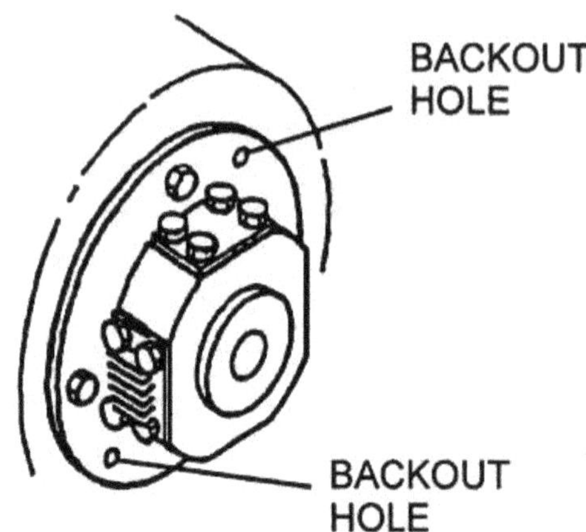

Figure 4-16. End Bell Removal – MEP-814A/MEP-814B.

4-27.2 Installation.

CAUTION

If bearing needs to be removed for any reason, always install new bearing. Main bearing is easily damaged when removed from rotor shaft. Damage to equipment could result.

a. a. Install new bearing (8, Figure 4-15). Ensure that bearing is seated squarely against shaft (9) shoulder by applying pressure to inner race only.

b. b. Position bearing (8) in end bell (5).

NOTE

It may be necessary to use a lifting device to raise and align end bell (5) with generator housing (10).

c. c. Position end bell (5), with exciter stator (17) attached, on generator housing (10) while pulling slack of wires F1 and F2 through side of generator housing. Secure end bell (5) with bolts (6) and lock-washers (7). Torque bolts 59 to 61 in.-lbs (6.7 to 7 Nm).

d. d. Tighten lead clamp assembly (37) at side of generator housing ensuring that generator leads are clamped securely.

e. e. Install screens (4) and plates (3) on end bell (5) with bolts (1) and new lock washers (2).

f. f. Install generator set housing rear panel (paragraph 2-18.4).

g. g. Connect negative battery c able. Close battery access door.

4-28 ROTATING RECTIFIER (MEP-814A/MEP-814B)

WARNING

High voltage is produced when the generator set is in operation. Never attempt to start the generator set unless it is properly grounded. Failure to comply with this warning can cause injury or death to personnel.

WARNING

Dangerous voltage exists on live circuits. Always observe precautions and never work alone. Failure to comply with this warning can cause injury or death to personnel.

WARNING

Ensure that the engine cannot be started while maintenance is being performed. (ENGINE CONTROL switch set to OFF/RESET; Battery Disconnect Switch is OFF; DEAD CRANK SWITCH is OFF). Failure to comply with this warning can cause injury or death to personnel.

WARNING

DC voltages are present at generator set electrical components even with generator set shut down. Avoid shorting any positive with ground/negative. Failure to comply with this warning can cause injury to personnel, and damage to equipment.

4-28.1 Removal.

WARNING

Many components require a two-person lift. Lifting heavy components can cause back strain. Ensure proper lifting techniques are used when lifting heavy components. Failure to comply with this warning can cause injury to personnel.

WARNING

The generator set, engine, and generator are extremely heavy and require an assistant and a lifting device (forklift, overhead lifting device) with sufficient capacity. Failure to comply with this warning can cause serious injury or death to personnel.

WARNING

All metal jewelry can conduct electricity and become entangled in generator set components. Remove all jewelry when working on generator set. Failure to comply with this warning can cause injury or death to personnel.

WARNING

DO NOT wear loose clothing when performing checks, services and maintenance. Failure to comply with this warning can cause injury or death to personnel.

a. a. Shut down generator set.

b. b. Open battery access door and disconnect negative battery cable.

c. c. Remove generator set housin g rear panel (paragraph 2-18.1).

d. d. Remove generator end bell and main bearing (paragraph 4-27.1).

e. e. Remove bolts (3, Figure 4-14) and lock washers (4) securing exciter rotor leads (3 places) to rotating rectifier. Tag and remove exciter rotor leads. Discard lock washers (4).

f. f. Remove nuts (1) and lock washers (2) securing generator rotor leads (2 places) to rotating rectifier. Tag and remove generator rotor leads. Discard lock washers (2).

g. g. Remove bolts (11, Figure 4-15), lock washers (12), and rotating rectifier (13) from rectifier hub (14).

Discard lock washers (12).

4-28.2 Installation.

a. a. Install rotating rectifier (13, Figure 4-15) on rectifier hub (14) with bolts (11) and new lock washers (12).

b. b. Connect two generator rotor leads to rotating rectifier with nuts (1, Figure 4-14) and new lock washers (2). Remove tags.

c. c. Connect three exciter rotor leads to rotating rectifier with bolts (3) and new lock washers (4).

d. d. Install main bearing and generat or end bell (paragraph 4-27.2).

e. e. Install generator set housin g rear panel (paragraph 2-18.4).

f. f. Connect negative battery cabl e. Close battery access door.

WARNING

High voltage is produced when the generator set is in operation. Never attempt to start the generator set unless it is properly grounded. Failure to comply with this warning can cause injury or death to personnel.

WARNING

Dangerous voltage exists on live circuits. Always observe precautions and never work alone. Failure to comply with this warning can cause injury or death to personnel.

WARNING

Ensure that the engine cannot be started while maintenance is being performed. (ENGINE CONTROL switch set to OFF/RESET; Battery Disconnect Switch is OFF; DEAD CRANK SWITCH is OFF). Failure to comply with this warning can cause injury or death to personnel.

WARNING

DC voltages are present at generator set electrical components even with generator set shut down. Avoid shorting any positive with ground/negative. Failure to comply with this warning can cause injury to personnel, and damage to equipment.

4-29.1 Removal.

WARNING

Many components require a two-person lift. Lifting heavy components can cause back strain. Ensure proper lifting techniques are used when lifting heavy components. Failure to comply with this warning can cause injury to personnel.

WARNING

The generator set, engine, and generator are extremely heavy and require an assistant and a lifting device (forklift, overhead lifting device) with sufficient capacity. Failure to comply with this warning can cause serious injury or death to personnel.

WARNING

All metal jewelry can conduct electricity and become entangled in generator set components. Remove all jewelry when working on generator set. Failure to comply with this warning can cause injury or death to personnel.

WARNING

DO NOT wear loose clothing when performing checks, services and maintenance. Failure to comply with this warning can cause injury or death to personnel.

a. a. Shut down generator set.

b. b. Open battery access door and disconnect negative battery cable.

c. c. Remove generator set housin g rear panel (paragraph 2-18.1).

d. d. Remove generator end bell (paragraph 4-27.1).

e. e. Open output box access door and remove exciter field leads F1 and F2 from terminals 1 and 2 of TB8.

f. f. Attach “fish wires” to disconnected leads F1 and F2 to aid in installation process.

g. g. Remove setscrews (15 and 16, Figure 4-15) and exciter stator (17) from end bell (5).

h. h. Detach “fish wires” once leads F1 and F2 clear generator housing (10).

4-29.2 Testing.

a. a. Shut down generator set. Allow generator to cool to ambient temperature.

b. b. Open battery access door and disconnect negative battery cable.

c. c. Open output box access door and disconnect exciter field leads F1 and F2 from terminals 1 and 2 of TB8.

d. d. Set multimeter for ohms and connect between disconnected exciter field leads. Multimeter reading should be as shown in Table 4-3.

e. e. Reading other than shown in Table 4-3 indicates that open or shorted windings and exciter stator must be replaced.

f. f. Connect multimeter between each exciter field lead and generator frame in turn.

g. Reading of less than infinity indicates defective ground insulation and exciter stator must be replaced.

h. h. Connect exciter field leads to terminals 1 and 2 of TB8. Close output box access door.

i. i. Connect negative battery c able. Close battery access door.

4-29.3 Installation.

a. a. Attach "fish wires" to leads F1 and F2 of exciter stator (17, Figure 4-15).

b. b. Gently pull on "fish wires" to pull leads F1 and F2 back through generator housing. Disconnect "fish wires" and position leads F1 and F2 in output box.

c. c. Position exciter stator (17) in end bell (5) and secure in place with setscrews (15 and 16).

d. d. Install generator end bell (paragraph 4-27.2).

e. e. Install generator set housing rear panel (paragraph 2-18.4).

f. f. Connect exciter field leads F1 and F2 to terminals 1 and 2 of TB8. Close output box access door.

g. g. Connect negative battery c able. Close battery access door.

4-30 EXCITER ROTOR (MEP-814A/MEP-814B)

WARNING

High voltage is produced when the generator set is in operation. Never attempt to start the generator set unless it is properly grounded. Failure to comply with this warning can cause injury or death to personnel.

WARNING

Dangerous voltage exists on live circuits. Always observe precautions and never work alone. Failure to comply with this warning can cause injury or death to personnel.

WARNING

Ensure that the engine cannot be started while maintenance is being performed. (ENGINE CONTROL switch set to OFF/RESET; Battery Disconnect Switch is OFF; DEAD CRANK SWITCH is OFF). Failure to comply with this warning can cause injury or death to personnel.

WARNING

DC voltages are present at generator set electrical components even with generator set shut down. Avoid shorting any positive with ground/negative. Failure to comply with this warning can cause injury to personnel, and damage to equipment.

4-30.1 Removal.

WARNING

Many components require a two-person lift. Lifting heavy components can cause back strain. Ensure proper lifting techniques are used when lifting heavy components. Failure to comply with this warning can cause injury to personnel.

WARNING

The generator set, engine, and generator are extremely heavy and require an assistant and a lifting device (forklift, overhead lifting device) with sufficient capacity. Failure to comply with this warning can cause serious injury or death to personnel.

WARNING

All metal jewelry can conduct electricity and become entangled in generator set components. Remove all jewelry when working on generator set. Failure to comply with this warning can cause injury or death to personnel.

WARNING

DO NOT wear loose clothing when performing checks, services and maintenance. Failure to comply with this warning can cause injury or death to personnel.

a. a. Shut down generator set.

b. b. Open battery access door and disconnect negative battery cable.

c. c. Remove generator set housin g rear panel (paragraph 2-18.1).

d. d. Remove generator end bell and main bearing (paragraph 4-27.1).

e. e. Remove rotating rectifier (paragraph 4-28.1).

WARNING

Use protective gloves when handling heated rectifier hub. Failure to comply

NOTE

Rectifier hub is heat shrunk onto shaft and will require the use of heat to remove.

f. f. Using bearing puller, remove rectifier hub (14, Figure 4-15) from main rotor shaft (9).

g. g. Pull two main rotor leads out of holes in exciter rotor (18).

h. h. Attach hub puller to two lead holes and re move exciter rotor (18) from shaft (9).

i. i. Remove key (20) and retaining ring (19) from shaft (9).

4-30.2 Testing.

a. a. Shut down generator set. Allow generator to cool to ambient temperature.

b. b. Open battery access door and disconnect negative battery cable.

c. c. Remove generator set housing rear panel (paragraph 2-18.1).

d. d. Remove generator end bell plates and screens (paragraph 4-27.1, step d).

NOTE

It will be necessary to bar (turn) the engine in order to position a specific area of the rotating rectifier at one of the end bell access holes.

e. e. Tag and disconnect exciter rotor leads (3 places) from rotating rectifier by removing bolts (3, Figure 4-14) and lock washers (4). Discard lock washers (4).

f. f. Connect resistance bridge between two exciter rotor leads and note resistance reading. Continue this procedure until readings are noted for each combination of leads (i.e., 1 and 2, 1 and 3, and 2 and 3).

g. g. Resistance readings should be as shown in Table 4-3 for each combination of leads. Reading other than shown in Table 4-3 indicates that open or shorted windings and exciter rotor must be replaced.

h. h. Set multimeter for ohms and connect between each exciter rotor lead and exciter stator mounting bolt in turn.

i. i. Reading of less than infinity indicates defective ground insulation and exciter rotor must be replaced.

j. j. Connect exciter rotor leads to rotating rectifier with new lock washers (4) and bolts (3). Remove tags.

k. k. Install end bell screens and plates (paragraph 4-27.2, step e).

l. l. Install generator set housing rear panel (paragraph 2-18.4).

m. m. Connect negative battery c able. Close battery access door.

4-30.3 Installation.

a. a. Install retaining ring (19, Figure 4-15) and key (20) on main rotor shaft (9).

b. b. Press exciter rotor (18) onto shaft (9) over key (20) and against retaining ring (19).

WARNING

Use protective gloves when handling heated rectifier hub. Failure to comply with this warning can cause injury to personnel.

c. c. Heat rectifier hub (14) until it slides into place on shaft (9) with slight pressure.

d. d. Once rectifier hub (14) has cooled, pull two main rotor leads through holes in exciter rotor (18).

e. e. Install rotating rectifier (paragraph 4-28.2).

f. f. Install main bearing and generator end bell (paragraph 4-27.2)

g. g. Install generator set housin g rear panel (paragraph 2-18.4)

h. h. Connect negative battery c able. Close battery access door.

4-31 MAIN ROTOR ASSEMBLY (MEP-814A/MEP-814B)

WARNING

High voltage is produced when the generator set is in operation. Never attempt to start the generator set unless it is properly grounded. Failure to comply with this warning can cause injury or death to personnel.

WARNING

Dangerous voltage exists on live circuits. Always observe precautions and never work alone. Failure to comply with this warning can cause injury or death to personnel.

WARNING

Ensure that the engine cannot be started while maintenance is being performed. (ENGINE CONTROL switch set to OFF/RESET; Battery Disconnect Switch is OFF; DEAD CRANK SWITCH is OFF). Failure to comply with this warning can cause injury or death to personnel.

WARNING

DC voltages are present at generator set electrical components even with generator set shut down. Avoid shorting any positive with ground/negative. Failure to comply with this warning can cause injury to personnel, and damage to equipment.

4-31.1 Removal.

WARNING

Many components require a two-person lift. Lifting heavy components can cause back strain. Ensure proper lifting techniques are used when lifting heavy components. Failure to comply with this warning can cause injury to personnel.

WARNING

The generator set, engine, and generator are extremely heavy and require an assistant and a lifting device (forklift, overhead lifting device) with sufficient capacity. Failure to comply with this warning can cause serious injury or death to personnel.

WARNING

All metal jewelry can conduct electricity and become entangled in generator set components. Remove all jewelry when working on generator set. Failure to comply with this warning can cause injury or death to personnel.

WARNING

DO NOT wear loose clothing when performing checks, services and maintenance. Failure to comply with this warning can cause injury or death to personnel.

a. a. Shut down generator set.

b. b. Remove generator assembly from generator set (paragraph 4-17.1)

NOTE

Mark location of two spacers, washers, and bolts to ensure correct positioning during installation procedures.

c. c. Remove bolts (21 and 23, Figure 4-15), washers (22 and 24), spacers (25), and drive disc (26) from drive hub (27).

d. d. Attach a suitable rotor lifting fixture to drive hub (27) and overhead hoist. Refer to Figure 4-13.

e. e. Remove generator end bell (paragraph 4-27.1).

CAUTION

Special care should be taken when removing main rotor. Winding damage could result if rotor is allowed to hit main stator.

f. f. Carefully remove main rotor (29) and attach ed components from generator housing (10).

g. g. Remove main bearing (paragraph 4-27.1).

h. h. Remove rotating rectifier (paragraph 4-28.1).

i. i. Remove exciter rotor (paragraph 4-30.1).

. Using bearing puller, remove drive hub (27) and key (28) from main rotor shaft (9).

k. Remove bolts (30), nuts (31), and fan halves (32) from main rotor shaft (9).

4-31.2 Testing.

a. Shut down generator set. Allow generator to cool to ambient temperature.

b. b. Open battery access door and disconnect negative battery cable.

c. c. Remove generator set housin g rear panel (paragraph 2-18.1).

d. d. Remove generator end bell plates and screens (paragraph 4-27.1, step d).

NOTE

It will be necessary to bar (turn) engine in order to position a specific area of the rotating rectifier at one of the end bell access holes.

e. e. Tag and disconnect main rotor leads (2 places) fr om rotating rectifier by removing screws (1, Figure 4-14) and lock washers (2). Discard lock washers (2).

f. f. Set multimeter for ohms and connect between disconnected main rotor leads. Multimeter reading should be as shown in Table 4-3.

g. g. Reading other than show in Table 4-3 indicates that shorted or open windings and main rotor must be replaced.

h. h. Connect multimeter between each main rotor lead and exciter stator mounting bolt in turn.

i. i. Reading of less than infinity indicates defective ground insulation and main rotor must be replaced.

j. j. Connect main rotor leads to rotating rect ifier with new lock washers (2) and nuts (1).

k. k. Install generator end bell plates and screens (paragraph 4-27.2, step e).

l. l. Install generator set housin g rear panel (paragraph 2-18.4).

m. m. Connect negative battery c able. Close battery access door.

4-31.3 Installation.

a. a. Place key (28, Figure 4-15) on main rotor shaft (9) and press drive hub (27) over key (28) and onto shaft (9).

NOTE

Ensure gap between fan halves are equal from side-to-side.

b. b. Install fan halves (32) onto shaft (9) with bolts (30) and nuts (31).

c. c. Install exciter rotor (paragraph 4-30.3).

d. d. Install rotating rectifier (paragraph 4-28.2).

e. e. Install main bearing (paragraph 4-27.2).

CAUTION

Special care must be taken when installing main rotor. Winding damage could result if rotor is allowed to hit main stator.

f. f. Attach a suitable rotor lifting figure to drive hub (27) and overhead hoist. Refer to Figure 4-13.

g. g. Carefully install rotor (29) and attached components into generator housing (10).

h. h. Install generator end bell (paragraph 4-27.2) and remove rotor lifting figure.

NOTE

Ensure correct positioning of spacers, washers, and bolts as marked on removal.

i. i. Install drive disc (26) on drive hub (27) with spacers (25), washers (22 and 24), and bolts (21and 23). Torque bolts (21) to 28 ft-lbs (38 Nm) and bolts (23) to 35 ft-lbs (47 Nm).

j. j. Install generator assembly in generator set (paragraph 4-17.2).

4-32 MAIN STATOR (MEP-814A/MEP-814B)

WARNING

High voltage is produced when the generator set is in operation. Never attempt to start the generator set unless it is properly grounded. Failure to comply with this warning can cause injury or death to personnel.

WARNING

Dangerous voltage exists on live circuits. Always observe precautions and never work alone. Failure to comply with this warning can cause injury or death to personnel.

WARNING

Ensure that the engine cannot be started while maintenance is being performed. (ENGINE CONTROL switch set to OFF/RESET; Battery Disconnect Switch is OFF; DEAD CRANK SWITCH is OFF). Failure to comply with this warning can cause injury or death to personnel.

WARNING

DC voltages are present at generator set electrical components even with generator set shut down. Avoid shorting any positive with ground/negative. Failure to comply with this warning can cause injury to personnel, and damage to equipment.

4-32.1 Removal.

WARNING

Many components require a two-person lift. Lifting heavy components can cause back strain. Ensure proper lifting techniques are used when lifting heavy components. Failure to comply with this warning can cause injury to personnel.

WARNING

The generator set, engine, and generator are extremely heavy and require an assistant and a lifting device (forklift, overhead lifting device) with sufficient capacity. Failure to comply with this warning can cause serious injury or death to personnel.

WARNING

All metal jewelry can conduct electricity and become entangled in generator set components. Remove all jewelry when working on generator set. Failure to comply with this warning can cause injury or death to personnel.

WARNING

DO NOT wear loose clothing when performing checks, services and maintenance. Failure to comply with this warning can cause injury or death to personnel.

a. a. Shut down generator set.

b. b. Remove generator assembly from generator set (paragraph 4-17.1)

c. c. Remove main rotor and attached rotating components (paragraph 4-31.1).

d. d. Remove bolts (35, Figure 4-15), lock washers (36), lead clamp assembly (37), and gasket (38) from generator housing (10). Discard lock washers (36).

e. e. Remove setscrews (33) and main stator (34) from housing assembly (10).

4-32.2 <u>Testing</u>.

a. a. Shut down generator set. Allow generator to cool to ambient temperature.

b. b. Open battery access door and disconnect negative battery cable.

c. c. Remove protective cover and moveable terminal board from voltage reconnection board (paragraph 2-107.2).

d. d. Disconnect two electrical connectors from bottom of control box assembly.

e. e. Tag and disconnect wires 107C and 109J fr om terminals 1 and 2 of power potential transformer (37, Figure 4-3).

f. f. Connect resistance bridge between terminals T1 and T4, T2 and T5, T3 and T6, T7 and T0, T8 and T0, and T9 and T0 of voltage reconnection board.

g. g. All resistance readings should be as shown in Table 4-3.

h. h. If resistance is low, there are shorted windings. If resistance is high, stator windings are open. In either case, stator must be replaced.

i. i. Disconnect removable grounding link (17, Figure 2-34) on output load terminal board.

j. j. Set multimeter for ohms and connect between each coil lead and ground in turn.

k. k. If multimeter indicates resistance on any connection, stator windings are grounded and stator must be replaced.

m. m. Connect wires 107C and 109J to terminals 1 and 2 of power potential transformer (37, Figure 4-3).

Remove tags.

n. n. Connect two electrical connectors to bottom of control box assembly.

o. o. Install moveable terminal board and protective cover on voltage reconnection board (paragraph 2-107.3).

p. p. Connect negative battery c able. Close battery access door.

4-32.3 Installation.

a. a. Install main stator (34, Figure 4-15) in housing assembly (10) with setscrews (33).

b. b. Pull main stator (34) leads through side of housing (10).

c. c. Position exciter stator (17) leads and main stator (34) leads through gasket (38) and in lead clamp assembly (37).

d. d. Install gasket (38) and lead clamp assembly (3 7) on housing (10) with bolts (35) and new lock washers (36).

e. e. Install main rotor and attached rotati ng components (paragraph 4-31.3).

f. f. Install generator assembly in generator set (paragraph 4-17.2).

4-33 GENERATOR HOUSING (MEP-814A/MEP-814B)

WARNING

High voltage is produced when the generator set is in operation. Never attempt to start the generator set unless it is properly grounded. Failure to comply with this warning can cause injury or death to personnel.

WARNING

Dangerous voltage exists on live circuits. Always observe precautions and never work alone. Failure to comply with this warning can cause injury or death to personnel.

WARNING

Ensure that the engine cannot be started while maintenance is being performed. (ENGINE CONTROL switch set to OFF/RESET; Battery Disconnect Switch is OFF; DEAD CRANK SWITCH is OFF). Failure to comply with this warning can cause injury or death to personnel.

WARNING

DC voltages are present at generator set electrical components even with generator set shut down. Avoid shorting any positive with ground/negative. Failure to comply with this warning can cause injury to personnel, and damage to equipment.

4-33.1 Removal.

WARNING

Many components require a two-person lift. Lifting heavy components can cause back strain. Ensure proper lifting techniques are used when lifting heavy components. Failure to comply with this warning can cause injury to personnel.

WARNING

The generator set, engine, and generator are extremely heavy and require an assistant and a lifting device (forklift, overhead lifting device) with sufficient capacity. Failure to comply with this warning can cause serious injury or death to personnel.

WARNING

All metal jewelry can conduct electricity and become entangled in generator set components. Remove all jewelry when working on generator set. Failure to comply with this warning can cause injury or death to personnel.

WARNING

DO NOT wear loose clothing when performing checks, services and maintenance. Failure to comply with this warning can cause injury or death to personnel.

a. a. Shut down generator set.

b. b. Remove generator assembly from generator set (paragraph 4-17.1)

c. c. Remove main rotor and attached rotating components (paragraph 4-31.1).

d. d. Remove exciter stator (paragraph 4-29.1).

e. e. Remove main stator (paragraph 4-32.1).

f. f. Remove bolts (39, Figure 4-15), lock washers (40), and plate (41) from housing assembly (10). Discard lock washers (40).

g. g. Remove bolts (42), lock washers (43), and screen (44) from housing (10). Discard lock washers (43).

h. h. Remove nuts (45 and 48), washer (46), lock washers (47 and 49), and grounding stud (50) from housing (10). Discard lock washers (47 and 49).

4-33.2 Installation.

a. a. Install grounding stud (50, Figure 4-15), new lock washers (47 and 49), washer (46), and nuts (45 and 48) on housing (10).

b. b. Install screen (44) on housing (10) with bolts (42) and new lock washers (44).

c. c. Install plate (41) on housing assembly (10) with screws (39) and new lock washers (40).

d. d. Install exciter stator (paragraph 4-29.3).

e. e. Install main stator (paragraph 4-32.3).

f. f. Install main rotor and attached rotati ng components (paragraph 4-31.3).

g. g. Install generator assembly in generator set (paragraph 4-17.2)

Section VII. MAINTENANCE OF SKID BASE

4-34 SKID BASE 4-34.1

Removal.

WARNING

Many components require a two-person lift. Lifting heavy components can cause back strain. Ensure proper lifting techniques are used when lifting heavy components. Failure to comply with this warning can cause injury to personnel.

WARNING

The generator set, engine, and generator are extremely heavy and require an assistant and a lifting device (forklift, overhead lifting device) with sufficient capacity. Failure to comply with this warning can cause serious injury or death to personnel.

WARNING

All metal jewelry can conduct electricity and become entangled in generator set components. Remove all jewelry when working on generator set. Failure to comply with this warning can cause injury or death to personnel.

WARNING

DO NOT wear loose clothing when performing checks, services and maintenance. Failure to comply with this warning can cause injury or death to personnel.

WARNING

Support components when removing attaching hardware or component may fall. Failure to comply with this warning can cause injury to personnel, and damage to equipment.

a. a. Shut down generator set.

b. b. Remove engine and generator assembly (paragraph 3-6.1 (MEP-804A/MEP-814A)/paragraph
3-7.1
(MEP-804B/MEP-814B)).

c. c. Remove fuel tank (paragraph 4-6.2).

d. d. Remove nuts (1, Figure 4-17), lock washers (2), bolts (3), washers (4), and forklift guides (5) from skid base. Discard lock washers (2).

4-34.2 Repair. Repair of skid base will be limited to corrosion control and spot welding minor cracks. If major

WARNING

CARC paint is a health hazard, and is irritating to eyes, skin, and respiratory system. Wear protective eyewear, mask, and gloves when applying or removing CARC paint. Failure to comply with this warning can cause injury to personnel.

WARNING

Solvent used to clean parts is potentially dangerous to personnel and property. Clean parts in a well-ventilated area. Avoid inhalation of solvent fumes. Wear goggles and rubber gloves to protect eyes and skin. Wash exposed skin thoroughly. DO NOT smoke or use near open flames or excessive heat. Failure to comply with this warning can cause injury to personnel, and damage to the equipment.

WARNING

**Eye protection is required when working with compressed air.
Compressed air can propel particles at high velocity and injure eyes. Do not exceed 15 psi pressure when using compressed air. Failure to comply with this warning can cause injury to personnel.**

WARNING

Avoid breathing fumes generated by soldering. Eye protection is required. Good general ventilation is normally adequate. Failure to comply with this warning can cause injury to personnel.

4-34.3 Installation.

a. a. Install forklift guides (5, Figure 4-17) in skid base with bolts (3), washers (4), new lock washers (2), and nuts (1).

b. b. Install fuel tank (paragraph 4-6.3).

c. c. Install engine and generator assembly (paragraph 3-6.2 (MEP-804A/MEP-814A)/paragraph 3-7.2

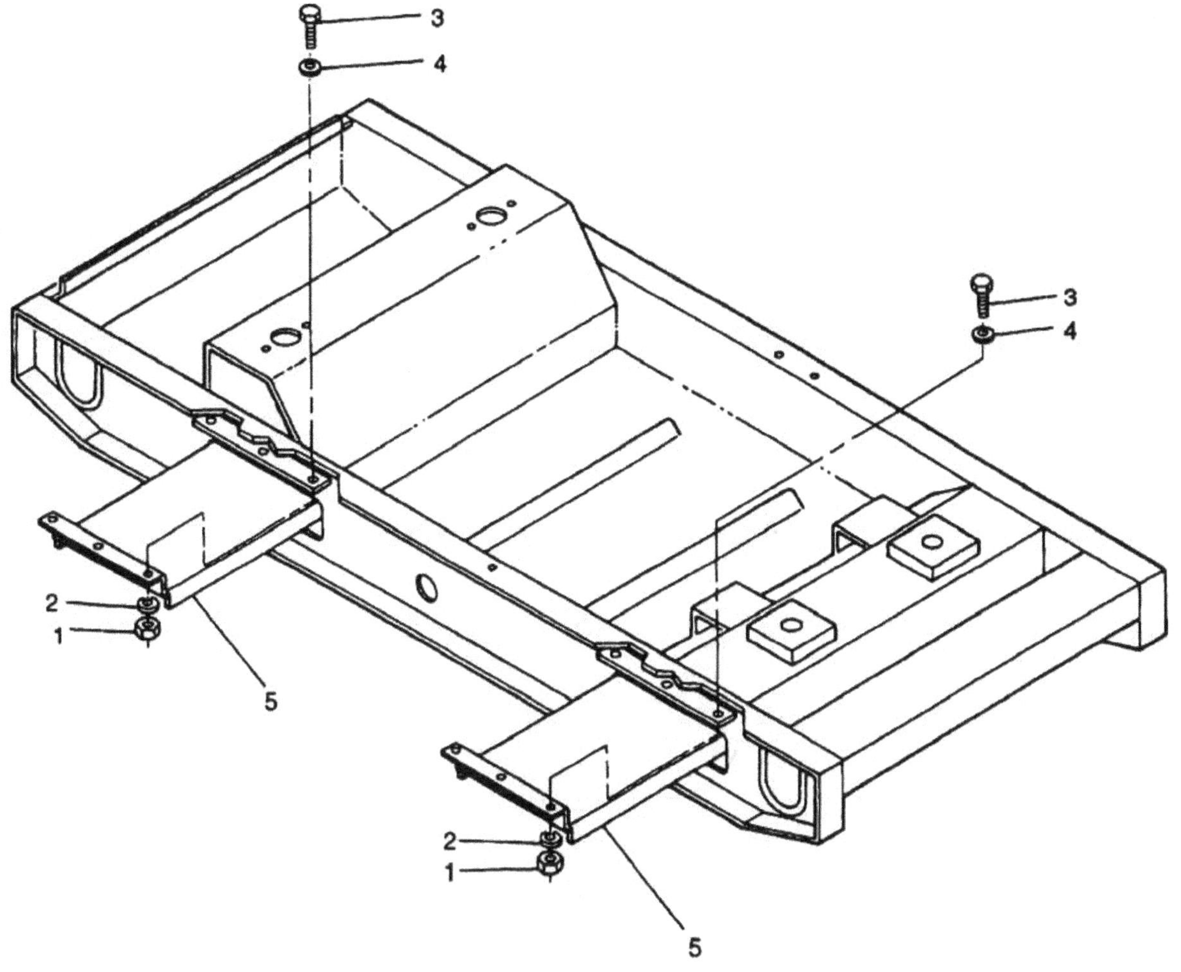

Figure 4-17. Skid Base.

CHAPTER 5

RE-ENGINE INSTRUCTIONS

Subject Index Page

Section I. INTRODUCTION

5-1 SCOPE

This section provides instructions for removing the Isuzu engine and associated components from an MEP-804A/MEP-814A Tactical Quiet (TQ) Generator Set and installing a Yanmar engine and associated components, converting the generator set to an MEP-804B/MEP-814B. The Isuzu C240 engine in the MEP-804A/MEP-814A is no longer manufactured. The replacement engine is a Yanmar engine model 4TNV84T-DFM. This chapter contains instructions to install a Yanmar engine into a MEP-804A/MEP-814A converting it to a MEP-804B/MEP-814B.

In order to perform the re-engine procedures, two items are required: One Yanmar model number 4TNV84T-DFM, turbocharged diesel engine, NSN 2815-01-538-4237, and one Field Installation Kit, Engine (FIKE), part number 97-24103-01 (MEP-804 Generator Set) or 97-24103-02 (MEP-814 Generator Set).

Section II. ENGINE REMOVAL

5-2 ENGINE REMOVAL PROCEDURES

WARNING

All metal jewelry can conduct electricity and become entangled in generator set components. Remove all jewelry when working on generator set. Failure to comply with this warning can cause injury or death to personnel.

WARNING

DO NOT wear loose clothing when performing checks, services and maintenance. Failure to comply with this warning can cause injury or death to personnel.

WARNING

Dangerous voltage exists on live circuits. Always observe precautions and never work alone. Failure to comply with this warning can cause injury or death to personnel.

WARNING

DC voltages are present at generator set electrical components even with generator set shut down. Avoid shorting any positive with ground/negative. Failure to comply with this warning can cause injury to personnel, and damage to equipment.

WARNING

Slave receptacle (NATO connector) is electrically live at all times and is unfused. The Battery Disconnect Switch does not remove power from the slave receptacle. NATO slave receptacle has 24 VDC even when Battery Disconnect Switch is set to OFF. This circuit is only dead when the batteries are fully disconnected. Disconnect the batteries before performing maintenance on the slave receptacle. Failure to comply with this warning can cause injury or death to personnel.

WARNING

Ensure that the engine cannot be started while maintenance is being performed. (ENGINE CONTROL switch set to OFF/RESET; Battery Disconnect Switch is OFF; DEAD CRANK SWITCH is OFF). Failure to comply with this warning can cause injury or death to personnel.

WARNING

When disconnecting or removing batteries, disconnect the negative lead that connects directly to the grounding stud first; disconnect the negative end of the interconnection cable next. When installing batteries, reverse the connection sequence. Failure to comply with this warning can cause injury to personnel.

WARNING

Many components require a two-person lift. Lifting heavy components can cause back strain. Ensure proper lifting techniques are used when lifting heavy components. Failure to comply with this warning can cause injury to personnel.

WARNING

Each battery weighs more than 70 pounds (32 kg) and requires a two-person lift. Lifting batteries can cause back strain. Ensure proper lifting techniques are used when lifting batteries. Failure to comply with this warning can cause injury to personnel.

WARNING

Support components when removing attaching hardware or component may fall. Failure to comply with this warning can cause injury to personnel, and damage to equipment.

WARNING

The generator set, engine, and generator are extremely heavy and require an assistant and a lifting device (forklift, overhead lifting device) with sufficient capacity. Failure to comply with this warning can cause injury or death to personnel.

WARNING

The connection of any electrical equipment and the disconnection of any electrical equipment may cause an explosion hazard. Do not connect any electrical equipment or disconnect any electrical equipment in an explosive atmosphere. Failure to comply with this warning can cause injury or death to personnel.

WARNING

Rated capacity of overhead hoist should be at least 1,500 pounds (680 kg). Do not use a hoist with less capacity. Failure to comply with this warning can cause injury or death to personnel, and damage to equipment.

WARNING

Always remove radiator cap slowly to permit any pressure to escape. Failure to comply with this warning can cause injury to personnel.

WARNING

The high pressure oil system operates at high temperature and pressure. Contact with hot oil can result in burns and scalding. Shut down generator set, and allow system to cool before performing checks, services, and maintenance. Wear heat resistant gloves and avoid contacting hot surfaces. Do not allow hot oil or components to contact skin or hands. Failure to comply with this warning can cause injury or death to personnel.

CAUTION

Plug/cap all fuel line connections during disassembly to prevent contamina-

NOTE

Unless otherwise directed, retain all removed parts. The retained parts should be stored, labeled and organized according to which subassembly they belong to in order to make the installation process manageable.

Throughout the removal procedure it is very important to tag wires and cables as they are disconnected. Identify each with an appropriate number and where and to which device it connects.

The winterization kit may have to be removed in order to install the new engine. Refer to Chapter 6 for removal instructions.

5-2.1 Removal Preparation Procedures.

WARNING

When disconnecting or removing batteries, disconnect the negative lead that connects directly to the grounding stud first; disconnect the negative end of the interconnection cable next. When installing batteries, reverse the connection sequence. Failure to comply with this warning can cause injury to personnel.

WARNING

Batteries give off a flammable gas. Do not smoke or use open flame when performing maintenance. Failure to comply can cause injury or death to personnel and equipment damage due to flames and explosion.

WARNING

Battery acid can cause burns to unprotected skin. Wear safety goggles and chemical gloves and avoid acid splash while working on batteries. Failure to comply with this warning can cause injury to personnel.

WARNING

The connection of any electrical equipment and the disconnection of any electrical equipment may cause an explosion hazard. Do not connect any electrical equipment or disconnect any electrical equipment in an explosive atmosphere. Failure to comply with this warning can cause injury or death to personnel.

a. a. Set MASTER SWITCH to OFF position.

b. b. Push in EMERGENCY STOP switch.

c. c. Open battery compartment access door.

d. d. Disconnect negative battery cable terminal lug (1, Figure 5-1).

e. e. Disconnect terminals (2 and 3) and remove interconnect battery cable.

f. f. Disconnect positive battery cable terminal lug (4).

g. g. Remove nuts (5), washers (6), lock washers (7), hook bolts (8), and retaining bar (9). Discard lock washers (7).

h. Remove batteries (10 and 11).

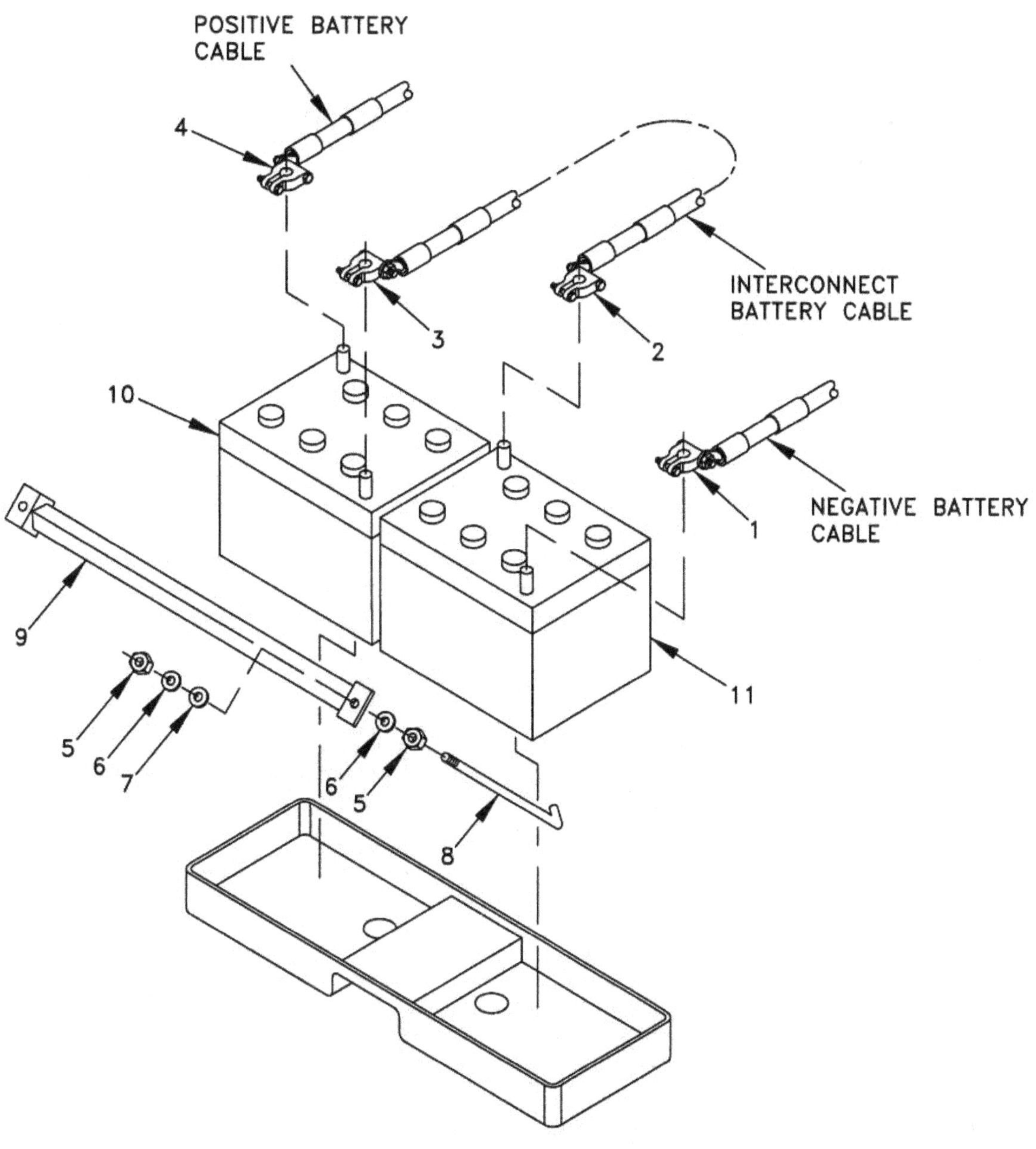

Figure 5-1. Batteries.

5-2.2 Drain Oil, Coolant, and Fuel.

WARNING

Cooling system operates at high temperature and pressure. Contact with high pressure steam and/or liquids can result in burns and scalding. Shut down generator set, and allow system to cool before performing checks, services and maintenance, or wear gloves and additional protective clothing and goggles as required. Failure to comply with this warning can cause injury or death to personnel.

WARNING

Always remove radiator cap slowly to permit pressure to escape. Failure to

WARNING

The high pressure oil system operates at high temperature and pressure. Contact with hot oil can result in burns and scalding. Shut down generator set, and allow system to cool before performing checks, services, and maintenance. Wear heat resistant gloves and avoid contacting hot surfaces. Do not allow hot oil or components to contact skin or hands. Failure to comply with this warning can cause injury or death to personnel.

WARNING

Wear heat resistant gloves and avoid contacting hot metal surfaces with your hands after components have been heated. Wear additional protective clothing as required. Failure to comply with this warning can cause injury to personnel.

WARNING

The connection of any electrical equipment and the disconnection of any electrical equipment may cause an explosion hazard. Do not connect any electrical equipment or disconnect any electrical equipment in an explosive atmosphere. Failure to comply with this warning can cause injury or death to personnel.

WARNING

Diesel fuel is flammable and toxic to eyes, skin, and respiratory tract. Skin and eye protection are required when working in contact with diesel fuel. Avoid repeated or prolonged contact. Provide adequate ventilation. Operators are to wash exposed skin and change chemical soaked clothing promptly if exposed to fuel. Failure to comply with this warning can cause injury or death to personnel.

WARNING

Fuels used in the generator set are flammable. Do not smoke or use open flames when performing maintenance. Failure to comply with this warning can cause injury or death to personnel, and damage to the generator set.

a. a. Using a suitable container, remove oil cap and drain engine oil from generator set (6 quarts (5.7 liters)) (TM 750-254).

b. b. Using a suitable container, drain engine coolant from radiator and engine block (13.5 quarts (12.8 liters)) (TM 750-254).

c. c. Using suitable container, drain fuel from fuel/water separator.

5-2.3 Control Panel Access Door Removal.

WARNING

DC voltages are present at generator set electrical components even with generator set shut down. Avoid shorting any positive with ground/negative. Failure to comply with this warning can cause injury to personnel, and damage to equipment.

WARNING

Dangerous voltage exists on live circuits. Always observe precautions and never work alone. Failure to comply with this warning can cause injury or death to personnel.

WARNING

The connection of any electrical equipment and the disconnection of any electrical equipment may cause an explosion hazard. Do not connect any electrical equipment or disconnect any electrical equipment in an explosive atmosphere. Failure to comply with this warning can cause injury or death to personnel.

a. a. Open control panel access door.

b. b. Remove seven nuts (1, Figure 5-2), screws (2), control panel access door (3), and hinge (4).

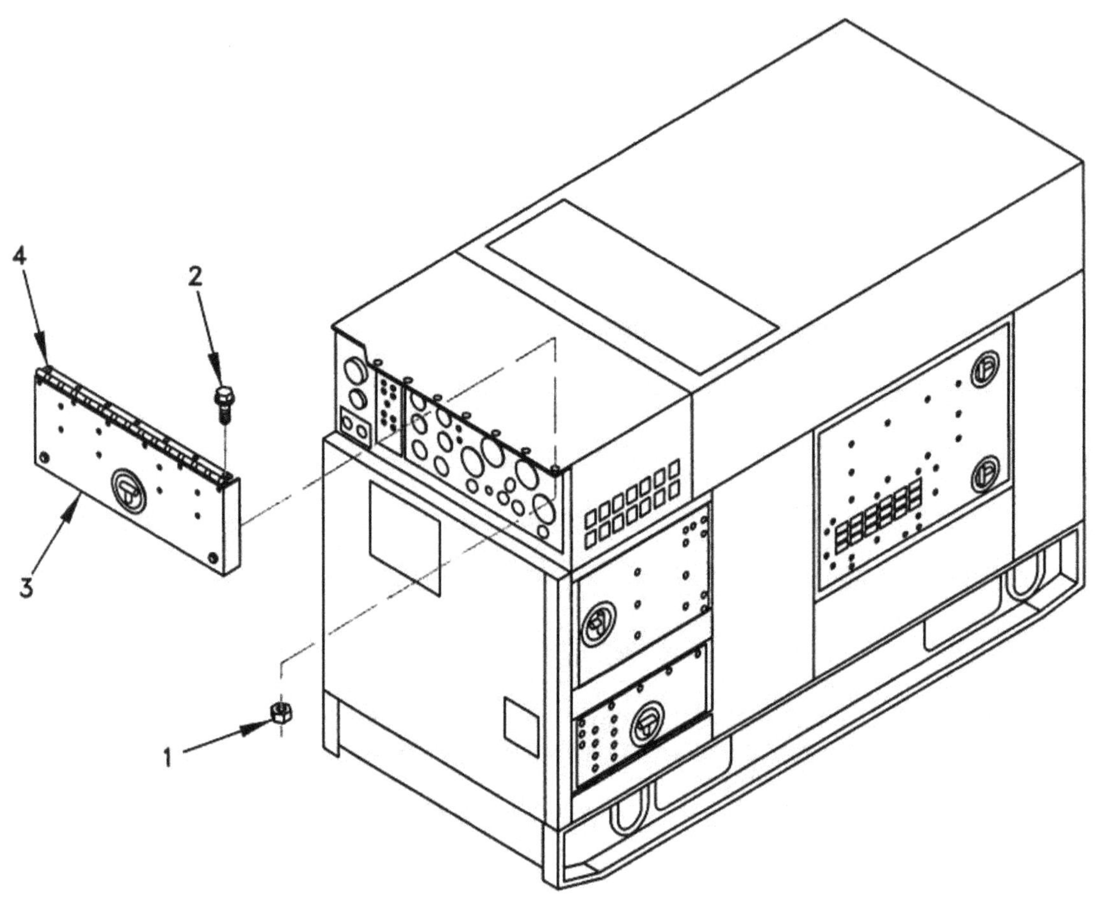

Figure 5-2. Control Panel Access Door.

5-2.4 Control Box Top Panel Removal.

WARNING

DC voltages are present at generator set electrical components even with generator set shut down. Avoid shorting any positive with ground/negative. Failure to comply with this warning can cause injury to personnel, and damage to equipment.

WARNING

Dangerous voltage exists on live circuits. Always observe precautions and never work alone. Failure to comply with this warning can cause injury or death to personnel.

WARNING

The connection of any electrical equipment and the disconnection of any electrical equipment may cause an explosion hazard. Do not connect any electrical equipment or disconnect any electrical equipment in an explosive atmosphere. Failure to comply with this warning can cause injury or death to personnel.

a. a. Release control panel by turning two fa steners and lower control panel carefully.

b. b. Remove 16 bolts (1, Figure 5-3), flat washers (2), lock washers (3), and two nuts (4) from control box top panel (5). Discard lock washers (3).

CAUTION

The control box top panel is attached to the generator set with a silicone sealant to prevent water from entering the control box assembly. Care must be taken not to bend or scratch the control box top panel when separating.

c. c. Separate and remove control box top panel (5) by prying to break loose sealant.

d. d. Inspect gasket (6). Remo ve and replace, as required.

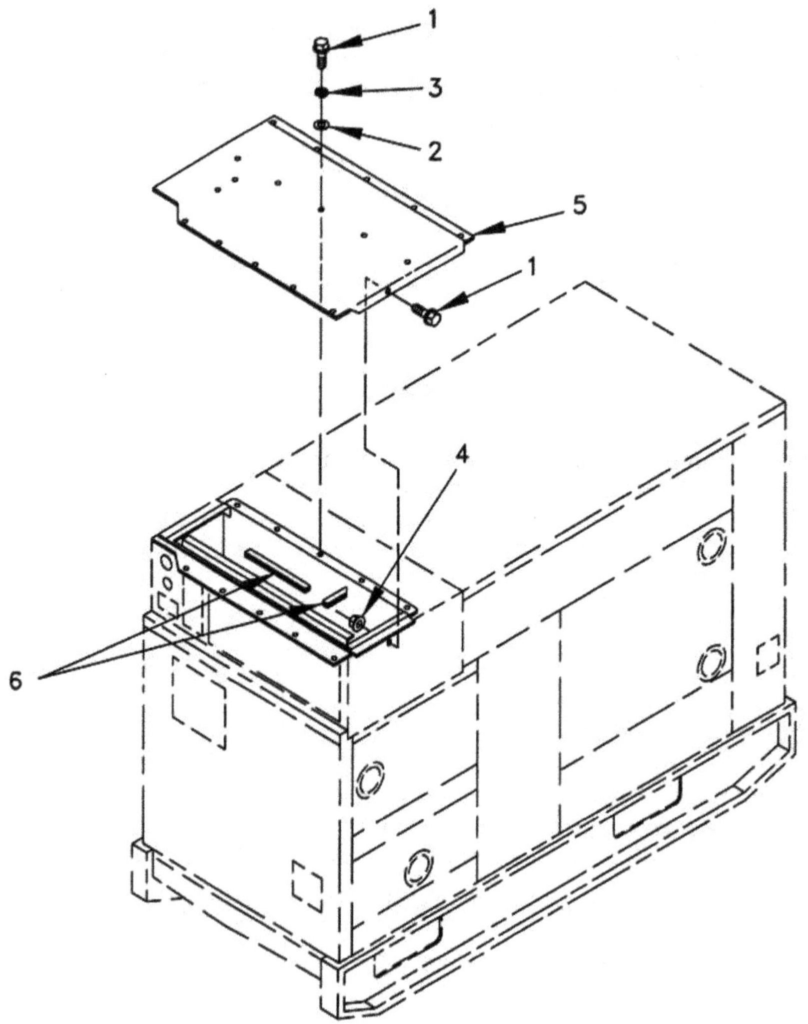

Figure 5-3. Control Box Top Panel.

5-2.5 Control Box Assembly Removal.

WARNING

DC voltages are present at generator set electrical components even with generator set shut down. Avoid shorting any positive with ground/negative. Failure to comply with this warning can cause injury to personnel, and damage to equipment.

WARNING

Dangerous voltage exists on live circuits. Always observe precautions and never work alone. Failure to comply with this warning can cause injury or death to personnel.

WARNING

The connection of any electrical equipment and the disconnection of any electrical equipment may cause an explosion hazard. Do not connect any electrical equipment or disconnect any electrical equipment in an explosive atmosphere. Failure to comply with this warning can cause injury or death to personnel.

a. a. Open output box access door.

b. b. Disconnect wire harness connectors P5 and P6 from control box assembly.

c. c. Remove 20 bolts (1, Figure 5-4), lock washers (2), washers (3), and nuts (4). Output box access door hinge will be loose from generator set. Discard lock washers (2).

d. Remove two bolts (5), lock washers (6), washers (7), nuts (8), and control box assembly (9). Discard lock washers (6).

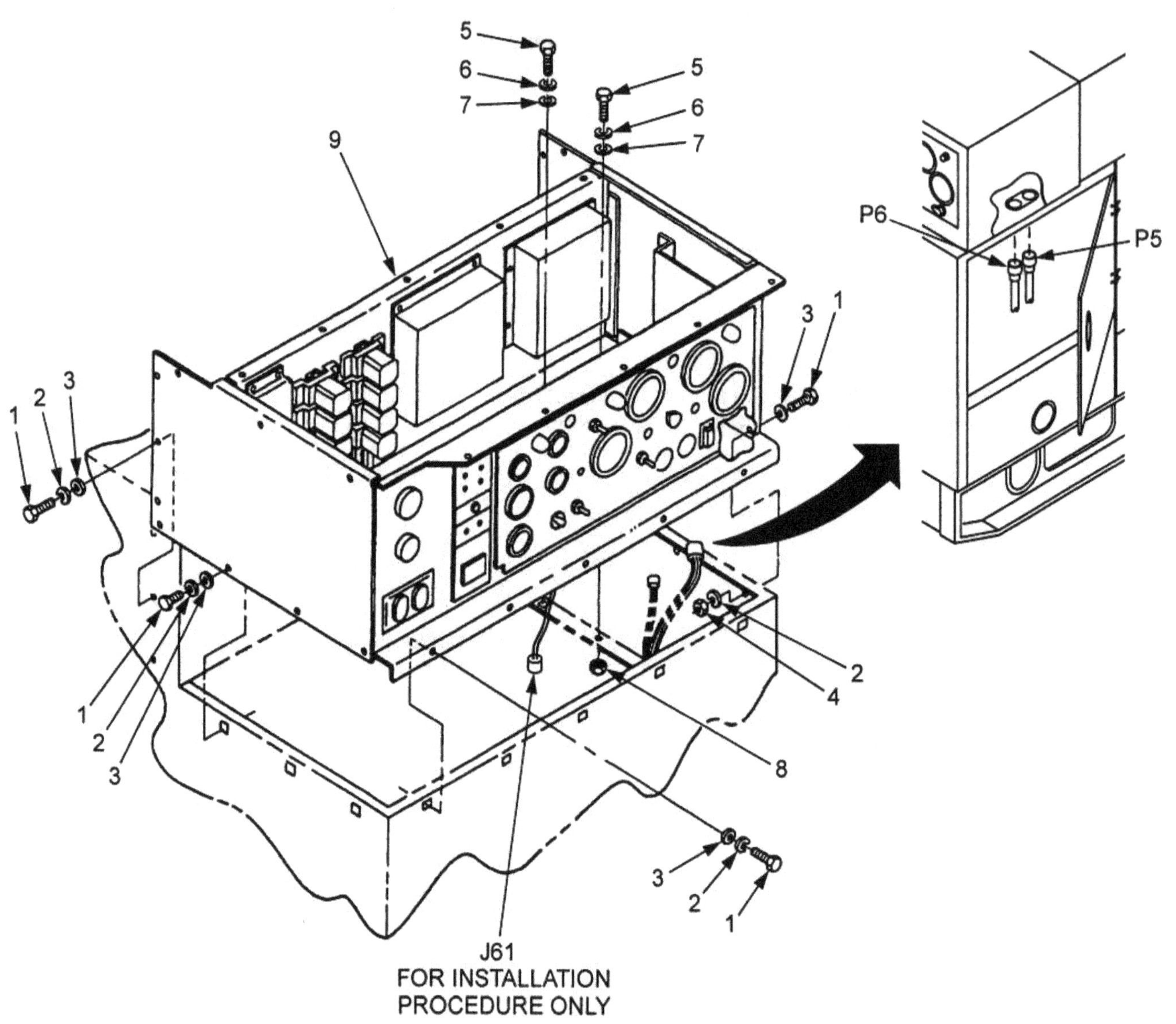

Figure 5-4. Control Box Assembly.

5-2.6 Top Housing Panel, Coolant Filler Hose, and Panel Assembly Removal.

WARNING

All metal jewelry can conduct electricity and become entangled in generator set components. Remove all jewelry when working on generator set. Failure to comply with this warning can cause injury or death to personnel.

WARNING

DO NOT wear loose clothing when performing checks, services and maintenance. Failure to comply with this warning can cause injury or death to personnel.

WARNING

Cooling system operates at high temperature and pressure. Contact with high pressure steam and/or liquids can result in burns and scalding. Shut down generator set, and allow system to cool before performing checks, services and maintenance, or wear gloves and additional protective clothing and goggles as required. Failure to comply with this warning can cause injury or death to personnel.

WARNING

When running, generator set engine has hot metal surfaces that will burn flesh on contact. Shut down generator set and allow engine to cool before performing checks, services, and maintenance. Wear gloves and additional protective clothing as required. Failure to comply with this warning can cause injury or death to personnel.

WARNING

Top housing panels and exhaust system can get very hot. Shut down generator set, and allow system to cool before performing checks, services, and maintenance. Failure to comply with this warning can cause severe burns and injury to personnel.

a. a. Remove 43 bolts (1 and 4, Figure 5-5), lock washers (2 and 5), washers (3 and 6), and top housing panel (7) from generator set. Discard lock washers (2 and 5).

b. b. Loosen clamp (1, Figure 5-6) at radiator (2) and disconnect coolant filler hose and panel assembly (5) from radiator.

c. c. Loosen clamp (3) and disconnect coolant overflow hose (4) from coolant filler hose and panel assembly (5).

d. d. Remove coolant filler hose and panel assembly (5) from generator set.

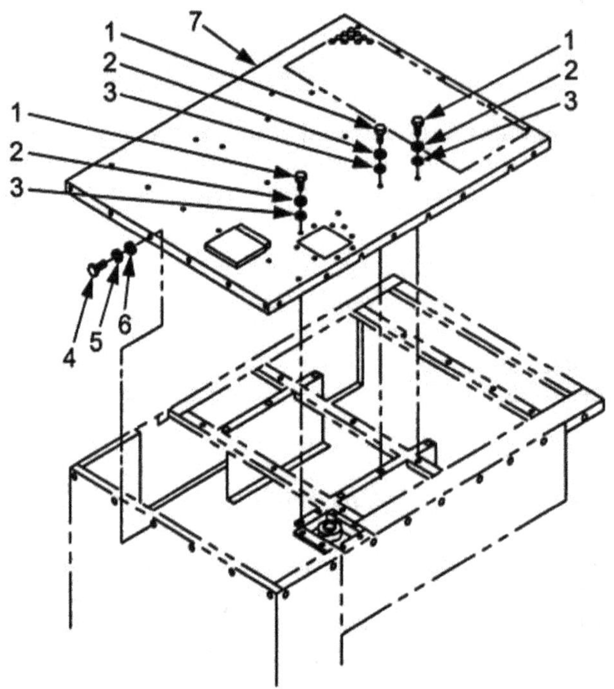

Figure 5-5. Top Housing Panel.

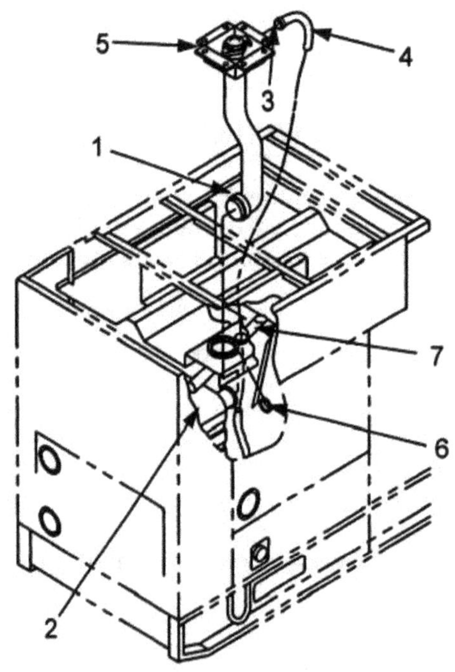

Figure 5-6. Coolant Filler Hose and Panel Assembly.

5-2.7 Coolant Overflow Bottle Removal.

WARNING

All metal jewelry can conduct electricity and become entangled in generator set components. Remove all jewelry when working on generator set. Failure to comply with this warning can cause injury or death to personnel.

WARNING

DO NOT wear loose clothing when performing checks, services and maintenance. Failure to comply with this warning can cause injury or death to personnel.

WARNING

Cooling system operates at high temperature and pressure. Contact with high pressure steam and/or liquids can result in burns and scalding. Shut down generator set, and allow system to cool before performing checks, services and maintenance, or wear gloves and additional protective clothing and goggles as required. Failure to comply with this warning can cause injury or death to personnel.

a. a. Remove clamp (1, Figure 5-7) and disconnect hose (2) from coolant overflow bottle (6).

b. b. Remove two bolts (3), nuts (4), lock washers (5), and coolant overflow bottle (6) and brackets from left side of engine. Discard lock washers (5).

c. Inspect coolant overflow bottle (6) and attaching hoses. Replace, as necessary.

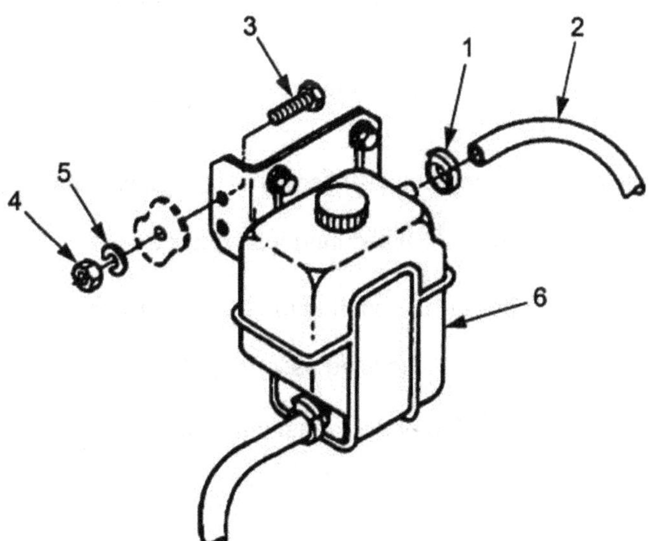

Figure 5-7. Coolant Overflow Bottle.

5-2.8 Top Housing Assembly Removal.

WARNING

All metal jewelry can conduct electricity and become entangled in generator set components. Remove all jewelry when working on generator set. Failure to comply with this warning can cause injury or death to personnel.

WARNING

DO NOT wear loose clothing when performing checks, services and maintenance. Failure to comply with this warning can cause injury or death to personnel.

WARNING

High voltage is produced when this generator set is in operation. Make sure generator set is completely shut down and free of any power source before attempting any repair or maintenance on the set, or when connecting or disconnecting load cables. Failure to comply with this warning can cause injury or death to personnel.

WARNING

Many components require a two-person lift. Lifting heavy components can cause back strain. Ensure proper lifting techniques are used when lifting heavy components. Failure to comply with this warning can cause injury to personnel.

a. a. Remove three nuts (1, Figure 5-8) securing exhaust manifold adapter (2) to engine.

b. b. Remove six nuts (3), lock washers (4), washers 6), and bolts (5) securing top housing assembly (11) to front housing assembly. Discard lock washers (4).

c. c. Remove four assembled nuts (7) and bolts (8) securing top housing assembly (11) to rear side panels.

d. d. Remove assembled nut (9) and bolt (10) securing top housing assembly (11) to output box angle.

e. e. Remove 12 nuts (12) and bolts (13) securing top housing assembly (11) to support bracket.

f. Remove two nuts (14) and bolts (15) securing top housing assembly (11) to end cover.

g. g. Use lifting device and remove top housing assembly (11).

h. h. Remove exhaust pipe clamp (16) and exhaust manifold adapter (2) from flexible muffler tube (17).

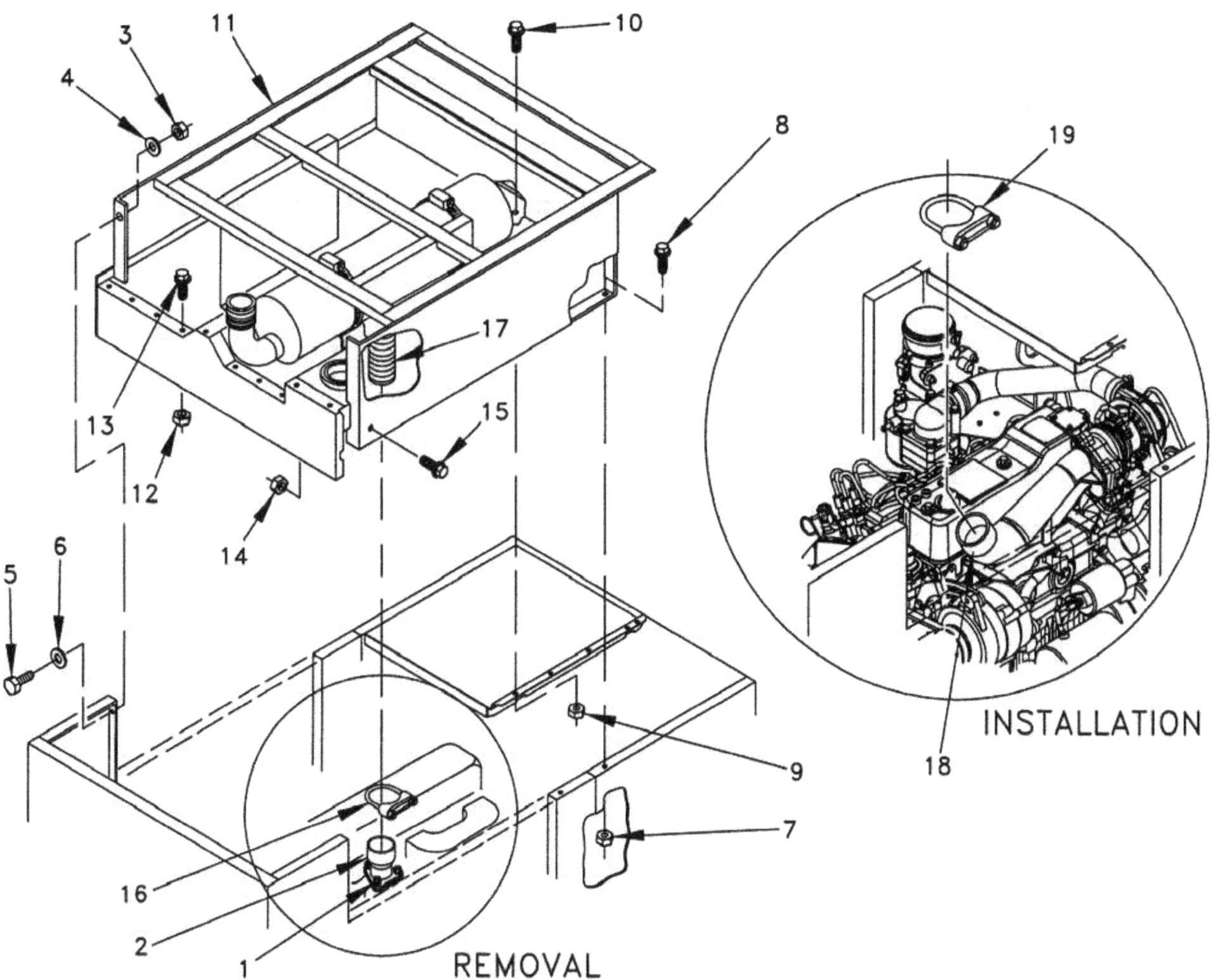

Figure 5-8. Top Housing Assembly.

5-2.9 Front Housing Assembly and Side Panel Removal.

WARNING

All metal jewelry can conduct electricity and become entangled in generator set components. Remove all jewelry when working on generator set. Failure to comply with this warning can cause injury or death to personnel.

WARNING

DO NOT wear loose clothing when performing checks, services and maintenance. Failure to comply with this warning can cause injury or death to personnel.

WARNING

High voltage is produced when this generator set is in operation. Make sure generator set is completely shut down and free of any power source before attempting any repair or maintenance on the set, or when connecting or disconnecting load cables. Failure to comply with this warning can cause injury or death to personnel.

WARNING

Many components require a two-person lift. Lifting heavy components can cause back strain. Ensure proper lifting techniques are used when lifting heavy components. Failure to comply with this warning can cause injury to personnel.

a. a. Remove bolt (1,Figure 5-9) and ground rods (2) from inside left side panel (3).

b. b. Remove six bolts (4), lock washers (5), washers (6), and nuts (7) securing front housing assembly (8) to side panels (3 and 9). Discard lock washers (5).

c. c. Remove two bolts (10), lock washers (11), washers (12), and nuts (13) securing the front housing assembly (8) to skid base (14). Discard lock washers (11).

d. d. Remove six bolts (15), lock washers (16), washers (17), and front housing assembly (8). Discard lock washers (16).

e. e. Remove three bolts (18), lock washers (19), washers (20), and nuts (21) securing right side panel (9) to right rear panel (22). Discard lock washers (19).

f. f. Remove seven bolts (23), lock washers (24), washers (25), and right side panel (9). Discard lock washers (24).

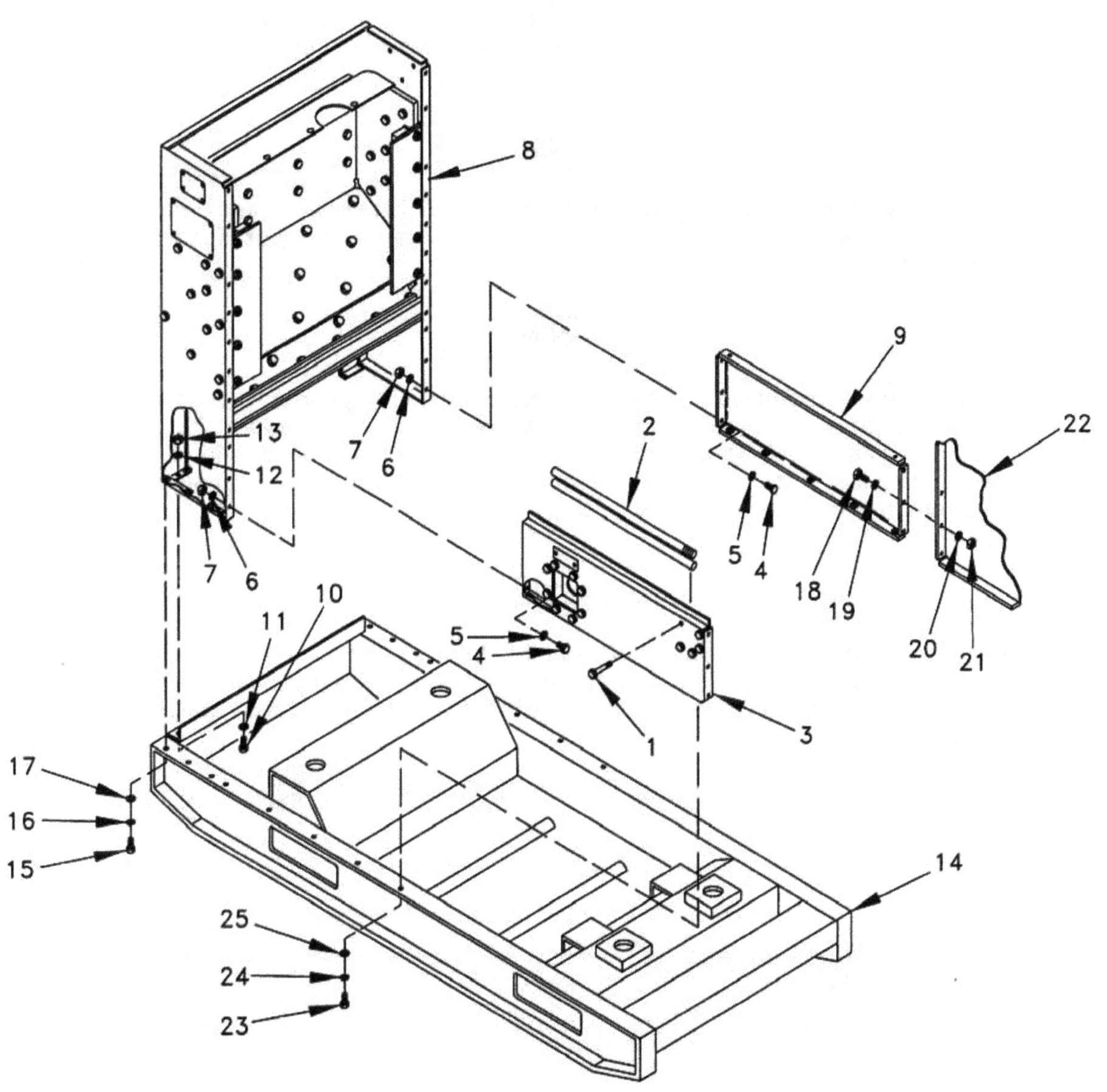

Figure 5-9. Front Housing Assembly and Side Panel.

5-2.10 Rear Housing Panel Removal.

WARNING

All metal jewelry can conduct electricity and become entangled in generator set components. Remove all jewelry when working on generator set. Failure to comply with this warning can cause injury or death to personnel.

WARNING

DO NOT wear loose clothing when performing checks, services and maintenance. Failure to comply with this warning can cause injury or death to personnel.

WARNING

High voltage is produced when this generator set is in operation. Make sure generator set is completely shut down and free of any power source before attempting any repair or maintenance on the set, or when connecting or disconnecting load cables. Failure to comply with this warning can cause injury or death to personnel.

WARNING

Many components require a two-person lift. Lifting heavy components can cause back strain. Ensure proper lifting techniques are used when lifting heavy components. Failure to comply with this warning can cause injury to personnel.

a. a. Remove six bolts (1, Figure 5-10), washers (2), lock washers (3), and nuts (4). Discard lock washers (3).

NOTE

Remove bolts from the bottom of the top.

b. b. Remove 12 bolts (5), lock washers (6), washers (7), and rear housing panel (8) from generator set. Discard lock washers (6).

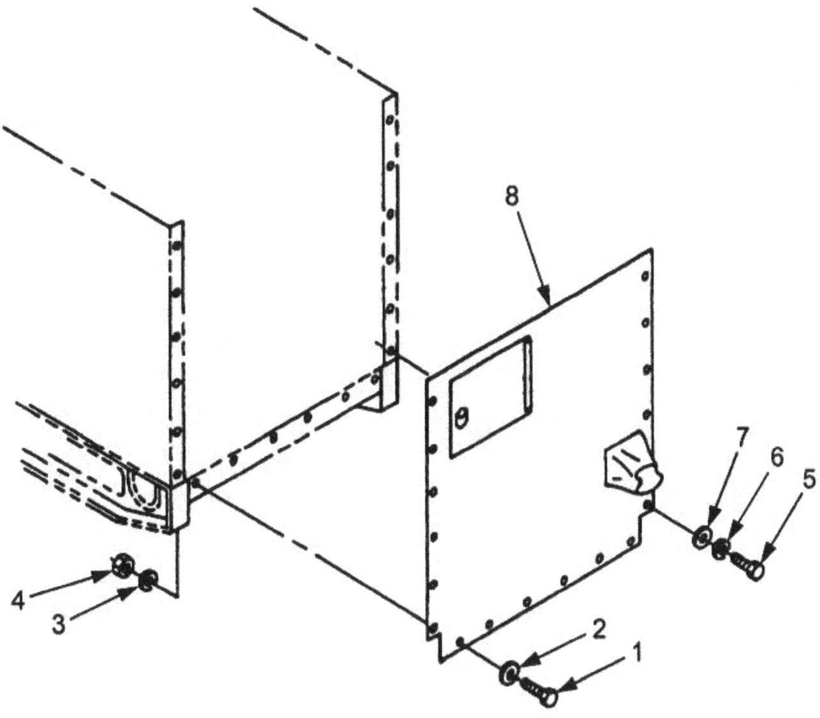

Figure 5-10. Rear Housing Panel.

5-2.11 Fuel Filler Panel Removal.

WARNING

All metal jewelry can conduct electricity and become entangled in generator set components. Remove all jewelry when working on generator set. Failure to comply with this warning can cause injury or death to personnel.

WARNING

DO NOT wear loose clothing when performing checks, services and maintenance. Failure to comply with this warning can cause injury or death to personnel.

WARNING

High voltage is produced when this generator set is in operation. Make sure generator set is completely shut down and free of any power source before attempting any repair or maintenance on the set, or when connecting or disconnecting load cables. Failure to comply with this warning can cause injury or death to personnel.

WARNING

Diesel fuel is flammable and toxic to eyes, skin, and respiratory tract. Skin and eye protection are required when working in contact with diesel fuel. Avoid repeated or prolonged contact. Provide adequate ventilation. Operators are to wash exposed skin and change chemical soaked clothing promptly if exposed to fuel. Failure to comply with this warning can cause injury or death to personnel.

WARNING

Fuels used in the generator set are flammable. Do not smoke or use open flames when performing maintenance. Failure to comply with this warning can cause injury or death to personnel, and damage to the generator set.

a. a. Loosen clamp (1, Figure 5-11) securing fuel filler panel (5) to fuel tank. Cap openings.

b. b. Loosen clamp (2) and disconnect hose (3) from fuel filler panel (5). Cap openings.

c. c. Disconnect fuel line (4) from fuel pump (6). Cap openings.

d. d. Tag and disconnect auxiliary fuel pump (6) electri cal connector P11 from harness connector J11.

e. Tag and disconnect fuel float module (7) electrical connectors J12 and P15 from harness connectors P12 and J15, respectively.

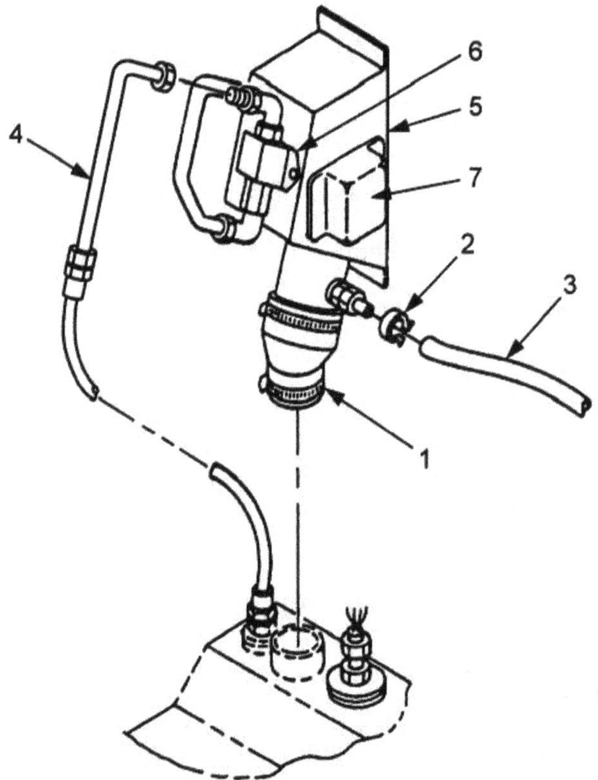

Figure 5-11. Fuel Filler Panel.

WARNING

All metal jewelry can conduct electricity and become entangled in generator set components. Remove all jewelry when working on generator set. Failure to comply with this warning can cause injury or death to personnel.

WARNING

DO NOT wear loose clothing when performing checks, services and maintenance. Failure to comply with this warning can cause injury or death to personnel.

WARNING

High voltage is produced when this generator set is in operation. Make sure generator set is completely shut down and free of any power source before attempting any repair or maintenance on the set, or when connecting or disconnecting load cables. Failure to comply with this warning can cause injury or death to personnel.

WARNING

Many components require a two-person lift. Lifting heavy components can cause back strain. Ensure proper lifting techniques are used when lifting heavy components. Failure to comply with this warning can cause injury to personnel.

WARNING

Slave receptacle (NATO connector) is electrically live at all times and is unfused. The Battery Disconnect Switch does not remove power from the slave receptacle. NATO slave receptacle has 24 VDC even when Battery Disconnect Switch is set to OFF. This circuit is only dead when the batteries are fully disconnected. Disconnect the batteries before performing maintenance on the slave receptacle. Failure to comply with this warning can cause injury or death to personnel.

a. a. Remove bolts (1, Figure 5-12), lock washers (2), battery cables (3), and slave cables (4) from NATO slave receptacle (5). Discard lock washers (2).

b. b. Remove four bolts (1, Figure 5-13), lock washers (2), and washers (3) securing left rear side panel (4) to skid base. Discard lock washers (2).

c. c. Remove seven bolts (5), lock washers (6), and washers (7) securing lower left side panel (8) to skid base. Discard lock washers (6).

d. d. With aid of an assistant, remove left rear side panel (4), lower left side panel (8), and door (9) as an assembly.

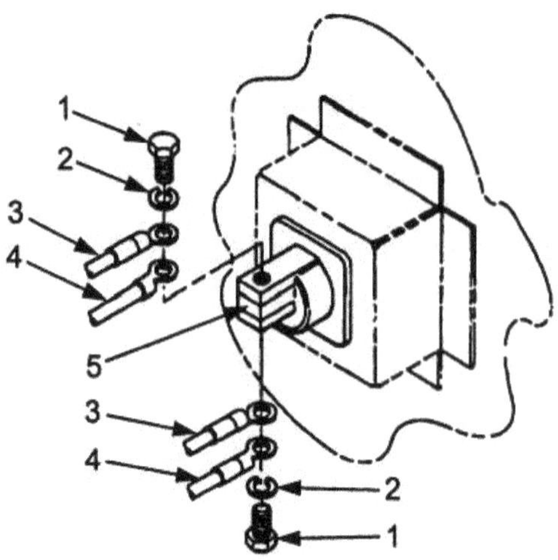

Figure 5-12. NATO Slave Receptacle Wiring.

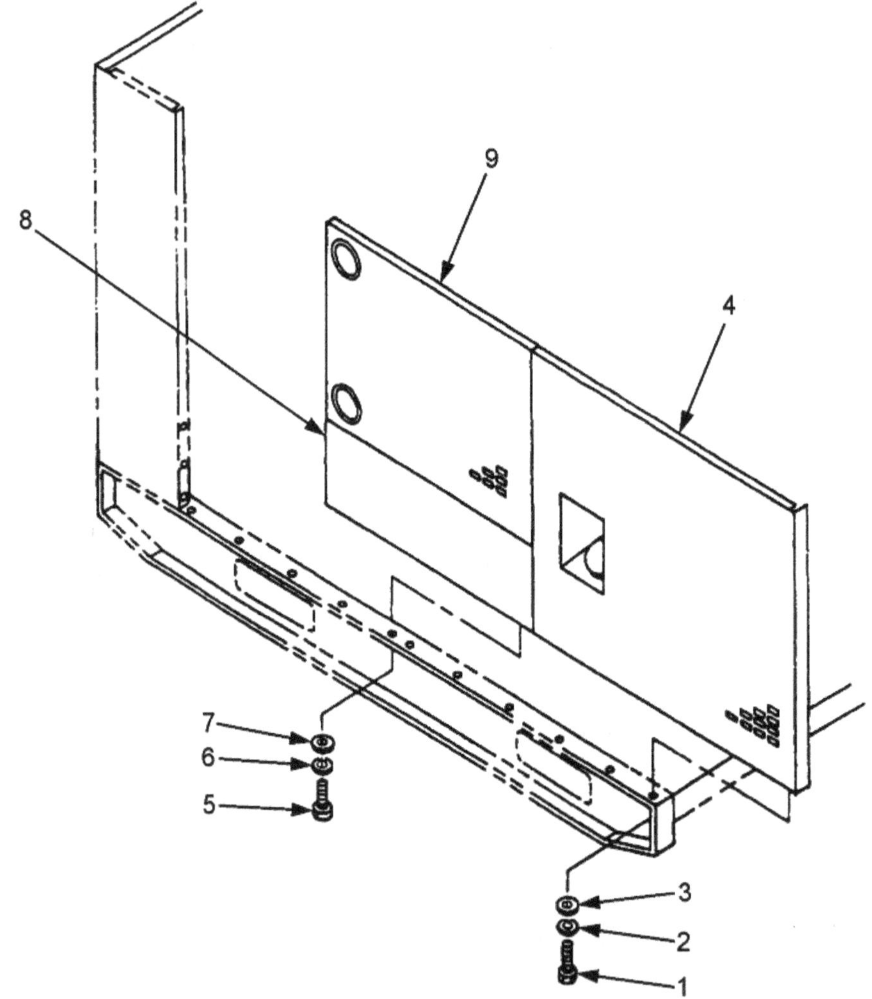

Figure 5-13. Left Rear Side Panel.

5-2.13 Fan Guards and Mounting Brackets Removal.

WARNING

All metal jewelry can conduct electricity and become entangled in generator set components. Remove all jewelry when working on generator set. Failure to comply with this warning can cause injury or death to personnel.

WARNING

DO NOT wear loose clothing when performing checks, services and maintenance. Failure to comply with this warning can cause injury or death to personnel.

WARNING

Cooling system operates at high temperature and pressure. Contact with high pressure steam and/or liquids can result in burns and scalding. Shut down generator set, and allow system to cool before performing checks, services and maintenance, or wear gloves and additional protective clothing and goggles as required. Failure to comply with this warning can cause injury or death to personnel.

WARNING

Fan has sharp blades. Use caution and wear gloves when removing or installing belts. Failure to comply with this warning can cause injury to personnel.

a. a. Remove bolts (1 and 5, Figure 5-14), washers (2 and 6), nuts (3 and 7), and lock washers (4 and 8) securing fans guards (9 and 10). Discard lock washers (4 and 8).

b. b. Remove bolts (11 and 15), lock washers (12 and 16), and washers (17) securing brackets (13, 14, and 18). Discard lock washers (12 and 16).

c. c. Discard fan guards and associated parts.

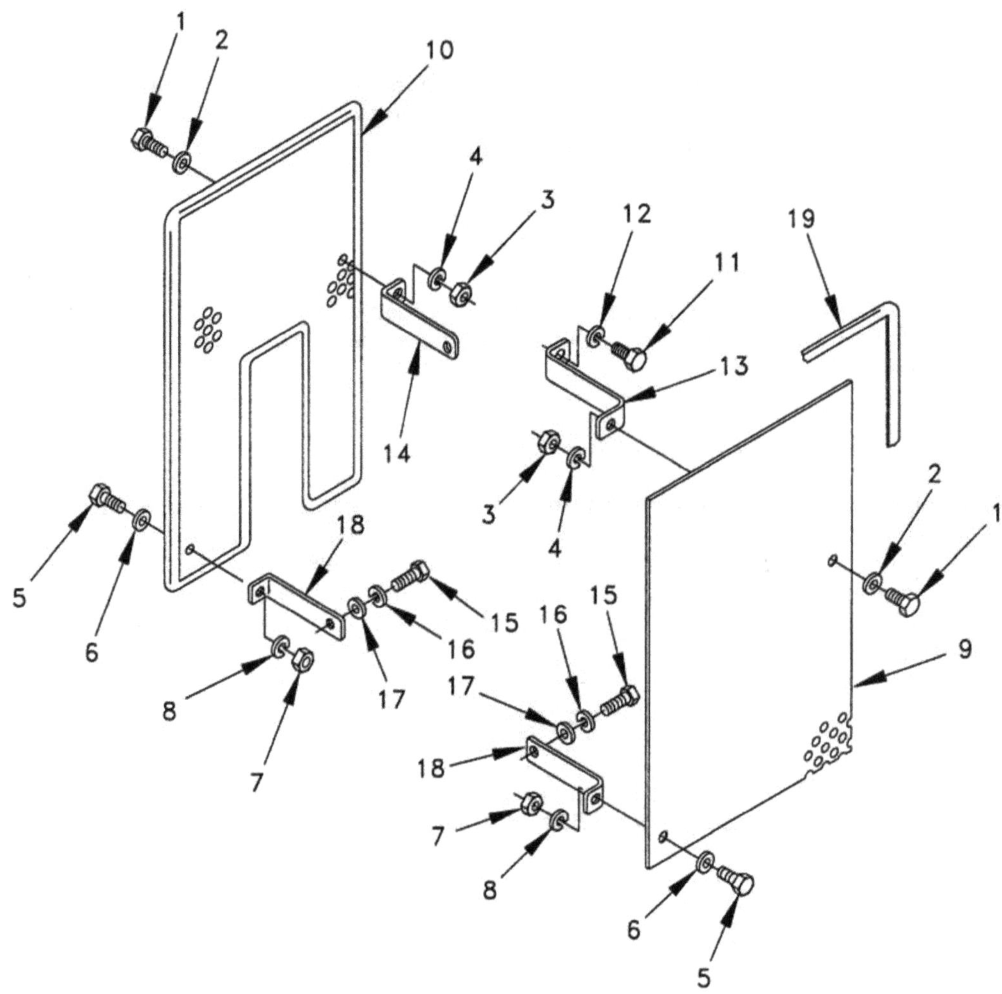

Figure 5-14. Fan Guards and Mounting Brackets.

5-2.14 Radiator Removal.

WARNING

All metal jewelry can conduct electricity and become entangled in generator set components. Remove all jewelry when working on generator set. Failure to comply with this warning can cause injury or death to personnel.

WARNING

DO NOT wear loose clothing when performing checks, services and maintenance. Failure to comply with this warning can cause injury or death to personnel.

WARNING

Cooling system operates at high temperature and pressure. Contact with high pressure steam and/or liquids can result in burns and scalding. Shut down generator set, and allow system to cool before performing checks, services and maintenance, or wear gloves and additional protective clothing and goggles as required. Failure to comply with this warning can cause injury or death to personnel.

WARNING

Fan has sharp blades. Use caution and wear gloves when removing or installing belts. Failure to comply with this warning can cause injury to personnel.

NOTE

Have container ready to capture any residual coolant from hoses during removal.

Some units may have a threaded rubber bushing securing the radiator. If so, remove and discard bushing.

a. a. Remove two nuts (1, Figure 5- 15), lock washers (2), bolts (3), and washers (4) securing support rods (5) to engine lifting bracket (6). Discard lock washers (2).

b. b. Remove two bolts (7), lock washers (8), and support rods (5). Discard lock washers (8).

c. c. Loosen two clamps (9) and remove hose (10).

d. d. Loosen two clamps (11) and remove disconnect hose (12).

e. e. Remove two bolts (13), washers (14), lock washers (15), and nuts (16). Discard lock washers (15).

f. f. Remove eight bolts (17), washers (18), and lock washers (19). Lean shroud halves (20) back against radiator (26). Discard lock washers (19).

g. g. Remove clamp (21) at radiator drain valve (22) and tag and disconnect hose (23).

h. h. Remove two nuts (24) and washers (25) securing radiator (26). With aid of an assistant, remove radiator (26) and shims (27).

i. Retain radiator and mounting hardware. Discard shroud halves, fan, and all other removed parts. Tie tagged hose out of the way. It will be connected to new radiator.

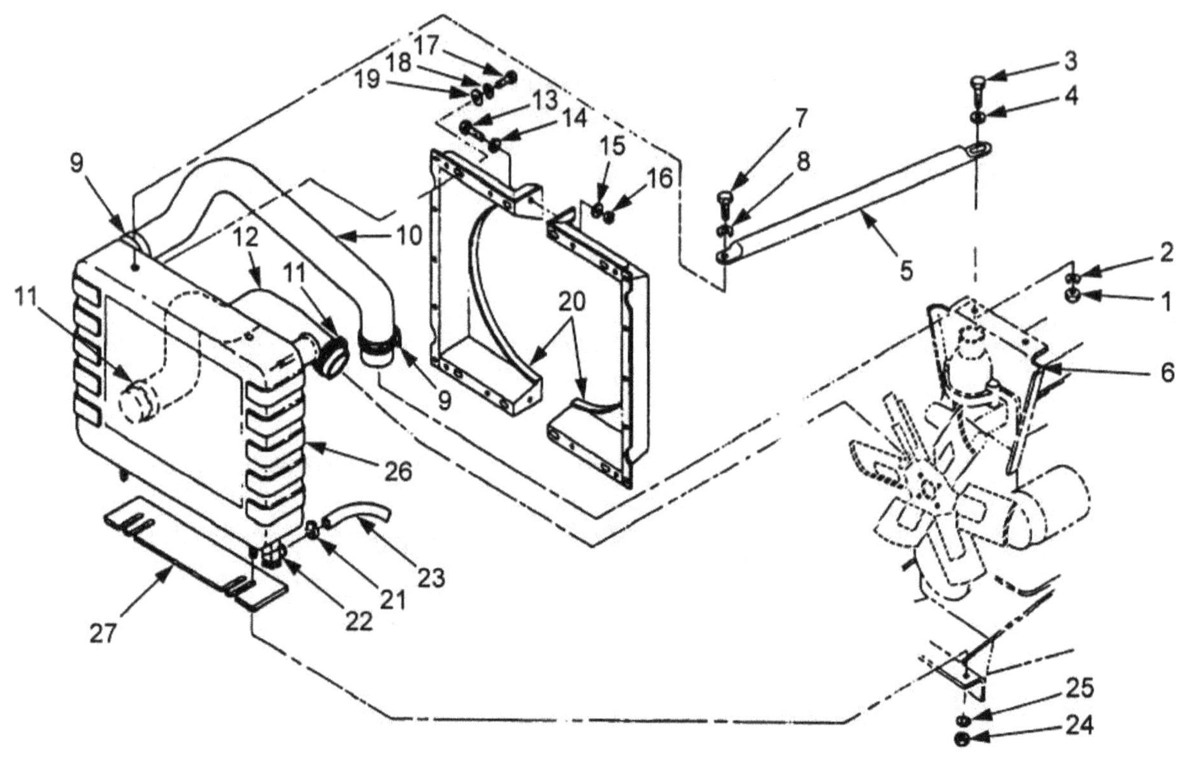

Figure 5-15. Radiator.

5-2.15 Air Intake Hose Removal.

WARNING

All metal jewelry can conduct electricity and become entangled in generator set components. Remove all jewelry when working on generator set. Failure to comply with this warning can cause injury or death to personnel.

WARNING

DO NOT wear loose clothing when performing checks, services and maintenance. Failure to comply with this warning can cause injury or death to personnel.

Loosen clamps (1, Figure 5-16) and remove air intake hoses (2, 3, and 4) as an assembly. Discard all removed items.

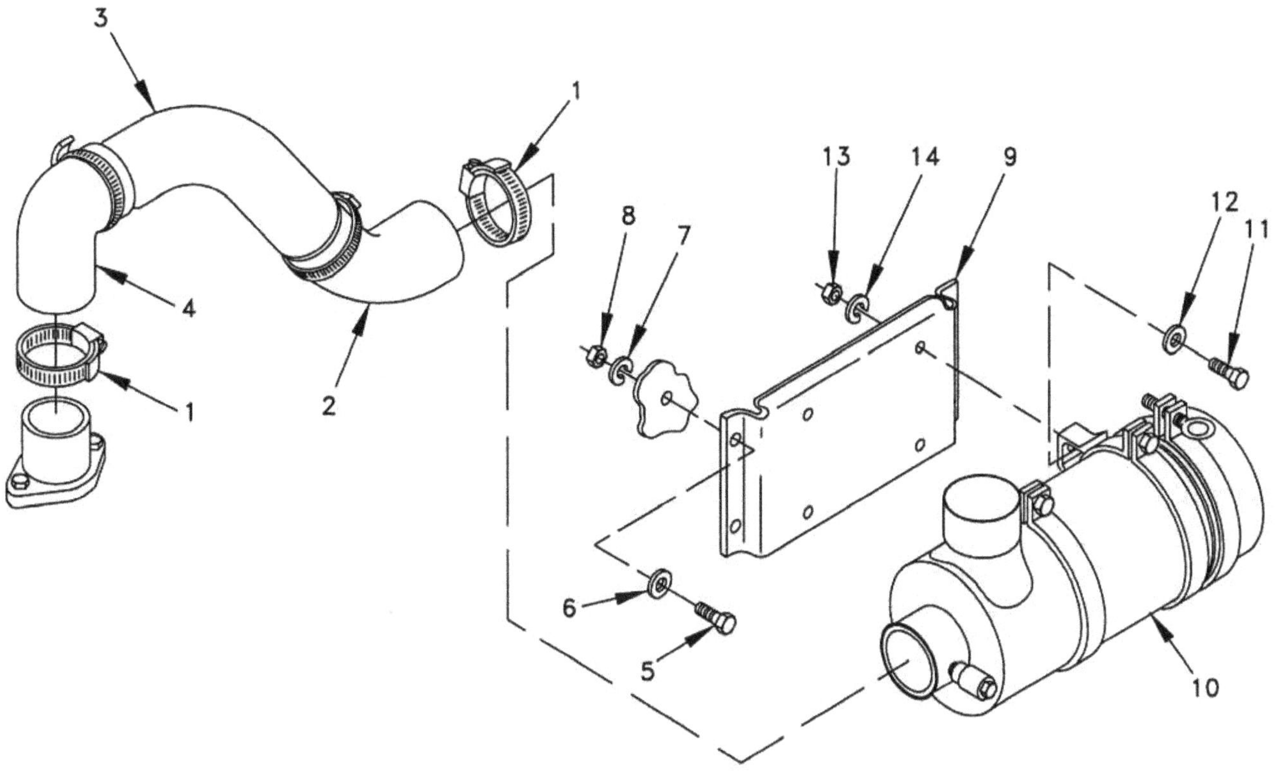

Figure 5-16. Air Intake Hose, Air Cleaner, and Bracket.

5-2.16 Air Cleaner and Bracket Removal.

WARNING

All metal jewelry can conduct electricity and become entangled in generator set components. Remove all jewelry when working on generator set. Failure to comply with this warning can cause injury or death to personnel.

WARNING

DO NOT wear loose clothing when performing checks, services and maintenance. Failure to comply with this warning can cause injury or death to personnel.

a. a. Remove bolts (5, Figure 5-16), washers (6), lock washers (7), nuts (8), mounting bracket (9), and air cleaner assembly (10) from generator set. Discard lock washers (7).

b. b. Remove bolts (11), washers (12), nuts (13), lock washers (14), and mounting bracket (9) from air cleaner assembly (10). Discard lock washers (14).

c. c. Discard mounting bracket. Retain air cleaner assembly and all mounting hardware for installation

5-2.17 Oil Drain Hose Disconnection.

WARNING

All metal jewelry can conduct electricity and become entangled in generator set components. Remove all jewelry when working on generator set. Failure to comply with this warning can cause injury or death to personnel.

WARNING

DO NOT wear loose clothing when performing checks, services and maintenance. Failure to comply with this warning can cause injury or death to personnel.

WARNING

The high pressure oil system operates at high temperature and pressure. Contact with hot oil can result in burns and scalding. Shut down generator set, and allow system to cool before performing checks, services, and maintenance. Wear heat resistant gloves and avoid contacting hot surfaces. Do not allow hot oil or components to contact skin or hands. Failure to comply with this warning can cause injury or death to personnel.

Loosen clamps (1, Figure 5-17) and disconnect oil drain hose (2) from engine oil pan.

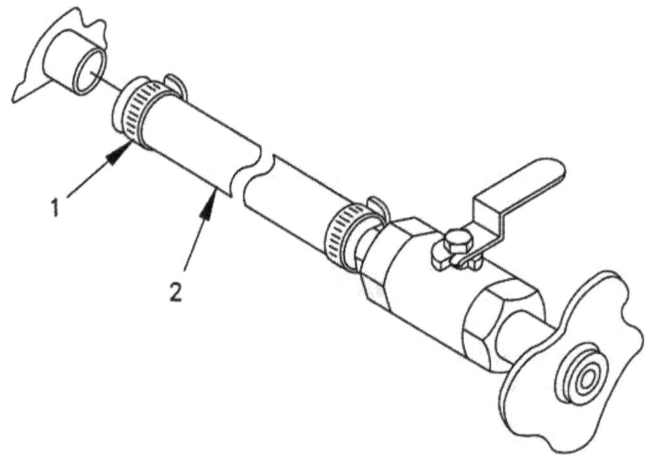

Figure 5-17. Oil Drain Hose.

WARNING

All metal jewelry can conduct electricity and become entangled in generator set components. Remove all jewelry when working on generator set. Failure to comply with this warning can cause injury or death to personnel.

WARNING

DO NOT wear loose clothing when performing checks, services and maintenance. Failure to comply with this warning can cause injury or death to personnel.

WARNING

The high pressure oil system operates at high temperature and pressure. Contact with hot oil can result in burns and scalding. Shut down generator set, and allow system to cool before performing checks, services, and maintenance. Wear heat resistant gloves and avoid contacting hot surfaces. Do not allow hot oil or components to contact skin or hands. Failure to comply with this warning can cause injury or death to personnel.

WARNING

Wear heat resistant gloves and avoid contacting hot metal surfaces with your hands after components have been heated. Wear additional protective clothing as required. Failure to comply with this warning can cause injury to personnel.

WARNING

The connection of any electrical equipment and the disconnection of any electrical equipment may cause an explosion hazard. Do not connect any electrical equipment or disconnect any electrical equipment in an explosive atmosphere. Failure to comply with this warning can cause injury or death to personnel.

WARNING

Diesel fuel is flammable and toxic to eyes, skin, and respiratory tract. Skin and eye protection are required when working in contact with diesel fuel. Avoid repeated or prolonged contact. Provide adequate ventilation. Operators are to wash exposed skin and change chemical soaked clothing promptly if exposed to fuel. Failure to comply with this warning can cause injury or death to personnel.

WARNING

Fuels used in the generator set are flammable. Do not smoke or use open flames when performing maintenance. Failure to comply with this warning can cause injury or death to personnel, and damage to the generator set.

a. a. On right side of engine, tag and disconnect electrical leads from low oil pressure switch (1, Figure 5-18), oil pressure sender (2), temperature sending unit (3), magnetic pickup (4), and fuel injection pump governor actuator (5).

b. b. Disconnect wire 243A fr om glow plug bus bar (6).

c. c. Remove two screws (7), lock nuts (8), and contactor (9) with wires connected. Tie contactor back to harness out of the way for engine removal.

d. d. Remove DEAD CRANK switch (10).

e. e. Disconnect fuel line (11) from engine fuel filter (12). Cap openings.

f. f. Remove clamp (13) and tag and disconnect hose (14) from fuel/water separator assembly (15). Cap openings. Discard hose and clamp.

g. g. Disconnect fuel line (16) from fuel transfer pump (17). Cap openings.

h. h. Remove nuts (18), lock washers (19), bolts (20) , washers (21), and fuel filter/water separator assembly (15) from engine. Discard fuel/water separator and mounting hardware.

i. i. Disconnect fuel line (22) from fuel transfer pump (17). Cap openings.

j. j. Loosen clamp (23) and remove fuel return line (24) from engine fuel filter (12). Cap openings.

k. k. Disconnect two clamps (25) (one on each end of hose) and fuel hose (26).

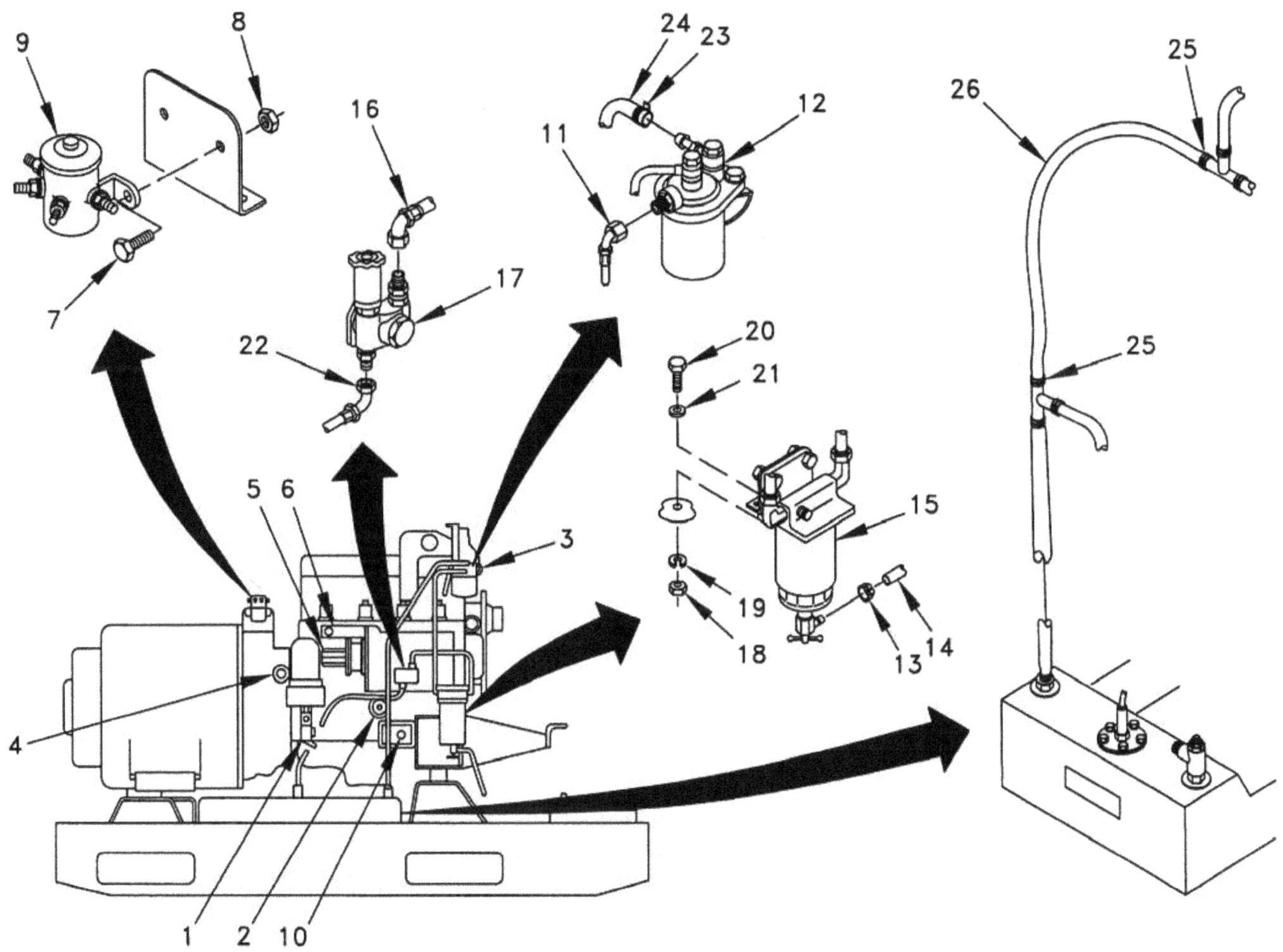

Figure 5-18. Right Side Engine Components.

5-2.19 Left Side Engine Components Removal.

WARNING

All metal jewelry can conduct electricity and become entangled in generator set components. Remove all jewelry when working on generator set. Failure to comply with this warning can cause injury or death to personnel.

WARNING

DO NOT wear loose clothing when performing checks, services and maintenance. Failure to comply with this warning can cause injury or death to personnel.

WARNING

The high pressure oil system operates at high temperature and pressure. Contact with hot oil can result in burns and scalding. Shut down generator set, and allow system to cool before performing checks, services, and maintenance. Wear heat resistant gloves and avoid contacting hot surfaces. Do not allow hot oil or components to contact skin or hands. Failure to comply with this warning can cause injury or death to personnel.

WARNING

Wear heat resistant gloves and avoid contacting hot metal surfaces with your hands after components have been heated. Wear additional protective clothing as required. Failure to comply with this warning can cause injury to personnel.

WARNING

The connection of any electrical equipment and the disconnection of any electrical equipment may cause an explosion hazard. Do not connect any electrical equipment or disconnect any electrical equipment in an explosive atmosphere. Failure to comply with this warning can cause injury or death to personnel.

WARNING

Diesel fuel is flammable and toxic to eyes, skin, and respiratory tract. Skin and eye protection are required when working in contact with diesel fuel. Avoid repeated or prolonged contact. Provide adequate ventilation.
Operators are to wash exposed skin and change chemical soaked clothing promptly if exposed to fuel. Failure to comply with this warning can cause injury or death to personnel.

WARNING

Fuels used in the generator set are flammable. Do not smoke or use open flames when performing maintenance. Failure to comply with this warning can cause injury or death to personnel, and damage to the generator set.

a. a. On left side of engine, remove three bolts (1, Figure 5-19), lock washers (2), washers (3), and clamps (4, 5, and 6) securing harness to engine. Discard lock washers (2).

b. b. On left side of engine, tag and disconnect elect rical leads from battery charging alternator (7), starter solenoid (8), starter motor (9), and coolant high temperature switch (10). Move engine electrical harness to rear of generator set and clear of engine.

c. c. Loosen clamp (11) and disconnect coolant drain hose (12) from engine block coolant drain valve (13).

d. Remove nut (14), lock washer (15), and wire (16) from center post of fuel sending unit (17). Remove screw (18), lock washer (19), flat washer (20), and wire (21) from fuel sending unit. Move wiring away from engine as much as practical for engine removal. Temporarily reinstall hardware to sending unit for wiring installation later. Discard lock washers (15 and 19).

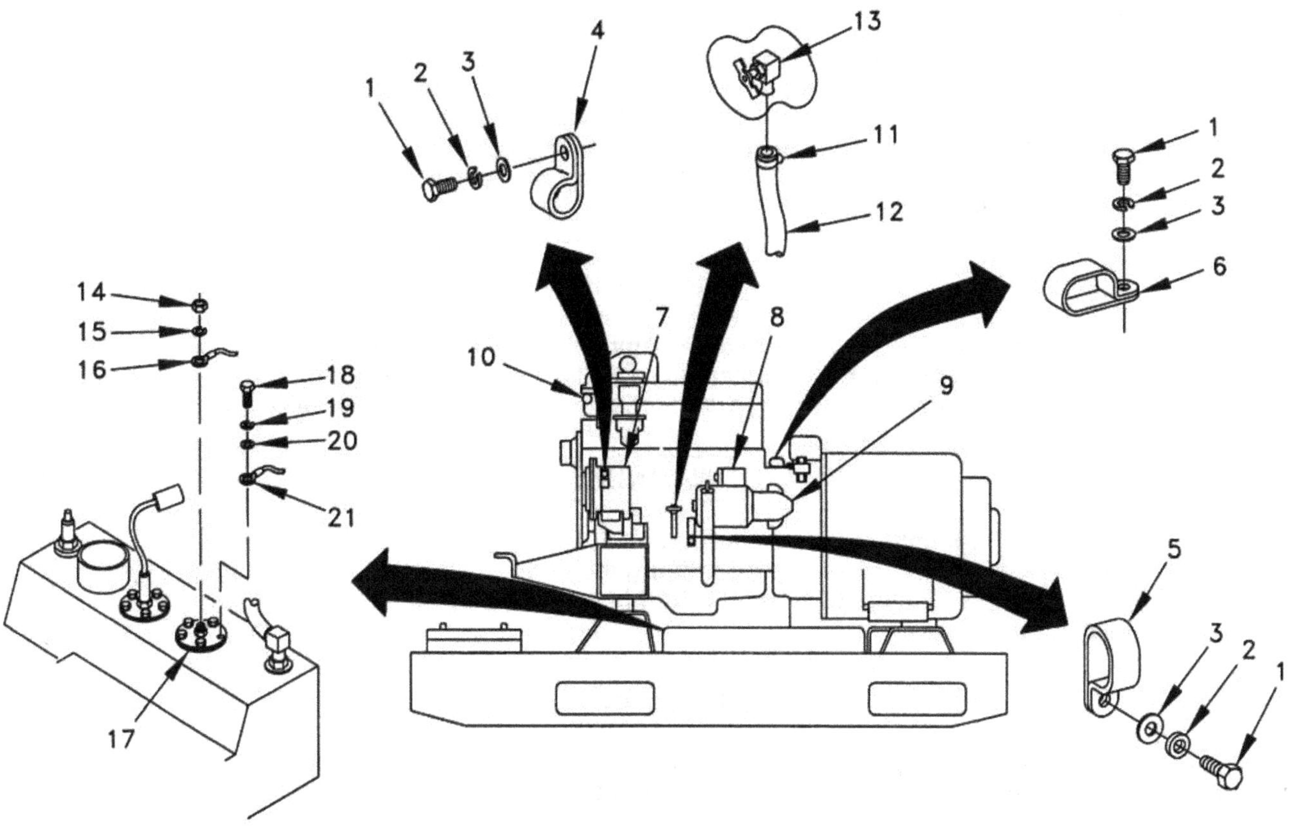

Figure 5-19. Left Side Engine Components.

5-2.20 Engine Removal.

WARNING

All metal jewelry can conduct electricity and become entangled in generator set components. Remove all jewelry when working on generator set. Failure to comply with this warning can cause injury or death to personnel.

WARNING

DO NOT wear loose clothing when performing checks, services and maintenance. Failure to comply with this warning can cause injury or death to personnel.

WARNING

The connection of any electrical equipment and the disconnection of any electrical equipment may cause an explosion hazard. Do not connect any electrical equipment or disconnect any electrical equipment in an explosive atmosphere. Failure to comply with this warning can cause injury or death to personnel.

WARNING

Diesel fuel is flammable and toxic to eyes, skin, and respiratory tract. Skin and eye protection are required when working in contact with diesel fuel. Avoid repeated or prolonged contact. Provide adequate ventilation. Operators are to wash exposed skin and change chemical soaked clothing promptly if exposed to fuel. Failure to comply with this warning can cause injury or death to personnel.

WARNING

Fuels used in the generator set are flammable. Do not smoke or use open flames when performing maintenance. Failure to comply with this warning can cause injury or death to personnel, and damage to the generator set.

WARNING

Many components require a two-person lift. Lifting heavy components can cause back strain. Ensure proper lifting techniques are used when lifting heavy components. Failure to comply with this warning can cause injury to personnel.

WARNING

Support components when removing attaching hardware or component may fall. Failure to comply with this warning can cause injury to personnel, and damage to equipment.

WARNING

The generator set, engine, and generator are extremely heavy and require an assistant and a lifting device (forklift, overhead lifting device) with sufficient capacity. Failure to comply with this warning can cause injury or death to personnel.

a. a. Loosen nuts (1 and 2, Figure 5-20), turn bolts (3) to contact skid base, and tighten nuts (1 and 2).

b. b. Remove screw (4), washer (5), and screen/cover (6) from generator case.

c. c. Remove eight bolts (7) and lock washers (8) (inner) securing generator drive disc to engine flywheel. Discard lock washers (8).

d. Remove two bolts (9), lock wa shers (10), washers (11), and right rear engine support (12). Discard lock washers (10).

WARNING

Rated capacity of overhead hoist should be at least 1,500 pounds (680 kg). Do not use a hoist with less capacity. Failure to comply with this warning can cause injury or death to personnel, and damage to equipment.

e. Attach lifting harness to engine and overhead hoist. Take up slack.

f. f. Remove 12 bolts (13) and lock washers (14) securing generator to flywheel housing. Discard hardware.

g. g. Place rope around generator rotor and tie rope to generator lifting eye to secure rotor from slipping or falling during engine removal.

h. h. Remove two nuts (15), washers (16), snubbing washers (17), and bolts (18) securing support bracket (19) to skid base.

WARNING

Keep hands and feet from underside of engine and generator while using lifting device to remove them from the skid base. Failure to comply with this warning can cause injury or death to personnel.

i. i. With aid of an assistant, slowly lift engine from skid base, ensuring that engine flywheel housing separates smoothly from generator without binding. Remove engine from generator set housing and place on engine stand.

j. j. Remove four nuts (20), lock washers (21), bolts (22), and two shock mounts (23) from skid base. Discard shock mounts (23) and lock washers (21).

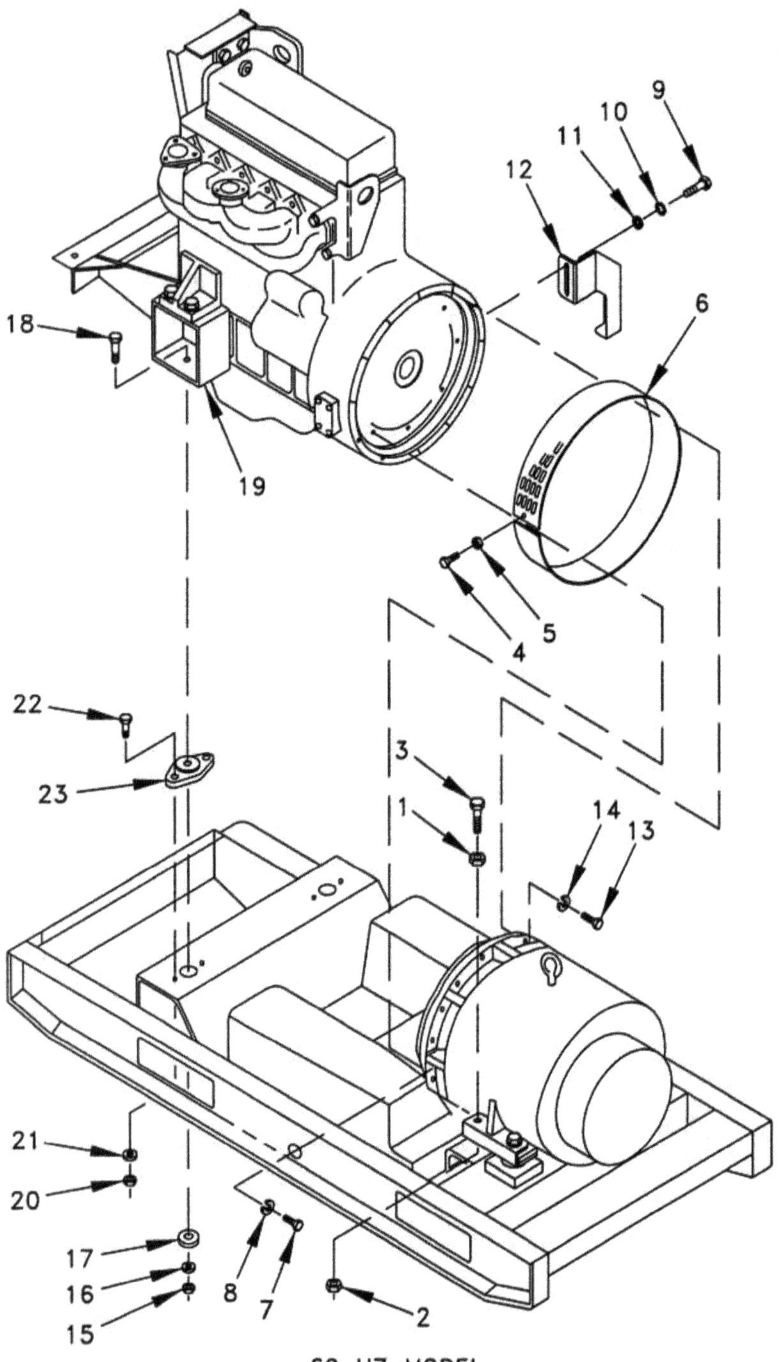

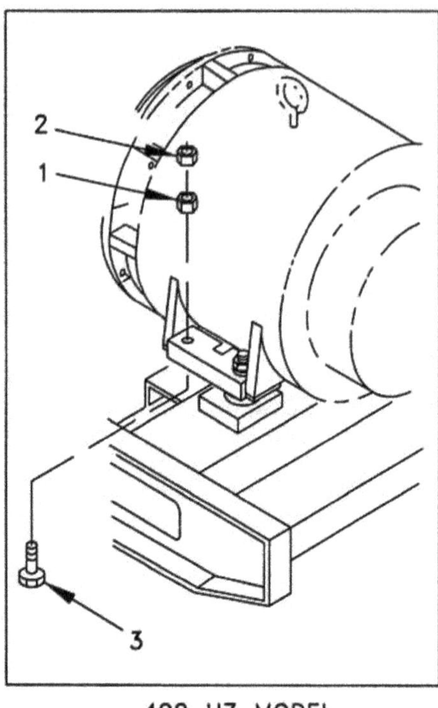

400 HZ MODEL

60 HZ MODEL

Figure 5-20. Engine.

Section III. PRE-ENGINE INSTALLATION MODIFICATION

5-3 NEW ENGINE PREPARATION

The FIKE parts list is identified in Table 5-1. Part number 97-24103-01 Kit is used for the MEP-804B Generator Set and for the MEP-814B Generator Set. Parts which are to be used for only one model of the generator set are identified in Table 5-1. Excess parts can be discarded. The expendable/durable supplies and materials required to perform the modification are listed in Table 5-2. One Yanmar model number 4TNV84T-DFM turbocharged diesel engine, NSN 2815-01-538-4237, is also required, but is not part of the kit.

Table 5-1. FIKE Parts List.

Item No.	Qty/Kit	Part Number	Nomenclature
1	REF	97-24010	Engine
2	1	97-24002	Closed Crankcase Ventilation (CCV) Filter
3	1	97-24011	Fitting
4	1	97-24013	Right Engine Support
5	1	97-24012	Left Engine Support
6	1	97-24014	Crossmember
7	1	97-24015	Manifold
8	1	97-24016	Manifold Bracket
9	1	88-21126	Oil Pressure Switch
10	1	88-21633	Oil Sampling Valve
11	1	88-22706	Oil Pressure Sender
12	1	97-24017	Magnetic Pickup Bracket
13	1	97-24114	Governor Actuator
14	1	97-24019	Upper Radiator Support
15	1	97-24020	Turbo Output Manifold Bracket
16	1	97-24021	Exhaust Manifold
17	4	88-20556-6	Lock Washer, 5/16 In.
18	1	97-24023	Rear Lifting Ring
19	1	88-20579-4	Hose, Nonmetallic, 0.312 ID, 11 In. (27.9 cm) Length
20	1	88-20579-4	Hose, Nonmetallic, 0.312 ID, 13 In. (33.0 cm) Length
21	8	88-20561-1	Hose Clamp
22	1	88-22550	Magnetic Pickup
23	1	88-20579-6	Hose, Nonmetallic, 0.50 ID, 27 In. (68.6 cm) Length
24	1	97-24028	Air Cleaner Reducer
25	1		Hose, 1.75 In. (4.4 cm) ID X 2 In. (5.1 cm) Length, Type R1, Class D2 per SAE J20

Table 5-1. FIKE Parts List – Continued.

Item No.	Qty/Kit	Part Number	Nomenclature
26	1	97-24031	Hose, 2.50 In. (6.4 cm) ID X 2 In. (5.1 cm) Length, Type R1, Class D2 per SAE J20
27	1	88-29544-1	T-Fitting
28	3	97-24032	Cable Clamp
29	1	97-24033	Right Fan Shroud
30	1	97-24034	Left Fan Shroud
31	2	97-24035	Radiator Tie Rod
32	1	97-24036	Lower Radiator Hose
33	1	00-24004	Upper Radiator Hose
34	2	88-21674-1	Radiator Mount
35	6	97-24039	Cage Nut, #10-32
36	1	Type M, #10	Fan Guard Support
37	2	B18241B080	Hose Clamp
38	7	88-20561-2	Nut, M8, CL 10 per ASTM B18.2.4.1M
39	2	0116-1828-01	Hose Clamp
40	5	88-20561-3	Screw, M8, 1.25 Pitch X 16 mm
41	8	88-20561-4	Hose Clamp
42	2	88-22546-2	Hose Clamp
43	1	00-24000	Fuel Pump
44	1	8-4-430160	Fuel Filter
45	1	97-24113	Fitting, 1/2 In. ID Hose Barb To 1/4 In. NPT Male Pipe
46	1	4-130109E	Fitting, 5/8 In. ID Hose Barb To 3/8 In. NPT Male Pipe
47	1		Pipe Plug, 1/4 In. NPT
48	–	2-2-140239	Deleted
49	2	88-22790-2	Street Elbow, 1/8 In. NPT
50	4	B18231A10020N	Nut, 5/16-18
51	14	0116-1828-02	Screw, M10, 1.5 Pitch X 20 mm
52	8	88-20260-45	Screw, M8, 1.25 Pitch, 20 mm
53	4	88-22331-2	Washerhead Screw, 5/16-18 X 1.25 In.
54	32	88-22331-1	Lock Washer, M10
55	19	88-20033-31A	Lock Washer, M8
56	14	88-20033-24A	Flat Washer, 13/32 In.
57	23		Flat Washer, 11/32 In.

Table 5-1. FIKE Parts List – Continued.

Item No.	Qty/Kit	Part Number	Nomenclature
58	23	88-20556-16	Lock Washer, 1/4 In.
59	20	88-20033-18A	Flat Washer, 1/4 In.
60	1	88-21101	Fuel/Water Separator
61	9	88-22790-1	Nut, 1/4-20
62	21	88-20260-32	Washerhead Screw, 1/4-20 X 0.75 In.
63	2	88-22790-3	Nut, 3/8-16
64	2	B1821BH038C62N	Screw, 3/8-16 X 5/8 In.
65	4	88-20556-18	Lock Washer, 3/8 In.
66	2	00-24003-1	Radiator Grommet
67	2	00-24003-2	Radiator Bushing
68	2	B1831BH060016	Socket Head Screw, M6, 1 Pitch X 16 mm
69	1	97-24061	Temperature Switch
70	1	88-22322	Temperature Sender
71	1	97-24044	Fuel Pump Bracket
72	3	4-4-070202	Elbow, Male, 90° per SAE J514
73	4	98-19744-03	Fitting, 1/4 In. ID Tube To 5/16 In. ID Hose
74	1	88-20579-7	Hose, Nonmetallic, 0.625 ID, 2.75 In. (7.0 cm) Length
75	1	97-24102-05	Wire Assembly (100HTR, plus terminal for wire 243A)
76	4	88-20564-2	Flat Washer
77	4	69-561-5	Nut With Captive Washer 1/4-20
78	1	88-20544-6	Cushioned Loop Clamp
79	2	88-20260-35	Washerhead Screw, 1/4-20 X 1.25 In.
80	1	97-24115	Oil Dip Stick Spacer
81	1	97-24102-04	Wire Assembly (JUMPER)
82	2	88-22331-3	Lock Washer, M6
83	3	88-20033-22A	Flat Washer, 9/32 In.
84	1	97-24102-03	Wire Assembly (P61/J13, plus terminal for wire 164D)
85	1	97-24102-01	Wire Assembly (163K16, plus terminal for wire 163B16)
86	2	88-21755-4	Fitting, 1/8 NPT To 5/16 ID Hose
87	1	88-21975	Muffler Clamp
88	1	97-24037	Generator Spacer
89	12	97-24116-2	Screw, M10, 1.5 Pitch X 40 mm (MEP-804B only)

Table 5-1. FIKE Parts List – Continued.

Item No.	Qty/Kit	Part Number	Nomenclature
90	1	97-24118	Copper Washer
91	1	97-24045	DEAD CRANK Switch Bracket
92	1	97-24050-01	Ratings Identification (ID) Plate (50/60 Hz) (MEP-804B only)
93	1	97-24051	Schematic Diagram Plate
94	1	97-24052	Wiring Diagram Plate
95	1	97-24053	Operating Instructions Plate
96	1	4-4-070102	Adapter
97	1	97-24042	Air Cleaner Bracket
98	1	88-22554-3	Hose Clamp
99	1	88-22752-7	Hose, .25 In. ID, 38 In. (96.5 cm) Length
100	36	13214E3789-2	Rivet, Blind
101	1	88-20579-4	Hose, Nonmetallic, 0.312 ID, 14 In. (35.6 cm) Length
102	1	88-20579-4	Hose, Nonmetallic, 0.312 ID, 36 In. (91.4 cm) Length
103	2	88-20561-5	Hose Clamp
104	20	97-24116-3	Screw, M10, 1.5 Pitch X 50 mm
105	2	97-24038	Rear Engine Support
106	1	97-24040	Left Fan Guard
107	1	97-24041	Right Fan Guard
108	4	B18231A10025N	Screw, M10, 1.5 Pitch X 25 mm
109	4	88-22331-2	Lock Washer, M10
110	4	88-20033-31	Flat Washer, 13/32 In.
111	7	88-20260-23	Washerhead Screw, 10-32 X 3/4 In.
112	7	88-20556-37	Lock Washer, #10
113	7	88-20033-8	Flat Washer, #10
114	1	88-21776	DEAD CRANK Switch Plate
115	1	97-24122	Inertia Ring
116	2	88-21071-3	Shock Mount
117	1	88-20260-23	Washerhead Screw, #10-32 X 3/4 In.
118	2	88-20559-1	Nut, #10-32
119	1	88-20033-11A	Flat Washer, #10
120	1	88-20556-15	Lock Washer, #10
121			Deleted

Table 5-1. FIKE Parts List – Continued.

Item No.	Qty/Kit	Part Number	Nomenclature
122	1	97-24050-02	Deleted
123	1	88-20557-4	Ratings ID Plate (400 Hz) (MEP-814B only)
124	1	97-24102-02	Grommet
125	1	97-24110	Wire Assembly (J61, Plus Terminal For Wire 164A16)
126			Circuit Breaker
127	1	97-24119-01	BATTERY CHARGER CIRCUIT BREAKER Label
128	1	97-24119-02	CB2 Label
129	1	97-24123-01	Cable Shielding
130	1	97-24123-02	Cable Shielding
131	1	97-24123-03	Cable Shielding
132	1	97-24130	Seal Retainer
133	1	97-24124	Seal
134	1	97-24131	Adhesive-backed Seal
135	1	97-24132	Seal
136			Ground Strap (12.0 In. Length)
137	2	97-24125	Ground Strap (22.0 In. Length)
138	1	97-24126	Ground Strap (8.0 In. Length)
139	2	97-24128	Ground Strap (26.0 In. Length)
140	2	97-24129	Governor Control Unit (MEP-804B Only)
141	1	97-24134	Elbow, 45° St., 3/4 Male NPT X 3/4 Female NPT per SAE J516
142	1	12-12-140330C	Nameplate, Oil Level Check Instructions
	1	97-24146	

Table 5-2. FIKE Expendable/Durable Supplies and Materials List.

Item Number	Level	National Stock Number	Item Name, Description, CAGEC, and Part Number	U/M
1	F	6850-00-181-7933	Antifreeze, 1 Gallon (81349) A-A-52624A	GL
2	F	5350-00-221-0872	Cloth, Abrasive, Crocus (58536) ANSI B74.18	PG
3	F	7920-01-338-3329	Cloth, Cleaning (21994) TX-1250	
3.1	F	8030-00-779-4699	Coating, Conversion, MIL-DTL-5541, Type 1, Class 3	KT
4	F	9150-00-186-6681	Lubricating Oil, Internal Combustion Engine, OE/HDO-30, 1 Quart (81349) MIL-PRF-2104H	QT
5	F	7920-00-205-3571	Rag, Wiping, Cotton and Cotton Synthetic, Grade B (81348) DDD-R-0030	BX
6	F	88-20593-2	Sealing Compound (30554) ASTM D5363, Grade HW	TU
7	F	6850-01-378-0679	Solvent, Cleaning Compound, Breakthrough, 5 Gallon Can (OK209)	CN
8	F	5975-00-899-4606	Strap, Tiedown, Electrical Component (96906) SAE-AS33671	HD
9	F	9905-00-537-8954	Tag, Marker, 50 Each Bundle (81349) A-A-52101	BD

5-3.1 Skid Base Preparation.

WARNING

All metal jewelry can conduct electricity and become entangled in generator set components. Remove all jewelry when working on generator set. Failure to comply with this warning can cause injury or death to personnel.

WARNING

DO NOT wear loose clothing when performing checks, services and maintenance. Failure to comply with this warning can cause injury or death to personnel.

WARNING

Solvent used to clean parts is potentially dangerous to personnel and property. Clean parts in a well-ventilated area. Avoid inhalation of solvent fumes. Wear goggles and rubber gloves to protect eyes and skin. Wash exposed skin thoroughly. DO NOT smoke or use near open flames or excessive heat. Failure to comply with this warning can cause injury to personnel, and damage to the equipment.

WARNING

High pressure steam can blow particles or chemicals into eyes, can cause severe burns, and creates hazardous noise levels. Wear protective eye, skin, and hearing protection when using high pressure steam. Failure to comply with this warning can cause injury to personnel.

WARNING

Eye protection is required when working with compressed air. Compressed air can propel particles at high velocity and injure eyes. Do not exceed 15 psi pressure when using compressed air. Failure to comply could cause serious injury to personnel.

WARNING

Cleaning solvent is flammable and toxic to eyes, skin, and respiratory tract. Skin and eye protection are required when working in contact with cleaning solvent. Avoid repeated or prolonged contact. Work in ventilated area only. Failure to comply with this warning can cause injury or death to personnel.

WARNING

Cleaning compound is toxic. Avoid prolonged breathing of vapors. Use only in a well-ventilated area. Failure to comply with this warning can cause injury to personnel.

WARNING

Avoid breathing fumes generated by soldering. Eye protection is required. Good general ventilation is normally adequate. Failure to comply with this warning can cause injury to personnel.

WARNING

CARC paint is a health hazard, and is irritating to eyes, skin, and respiratory system. Wear protective eyewear, mask, and gloves when applying or removing CARC paint. Failure to comply with this warning can cause injury to personnel.

a. a. Clean skid base with dry cleaning solvent (Item 7, Table 5-2) and cleaning cloth (Item 3, Table 5-2).

b. b. Repair/replace any damaged panels.

c. Inspect and repair/replace any insulation on panels.

5-3.2 Engine Support Assembly.

WARNING

All metal jewelry can conduct electricity and become entangled in generator set components. Remove all jewelry when working on generator set. Failure to comply with this warning can cause injury or death to personnel.

WARNING

DO NOT wear loose clothing when performing checks, services and maintenance. Failure to comply with this warning can cause injury or death to personnel.

NOTE

All pipe thread connections shall be made liquid tight by applying sealing compound to the male threads of each joint.

New engine is shipped with a box of parts. A new gasket for the turbocharger connection is in the box. Remove this gasket and discard remainder of parts.

a. a. If supplied, remove lifting rings from ne w engine. Discard lifting rings and hardware.

b. b. Install upper radiator support (1, Figure 5-21) (Item 14, Table 5-1) using three screws (2) (Item 52, Table 5-1), lock washers (3) (Item 55, Table 5-1), and flat washers (4) (Item 57, Table 5-1).

c. Install rear lifting ring (5) (Item 18, Table 5-1) using three screws (6) (Item 52, Table 5-1), lock washers (7) (Item 55, Table 5-1), and flat washers (8) (Item 57, Table 5-1).

WARNING

Rated capacity of overhead hoist should be at least 1,500 lbs (680 kg). Do not use a hoist with less capacity. Failure to comply with this warning can cause injury or death to personnel, and damage to equipment.

d. d. Attach lifting harness to engine (9) and overhead hoist. Take up slack.

e. e. Loosely assemble engine support assembly, consisting of left engine support (10) (Item 5, Table 5-1), right engine support (11) (Item 4, Table 5-1), and crossmember (12) (Item 6, Table 5-1), using four screws (13) (Item 53, Table 5-1), lock washers (14) (Item 17, Table 5-1), flat washers (15) (Item 57, Table 5-1), and nuts (16) (Item 50, Table 5-1).

f. f. Remove brackets, or equivalent, securing new engine to shipping pallet.

g. g. With aid of an assistant, lift engine and position engine on engine support assembly (assembled in step e above). Attach engine to support assembly using ten screws (17) (Item 51, Table 5-1), lock washers (18) (Item 54, Table 5-1), and flat washers (19) (Item 56, Table 5-1). Tighten all hardware.

h. h. Remove shipping plug from engine oil pan, if necessary, and install fitting (20) (Item 3, Table 5-1) using sealing compound (Item 6, Table 5-2).

i. Install fan guard support (21) (Item 36, Table 5-1) using two screws (22) (Item 62, Table 5-1), lock washers (23) (Item 58, Table 5-1), flat washers (24) (Item 59, Table 5-1), and nuts (25) (Item 61, Table 5-1).

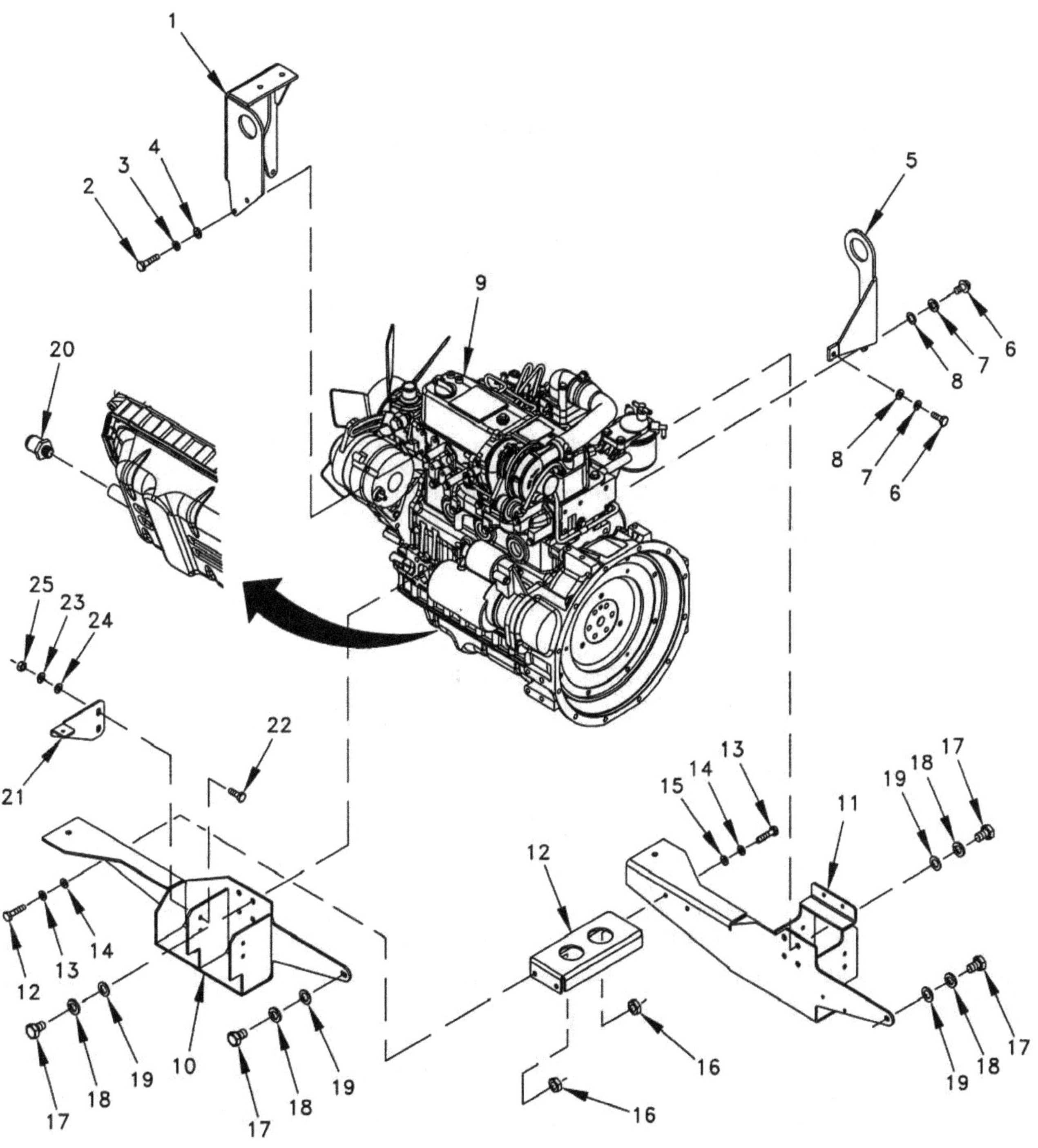

Figure 5-21. Engine Support Assembly.

5-3.3 Left Side Engine Components Installation.

WARNING

All metal jewelry can conduct electricity and become entangled in generator set components. Remove all jewelry when working on generator set. Failure to comply with this warning can cause injury or death to personnel.

WARNING

DO NOT wear loose clothing when performing checks, services and maintenance. Failure to comply with this warning can cause injury or death to personnel.

WARNING

The connection of any electrical equipment and the disconnection of any electrical equipment may cause an explosion hazard. Do not connect any electrical equipment or disconnect any electrical equipment in an explosive atmosphere. Failure to comply with this warning can cause injury or death to personnel.

WARNING

Diesel fuel is flammable and toxic to eyes, skin, and respiratory tract. Skin and eye protection are required when working in contact with diesel fuel. Avoid repeated or prolonged contact. Provide adequate ventilation. Operators are to wash exposed skin and change chemical soaked clothing promptly if exposed to fuel. Failure to comply with this warning can cause injury or death to personnel.

WARNING

Fuels used in the generator set are flammable. Do not smoke or use open flames when performing maintenance. Failure to comply with this warning can cause injury or death to personnel, and damage to the generator set.

a. a. Remove shipping plug, if necessary, and install T-fitt ing (1, Figure 5-22) (Item 27, Table 5-1) using sealing compound (Item 6, Table 5-2).

b. b. Install fitting (2) (Item 46, Table 5-1) to T-fitting (1) using sealing compound (Item 6, Table 5-2).

c. c. Install hose (3) (Item 74, Table 5-1) using two hose clamps (4) (Item 39, Table 5-1).

d. d. Install fitting (5) (Item 45, Table 5-1) to T-fitting (1) using sealing compound (Item 6, Table 5-2).

e. e. Connect one end of 27-inch (68.6 cm) hose (6) (Item 23, Table 5-1) to fitting (5) using hose clamp (7) (Item 37, Table 5-1). Other end of hose will connect to closed CCV filter after it is installed.

f. f. Install turbo output manifold bracket (8) (Item 15, Table 5-1) using two screws (9) (Item 40, Table 5-1), lock washers (10) (Item 55, Table 5-1), and flat washers (11) (Item 57, Table 5-1).

g. g. Install pipe plug (12) (Item 47, Table 5-1) to exhaust manifold (13) (Item 16, Table 5-1) using

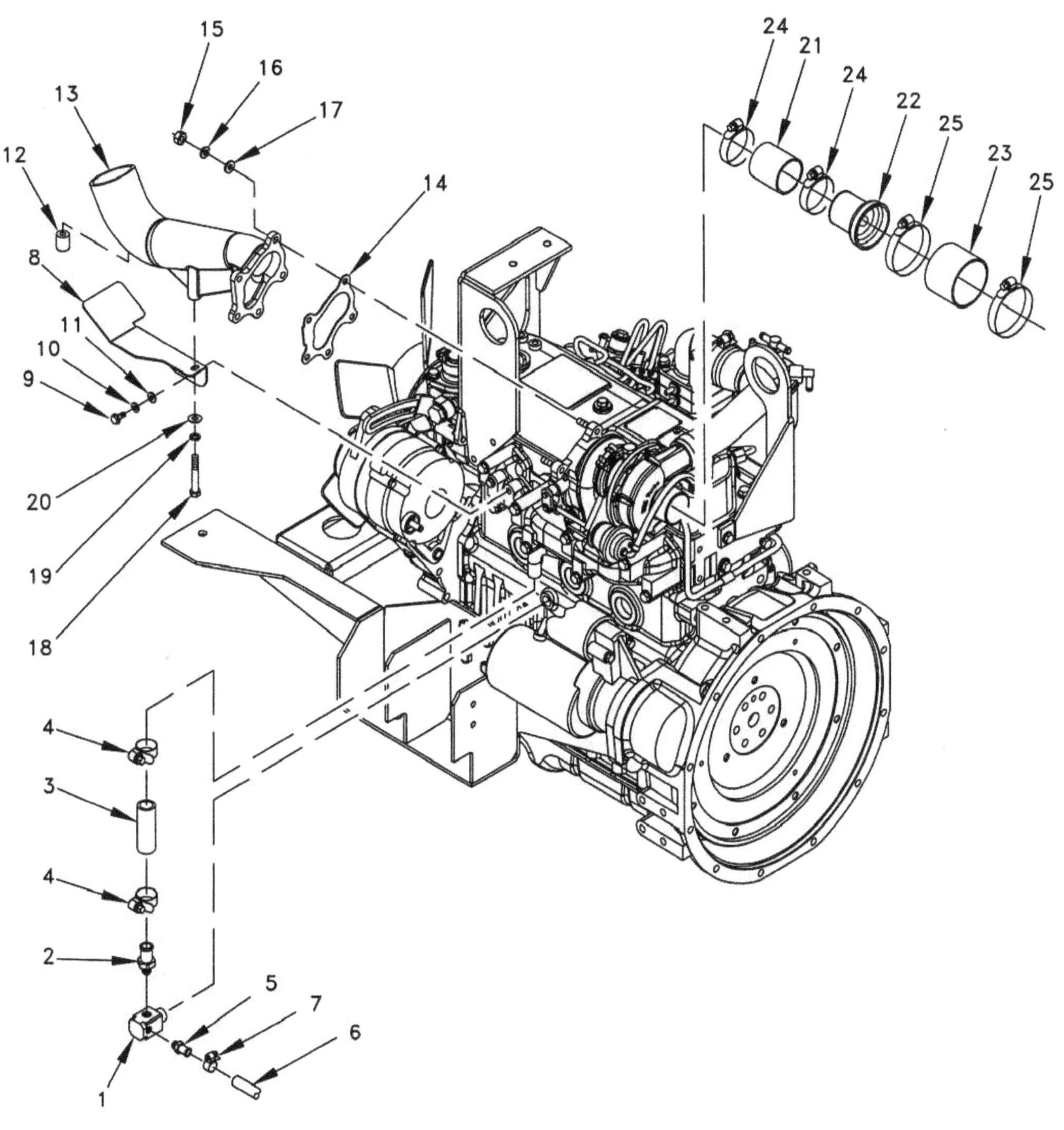

Figure 5-22. Left Side Engine Components.

NOTE

Install all hardware before tightening.

h. h. Install exhaust manifold (13) and gasket (14) (supplied with engine) using five nuts (15) (Item 38, Table 5-1), lock washers (16) (Item 55, Table 5-1), flat washers (17) (Item 57, Table 5-1), one screw (18) (Item 62, Table 5-1), lock washer (19) (Item 58, Table 5-1), and flat washer (20) (Item 59, Table 5-1).

i. i. Install hose (21) (Item 25, Table 5-1), air cleaner reducer (22) (Item 24, Table 5-1), and hose (23) (Item 26, Table 5-1) using two hose clamps (24) (Item 42, Table 5-1) and two hose clamps (25) (Item 103, Table 5-1). Orient small tube on side of air cleaner reducer (22) 45 degrees down from horizontal toward engine. Do not tighten clamps for air cleaner reducer at this time.

5-3.4 Right Side Engine Fuel Components Installation.

WARNING

All metal jewelry can conduct electricity and become entangled in generator set components. Remove all jewelry when working on generator set. Failure to comply with this warning can cause injury or death to personnel.

WARNING

DO NOT wear loose clothing when performing checks, services and maintenance. Failure to comply with this warning can cause injury or death to personnel.

WARNING

The connection of any electrical equipment and the disconnection of any electrical equipment may cause an explosion hazard. Do not connect any electrical equipment or disconnect any electrical equipment in an explosive atmosphere. Failure to comply with this warning can cause injury or death to personnel.

WARNING

Diesel fuel is flammable and to~~xic to eyes, s~~kin, and respiratory tract. Skin and eye protection are required when working in contact with diesel fuel. Avoid repeated or prolonged contact. Provide adequate ventilation.
Operators are to wash exposed skin and change chemical soaked clothing promptly if exposed to fuel. Failure to comply with this warning can cause injury or death to personnel.

WARNING

Fuels used in the generator set~~ are flammable~~. Do not smoke or use open flames when performing maintenance. Failure to comply with this warning can cause injury or death to personnel, and damage to the generator set.

a. Install elbow (1, Figure 5-23) (Item 49, Table 5- 1) to fuel pump (2) (Item 43, Table 5-1) using sealing compound (Item 6, Table 5-2).

b. b. Install fuel filter (3) (Item 44, Table 5-1) to elbow (1) using sealing compound (Item 6, Table 5-2). Orient filter approximately 20 degrees to left so that hose connected later will not interfere with fuel/ water separator (4).

c. c. Install fitting (5) (Item 86, Table 5-1) to fuel filter (3) using sealing compound (Item 6, Table 5-2).

d. d. Install elbow (6) (Item 49, Table 5-1) to fuel pump (2) using sealing compound (Item 6, Table 5-2).

e. e. Install fitting (7) (Item 86, Table 5-1) to elbow (6) using sealing compound (Item 6, Table 5-2).

f. f. Install fuel pump (2) to bracket (8) (Item 71, Table 5-1) using two nuts (9) (Item 61, Table 5-1), lock washers (10) (Item 58, Table 5-1), and flat washers (11) (Item 59, Table 5-1).

g. g. Install fuel pump bracket (8) using three screws (12) (Item 62, Table 5-1), lock washers (13) (Item 58, Table 5-1), and flat washers (14) (Item 59, Table 5-1).

h. h. Attach bracket on oil dipstick extension (15) using spacer (16) (Item 80, Table 5-1) and two screws (17) (Item 62, Table 5-1).

i. i. Install 11-inch (27.9 cm) hose (18) (Item 19, Table 5-1) to fuel pump (2) using hose clamp (19) (Item 21, Table 5-1).

j. j. Install fuel/water separator (4) (Item 60, Table 5-1) using two screws (20) (Item 79, Table 5-1), lock washers (21) (Item 58, Table 5-1), flat washers (22) (Item 59, Table 5-1), and nuts (23) (Item 61, Table 5-1).

k. k. Install two elbows (24) (Item 72, Table 5-1) and fittings (25) (Item 73, Table 5-1) to fuel/water separator (4) using sealing compound (Item 6, Table 5-2).

l. l. Connect hose (18) from fuel pump (2) to fitting (25) with hose clamp (26) (Item 21, Table 5-1).

m. m. Install hose (27) (Item 99, Table 5-1) to port on bottom of fuel/water separator (4) and secure with hose clamp (28) (Item 98, Table 5-1). Route other end of hose through drain hole in skid base.

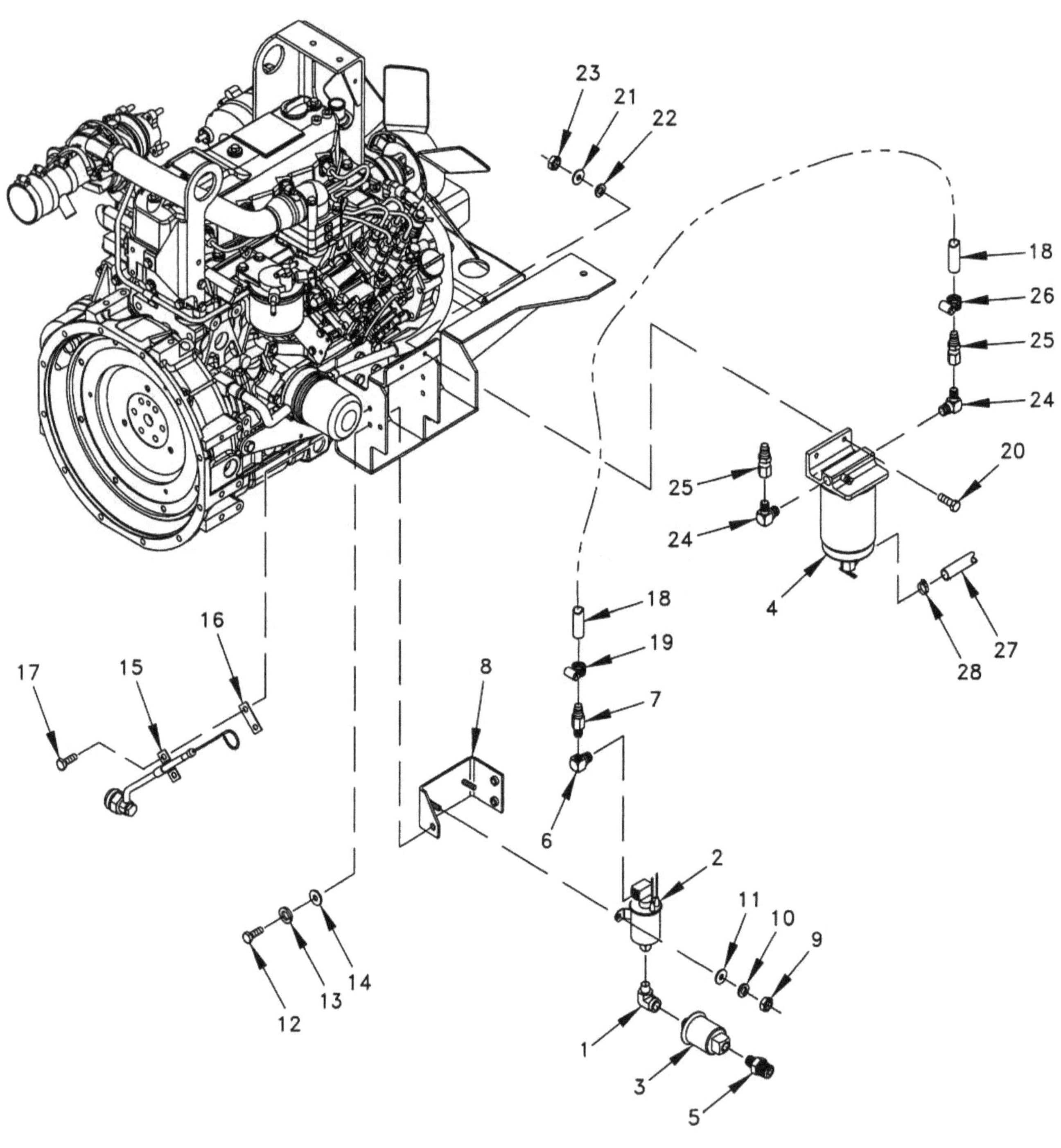

Figure 5-23. Right Side Engine Fuel Components.

5-3.5 Right Side Engine Sensors and Components Installation.

WARNING

All metal jewelry can conduct electricity and become entangled in generator set components. Remove all jewelry when working on generator set. Failure to comply with this warning can cause injury or death to personnel.

WARNING

DO NOT wear loose clothing when performing checks, services and maintenance. Failure to comply with this warning can cause injury or death to personnel.

WARNING

The connection of any electrical equipment and the disconnection of any electrical equipment may cause an explosion hazard. Do not connect any electrical equipment or disconnect any electrical equipment in an explosive atmosphere. Failure to comply with this warning can cause injury or death to personnel.

a. a. Remove oil pressure switch (1) from engine block and install manifold (2, Figure 5-24) (Item 7, Table 5-1) and manifold bracket (3) (Item 8, Table 5-1) using two screws (4) (Item 51, Table 5-1), lock washers (5) (Item 54, Table 5-1), and flat washers (6) (Item 56, Table 5-1). Be sure holes in sides of manifold are oriented correctly for installation of items 7 and 8.

b. b. Install new oil pressure switch (7) (Item 9, Table 5-1), oil pressure sender (8) (Item 11, Table 5-1), and oil sampling valve (9) (Item 10, Table 5-1) using sealing compound (Item 6, Table 5-2).

c. c. Remove shutdown solenoid, O-ring, and mounting hardware. Discard solenoid and hardware. Retain O-ring for installation of governor actuator.

d. d. Install governor actuator (10) (Item 13, Table 5-1) and O-ring (11) (from shutdown solenoid) using two screws (12) (Item 68, Table 5-1) and lock washers (13) (Item 82, Table 5-1).

e. e. Remove rubber cover (14) from engine bell housing.

f. f. Install magnetic pickup bracket (15) (Item 12, Table 5-1) using two screws (16) (Item 40, Table 5-1), lock washers (17) (Item 55, Table 5-1), and flat washers (18) (Item 57, Table 5-1).

g. g. Install magnetic pickup (19) (Item 22, Table 5-1) and lock nut (20) (supplied with transducer).

h. h. Screw magnetic pickup (19) into bracket (15) until pickup contacts top surface of gear tooth on flywheel. Back out one turn and tighten lock nut (20).

i. Remove shipping plug and temperature switch and install temperature sender (21) (Item 70, Table 5-1), copper washer (22) (Item 90, Table 5-1), and temperature switch (23) (Item 69, Table 5-1) using sealing compound (Item 6, Table 5-2) on threads.

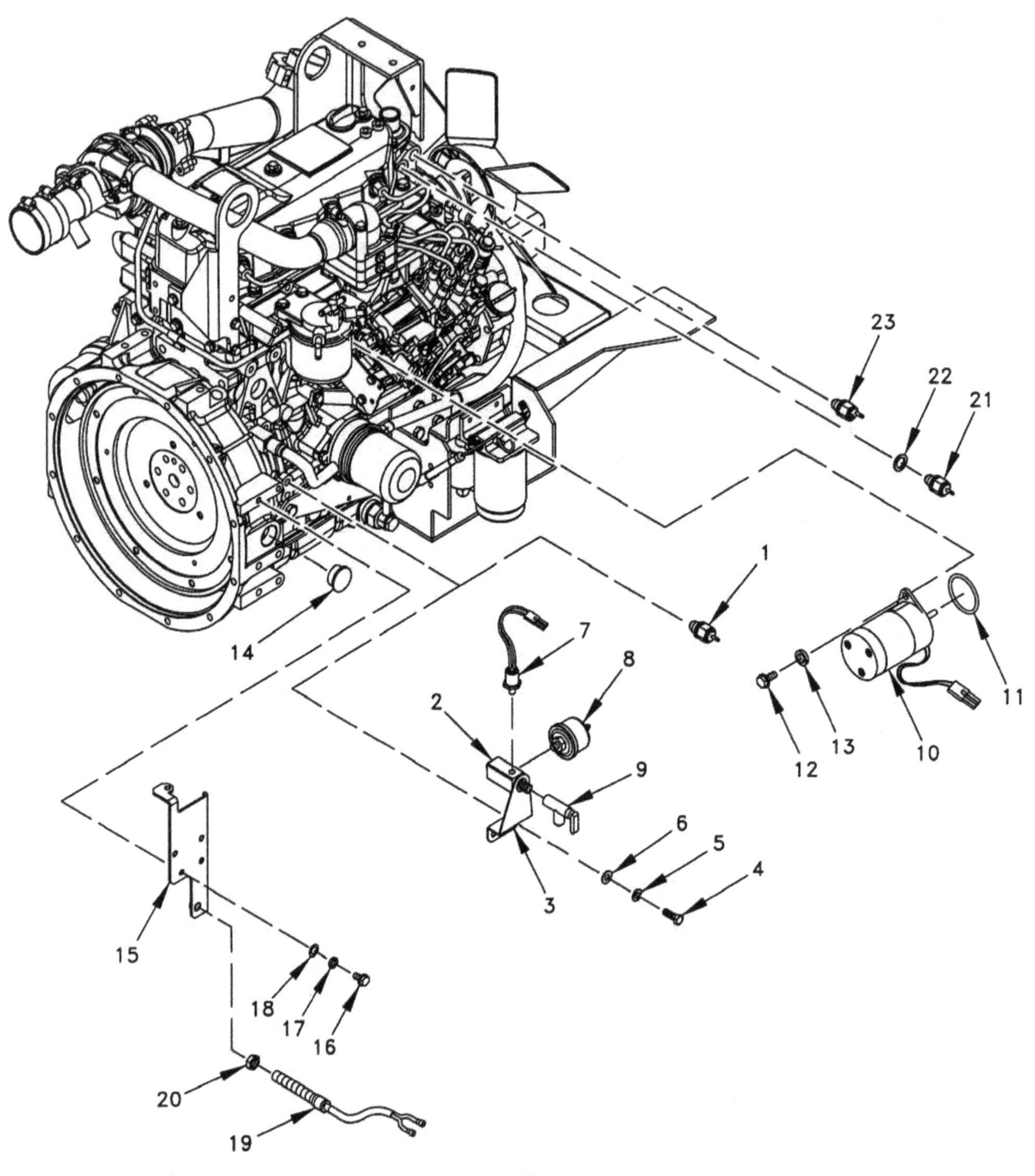

Figure 5-24. Right Side Engine Sensors and Components.

5-3.6 CCV Filter Installation.

WARNING

All metal jewelry can conduct electricity and become entangled in generator set components. Remove all jewelry when working on generator set. Failure to comply with this warning can cause injury or death to personnel.

WARNING

DO NOT wear loose clothing when performing checks, services and maintenance. Failure to comply with this warning can cause injury or death to personnel.

WARNING

The connection of any electrical equipment and the disconnection of any electrical equipment may cause an explosion hazard. Do not connect any electrical equipment or disconnect any electrical equipment in an explosive atmosphere. Failure to comply with this warning can cause injury or death to personnel.

a. a. Remove fuel filter (1, Figure 5-25) from engine (2). Retain mounting hardware for reinstallation.

b. b. Install CCV filter (3) (Item 2, Table 5-1) using two screws (4) (Item 52, Table 5-1), lock washers (5) (Item 55, Table 5-1), and flat washers (6) (Item 57, Table 5-1).

c. c. Install screw (7) (Item 62, Table 5-1), lock washer (8) (Item 58, Table 5-1), flat washer (9) (Item 83, Table 5-1), and nut (10) (Item 61, Table 5-1).

d. d. Install fuel filter (1) next to CCV filter (3) on same mounting bracket using original fuel filter mounting hardware plus two washers (item 57, Table 5-1), lock washers (item 55, Table 5-1), and nuts (item 38, Table 5-1).

e. e. Locate hose (6, Figure 5-22) connected to T-fitting on engine oil port. Attach loose end of this hose (11, Figure 5-25) to CCV filter (3). Secure with hose clamp (12) (Item 37, Table 5-1).

NOTE

If hose is attached to engine port (15), remove and discard. Hose may require lubrication for installation.

f. f. CCV filter (3) comes with a piece of hose. Cut two pieces from hose to connect filter assembly to engine.

g. g. Connect one piece of hose (13) to port on bracket (14) and to port on engine (15). Secure with two hose clamps (16) (Item 41, Table 5-1).

h. h. Connect second piece of hose (17) to port (18) on CCV filter (3) and to port on reducer (19). Secure with two hose clamps (20) (Item 41, Table 5-1).

i. i. Install 13-inch (33.0 cm) hose (21) (Item 20, Table 5-1) to fuel/water separator (22). Secure with hose clamp (23) (Item 21, Table 5-1).

j. j. Connect loose end of hose (21) to fuel filter (1). Secure with hose clamp (24) (Item 21, Table 5-1).

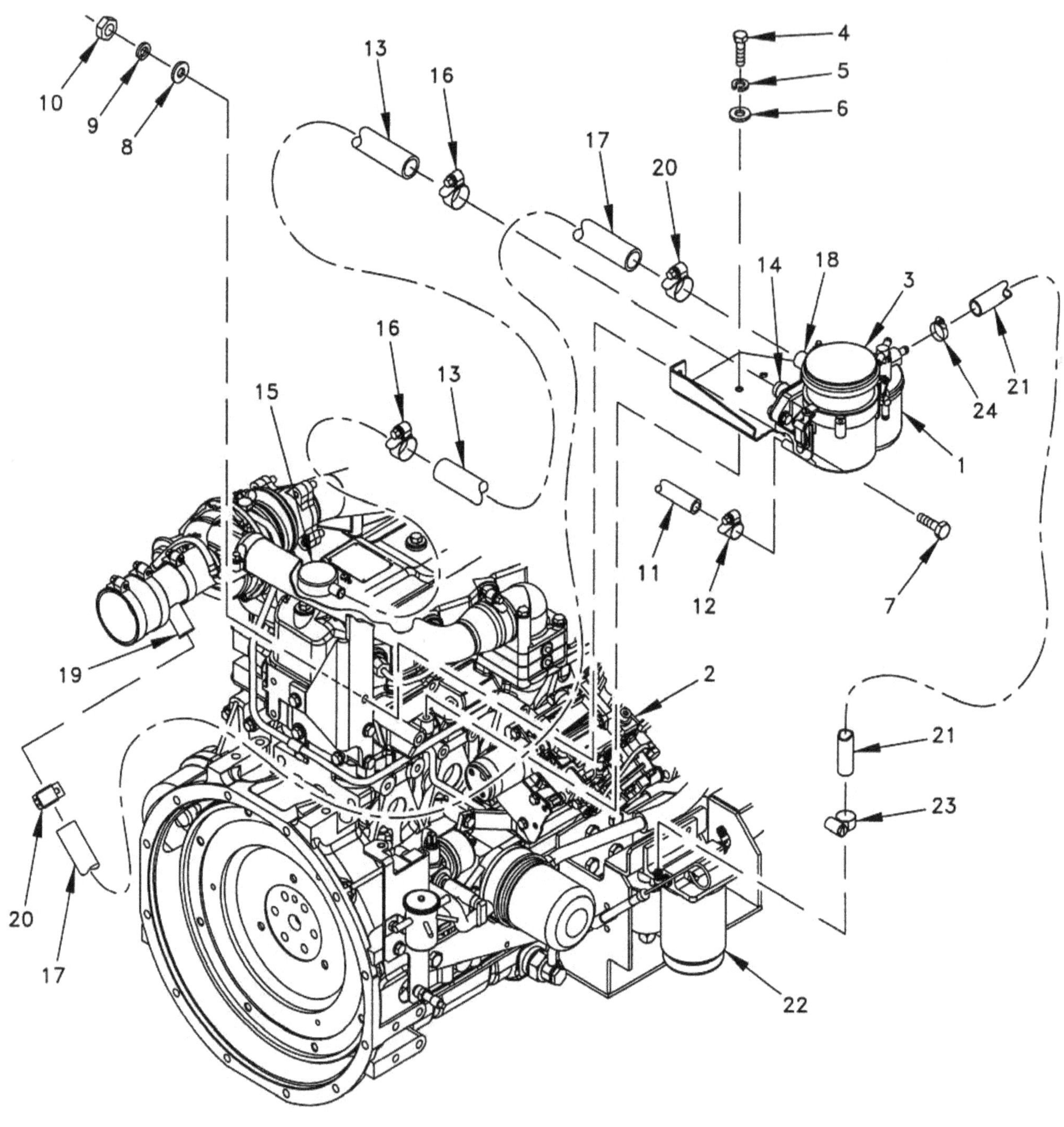

Figure 5-25. CCV Filter.

5-4 PRE-ENGINE INSTALLATION MODIFICATION REQUIREMENTS

WARNING

All metal jewelry can conduct electricity and become entangled in generator set components. Remove all jewelry when working on generator set. Failure to comply with this warning can cause injury or death to personnel.

WARNING

DO NOT wear loose clothing when performing checks, services and maintenance. Failure to comply with this warning can cause injury or death to personnel.

WARNING

The connection of any electrical equipment and the disconnection of any electrical equipment may cause an explosion hazard. Do not connect any electrical equipment or disconnect any electrical equipment in an explosive atmosphere. Failure to comply with this warning can cause injury or death to personnel.

5-4.1 Fuel Line Upgrade.

WARNING

Diesel fuel is flammable and toxic to eyes, skin, and respiratory tract. Skin and eye protection are required when working in contact with diesel fuel. Avoid repeated or prolonged contact. Provide adequate ventilation.
Operators are to wash exposed skin and change chemical soaked clothing promptly if exposed to fuel. Failure to comply with this warning can cause injury or death to personnel.

WARNING

Fuels used in the generator set are flammable. Do not smoke or use open flames when performing maintenance. Failure to comply with this warning can cause injury or death to personnel, and damage to the generator set.

NOTE

All pipe thread connections shall be made liquid tight by applying sealing compound to the male threads of each joint.

a. a. Remove hose (1, Figure 5-26) and hose fitting (2) from fuel supply line.

b. b. Apply sealing compound (Item 6, Table 5-2) and install elbow (3) (Item 72, Table 5-1), fitting (4) (Item 73, Table 5-1), 14-inch (35.6 cm) hose (5) (Item 101, Table 5-1), and hose clamp (6) (Item 21, Table 5-1).

c. c. Tag new hose "To fuel filter inlet".

d. d. Remove fitting (7) and hose (8) above T fitting (9) on fuel supply line.

e. e. Apply sealing compound (Item 6, T able 5-2) and install adapter (10) (Item 96, Table 5-1), fitting

(Item 73, Table 5-1), 36-inch (91.4 cm) hose (12) (Item 102, Table 5-1), and hose clamp (13) (Item 21, Table 5-1).

f. Tag new hose "To fuel return from secondary fuel filter".

5-4.2 Shock Mount Installation.

WARNING

All metal jewelry can conduct electricity and become entangled in generator set components. Remove all jewelry when working on generator set. Failure to comply with this warning can cause injury or death to personnel.

WARNING

DO NOT wear loose clothing when performing checks, services and maintenance. Failure to comply with this warning can cause injury or death to personnel.

Install two shock mounts (23, Figure 5-20) (Item 116, Table 5-1) with four bolts (22), new lock washers (21), and nuts (20).

5-4.3 Inertia Ring Installation.

NOTE

Mark location of spacers, washers, and bolts to ensure correct positioning during installation.

a. a. Remove bolts (1 and 2, Figure 5-26.1), washers (3), spacers (4), and drive discs (5) from generator drive hub (6).

b. b. Place inertia ring (7) (Item 115, Table 5-1) on generator shaft just inside drive hub (6).

NOTE

Ensure correct positioning of spacers, washers, and bolts as marked on removal.

c. c. Install bolts (1 and 2), washers (3), spacers (4), and drive discs (5) to drive hub (6).

d. d. Torque bolts (1 and 2) to 35 ft-lbs (47 Nm).

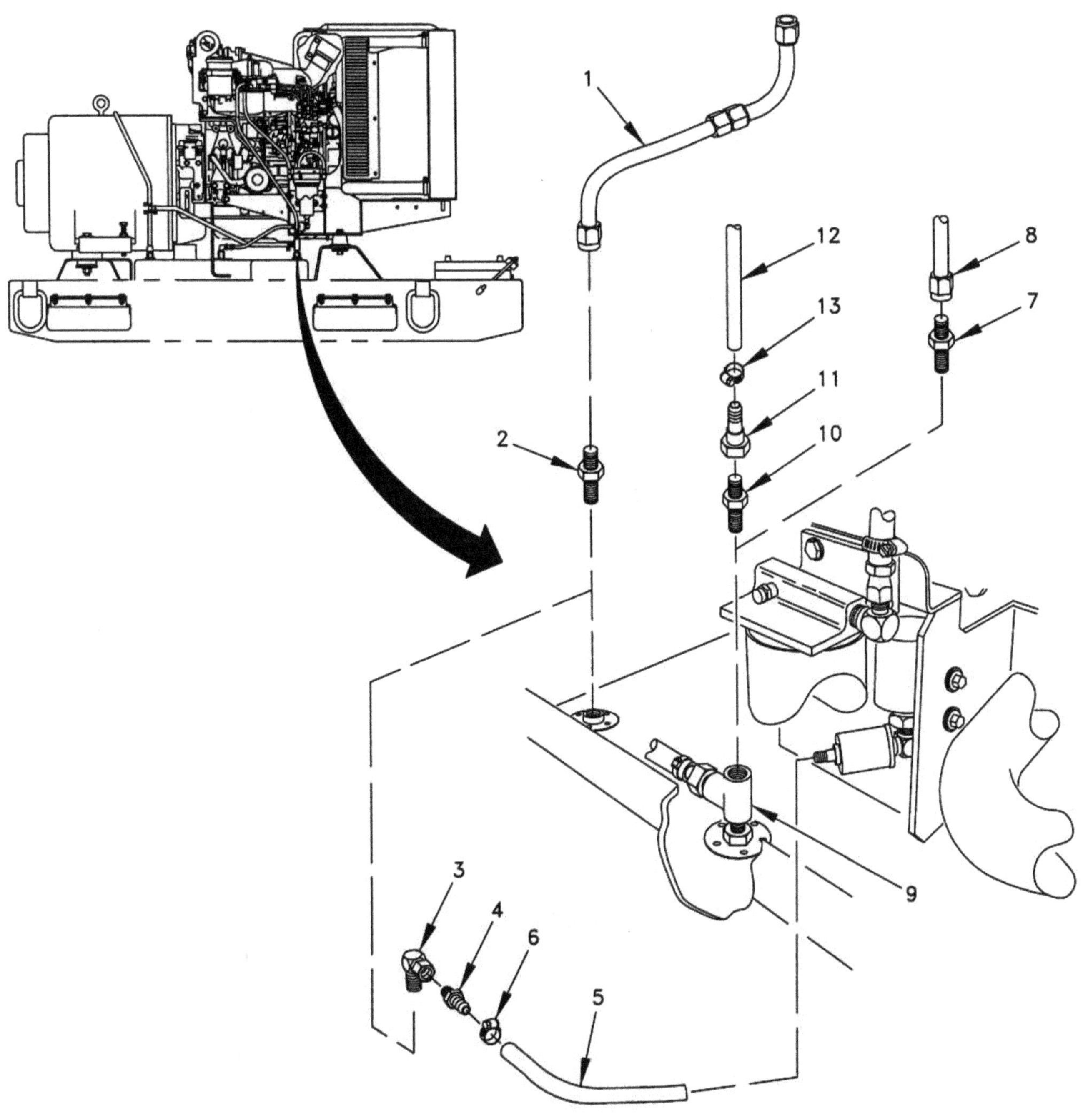

Figure 5-26. Fuel Line.

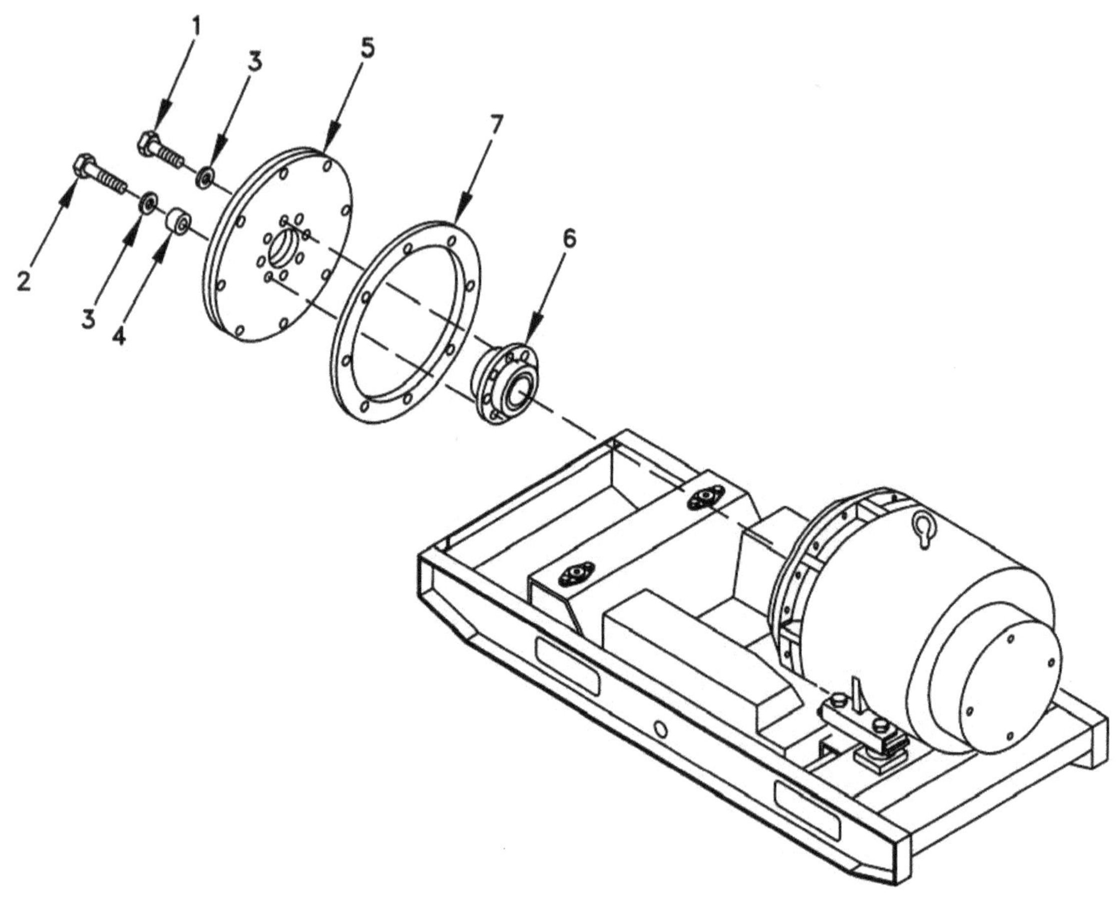

Figure 5-26.1. Inertia Ring.

Section IV. ENGINE INSTALLATION

5-5 NEW ENGINE INSTALLATION PROCEDURES

5-5.1 Engine Installation.

WARNING

All metal jewelry can conduct electricity and become entangled in generator set components. Remove all jewelry when working on generator set. Failure to comply with this warning can cause injury or death to personnel.

WARNING

DO NOT wear loose clothing when performing checks, services and maintenance. Failure to comply with this warning can cause injury or death to personnel.

WARNING

Rated capacity of overhead hoist should be at least 1,500 pounds (680 kg). Do not use a hoist with less capacity. Failure to comply with this warning can cause injury or death to personnel, and damage to equipment.

WARNING

Keep hands and feet from underside of engine and generator while using lifting device to remove them from the skid base. Failure to comply with this warning can cause injury or death to personnel.

NOTE

Winterization kit may have to be installed after installation of new engine. Refer to Chapter 6 for installation instructions.

a. a. Attach lifting harness to engine (1, Figure 5-27) and overhead hoist. Take up slack.

b. b. With aid of an assistant, lift engine from engine stand and position engine on skid base (2), aligning flywheel housing to mounting holes, brackets, and generator. Insert generator spacer (3) (Item 88, Table 5-1) between engine and generator.

c. Locate two spacers on generator coupling disk 45 degrees off from bolts in engine flywheel (Figure 5-27).

WARNING

Do not use the engine starter to turn the flywheel. Failure to comply with this warning can cause injury to personnel.

WARNING

Support components when removing attaching hardware or component may fall. Failure to comply with this warning can cause injury to personnel, and damage to equipment.

NOTE

Align flywheel and generator coupling disk at top dead center.

d. d. Remove four screws (4), washers (5), and cover plate (6) from end of generator.

e. e. Slide generator rotor toward engine 2 or 3 inches in order to have enough room to install bolts securing engine flywheel to disks on generator.

f. f. Secure generator drive disc and inertia ring to engine flywheel using eight lock washers (8) (Item 54, Table 5-1) and screws (7) (Item 104, Table 5-1). Torque screws to 35 ft-lbs (47 Nm).

g. g. Install cover plate (6) to end of generator with four screws (4) and washers (5).

h. h. 50/60 Hz generator set (MEP-804): Secure engine flywheel housing to generator housing lip using 12 lock washers (10) (Item 54, Table 5-1) and screws (9) (Item 89, Table 5-1). Tighten screws slowly to ensure even and proper seating of generator housing lip to flywheel housing. Torque screws to 31 ft-lbs (42 Nm) (flywheel housing).

i. i. 400 Hz generator set (MEP-814): Secure engine flywheel housing to generator housing lip using 12 lock washers (10) (Item 54, Table 5-1) and screws (9) (Item 104, Table 5-1). Tighten screws slowly to ensure even and proper seating of generator housing lip to flywheel housing. Torque screws to 31 ft-lbs (42 Nm) (flywheel housing).

j. j. Secure engine to skid base using two bolts (1 1), snubbing washers (12), washers (13), and nuts (14) removed from old engine. Torque screws to 75 ft-lbs (102 Nm) (engine to skid base).

k. k. Install removed screen/cover (15) to generator case with removed washer (17) and screw (16).

l. l. Loosen nuts (18) (one on each side of generat or) and adjust bolts (19) to approximately 1/2 inch (1.2 cm) from frame. Tighten nuts.

m. Untie rope from generator lifting eye and remove.

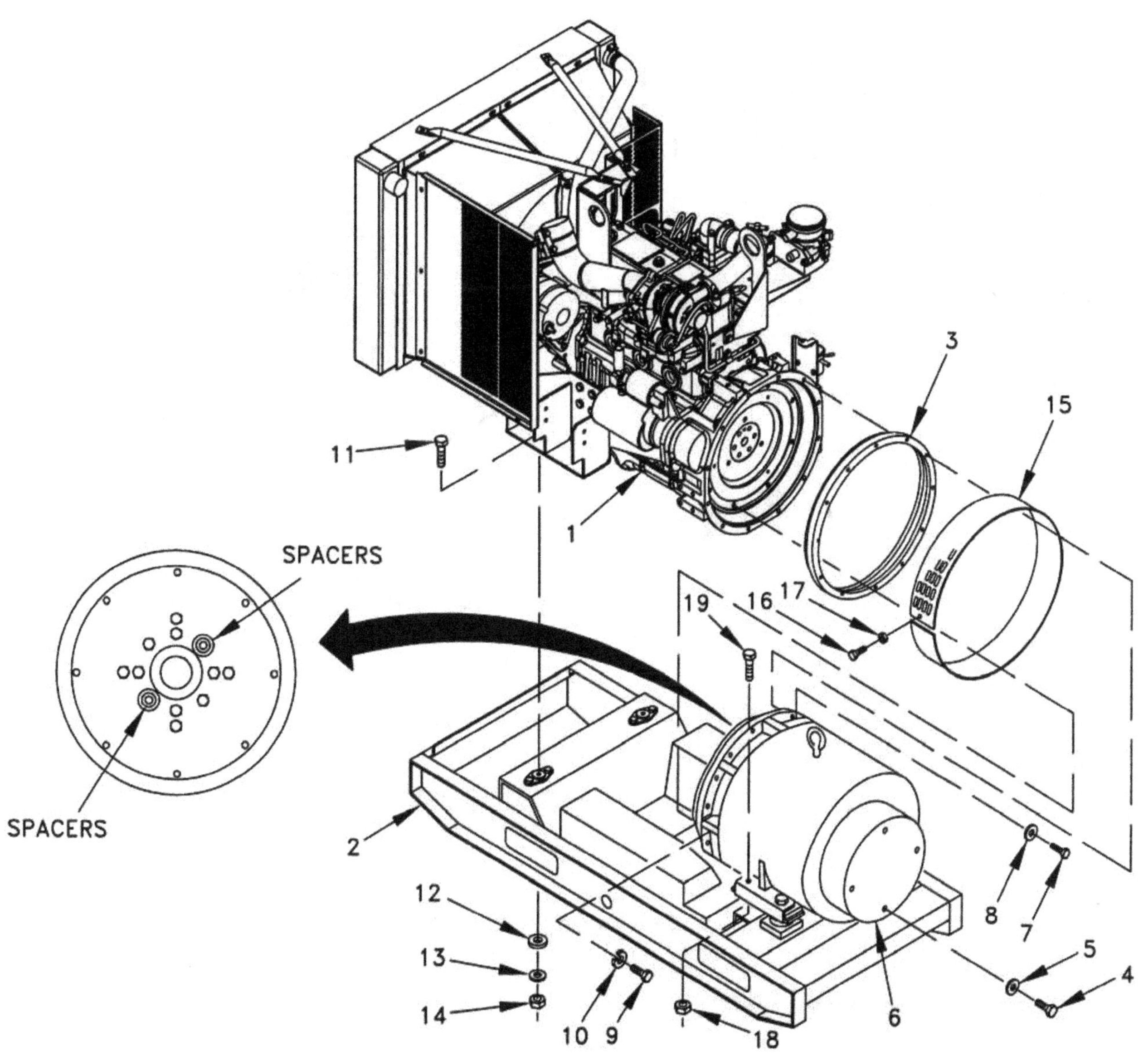

Figure 5-27. Engine.

n. Install two rear engine supports (1, Figure 5-28) (Item 105, Table 5-1) (one on each side of engine) with four screws (2) (Item 108, Table 5-1), lock washers (3) (Item 109, Table 5-1), and flat washers (4) (Item 110, Table 5-1).

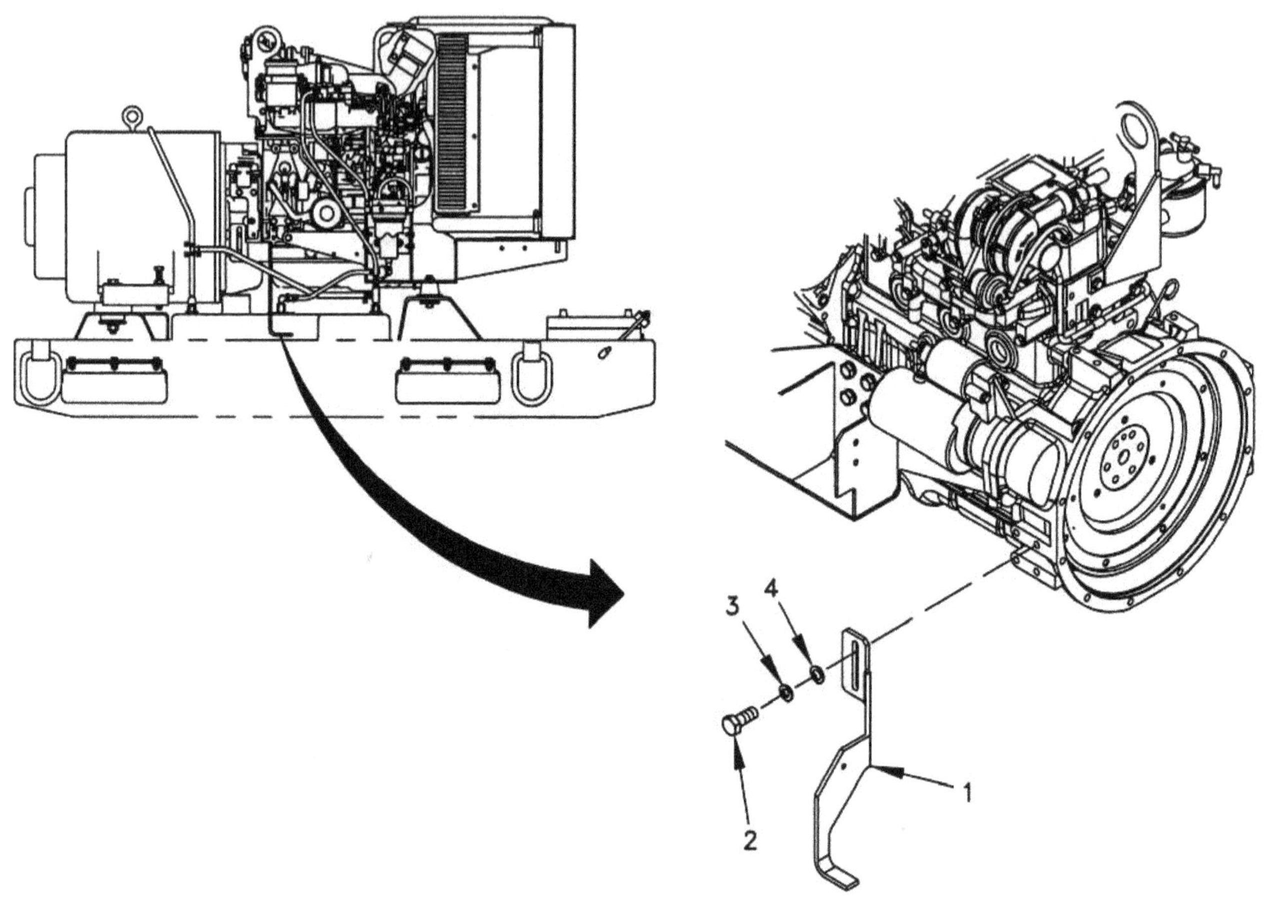

Figure 5-28. Rear Engine Supports.

5-5.2 Air Cleaner and Bracket Installation.

WARNING

All metal jewelry can conduct electricity and become entangled in generator set components. Remove all jewelry when working on generator set. Failure to comply with this warning can cause injury or death to personnel.

WARNING

DO NOT wear loose clothing when performing checks, services and maintenance. Failure to comply with this warning can cause injury or death to personnel.

a. a. Install air cleaner (1, Figure 5-29) to new air cleaner bracket (2) (Item 97, Table 5-1) with four bolts (3), washers (4), new lock washers (5), and nuts (6) retained from removal.

b. b. Loosen two clamps (7) holding air cleaner (2) in position.

c. c. Slide air cleaner (1) away from engine approximately 2 inches (5.1 cm). Do not tighten clamps (7).

d. d. Install air cleaner and bracket with four bolts (8), washers (9), new lock washers (10), and nuts (11) retained from removal of old bracket. Ensure hose (12) mates with opening on air cleaner (1).

e. Tighten clamps (13 and 7), as required.

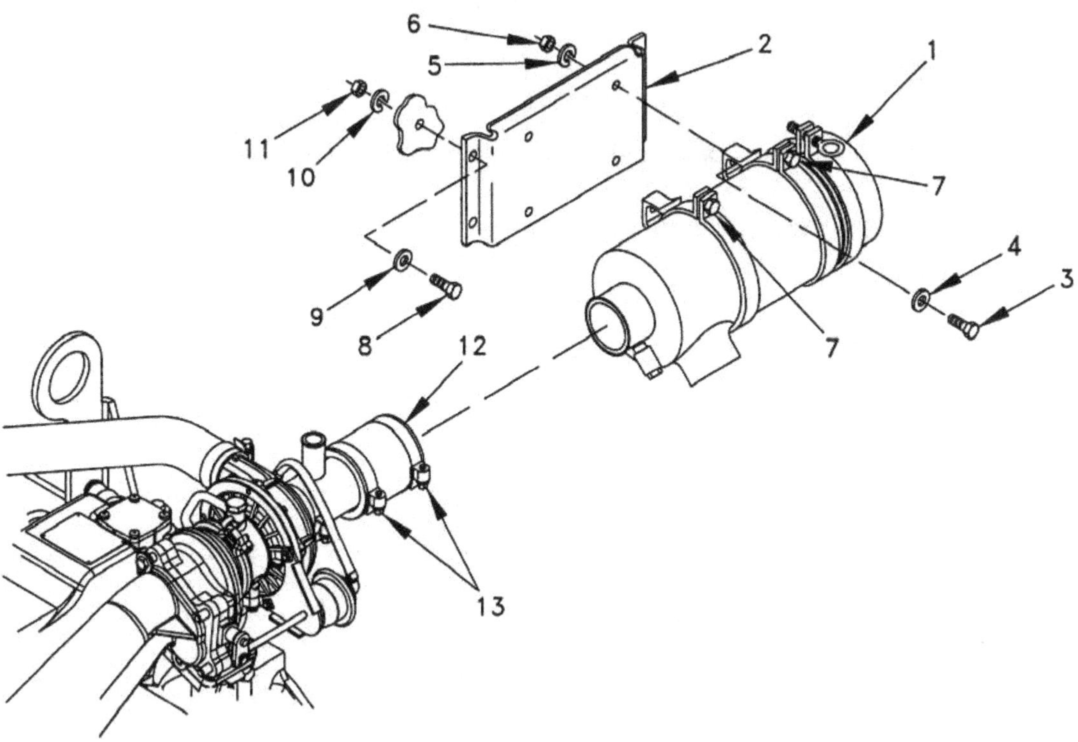

Figure 5-29. Air Cleaner and Bracket.

5-5.3 Fuel Line Installation.

WARNING

All metal jewelry can conduct electricity and become entangled in generator set components. Remove all jewelry when working on generator set. Failure to comply with this warning can cause injury or death to personnel.

WARNING

DO NOT wear loose clothing when performing checks, services and maintenance. Failure to comply with this warning can cause injury or death to personnel.

WARNING

Diesel fuel is flammable and toxic to eyes, skin, and respiratory tract. Skin and eye protection are required when working in contact with diesel fuel. Avoid repeated or prolonged contact. Provide adequate ventilation. Operators are to wash exposed skin and change chemical soaked clothing promptly if exposed to fuel. Failure to comply with this warning can cause injury or death to personnel.

WARNING

Fuels used in the generator set are flammable. Do not smoke or use open flames when performing maintenance. Failure to comply with this warning can cause injury or death to personnel, and damage to the generator set.

NOTE

All pipe thread connections shall be made liquid tight by applying sealing compound to the male threads of each joint.

a. a. Attach new hose tagged "To Fuel Filter Inlet" to fuel filter (Figure 5-30). Secure with hose clamp (Item 21, Table 5-1).

b. b. Attach new hose tagged "To Fuel Return from Secondary Fuel Filter" to secondary fuel filter.

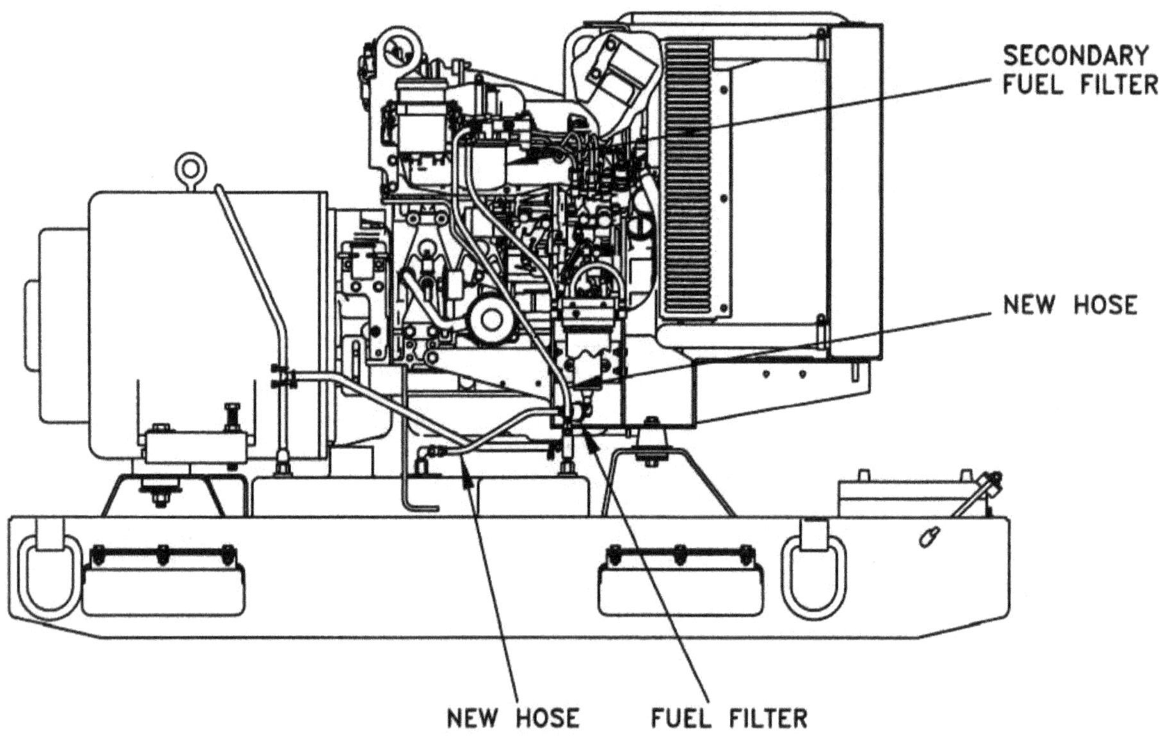

Figure 5-30. Fuel Line.

5-5.4 Radiator Installation.

WARNING

All metal jewelry can conduct electricity and become entangled in generator set components. Remove all jewelry when working on generator set. Failure to comply with this warning can cause injury or death to personnel.

WARNING

DO NOT wear loose clothing when performing checks, services and maintenance. Failure to comply with this warning can cause injury or death to personnel.

NOTE

Grommets between the radiator and the mounting surface set the clearance between the fan blade and the fan shroud.

Mount bushing with the threaded end first and secure with washers and nut.

a. a. Install original radiator (1, Figure 5-31) and two grommets (2) (Item 66, Table 5-1) with two bushings (3) (Item 67, Table 5-1), flat washers (4) (Item 56, Table 5-1), lock washers (5) (Item 65, Table 5-1), and nuts (6) (Item 63, Table 5-1).

b. b. Install two radiator mounts (7) (Item 34, Table 5-1) using two flat washers (8) (Item 76, Table 5-1) and nuts (9) (Item 77, Table 5-1).

c. c. Install two radiator tie rods (10) (Item 31, Table 5-1) using two screws (11) (Item 64, Table 5-1), lock washers (12) (Item 65, Table 5-1), flat washers (13) (Item 76, Table 5-1), and nuts (14) (Item 77, Table 5-1).

d. d. Install three cage nuts (15) (Item 35, Table 5-1) to left fan shroud (16) (Item 30 Table 5-1).

e. e. Install three cage nuts (17) (Item 35, Table 5-1) to right fan shroud (18) (Item 29, Table 5-1).

f. f. Install left fan shroud (16) and right fan shroud (18) using eight screws (19) (Item 62, Table 5-1), lock washers (20) (Item 58, Table 5-1), and flat washers (21) (Item 59, Table 5-1).

g. g. Install lower radiator hose (22) (Item 32, Table 5-1). Secure with two hose clamps (23) (Item 41, Table 5-1).

h. h. Install upper radiator hose (24) (Item 33, Table 5-1). Secure with two hose clamps (25) (Item 41, Table 5-1).

i. i. Install two screws (26) (Item 62, Table 5-1), flat washers (27) (Item 59, Table 5-1), lock washers (28) (Item 58, Table 5-1), and nuts (29) (Item 61, Table 5-1) to top and bottom of fan guard.

j. j. Install identification plate (Item 142, Table 5-1) to right shroud (Item 29, Table 5-1) using 4 rivets (Item 100, Table 5-1).

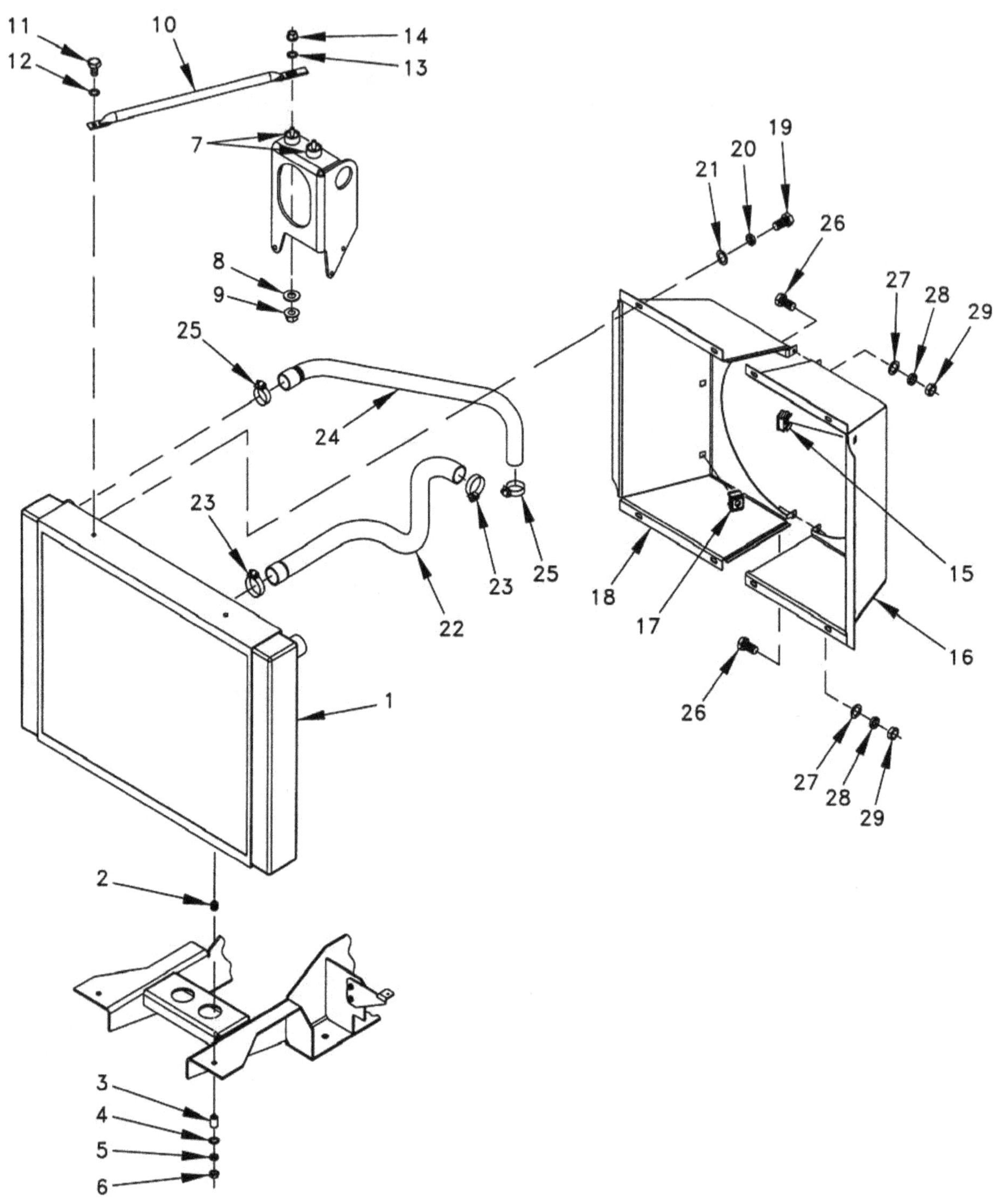

Figure 5-31. Radiator.

5-5.5 Coolant Overflow Bottle and Hose Installation.

WARNING

All metal jewelry can conduct electricity and become entangled in generator set components. Remove all jewelry when working on generator set. Failure to comply with this warning can cause injury or death to personnel.

WARNING

DO NOT wear loose clothing when performing checks, services and maintenance. Failure to comply with this warning can cause injury or death to personnel.

a. a. Install removed coolant overflow bottle (6, Figure 5-7) and brackets and secure with two new lock washers (5) and original nuts (4) and bolts (3).

b. b. Hose (1, Figure 5-32) connected to bottom of coolant overflow bottle will be connected to spout on coolant filler hose and panel assembly to be installed later.

c. c. Connect hose (2) retained from removal to radiator drain valve (3) using existing clamp.

d. d. From old engine configuration there are a pipe T and length of hose coming off water drain line at approximately index number (4). This was connected to an engine drain port which is not needed on new engine.

e. e. Remove pipe T (4) and hose branch plus short hose piece connected to T (5).

f. f. Connect hose (6) to top of coolant overflow bottle and to T at hose (2).

g. g. Connect fuel/water separator drain hose to T on coolant drain line.

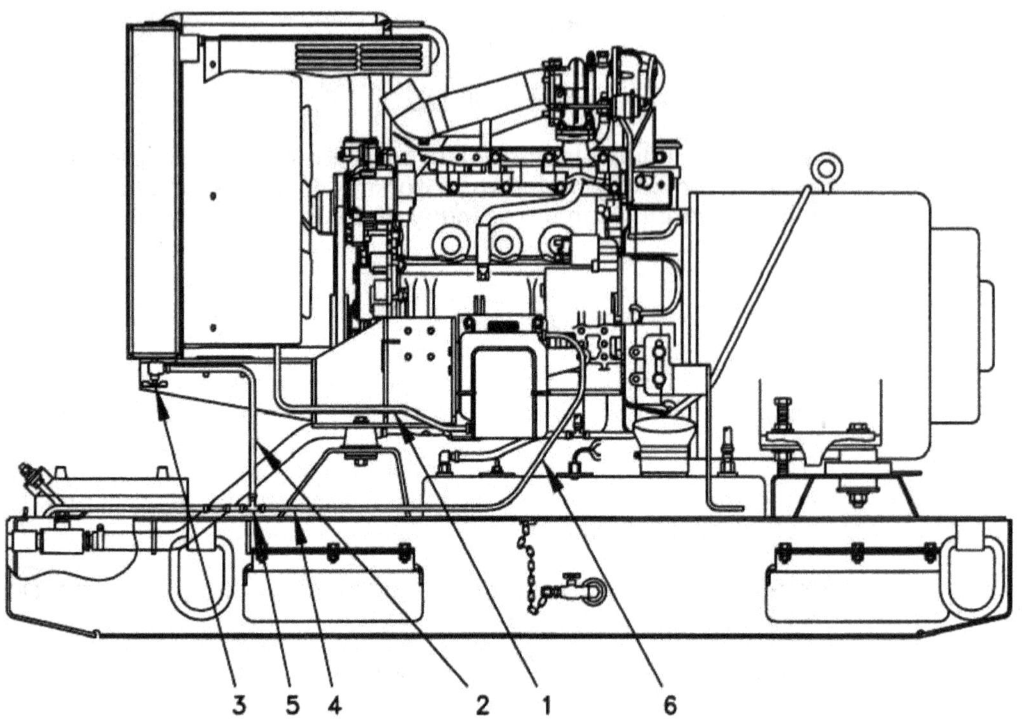

Figure 5-32. Coolant Hose.

5-5.6 Oil Drain Hose Connection.

WARNING

All metal jewelry can conduct electricity and become entangled in generator set components. Remove all jewelry when working on generator set. Failure to comply with this warning can cause injury or death to personnel.

WARNING

DO NOT wear loose clothing when performing checks, services and maintenance. Failure to comply with this warning can cause injury or death to personnel.

Connect elbow (Item 141, Table 5-1) to oil drain hose (2, Figure 5-17) and oil drain valve. Connect oil drain hose (2) to engine oil pan and tighten clamps (1).

5-5.7 Engine Wire Harness Installation.

WARNING

All metal jewelry can conduct electricity and become entangled in generator set components. Remove all jewelry when working on generator set. Failure to comply with this warning can cause injury or death to personnel.

WARNING

DO NOT wear loose clothing when performing checks, services and maintenance. Failure to comply with this warning can cause injury or death to personnel.

Attach engine wire harness to new engine as follows:

NOTE

For wire numbers, the numbers before the letter and the letter constitute the wire number. The numbers after the letter identify the wire gauge.

a. a. Locate branch of wire harness with wires marked 168B20 and 223B20.

b. b. Connect wire 168B20 to C contact of coolant high temperature switch HT (Figure 5-33).

c. c. Connect wire 223B20 to NO contact of coolant high temperature switch HT.

d. d. Locate branch of wire harness with wire marked 176B20.

e. e. Connect wire 176B20 to contact of coolant temperature sending unit MT6.

f. f. Locate branch of wire harness with wires marked 100R16 and a shielded cable.

g. g. Connect wire 100R16 and shield (blk/wht) of shielded cable to ground lug on body of battery charging alternator G2 per Figure 5-34 and wire markings.

h. h. Locate wire 164D16 and cut terminal from end of wire. Install new terminal (loose packed with Item 84, Table 5-1).

i. i. Connect wire 164D16 to large electrical stud on rear of battery charging alternator G2.

j. j. Connect wire 248A16 to IGN spade terminal on battery charging alternator G2.

k. k. Locate branch of wire harness with wires marked 147B16 and 148B16.

l. l. Connect wires 147B16 and 148B16 to magnetic pickup unit (MPU) (Figure 5-35). Polarity is not important.

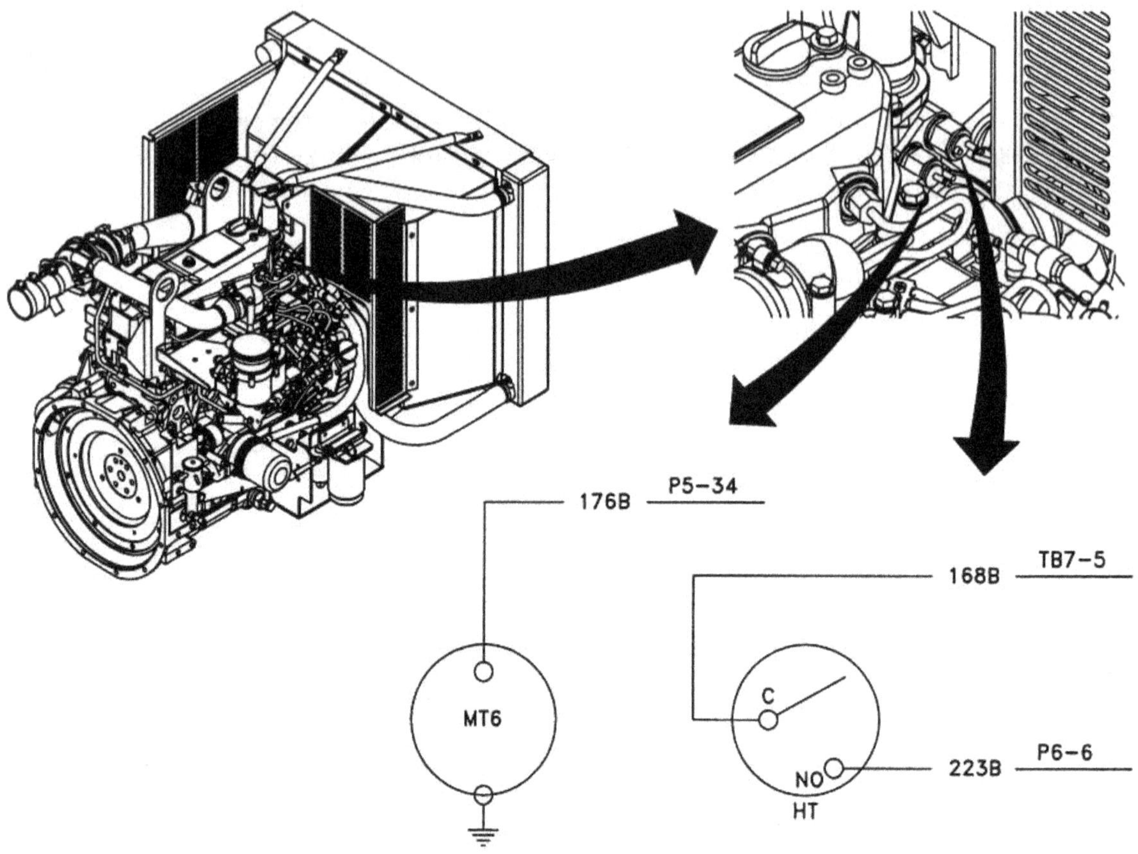

Figure 5-33. Coolant Sensor Wire Harness.

m. m. Locate branch of wire harness with wire marked 166A16.

n. n. Connect wire 166A16 to + terminal of starter solenoid L4 (Figure 5-36) (small electrical terminal on rear of starter solenoid).

o. o. Locate branch of wire harness with wire marked 165F8.

p. p. Connect wire 165F8 to terminal 1 of starter solenoid L4 (large electrical terminal on starter solenoid).

q. q. Locate branch of wire harness with wire marked 125A20.

r. r. Connect wire 125A20 to + terminal of starter B1. Peel back rubber boot to expose electrical connection.

s. s. Locate branch of wire harness with wire marked 100L16.

t. t. Connect wire 100L16 to - terminal of starter B1. Do not tighten.

u. u. Locate branch of wire harness with shielded cable with wires marked 100D16 and 165C16.

v. v. Connect wire 165C16 to terminal 1 of starter solenoid L4.

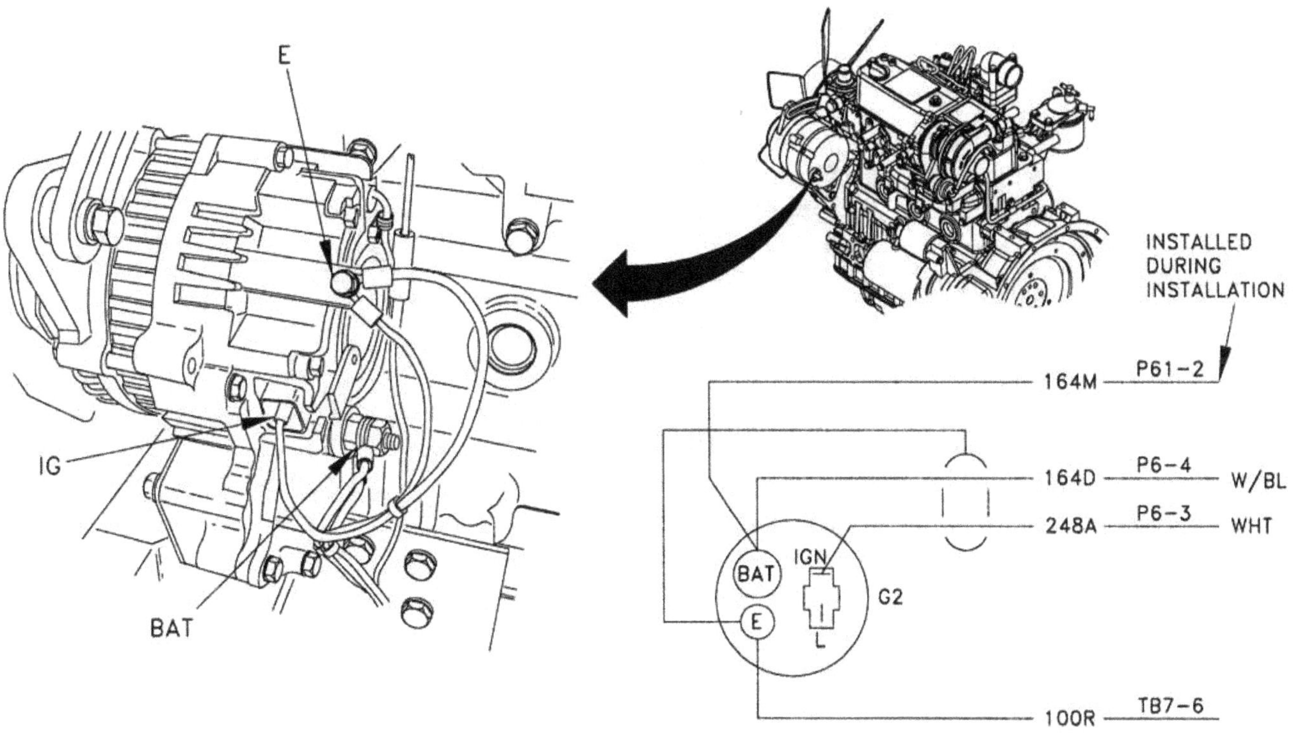

Figure 5-34. Battery Charging Alternator Wire Harness.

w. w. Connect wire 100D16 and shield for this wire (blk/wht) to - terminal of starter B1. Do not tighten.

x. x. Locate 6-inch (15.2 cm) branch of wire harness with wire marked 100C10.

y. y. Connect wire 100C10 to - terminal of starter B1. Do not tighten. (There is also an 11.5-inch (29.2 cm) branch containing other end of this wire. It will be connected later.)

z. Locate branch of wire harness with connector J10.

aa. Connect connector J1 0 to governor actuator A6 (1, Figure 5-37).

ab. Locate branch of wire harness with wire marked 177B20. ac. Con-

nect wire 177B20 to oil pressure sending unit MT7 (2).

ad. Locate branch of wire harness with wires marked 168A20 and 169A20. ae. Con-

nect wire 168A20 (black wire) to C contact of low oil pressure switch OP (3). af. Con-

nect wire 169A20 (red wire) to NC contact of low oil pressure switch OP (3).

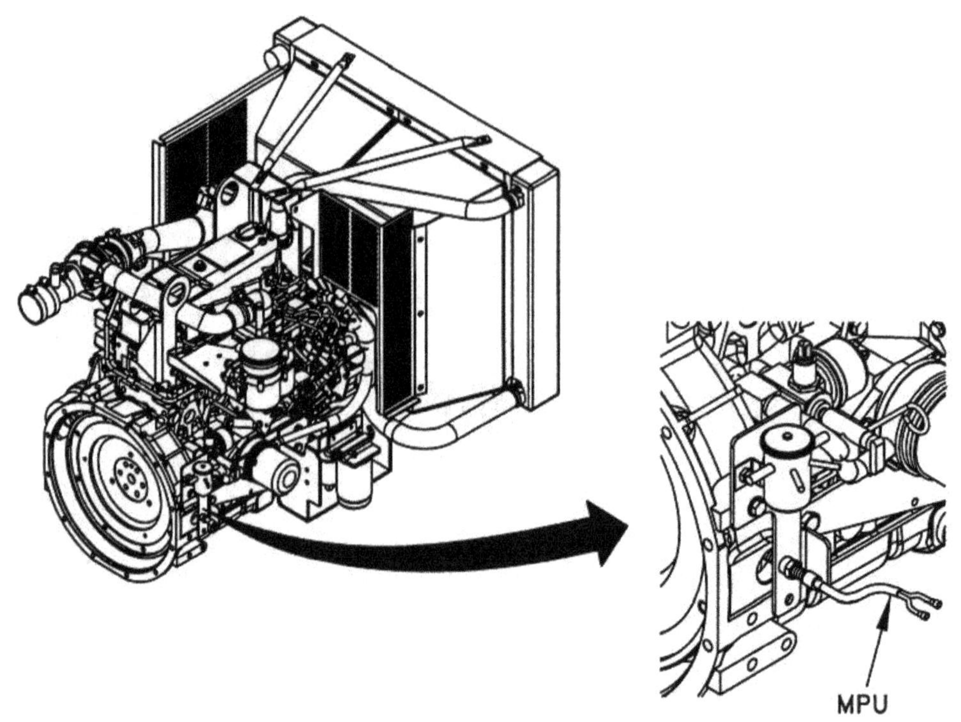

Figure 5-35. MPU Wire Harness.

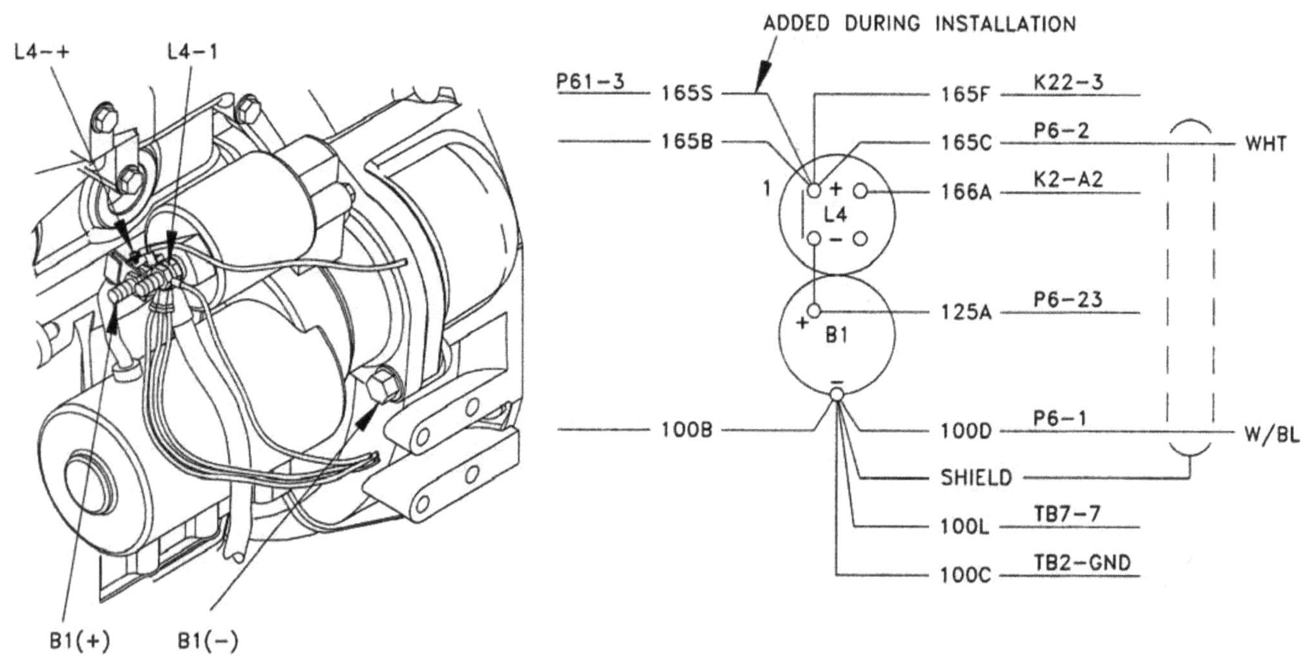

Figure 5-36. Starter and Starter Solenoid Wire Harness.

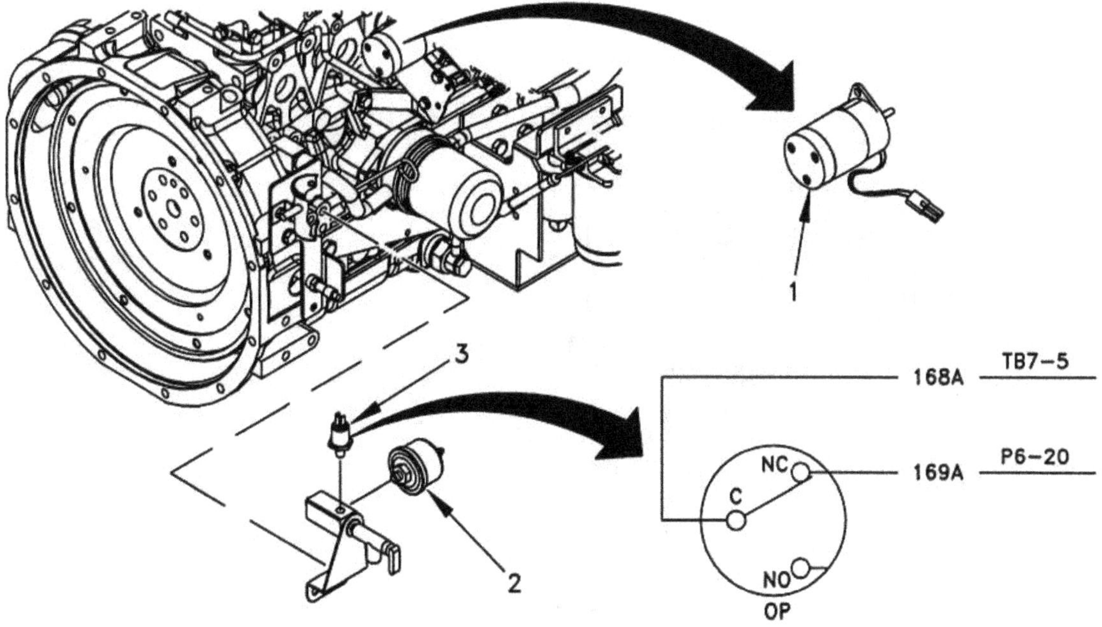

Figure 5-37. Lubrication System Wire Harness.

ag. Locate wire assembly with connector J13 (Item 84, Table 5-1).

ah. Connect connector J13 to fuel transfer pump E2 connector P13 (Figure 5-38).

ai Harness containing connector J13 also contains connector P61. Lay harness containing connector P61 over air cleaner bracket and position connector P61 near air inlet access door.

aj Locate wire 164M16 (in P61) and route loose end to battery charging alternator G2. Connect wire 164M16 to G2 BAT connection (Figure 5-34).

ak. Locate wire 165S16 (in P61) and route loose end to starter solenoid L4. Connect wire 165S16 to terminal 1 of starter solenoid L4 (Figure 5-36).

al. Install wire (16, Figure 5-19) to fuel sending unit (17) using nut (14) and new lock washer (15).

am. Install wire (21) to fuel sending unit (17) using screw (18), new lock washer (19), and flat washer (20).

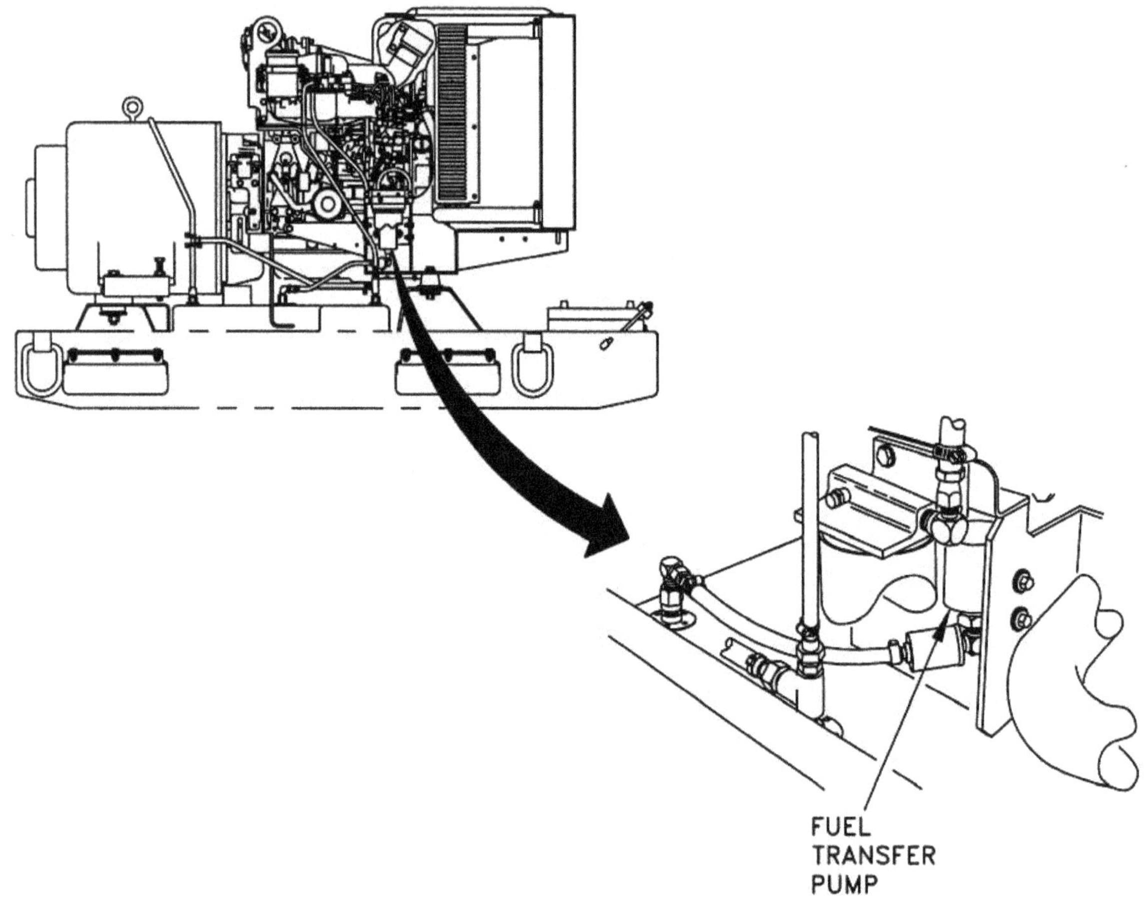

Figure 5-38. Fuel Transfer Pump Wire Harness.

an. Install contactor K22 (1, Figure 5-39) using two screws (2), (Item 62, Table 5-1), lock washers (3) (Item 58, Table 5-1), and washers (4) (Item 83, Table 5-1).

ao. Locate wire 243A8. One end is connected to contactor K22, pin 4; the other end is loose. Cut terminal off of loose end of this wire.

ap. Strip loose end of this wire and crimp on new terminal (loose packed with Item 75, Table 5-1).

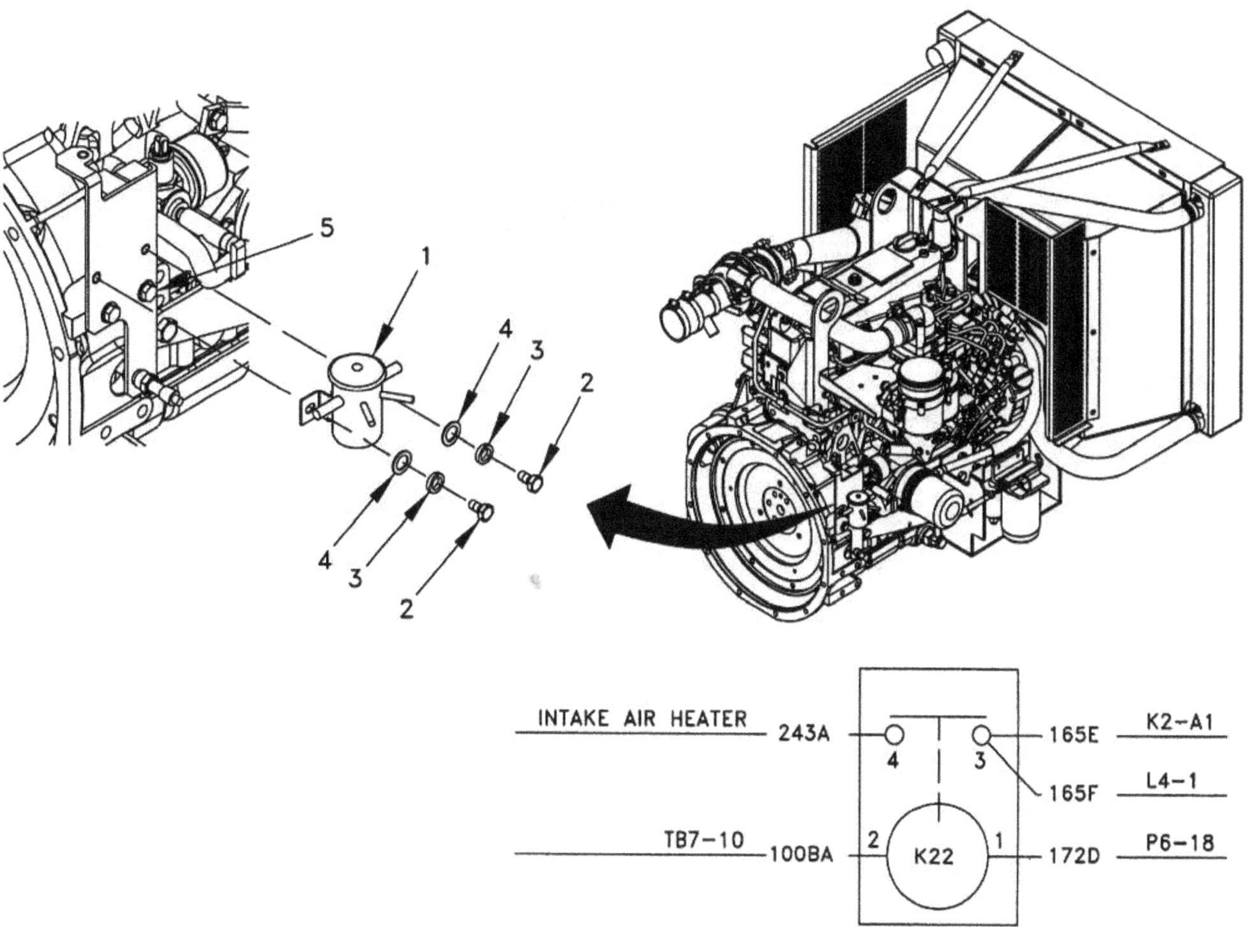

Figure 5-39. Engine Ground Stud.

aq. Connect end of wire 243A8 to lower intak e air preheater terminal 1 (Figure 5-40).

ar. Locate single wire assembly 100HTR (Item 75, Table 5-1) and connect to upper intake air preheater terminal 1 (Figure 5-40).

as. Remove mounting bolt (5, Figure 5-39) from magnetic pickup bracket and install ground wire 100FP16 (from connector P13) and wire 100HTR and tighten.

at. Locate wire labeled JUMPER (Item 81, Table 5-1).

au. Connect wire labeled JUMPER between terminal 2 of upper intake air preheater and terminal 2 of lower air intake preheater (Figure 5-40).

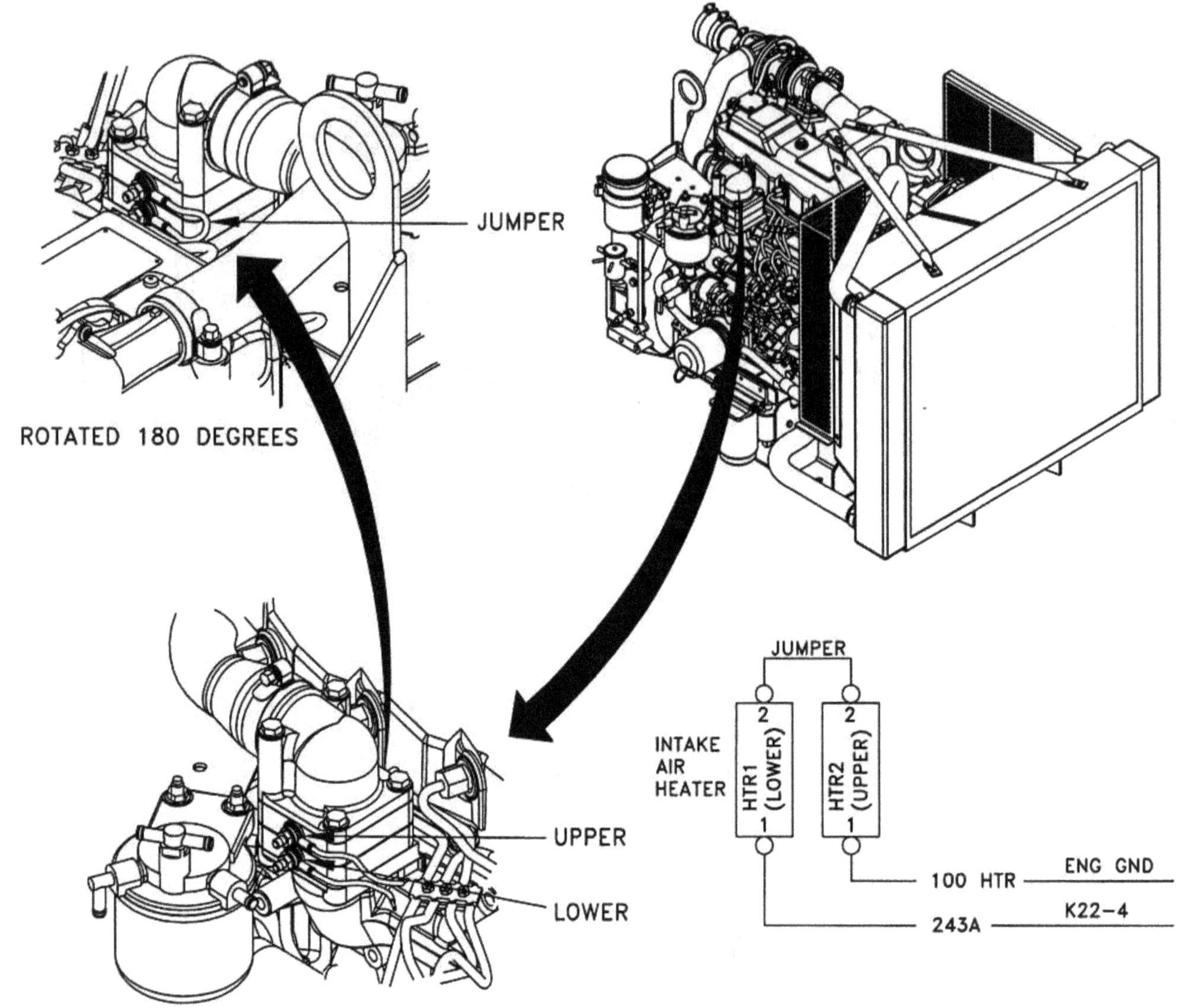

Figure 5-40. Intake Air Preheater Wire Harness.

5-5.7A Harness Shielding Installation.

a. a. Install EMI/RFI shielding (1, Figure 5-40.1) (Item 130, Table 5-1) and ground strap (2) (Item 137, Table 5-1) to harness.

b. b. Attach ground strap (2) to ground stud (3) on generator using existing hardware.

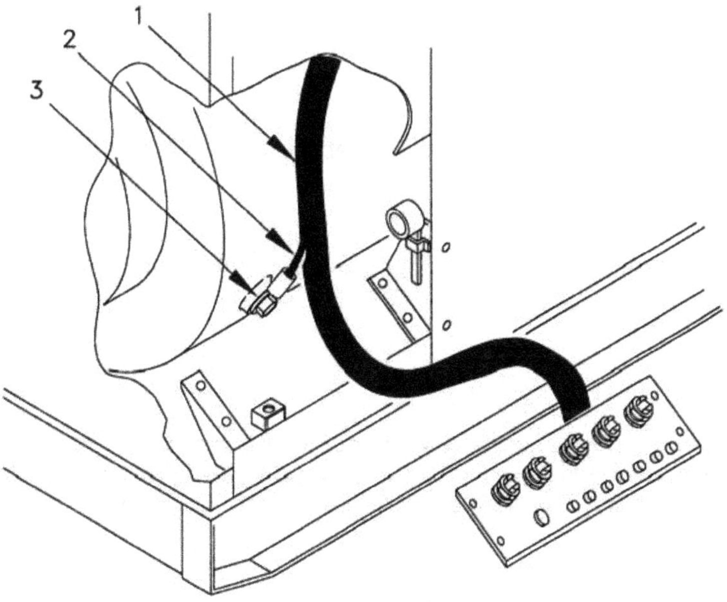

Figure 5-40.1. Harness Shielding.

5-5.8 Left Rear Side Panel Installation.

WARNING

All metal jewelry can conduct electricity and become entangled in generator set components. Remove all jewelry when working on generator set. Failure to comply with this warning can cause injury or death to personnel.

WARNING

DO NOT wear loose clothing when performing checks, services and maintenance. Failure to comply with this warning can cause injury or death to personnel.

WARNING

Many components require a two-person lift. Lifting heavy components can cause back strain. Ensure proper lifting techniques are used when lifting heavy components. Failure to comply with this warning can cause injury to personnel.

a. a. With aid of an assistant, position left rear side housing panel (4, Figure 5-13), lower side panel (8), and door (9) on generator set.

b. b. Secure left rear side panel (4) to skid base with four washers (3), new lock washers (2), and bolts (1).

c. c. Secure lower left side panel (8) to skid base with seven washers (7), new lock washers (6), and bolts (5).

d. d. Secure fuel drain cap chain to bolt on lower left side panel (8).

e. e. Connect battery cables (3, Figure 5-12) and slave cables (4) to NATO slave receptacle (5) with new lock washers (2) and bolts (1).

f. Connect wire 100B to - terminal of star ter B1 (Figure 5-36). Tighten bolt.

WARNING

All metal jewelry can conduct electricity and become entangled in generator set components. Remove all jewelry when working on generator set. Failure to comply with this warning can cause injury or death to personnel.

WARNING

DO NOT wear loose clothing when performing checks, services and maintenance. Failure to comply with this warning can cause injury or death to personnel.

WARNING

Many components require a two-person lift. Lifting heavy components can cause back strain. Ensure proper lifting techniques are used when lifting heavy components. Failure to comply with this warning can cause injury to personnel.

WARNING

Diesel fuel is flammable and toxic to eyes, skin, and respiratory tract. Skin and eye protection are required when working in contact with diesel fuel. Avoid repeated or prolonged contact. Provide adequate ventilation. Operators are to wash exposed skin and change chemical soaked clothing promptly if exposed to fuel. Failure to comply with this warning can cause injury or death to personnel.

WARNING

Fuels used in the generator set are flammable. Do not smoke or use open flames when performing maintenance. Failure to comply with this warning can cause injury or death to personnel, and damage to the generator set.

a. Connect auxiliary fuel pump (6, Figure 5-11) and fuel float module (7) electrical connectors. Remove tags.

b. b. Remove caps and connect fuel line (4) to fuel pump (6).

c. c. Remove caps and connect hose (3) to fuel filler panel (5). Tighten clamp (2).

d. Remove caps and connect fuel filler panel (5) to fuel tank. Tighten clamp (1).

5-5.10 Rear Housing Panel Installation.

WARNING

All metal jewelry can conduct electricity and become entangled in generator set components. Remove all jewelry when working on generator set. Failure to comply with this warning can cause injury or death to personnel.

WARNING

DO NOT wear loose clothing when performing checks, services and maintenance. Failure to comply with this warning can cause injury or death to personnel.

WARNING

Many components require a two-person lift. Lifting heavy components can cause back strain. Ensure proper lifting techniques are used when lifting heavy components. Failure to comply with this warning can cause injury to personnel.

a. Install rear housing panel (8, Figure 5-10) on generator set with 12 washers (7), new lock washers (6), and bolts (5).

b. Install six nuts (4), new lock washers (3), washers (2), and bolts (1).

5-5.11 Front Housing Assembly and Side Panel Installation.

WARNING

All metal jewelry can conduct electricity and become entangled in generator set components. Remove all jewelry when working on generator set. Failure to comply with this warning can cause injury or death to personnel.

WARNING

DO NOT wear loose clothing when performing checks, services and maintenance. Failure to comply with this warning can cause injury or death to personnel.

WARNING

Many components require a two-person lift. Lifting heavy components can cause back strain. Ensure proper lifting techniques are used when lifting heavy components. Failure to comply with this warning can cause injury to personnel.

a. Install right side panel (9, Figure 5-9) using seven washers (25), new lock washers (24), and bolts (23).

b. b. Install three nuts (21), washers (20), new lock washers (19), and bolts (18) securing right side panel (9) to right rear panel (22).

c. c. Install front housing assembly (8) with six washers (17), new lock washers (16), and bolts (15).

d. d. Install two nuts (13), washers (12), new lock washers (11), and bolts (10) securing front housing assembly (8) to skid base (14).

e. Install six nuts (7), washers (6), new lock wa shers (5), and bolts (4) securing front housing assembly (8) to side panels (3 and 9).

f. Install ground rods (2) inside left side panel (3) and secure with bolt (1).

5-5.12 Fan Guard Installation.

WARNING

All metal jewelry can conduct electricity and become entangled in generator set components. Remove all jewelry when working on generator set. Failure to comply with this warning can cause injury or death to personnel.

WARNING

DO NOT wear loose clothing when performing checks, services and maintenance. Failure to comply with this warning can cause injury or death to personnel.

WARNING

Many components require a two-person lift. Lifting heavy components can cause back strain. Ensure proper lifting techniques are used when lifting heavy components. Failure to comply with this warning can cause injury to personnel.

WARNING

Fan has sharp blades. Use caution and wear gloves when removing or installing belts. Failure to comply with this warning can cause injury to personnel.

a. a. Install left fan guard (8, Figure 5-41) (Item 106, Table 5-1) using two screws (1) (Item 111, Table 5-1), lock washers (2) (Item 112, Table 5-1), flat washers (3) (Item 113, Table 5-1), screw (4) (Item 117, Table 5-1), flat washer (5) (Item 119, Table 5-1), lock washer (6) (Item 120, Table 5-1), and nut (7) (Item 118, Table 5-1).

b. b. Locate hose (9) connected to bottom of overfl ow bottle (10), route along left fan guard (8), and secure with screw (11) (Item 111, Table 5-1), lock washer (12) (Item 112, Table 5-1), flat washer (13) (Item 113, Table 5-1), and clamp (14) (Item 78, Table 5-1). Hose eventually goes to overflow tube on coolant filler hose and panel assembly.

c. c. Install right fan guard (19) (Item 107, Table 5-1) and chain (20) from oil fill cap using four screws (15) (Item 111, Table 5-1), lock washers (16) (Item 112, Table 5-1), flat washers (17) (Item 113, Table 5-1), and nut (18) (Item 118, Table 5-1).

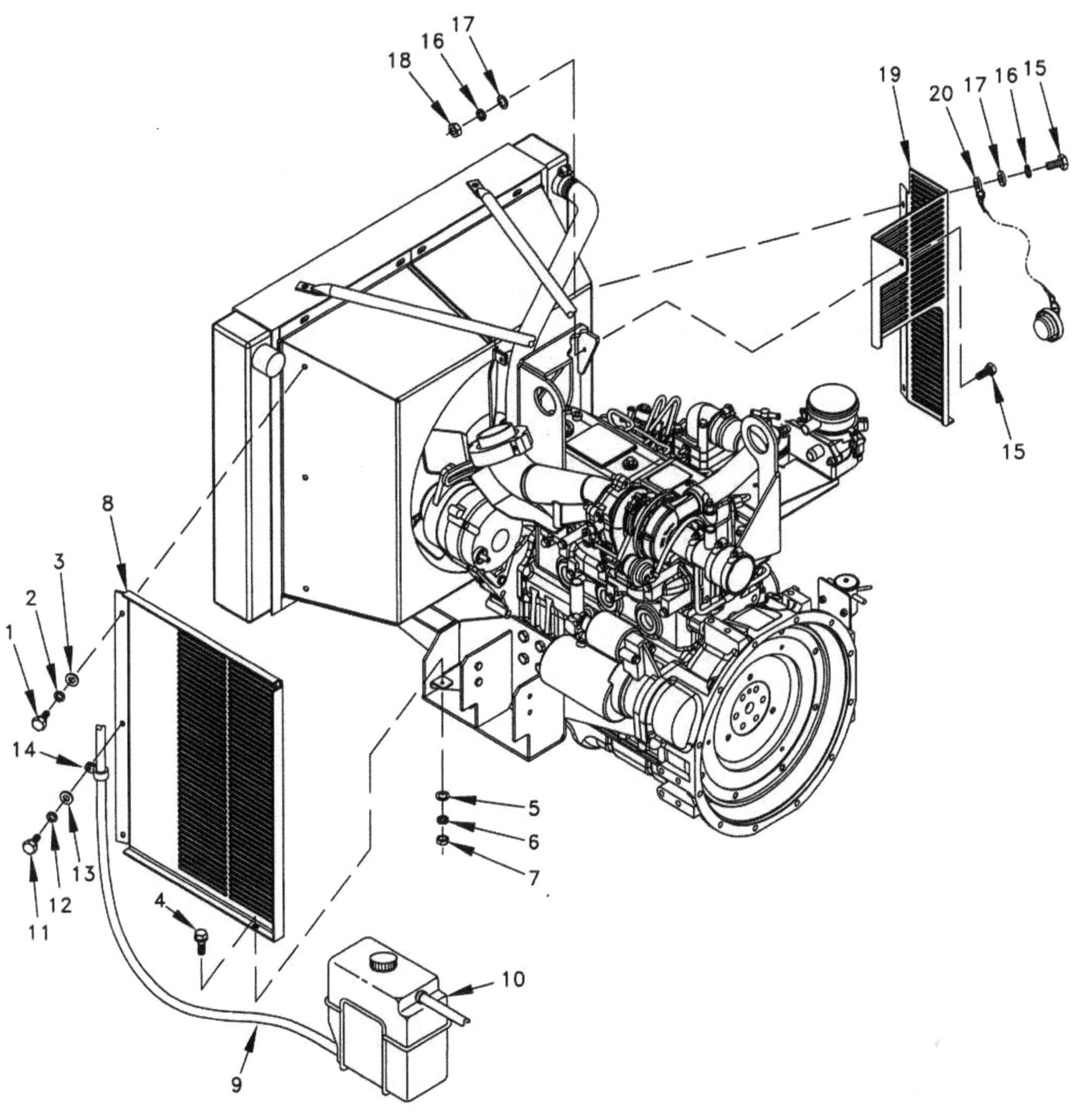

Figure 5-41. Fan Guard.

5-5.13 DEAD CRANK Switch Installation.

WARNING

All metal jewelry can conduct electricity and become entangled in generator set components. Remove all jewelry when working on generator set. Failure to comply with this warning can cause injury or death to personnel.

WARNING

DO NOT wear loose clothing when performing checks, services and maintenance. Failure to comply with this warning can cause injury or death to personnel.

NOTE

Carefully note orientation of plate to ensure text is oriented correctly after bracket is installed.

a. a. Install switch plate (1, Figure 5-42) (Item 114, Table 5-1) to DEAD CRANK switch bracket (2) (Item 91, Table 5-1) using two rivets (3) (Item 100, Table 5-1).

b. b. Install DEAD CRANK switch S10 (4) to bracket (2) using hardware removed with switch.

c. c. Remove existing hardware (5) and install new DEAD CRANK switch bracket (2) using removed

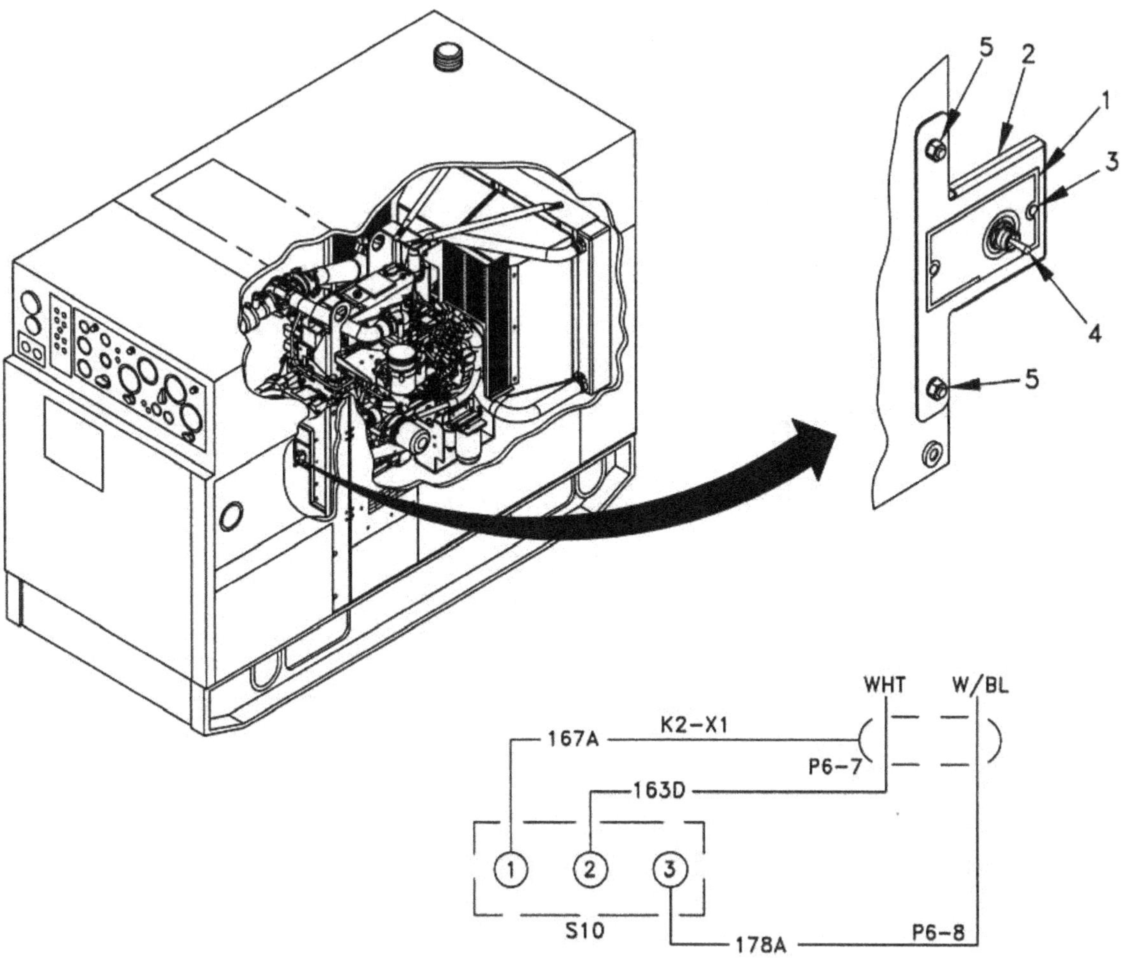

Figure 5-42. DEAD CRANK Switch.

5-5.14 Top Housing Assembly Installation.

WARNING

All metal jewelry can conduct electricity and become entangled in generator set components. Remove all jewelry when working on generator set. Failure to comply with this warning can cause injury or death to personnel.

WARNING

DO NOT wear loose clothing when performing checks, services and maintenance. Failure to comply with this warning can cause injury or death to personnel.

WARNING

Many components require a two-person lift. Lifting heavy components can cause back strain. Ensure proper lifting techniques are used when lifting heavy components. Failure to comply with this warning can cause injury to personnel.

a. Use lifting device and position top housing assembly (11, Figure 5-8) on generator set.

b. Secure top housing assembly (11) to output box angle with bolt (10) and assembled nut (9).

c. Secure top housing assembly (11) to rear side panels with four bolts (8) and assembled nuts (7).

d. Secure top housing assembly (11) to front housing assembly with six bolts (5), washers (6), new lock washers (4), and nuts (3).

e. Connect flexible muffler pipe (17) to engine exhaust (18) using clamp (19) (Item 87, Table 5-1). Tighten clamp.

5-5.15 Coolant Filler Hose and Panel Assembly Installation.

WARNING

All metal jewelry can conduct electricity and become entangled in generator set components. Remove all jewelry when working on generator set. Failure to comply with this warning can cause injury or death to personnel.

WARNING

DO NOT wear loose clothing when performing checks, services and maintenance. Failure to comply with this warning can cause injury or death to personnel.

WARNING

Many components require a two-person lift. Lifting heavy components can cause back strain. Ensure proper lifting techniques are used when lifting heavy components. Failure to comply with this warning can cause injury to personnel.

a. Position coolant filler hose and panel assembly (5, Figure 5-6) in generator set and connect to radiator (2). Tighten clamp (1).

b. Connect overflow hose (4) to coolant filler hose and panel assembly (5). Tighten clamp (3).

5-5.16 Top Housing Panel Installation.

WARNING

All metal jewelry can conduct electricity and become entangled in generator set components. Remove all jewelry when working on generator set. Failure to comply with this warning can cause injury or death to personnel.

WARNING

DO NOT wear loose clothing when performing checks, services and maintenance. Failure to comply with this warning can cause injury or death to personnel.

WARNING

Many components require a two-person lift. Lifting heavy components can cause back strain. Ensure proper lifting techniques are used when lifting heavy components. Failure to comply with this warning can cause injury to personnel.

Install top housing panel (7, Figure 5-5) with 43 washers (3 and 6), new lock washers (2 and 5), and bolts (1 and

5-5.17 Control Box Assembly Wire Harness Installation.

WARNING

All metal jewelry can conduct electricity and become entangled in generator set components. Remove all jewelry when working on generator set. Failure to comply with this warning can cause injury or death to personnel.

WARNING

DO NOT wear loose clothing when performing checks, services and maintenance. Failure to comply with this warning can cause injury or death to personnel.

WARNING

Many components require a two-person lift. Lifting heavy components can cause back strain. Ensure proper lifting techniques are used when lifting heavy components. Failure to comply with this warning can cause injury to personnel.

a. a. Place control box assembly on a clean workbench with front panel lowered.

b. b. Drill 0.56 inch (9/16 in.) hole in control box bottom panel approximately 1.38 inches (3.5 cm) back from hinge and 4.75 inches (12.1 cm) from left edge of front panel opening (Figure 5-43). Deburr edge of hole.

c. c. Insert grommet (Item 124, Table 5-1) in hole.

d. d. Position wire assembly (Item 125, Table 5-1) connector J61 under control box assembly and route wires through new grommet into inside of control box assembly.

e. e. Tag and cut wires 163B16 and 164A16 from rear of fuse FU1 (Figure 5-44).

f. f. Remove nut (1) and fuse FU1 (2).

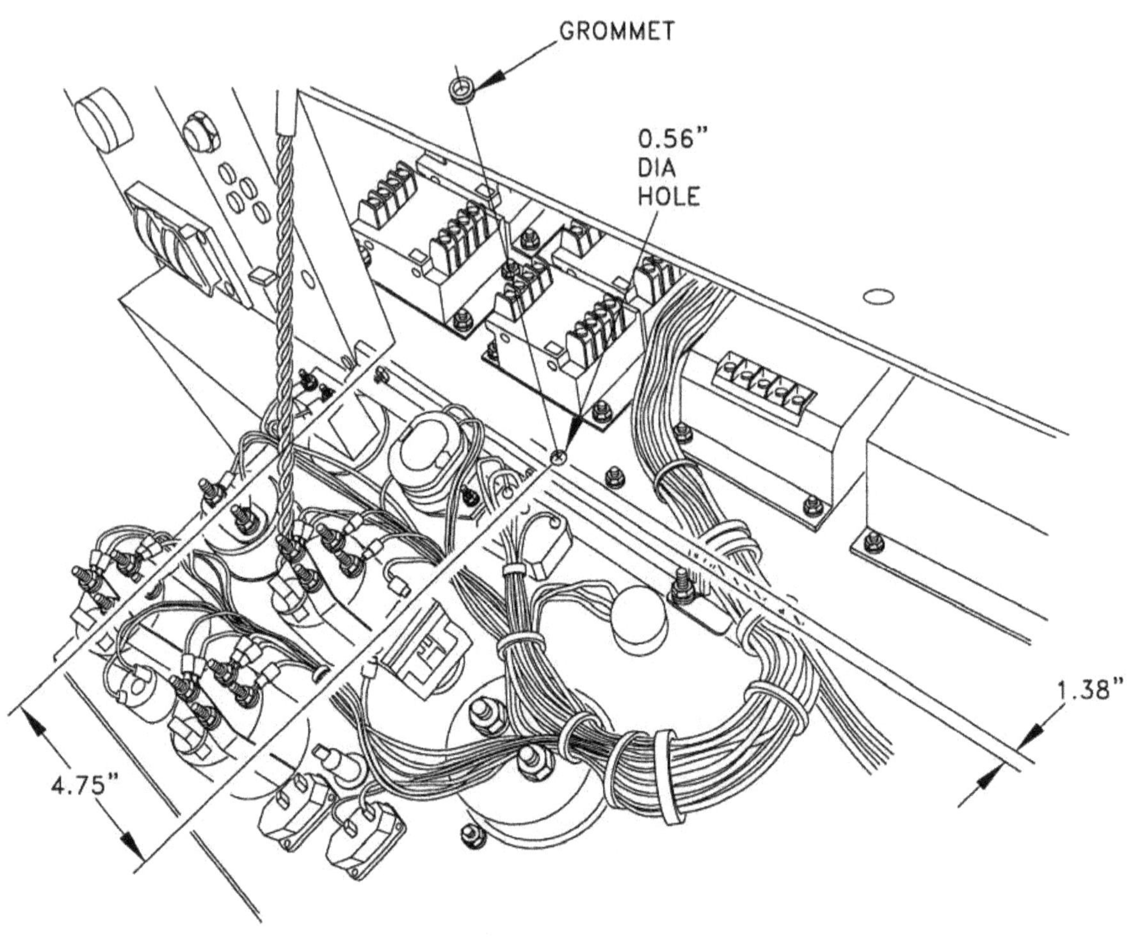

Figure 5-43. Control Panel Modification.

g. g. Install circuit breaker CB2 (3) (Item 126, Table 5-1) where fuse FU1 was located. Secure with nut and washer supplied with circuit breaker CB2.

h. h. Strip wires 163B16 and 164A16 that were cut from fuse FU1 and crimp new terminals (loose packed with Items 85 and 125, Table 5-1) on ends of wires.

i. i. Connect wire 163B16 to LOAD terminal on circuit breaker CB2. Remove tag. Connect wire 164A16 to LINE terminal on circuit breaker CB2. Remove tag.

CAUTION

Match wire numbers of new wires with existing wiring.

j. j. Connect single wire assembly labeled 163K16 (Item 85, Table 5-1) between ammeter shunt MT4-1 (4) and circuit breaker CB2 LOAD terminal. (Ammeter shunt MT4 is located behind bracket that holds circuit breaker CB2 in place.)

k. k. Connect wire labeled 225N16 (J61-1) to terminal board TB5 terminal 17.

l. l. Connect wire labeled 164N16 (J61-2) to circuit breaker CB2 LINE terminal.

CAUTION

Match wire numbers of new wires with existing wiring.

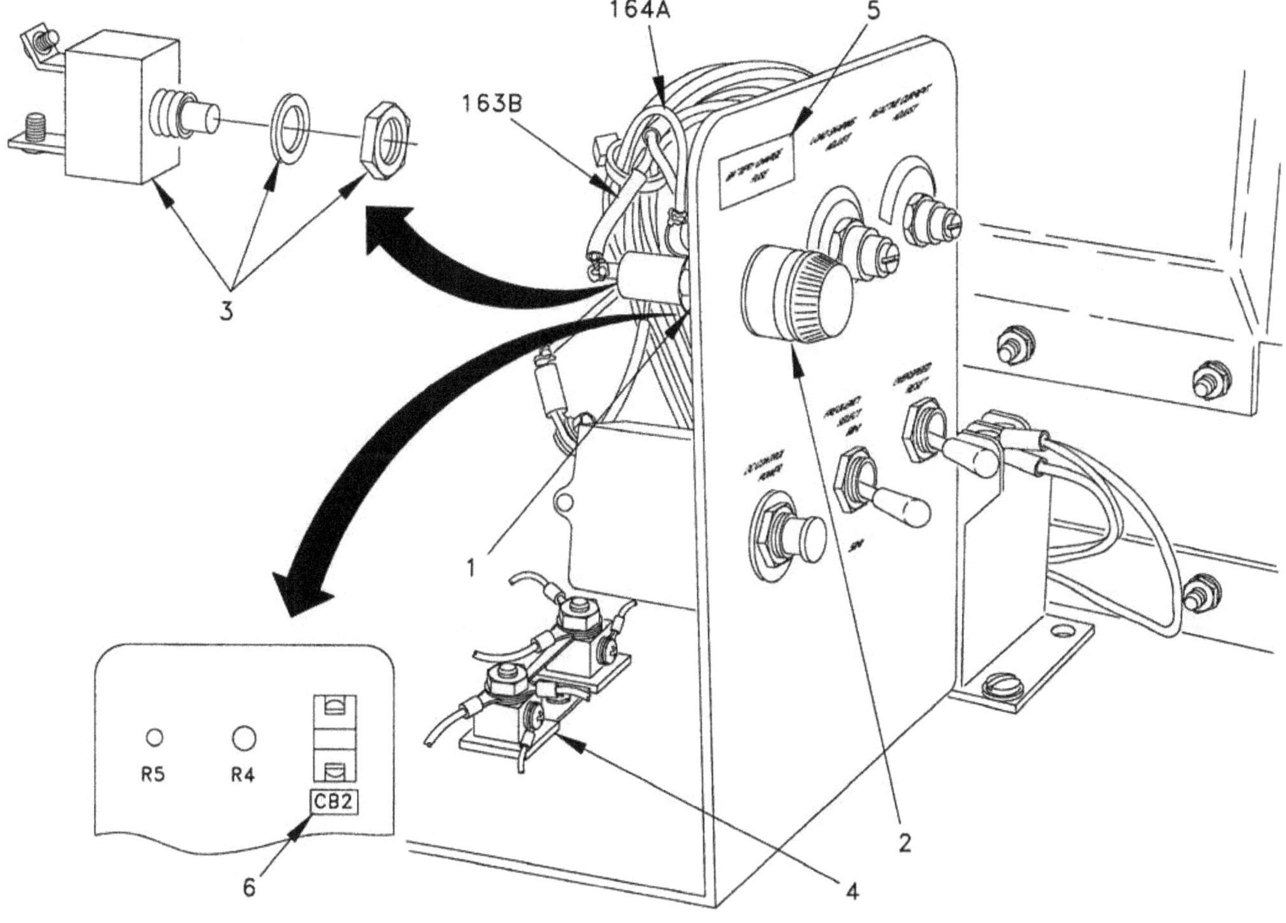

Figure 5-44. Fuse FU1 Replacement with Circuit Breaker CB2.

m. m. Connect wire labeled 165R16 (J61-3) to ammeter shunt MT4 (4) terminal 4.

n. n. Tie wires to existing harness with tiedown straps (Item 8, Table 5-2).

o. o. Clean surfaces with isopropyl alcohol and install new BATTERY CHARGER CIRCUIT BREAKER label (5) (Item 127, Table 5-1) to cover reference to BATTERY CHARGER FUSE and new CB2 label (6) (Item 128, Table 5-1) to cover reference to fuse FU1 (rear of bracket).

5-5.17A Governor Control Unit Replacement (MEP-804B).

a. a. Obtain new governor control unit from kit (Item 140, Table 5-1).

b. b. Carefully, one terminal at a time, remove wiring from old governor control unit and connect wire to new governor control unit.

c. Remove old governor control unit. Retain hardware.

d. d. Install new governor control unit using original hardware.

5-5.17B Control Box Shielding Installation.

a. a. Loosen fasteners and lower front of control box assembly.

b. b. Remove bolts (1, Figure 5-44.1) and nuts (2) securing hinge (3) to control panel frame
(4).

WARNING

CARC paint is a health hazard, and is irritating to eyes, skin, and respiratory system. Wear protective eyewear, mask, and gloves when applying or removing CARC paint. Failure to comply with this warning can cause injury to personnel.

d. d. Remove paint from inside of hinge (3) where seal (8) (Item 133, Table 5-1) will contact hinge.

e. e. Remove paint from edges of control panel frame (4) where adhesive-backed seal (10) (Item
134,

WARNING

Conversion coating material is toxic to eyes, skin, and respiratory tract. Skin and eye protection are required when working in contact with conversion coating material. Avoid repeated or prolonged contact. Work in ventilated area only. Failure to comply with this warning can cause injury or death to personnel.

f. f. Coat bare metal with conversion coating (Item 3.1, Table 5-2). Allow to dry.

g. g. Remove protective paper and install adhesive-backed seal (10) to control panel frame (4).

h. h. Slip seal (11) over control panel frame (4) edge.

i. i. Install hinge (3) to control panel (7) with bolts (5) and nuts (6).

j. j. Install seal (8) and secure with seal retainer (9) (Item 132, Table 5-1), bolts (1), and nuts

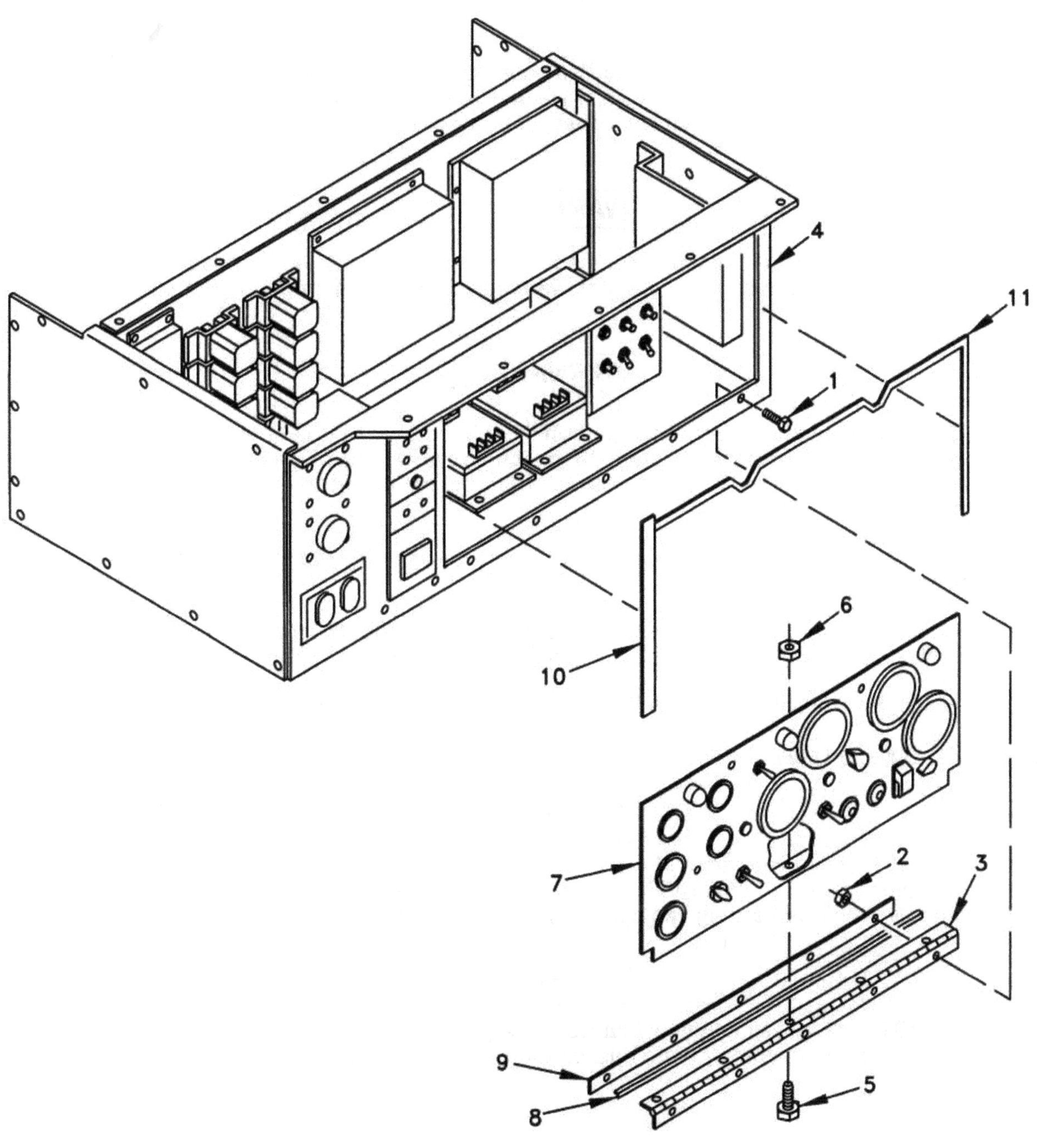

Figure 5-44.1. Control Box Shielding.

5-5.18 Control Box Assembly Installation.

WARNING

All metal jewelry can conduct electricity and become entangled in generator set components. Remove all jewelry when working on generator set. Failure to comply with this warning can cause injury or death to personnel.

WARNING

DO NOT wear loose clothing when performing checks, services and maintenance. Failure to comply with this warning can cause injury or death to personnel.

WARNING

Many components require a two-person lift. Lifting heavy components can cause back strain. Ensure proper lifting techniques are used when lifting heavy components. Failure to comply with this warning can cause injury to personnel.

a. a. Install control box assembly (9, Figure 5-4) on generator set with two nuts (8), washers (7), new lock washers (6), and bolts (5).

b. b. Install 20 nuts (4), washers (3), new lock washers (2), and bolts (1). Hardware on right side of control box assembly also secures output box access door hinge in place.

c. c. Connect wire harness connectors P5, P6, and P61 to control box connectors J5, J6, and J61.

5-5.19 Control Box Top Panel Installation.

WARNING

All metal jewelry can conduct electricity and become entangled in generator set components. Remove all jewelry when working on generator set. Failure to comply with this warning can cause injury or death to personnel.

WARNING

DO NOT wear loose clothing when performing checks, services and maintenance. Failure to comply with this warning can cause injury or death to personnel.

WARNING

Many components require a two-person lift. Lifting heavy components can cause back strain. Ensure proper lifting techniques are used when lifting heavy components. Failure to comply with this warning can cause injury to personnel.

Install control box top panel (5, Figure 5-3) with two nuts (4), new 16 lock washers (3), flat washers (2), and bolts (l).

WARNING

All metal jewelry can conduct electricity and become entangled in generator set components. Remove all jewelry when working on generator set. Failure to comply with this warning can cause injury or death to personnel.

WARNING

DO NOT wear loose clothing when performing checks, services and maintenance. Failure to comply with this warning can cause injury or death to personnel.

WARNING

Many components require a two-person lift. Lifting heavy components can cause back strain. Ensure proper lifting techniques are used when lifting heavy components. Failure to comply with this warning can cause injury to personnel.

a. Install control panel access door (3, Figure 5-2) and hinge (4) on generator set with seven screws (2) and nuts (1).

b. Close control panel access door.

5-5.21 Final Generator Set Wire Harness Connection and Installation.

WARNING

All metal jewelry can conduct electricity and become entangled in generator set components. Remove all jewelry when working on generator set. Failure to comply with this warning can cause injury or death to personnel.

WARNING

DO NOT wear loose clothing when performing checks, services and maintenance. Failure to comply with this warning can cause injury or death to personnel.

a. a. Remove left and right engine access doors (paragraph 2-14.1a thru d).

b. b. Remove hinges for access doors (paragraph 2-14.1c).

WARNING

CARC paint is a health hazard, and is irritating to eyes, skin, and respiratory system. Wear protective eyewear, mask, and gloves when applying or removing CARC paint. Failure to comply with this warning can cause injury to personnel.

c. c. Remove paint from hinge and door at areas of contact.

WARNING

Conversion coating material is toxic to eyes, skin, and respiratory tract. Skin and eye protection are required when working in contact with conversion coating material. Avoid repeated or prolonged contact. Work in ventilated area only. Failure to comply with this warning can cause injury or death to personnel.

e. e. Coat bare metal with conversion coating (Item 3.1, Table 5-2). Allow to dry.

f. f. Install doors and hinges with two ground straps (1 and 2, Figure 5-45) (Item 138 and Item 139, Table 5-1).

g. g. Wrap harness with EMI/RFI shielding (3) (Item 129, Table 5-1). Secure ground strap (4) (Item 136, Table 5-1) to shielding (3).

h. h. Wrap harness with EMI/RFI shielding (5) (Item 131, Table 5-1). Secure ground strap (6) (Item 137, Table 5-1) to shielding (3).

i. i. Secure harness and ground straps (2, 4, and 6) to engine using screw (5) (Item 40, Table 5-1), cable clamp (8) (Item 28, Table 5-1), two screws (7) (Item 51, Table 5-1), and cable clamps (8) (Item 28, Table 5-1).

j. j. Tie new harnesses to engine brackets and loose hoses to adjacent components using tiedown straps (Item 8, Table 5-2).

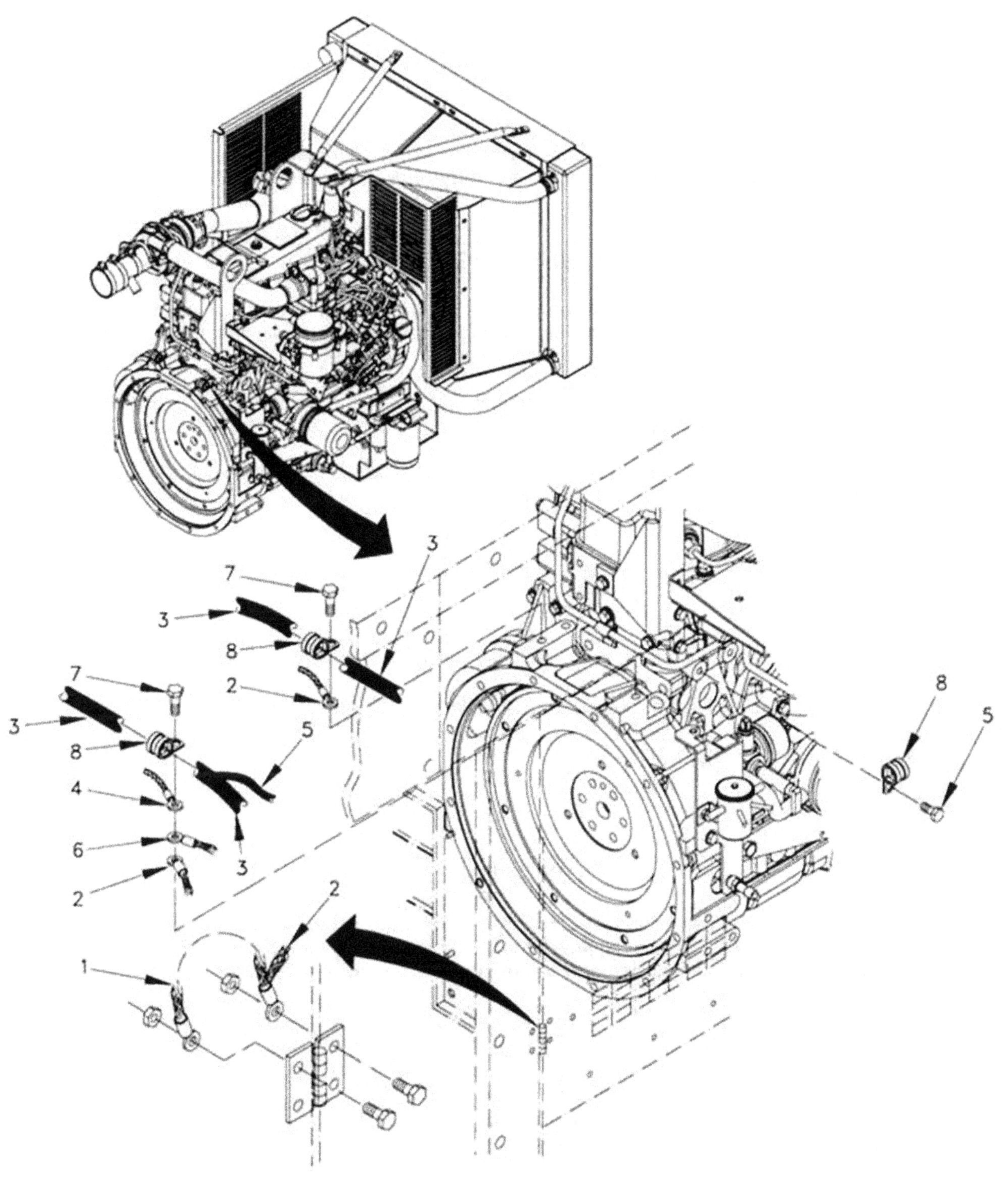

Figure 5-45. Engine Wire Harness Clamps and Shielding.

5-5.22 Generator Set Data Plate Installation.

WARNING

All metal jewelry can conduct electricity and become entangled in generator set components. Remove all jewelry when working on generator set. Failure to comply with this warning can cause injury or death to personnel.

WARNING

DO NOT wear loose clothing when performing checks, services and maintenance. Failure to comply with this warning can cause injury or death to personnel.

a. a. Remove the following plates and attaching hardware: ID plate, schematic diagram plate, wiring diagram plate, operating instructions plate.

b. b. 50/60 Hz generator set (MEP-804): Install ratings ID plate (Item 92, Table 5-1) using four rivets (Item 100, Table 5-1).

c. c. 400 Hz generator set (MEP-814): Install ratings ID plate (Item 123, Table 5-1) using four rivets (Item 100, Table 5-1).

d. d. Install schematic diagram plate (Item 93, Table 5-1) using four rivets (Item 100, Table 5-1).

e. e. Install wiring diagram plate (Item 94, Table 5-1) using 10 rivets (Item 100, Table 5-1).

f. f. Install operating instructions plate (Item 95, Table 5-1) using eight rivets (Item 100, Table 5-1).

5-6 COOLANT, OIL, AND FUEL SERVICE

5-6.1 Coolant Service.

WARNING

All metal jewelry can conduct electricity and become entangled in generator set components. Remove all jewelry when working on generator set. Failure to comply with this warning can cause injury or death to personnel.

WARNING

DO NOT wear loose clothing when performing checks, services and maintenance. Failure to comply with this warning can cause injury or death to personnel.

a. a. Remove radiator cap.

b. b. Check that radiator drain valve (3, Figure 5-32) is closed.

c. c. Fill radiator with proper coolant/antifreeze (Item 1, Table 5-2) per Table 5-3. Fill radiator to a level
2
inches (5.1 cm) below fill opening.

e. e. Fill coolant overflow bottle to COLD level.

f. f. Install coolant overflow bottle cap and radiator cap.

Table 5-3. Coolant.

Ambient Temperature	Radiator Coolant	Ratio
+40°F to +120°F (4°C to 49°C)	Water: MIL-A-53009A(1) Inhibitor, Corrosion	35:1
-25°F to +120°F (-32°C to 49°C)	Water: A-A-52624A Antifreeze	1:1
-25°F to +120°F (-32°C to 49°C)	A-A-52624A Antifreeze	N/A

5-6.2 Oil Service.

WARNING

All metal jewelry can conduct electricity and become entangled in generator set components. Remove all jewelry when working on generator set. Failure to comply with this warning can cause injury or death to personnel.

WARNING

DO NOT wear loose clothing when performing checks, services and maintenance. Failure to comply with this warning can cause injury or death to personnel.

NOTE

Dipstick is marked indicating that oil level can be checked and oil added when engine is running or stopped. Make sure the correct side of dipstick is checked.

a. a. Fill engine with proper lubricating oil (Item 4, Table 5-2) per Table 5-4 to FULL mark on dipstick. Engine oil capacity is 6 quarts (5.7 liters).

b. b. Install oil fill cap.

Table 5-4. Lubricating Oil.

Ambient Temperature	Lubricating Oil
+20°F to +120°F (-7°C to 49°C)	MIL-PRF-2104H OE HDO-30 or OE HDO-15/40
0°F to +20°F (-17°C to –7°C)	MIL-PRF-2104H OE HDO-10
-25°F to 0°F (-32°C to –17°C)	MIL-PRF-46167C

5-6.3 Fuel Service.

WARNING

All metal jewelry can conduct electricity and become entangled in generator set components. Remove all jewelry when working on generator set. Failure to comply with this warning can cause injury or death to personnel.

WARNING

DO NOT wear loose clothing when performing checks, services and maintenance. Failure to comply with this warning can cause injury or death to personnel.

WARNING

Diesel fuel is flammable and toxic to eyes, skin, and respiratory tract. Skin and eye protection are required when working in contact with diesel fuel. Avoid repeated or prolonged contact. Provide adequate ventilation. Operators are to wash exposed skin and change chemical soaked clothing promptly if exposed to fuel. Failure to comply with this warning can cause injury or death to personnel.

WARNING

Fuels used in the generator set are flammable. Do not smoke or use open flames when performing maintenance. Failure to comply with this warning can cause injury or death to personnel, and damage to the generator set.

WARNING

Fuels used in the generator set are flammable. When filling the fuel tank, maintain metal-to-metal contact between filler nozzle and fuel tank opening to eliminate static electrical discharge. Failure to comply with this warning can cause injury or death to personnel, and damage to the generator set.

a. a. Remove fuel tank filler cap.

b. b. Fill fuel tank with proper fuel per Table 5-5.

c. c. Install fuel tank filler cap.

Table 5-5. Fuel.

Ambient Temperature	Lubricating Oil
+20°F to +120°F (-7°C to 49°C)	A-A-52557A, Grade 2-D MIL-DTL-83133E, JP-8
-25°F to +20°F (-32°C to -7°C)	A-A-52557A, Grade 1-D MIL-DTL-5624T, JP-5

5-7 FINISH ENGINE REHOST

WARNING

All metal jewelry can conduct electricity and become entangled in generator set components. Remove all jewelry when working on generator set. Failure to comply with this warning can cause injury or death to personnel.

WARNING

DO NOT wear loose clothing when performing checks, services and maintenance. Failure to comply with this warning can cause injury or death to personnel.

WARNING

Batteries give off a flammable gas. Do not smoke or use open flame when performing maintenance. Failure to comply can cause injury or death to personnel and equipment damage due to flames and explosion.

WARNING

Battery acid can cause burns to unprotected skin. Wear safety goggles and chemical gloves and avoid acid splash while working on batteries. Failure to comply with this warning can cause injury to personnel.

WARNING

When disconnecting or removing batteries, disconnect the negative lead that connects directly to the grounding stud first; disconnect the negative end of the interconnection cable next. When installing batteries, reverse the connection sequence. Failure to comply with this warning can cause injury to personnel.

WARNING

Many components require a two-person lift. Lifting heavy components can cause back strain. Ensure proper lifting techniques are used when lifting heavy components. Failure to comply with this warning can cause injury to personnel.

WARNING

Each battery weighs more than 70 pounds (32 kg) and requires a two-person lift. Lifting batteries can cause back strain. Ensure proper lifting techniques are used when lifting batteries. Failure to comply with this warning can cause injury to personnel.

a. a. Install batteries (10 and 11, Figure 5-1).

b. b. Install retaining bar (9), hook bolts (8), new lock washers (7), washers (6), and nuts

c. c. Connect positive ba ttery cable terminal lug (4).

d. d. Install interconnect battery cable and connect terminals (3 and 2).

e. e. Connect negative battery terminal lug (1) and close battery compartment access door.

f. f. Start generator set and check for leaks and proper operation.

Section V. FINAL ADJUSTMENTS

5-8 ENGINE ADJUSTMENT PROCEDURES 5-8.1

Magnetic Pickup Adjustment Procedure.

WARNING

All metal jewelry can conduct electricity and become entangled in generator set components. Remove all jewelry when working on generator set. Failure to comply with this warning can cause injury or death to personnel.

WARNING

DO NOT wear loose clothing when performing checks, services and maintenance. Failure to comply with this warning can cause injury or death to personnel.

WARNING

High voltage is produced when the generator set is in operation. Never attempt to start the generator set unless it is properly grounded. Failure to comply with this warning can cause injury or death to personnel.

WARNING

Dangerous voltage exists on live circuits. Always observe precautions and never work alone. Failure to comply with this warning can cause injury or death to personnel.

WARNING

DC voltages are present at generator set electrical components even with generator set shut down. Avoid shorting any positive with ground/negative. Failure to comply with this warning can cause injury to personnel, and damage to equipment.

a. a. Release control panel by turning two fasteners and lower control panel carefully.

b. b. Disconnect wire 147C from terminal 16 and wire 148C from terminal 17 of governor control unit.

c. c. Set multimeter to ohms function and connect lead s to ends of disconnected wires 147C and 148C. Multimeter should indicate between 800 and 900 ohms.

d. d. Leave multimeter connected to wires and set multimeter for AC volts.

e. e. Crank engine with DEAD CRANK switch and observe multimeter. Multimeter indication should be between 2.0 and 3.0 VAC.

CAUTION

Do not adjust magnetic pickup inward more than one eighth turn each time or damage to magnetic pickup may result.

f. f. To adjust output voltage measured in step e above, loosen jam nut and turn magnetic pickup in no more than 1/8 turn at a time to increase output voltage and out no more than 1/8 turn at a time to decrease output voltage. Tighten jam nut.

g. g. Repeat steps e and f above until proper output voltage is achieved.

h. h. Remove multimeter and connect wires to governor control unit.

i. i. Raise and secure control panel.

5-8.2 Fan Belt Adjustment Procedure.

NOTE

Run engine for 5 minutes if belt is cold. If belt is hot, let cool for 10 to 15 minutes.

a. Shut down generator set.

WARNING

Batteries give off a flammable gas. Do not smoke or use open flame when performing maintenance. Failure to comply can cause injury or death to personnel and equipment damage due to flames and explosion.

WARNING

When disconnecting or removing batteries, disconnect the negative lead that connects directly to the grounding stud first; disconnect the negative end of the interconnection cable next. When installing batteries, reverse the connection sequence. Failure to comply with this warning can cause injury to personnel.

b. b. Open battery access door, disconnect negative battery cable, and open engine access doors.

c. c. Check fan belt (6, Figure 5-46) for proper tension using a suitable belt tension gauge. Belt tension shall be 70 pounds.

d. d. If fan belt needs adjustment, loosen alternator mounting bolt (1) and nut (2).

CAUTION

Do not pry against alternator rear frame. Damage to alternator or mounting brackets could occur.

e. e. Apply outward pressure to alternator front frame until belt tension is correct.

f. f. Tighten alternator mounting bolt (1) and nut (2).

g. g. Close engine access doors, connect negative battery cable, and close battery access door.

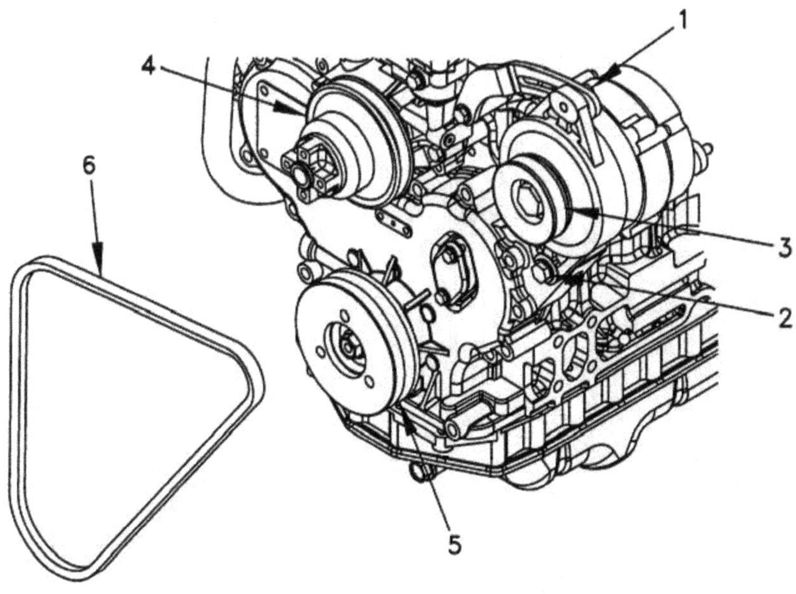

Figure 5-46. Fan Belt.

5-8.3 Governor Actuator Adjustment Procedure.

NOTE

Governor actuator is calibrated at the factory.

Loosen lock nut (1, Figure 5-47) and adjust throttle minimum adjust screw (2) to where throttle (3) is within 3/16 inch (0.47 cm) of throttle maximum fuel stop (4).

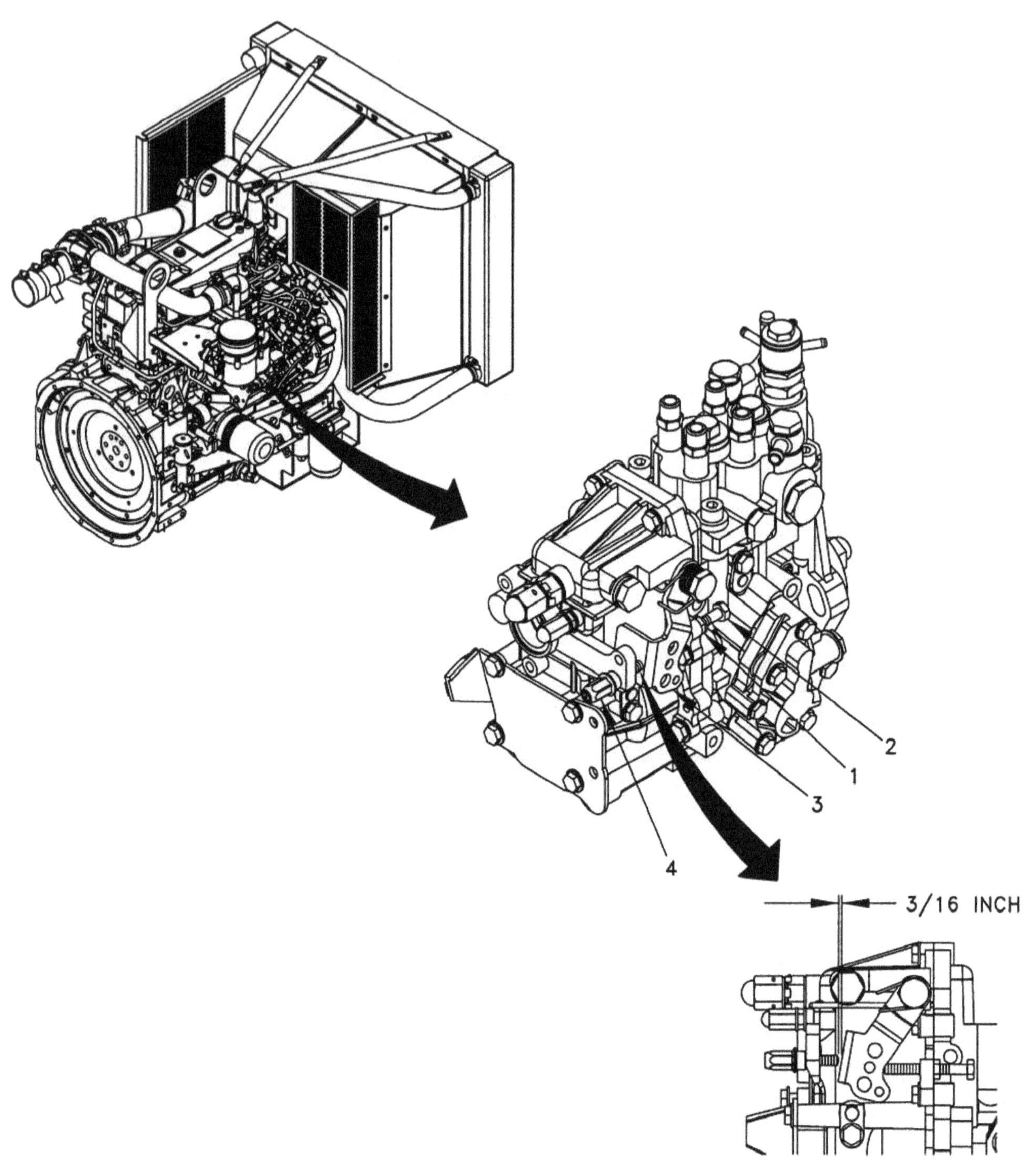

Figure 5-47. Governor Actuator.

5-9 GENERATOR SET ADJUSTMENT PROCEDURES

NOTE

The following steps will place the generator set in a functional condition. The generator set will not necessarily function to its maximum capacity without adjusting the governor control unit.

a. a. With engine stopped, turn idle speed screw (1, Figure 5-48) on engine all the way in and tighten jam nut (2).

b. b. End of idle speed screw must move throttle bracket (3) within 3/16 inch (0.47 cm) of maximum speed screw (4). If throttle speed screw does not push throttle bracket far enough, throttle speed screw must be replaced with a slightly longer screw.

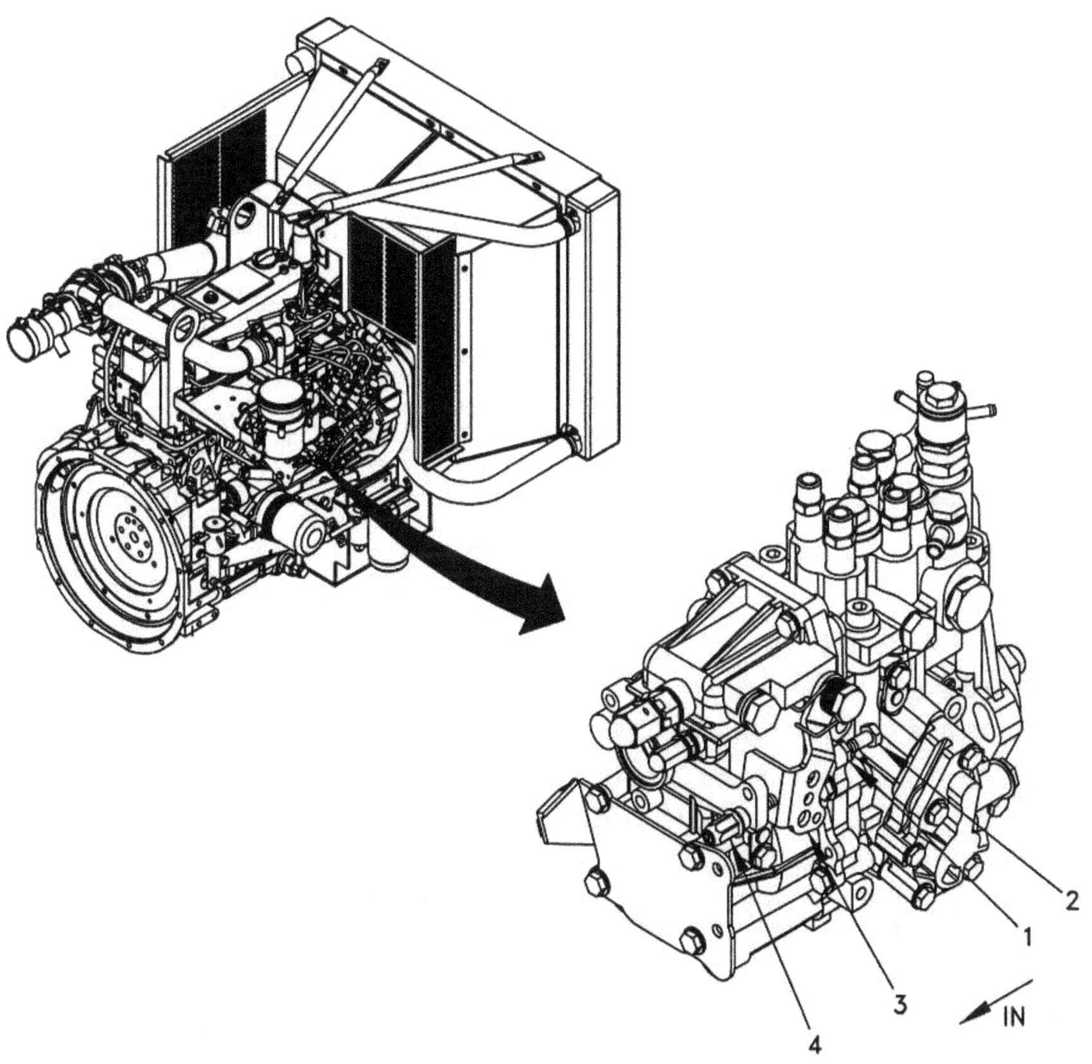

Figure 5-48. Fuel Injection Pump Idle Speed Screw.

c. Turn OVER SPD. potentiometer on governor control unit in control box assembly clockwise (CW) two turns (Figure 5-49).

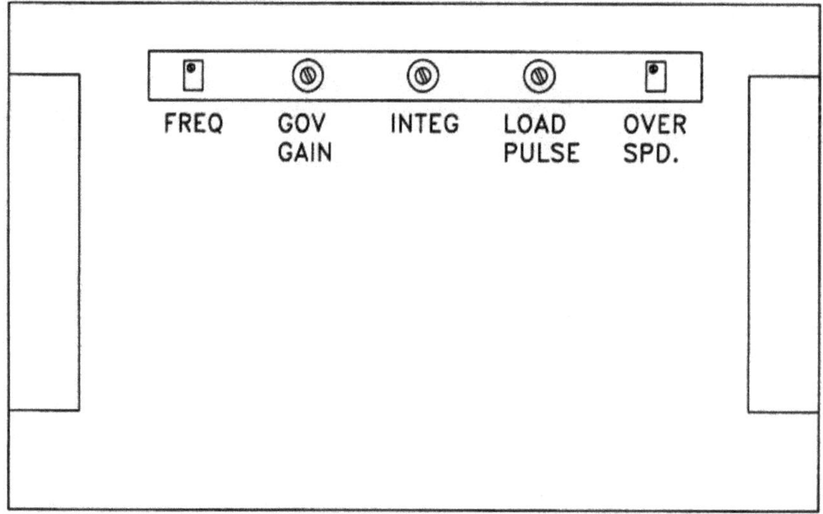

Figure 5-49. Governor Control Unit.

d. d. MEP-804B: Set frequency selector switch inside control box assembly to 60 Hz.

e. e. Turn FREQUENCY adjust knob on control panel fully counterclockwise (CCW).

f. f. Start engine.

g. g. Turn FREQ potentiometer on governor control unit located in control box assembly CW until frequency reads 62 Hz (MEP-804B)/412 Hz (MEP-814B) on FREQUENCY meter on control panel.

h. h. If generator set shuts down because of overspeed (overspeed fault lights illuminate), go to steps

(1)

thru (5) below. If generator set does not shut down, go to step i below.

(1) Operate OVERSPEED RESET switch momentarily inside control panel.

(2) Turn OVER SPD. potentiometer on governor control unit CW one turn.

(3) Press PUSH TEST switch on control panel to reset overspeed fault lamps.

(4) Rotate MASTER switch to OFF and back to START in order to start engine.

(5) Repeat from step f above until generator set runs continuously at 62 Hz (MEP-804B)/412 Hz (MEP-814B).

i. i. Turn FREQUENCY adjust potentiometer on control panel slowly CW until generator set either shuts down or knob reaches its maximum limit. If generator set shuts down before knob reaches its limit, go to step j below. If knob reaches its limit, go to step k below.

j. j. If generator set shuts down because of overspeed, go to steps (1) thru (6) below:

(1) Operate OVERSPEED RESET switch momentarily inside control panel.

(2) Turn OVER SPD. potentiometer on governor control unit CW 1/2 turn.

(3) Press PUSH TEST switch on control panel to reset overspeed fault lamps.

(4) Rotate MASTER switch to OFF and back to START in order to start generator set.

(5) If generator set shuts down upon startup, repeat steps j(1) thru j(4) above.

(6) If generator set does not shut down upon startup, repeat from step i above.

NOTE

At this point, the overspeed shutdown circuit is set for approximately 68 to 69 Hz (MEP-804B/455Hz (MEP-814B).

k. Adjust FREQ potentiometer on governor control unit CCW until 62 Hz (MEP-804B)/412 Hz (MEP-814B) is read on FREQ meter on control panel.

5-9.2 Governor Control Unit Adjustment Procedure.

a. a. Shut down generator set.

b. b. Open output box access door, remove cover from voltage reconnection terminal board, and attach voltage and frequency recorder to terminals 9 and 12 of voltage reconnection terminal board.

NOTE

The following steps require monitoring frequency, voltage, current, and power. All readings except for frequencies will utilize the generator set control panel gauges. However, a frequency counter will be required because the designed overspeed trip frequency is greater than the range of the control panel FREQUENCY meter (HERTZ).

c. c. Attach frequency counter to voltage reconnection terminal board terminals 9 and 12.

d. d. Open output load terminal door and attach load bank to generator set (four wire connection).

NOTE

Ensure load bank and generator set voltage reconnection are set for same configuration (i.e., 120/208 or 240/416 VAC).

e. e. Check for proper adjustment of governor actuator (paragraph 5-8.3). Adjust, as necessary.

f. f. Check for proper adjustment of magnetic pickup (paragraph 5-8.1). Adjust, as necessary.

g. g. Lower generator set control panel and turn INTEG, GOV GAIN, and LOAD PULSE potentiome- ter

on governor control unit to their full CCW positions.

h. h. Start generator set and operate at rated voltage and frequency.

i. i. Observe frequency counter and slowly increase operating frequency of generator set by turning governor control unit FREQ range potentiometer CW until frequency counter indicates between 65.94 and 66.06 Hz (MEP-804A)/436 and 444 Hz (MEP-814A). At this point, generator set has reached overspeed trip frequency and should shut down.

NOTE

Perform steps j and k below if generator set does not shut down within limits noted in step i above. Otherwise, proceed to step l below.

j. If generator set has not shut down at upper limit of frequency noted in step i above, proceed as follows:

(1) Turn FREQ range potentiometer CCW until frequency counter indicates midrange of overspeed trip frequency. (Example: 66 Hz for MEP-804A).

(2) Turn OVERSPD control potentiometer on governor control unit CCW until generator set shuts down.

(3) Activate OVERSPD RESET switch.

(4) Repeat steps h and i above.

k. k. If generator set shuts down prior to reaching lowe r limit of frequency noted in step i above, proceed as follows:

(1) Turn OVERSPD control potentiometer CW one turn for each hertz generator set shut down prior to lower frequency limit.

(2) Activate OVERSPD RESET switch.

(3) Repeat steps h and i above.

l. l. Actuate OVERSPD RESET switch.

m. m. Turn FREQ range potentiometer on governor control unit two turns CCW.

n. n. Start and operate generator set at rated volt age and turn FREQUENCY adjust potentiometer on control panel to midrange.

o. o. Turn FREQ range potentiometer until rated frequency (50, 60, or 400 Hz) is indicated on control panel FREQUENCY meter (HERTZ).

p. p. Set load bank for generator set rated load and apply load. Observe generator set instruments and adjust load as needed to ensure rated load is applied.

q. q. Set multimeter for DC volts and connect to terminals 11 and 12 of governor control unit.

r. r. Adjust LOAD SHARING ADJUST rheostat until multimeter indicates 6 VDC. Disconnect multimeter.

s. s. Remove load.

t. t. Turn on voltage and frequency recorder and operate at minimum chart speed of 5 mrn/sec (chart resolution of 0.2 mm/sec). Adjust recorder voltage amplifier for minimum chart resolution of 1.0 volt/mm and frequency deviation of 0.2 Hz/mm.

u. u. Adjust GOV GAIN potentiometer on governor control unit as follows:

(1) Turn GOV GAIN potentiometer to its full CW position.

(2) Momentarily actuate and turn off LOAD switch on control panel.

(3) Observe strip chart on recorder for frequency oscillation (hunting). If required, slowly turn GOV GAIN CCW until frequency oscillation disappears.

v. v. Apply and remove rated load to generator set at 40-second intervals. Repeat this step two more times.

w. w. Shut down generator set and turn off recorder.

x. x. Examine voltage and frequency strip chart for the following performance criteria:

(1) Frequency regulation (shall not exceed 1/4 of 1 percent of rated frequency).

(2) Frequency short-term stability (30 seconds) (will remain constant within a bandwidth equal to 1/2 of 1 percent rated frequency without repetitive frequency variations (hunting)).

(3) Generator set will reestablish stable engine operation within 2 seconds after a sudden load change (i.e., from a load to no-load condition) (within one second for 400 Hz units).

y. y. Maximum transient frequency change above (overshoot) and below (undershoot) new steady state frequency shall not be more than 4 percent of rated frequency (not more than 1-1/2 percent for 400 Hz units).

NOTE

All required INTEG and LOAD PULSE potentiometer adjustments will be in 10 percent increments.

z. z. Adjust INTEG potentiometer on governor control unit CW to decrease recovery time of load transients.

aa. Adjust LOAD PULSE potentiometer on governor control unit CW to decrease frequency overshoot and undershoot and to decrease recovery time of overshoot/undershoot transients.

NOTE

Steps u(1) and u(2) above are not required when doing step ab below.

ab. Start generator set.

ac. Repeat steps t thru v above until generator set meets performance requirements stated in step x above.

ad. Apply and remove 75 percent rated load to ge nerator set at 40-second intervals. Repeat this step two more times.

ae. Apply and remove 50 percent rated load to generator set at 40-second intervals. Repeat this step two more times.

af. Apply and remove 25 percent rated load to generator set at 40-second intervals. Repeat this step two more times.

ag. Shut down generator set and turn off strip chart recorder.

ah. Examine voltage and frequency strip chart for the following performance criteria:

(1) Frequency regulation (shall not exceed 1/4 of 1 percent of rated frequency).

(2) Frequency short-term stability (30 seconds) (will remain constant within a bandwidth equal to 1/2 of 1 percent rated frequency without repetitive frequency variations (hunting)).

(3) Generator set will reestablish stable engine operation within 2 seconds after a sudden load change (i.e., from a load to no-load condition) (within one second for 400 Hz units).

ai. Maximum transient frequency change above (overshoot) and below (undershoot) new steady state frequency shall not be more than 4 percent of rated frequency (not more than 1-1/2 percent for 400 Hz units).

aj. Disconnect load bank and clos e load terminal board access door.

ak. Disconnect frequency counter and voltage and frequency recorder from voltage reconnection terminal board.

al. Install voltage reconnection terminal boar d cover and close output box access door.

5-9.3 Load Sharing Rheostat Adjustment Procedure.

While generator set is operating at maximum load, adjust rheostat inside control box assembly for 6 VDC on terminals 11 and 12 on governor control unit.

5-10 GENERATOR SET PLACEMENT INTO SERVICE

Perform Preventive Maintenance Checks and Services (PMCS), as required, and generator set is ready to be placed into service.

CHAPTER 6

WINTERIZATION KIT

Subject Index Page

Subject Index Page

Section I. INTRODUCTION

6-1 SCOPE

This section is for your use in operating and maintaining the Winterization Kit installed on the 15kW Tactical Quiet Generator Sets. The chapter covers unit maintenance, direct support maintenance, installation instructions, and removal instructions as well as troubleshooting procedures for the kit.

6-1.1 Levels of Maintenance.

Army users shall refer to the Maintenance Allocation Chart (MAC) for tasks and levels of maintenance to be performed.

Section II. EQUIPMENT DESCRIPTION AND DATA

6-2 GENERAL

The winterization kit is designed to be mounted in generator sets where extreme cold temperatures are anticipated. The kit consists of a coolant heater, which allows the generator set to operate to -50□F (-45.6□C). The coolant heater circulates the coolant from the generator set through the heater pump and heats the coolant by the heater and returns it back through the radiator of the generator set. The cycle continues until the temperature reaches 176□F (80□C). The heater then goes into low-heat mode; if the coolant temperature drops to 158□F (70□C), the heater will switch to high-heat mode.

6-2.1 Equipment Characteristics, Capabilities, and Features.

a. a. Characteristics. The Winterization Kit consists of a coolant heater, which allows the generator set to operate to -50□F (-45.6□C).

b. b. Capabilities and Features. The heater burns fuel from the generator fuel tank to heat the coolant, which is pumped through the engine block. The kit consists of a heater and coolant pump, a control unit, an ON-OFF switch, a fuel pump and line, coolant circulating lines, a wiring harness and mounting hardware to ensure operation to -50□F (-45.6□C).

c. Location and Description of Major Components . Figure 6-1 illustrates the major components of the kit and shows their locations on the 15kW Tactical Quiet Generator Set. (Refer to Table 6-1 for item names).

Table 6-1. Description of Major Components.

Item No.	Item Name	Description
	Winterization Kit	A fuel-burning heater, pre-heats engine coolant permitting generator set Operation to -50°F (-45.6°C).
1	Control Unit	Controls heater operations.
2	Heater	Heats coolant for operation in extreme cold temperatures.
3	Fuel Pump	Pumps fuel from the generator set fuel tank to the heater.
4	Fuel Lines	Provides a means of transporting fuel to heater.
5	Coolant Pump	Circulates coolant from generator set through the heater.
6	Coolant Lines	Provides a means of transporting coolant for circulation.
7	Bypass Valve	Allows coolant to bypass heater when Winterization Kit is not in use. (MEP-804B/MEP-814B Only)
8	Switch/Lamp	Switches heater on or off and lamp indicates heater function codes.
9	Wiring Harness	Electrically connects Winterization Kit components.
10	Exhaust Hose	Provides a means of exhausting combustion gases from heater.
11	Air Inlet Hose	Provides intake air to winterization heater.

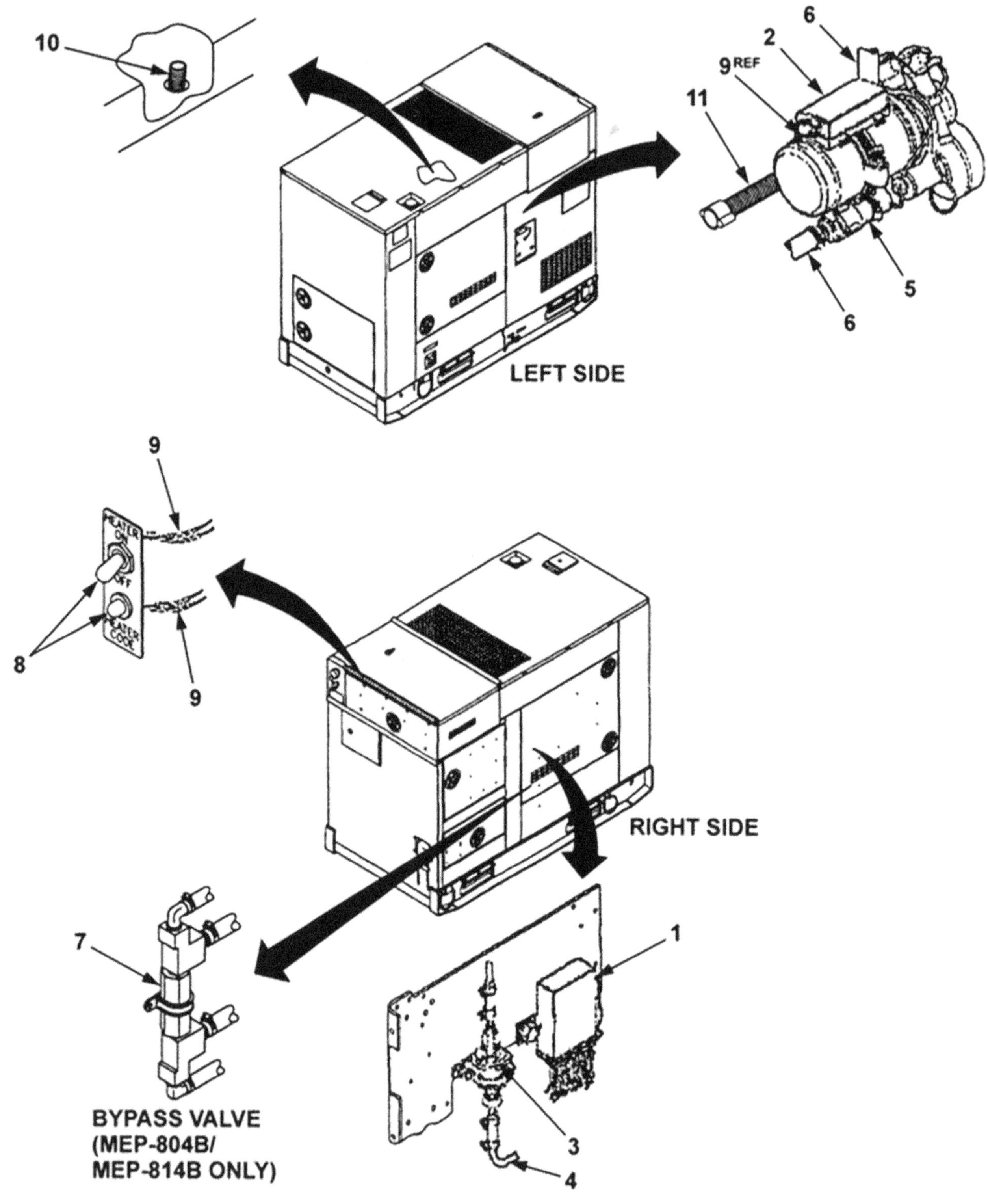

Figure 6-1. Location of Major Components.

6-2.2 Tabulated/Illustrated Data.

Tabulated data for the heater is located in Table 6-2.

Table 6-2. Tabulated Data for the Heater.

Item Name	Data
1. Winterization Kit	
a. National Stock Number	6115-01-477-0566
b. Overall Length	10.787 inches
c. Overall Width	5.984 inches
d. Overall Height	7.815 inches
e. Weight	25 lbs
2. Heater:	
a. Manufacturer	Active Gear
b. Model	D5W
3. Heating	Water Coolant
Capacity	High: 17,000 BTU/Hr. Low: 4250 BTU/Hr.
4. Rated Voltage	24 Vdc
a. Operating Voltage Range	20 to 28 Vdc
b. Current at 24 Vdc	Start: 20 Amps/Hr. Running High: 1.8 Amps/Hr. Running Low: 1.2 Amps/Hr.
5. Fuel	Diesel
Fuel Consumption	High: 0.16 Gal/Hr. Low: 0.04 Gal/Hr.
6. Coolant Pump Flow	250 Gal/Hr.

Section III. PRINCIPLES OF OPERATION

6-3 GENERAL

This section contains a functional description of the winterization kit.

6-3.1 Functional Description.

When the heater is switched on, the indicator lamp comes on, the combustion air fan blower comes on high, then low, and the start cycle begins. The water-circulating pump begins to run, and after a short time the fuel pump begins to operate. The fuel is ignited by the ignition element. A flame detector turns off the ignition element when combustion is established. The heater runs in high-heat mode. When the coolant temperature has reached the operating point, the temperature sensor sends a signal to the control unit to reduce the heat output. If the temperature remains at the upper limit, the heater turns off, but the coolant pump continues to run. The heater will then operate in the low-heat mode. The heater will automatically restart once the system's temperature has dropped to the lower temperature switch point of the sensor. If the coolant temperature drops, the heater automatically switches back to high-heat mode. The heater continues to cycle itself between high- and low-heat modes until the ON/OFF switch is placed in the OFF position. At this point the indicator lamp will go off, but the combustion air blower and the circulating pump will continue to run for several minutes, and then will shut off.

NOTE

Heater warms generator engine block sufficiently to start in below zero temperatures.

a. The circulation pump, ceramic glow plug, and combustion air fan start operation after the heater is turned on.

b. After approximately 50 seconds, the fuel pump starts, combustion starts, and the ceramic igniter/glow plug is turned off.

c. c. When the coolant temperature reaches 176□F (80□C), the heater switches to low-heat mode.

d. d. If the coolant temperature drops to 158□F (70□C), the heater switches to high-heat mode.

e. e. If the coolant temperature rises to 185□F (85□C), the heater will switch off. The heater will automatically restart in high-heat mode when the coolant temperature reaches 158□F (70□C).

f. f. When switched off, the fuel pump stops and the flame is extinguished. The combustion air fan blower and coolant circulating pump continue to run for a 3-minute cool-down cycle.

9. If the voltage drops below 21.0 VDC or rises above 30.0 VDC at any time, the heater will shut down after a 20 second delay.

Section IV. DESCRIPTION AND USE OF CONTROLS AND INDICATORS

6-4 GENERAL

This section describes and illustrates the winterization kit controls and indicators to ensure proper operations.

a. Controls/Indicators. There are three controls/indicators; the heater indicator light, a control unit, and the ON-OFF switch.

b. Heater Indicator Light . A light at the heater ON-OFF switch (item 8, Figure 6-1) lights when the heater is operating. The light also serves as a troubleshooting code light (see Section X, Troubleshooting Procedures).

c. Control Unit. **The control unit is a sealed unit, mounted on the generator wall, which controls heater** operation.

d. d. Heater ON-OFF Switch . The ON-OFF switch is a single-pole, single-throw toggle switch. Placed in the ON position, the switch closes the 24 VDC circuit to the control unit and illuminates the heater indicator light.

e. e. Function and Identification Plates . There are three data plates pertaining to the winterization kit.

Table 6-3. Data Plates.

Item No.	Item Name/Description	Function/Location
1.	Heater Operating Instructions	Mounted on generator control panel access door that describes heater operating procedures.

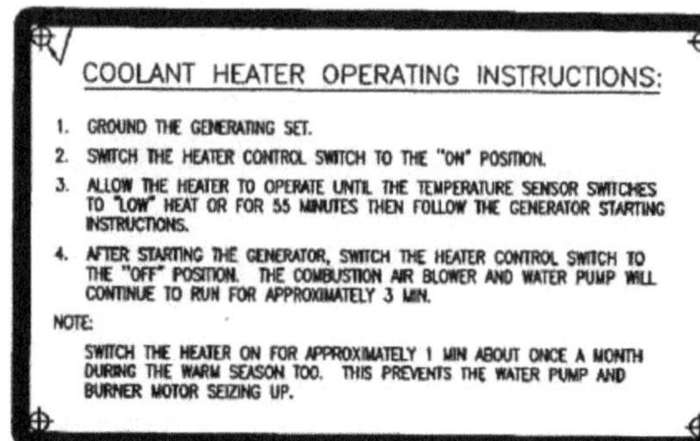

COOLANT HEATER OPERATING INSTRUCTIONS:

1. GROUND THE GENERATING SET.
2. SWITCH THE HEATER CONTROL SWITCH TO THE "ON" POSITION.
3. ALLOW THE HEATER TO OPERATE UNTIL THE TEMPERATURE SENSOR SWITCHES TO "LOW" HEAT OR FOR 55 MINUTES THEN FOLLOW THE GENERATOR STARTING INSTRUCTIONS.
4. AFTER STARTING THE GENERATOR, SWITCH THE HEATER CONTROL SWITCH TO THE "OFF" POSITION. THE COMBUSTION AIR BLOWER AND WATER PUMP WILL CONTINUE TO RUN FOR APPROXIMATELY 3 MIN.

NOTE:

SWITCH THE HEATER ON FOR APPROXIMATELY 1 MIN ABOUT ONCE A MONTH DURING THE WARM SEASON TOO. THIS PREVENTS THE WATER PUMP AND BURNER MOTOR SEIZING UP.

"Operating Instruction Plate

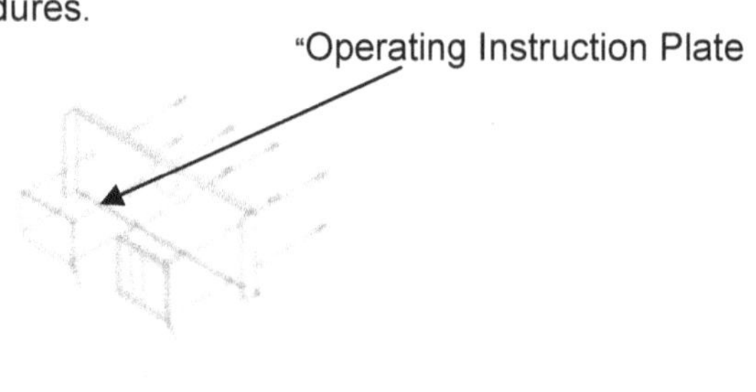

2. Function Code Plate Mounted on the generator control panel access door and lists the sequence of pulses

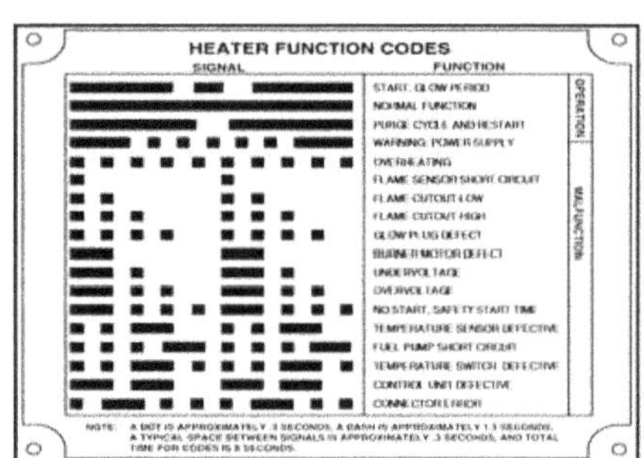

HEATER FUNCTION CODES

SIGNAL	FUNCTION	
	START, GLOW PERIOD	OPERATION
	NORMAL FUNCTION	
	PURGE CYCLE AND RESTART	
	WARNING: POWER SUPPLY	
	OVERHEATING	MALFUNCTION
	FLAME SENSOR SHORT CIRCUIT	
	FLAME CUTOUT LOW	
	FLAME CUTOUT HIGH	
	GLOW PLUG DEFECT	
	BURNER MOTOR DEFECT	
	UNDERVOLTAGE	
	OVERVOLTAGE	
	NO START, SAFETY START TIME	
	TEMPERATURE SENSOR DEFECTIVE	
	FUEL PUMP SHORT CIRCUIT	
	TEMPERATURE SWITCH DEFECTIVE	
	CONTROL UNIT DEFECTIVE	
	CONNECTOR ERROR	

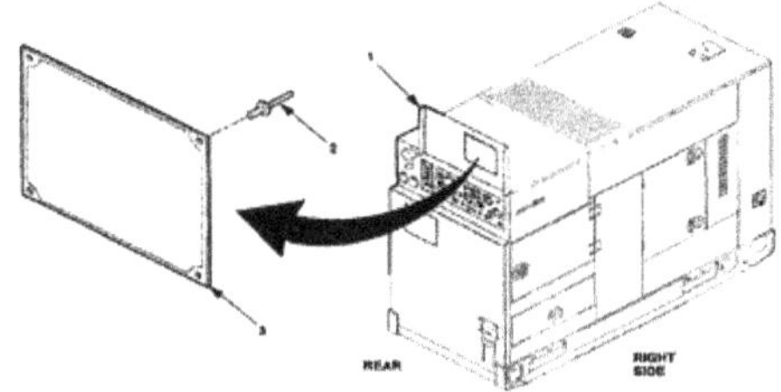

3. Identification Data Plate Mounted on the generator access door and identifies winterization kit characteristics

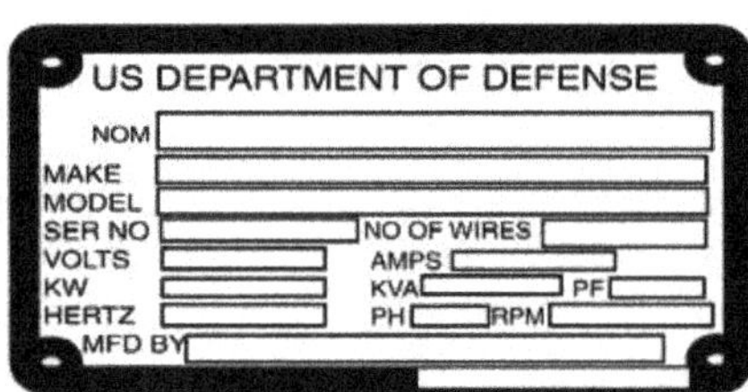

US DEPARTMENT OF DEFENSE

NOM
MAKE
MODEL
SER NO | NO OF WIRES
VOLTS | AMPS
KW | KVA | PF
HERTZ | PH | RPM
MFD BY

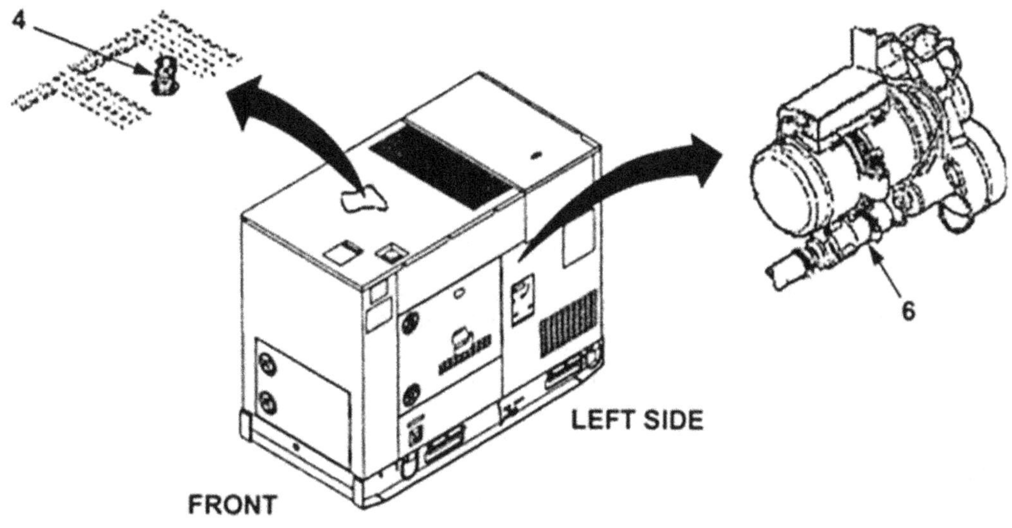

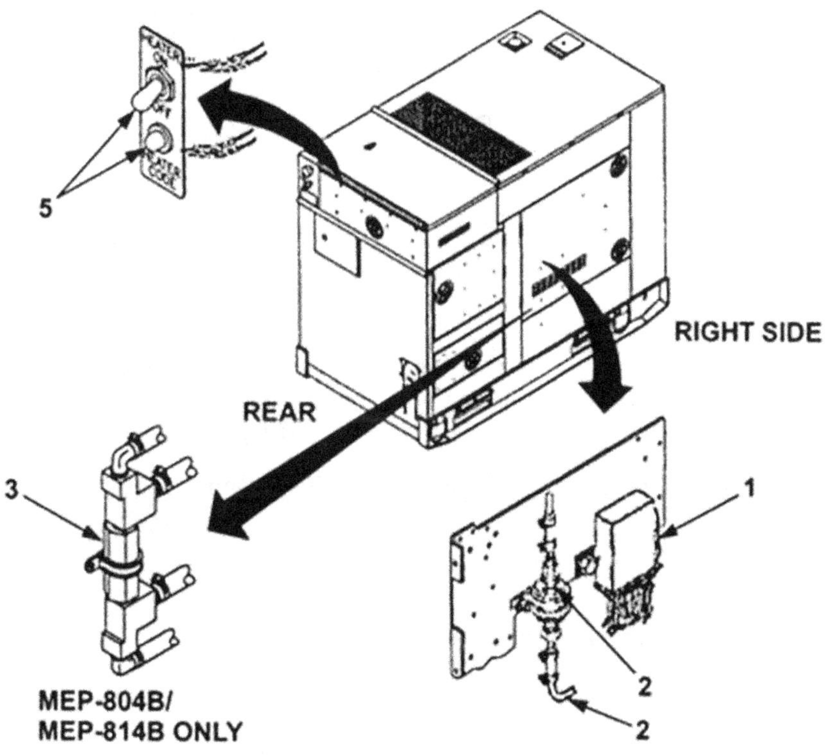

Figure 6-2. PMCS Routing Diagram.

Section V. PREVENTIVE MAINTENANCE CHECKS AND SERVICES (PMCS)

6-5 GENERAL

Table 6-4, (PMCS Table) has been provided so you can keep your equipment in good operating condition and ready for its primary mission.

6-5.1 Warnings, Cautions, and Notes.

Always observe the ***WARNINGS, CAUTIONS,*** and ***NOTES*** appearing in your PMCS Table. Warnings and Cautions appear before applicable procedures. You must observe ***WARNINGS*** to prevent serious injury to yourself and to others. You must observe ***CAUTIONS*** to prevent your equipment from being damaged. You must observe ***NOTES*** to ensure procedures are performed properly.

6-5.2 Explanation of Table Entries.

a. Item No. Column . Numbers in this column are for reference. When completing DA Form 2404 (Equipment/Inspection and Maintenance Worksheet) or DD Form 5988E, include the item number for the checks/service indicating a fault. Item numbers also appear in the order that you must do checks and services for the intervals listed.

b. b. Interval Column. This column tells you when you must do the procedure in the procedure column. "BEFORE" procedures must be done before you operate the generator set with modification kit installed for its intended mission. "DURING" procedures must be done during the time you are operating the generator set for its intended mission. "AFTER" procedures must be done immediately after shutting down the generator set. Perform "WEEKLY" procedures at the listed interval.

c. c. Location, Item to Check/Service Column . This column lists the location and the item to be checked or serviced. The item location is underlined.

d. Procedure Column. This column gives the procedure for checking or servicing the item listed in the location, item to check/service column. You must perform the procedure to know if the generator set is ready or available for its intended mission or operation. You must do the procedure at the time stated in the interval column.

. Not Fully Mission Capable if: Column. Information in this column tells you what faults will keep your modified generator set from being capable of performing its primary mission. If you make checks or services that show faults listed in this column, do not operate the generator set.

f. Reporting and Correcting Deficiencies . If the Winterization Kit does not perform as required, refer to Section XI of this Chapter.

6-5.3 Other Table Entries.

Be sure to observe all special information and notes that appear in your table.

6-5.4 Special Instructions.

Preventive maintenance is not limited to performing the checks and services listed in the PMCS Table. Figure 6-2 depicts general Winterization Kit PMCS routing. Covering unused receptacles, stowing unused accessories and other routine procedures such as equipment inventory, cleaning components, and touch up painting are not listed in the table. These are things you should do any time you see that they need to be done. If a routine check is listed in the PMCS Table, it is because experience has shown that problems may occur with this item. Take along tools and cleaning cloths needed to perform the required checks and services. Use the following information to help identify potential problems before and during checks and services. Use the information in the following paragraphs to help you identify problems at any time.

WARNING

Solvent used to clean parts is potentially dangerous to personnel and property. Clean parts in a well-ventilated area. Avoid inhalation of solvent fumes. Wear goggles and rubber gloves to protect eyes and skin. Wash exposed skin thoroughly. Do not smoke or use near open flame or excessive heat. Failure to comply with this warning can cause injury to personnel, and damage to the equipment.

CAUTION

Keep cleaning solvents, gasoline and lubricants away from rubber or soft plastic parts. They will deteriorate material.

a. Keep it clean. Dirt, grease, and oil get in the way and may cover up a serious problem. Use a dry cleaning solvent to clean metal surfaces.

. Use soap and water to clean rubber or plastic parts and material.

c

. Check all bolts, nuts, and screws to make sure they are not loose, missing, bent, or broken. Do not try to check them all with a tool, but look for chipped paint, bare metal, or rust around bolt heads. If you find one loose, tighten it or report it to unit level of maintenance.

d. d. Inspect welds. Look for loose or chipped paint, rust, or gaps where parts are welded together. If a broken weld is found, report it to unit level of maintenance.

e. e. Inspect electrical wires, connectors, terminals , and receptacles. Look for cracked or broken insulation, bare wires, and loose or broken connectors. Tighten loose connectors and make sure wires are in good condition. Examine terminals and receptacles for serviceability. If deficiencies are found, report them to unit level of maintenance.

f. f. Inspect hoses and fluid lines. Look for wear, damage, and leaks. Make sure that clamps and fittings are tight. Wet spots and stains around a fitting or connector can mean a leak. If a leak comes from a loose connector, or if something is broken or worn out, report it to unit level of maintenance.

6-5.5 Leakage Definitions.

You must know how fluid leakage affects the status of your equipment. The following are definitions of the types/ classes of leakage you need to know to be able to determine the status of your equipment. Learn and be familiar with them. When in doubt, notify your supervisor.

Leakage Class	Leakage Definition
Class I	Seepage of fluid (as indicated by wetness or discoloration) not great enough to form drops.
Class II	Leakage of fluid great enough to form drops, but not enough to cause drops to drip from the item being checked/inspected.
Class III	Leakage of fluid (other than fuel) greater than three drops per minute that fall from the item being checked/inspected.

Table 6-4. Preventive Maintenance Checks and Services.

Item No.	Interval	Location Item to Check/Service	Procedure	Not Fully Mission Capable if:
			NOTE **Be sure that Generator Set PMCS is completed first in accordance with Chapter 2 and TM 9-6115-643-10.**	
		VISUAL INSPECTION		
1	Before	HEATER ASSEMBLY	a. Check for damage.	Damage that renders equipment unsafe.
			b. Ensure that heater assembly is mounted securely.	Heater not mounted securely.
		CONTROL UNIT	Check for loose or broken wires or damage.	Wires loose or broken or control unit damaged.
		FUEL AND COOLANT LINES	Check on, around, and under equipment for fuel, oil, or coolant leaks.	Class III coolant or any class fuel leak is detected.
2	Before	HEATER	Inspect heater for signs of leaks.	Class III coolant or any class fuel leak is detected.
		FUEL LINES	Inspect winterization kit fuel lines for kinks, leaks, loose or damaged clamps.	Fuel lines damaged; clamps missing.
		FUEL PUMP	Inspect fuel pump for leaks.	Any fuel leak.
		EXHAUST HOSE	Inspect for obstruction, missing or damaged mounting clamp.	Hose obstructed; hose or clamp missing or damaged.
		AIR INLET HOSE	Inspect for obstruction, missing or damaged mounting clamp.	Inlet hose obstructed.

WARNING

Cooling system operates at high temperatures and pressure. Contact with high pressure steam and/or liquids can result in burns and scalding. Shut down generator set, and allow system to cool before performing checks, services, and maintenance, or wear gloves and additional protective clothing and goggles as required. Failure to comply with this warning can cause injury or death to personnel.

Table 6-3. Preventive Maintenance Checks and Services - Continued.

Item No.	Interval	Location Item to Check/Service	Procedure	Not Fully Mission Capable if:
3	Before	WINTERIZATION KIT COOLANT LINES	Inspect for loose, damaged or missing clamps.	Class III leaks or missing clamps or hoses.
			Inspect for leaks.	Class III leaks or missing clamps or hoses.
		COOLANT PUMP	Inspect for leaks.	Class III leaks or missing clamps or hoses.
		BYPASS VALVE	Inspect for leaks, damage, loose clamps, or other damage.	Class III leaks or missing clamps or hoses.
4	Before	WIRE HARNESS	Inspect wiring for burned or frayed insulation or loose terminals. a. Check that indicator light is on when heater is operating.	Wiring is loose or burned.
5	During	HEATER CONTROL AND SWITCH LAMP	b. Check Heater function Code	Lamp blinks showing failure in accordance with Heater Function Code Plate.
			Plate.a. Check for damage.	
6	After	HEATER ASSEMBLY	b. Ensure that heater assembly is mounted securely.	Damage that renders equipment unsafe.
			Loose or broken wires or damage.	Heater not mounted securely.
		CONTROL UNIT		Wires loose or broken or control unit damaged.
		FUEL AND COOLANT LINES	Check on, around, and under equipment for fuel, oil, or coolant leaks.	Class III coolant or any class fuel leak is detected
7	After	HEATER	Inspect heater for signs of leaks.	
		FUEL LINES	Inspect winterization kit fuel lines for kinks, leaks, loose or damaged clamps.	Fuel lines damaged, clamps missing, or any leaks.

Table 6-3. Preventive Maintenance Checks and Services - Continued.

Item No.	Interval	Location Item to Check/Service	Procedure	Not Fully Mission Capable if:
7-cont.		FUEL PUMP	Inspect fuel pump for leaks.	Any fuel leak.
		EXHAUST HOSE	Inspect for obstruction, missing or damaged mounting clamp.	Obstructed exhaust.
		AIR INLET HOSE	Inspect for obstruction, missing or damaged mounting clamp.	Air inlet obstructed.

WARNING

Cooling system operates at high temperatures and pressure. Contact with high pressure steam and/or liquids can result in burns and scalding. Shut down generator set, and allow system to cool before performing checks, services, and maintenance, or wear gloves and additional protective clothing and goggles as required. Failure to comply with this warning can cause injury or death to personnel.

Item No.	Interval	Location Item to Check/Service	Procedure	Not Fully Mission Capable if:
8	After	WINTERIZATION KIT COOLANT LINES	Inspect for loose, damaged, or missing clamps.	Class III leaks or missing clamps or hoses.
			Inspect for leaks.	Class III leaks or missing clamps or hoses.
		COOLANT PUMP	Inspect for leaks.	Class III leaks or missing clamps or hoses.
		BYPASS VALVE	Inspect for leaks, damage, loose clamps, or other damage.	Class III leaks or missing clamps or hoses.
9	After	WIRE HARNESS	Inspect wiring for burned or frayed insulation or loose terminals.	Wiring is loose or damaged.
10	After	HEATER CONTROL AND SWITCH LAMP	Check that indicator light is operable.	Indicator light not operable.
			Check Heater Function Code Plate.	Heater Function Code Plate missing.

Section VI. REPAIR PARTS; SPECIAL TOOLS; TEST, MEASUREMENT, AND DIAGNOSTIC EQUIPMENT (TMDE); AND SPECIAL SUPPORT EQUIPMENT

6-6 COMMON TOOLS AND EQUIPMENT

For authorized common tools and equipment refer to the Modified Table of Organization and Equipment (MTOE) applicable to your equipment.

6-6.1 Special Tools, TMDE, and Special Support Equipment.

No special tools or special support equipment are required for maintenance of the modification kit.

6-6.2 Repair Parts.

Repair parts for the modification kit are listed and illustrated in Appendix C.

Section VII. SERVICE UPON RECEIPT OF MATERIEL

If the winterization kit is already installed on the generator set, the normal service on receipt of the generator set is sufficient, and no separate service on receipt is required by unit for the winterization kit. If the kit is not installed, refer to Section IX of this Chapter.

6-7 REMOVAL INSTRUCTIONS

Refer to Section XIII of this Chapter.

Section VIII. UNIT LUBRICATION

6-8 LUBRICATION (NOT APPLICABLE TO MODIFICATION KITS)

No lubrication is required on the winterization kit.

Section IX. INSTALLATION INSTRUCTIONS

The following instructions have been provided so you can install the Winterization Kit, NSN: 6115-01-477-0566 on your 15kW Generator Set.

WARNING

Disconnect negative battery cable from right battery and the positive battery cable from the left battery before doing the following procedures. Reconnect cables in reverse order. Failure to comply with this warning can cause injury or death to personnel.

WARNING

Cooling system operates at high temperatures and pressure. Contact with high pressure steam and/or liquids can result in burns and scalding. Shut down generator set, and allow system to cool before performing checks, services, and maintenance, or wear gloves and additional protective clothing and goggles as required. Failure to comply with this warning can cause injury or death to personnel.

WARNING

Always remove radiator cap slowly to permit any pressure to escape. Failure to comply with this warning can cause injury to personnel.

WARNING

Diesel fuel is flammable and toxic to eyes, skin, and respiratory tract. Skin and eye protection are required when working in contact with diesel fuel. Avoid repeated or prolonged contact. Provide adequate ventilation.
Operators are to wash exposed skin and change chemical soaked clothing promptly if exposed to fuel. Failure to comply with this warning can cause injury or death to personnel.

WARNING

Catch fuel in suitable container. Keep spilled fuel away from hot engine and all fires. Failure to comply with this warning can cause injury or death to personnel.

6-9 INSTALLATION PROCEDURES

NOTE

Use packing list provided with kit.

6-9.1 Open battery access door and disconnect negative battery cable from right battery and the positive battery cable from the left battery per paragraph 2-12, before doing the following procedures.

6-9.2 Using the heater mounting plate (5, Figure 6-3) as a template, drill holes from the outside on left side panel. Before drilling holes, open left side rear door and look inside generator set to verify proper positioning and clearance of mounting plate. To install the heater mounting plate, perform the following steps: using the plate as a template, place it 3.5" from the top left access door flange and 15" across from the door hinge flange to locate the edge of the plate. For the B model, install heater mounting plate by using plate as a template and place it 3.5" from the forward edge and 2.5" from the top of panel to locate edge of plate.

NOTE

When mounting the heater (7), to the heater mounting plate, the burner end of the heater shall be higher than the coolant intake end.

6-9.3 Open left side rear door (8) and install heater mounting plate (6) using bolts (4), flat washers (3), lock washers (2), and nuts (1). Attach heater (7) with bracket (part of 7) to the mounting plate (5) and to the bulkhead (6) with bolts (4), flat washers (3), lock washers (2) and nuts (1). The bolts come through the mounting plates from outside of the generator set.

6-9.4 Approximately 10" from inside of left door hinge (Figure 6-3) drill 1 7/16" diameter hole in top (left side) of bulkhead housing.

6.9.5 Remove top panel from Generator Set in accordance with maintenance procedure at paragraph 2-16.

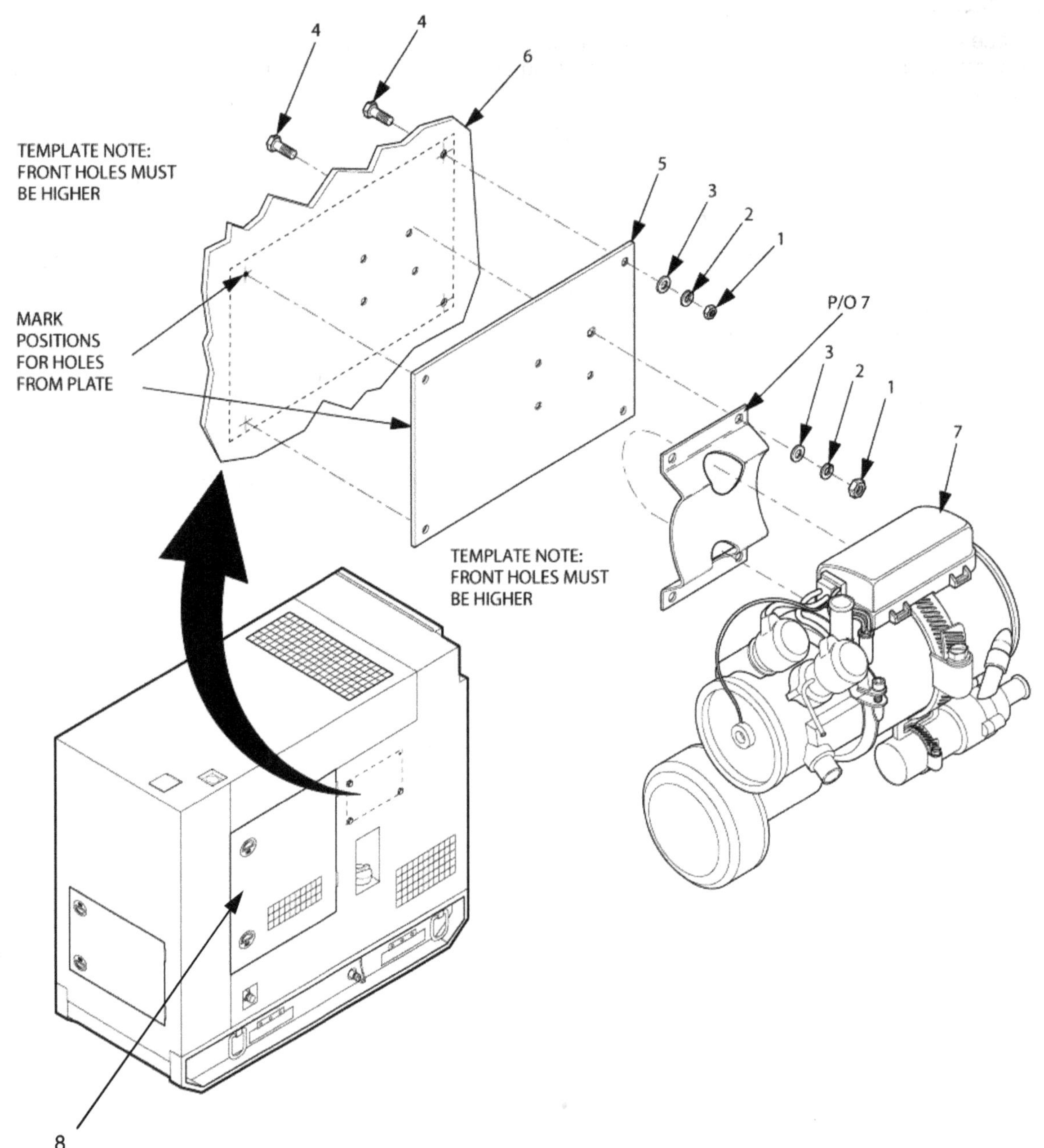

Figure 6-3. Heater Mounting.

6-9.6 Attach one end of exhaust hose (2, Figure 6-4) to exhaust elbow (5), using adapter (4) with clamp (3). Attach it to exhaust port on heater. Apply 2 thick beads of RTV around exhaust port of heater and attach exhaust elbow to heater. Move elbow in a position that doesn't interfere with anything in the route to the hole drilled above in paragraph 6-9.4.

6-9.7 Route hose through hole drilled (paragraph 6-9.4) and use sealant provided in kit around hose. Install clamp (9) to middle of exhaust hose and attach at to the top of the bulkhead at first screw closest to the clamp. Secure using bolt (10), spacer (1), nut (6), lock washer (7), and washer (8). Reinstall top panel per paragraph 2-16.

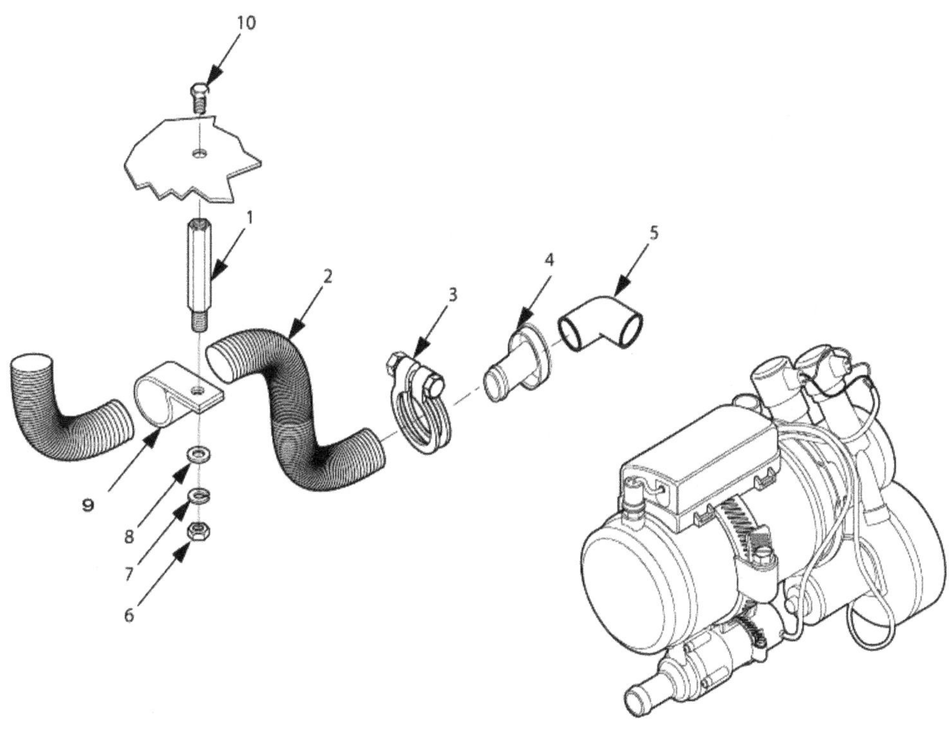

Figure 6-4. Exhaust Hose Assembly.

6-9.8 Drain coolant system in accordance with maintenance procedure at paragraph 2-79.

NOTE

(MEP-804A/MEP- 814A Only) Perform steps a thru d to install coolant lines to heater.

a. a. Bottom radiator hose (8, Figure 6-5) must be cut into two pieces. The cut should be about half-way. After the cut is made, a tee valve (2) is inserted between the two halves and secured with two clamps (1). The tee is orientated toward the engine.

b. b. Attach a preformed non-metallic hose (4) to the tee valve (2) and secure one end with hose clamp (3) to lower radiator hose (8) as shown.

c. Run the other end of this coolant line under engine and attach to output end of heater with clamp (3).

d. Attach new coolant hose (7) to intake side of heater with clamp (6) and secure opposite end of coolant hose (7) to engine port adapter (out-take) (5). Secure with clamp (6).

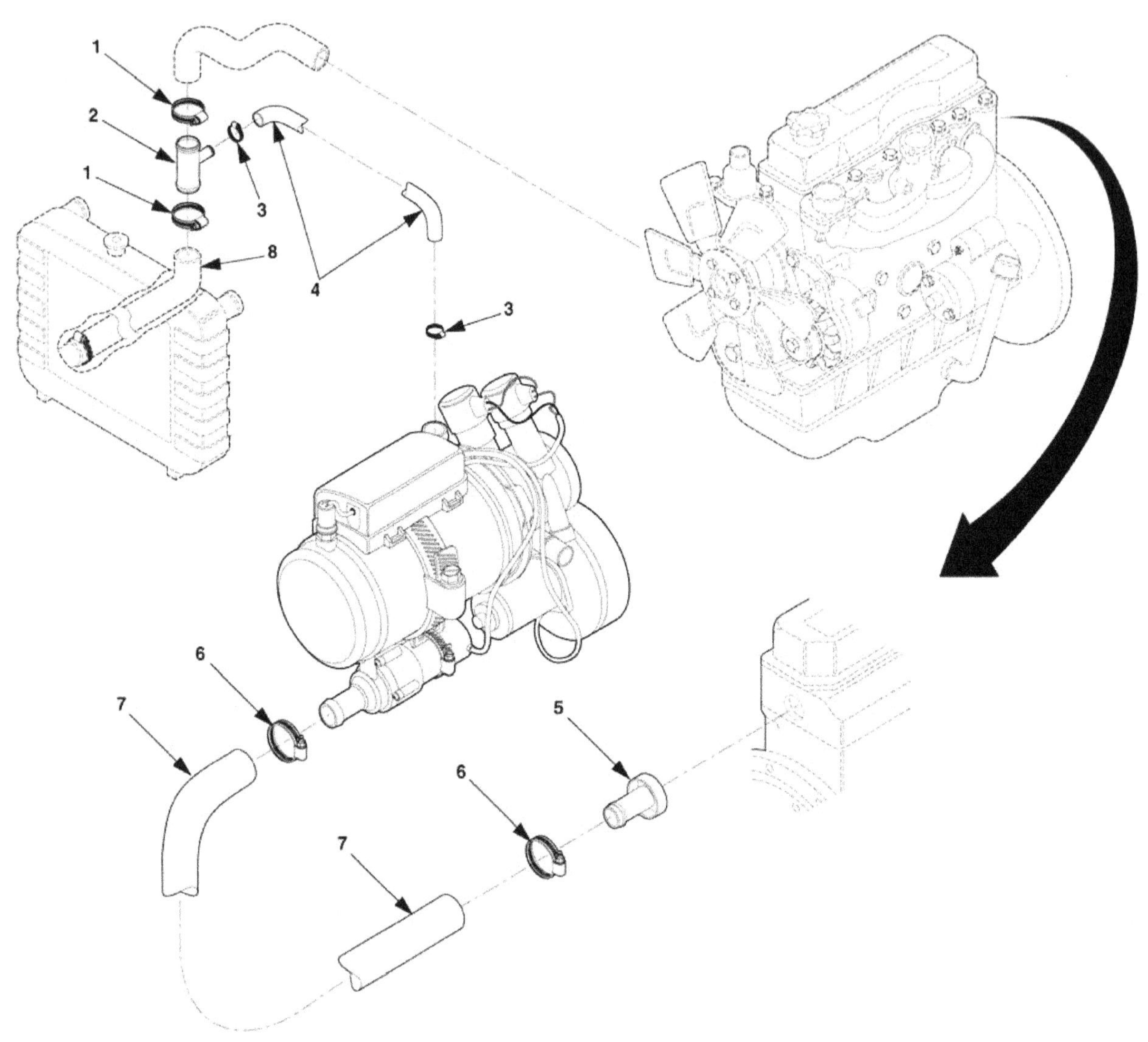

Figure 6-5. Coolant Hoses.

NOTE

(MEP-804B/MEP-814B Only) Install coolant lines to heater as follows:

e. e. Assemble bypass valve assembly (5, Figure 6-6 (Sheet 1)), consisting of two elbow fittings (1), two straight fittings (2) two adapter fittings (3) and check valve (4), and clamp (6). Direction arrow on check valve should point down.

f. f. Attach bypass valve assembly (5) and secure to rear engine support (7) with screw (8), two lock washers (9), flat washer (10), and nut (11).

g. g. Remove hose (12, Figure 6-6 (Sheet 2)) between engine port (13) and oil cooler (14). Be sure to use a suitable container to catch coolant from hoses.

h. h. Install coolant hose (15), with clamps (17), fr om bottom adapter fitting on bypass valve assemble (ref 5) to the top fitting on the heater (18).

i. i. Install coolant hose (16), with clamps (17), from top adapter fitting on bypass valve assembly (ref 5) to the bottom fitting on the heater (18).

j. j. Install coolant hose (19, Figure 6-6 (Sheet 3)), between bottom elbow fitting on bypass valve assembly (ref 5) and oil cooler (ref 14) with two clamps (20).

k. Install coolant hose (21) to top elbow fitting on bypass valve assembly (ref 5) with clamp (22) and to engine port (ref 13) with pressure clamp (23).

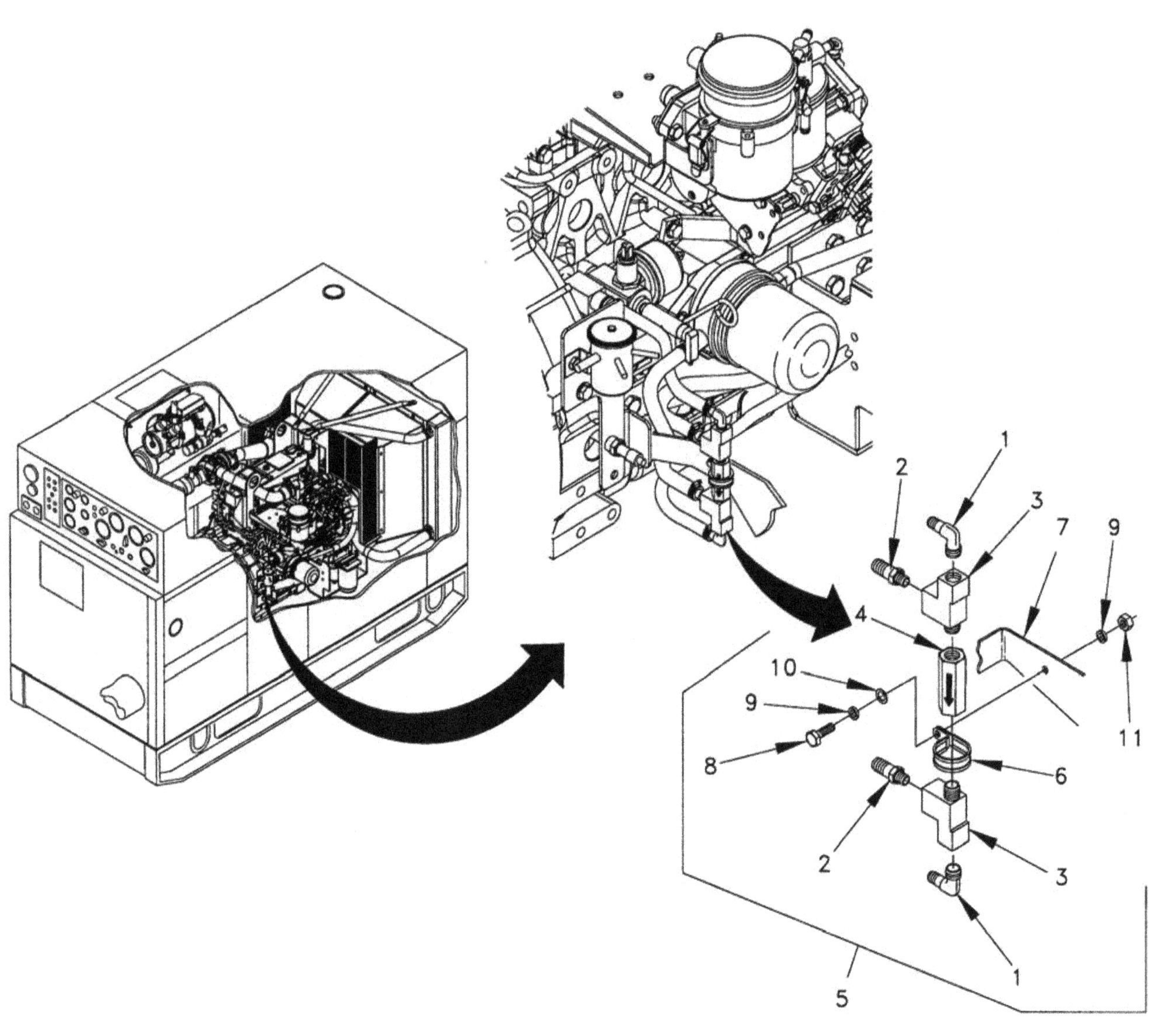

Figure 6-6. Coolant Hoses and Bypass Valve (MEP-804B/MEP-814B) (Sheet 1 of 3).

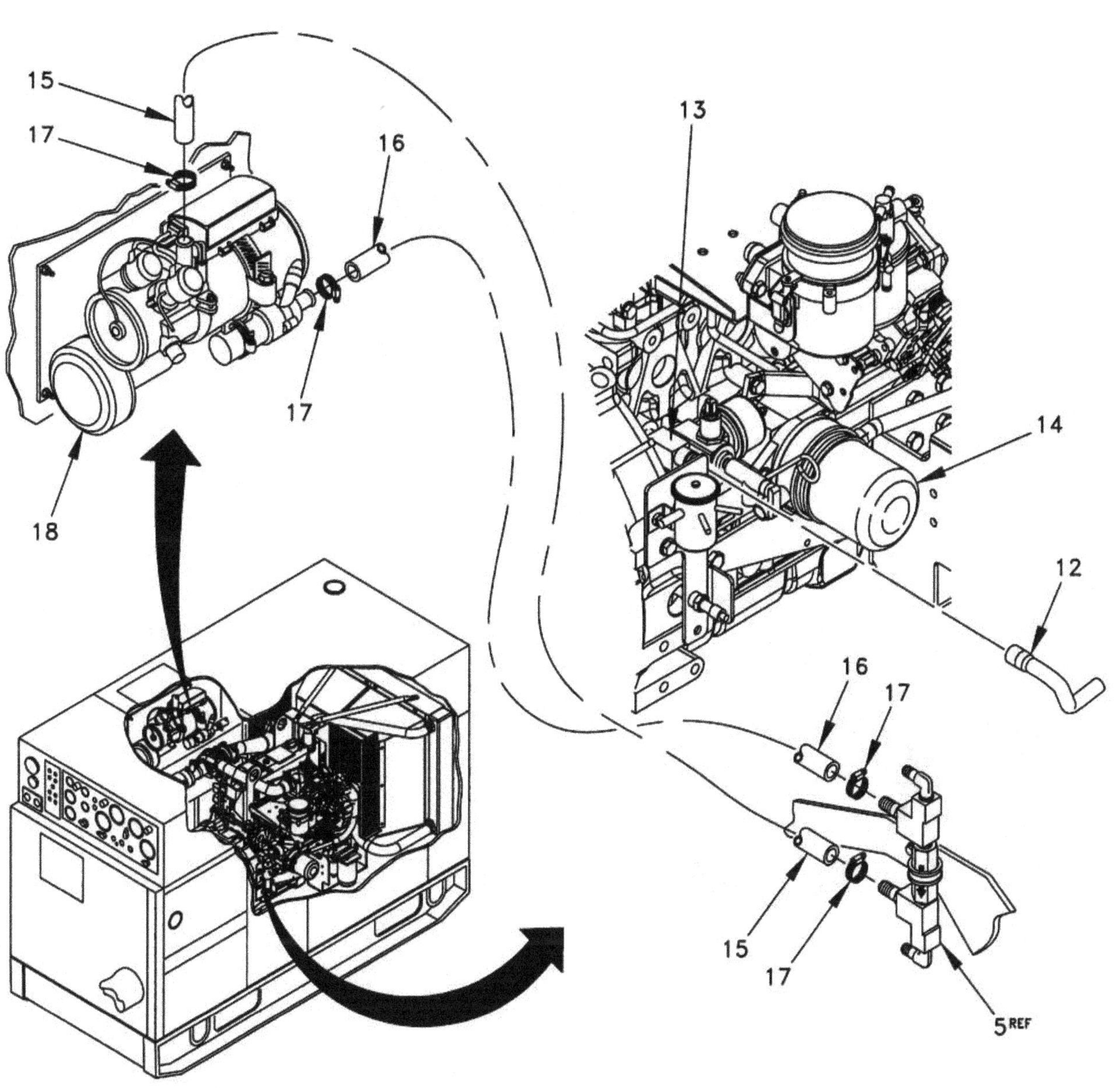

Figure 6-6. Coolant Hoses and Bypass Valve (MEP-804B/MEP-814B) (Sheet 2 of 3).

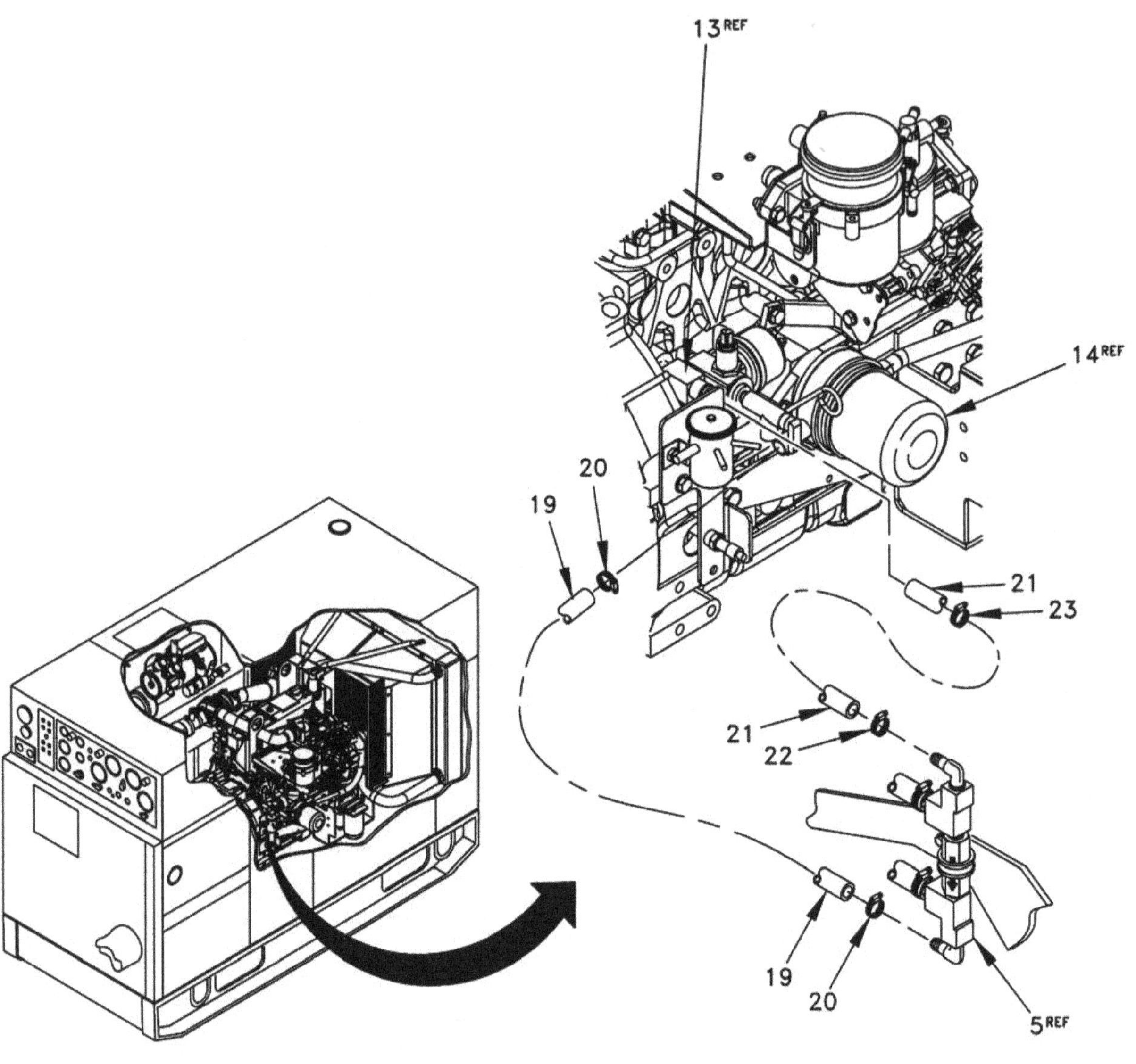

Figure 6-6. Coolant Hoses and Bypass Valve (MEP-804B/MEP-814B) (Sheet 3 of 3).

l. l. Install control unit (3, Figure 6-7), with bolt (4) and nut (1) to the engine side of the output box (2).

m. m. Install fuel pump (5) with clamp (6) using bolt found on the engine side of the output box and nut

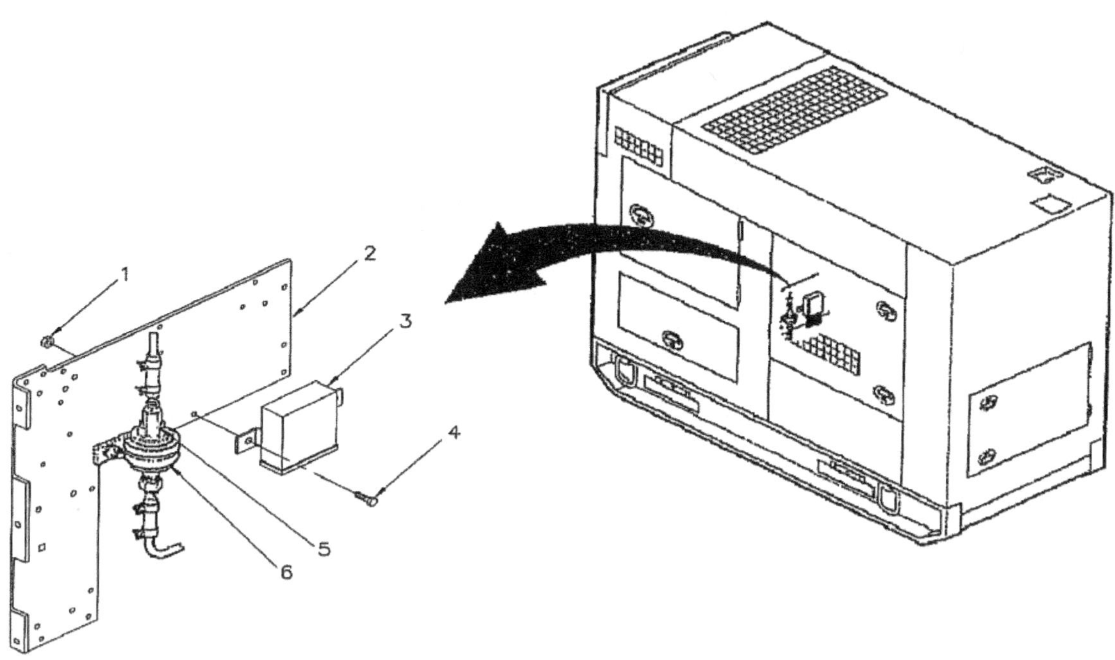

Figure 6-7. Control Unit Assembly.

n. Disconnect generator fuel supply line (12, Figure 6-8) and adapter (11) from fuel pickup tube assembly (13). Attach tee (10) to fuel pickup tube assembly (13) re-attach generator fuel supply line (12) to top of tee (10) using adapter (11).

NOTE

The proper connection of fuel lines is with use of a butt joint.

o. o. Install barbed adapter (9) to tee fitting (10) along with butt splice (7), fuel tubing (8), securing with clamps (6).

p. p. Install butt splice (7) with fuel tubing (8) to inlet of pump (4), securing with two clamps (6). Attach butt splices (2) and tubing (3) to outlet from pump (4) and to heater inlet tube using clamps (1).

NOTE

Seal all pipe threads with sealing compound contained in kit.

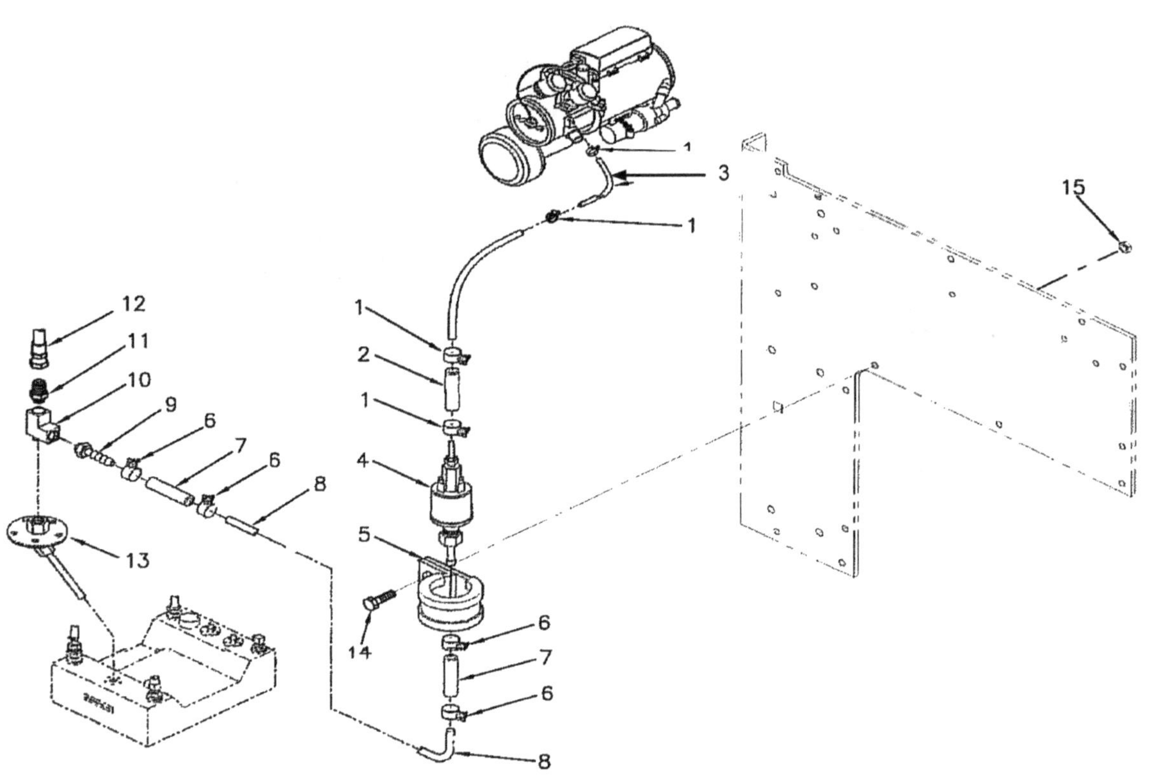

Figure 6-8. Fuel System Assembly.

q. q. Open right engine compartment panel door.

r. r. Locate wiring harness and remove and discard pins from P6 (16, 17, and 24), Figure 6-9.

s. s. Install wiring harness (2, Figure 6-9) in accordance with wiring diagram (see Figure 6-9). Secure harness with tie wraps contained in kit.

NOTE

Install fuse (25 AMP, 32 V) if not already installed.

t. t. Install fuse by connecting (+) terminal wire of fuse holder (1) to the slave receptacle. Then install fuse XF1 fuse holder (bottom left radiator housing).

u. u. Drill a 13/64th hole at lower left hand radiator side stiffener and mount fuse holder (1) using screw, washer, and nut provided in kit.

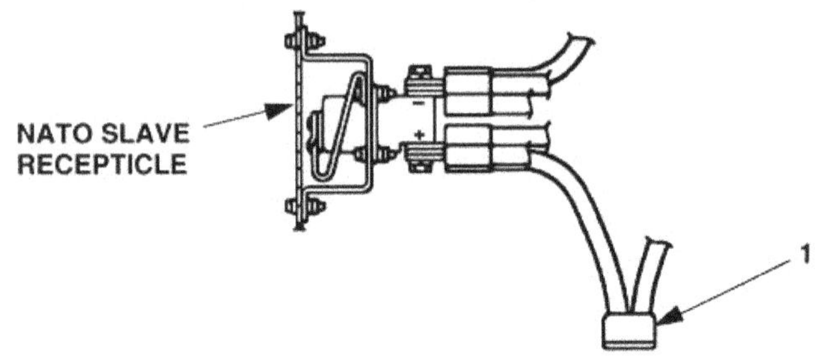

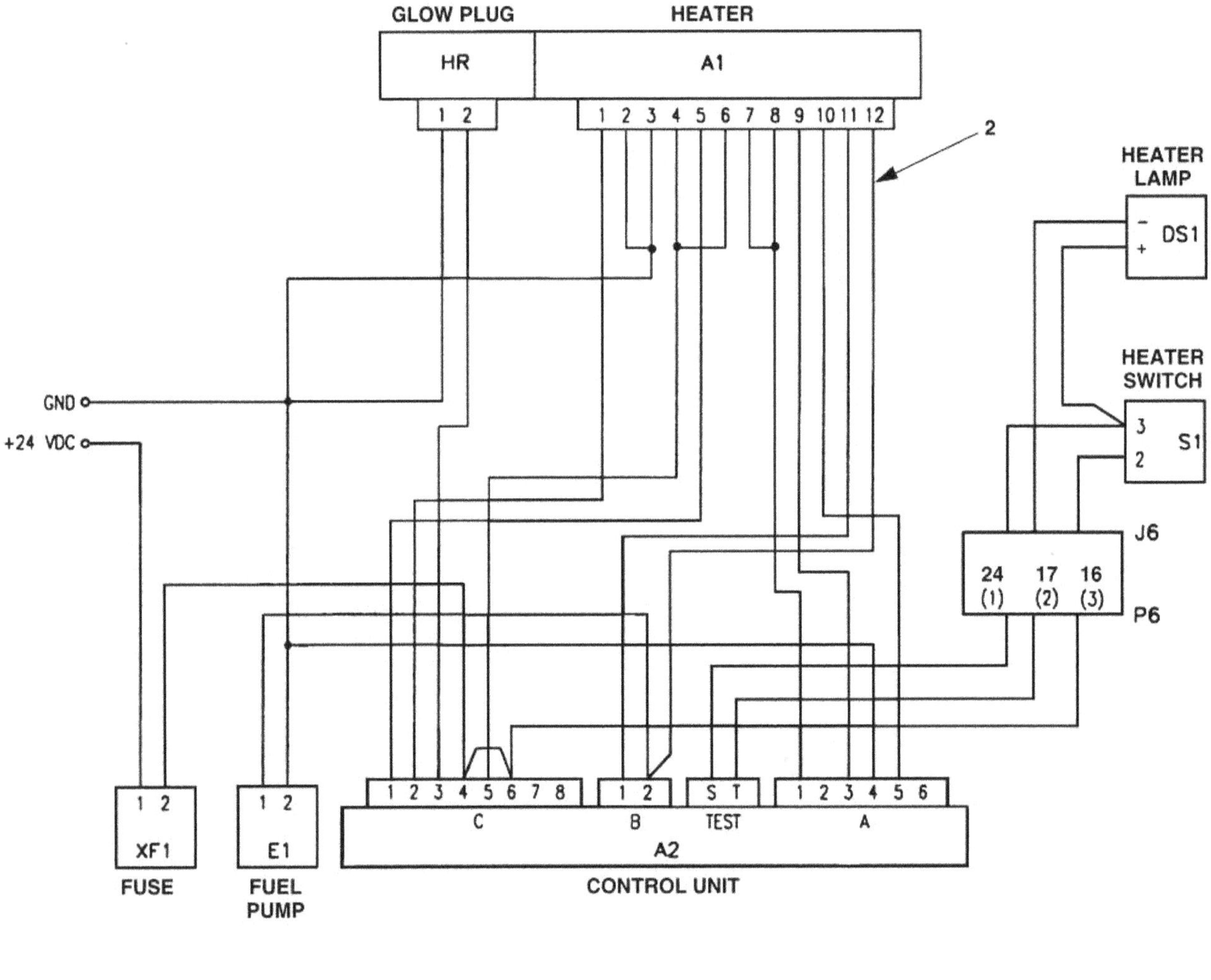

GLOW PLUG
HEATER
HR
A1
2
HEATER LAMP
DS1
HEATER SWITCH
S1
GND
+24 VDC
J6
24 (1)
17 (2)
16 (3)
P6
XF1
FUSE
E1
FUEL PUMP
C
B
TEST
A
A2
CONTROL UNIT

Figure 6-9. Fuse and Wiring Harness.

v. v. Open control panel access door (1, Figure 6-10).

w. w. Locate operating instruction plate (11) and heater function codes plate (10), and match drill 3/16" holes in control panel access door (1). Plates shall be readable when the door is open.

x. x. Install rivets (2) Figure 6-10. Rivet head shall be to the outside of the unit.

y. y. Locate heater switch label (6) Figure 6-10 and drill a 5/16" hole for the light and a 1/2" hole for the switch, in control panel (9).

z. Install heater switch label (6), indicator light (7), and toggle switch (3).

aa. Remove and maintain (for removal purposes) pins from J6-16, 17, and 24 (see Figure 6-9).

ab. Install electrical leads in accordance with wiring diagram (see Figure 6-9). Secure leads with tie wraps contained in kit.

CAUTION

The coolant in the system shall contain the proper mixture of water and antifreeze to prevent coolant from freezing or slushing. Failure to observe this caution could cause engine damage.

ac. Add coolant to proper level.

ad. Connect positive and negative battery cables.

ae. Start and run engine until the r adiator thermostat has opened.

af. Start the heater coolant pump by connecting a jumper from positive power to pin A6 of the A2 control unit. This will start the coolant pump only. Continue until pitch sound changes.

ag. If necessary, top off coolant.

ah. Disconnect jumper lead from pin A6 of the A2. ai.

Double check all hose connections for leaks.

aj. With engine running, start heater. Check the indi cator light (3, Figure 6-10) per the heater function codes plate (10) for heater operation.

ak. Turn off the engine and allow heater to run until the coolant reaches temperature at which the heater will cycle to low-heat mode.

al. Switch heater off.

NOTE

The water pump and combustion air blower will continue to run for approximately 3 minutes.

am. Locate identification plate (see Table 6-3) below existing plates on left side of housing. Match and drill four 3/16" holes in housing. Install four rivets.

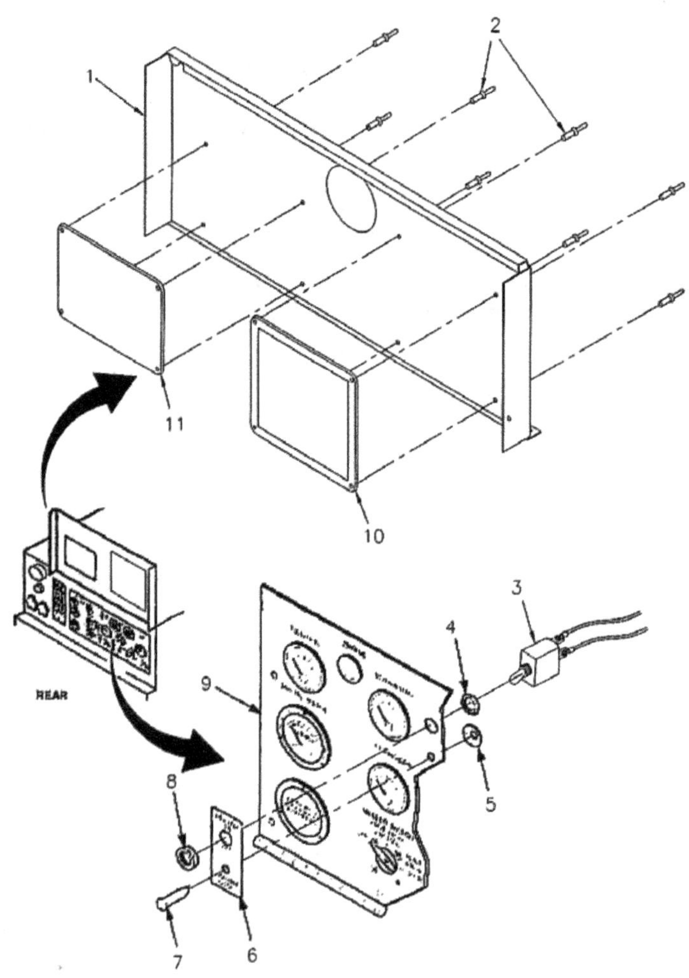

Figure 6-10. Cover and Instrument Panel

Section X. TROUBLESHOOTING PROCEDURES

6-10 GENERAL

Refer to Chapters 2 and 3 for generator set troubleshooting procedures (TM 9-2815-538-24&P for engine troubleshooting procedures). This section lists diagnostic and symptom related malfunctions you may find during operation of the generator set with the Winterization Kit installed and the generator set running. These pulses are shown visually on the Function Codes plate mounted inside the generator set control panel cover. Each malfunction is listed individually in the Symptom Index (Section XI), and helps determine probable causes and corrective actions to take. The troubleshooting symptom or diagnostic fault index cannot list all faults that may occur, nor all the tests, inspections, and/or corrective actions. If a malfunction is not listed or cannot be corrected by listed corrective actions, notify your supervisor. There are two types of faults associated with troubleshooting: *Symptom* related Faults and *Diagnostic* Faults.

a. Symptom Related Faults . This is the soldier's ability to recognize faults occurring during normal operation that may or may not prompt a diagnostic/lamp fault.

Basic items of concern would include:

- Fuses
- Proper Electrical Connections
- Air Inlet and Exhaust Pipe clearance
- Fuel in the tank
- Correct Battery Voltage
- Coolant Flow

b. Diagnostic Faults. The indicator light near the heater switch is designed to blink code sequences to signal malfunctions in the system. See paragraph 6-11.

6-11 HEATER FUNCTION CODES

Code Light Pulses. The indicator light near the heater ON-OFF switch will blink in different sequences of long and short pulses to indicate malfunctions. A plate (Figure 6-11) mounted on the generator control panel access door (Figure 6-10) lists the malfunctions and shows each sequence of pulses for diagnostic faults given. Figure 6-12 provides an example of the Symptom Index Troubleshooting Table found in Section XI of this Chapter.

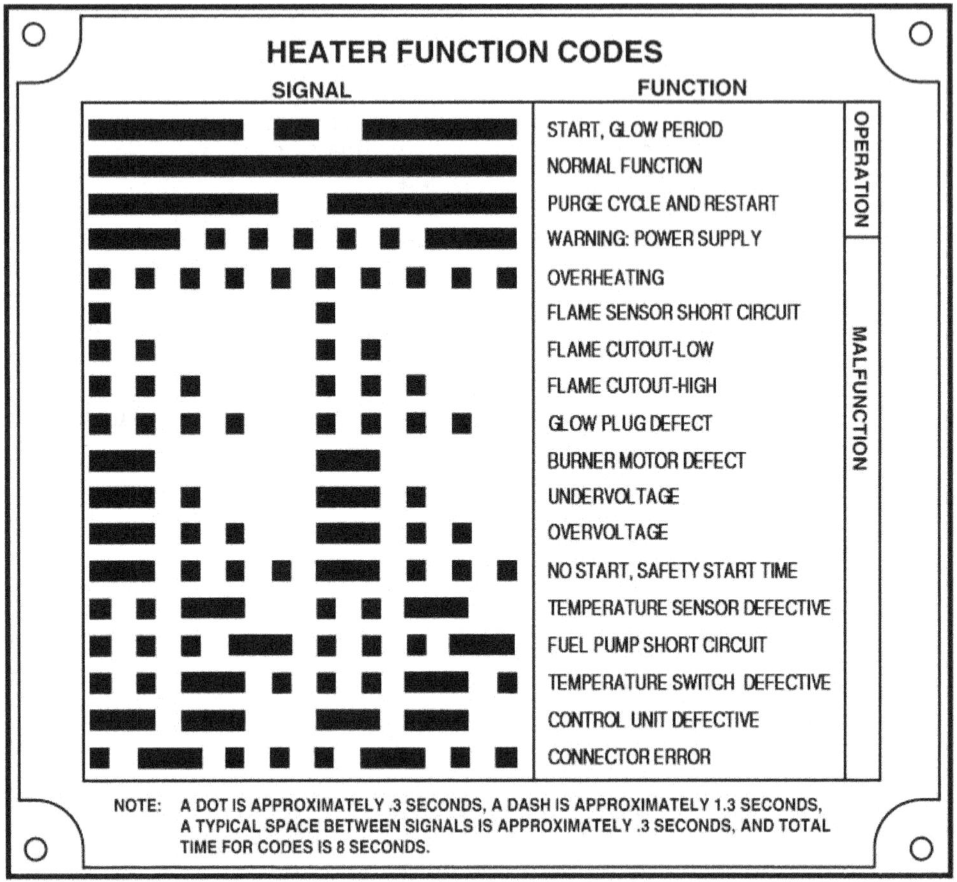

Figure 6-11. Heater Function Codes Plate.

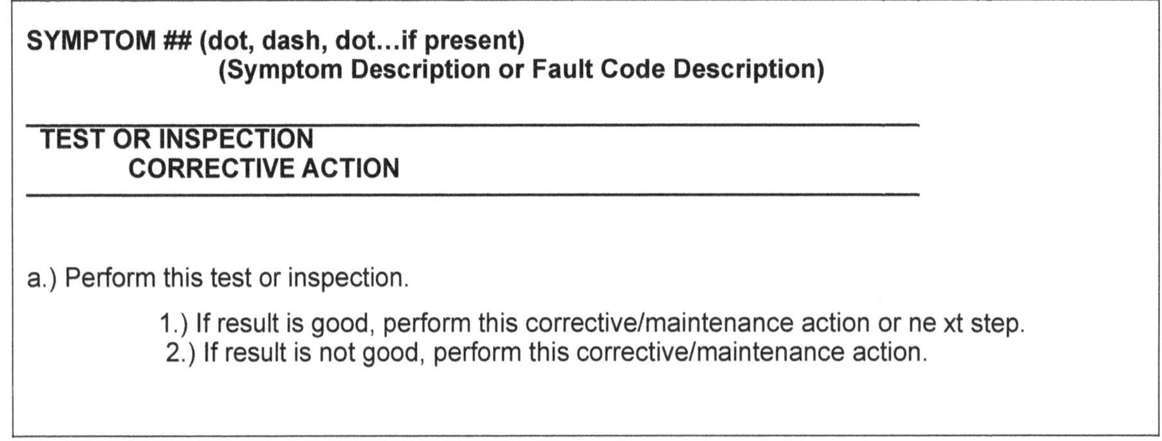

SYMPTOM ## (dot, dash, dot...if present)
(Symptom Description or Fault Code Description)

TEST OR INSPECTION
CORRECTIVE ACTION

a.) Perform this test or inspection.

1.) If result is good, perform this corrective/maintenance action or ne xt step.
2.) If result is not good, perform this corrective/maintenance action.

Figure 6-12. Example Troubleshooting Table.

Section XI. SYMPTOM INDEX, WINTERIZATION KIT

NOTE

When the heater is switched on, the light will perform one of the sequences of light pulses shown visually on the Heater Function Codes Plate (Figure 6-11) mounted inside the generator set control panel cover. Before each symptom, this index lists in parentheses the light sequence associated with it.

Troubleshooting

SYMPTOM 1

(dash, dash) Purge Cycle and Heater Restart (Starts while heater still NOT shut OFF).

TEST OR INSPECTION CORRECTIVE ACTION

If the heater fails to start the first time, it will automatically attempt a second start as long as coolant temperature is 70° (158°C). If that attempt fails, the heater will automatically shut down.

Do not attempt to perform any maintenance tasks on the Winterization Kit while the generator set is operating. Failure to comply with this warning can cause injury or death to personnel.

a. Shut down generator set.

b. Make sure battery power is still applied and EMERGENCY STOP SWITCH is pushed IN.

c. Release DCS control panel by turning two fasteners and lowering panel.

NOTE

See Figure 6-9 for all Wiring Harness troubleshooting.

d. Check fuel level, check for leaks and that fuel pump is properly working. Place a container under the output end of the fuel pump line and remove the line. Turn the pump on and observe the fuel flow.

(1) If no fuel leaks and pump IS working properly, go to Step e.

(2) If fuel pump is NOT working properly, continue to SYMPTOM 13 (Fuel Pump Short Circuit).

e. Ensure purge cycle of at least 90 seconds has elapsed and coolant temperature falls below the triggering point of 70°C (158°F).

(1) If temperature is as specified and flame is NOT stable, inspect glow plug for proper seating. See paragraph 6-22.

(2) If temperature is as specified and flame IS stable, replace Heater Assembly.

END OF TASK

SYMPTOM 2

(dash, 5 dots, dash) Warning: Power Supply

TEST OR INSPECTION CORRECTIVE ACTION

NOTE

This fault will more than likely accompany SYMPTOMS 9/10 and a safety shut down will occur.

Do not attempt to perform any maintenance tasks on the Winterization Kit while the generator set is operating. Failure to comply with this warning can cause injury or death to personnel.

a. Shut down generator set.

b. Make sure battery power is still applied and EMERGENCY STOP SWITCH is pushed IN.

c. Release DCS control panel by turning two fasteners and lowering panel.

d. Record any accompanying fault/s and check voltage output of batteries.

(1) If voltage is less than 21 VDC, check electrolyte levels as per paragraph 2-11. Charge batteries as necessary.

(2) If voltage is greater than 30 VDC, troubleshoot 25A main fuse/s as per wiring diagram Figure 6-9.

(3) If voltage is within normal range between 21and 30 VDC, replace control unit per paragraph 6-12.

END OF TASK

SYMPTOM 3

(10 dots) Overheating

TEST OR INSPECTION
CORRECTIVE ACTION

If the temperature of the heat exchanger or the coolant rises above 85° C (185° F) at any given time, this fault will more than likely produce a safety cut out and malfunction shutdown. Fault is normally due to a lack of water/coolant, air restriction, or poorly bled coolant system.

Do not attempt to perform any maintenance tasks on the Winterization Kit while the generator set is operating. Failure to comply with this warning can cause injury or death to personnel.

a. Shut down generator set.

b. Make sure EMERGENCY STOP SWITCH is pushed IN on generator set.

c. Release DCS control panel by turning two fasteners and lowering panel.

CAUTION

Coolant must contain a minimum of 10% antifreeze at all times to protect against corrosion. Fresh water will corrode internal heater parts.

d. Check coolant level and add coolant as required per paragraph 2-75.

e. Check to ensure coolant pump is working properly. Measure continuity of coolant pump wiring: A2-C2 and A1-1 (A1-2 GND).

(1) If continuity is present, go to Step f.

(2) If continuity is not present, repair/replace wiring harness as required per paragraph 6-21.

f. Check to ensure voltage is present at positive end of coolant pump with ON/OFF heater switch in the ON position.

(1) If voltage is as specified, replace heater per paragraph 6-16.

(2) If voltage is NOT as specified, go to SYMPTOM 2 (Warning: Power Supply).

END OF TASK

SYMPTOM 4

(dot, dot) Flame Sensor Short Cir-

TEST OR INSPECTION CORRECTIVE ACTION

The flame is monitored by the flame sensor and at the point of ignition, will automatically switch the glow plug OFF.

Do not attempt to perform any maintenance tasks on the Winterization Kit while the generator set is operating. Failure to comply with this warning can cause injury or death to personnel.

NOTE

See Figure 6-9 for Wiring Harness troubleshooting.

a. Shut down generator set.

b. Make sure EMERGENCY STOP SWITCH is pushed IN on generator set.

c. Remove heater assembly and test according to paragraph 6-13 (Heater Assembly Maintenance).

   (1) If flame sensor is good, repair/replace wiring harness as required (paragraph 6-21).

   (2) If continuity is NOT good, replace heater assembly (paragraph 6-13, Heater Assembly Maintenance).

END OF TASK

SYMPTOM 5

(2 dots, 2 dots) Flame Cutout – Low

~~TEST OR INSPECTION~~

CORRECTIVE ACTION

The flame is monitored by the flame sensor and at the point of ignition, will automatically switch the glow plug OFF. This fault will typically present itself when there is NOT a sufficient fuel supply or fuel pump operation is suspect. A higher than normal blower motor RPM (greater than 5500) may also prevent flame from sustaining itself. A fault shutdown is imminent.

Do not attempt to perform any maintenance tasks on the Winterization Kit while the generator set is operating. Failure to comply with this warning can cause injury or death to personnel.

NOTE

See Figure 6-9 for Wiring Harness troubleshooting.

a. Shut down generator set.

b. Make sure EMERGENCY STOP SWITCH is pushed IN on generator set.

c. Turn heater switch momentarily OFF and then back ON to reset the fault shutdown. If heater does not start, it will continue to shut down.

d. Check fuel level and/or fuel line restrictions. Refill if needed.

e. Test fuel metering pump per paragraph 6-14 for proper operation.

(1) If fuel metering pump works as specified, continue to step f.

(2) If fuel metering pump does NOT work as specified, repair/replace fuel metering pump per paragraph 6-14.

f. Measure blower motor voltage at A2-Pin C5.

If voltage is not as specified, verify 24 VDC power source and/or replace heater control unit per paragraph 6-12.

END OF TASK

SYMPTOM 6

(3 dots, 3 dots) Flame Cutout - High

TEST OR INSPECTION CORRECTIVE ACTION

The flame is monitored by the flame sensor and at the point of ignition, will automatically switch the glow plug OFF. This fault will typically present itself when the heater and/or fuel system is under extreme use and develops vapor lock in the fuel line.

Do not attempt to perform any maintenance tasks on the Winterization Kit while the generator set is operating. Failure to comply with this warning can cause injury or death to personnel.

NOTE

See Figure 6-9 for J7 Wiring Harness troubleshooting.

a. Shut down generator set and wait for a period of 30 minutes.

b. Make sure EMERGENCY STOP SWITCH is pushed IN on generator set.

c. Turn heater switch momentarily OFF and then back ON to reset the fault shutdown. If heater does not start, it will continue to shut down.

d. Check for use of proper fuel and fill level. Refill if needed.

NOTE

Ensure that fuel line is routed away from the heater assembly and exhaust.

e. Check installation position of fuel metering pump per paragraph 6-9.8 (m).

NOTE

Variances greater than 15° from vertical could cause fuel blockage.

f. Remove possible vapor lock from fuel line. Reduced suction will caution vapor in the line. See testing per paragraph 6-14.

END OF TASK

SYMPTOM 7

(4 dots, 4 dots) Glow Plug Defect

TEST OR INSPECTION
CORRECTIVE ACTION

The flame is monitored by the flame sensor and at the point of ignition, will automatically switch the glow plug OFF.

Do not attempt to perform any maintenance tasks on the Winterization Kit while the generator set is operating. Failure to comply with this warning can cause injury or death to personnel.

NOTE

See Figure 6-9 for Wiring Harness troubleshooting.

a. Shut down generator set and wait for a period of 30 minutes.

b. Make sure EMERGENCY STOP SWITCH is pushed IN on generator set.

c. Turn heater switch momentarily OFF and then back ON to reset the fault shutdown. If heater does not start, it will continue to shut down.

d. Perform TEST procedures per paragraph 6-22, Igniter/Glow Plug and Resister Maintenance) to determine if glow plug or resister is faulty.

(1) If glow plug and resistor continuity is good, continue to step e.

(2) If glow plug or resistor is bad, replace as needed.

e. Measure continuity between A2-C Pin 3 and HR Pin 2 and HR Pin 1 and HR Pin 2. If continuity is NOT good, repair/replace wiring harness as required per paragraph 6-21.

END OF TASK

SYMPTOM 8

(dash, dash) Burner Motor Defect

TEST OR INSPECTION CORRECTIVE ACTION

When the heater is started, the functioning of the blower motor is checked once. If it does not start, the heater will undergo a fault shutdown. During operation, the blower motor is monitored in cycles. If the motor speed is below the permitted limit, a fault shutdown will also occur. A fault shutdown may or may not occur during rough idling. This is typically due to a fuel, air mixture, or low voltage related problem.

Do not attempt to perform any maintenance tasks on the Winterization Kit while the generator set is operating. Failure to comply with this warning can cause injury or death to personnel.

NOTE

See Figure 6-9 for Wiring Harness troubleshooting.

a. Shut down generator set.

b. Make sure EMERGENCY STOP SWITCH is pushed IN on generator set.

c. Turn heater switch momentarily OFF and then back ON to reset the fault shutdown. If heater does not start, it will continue to shut down.

d. Clear any obstructions or debris from air intake or exhaust lines.

e. Check 7.5 AMP fuse.

  (1) If fuse is bad, replace fuse as required.

  (2) If fuse is good, continue to step f.

f. Measure voltage at A1- Pin 4 and GND to determine correct voltage is applied.

  If voltage is NOT as specified, replace heater assembly per paragraph 6-13..

END OF TASK

SYMPTOMS 9 and 10

(dash, dot, dash, dot) Under Voltage

(dash, 2 dots, dash, 2 dots) Over Voltage

TEST OR INSPECTION
CORRECTIVE ACTION

SYMPTOM 2 may or may not illuminate simultaneously with a SYMPTOM 9 or 10 fault. Troubleshooting for all three should be carried out in the same procedure as directed below. A fault shutdown will occur in the event the voltage is less than 21 VDC or greater than 30 VDC.

NOTE

See SYMPTOM 2 (Warning: Power Supply).

END OF TASK

SYMPTOM 11

(dash, 3 dots, dash, 3 dots) Non-Start, Automatic Cut-Out

TEST OR INSPECTION CORRECTIVE ACTION

This fault is the result of having one or more faults causing multiple indications simultaneously. As described below, this fault should be the LAST malfunction to troubleshoot if there are a series of malfunctions in order to find the root of the problem. Some of these symptoms may be eliminated due to proper PMCS being performed. Those include but not limited to:

- Fuel/Ignition system (SYMPTOM 5)
- Power requirements (SYMPTOM 2)
- Intermittent wiring or cable connections (SYMPTOM 15)

NOTE

If the heater fails to start the first time, it will automatically attempt a second start as long as coolant temperature is 70° (158°C). If that attempt fails, the heater will automatically shut down.

Do not attempt to perform any maintenance tasks on the Winterization Kit while the generator set is operating. Failure to comply with this warning can cause injury or death to personnel.

a. Shut down generator set.

b. Make sure battery power is still applied and EMERGENCY STOP SWITCH is pushed IN.

c. Release DCS control panel by turning two fasteners and lowering panel.

NOTE

See Figure 6-9 for all Wiring Harness troubleshooting.

d. Have all other SYMPTOMS been extinguished?

(1) If other SYMPTOMS exist, go to those respective SYMPTOMS.

(2) If there are no other SYMPTOMS, and fault still exist, replace control unit per paragraph 6-12.

END OF TASK

SYMPTOM 12

(2 dots, dash, 2 dots, dash) Temperature Sensor Defective

TEST OR INSPECTION CORRECTIVE ACTION

The temperature sensor is a device to acknowledge temperature change. If the temperature of the heat exchanger wall or the coolant rises above the maximum permissible temperature, the safety thermal cutout fuse (switch) is blown and initiates a malfunction shutdown. The inability to switch between LOW- and HIGH-Heat modes presents the possibility of a SYMPTOM 12 or 13 fault.

NOTE

If the heater fails to start the first time, it will automatically attempt a second start as long as coolant temperature is 70° (158°C). If that attempt fails, the heater will automatically shut down.

Do not attempt to perform any maintenance tasks on the Winterization Kit while the generator set is operating. Failure to comply with this warning can cause injury or death to personnel.

a. Shut down generator set.

b. Make sure battery power is NOT applied and EMERGENCY STOP SWITCH is pushed IN.

c. Remove Heater Assembly and test according to paragraph 6-13.

d. Turn heater switch momentarily to OFF and then back ON to reset the fault shutdown.

(1) If sensor is bad, replace heater assembly (paragraph 6-13, Heater Assembly Maintenance)

(2) If sensor is good and fault still remains, repair/replace wiring harness per paragraph 6-21.

END OF TASK

SYMPTOMS: 13, 14, 15, 16

13 (3 dots, dash, 3 dots, dash) Fuel Pump Short Circuit

14 (2 dots, dash, 3 dots, dash, dot) Temperature/Thermal Cutout Switch (Fuse Defective)

15 (4 dashes) Control Unit Defective

16 (dot, dash, 3 dots, dash, 2 dots) Connection Error

TEST OR INSPECTION CORRECTIVE ACTION

The above faults are usually as a result of a bad connection or considered an intermittent fault. Follow the procedures below as the corrective action is the same for each.

Do not attempt to perform any maintenance tasks on the Winterization Kit while the generator set is operating. Failure to comply with this warning can cause injury or death to personnel.

a. Troubleshoot the above SYMPTOM fault conditions by checking for possible wiring harness disconnects.

b. Turn heater switch momentarily to OFF and then back ON to reset the fault shutdown.

If wiring harness and/or connections are good, replace the respective faulty component:

- Fuel Pump Short Circuit per paragraph 6-14.
- Temperature/Thermal Cutout Switch (heater assembly per paragraph 6-13).
- Control Unit per paragraph 6-12.
- Wiring Harness per paragraph 6-21.

END OF TASK

SYMPTOM 17

(No Indication of Heater Operation When Switch is in ON Position)

TEST OR INSPECTION CORRECTIVE ACTION

Some of these symptoms may be eliminated due to proper PMCS being performed. Those include but not limited to:

- Power requirements (SYMPTOM 2).
- Intermittent wiring or cable connections (SYMPTOM 16).

NOTE

If the heater fails to start the first time, it will automatically attempt a second start as long as coolant temperature is 70° (158°C). If that attempt fails, the heater will automatically shut down.

Do not attempt to perform any maintenance tasks on the Winterization Kit while the generator set is operating. Failure to comply with this warning can cause injury or death to personnel.

a. Shut down generator set.

b. Make sure 24 ±2VDC battery power is still applied and EMERGENCY STOP SWITCH is pushed **IN.**

c. Release DCS control panel by turning two fasteners and lowering panel.

NOTE

See Figure 6-9 for all Wiring Harness troubleshooting.

d. Check 7.5A fuse (XF1) supplying power to blower motor.

(1) If fuse is bad, replace if needed.

(2) If fuse is good, continue to step e.

e. Check for +24 VDC at S1-2 on heater switch.

(1) If voltage is as specified, continue to step f.

(2) If voltage is not as specified, continue checking for +24 VDC between power source and heater switch according to Figure 6-9.

f. Place S1 switch to ON position and check for +24 VDC at S1-3.

(1) If voltage is as specified, continue to step g.

(2) If voltage is not as specified, replace S1 per paragraph 6-17.

g. Remove A2-Test connector from control unit. Check for +24 VDC at Pin S with S1 in ON position.

(1) If voltage is as specified, continue to step h.

(2) If voltage is not as specified, repair/replace wiring harness (paragraph 6-21).

h. Remove A2-A connector (6 pin) from control unit. Check continuity from Pin 3 to GND.

(1) If circuit indicates short, replace control unit per paragraph 6-12.

(2) If circuit indicates an open, repair/replace wiring harness (paragraph 6-21).

END OF TASK

Section XII. MAINTENANCE PROCEDURES

WARNING

Do not attempt to perform any maintenance tasks on the Winterization Kit while the generator set is operating. Failure to comply with this warning can cause injury or death to personnel.

6-12 CONTROL UNIT MAINTENANCE

6-12.1 Test.

Refer to troubleshooting procedures, symptom index 13/14/15/16.

6-12.2 Inspection.

a. a. Open right side engine access door.

b. b. Inspect control unit for loose or missing wires and damaged parts.

c. c. Close engine access door.

6-12.3 Removal – Control Unit Maintenance.

WARNING

Disconnect negative battery cable from right battery and the positive battery cable from the left battery before doing the following procedures. Reconnect cables in reverse order. Failure to comply with this warning can cause injury or death to personnel.

WARNING

Cooling system operates at high temperatures and pressure. Contact with high pressure steam and/or liquids can result in burns and scalding. Shut down generator set, and allow system to cool before performing checks, services, and maintenance, or wear gloves and additional protective clothing and goggles as required. Failure to comply with this warning can cause injury or death to personnel.

WARNING

Always remove radiator cap slowly to permit any pressure to escape. Failure to comply with this warning can cause injury to personnel.

a. a. Disconnect negative battery cable and then positive battery cable. b .

Disconnect electrical connectors.

b. c. Remove two nuts (5, Figure 6-13), two screws (2), lock washers (3), flat washers (4) and control unit (1). Discard lock washers (3).

6-12.4 Replacement.

a. a. Perform steps 1 thru 3 of removal.

b. b. Perform steps 1 thru 3 of installation, substituting new control unit for the defective one.

6-12.5 Installation.

a. a. Install control unit (1), securing with screws (2), new lock washers (3), flat washers (4) and nuts (5).

b. b. Reconnect wiring harness to control unit.

c. c. Reconnect positive battery cable and then negative battery cable.

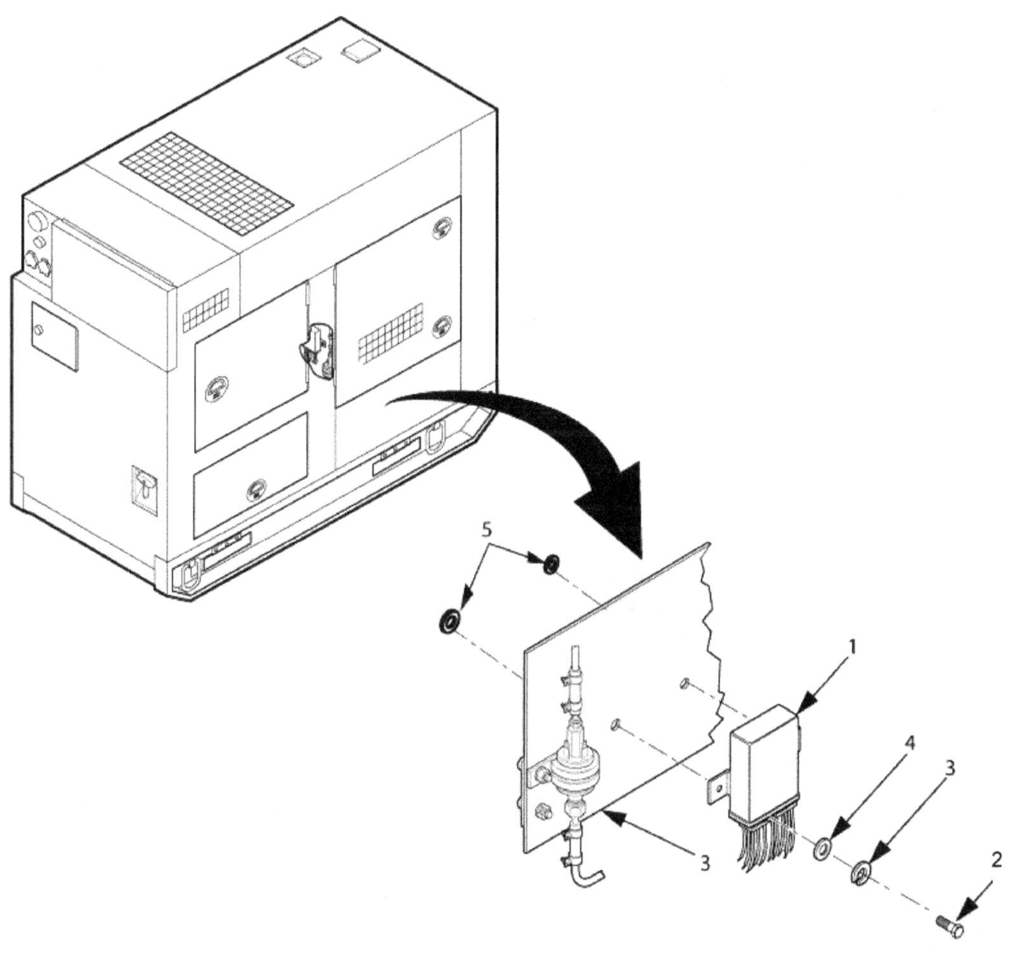

Figure 6-13. Control Unit Maintenance.

6-13 HEATER ASSEMBLY MAINTENANCE 6-13.1

Inspection.

Inspection is limited to visual inspection of components. Check for damaged and frayed wiring.

6-13.2 Removal.

WARNING

Disconnect negative battery cable from right battery and the positive battery cable from the left battery before doing the following procedures. Reconnect cables in reverse order. Failure to comply with this warning can cause injury or death to personnel.

WARNING

Cooling system operates at high temperatures and pressure. Contact with high pressure steam and/or liquids can result in burns and scalding. Shut down generator set, and allow system to cool before performing checks, services, and maintenance, or wear gloves and additional protective clothing and goggles as required. Failure to comply with this warning can cause injury or death to personnel.

WARNING

Always remove radiator cap slowly to permit any pressure to escape. Failure to comply with this warning can cause injury to personnel.

WARNING

Catch fuel in suitable container. Keep spilled fuel away from hot engine and all fires. Failure to comply with this warning can cause injury or death to personnel.

a. a. Disconnect negative battery cable and then positive battery cable.

b. b. Drain coolant system.

c. c. Disconnect fuel clamp (15, Figure 6-14) and fuel line (16) from heater and drain any fuel from hose into a suitable container.

d. d. Disconnect wiring harness connector (9) from heater.

e. e. Remove screw clamp (13), elbow (14), and exhaust hose (12) from heater (10).

f. f. Position a suitable container under heater to catch coolant from hoses, and loosen screw clamps (8) to disconnect coolant inlet hose (11) and outlet hose(7) from heater (10).

g. Remove four screws (1), flat washers (4), lock washers (5), nuts (6), and heater (10) with mounting plate (3) from bulkhead housing (2). Discard lock washers (5).

6-13.3 Repair.

Repair of the Heater Assembly is limited to the replacement of the igniter/glow plug assembly, resistor, and the heater pump.

6-13.4 Replacement.

a. a. Perform removal steps.

b. b. Perform installation steps substituting new heater assembly for the defective one.

6-13.5 Installation.

a. Install heater (10) with attached mounting plate (3) on bulkhead housing (2). Secure with screws (1), flat washers (4), new lock washers (5), and nuts (6).

b. Connect coolant inlet hose (11) and outlet hose (7) to heater (10), using screw clamps (8). Top off radiator to replace any coolant lost in removing hoses.

c. c. Attach elbow (14) to heater unit (10) and exhaust flex hose (12) to elbow using screw clamp (13).

d. d. Connect wiring har ness connector (9) to the heater (10).

e. e. Connect fuel line (16) to heater assembly (10) using clamp (15).

f. f. Top off coolant system.

g. g. Reconnect positive battery cable and then negative battery cable.

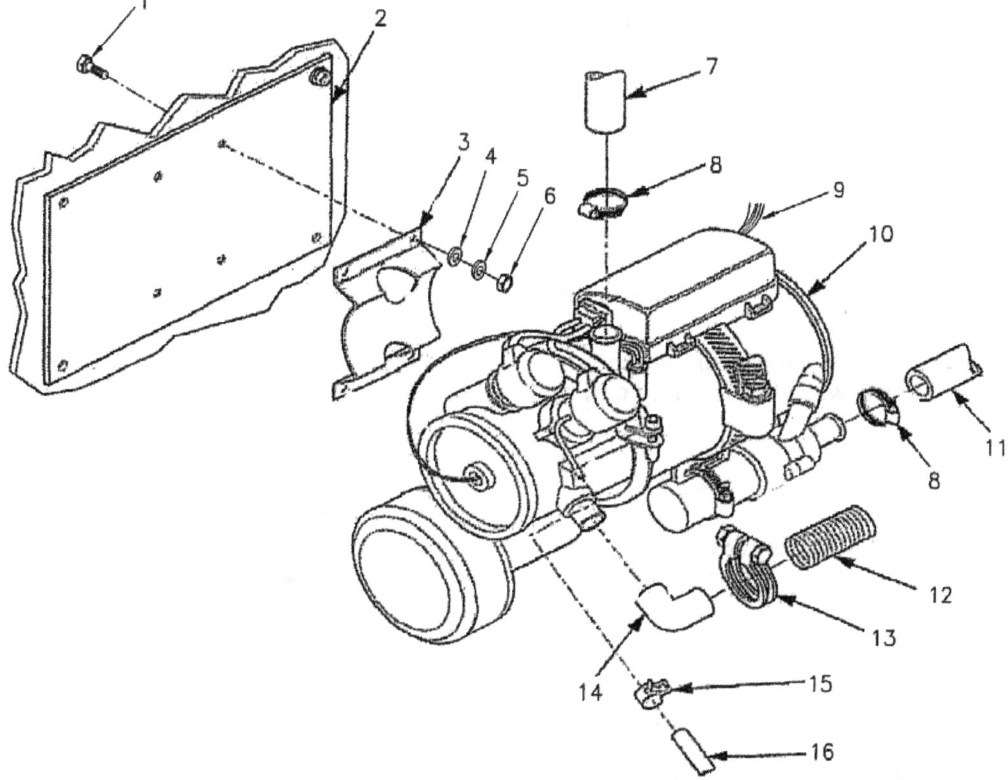

Figure 6-14. Heater Assembly Maintenance.

6-14 FUEL PUMP MAINTENANCE 6-14.1

Test.

a. Disconnect fuel line at heater assembly.

WARNING

Catch fuel in suitable container. Keep spilled fuel away from hot engine and all fires. Failure to comply with this warning can cause injury or death to personnel.

b. b. Remove fuel line from output end of fuel pump and place into container suitable for catching fuel.

c. c. Apply 24 VDC to electrical input of pump. Pu mp should hum. Fuel will then start flowing into the container.

6-14.2 Removal.

**WARN-
ING**

**Disconnect negative battery cab attery and the positive
Cooling system operates at high and pressure. Contact with
high pressure steam and/or liqu ttery before n burns and scalding. Shut
down generator set, and allow sdoinder. Fail- before performing checks,
ure to coel.**

**WARN-
ING**

services, and maintenance, or w additional protective clothing and goggles as required. Failure to comply with this warning can cause injury or death to personnel.

WARNING

Always remove radiator cap slowly to permit any pressure to escape. Failure to comply with this warning can cause injury to personnel.

WARNING

Catch fuel in suitable container. Keep spilled fuel away from hot engine and all fires. Failure to comply with this warning can cause injury or death to personnel.

a. a. Disconnect negative battery cable and then positive battery cable.

b. b. Disconnect electrical leads (6, Figure 6-15) from connector (7) on fuel pump (13).

c. c. Loosen screw clamp (9) to disconnect fuel line (11) and butt splice (10) from input end (8) of fuel pump (13). Drain any fuel from line into a suitable container.

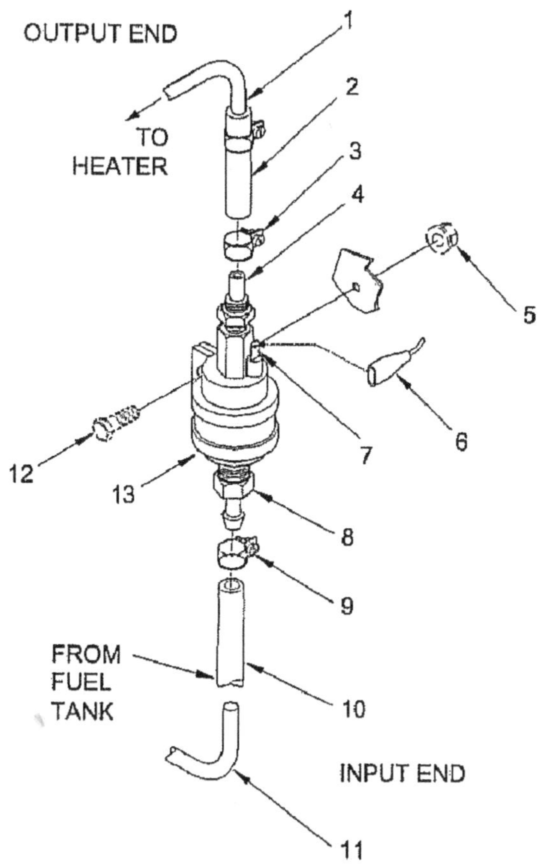

Figure 6-15. Fuel Pump Maintenance.

d. d. Loosen screw clamp (3) to disconnect butt splice (2) and fuel line (1) from output end (4) of fuel pump (13). Drain any fuel from line into a suitable container.

e. e. Remove nut (5) from bolt (12), then remove bolt (12).

f. f. Remove fuel pump (13).

6-14.3 <u>Replacement.</u>

a. a. Perform removal steps.

b. b. Perform installation steps substituting new fuel pump for the defective one.

6-14.4 Installation.

a. Install new fuel pump (13) and secure to housing with bolt (12) and nut (5).

b. b. Install butt splice (2) and fuel line (1) on outlet end (4) of fuel pump (13) and secure with screw clamps (3).

c. c. Install fuel line (11) and butt splice (10) on inlet end (8) of fuel pump (13) and secure with screw clamps (9).

d. d. Plug in electrical leads (6) at connector (7) on fuel pump (13).

e. Reconnect positive battery cable and then negative battery cable.

6-15 FUEL LINE MAINTENANCE 6-15.1

Inspection.

Perform a visual inspection for deterioration leakage, dry rot, etc.

6-15.2 Removal.

WARNING

Catch fuel in suitable container. Keep spilled fuel away from hot engine and all fires. Failure to comply with this warning can cause injury or death to personnel.

a. a. Loosen screw clamps (1, Figure 6-16), to remove butt splices (2) and fuel line (9) between primary fuel tank (3) and fuel pump (4). Drain any fuel from line into a suitable container.

b. b. Loosen screw clamps (5) to remove butt splices (6) and fuel line (7) between fuel pump (4) and heater (8).

6-15.3 Replacement.

a. Perform removal steps.

b. b. Perform installation steps substituting new fuel line for the defective one.

6-15.4 Installation.

NOTE

Ensure that fuel line is routed away from the engine and heater exhaust.

a. a. Install a new fuel line (9) and butt splices (2) between primary fuel tank (3) and fuel pump (4), securing with screw clamps (1).

b. b. Install a new fuel line (7) and butt splices (6) between fuel pump (4) and heater (8), securing with screw clamps (5).

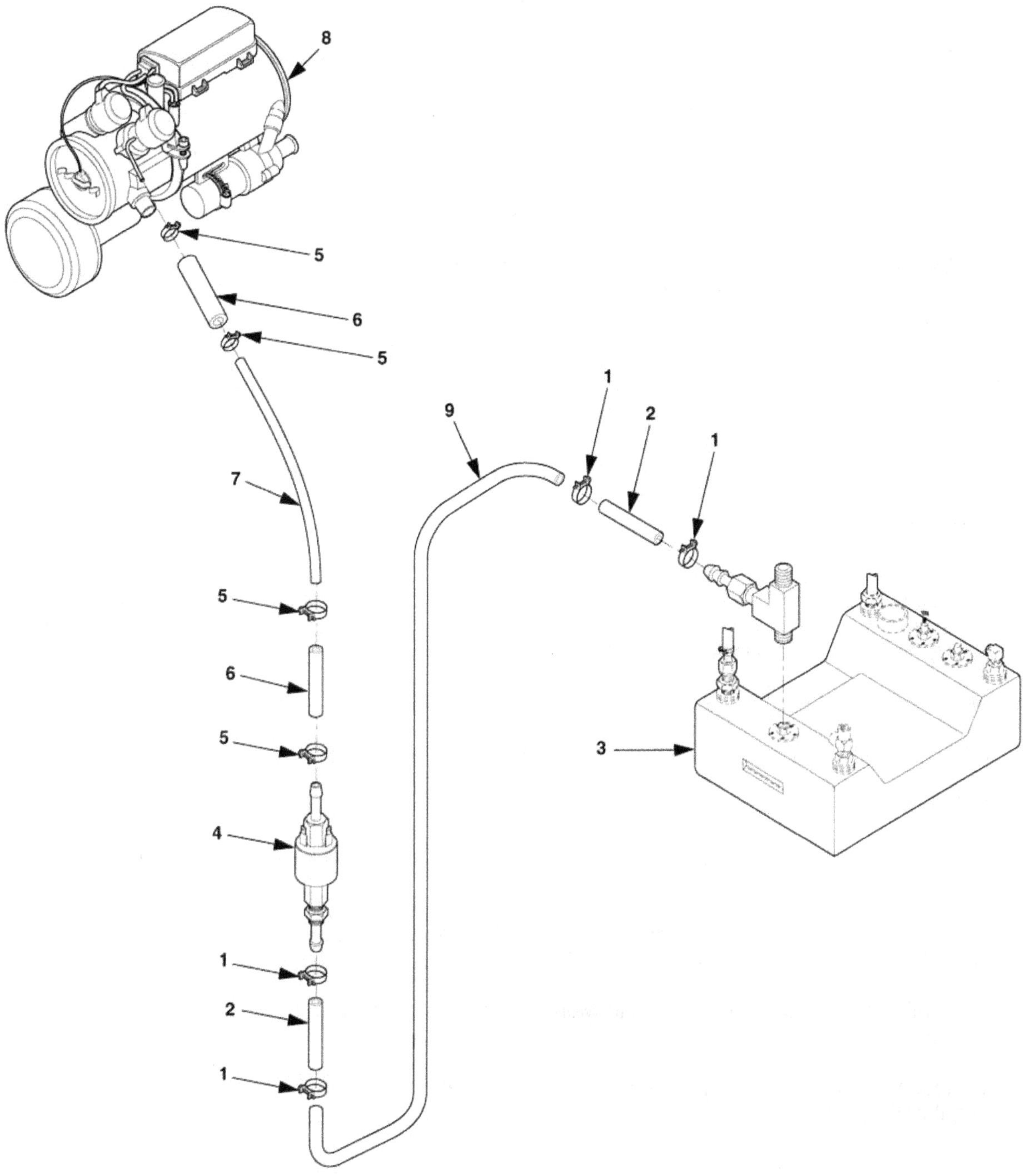

Figure 6-16. Fuel Line Maintenance.

6-16 HEATER ATTACHMENT MAINTENANCE

WARNING

Disconnect negative battery cable from right battery and the positive battery cable from the left battery before doing the following procedures. Reconnect cables in reverse order. Failure to comply with this warning can cause injury or death to personnel.

WARNING

Catch fuel in suitable container. Keep spilled fuel away from hot engine and all fires. Failure to comply with this warning can cause injury or death to personnel.

6-16.1 Removal.

a. a. Turn heater and generator off.

b. b. Open left and right access doors. Disconnect negative battery cable from the right battery and then disconnect positive battery cable from the left battery.

c. Support heater. Remove screws (1, 3, and 4, Figure 6-17), washers (2, 7, and 8), nuts (11), lock washers (5, 6, and 9) and heater from heater-mounting brace (10), then heater-mounting brace (10) from wall-mounting bracket (12), and wall-mounting bracket (12) and screws (13) from generator housing (14). Discard lock washers (5, 6, and 9).

6-16.2 Inspection.

Inspect all hardware (1-13) for damage and replace as required.

6-16.3 Installation.

a. a. Install wall-mounting bracket (12) with screws (4 and 13), new lock washers (6) and washers (8) to generator housing (14).

b. b. Install heater to heater-mounting brace (10) with screws (1), washers (2), new lock washers (9) and nuts (11).

c. c. Install heater-mounting brace (10) to wall-moun ting bracket (12) with screws (3), new lock washers (5), and washers (7).

d. d. Reconnect positive battery cable to the left battery and then reconnect negative battery cable to the right battery. Close access doors.

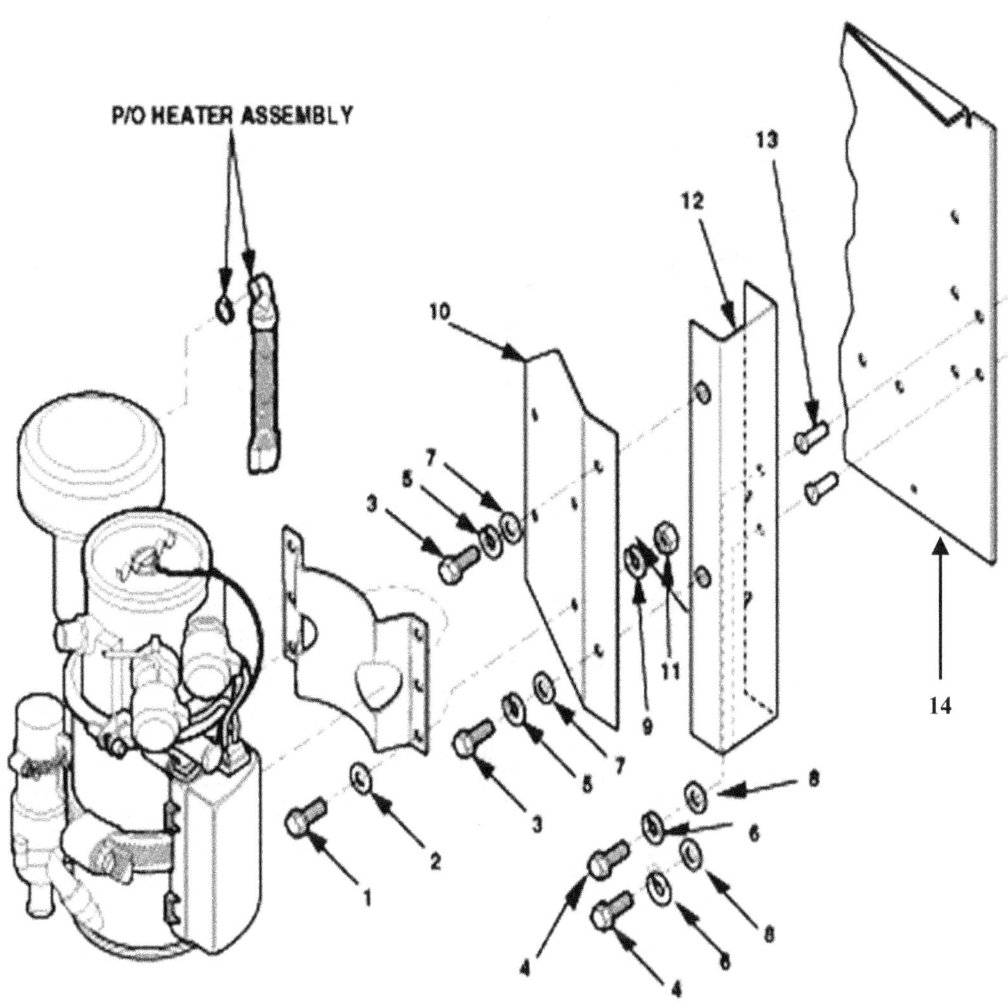

Figure 6-17. Heater Attachment Maintenance.

6-17 HEATER SWITCH/HEATER SWITCH LABEL/INDICATOR LIGHT MAINTENANCE 6-17.1

Installation – Heater Switch Plate.

a. Using the heater switch plate (4, Figure 6-18) as a guide, drill a 5/16" and a 1/2" hole in control panel (1).

b. Align and attach heater switch plate, using blind rivets.

6-17.2 Inspection – Heater Switch Plate.

Inspection is limited to visual inspection of components. Check for damaged and frayed wiring.

6-17.3 Test – Heater Switch Plate.

a. a. Disconnect lead from terminal of switch (2) and use multimeter to measure for continuity across switch terminals with switch in on position. With switch in off position, there should be an open circuit.

b. b. If readings are not as above, replace switch.

6-17.4 Removal – Heater Switch Plate.

WARNING

Disconnect negative battery cable from right battery and the positive battery cable from the left battery before doing the following procedures. Reconnect cables in reverse order. Failure to comply with this warning can cause injury or death to personnel.

a. a. Disconnect negative battery cable and then positive battery cable.

b. b. Open control box door and release control panel by turning two fasteners and carefully lowering control panel.

c. At generator control panel (1), tag and disconnect all electrical leads to switch (2).

d. d. Remove knurled nut (5) and mounting nut (6) from control panel (1).

6-17.5 Replacement – Heater Switch.

a. a. Perform steps a thru d of removal.

b. b. Perform steps a thru c of installation, substituting new heater switch for the defective one.

6.17.6 Installation – Heater Switch.

a. a. Install switch (2), mounting nut (6), knurled nut (5), and reconnect electrical leads to switch (2).

b. b. Raise and secure control panel (1) and close control box door.

c. Reconnect positive battery cable and then negative battery cable.

6-17.7 Inspection – Indicator Light.

Inspection is limited to visual inspection of components. Check for defects.

6-17.8 Installation – Indicator Light.

WARNING

Disconnect negative battery cable from right battery and the positive battery cable from the left battery before doing the following procedures. Reconnect cables in reverse order. Failure to comply with this warning can cause injury or death to personnel.

a. a. Disconnect negative battery cable and then positive battery cable.

b. b. Open control box door and release control panel by turning two fasteners and carefully lowering control panel.

c. At generator control panel (1), tag and disc onnect all electrical leads to switch (2).

d. d. Install indicator light (3) with push on nut (7).

6-17.9 Removal – Indicator Light. Remove push on nut (7)

from indicator light (3). **6-17.10 Replacement – Indicator**

Light.

a. a. Complete steps a through d of installation.

b. b. Complete step of removal.

c. c. Replace defective indicator light with new indicator light.

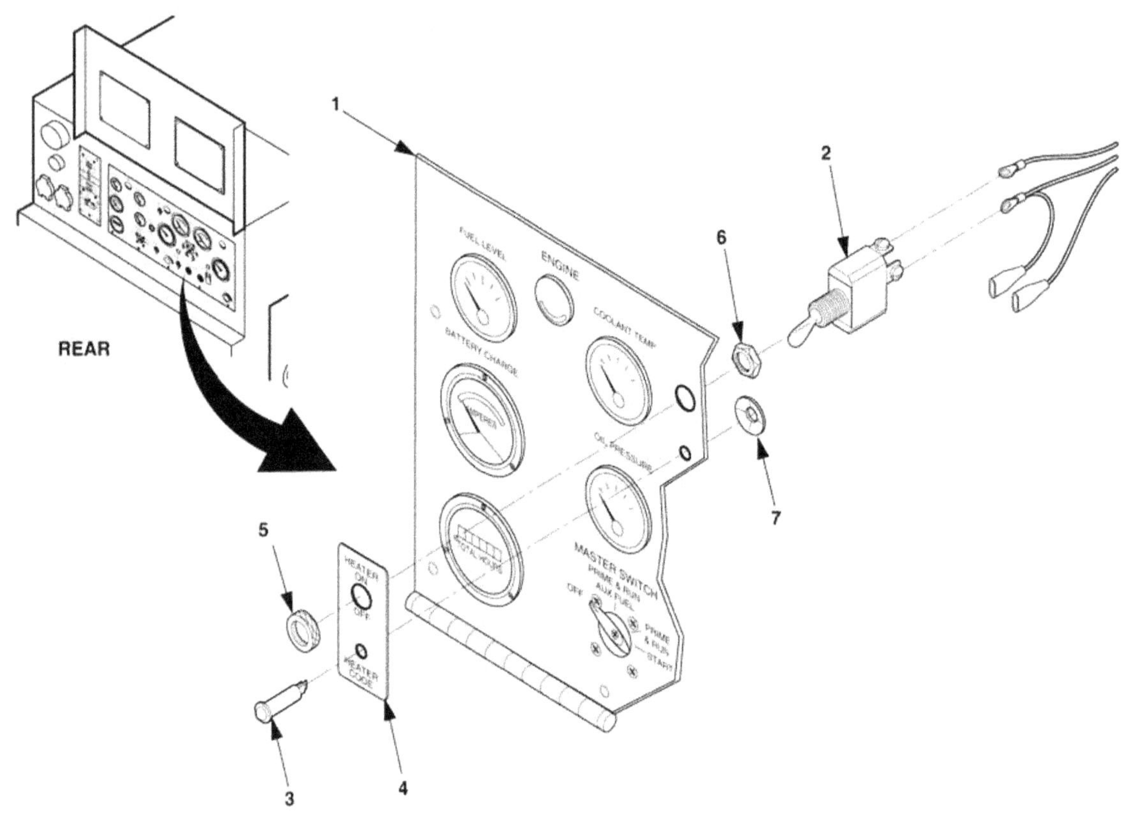

Figure 6-18. Heater Switch/Heater Switch Label//Indicator Light Maintenance

6-18 COOLANT HOSE MAINTENANCE 6-18.1 Inspection.

Inspection is limited to visual inspection of components. Check for leaks and cuts.

6-18.2 Removal.

WARNING

Cooling system operates at high temperatures and pressure. Contact with high pressure steam and/or liquids can result in burns and scalding. Shut down generator set, and allow system to cool before performing checks, services, and maintenance, or wear gloves and additional protective clothing and goggles as required. Failure to comply with this warning can cause injury or death to personnel.

WARNING

Always remove radiator cap slowly to permit any pressure to escape. Failure to comply with this warning can cause injury to personnel.

a. a. Drain coolant (see TM 750-254).

b. b. Loosen screw clamps (1, Figure 6-19) (MEP-804A/MEP-814A) to disconnect either or both coolant hoses, inlet (2) or outlet (3) from the heater. Drain any coolant from the hoses into a suitable container.

c. c. Loosen screw clamps (1, Figure 6-20) (MEP-804B/MEP-814B) to disconnect hoses (2 and 3) from the heater (4) and top and bottom hoses of the bypass valve (5) from the oil cooler (9) and from engine block (8).

6-18.3 Replacement.

a. Perform removal steps.

b. b. Perform installation steps substituting new coolant hose for defective one.

6-18.4 Installation.

a. a. Install new hose (2 and/or 3) securing with screw clamps (1).

b. b. Top off radiator coolant level to replace coolant lost in hose removal.

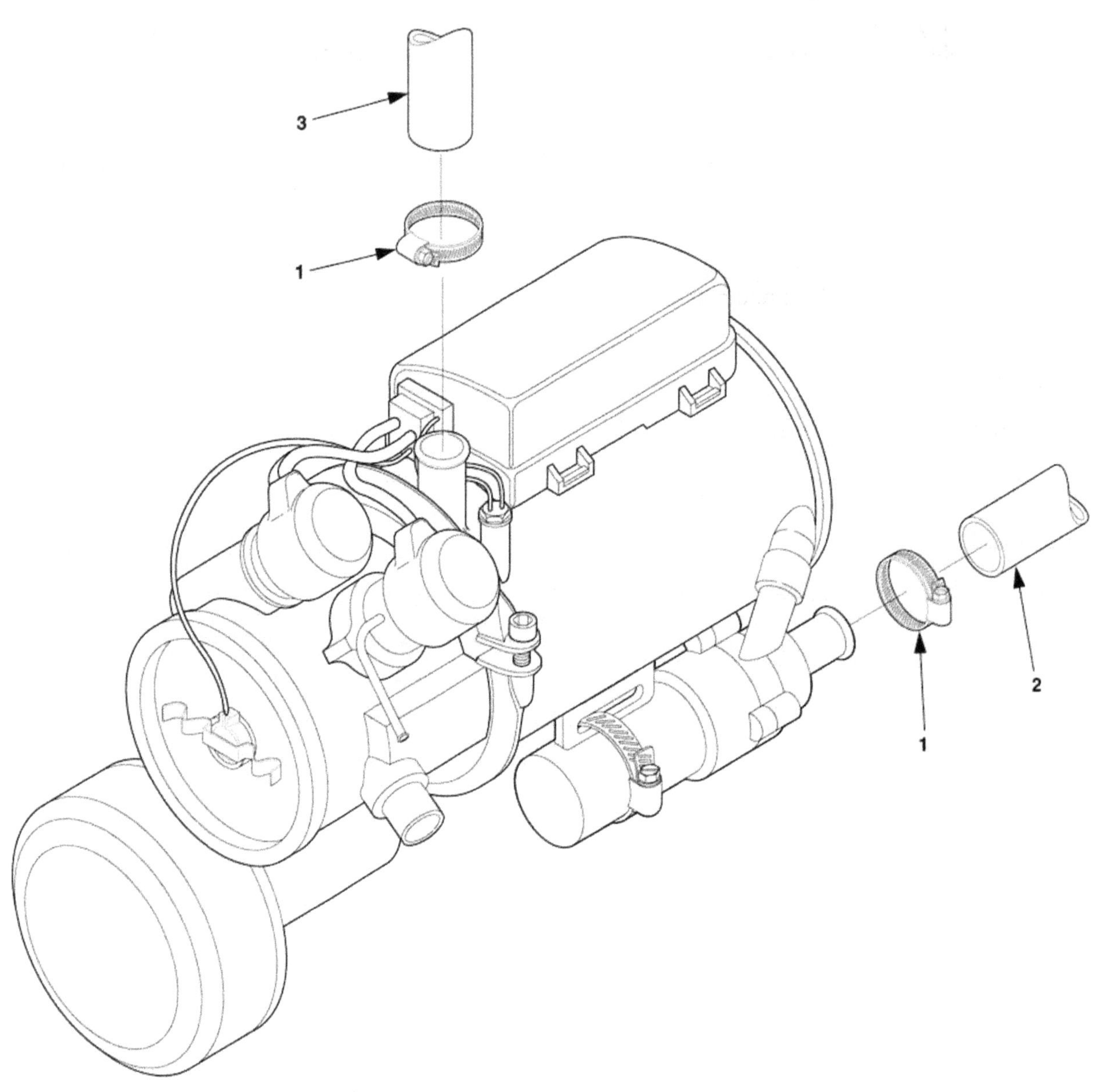

Figure 6-19. Coolant Hose Maintenance (MEP-804A/MEP-814A).

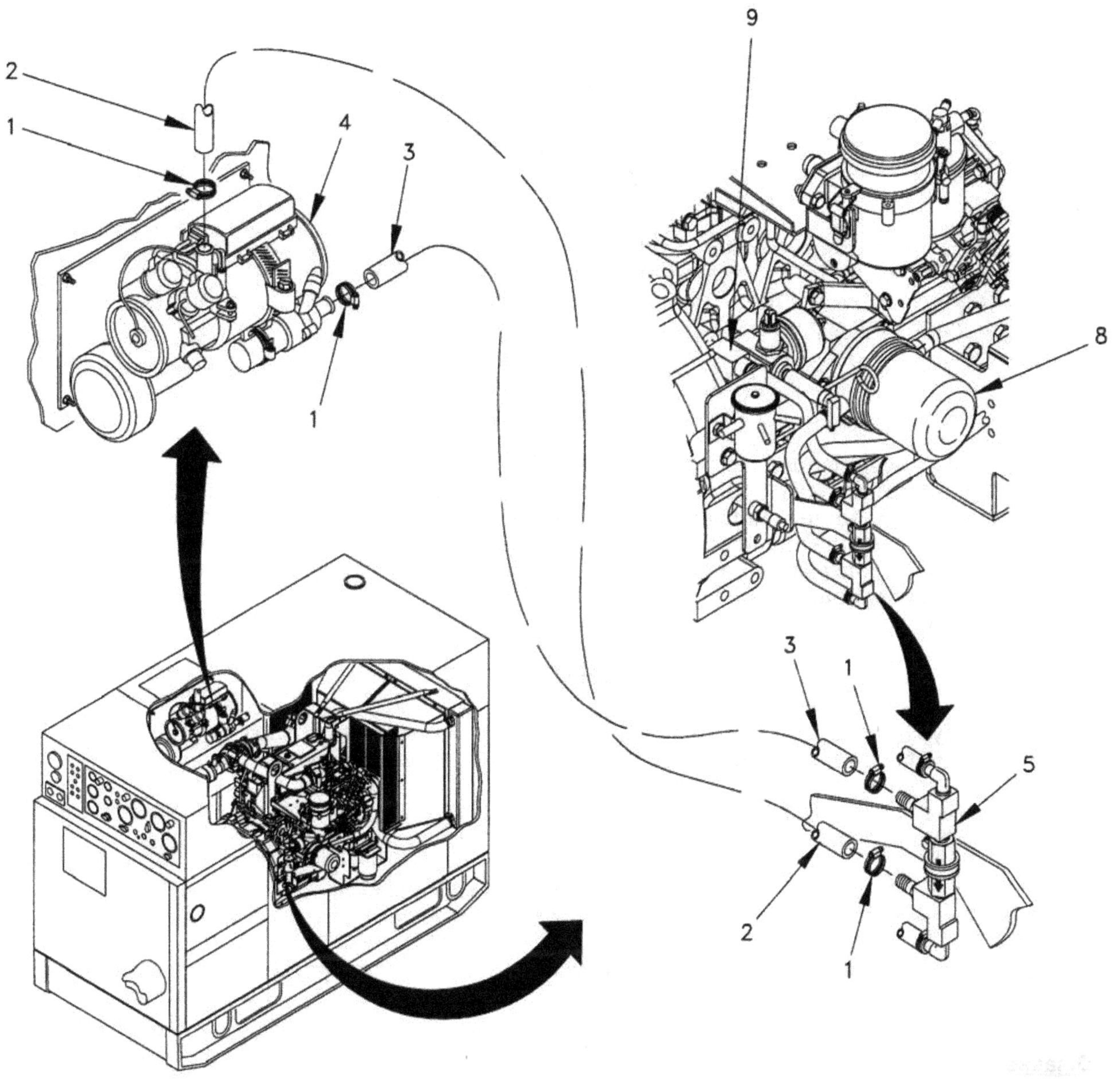

Figure 6-20. Coolant Hose Maintenance (MEP-804B/MEP-814B).

6-19 BYPASS VALVE MAINTENANCE 6-19.1

Test.

Look for cracks and\or class III leaks at the bypass valve and four coolant lines.

6-19.2 Removal.

WARNING

Disconnect negative battery cable from right battery and the positive battery cable from the left battery before doing the following procedures. Reconnect cables in reverse order. Failure to comply with this warning can cause injury or death to personnel.

WARNING

Cooling system operates at high temperatures and pressure. Contact with high pressure steam and/or liquids can result in burns and scalding. Shut down generator set, and allow system to cool before performing checks, services, and maintenance, or wear gloves and additional protective clothing and goggles as required. Failure to comply with this warning can cause injury or death to personnel.

WARNING

Always remove radiator cap slowly to permit any pressure to escape. Failure to comply with this warning can cause injury to personnel.

a. Open battery access door and disconnect cable labeled NEGATIVE; then, disconnect cable labeled POSITIVE from batteries.

b. b. b. Loosen hose clamps (1, Figure 6-21) and tag and disconnect four coolant hoses (2) from bypass valve assembly (8). Be sure to use a suitable container to catch coolant from hoses.

c. c. Remove screw (3), two lock washers (4), washer (5), nut (6), clamp (7), and bypass valve assembly

6-19.3 Disassembly.

a. a. Remove clamp (7) from bypass valve assembly (8).

b. b. Remove two elbow fittings (9), and two straight fittings (10) from two adapter fittings (11).

c. Remove two adapter fittings (11) from check valve (12). Observe direction of arrow on check valve for installation.

d. Inspect components of bypass valve assembly and replace damage components as necessary.

6-19.4 Assembly.

a. a. Install two elbow fittings (9) and two straight fittings (10) to adapter fittings (11).

c. With direction arrow on check valve (12) pointing down, install clamp (7) to bypass valve assembly (8).

6-19.5 Replacement.

a. a. Perform steps a thru c of removal.

b. b. Disassemble and inspect bypass valve assembly (8).

c. c. Replace defective parts as necessary and assemble.

d. Install bypass valve assembly.

6-19.6 Installation.

a. a. Install new bypass valve assembly (8) in clamp (7) with direction arrow on check valve pointing down and secure with screw (3), washer (5), two new lock washers (4), and nut (6).

b. b. Remove tags and attach four coolant hoses (2) to bypass valve assembly (8) and secure with four clamps (1).

c. c. Reconnect POSITIVE battery cable and then NEGATIVE battery cable.

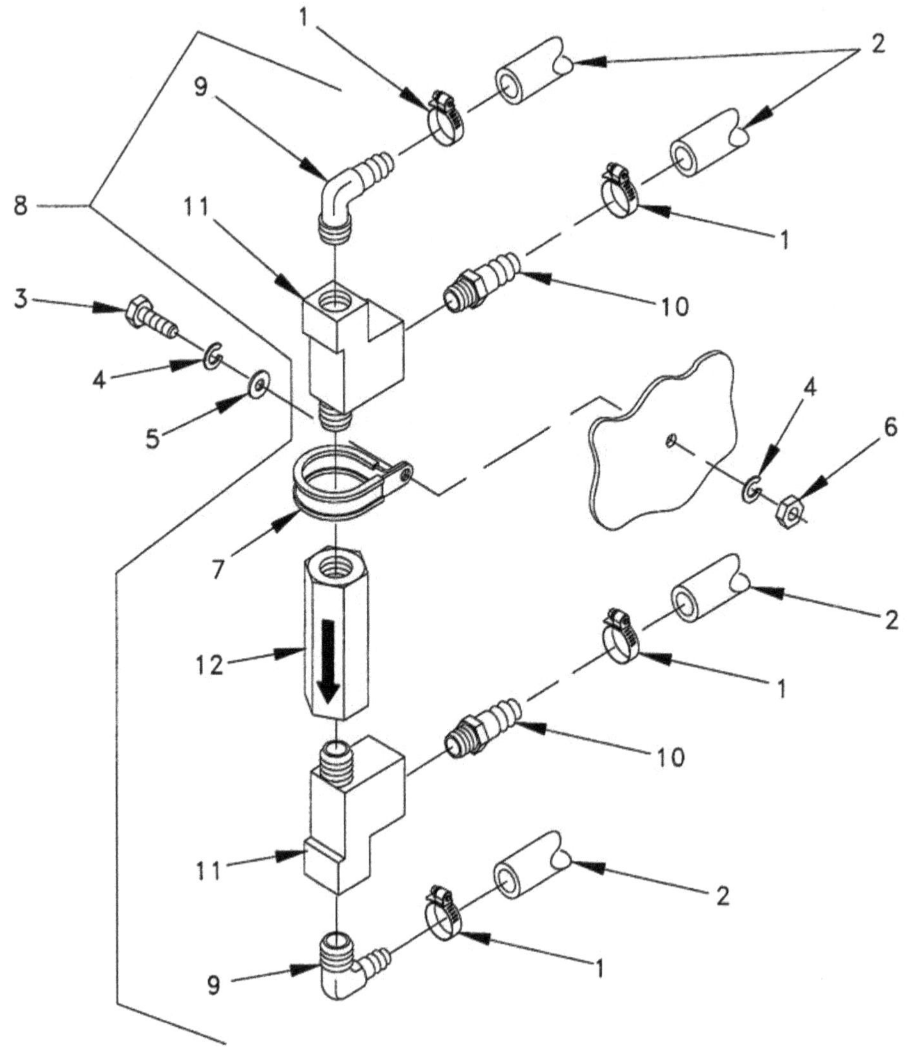

Figure 6-21. Bypass Valve Maintenance (MEP-804B/MEP-814B).

6-20 AIR INLET AND EXHAUST HOSE MAINTENANCE 6-20.1

Inspection.

Inspection is limited to visual inspection of components. Check for cuts, air leaks, and crushed hoses.

6-20.2 Removal.

WARNING

Muffler and flex hoses can get very hot. Allow them to cool before touching them. Failure to comply with this warning can cause severe burns and injury to personnel.

a. a. Loosen screw clamp (7, Figure 6-22) to remove air inlet hose (1) from heater (6).

b. b. Loosen screw clamp (4) to remove air exhaust hose (3) and elbow (5) from heater (6).

c. c. Remove bushing (8) as required.

d. d. Remove RTV (high heat sealing compound), air exhaust hose (3) and clamp (2) from standoff (9) by removing nut (10), lock washer (11) and flat washer (12). Discard lock washer (11).

6-20.3 Replacement.

a. Perform removal steps.

b. b. Perform installation steps substituting new air inlet and exhaust hose for the defective one.

6-20.4 Installation.

a. a. Install bushing (8), if removed.

b. b. Attach elbow (5), using RTV (high heat sealing compound), to heater exhaust port. Attach air exhaust hose (3) to elbow (5) with screw clamp (4).

c. c. Install exhaust hose (3) through clamp (2) and secure. Guide exhaust hose (3) through hole in bulkhead and secure with RTV (high heat sealing compound).

d. d. Attach clamp (2) to standoff (9) with nut (10), new lock washer (11) and flat washer (12).

e. e. Attach air inlet hose (1) to heater (6) with screw clamp (7).

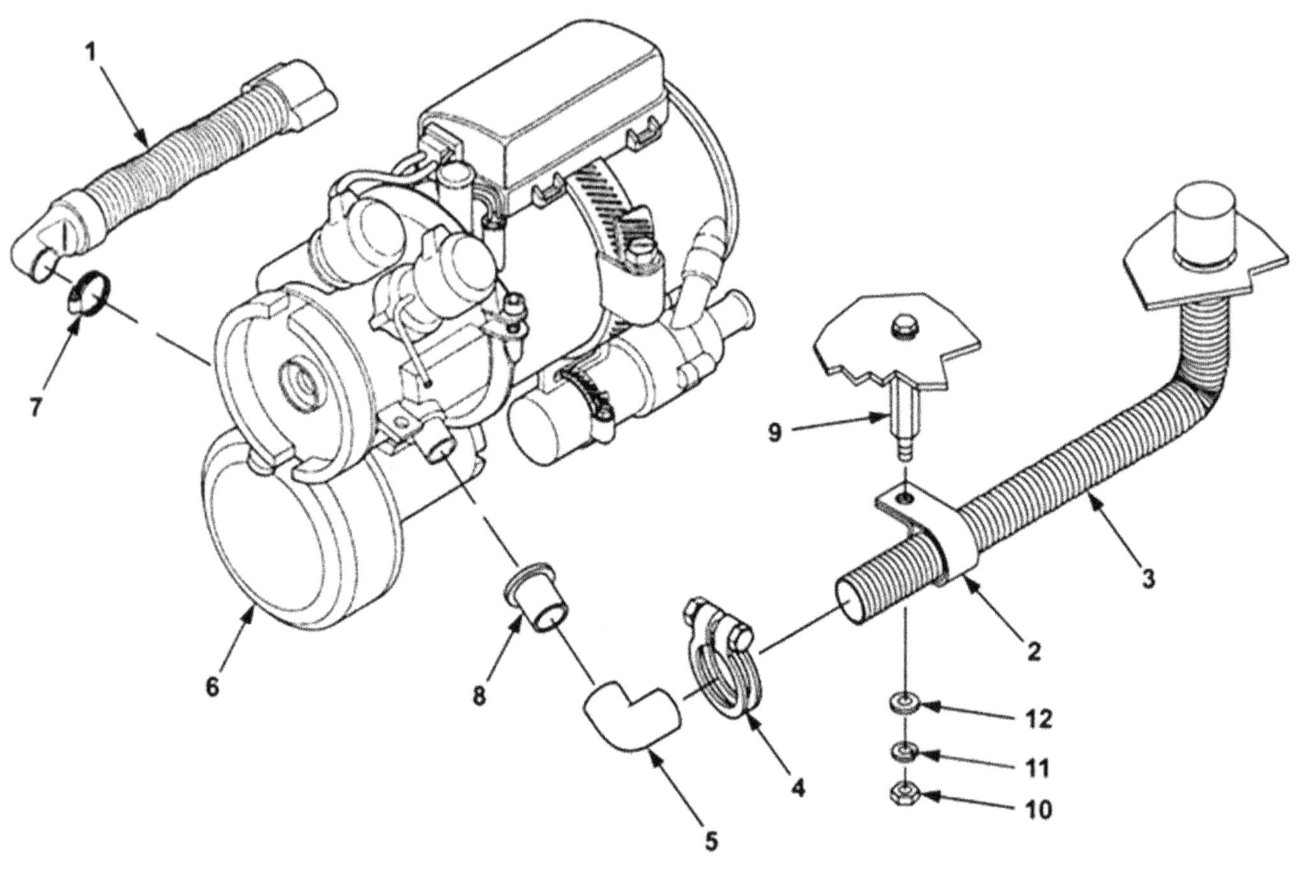

Figure 6-22. Air Inlet and Exhaust Hose Maintenance.

6-21 WIRING HARNESS MAINTENANCE

WARNING

High voltage is produced when this generator set is in operation. SHUT DOWN generator set and make sure it is free of any power source before attempting any repair or maintenance on the set, or when connecting or disconnecting load cables. Failure to comply with this warning can cause injury or death to personnel.

6-21.1 Inspection.

Inspection is limited to visual inspection of components. Check for damaged and frayed wiring.

6-21.2 Test.

Test wiring for continuity. Replace any open or broken wires or connectors.

6-21.3 Removal.

a. a. Prior to removal, refer to Figure 6-23 and tag all leads on wiring harness.

b. b. Disconnect all leads and plugs. Remove wiring harness.

6-21.4 Repair.

Repair consists of replacing wires, connectors, and terminals. Follow standard shop practice when performing all repairs.

6-21.5 Replacement.

a. Perform removal steps.

b. b. Perform installation steps substituting new wiring harness for the defective one.

6-21.6 Installation.

a. a. Position wiring harness in place.

b. b. Refer to Figure 6-23 and connect leads and plugs.

NATO SLAVE
RECEPTICLE

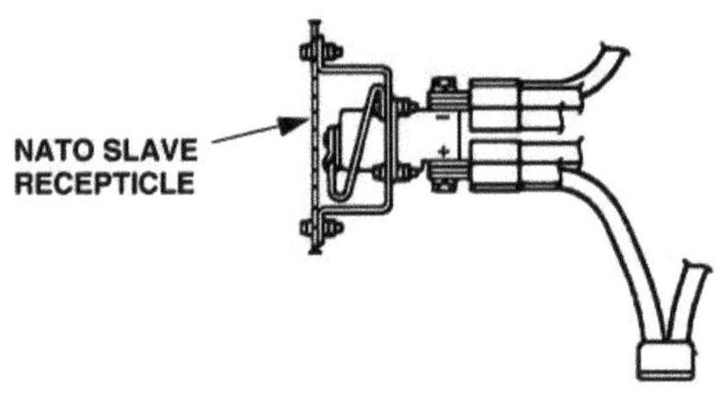

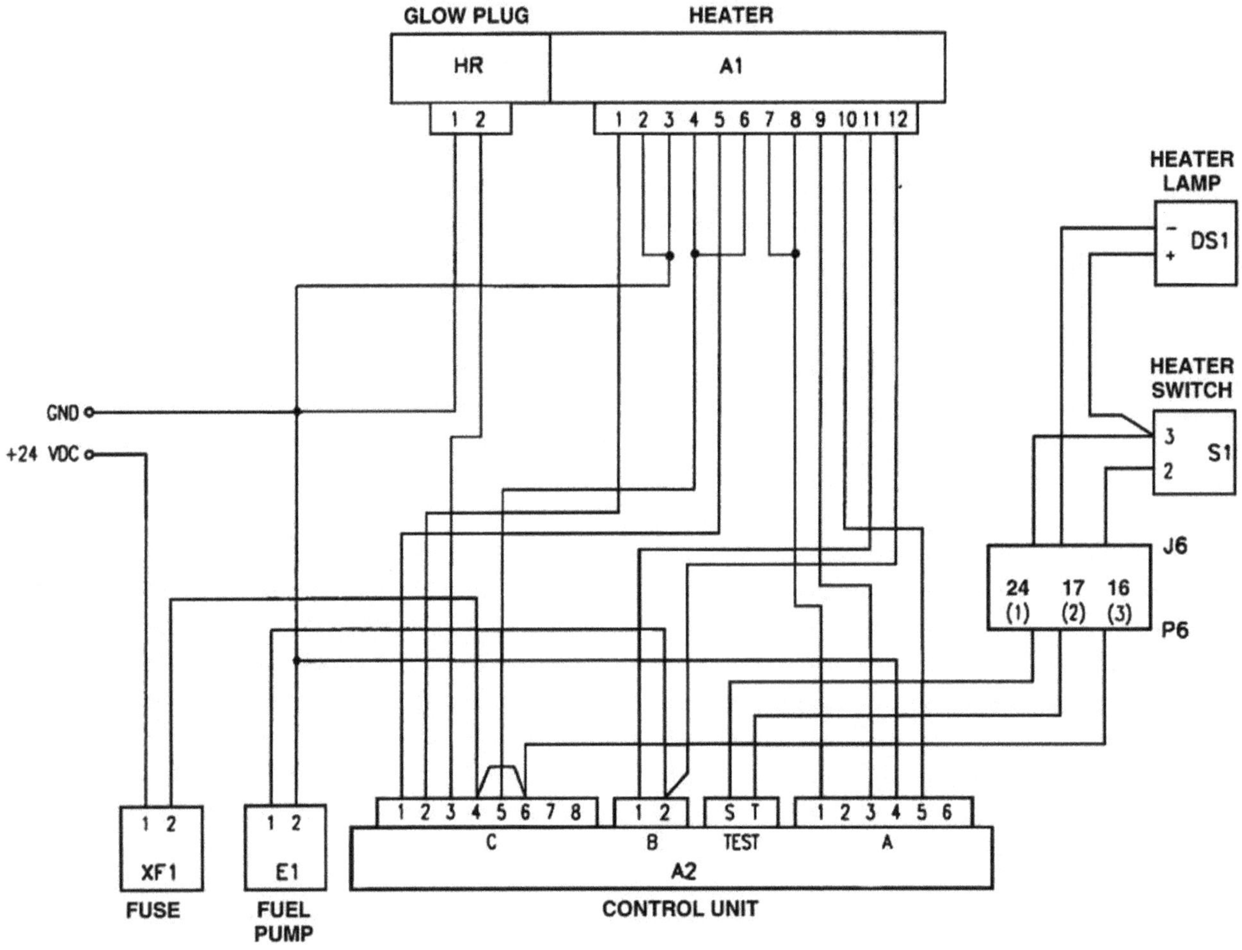

GLOW PLUG
HEATER
HR
A1
1 2
1 2 3 4 5 6 7 8 9 10 11 12
HEATER
LAMP
DS1
HEATER
SWITCH
3
S1
2
GND
+24 VDC
J6
24 17 16
(1) (2) (3)
P6
1 2
XF1
FUSE
1 2
E1
FUEL
PUMP
1 2 3 4 5 6 7 8
C
1 2
B
S T
TEST
1 2 3 4 5 6
A
A2
CONTROL UNIT

Figure 6-23. Wiring Harness Maintenance.

6-22 IGNITER/GLOW PLUG AND RESISTOR MAINTENANCE

WARNING

Disconnect negative battery cable from right battery and the positive battery cable from the left battery before doing the following procedures. Reconnect cables in reverse order. Failure to comply with this warning can cause injury or death to personnel.

6-22.1 Test.

a. a. Disconnect negative battery cable, and then disconnect positive battery cable.

b. b. Remove caps (5, Figure 6-24) from igniter/glow plug (4) and resistor (1).

c. c. Loosen hex nuts (2) and disconnect cable (3) from igniter/glow plug (4) and resistor (1).

d. d. Check continuity between igniter/glow plug (4) and ground and the resistor (1) and ground.

e. e. Reconnect cable (3) and tighten hex nuts (2) on igniter/glow plug (4) and resistor (1).

f. f. Replace caps (5) on igniter/ glow plug (4) and resistor (1).

g. g. Reconnect positive battery cable, and then reconnect negative battery cable.

6-22.2 Replacement.

a. a. Disconnect negative battery cable, and then disconnect positive battery cable.

b. b. Remove caps (5) from igniter/glow plug (4) and/or resistor (1). Loosen hex nuts (2) on igniter/glow plug (4) and/or resistor (1). Remove cable (3).

c. c. Unscrew igniter/glow plug (4) and/or resistor (1), as required, and remove it. Use an angled hook to clean the igniter/glow plug hole.

d. d. Install igniter/glow plug (4) and/or resistor (1).

e. e. Install cable (3), hex nuts (2), and caps (5).

f. f. Reconnect positive battery cable, and then reconnect negative battery cable.

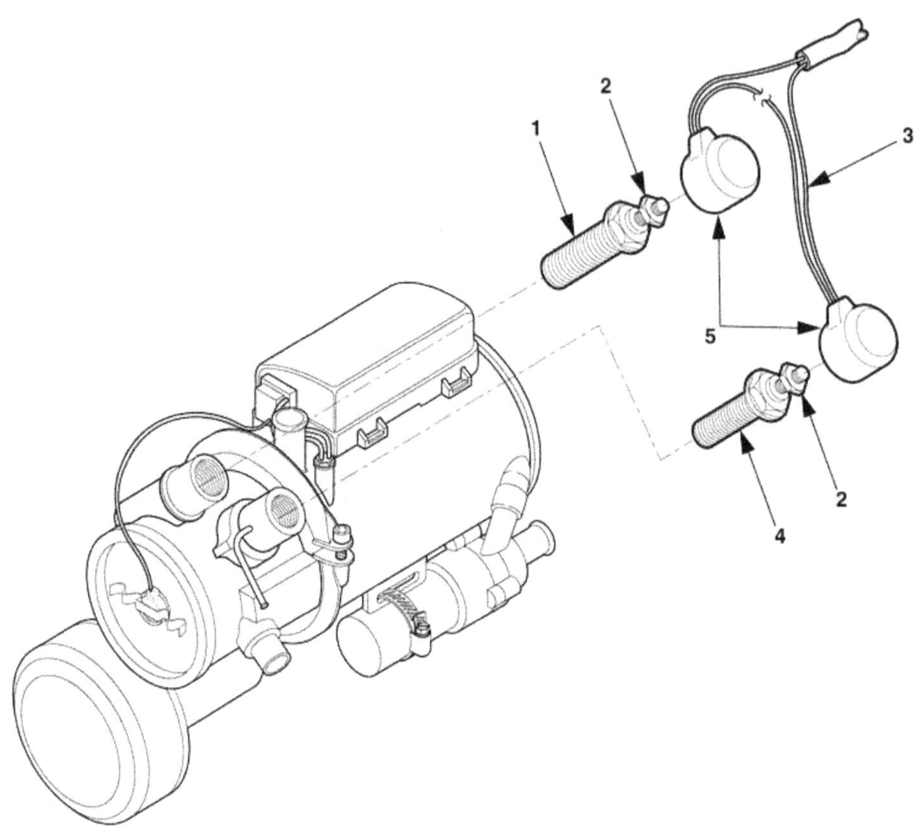

Figure 6-24. Igniter/Glow Plug And Resistor Maintenance.

6-23 FUNCTION CODES/OPERATING ID PLATE MAINTENANCE

WARNING

Disconnect negative battery cable from right battery and the positive battery cable from the left battery before doing the following procedures. Reconnect cables in reverse order. Failure to comply with this warning can cause injury or death to personnel.

NOTE

Placement of function code plate: Function code plate edge should be placed approximately .50" from top of control panel door when opened and approximately 11" from right edge of control panel door. Plate shall be readable when door is opened.

6-23.1 Inspection.

Inspect plate for illegible instructions, dents, cracks, etc.

6-23.2 Installation – Function Codes Plate.

a. a. Lift generator control box door (1, Figure 6-25).

b. b. Using function code plate (4) as a template, drill four 1/8" inch holes.

c. c. Install function codes plate (4) using four blind rivets (2). Rivet head shall be on the outside of unit.

6-23.3 Removal – Function Codes Plate.

a. a. Lift generator control box door (1).

b. b. Using electric drill with 1/8" drill bit, drill each of the rivets and punch out.

c. c. Remove four rivets (2) and function codes plate (4).

d. d. Close control box door (1).

6-23.4 Replacement – Function Codes Plate.

a. a. Lift generator control panel cover (1).

b. b. Perform steps b and c of removal.

c. c. Install new function codes plate (4) using four blind rivets (2). Rivet head shall be on outside of unit.

d. d. Close control box door (1).

6-23.5 Inspect – Operating Instruction Plate.

Inspect plate for illegible instructions, dents, cracks, etc.

WARNING

Disconnect negative battery cable from right battery and the positive battery cable from the left battery before doing the following procedures. Reconnect cables in reverse order. Failure to comply with this warning can cause injury or death to personnel.

NOTE

Placement of operation instruction code plate: Operating instruction plate edge should be placed approximately 1" from top of control panel door when opened and approximately 5" from left edge of control panel door. Plate shall be read-able when door is opened.

6-23.6 Install – Operating Instruction Plate.

a. a. Lift generator control box door (1).

b. b. Using operation instruction plate (3) as a template, drill four 1/8" inch holes.

c. c. Install operating instruction plate (3) using four blind rivets (2). Rivet head shall be on the outside of

6-23.7 Removal – Operating Instruction Plate.

a. a. Lift generator control box door (1).

b. b. Using electric drill with 1/8" drill bit, drill each of the rivets and punch out.

c. c. Remove four rivets (2) and operating instruction plate (3).

d. d. Close control box door (1).

6-23.8 Replacement – Operating Instruction Plate.

a. a. Perform steps a through c of removal.

b. b. Install new operating instruction plate (3) and using four blind rivets (2). Rivet head shall be on outside of unit.

c. Close control box door (1).

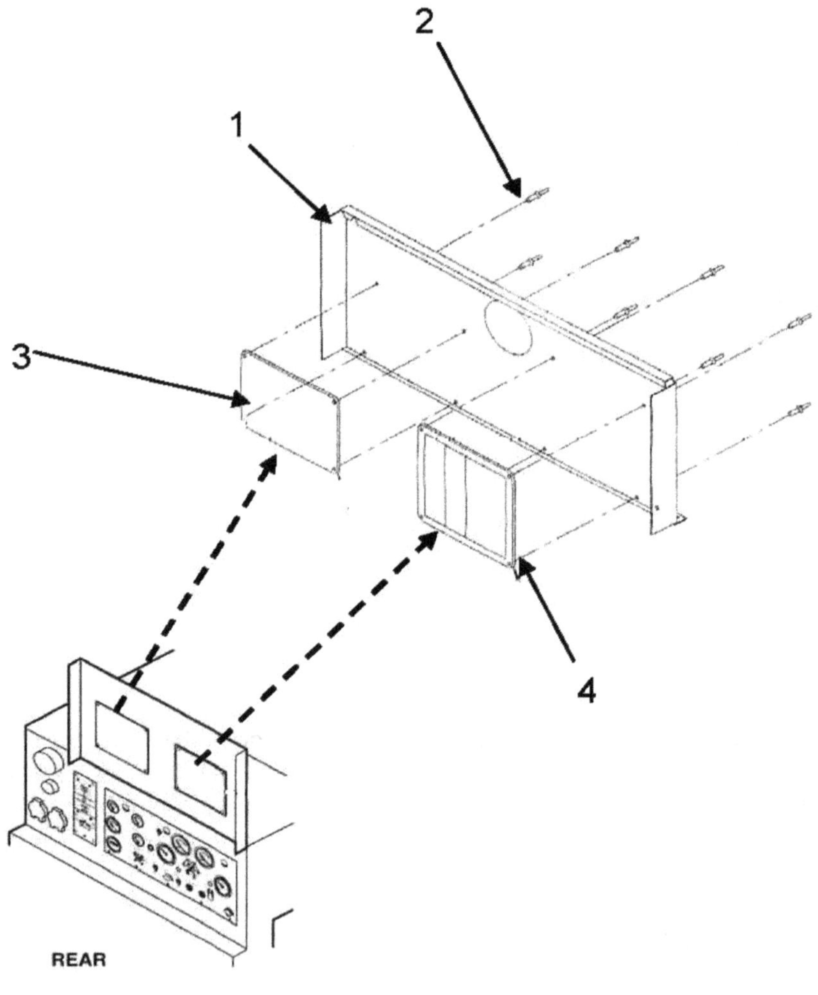

Figure 6-25. Function Codes Plate/Operating Instruction Plate Maintenance.

Section XIII. REMOVAL INSTRUCTIONS

The following instructions have been provided so you can remove Heater Kit, 15kW Generator Set NSN: 6115-01-477-0566 from your Generator Set.

WARNING

Disconnect negative battery cable from right battery and the positive battery cable from the left battery before doing the following procedures. Reconnect cables in reverse order. Failure to comply with this warning can cause injury or death to personnel.

WARNING

Cooling system operates at high temperatures and pressure. Contact with high pressure steam and/or liquids can result in burns and scalding. Shut down generator set, and allow system to cool before performing checks, services, and maintenance, or wear gloves and additional protective clothing and goggles as required. Failure to comply with this warning can cause injury or death to personnel.

WARNING

Always remove radiator cap slowly to permit any pressure to escape. Failure to comply with this warning can cause injury to personnel.

WARNING

Catch fuel in suitable container. Keep spilled fuel away from hot engine and all fires. Failure to comply with this warning can cause injury or death to personnel.

6-24 REMOVAL PROCEDURES

a. a. Turn heater and generator set off.

b. b. Open left and right access doors and disconnect negative battery cable from right battery and the positive battery cable from the left battery per paragraph 2-12.

c. c. Drain generator coolant system per paragraph 2-75.

d. d. Open control panel access door (1, Figure 6-10).

e. e. Remove electrical leads from J6-16, 17, and 24 (see Figure 6-30).

f. Remove toggle switch (3, Figure 6-10), indicato r light (7), and heater switch label (6) from instrument panel (9). Remove rivets (2) from function code plate (10) and operating plate (11). Remove function code and operating plates (10 and 11) from control panel door (1).

g. Remove two nuts (5, Figure 6-26), bolts (2), flat washers (3), lock washers (4) and control unit (1) from engine access side of output box. Replace existing hardware.

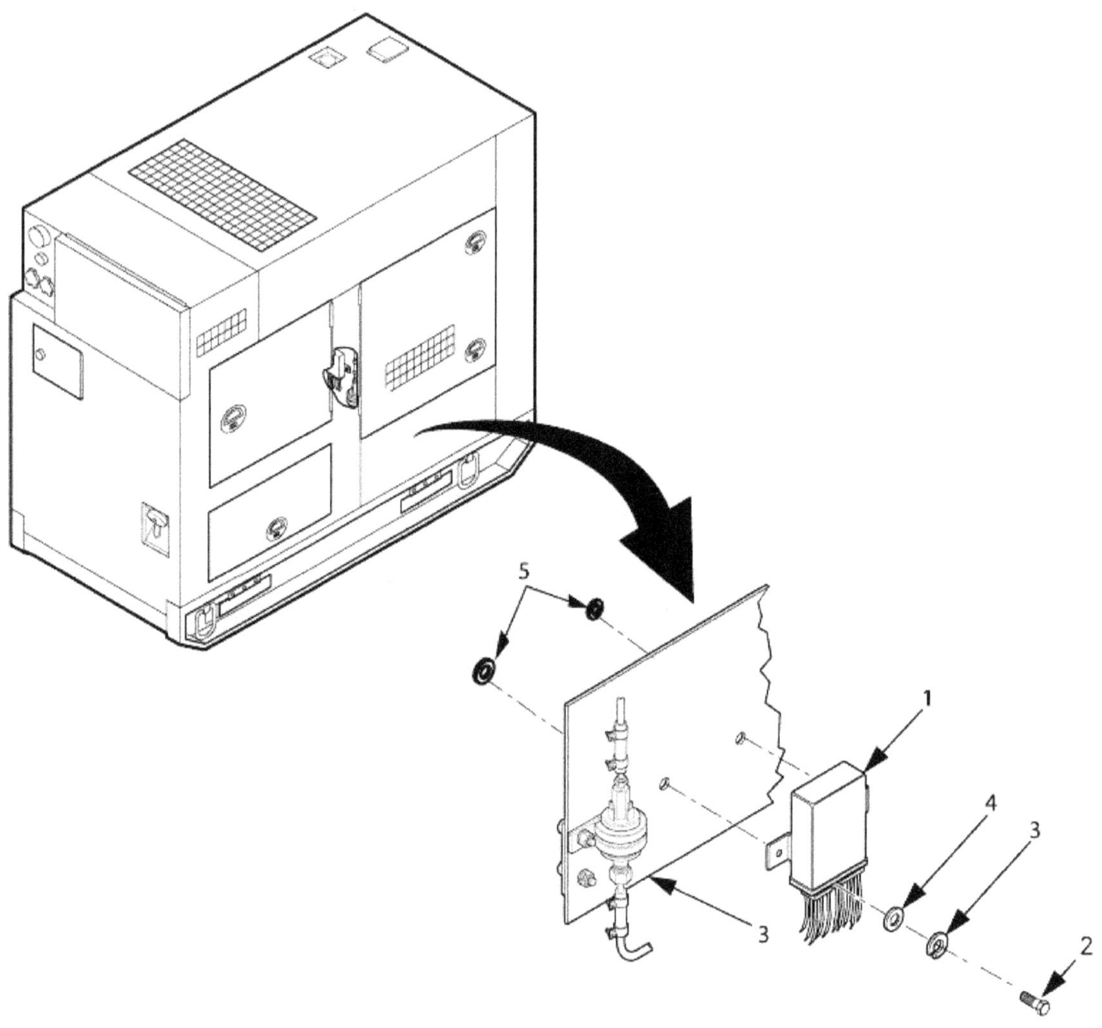

Figure 6-26. Control Unit Removal.

h. Remove Bolt (14, Figure 6-27), nut (15), clamp (5), fuel pump (4), butt splices (7), clamps (6 and 1), fuel lines (2, 3 and 8), barbed adapter (9) and fuel line tee (10) from fuel pickup tube assembly (13). Be sure to use a suitable container to catch any fuel from lines. Install fuel line adapter (11) back into fuel pickup tube assembly (13).

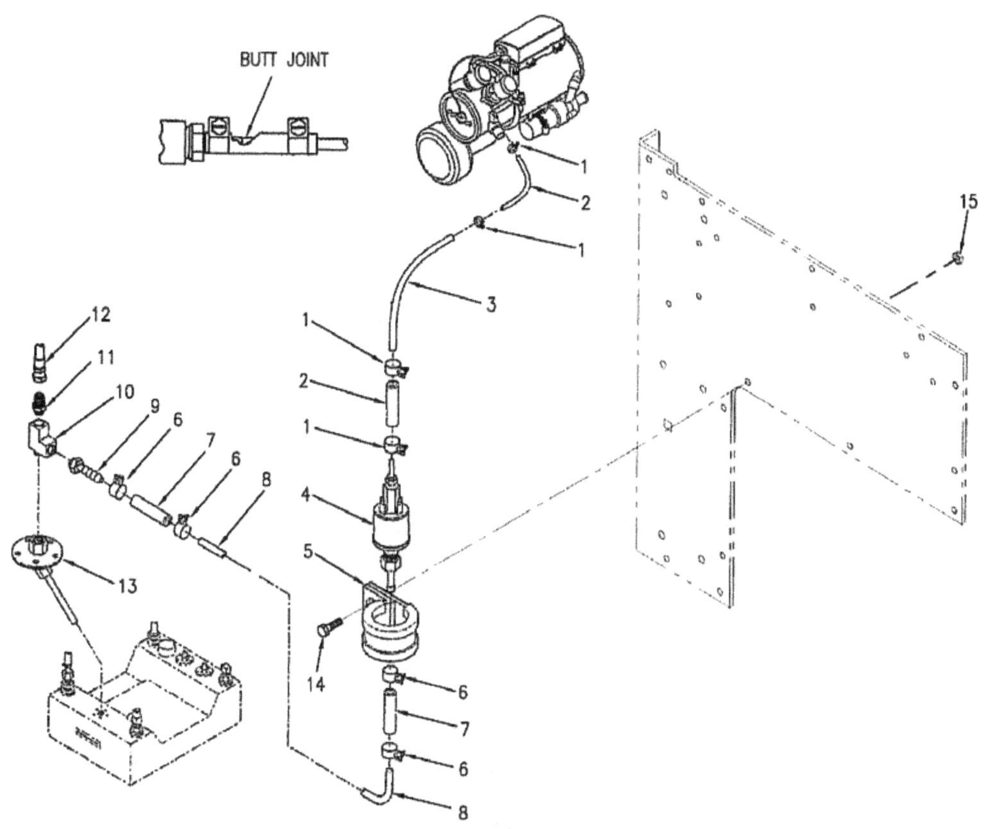

Figure 6-27. Fuel Pump and Lines Removal.

i. Loosen clamps (6, Figure 6-28) and remove coolant hose (7) and expansion plug (5). Replace the expansion plug with the correct type according to the engine manual. Be sure to use a suitable container to catch coolant from hoses. Refer to engine TM 9-2815-254-24 (Isuzu) or TM 9-2815-538-24&P (Yanmar). Remove clamps (1 and 3), coolant hose (4) and coolant hose tee fitting (2) from lower radiator hose (8) and replace with new radiator hose.

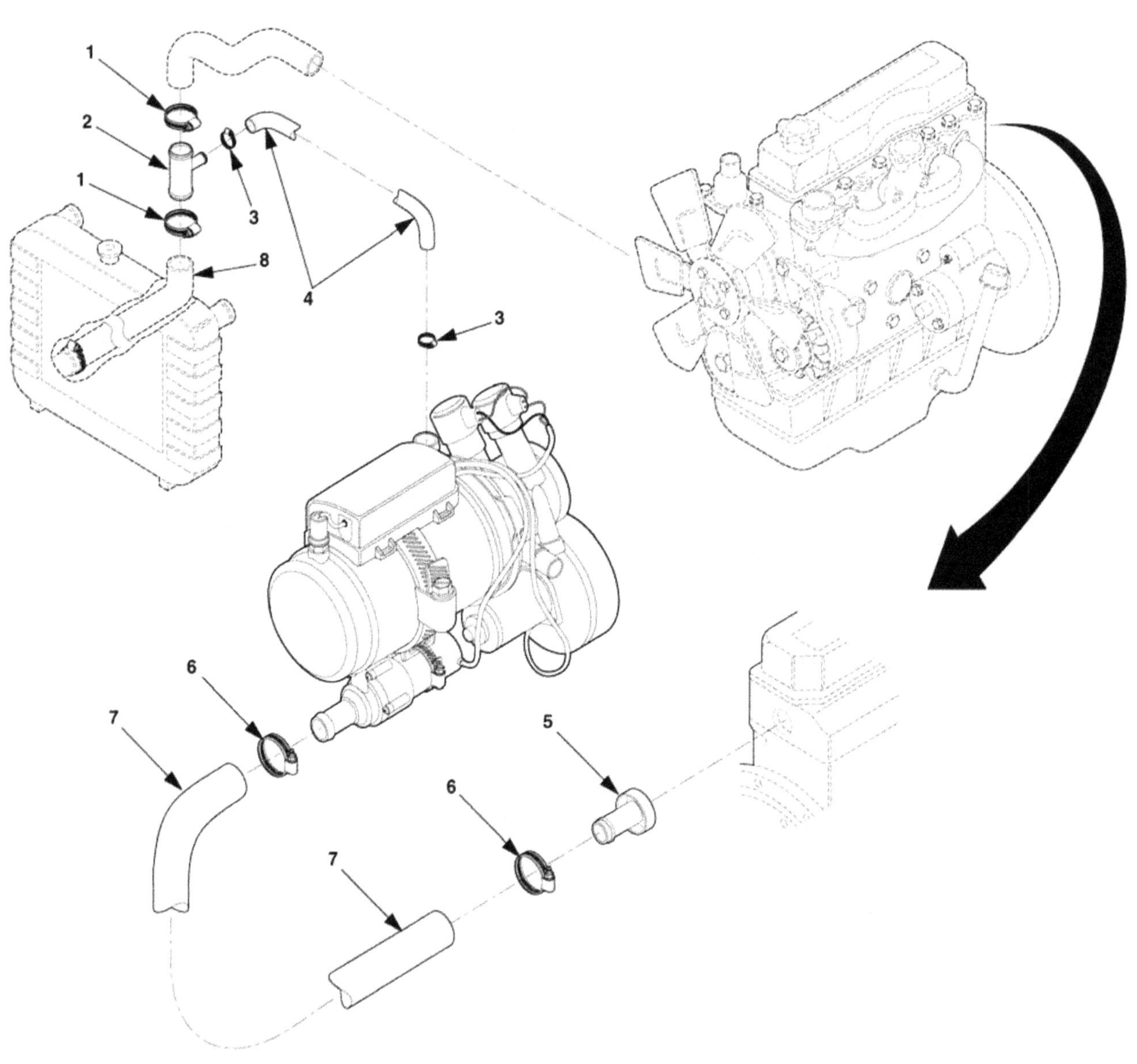

Figure 6-28. Coolant Hoses and Expansion Plug.

j. For the B model, remove clamps (17, Figure 6-29) and hoses (15 and 16) from heater (18) and from the bypass valve (ref 5). Replace hose (12) between engine port (13) and oil cooler (14).

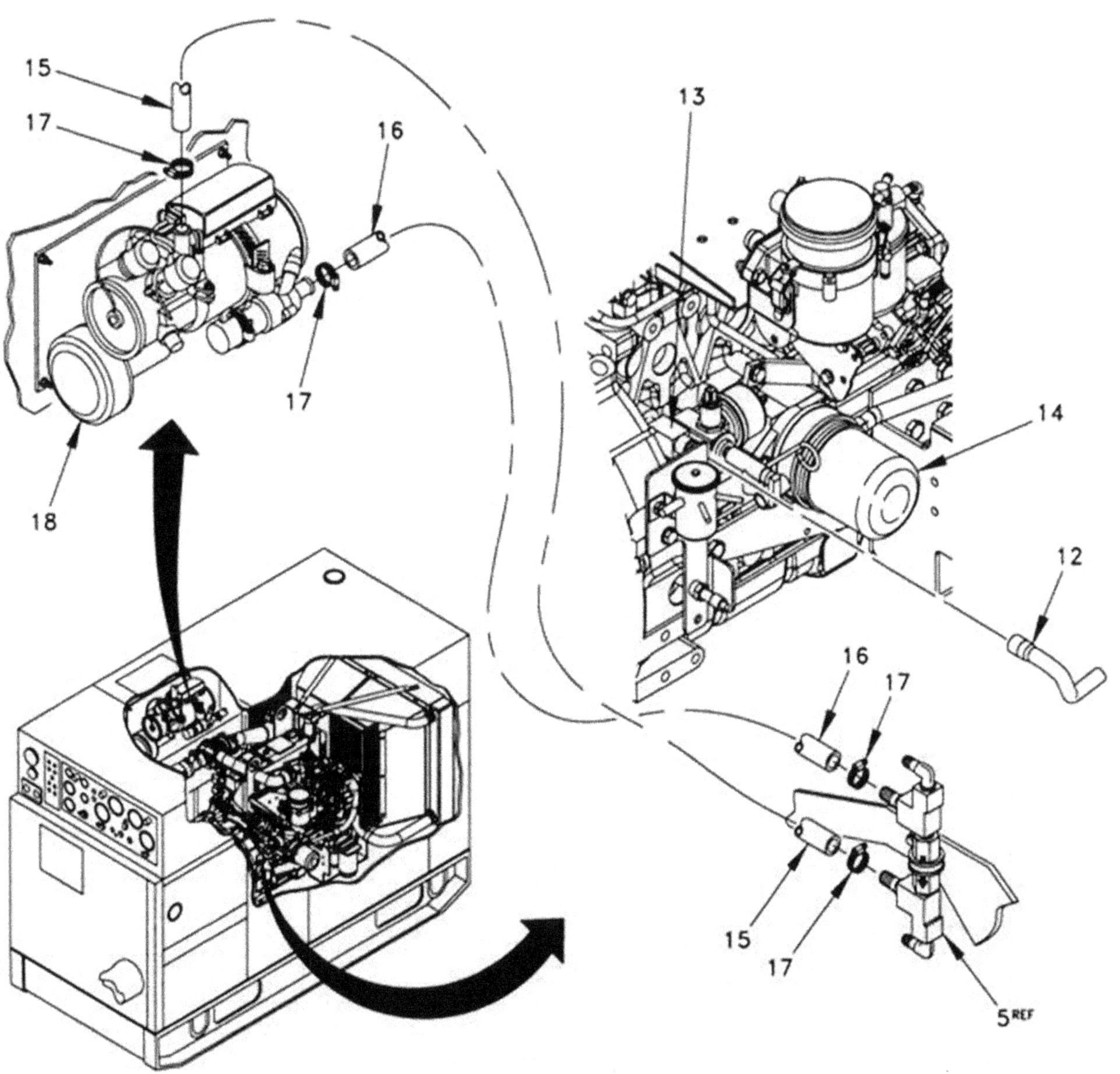

Figure 6-29. Coolant Hoses and Bypass Valve (MEP-804B/MEP-814B).

k. Remove wiring harness from generator set per Figure 6-30. Reinsert pins 16, 17, and 24 into J6 .

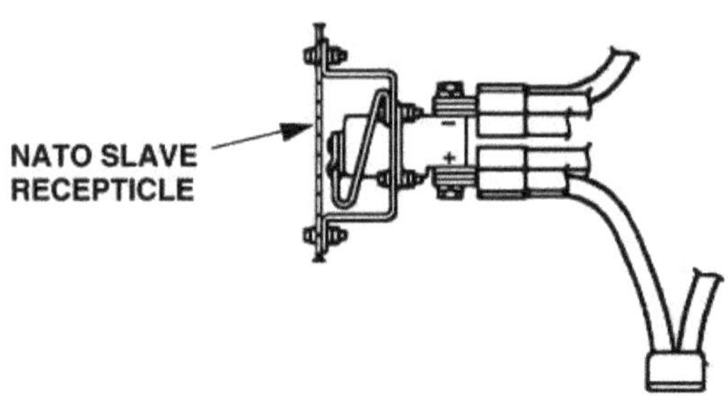

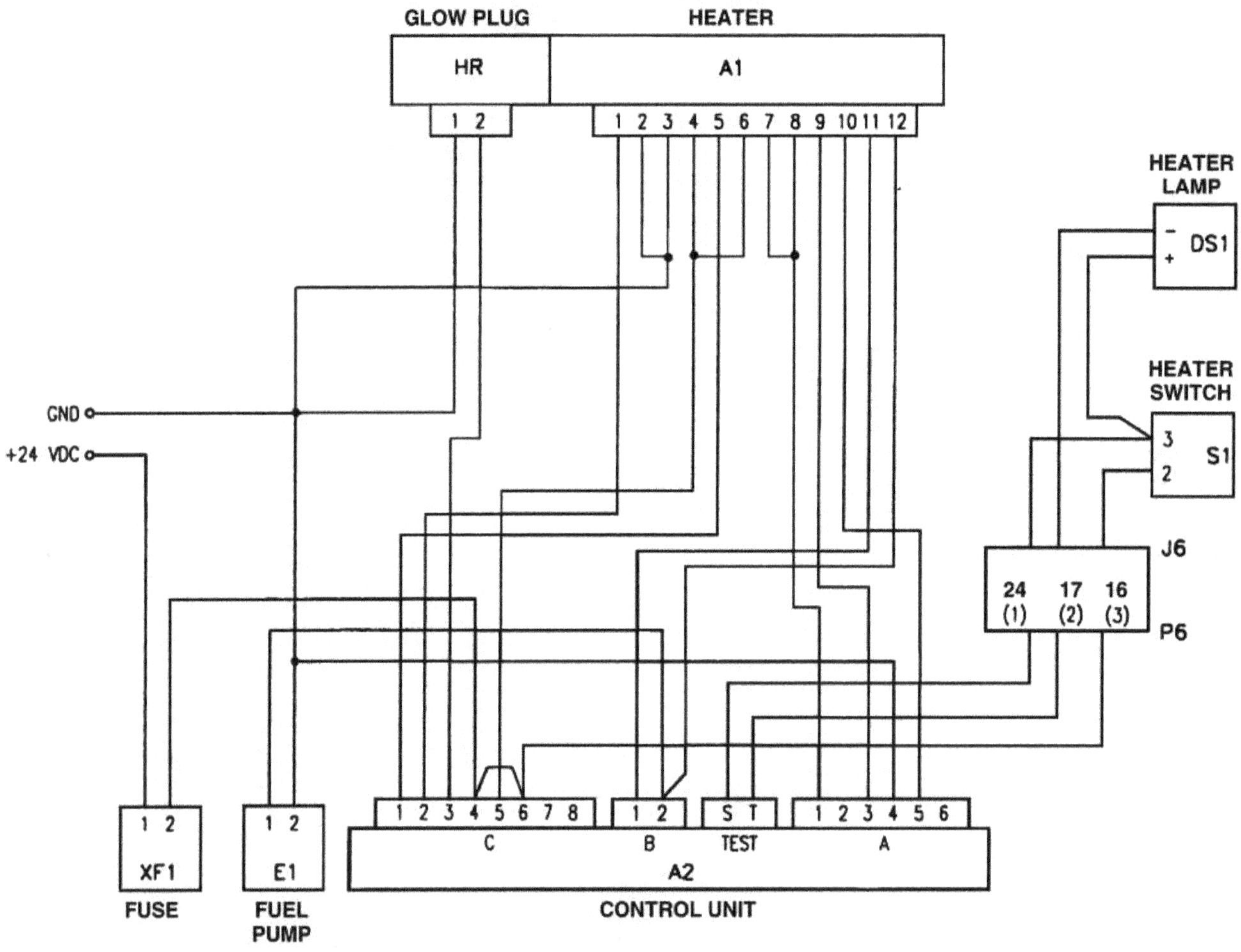

6-30. Wiring Harness Removal.

l. Remove heater assembly by removing bolts (1 , 3, 4 and 13, Figure 6-31), lock washers (5, 6 and 9), flat washers (8, 2, 7 and 11) and wall mounting bracket (10) from wall mounting brace (12) and from bulkhead (14). Discard lock washers (5, 6 and 9).

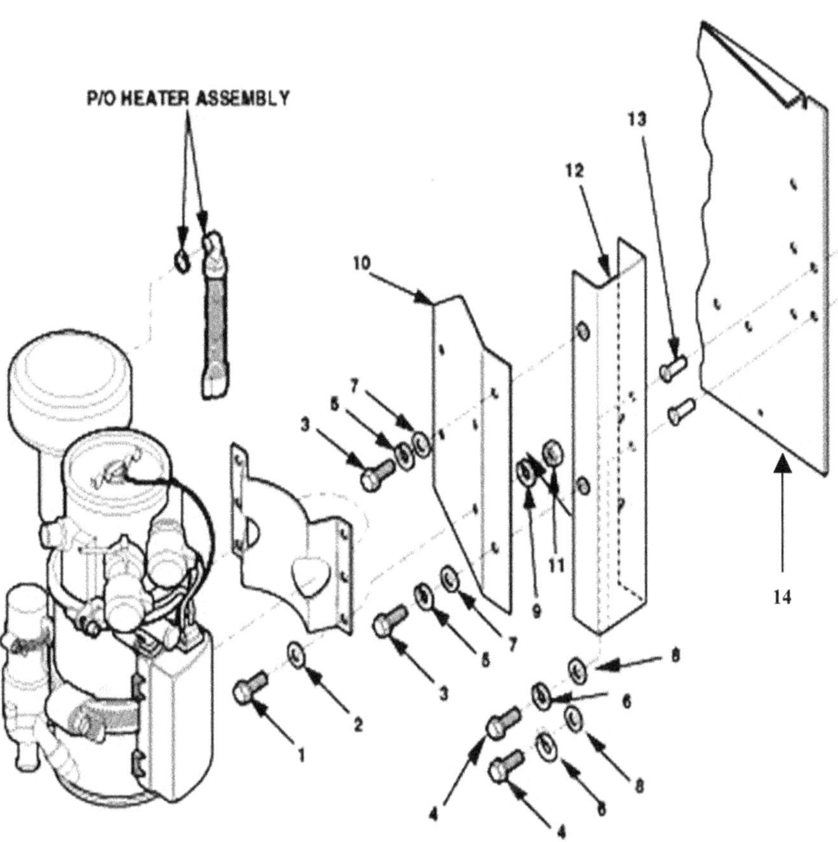

Figure 6-31. Heater Assembly Removal.

m. Remove exhaust hose (2, Figure 6-32) by loosening clamp (3), and removing adapter (4) from exhaust elbow (5). Remove bolt (10), spacer (1), nut (6), lock washer (7), and flat washer (8). Discard lock washer (7). Remove clamp (9). Be sure to replace bolt (10), flat washer (8), new lock washer (7) and nut (6) back in existing hole after removing hose.

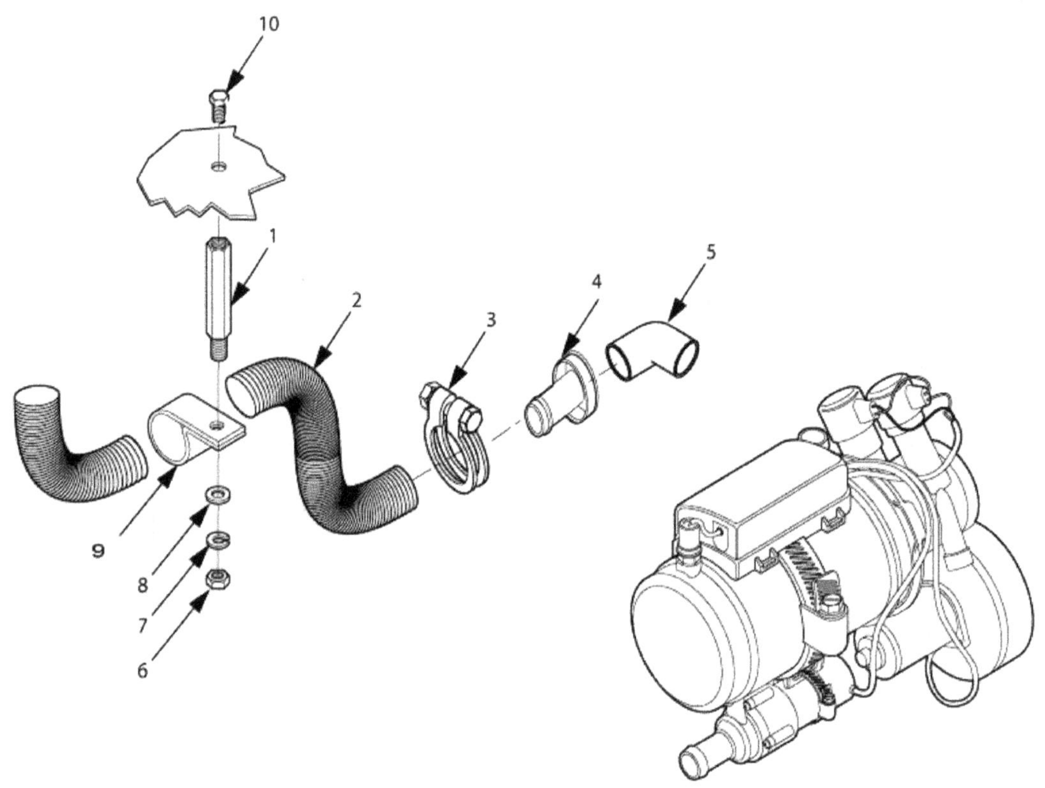

Figure 6-32. Exhaust Hose Removal.

APPENDIX A

REFERENCES

A-1 SCOPE

This appendix lists all forms, regulations, pamphlets, specifications, standards, technical manuals, technical bulletins, lubrication orders, field manuals, and miscellaneous publications referenced in this TM.

A-2 FORMS

Air Force Reporting of Errors Form .... AFTO Form 22
Recommended Changes to Publications and Blank Forms .... DA Form 2028
Recommended Changes to Equipment Technical Publications .... DA Form 2028-2
Equipment Inspection and Maintenance Worksheet .... DA Form 2404
Equipment Inspection and Maintenance Worksheet .... DA Form 5988-E
Preventive Maintenance Schedule and Record .... DD Form 314
Report of Discrepancy .... SF Form 364
Product Quality Deficiency Report .... SF Form 368

A-3 ARMY REGULATIONS

Dictionary of United States Army Terms .... AR 310-25

A-4 DEPARTMENT OF THE ARMY PAMPHLETS

The Army Maintenance Management System (TAMMS) .... DA PAM 738-50

A-5 MILITARY SPECIFICATIONS

Lubricating Oil, Internal Combustion Engine, Combat/Tactical Service.......... MIL-PRF-2104H
Turbine Fuel, Aviation, Grades JP-4 and JP-5 .... MIL-DTL-5624U
Lubricating Oil, Internal Combustion Engine, Arctic MIL-PRF-46167D
Additive, Antifreeze Extender, Liquid Cooling Systems .... MIL-A-53009A(1)
Turbine Fuels, Aviation, Kerosene Types, NATO F-34 (JP-8), NATO F-35, and JP-8+100.......... MIL-DTL-83133E

A-6 COMMERCIAL ITEM DESCRIPTIONS

Fuel Oil, Diesel; for Posts, Camps, and Stations.......... A-A-52557A
Antifreeze, Multi Engine Type.......... A-A-52624A
Generator Sets, Mobile Electric Power and Supplemental Equipment, Packaging of PPP-G-2919
Abbreviations for Use on Drawings, and in Specifications, Standards and Technical Documents ASME-Y14.38M

A-7 MILITARY STANDARDS

Corrosion and Corrosion Prevention Metals MIL-HDBK-729NOT1
Military Marking for Shipment and Storage .... MIL-STD-129P(4)

APPENDIX A

REFERENCES (CONT'D)

A-8 TECHNICAL MANUALS

Unit, Direct Support and General Support Maintenance Instructions,
Diesel Engine, Model C-240PW-8, 4-Cylinder, 2.4 Liter................................ TM 9-2815-254-24
Field Level Maintenance Manual Including Field and Sustainment Level Repair
Parts and Special Tools List, Yanmar Diesel Engine, 4TNV84T-DFM........... TM 9-2815-538-24&P
Operator's Manual, Generator Set, Skid Mounted, Tactical Quiet,
15kW, 50/60 and 400 Hz.............................. .... TM 9-6115-643-10
Unit, Direct Support and General Support Maintenance Repair Parts and Special
Tools List, Generator Set, Tactical Quiet, 15kW, 50/60 and 400 Hz.............. TM 9-6115-643-24P
Operator's, Unit, Direct Support and General Support Maintenance Manual for
Lead Storage Batteries................... .... TM 9-6140-200-14
Painting Instructions for Army Material .... TM 43-0139 Pro-
cedures for Destruction of Equipment to Prevent Enemy Use
(Mobility Equipment Command) .... TM 750-244-3
Cooling Systems: Tactical Vehicles .... TM 750-254

A-9 TECHNICAL BULLETINS

Preservation of USAMECOM Mechanical Equipment for Shipment
and Storage .... TB 740-97-2

A-10 LUBE ORDERS

Generator Set, Skid Mounted, Tactical Quiet 15kW, 50/60 and 400Hz................ LO 9-6115-643-12

A-11 FIELD MANUALS

Chemical and Biological Contamination Avoidance... FM 3-3
NBC Protection.. FM 3-4
NBC Decontamination.. FM 3-5
First Aid .... FM 4-25.11
Theater of Operations, Electrical Systems... FM 5-424
Operation and Maintenance of Ordnance Materiel in Cold Weather
(0□ to –65□)... FM 9-207
Military Symbols.. FM 21-30
Chemical, Biological, Radiological, and Nuclear Defense .... FM 21-40
Basic Cold Weather Manual.. FM 31-70

A-12 MISCELLANEOUS PUBLICATIONS

Air Force Maintenance Forms and Records... AFR 66-1
Army Logistics Readiness and Sustainability... AR 700-138
Reporting of Supply Discrepancies .... AR 735-11-2
Army Materiel Maintenance Policy and Retail Maintenance Operations .... AR 750-1
Procedures for Destruction of Electronics Materiel to Prevent Enemy Use.......... AR 750-244-2
Engines: Preparation for Storage and Shipment of, Purchase Description......... ATPD 2232
Army Medical Department Expendable/Durable Items .... CTA 8-100
Expendable Items (Except Medical Class V, Repair Parts and Heraldic Items)... CTA 50-970 Pro-
cessing and Inspection of Nonmounted, Nonaircraft Gasoline and
Diesel Engines for Storag e and Shipment .... TO 38-1-5

APPENDIX B

MAINTENANCE ALLOCATION CHART (MAC)

Section I. INTRODUCTION

B-1 GENERAL

a. This section provides a general explanation of all maintenance and repair functions authorized at various maintenance categories.

b. The Maintenance Allocation Chart (MAC) in Section II designates overall authority and responsibility for the performance of maintenance functions on the 15 kW 50/60 Hz and 400 Hz Tactical Quiet (TQ) Generator Sets and its components. The application of the maintenance functions to the generator sets or components will be consistent with the capacities and capabilities of the designated maintenance categories.

c. Section III lists the tools and test equipment (both special tools and common tool sets) required for each mainte-nance function as referenced from Section II.

d. Section IV contains supplemental instructions and explanatory notes for particular maintenance functions.

B-2 MAINTENANCE FUNCTION

Maintenance functions will be limited to and defined as follows:

a. Inspect. To determine the serviceability of an item by comparing its physical, mechanical, and/or electrical characteristics with established standards through examination (e.g., by sight, sound, or feel).

b. Test. To verify serviceability by measuring the mechanical, pneumatic, or electrical characteristics of an item and comparing those characteristics with prescribed standards.

c. Service. Operations required periodically to keep an item in proper operating condition, i.e., to clean (include decontaminate, when required), to preserve, to drain, to paint, or to replenish fuel, lubricants, chemical fluids, or gases.

d. Adjust. To maintain or regulate, within prescribed limits, by bringing into proper or exact position, or by setting the operating characteristics to specified parameters.

e. Align. To adjust specified variable elements of an item to bring about optimum or desired performance.

f. Calibrate. To determine and cause corrections to be made or to be adjusted on instruments or Test, Measurement, and Diagnostic Equipment (TMDE) used in precision measurement. Consists of comparisons of two instruments, one of which is a certified standard of known accuracy, to detect and adjust any discrepancy in the accuracy of the instrument being compared.

g. Remove/Install. To remove and install the same item when required to perform service or other maintenance functions. Install may be the act of emplacing, seating, or fixing into position a spare, repair part, or module (component or assembly) in a manner to allow the proper functioning of an equipment or system.

h. Replace. To remove an unserviceable item and install a serviceable counterpart in its place. “Replace” is authorized by the MAC and is shown as the 3rd position code of the SMR code.

i. Repair. The application of maintenance services, including fault location/troubleshooting, removal/installation and disassembly/assembly procedures, and maintenance actions to identify troubles and restore serviceability to an item by correcting specific damage, fault, malfunction, or failure in a part, subassembly, module (component or assembly), end item, or system.

j. Overhaul. That maintenance effort (service/action) prescribed to restore an item to a completely serviceable/operational condition as required by maintenance standards in appropriate technical publications (i.e., Depot Maintenance Work Requirement (DMWR)). Overhaul is normally the highest degree of maintenance performed by the Army. Overhaul does not normally return an item to like new condition.

k. Rebuild. Consists of those services/actions necessary for the restoration of unserviceable equipment to a like new condition in accordance with original manufacturing standards. Rebuild is the highest degree of material maintenance applied to Army equipment. The rebuild operation includes the act of returning to zero those age measurements (hours/miles, etc.) considered in classifying Army equipment/components.

B-3 EXPLANATION OF COLUMNS IN THE MAC, SECTION II

a. Column (1), Group Number. Column (1) lists functional group code numbers, the purpose of which are to identify maintenance significant components, assemblies, subassemblies, and modules with the next higher assembly. End item group number shall be "00".

b. Column (2), Component/Assembly. Column (2) contains the names of components, assemblies, sub-assemblies, and modules for which maintenance is authorized.

c. Column (3), Maintenance Function. Column (3) lists the function to be performed on the item listed in column (2). (For detailed explanation of these functions, see paragraph B-2.)

d. Column (4), Maintenance Category. Column (4) specifies, by the listing of a work time figure in the appropriate subcolumn(s), the category of maintenance authorized to perform the function listed in column (3). This figure represents the active time required to perform that maintenance function at the indicated category of maintenance. If the number or complexity of the tasks within the listed maintenance function vary at different maintenance categories, appropriate work time figures will be shown for each category. The work time figure represents the average time required to restore an item (assembly, subassembly, component, module, end item, or system) to a serviceable condition under typical field operating conditions, This time includes preparation time (including any necessary disassembly/assembly time), troubleshooting/fault location time, and quality assurance/quality control time in addition to the time required to perform the specific tasks identified for the maintenance functions authorized in the Maintenance Allocation Chart. The symbol designations for the various maintenance categories are as follows:

C .... Operator or crew
O .... Organization Maintenance
F.. Direct Support Maintenance
H .... General Support Maintenance
D .... Depot Maintenance

e. Column (5), Tools and Equipment. Column (5) specifies, by code, those common tool sets (not individual tools) and special tools, TMDE, and support equipment required to perform the designated function.

f. Column (6), Remarks. This column shall, when applicable, contain a letter code, in alphabetical order, which shall be keyed to the remarks contained in Section IV.

B-4 EXPLANATION OF COLUMNS IN TOOL AND TEST EQUIPMENT REQUIREMENTS, SECTION III

a. Column (1), Reference Code. The tool and test equipment reference code correlates with a code used in the MAC, Section II, Column (5).

b. Column (2), Maintenance Category. The lowest category of maintenance authorized to use the tool or test equipment.

c. Column (3), Nomenclature. Name or identification of the tool or test equipment.

d. Column (4), National Stock Number. The National Stock Number of the tool or test equipment.

e. Column (5), Tool Number. The manufacturer's part number of the tool or test equipment.

B-5 EXPLANATION OF COLUMNS IN REMARKS, SECTION IV

a. Column (1), Reference Code. The code recorded in column (6), Section II.

b. Column (2), Remarks. This column lists information pertinent to the maintenance function being performed

Section II. MAINTENANCE ALLOCATION CHART Table B-1.

(1) GROUP NUMBER	(2) COMPONENT/ASSEMBLY	(3) MAINTENANCE FUNCTION	(4) MAINTENANCE CATEGORY					(5) TOOLS AND EQUIP.	(6) REMARKS
			C	O	F	H	D		
00	GENERATOR SET 15KW (LESS ENGINE)	INSPECT	.2	.5					
		TEST		1.0	1.0			1,2,3,4	
		SERVICE	.3	.3				4	
		ADJUST		.3	1.0			1,2,4	
		REPAIR		2.5	3.5			1,3,3,4	
01	DC ELECTRICAL SYSTEM	INSPECT	.1	.1					
		TEST		.2				1,4	
		REPAIR		.3				1,4	
0101	BATTERY AND SLAVE RECEPTACLE CABLES	INSPECT	.1	.1					
		REPAIR		.3				1,4	
		REM/INST		.2				4	
		REPLACE		.3				1,4	
0102	BATTERIES	INSPECT	.1	.1					
		TEST		.1				1,4	
		SERVICE	.1	.1				4	
		REM/INST		.2				4	
		REPLACE		.2				4	B
0103	SLAVE RECEPTACLE	INSPECT	.1	.1					
		REM/INST		.1				4	
		REPLACE		.1				4	B
02	HOUSING	INSPECT	.2	.3					
		REPAIR		1.0				4	
		REM/INST		2.0				4	
0201	ACCESS DOORS	INSPECT	.1	.1					
		REPAIR		.5				4	
		REM/INST		.5				4	
		REPLACE		.5				4	B
0202	TOP HOUSING SECTION	INSPECT	.1	.2					
		REPAIR		1.0				4	
		REM/INST		1.0				4	
		REPLACE		1.0				4	B
0203	FRONT HOUSING SECTION	INSPECT	.1	.2					
		REPAIR		1.0				4	
		REM/INST		.6				4	
		REPLACE		.6				4	B
0204	REAR HOUSING SECTION	INSPECT	.1	.2					
		REPAIR		1.0				4	
		REM/INST		1.0				4	
		REPLACE		1.0				4	B

Section II. MAINTENANCE ALLOCATION CHART

Table B-1. MAC for MEP-804A/MEP-804B and MEP-814A/MEP-814B – Continued.

(1) GROUP NUMBER	(2) COMPONENT/ASSEMBLY	(3) MAINTENANCE FUNCTION	(4) MAINTENANCE CATEGORY					(5) TOOLS AND EQUIP.	(6) REMARKS
			C	O	F	H	D		
0205	DECALS AND PLATES	INSPECT	.1	.1					
		REM/INST		.3				1,5	
		REPLACE		.3				1,5	B
03	CONTROL BOX ASSEMBLY	INSPECT	.1	.2					
		TEST		1.0	.5			1,2,3,4, 6,7	
		REPAIR		1 .0	2.0			1,3,4	
		REM/INST		.2				4	
		REPLACE			.2			4	B
0301	PANEL LIGHTS	INSPECT	.1	.1					
		REM/INST		.3				4	
		REPLACE		.3				4	B
		REPAIR		.2					C
0302	INDICATORS	INSPECT	.1	.1					
		TEST		.3				1,4	
		REM/INST		.2				4	
		REPLACE		.2				4	
0303	SWITCHES	INSPECT	.1	.1					
		TEST		.2				1,4	
		REM/INST		.2				1,4	
		REPLACE		.2				1,4	
0304	CONVENIENCE RECEPTACLE	INSPECT	.1	.1					
		TEST		.2				1,4	
		REM/INST		.5				4	
		REPLACE		.5				4	B
0305	GROUND FAULT INTERRUPTER	INSPECT	.1	.1					
		TEST	.1	.1				1,4	
		REM/INST		.5				4	
		REPLACE		.5				4	B
0306	MALFUNCTION INDICATOR PANEL	INSPECT	.1	.1					
		TEST	.1	.1					
		REM/INST		.5				4	
		REPLACE		.5				4	
0307	FUSE AND CIRCUIT BREAKER	INSPECT	.1	.1					
		TEST		.2				1,4	
		REM/INST		.5 .5				1,4	
		REPLACE						1,4	

Section II. MAINTENANCE ALLOCATION CHART

Table B-1. MAC for MEP-804A/MEP-804B and MEP-814A/MEP-814B – Continued.

(1) GROUP NUMBER	(2) COMPONENT/ASSEMBLY	(3) MAINTENANCE FUNCTION	(4) MAINTENANCE CATEGORY					(5) TOOLS AND EQUIP.	(6) REMARKS
			C	O	F	H	D		
0308	AC VOLTAGE REGULATOR	INSPECT		.1					
		TEST			.5			1,4	
		REM/INST			.5			4	
		REPLACE			.5			4	B
0309	RELAYS AND TRANSDUCERS	INSPECT		.1					
		TEST		.3				1,4,6,7	
		REM/INST		.4				4	
		REPLACE		.4				4	
0310	GOVERNOR CONTROL UNIT	INSPECT	.1	.1					
		TEST			1.0			2,4	
		ADJUST			1.0			2,4	
		REM/INST		.2				4	
		REPLACE			.2			4	
0311	CONTROL BOX HARNESS	INSPECT	.1	.2	.2				
		TEST		1.0	1.0			1,3,4	
		REPAIR		.5	1.0			1,3,4	
		REM/INST			1.5			1,4	
		REPLACE			1.5			3,4	
0312	LOAD MEASURING UNIT	INSPECT	.1	.1					
		TEST			.5			3,4	
		REM/INST		.5				4	
		REPLACE			.5			4	B
0313	RESISTOR-DIODE ASSEMBLY	INSPECT	.1	.1					
		TEST		.5				1,4	
		REPAIR		1.0				1,4	
		REM/INST		1.0				1,4	
		REPLACE		1.0				1,4	B
0314	CONTROL BOX PANELS	INSPECT	.1	.1					
		REPAIR		.2				4	
		REM/INST		3.0				4	
		REPLACE		3.0				4	B
0315	DECALS AND PLATES	INSPECT	.1	.1					
		REM/INST		.3				1,5	
		REPLACE		.3				1,5	B
04	AIR INTAKE/EXHAUST SYSTEM	INSPECT	.2	.2					
		REPAIR		1.5				4	

Section II. MAINTENANCE ALLOCATION CHART

Table B-1. MAC for MEP-804A/MEP-804B and MEP-814A/MEP-814B – Continued.

(1) GROUP NUMBER	(2) COMPONENT/ASSEMBLY	(3) MAINTENANCE FUNCTION	(4) MAINTENANCE CATEGORY					(5) TOOLS AND EQUIP.	(6) REMARKS
			C	O	F	H	D		
0401	MUFFLER AND PIPES	INSPECT	.1	.5					
		REM/INST		.7				4	
		REPLACE		.7				4	B
0402	AIR CLEANER ASSEMBLY	INSPECT	.1	.2					
		SERVICE	.2	.2					
		REPAIR		.5				4	
		REM/INST		.5				4	
		REPLACE		.5				4	B
05	COOLANT SYSTEM	INSPECT	.1	.2					
		TEST		.2				1,8	
		SERVICE	.1	.5				4	
		REPAIR		1.0	2.0			1,3,4	B
0501	COOLANT HOSES	INSPECT	.1	.1					
		REM/INST		.5				4	
		REPLACE		.5				4	B
0502	RADIATOR	INSPECT	.1	.2					
		REPAIR		1.0	2.0			1,3,4	
		REM/INST		1.0				4	
		REPLACE		1.0				4	
0503	COOLING FAN	INSPECT	.1	.1				4	
		REM/INST		.8				4	
		REPLACE		.8				4	
0504	FAN BELT	INSPECT	.1	.1					
		TEST		.1					
		ADJUST		.5				4	
		REM/INST		.5				4	
		REPLACE		.5				4	B
0505	COOLANT RECOVERY SYSTEM	INSPECT	.1	.1					
		REM/INST		.5				4	
		REPLACE		.5				4	
06	FUEL SYSTEM	INSPECT	.1	.2					
		REPAIR		1.0	1.5			4	
0601	LOW PRESSURE FUEL LINES	INSPECT	.1	.2					
		REM/INST		.5				4	
		REPLACE		.5				4	
0602	AUXILIARY FUEL PUMP	INSPECT		.1					
		TEST		.5				4	
		REPAIR		.5				4	
		REM/INST		.5 .5				4	
		REPLACE						1,4	

Section II. MAINTENANCE ALLOCATION CHART

Table B-1. MAC for MEP-804A/MEP-804B and MEP-814A/MEP-814B – Continued.

(1) GROUP NUMBER	(2) COMPONENT/ASSEMBLY	(3) MAINTENANCE FUNCTION	(4) MAINTENANCE CATEGORY					(5) TOOLS AND EQUIP.	(6) REMARKS
			C	O	F	H	D		
0603	FUEL TANK	INSPECT		.2					
		SERVICE	.3						
		REM/INST			2.5			4	
		REPLACE			2.5			4	
0604	FUEL TANK FLOATS AND SWITCHES	INSPECT		.2					
		TEST		.3				1,4	
		REM/INST		.5				4	
		REPLACE		.5				4	
0605	FUEL FILTER/WATER SEPARATOR	INSPECT	.1	.2					
		SERVICE	.1	.4				4	
		REPAIR		.5				4	
		REM/INST		1.0				4	
		REPLACE		1.0				4	B
0606	AUXILIARY FUEL FILTER	INSPECT	.1						
		REM/INST		.2				4	
		REPLACE		.2				4	
07	OUTPUT BOX ASSEMBLY	INSPECT	.2						
		TEST	1.0	2.0				1,3,4	
		REPAIR	2.0	3.0				1,3,4	
		REM/INST		2.0				4	
		REPLACE		2.0				4	B
0701	VOLTAGE RECONNECTION BOARD	INSPECT		.2					
		REM/INST			.4			4	
		REPLACE			.4			4	B
0702	OUTPUT BOX HARNESS AND CABLES	INSPECT	.1	.2					
		TEST		.6				1,2,4	
		REPAIR		.5	1.0			1,3,4	
		REM/INST			2.0			4	
		REPLACE			2.0			4	B
0703	TRANSFORMERS	INSPECT	.1	.2					
		TEST			1.0			2,4	
		REM/INST			1.3			4	
		REPLACE			1.3			4	B
0704	AC CIRCUIT INTERRUPTER	INSPECT		.2					
		TEST		.5				1,4	
		REM/INST		.5				4	
		REPLACE		.5				4	B
0705	START RELAY	INSPECT		.2					
		TEST		.5				1,4	
		REM/INST		.5 .5				4	
		REPLACE						4	B

Section II. MAINTENANCE ALLOCATION CHART

Table B-1. MAC for MEP-804A/MEP-804B and MEP-814A/MEP-814B – Continued.

(1) GROUP NUMBER	(2) COMPONENT/ASSEMBLY	(3) MAINTENANCE FUNCTION	(4) MAINTENANCE CATEGORY					(5) TOOLS AND EQUIP.	(6) REMARKS
			C	O	F	H	D		
0706	OUTPUT BOX PANELS	INSPECT	.1	.1					
		REPAIR		.2	1.0			4	
		REM/INST			2.0			4	
		REPLACE			2.0			4	B
08	OUTPUT LOAD TERMINAL BOARD ASSEMBLY	INSPECT	.1	.1					
		TEST		.5				1	
		REPAIR		1.0				4	
		REM/INST		1.5				4	
		REPLACE		1.5				4	
0801	LOAD TERMINALS	INSPECT		.1					
		REPAIR		.5				4	
		REM/INST		.5				4	
		REPLACE		.5				4	
0802	VARISTORS	INSPECT		1.0					
		TEST		1.1				1,4	
		REM/INST		1.0				4	
		REPLACE		1.0				4	B
0803	LOAD TERMINAL BOARD	INSPECT	.1	.1					
		REM/INST		1.5				4	
		REPLACE		1.5				4	B
09	ENGINE ACCESSORIES	INSPECT	.1	.1					
		TEST		.5				1,4	
		REPAIR		.5				4	
0901	SENDERS AND SWITCHES	INSPECT	.1	.1					
		TEST		.5				1,4	
		REPAIR		.5				4	
		REM/INST		.5				4	
		REPLACE		.5				4	B
0902	DEAD CRANK SWITCH	INSPECT	.1	.1					
		TEST		.5				1,4	
		REM/INST		.5				4	
		REPLACE		.5				4	B
0903	GOVERNOR ACTUATOR	INSPECT	.1	.1					
		TEST			.3			4	
		ADJUST			.3			4	
		REPAIR		.5	.5			1,4	
		REM/INST			.5 .5			4	
		REPLACE						4	B

Section II. MAINTENANCE ALLOCATION CHART

Table B-1. MAC for MEP-804A/MEP-804B and MEP-814A/MEP-814B – Continued.

(1) GROUP NUMBER	(2) COMPONENT/ASSEMBLY	(3) MAINTENANCE FUNCTION	(4) MAINTENANCE CATEGORY					(5) TOOLS AND EQUIP.	(6) REMARKS
			C	O	F	H	D		
0904	DECALS AND PLATES	INSPECT	.1	.1					
		REM/INST		.3				1,5	
		REPLACE		.3				1,5	B
10	LUBRICATION SYSTEM	INSPECT	.1	.2					
		SERVICE	.2						A, D
		REPAIR		.5				4	A, D
1001	OIL DRAIN LINE	IN SPECT	.1	.2					
		REPAIR		.5				4	
		REM/INST		.5				4	
		REPLACE		.5				4	B
11	GENERATOR ASSEMBLY	INSPECT		.1	.1				
		TEST		1.0	1.0			1,2,3,4	
		REPAIR			4.0			4	
		REM/INST			4.0			1,3,4	
		REPLACE			4.0			1,3,4	B
12	ENGINE ASSEMBLY	INSPECT	.2						A, D
		SERVICE							A, D
		ADJUST							A, D
		REPAIR							A, D
		REM/INST			4.0			1,3,4	
		REPLACE			4.0			1,3,4	
13	SKID BASE	INSPECT	.1	.1					
		REPAIR			1.0			4	
		REM/INST			3.0			4	
		REPLACE			3.0			4	B

Section III. TOOL AND TEST EQUIPMENT REQUIREMENTS

Table B-2. Tool and Test Equipment Requirements for MEP-804A/MEP-804B and MEP-814A/MEP-814B.

TOOL OR TEST EQUIPMENT REF CODE	MAINTENANCE CATEGORY	NOMENCLATURE	NATIONAL/ NATO STOCK NUMBER	TOOL NUMBER
1	O	SHOP EQUIPMENT, AUTOMOTIVE MAINT AND REPAIR	4910-00-754-0654	SC4910-95-CL-A74
2	F	SHOP EQUIPMENT, ELECTRICAL REPAIR	4910-01-096-4475	SC4940-95-CL-B05
3	F	SHOP EQUIPMENT, AUTOMOTIVE MAINT AND REPAIR, FIELD	4910-00-348-7696	SC4910-95-CL-A02
4	O,F	TOOL KIT, GENERAL MECHANIC	5180-00-177-7033	SC5180-90-CL-N26
5	O	POP RIVET GUN	5120-00-508-1588	GGG-R-00395
6	O	POTENTIOMETER, 5,000 OHMS	–	–
7	O	POTENTIOMETER, 10,000 OHMS	–	–
8	F	RESISTOR, FIXED, 5 OHM, 25 WATT		

Section IV. REMARKS

Table B-3. Remarks for MEP-804A/MEP-804B and MEP-814A/MEP-814B.

REFERENCE CODE	REMARKS
A	Refer to TM 9-2815-254-24.
B	Replace function identical to removal/install function.
C	Repair is limited to replacement of bulb.
D	Refer to TM 9-2815-538-24&P.

APPENDIX C

EXPENDABLE/DURABLE SUPPLIES AND MATERIALS LIST

Section I. INTRODUCTION

C-1 SCOPE

This appendix lists expendable supplies and materials you will need to operate and maintain the generator set. These items are authorized to you by CTA 50-970, Expendable Items (except medical, class V, repair parts, and heraldic items).

C-2 EXPLANATION OF COLUMNS

a. Column (1) – Item Number. This number is assigned to the entry in the listing and is referenced in the narrative instructions to identify the material (e.g., "Use adhesive, Item 1, Appendix C").

b. Column (2) – Level. This column identifies the lowest level of maintenance that requires the listed item.

c. Column (3) – National Stock Number. This is the National Stock Number assigned to the item; use it to request or requisition the item.

d. Column (4) – Description. Indicates the Federal item name and, if required, a description to identify the item. The last line for each item indicates the Commercial and Government Entity (CAGE) code in parentheses followed by the part number.

e. Column (5) – Unit of Measure (U/M). Indicates the measure used in performing the actual maintenance function. This measure is expressed by a two-character alphabetical abbreviation (e.g., ea, in, pr). If the unit of measure differs from the unit of issue, requisition the lowest unit of issue that will satisfy your requirements.

Section II. TABULAR LIST

Table C-1. Expendable/Durable Supplies and Materials List.

(1) Item No.	(2) Level	(3) National Stock Number	(4) Description	(5) U/M
1	O,F	8040-00-380-7959	Adhesive, Seal, EC847	QT
2	O,F	6650-00-181-7929	Antifreeze, A-A-52624, 1 Gal. Can	GL
3	O,F	6650-00-181-7933	Antifreeze, A-A-52624, 5 Gal. Can	GL
4	O,F	6850-00-181-7940	Antifreeze, A-A-52624, 55 Gal. Drum	GL
5	O,F	6650-00-174-1806	Antifreeze, A-A-52624, 1 Gal. Can	GL
6	O,F	8030-01-234-2782	Antiseize, Compound, CP-8, 1/2 Lb. Can	OZ
7	O,F	7920-01-338-3329	Cloth, Cleaning, TX-1250	EA
8	F	8030-00-056-8673	Compound, Therm, Pentrox A	OZ
9	O,F	9150-00-663-1770	Grease, General Purpose, 630AA, 6 Lb. Can	LB
10	O,F	6850-01-160-3866	Inhibitor, Corrosion, MIL-A-53009A(1)	QT
11	O,F	9150-00-152-4117	Lubricating Oil, Engine, MIL-PRF-2104H, 15/40W	QT
12	O,F	9150-00-189-6727	Lubrication Oil, Engine, BRAYC0421C, 10W	QT
13	O,F	9150-00-186-6681	Lubricating Oil, Engine, ALIEDC030, 30W	QT
14	O,F	9150-00-402-2372	Lubricating Oil, Engine, MIL-PRF-46167D, OEA	QT
15	O,F	5330-00-543-3600	Paper, Abrasive, ALOXGRIT 80	SH
16	O,F	8040-00-643-0802	Sealant, RTV 108	OZ
17	O,F	8030-00-849-0071	Sealing Compound, FORM GASKET 2	TU
18	O,F	8030-01-025-1692	Sealing Compound, LOCKTITE 242	OZ
19	O,F	3439-00-974-1873	Solder, Tin Alloy, SN60WRAP2, 1 Lb. Spool	OZ
20	O,F	6850-00-264-8036	Solvent, Dry Cleaning, P-D-660, 5 Gal. Can	GL

APPENDIX D

FABRICATION/ASSEMBLY OF PARTS

D-1 INTRODUCTION

This appendix includes complete instructions for fabricating or assembling parts as required on this generator set.

NOTE

All dimensions are expressed in inches. Refer to Table D-1 for inches to metric conversions.

D-2 ILLUSTRATIONS

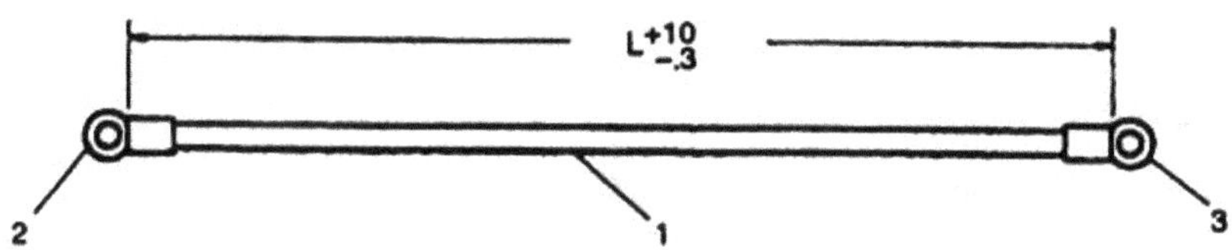

NOTES:

1. Dimensions shown are in inches.
2. Refer to TM 9-6115-643-24P for materials required and length (L) of wire.

PROCEDURES:

1. Cut wire (1) to length indicated.
2. Strip 0.75 inch of insulation from each end of wire (1).
3. Crimp terminal (2) on one end of wire (1) and terminal (3) on other end.

Figure D-1. AC Power Cable Assemblies (P/Ns 88-22126-1 thru 88-22126-

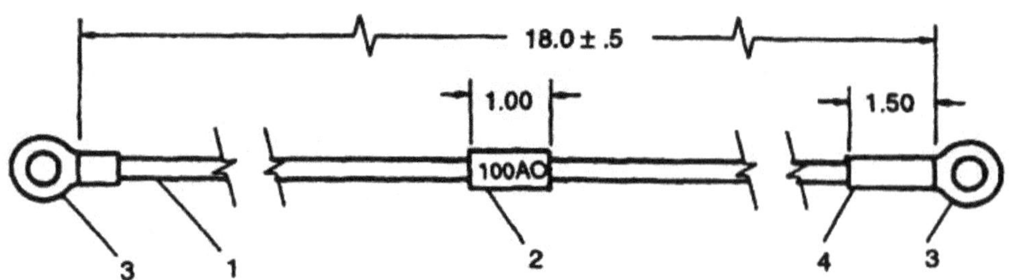

NOTES:

1. Dimensions shown are in inches.
2. Refer to TM 9-6115-643-24P for materials required.

PROCEDURES:

1. Cut wire (1) to length indicated.
2. Strip 0.75 inch of insulation from each end of wire (1).
3. Position insulation sleeving (2) on center of wire, mark with wire number "100AO" and shrink to fit.
4. Mark insulation sleeving (4) with "NEGATIVE" and slide over one end of wire (1).
5. Crimp terminal (3) on each end of wire (1).
6. Position insulation sleeving (4) as shown and shrink to fit.

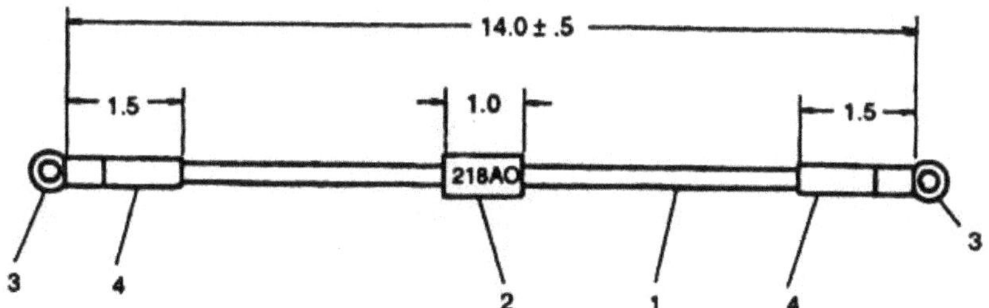

NOTES:

1. Dimensions shown are in inches.
2. Refer to TM 9-6115-643-24P for materials required.

PROCEDURES:

1. Cut wire (1) to length indicated.
2. Strip 0.75 inch of insulation from each end of wire (1).
3. Mark insulation sleeving (2) with wire number "218AO", position on center of wire and shrink to fit.
4. Mark one insulation sleeving (4) with "NEGATIVE", the other with "POSITIVE" and slide over each end of wire (1).
5. Crimp terminal (3) on each end of wire (1).
6. Position insulation sleeving (4) as shown and shrink to fit

Figure D-3. Battery Cable Assembly (P/N 88-22179).

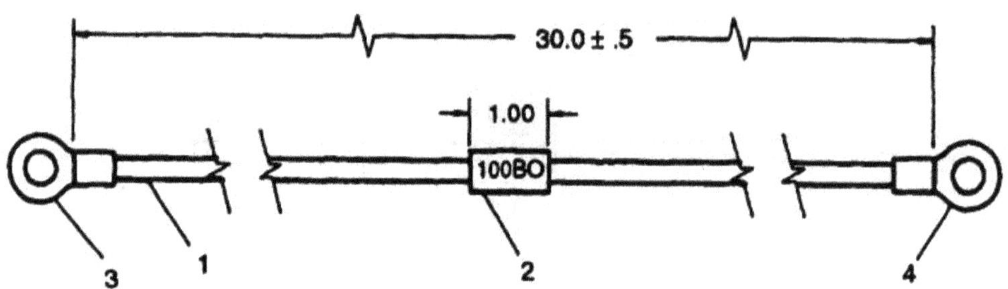

NOTES:

1. Dimensions shown are in inches.

2. Refer to TM 9-6115-643-24P for materials required.

PROCEDURES:

1. Cut wire (1) to length indicated.

2. Strip 0.75 inch of insulation from each end of wire (1).

3. Position insulation sleeving (2) on center of wire, mark with wire number "100BO" and shrink to fit.

4. Crimp terminal (3) on one end of wire (1) and terminal (4) on other end.

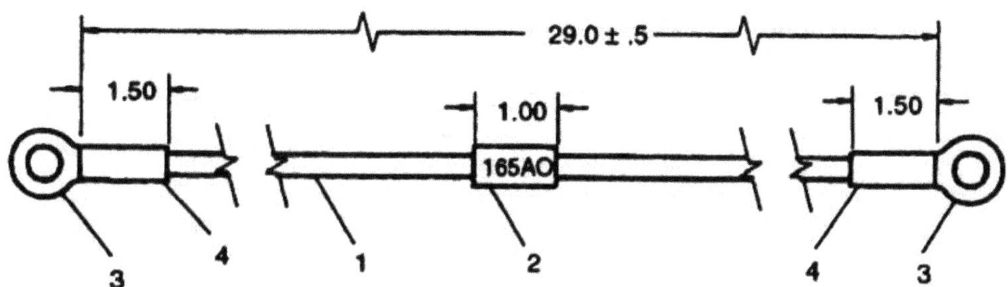

NOTES:

1. Dimensions shown are in inches.
2. Refer to TM 9-6115-643-24P for materials required.

PROCEDURES:

1. Cut wire (1) to length indicated.
2. Strip 0.75 inch of insulation sleeving from each end of wire (1).
3. Mark insulation sleeving (2) with wire number "165AO", position on center of wire and shrink to fit.
4. Mark insulation sleeving (4) with "POSITIVE" and slide over each end of wire (1).
5. Crimp terminal (3) on each end of wire (1).
6. Position insulation sleeving (4) as shown and shrink to fit.

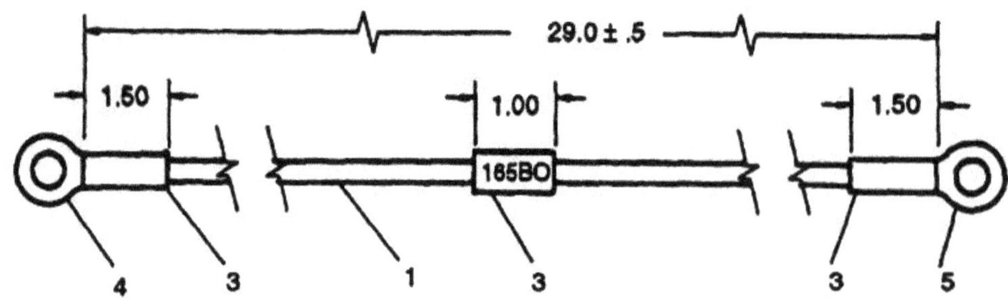

NOTES:

1. Dimensions shown are in inches.

2. Refer to TM 9-6115-643-24P for materials required.

PROCEDURES:

1. Cut wire (1) to length indicated.

2. Strip 0.75 inch of insulation sleeving from each end of wire (1).

3. Mark insulation sleeving (3) with wire number "165BO", position on center of wire and shrink to fit.

4. Mark insulation sleeving (3) with "POSITIVE" and slide over each end of wire (1).

5. Crimp terminal (4) on one end of wire (1) and terminal (5) on other end.

6. Position insulation sleeving (3) as shown and shrink to fit.

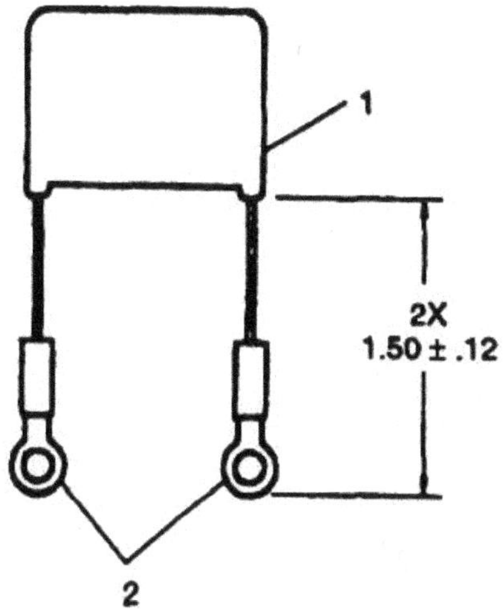

NOTES:

1. Dimensions shown are in inches.
2. Refer to TM 9-6115-643-24P for terminals required.

PROCEDURES:

1. Cut each lead of capacitor (1) to obtain dimension shown with terminals (2) installed.
2. Strip 0.25 inch from each lead of capacitor (1).
3. Crimp and solder terminals (2) on end of each lead.

Figure D-7. Electromagnetic Interference (EMI) Capacitor Assembly (P/N 88-

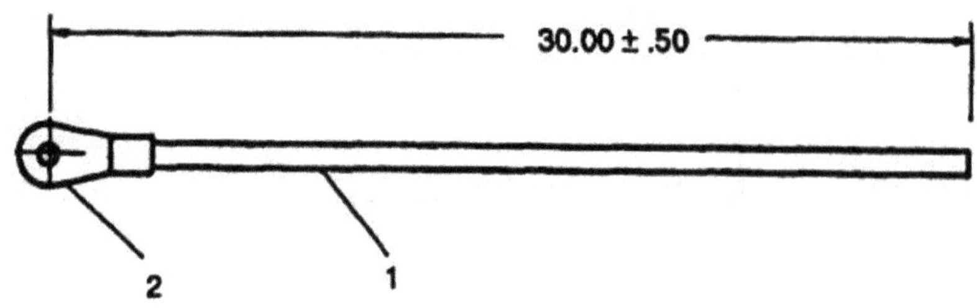

NOTES:

1. Dimensions shown are in inches.

2. Refer to TM 9-6115-643-24P for materials required.

PROCEDURES:

1. Cut rope (1) to length indicated.

2. Crimp terminal (2) on one end of rope (1).

Figure D-8. Load Wrench Cord (P/N 88-

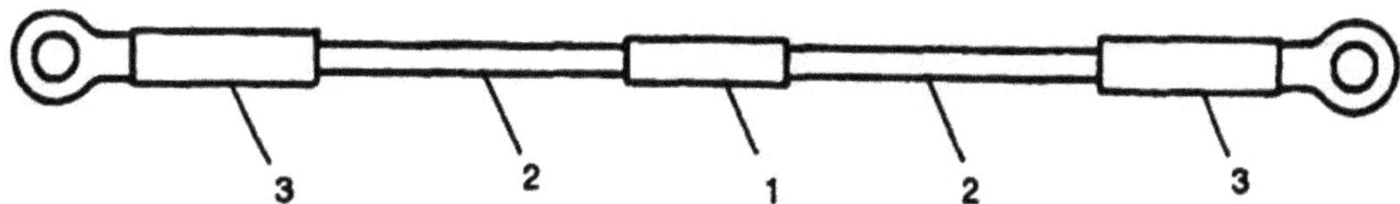

NOTES:

Refer to TM 9-6115-643-24P for materials required.

PROCEDURES:

1. Position 0.75 inch of insulation sleeving (2) on each lead of diode (1), leaving 0.25 inch of bare wire on each lead. Shrink sleeving to fit.

2. Crimp and solder terminals (3) on end of each diode (1) lead.

Figure D-9. Diode Assembly (P/N 88-22418-2).

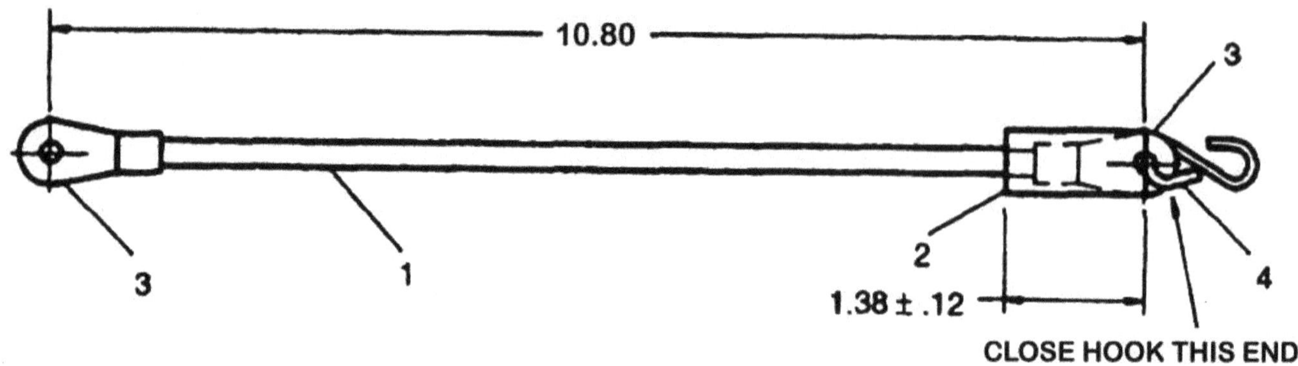

NOTES:

1. Dimensions shown are in inches.
2. Refer to TM 9-6115-643-24P for materials required.

PROCEDURES:

1. Cut rope (1) to length indicated.
2. Slide insulation sleeving (2) over one end of rope (1).
3. Crimp terminal (3) on each end of rope (1).
4. Install hook (4) in one terminal (3) and close hook end securing it to terminal.
5. Position insulation sleeving (2) as shown and shrink to fit.

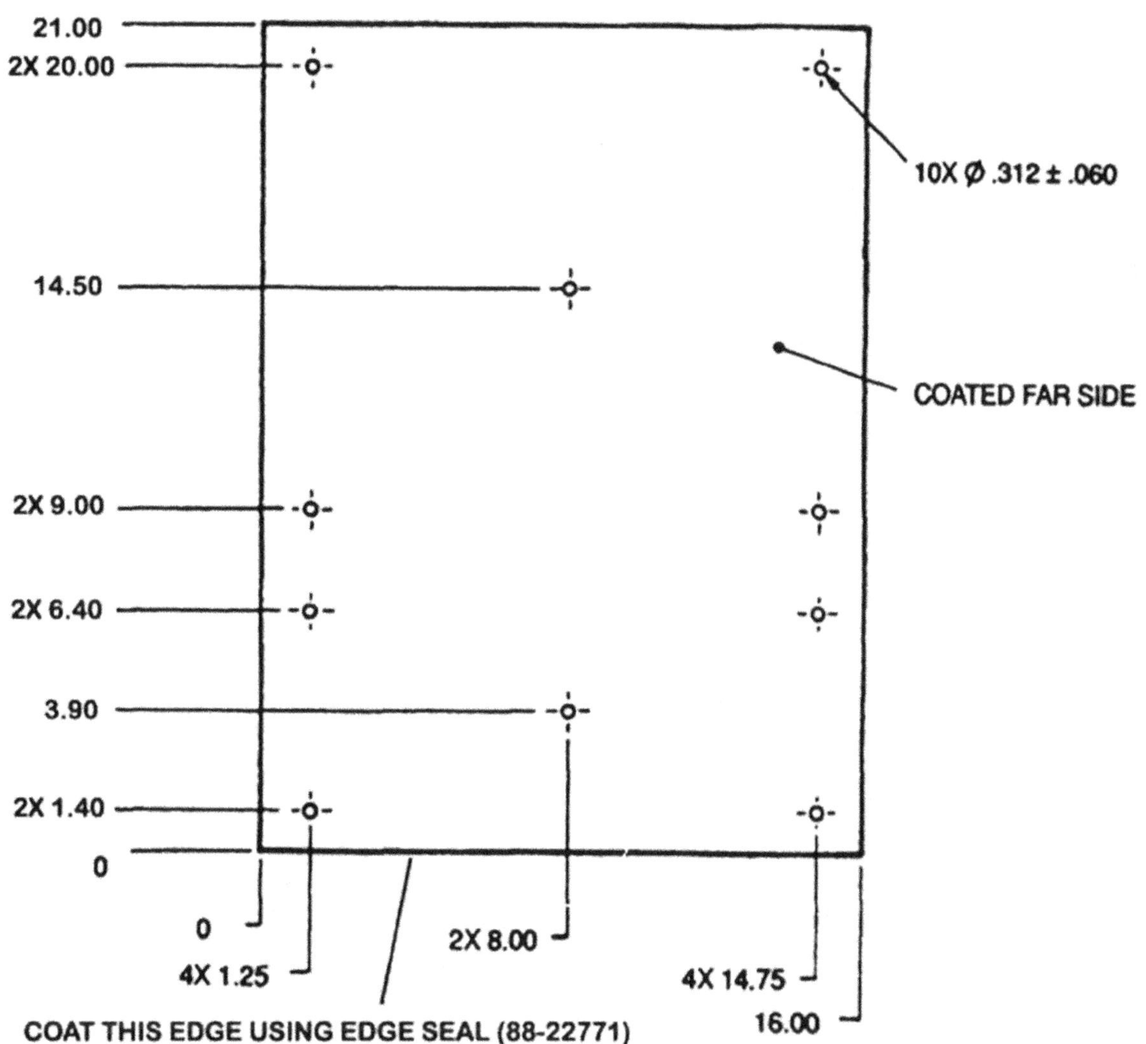

MATERIALS	
Description	Part Number
Foam, Sound Absorbing	88-21110

NOTES:

1. Dimensions shown are in inches.
2. Tolerances are 0.1 inch unless otherwise stated.

PROCEDURES:

1. Cut foam to dimensions shown.
2. Drill holes as shown.

Figure D-11. Baffle Insulation (P/N 88-

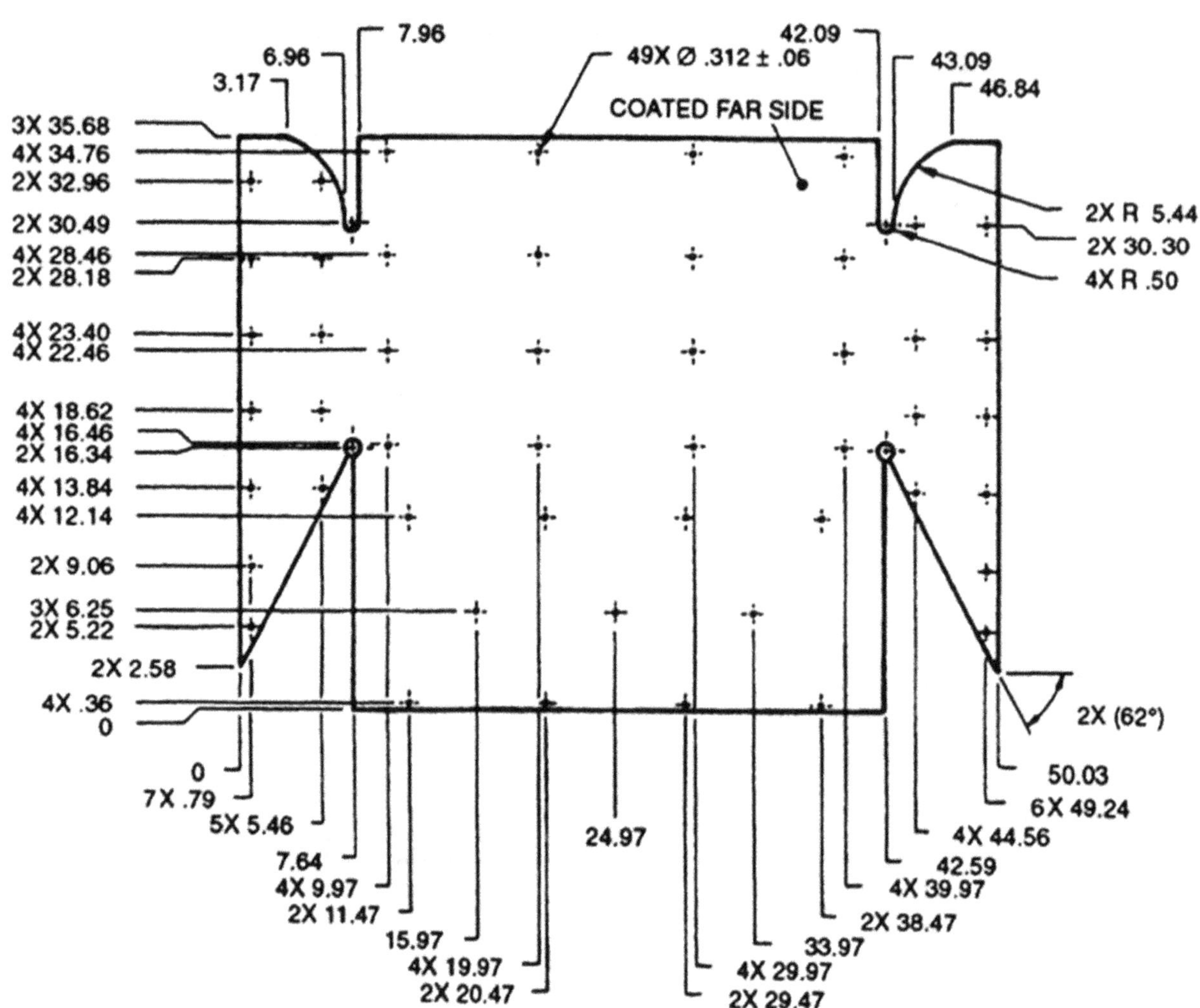

MATERIALS	
Description	**Part Number**
Foam, Sound Absorbing	FF40JM02

NOTES:

1. Dimensions shown are in inches.
2. Tolerances are 0.1 inch unless otherwise stated.

PROCEDURES:

1. Cut foam to dimensions shown.
2. Drill holes as shown.

Figure D-12. Front Housing Insulation (P/N 88-

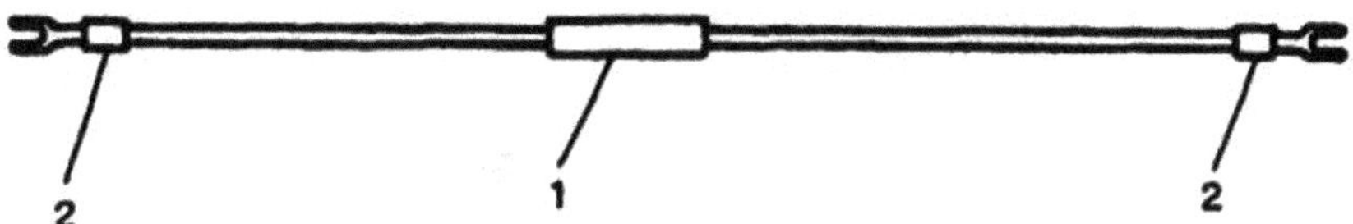

NOTES:

1. Refer to TM 9-6115-643-24P for materials required.
2. Resistors (1) are different depending on dash number being assembled.

PROCEDURES:

Crimp terminal (2) on end of each resistor (1) lead.

Figure D-13. Resistor Assemblies (P/Ns 01-21506-1 and 01-21506-

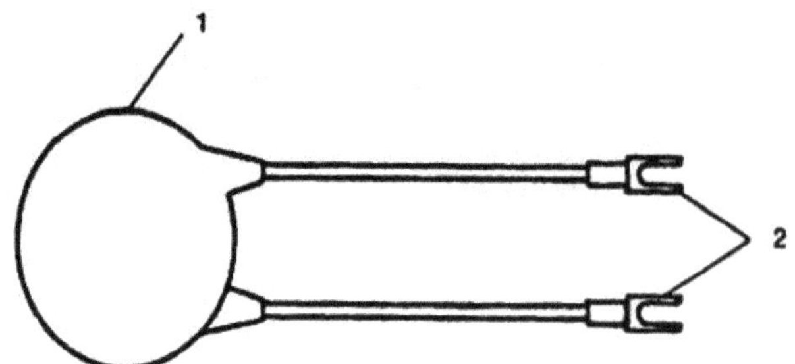

NOTES:

Refer to TM 9-6115-643-24P for materials required.

PROCEDURES:

Crimp and solder terminals (2) on end of each resistor (1) lead.

Figure D-14. Volt Resistor Assembly (P/N 88-

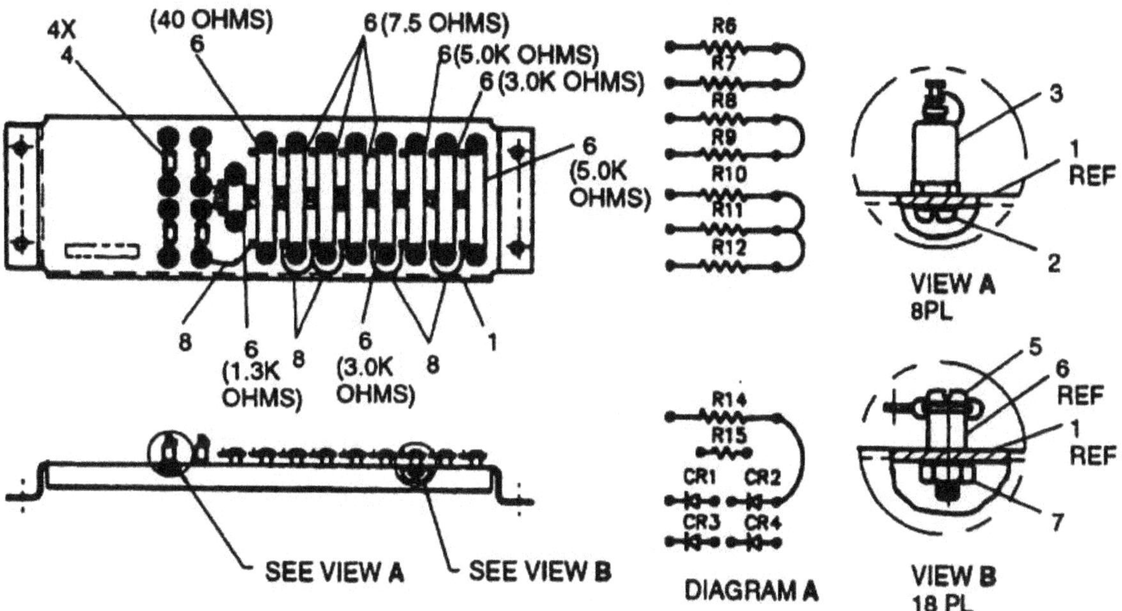

NOTES:

Refer to TM 9-6115-643-24P for materials required and positioning of resistors by ohm rating.

PROCEDURES:

1. Install screws (2) and insulated terminals (3) on bracket (1) as shown.
2. Solder diodes (4) to insulated terminals (3) as shown.
3. Install screws (5), resistors (6), and nuts (7) on bracket (1) as shown.
4. Position wires (8) and solder connections as shown.

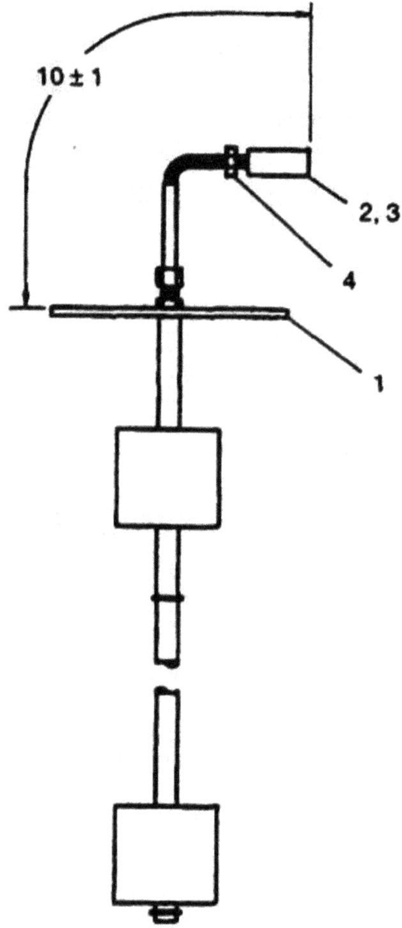

NOTES:

1. Dimensions shown are in inches.
2. Refer to TM 9-6115-643-24P for materials required.

PROCEDURES:

1. Strip 0.125 inch of insulation from end of each switch (1) lead.
2. Crimp pin (2) on end of each lead.
3. Insert pins into housing (3) with lead A in position 1, lead B in position 2, lead C in position 3, and lead D in position 4.
4. Mark "P12" on strap (4) and install in position shown.

Figure D-16. Fuel Level Switch Assembly (P/N 88-22792).

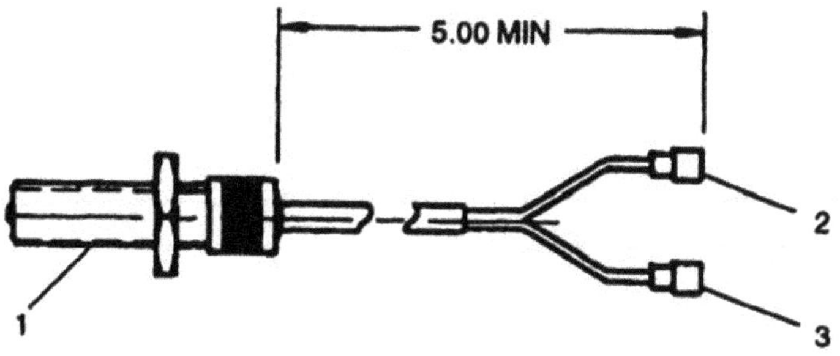

NOTES:

1. Dimensions shown are in inches.

2. Refer to TM 9-6115-643-24P for materials required.

PROCEDURES:

1. Strip 0.25 inch of insulation from end of each transducer (1) lead.

2. Crimp male terminal (2) on red wire and female terminal (3) on black wire.

Figure D-17. Transducer Assembly (P/N 88-

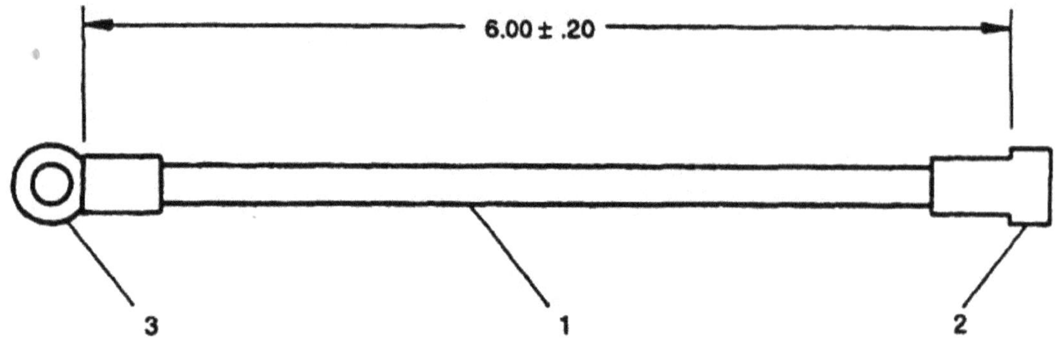

NOTES:

1. Dimensions shown are in inches.
2. Refer to TM 9-6115-643-24P for materials required.

PROCEDURES:

1. Cut wire (1) to length indicated.
2. Strip 0.25 inch from each end of wire (1).
3. Crimp terminal (2) on one end of wire (1) and terminal (3) on other end.

Figure D-18. Varistor Wires L1 thru L3 and L0 (P/Ns 88-20305-1 thru 88-20305-3 and 88-

Table D-1. Inches to Metric Conversion.

Part I. Fractional Equivalent.

Fractional Inches	Decimal Inches	mm
1/16	.0625	1.587
1/8	.1250	3.175
3/16	.1875	4.762
1/4	.2500	6.350
5/16	.3125	7.937
3/8	.3750	9.525
7/16	.4375	11.112
1/2	.5000	12.700
9/16	.5625	14.287
5/8	.6250	15.875
11/16	.6875	17.462
3/4	.7500	19.050
13/16	.8125	20.637
7/8	.8750	22.225
15/16	.9375	23.812
1	1	25.400

Part II. Inches to Centimeters.

Inches	cm
1	2.540
2	5.080
3	7.620
4	10.16
5	12.70
6	15.24
7	17.78
8	20.32
9	22.86
10	25.40
20	50.80
30	76.20
40	101.6
50	127.0
60	152.4
70	177.8
80	203.2
90	228.6
100	254.0

INDEX

Subject Figure, Table — Paragraph, Number

A

INDEX – Continued

Subject Figure, Table **Paragraph, Number**

A – Continued

B

INDEX – Continued

Subject Figure, Table **Paragraph, Number**

B – Continued

C

INDEX – Continued

Subject Figure, Table **Paragraph, Number**

C – Continued

INDEX – Continued

Subject Figure, Table **Paragraph, Number**

C – Continued

INDEX – Continued

Subject Figure, Table — Paragraph, Number

C – Continued

D

INDEX – Continued

Subject Figure, Table **Paragraph, Number**

D – Continued

E

INDEX – Continued

Subject Figure, Table **Paragraph, Number**

E – Continued

INDEX – Continued

Subject Figure, Table **Paragraph, Number**

E – Continued

F

INDEX – Continued

Subject Figure, Table **Paragraph, Number**

F – Continued

INDEX – Continued

Subject Figure, Table **Paragraph, Number**

F – Continued

INDEX – Continued

Subject Figure, Table **Paragraph, Number**

F – Continued

G

INDEX – Continued

Subject Figure, Table **Paragraph, Number**

G – Continued

INDEX – Continued

Subject Figure, Table **Paragraph, Number**

INDEX – Continued

Subject Figure, Table **Paragraph, Number**

L

INDEX – Continued

Subject Figure, Table **Paragraph, Number**

L – Continued

INDEX – Continued

Subject Figure, Table **Paragraph, Number**

M

INDEX – Continued

Subject Figure, Table **Paragraph, Number**

M – Continued

INDEX – Continued

Subject Figure, Table **Paragraph, Number**

O – Continued

INDEX – Continued

Subject Figure, Table **Paragraph, Number**

P, Q

INDEX – Continued

Subject Figure, Table **Paragraph, Number**

P, Q – Continued

R

INDEX – Continued

Subject Figure, Table **Paragraph, Number**

R – Continued

INDEX – Continued

Subject Figure, Table — Paragraph, Number

R – Continued

INDEX – Continued

Subject Figure, Table **Paragraph, Number**

R – Continued

INDEX – Continued

Subject Figure, Table **Paragraph, Number**

R – Continued

S

INDEX – Continued

Subject Figure, Table **Paragraph, Number**

S – Continued

INDEX – Continued

Subject Figure, Table **Paragraph, Number**

S – Continued

INDEX – Continued

Subject Figure, Table **Paragraph, Number**

V

INDEX – Continued

Subject Figure, Table **Paragraph, Number**

W, X, Y, Z

INDEX – Continued

Subject Figure, Table **Paragraph, Number**

W, X, Y, Z – Continued

By Order of the Secretary of the Army:

Official:

JOYCE E. MORROW
Administrative Assistant to the Secretary of the Army
0813605

GEORGE W. CASEY, JR
General, United States Army
Chief of Staff

By Order of the Secretary of the Air Force:

T. MICHAEL MOSELEY
General, United States Air Force
Chief of Staff

Official:

BRUCE CARLSON
General, United States Air Force
Commander, AFMC

DISTRIBUTION:

To be distributed in accordance with the initial distribution number (IDN) 255266 requirements for TM 9-6115-643-24.

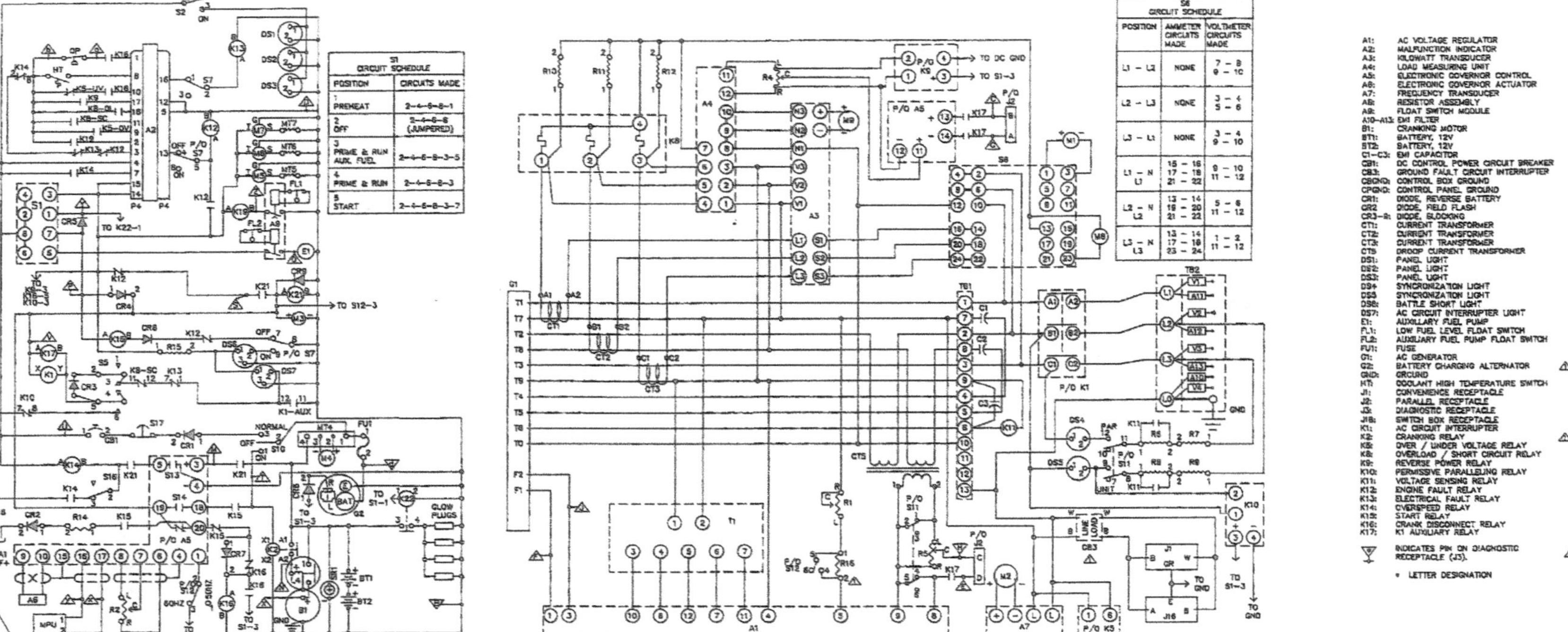

S1 CIRCUIT SCHEDULE	
POSITION	CIRCUITS MADE
1 PREHEAT	2-4-6-8-1
2 OFF	2-4-6-8 (JUMPERED)
3 PRIME & RUN AUX. FUEL	2-4-6-8-3-5
4 PRIME & RUN	2-4-6-8-3
5 START	2-4-6-8-3-7

S6 CIRCUIT SCHEDULE		
POSITION	AMMETER CIRCUITS MADE	VOLTMETER CIRCUITS MADE
L1 - L2	NONE	7 - 8 9 - 10
L2 - L3	NONE	3 - 4 5 - 6
L3 - L1	NONE	3 - 4 9 - 10
L1 - N L1	15 - 16 17 - 18 21 - 22	9 - 10 11 - 12
L2 - N L2	13 - 14 19 - 20 21 - 22	5 - 6 11 - 12
L3 - N L3	13 - 14 17 - 18 23 - 24	1 - 2 11 - 12

A1: AC VOLTAGE REGULATOR
A2: MALFUNCTION INDICATOR
A3: KILOWATT TRANSDUCER
A4: LOAD MEASURING UNIT
A5: ELECTRONIC GOVERNOR CONTROL
A6: ELECTRONIC GOVERNOR ACTUATOR
A7: FREQUENCY TRANSDUCER
A8: RESISTOR ASSEMBLY
A9: FLOAT SWITCH MODULE
A10–A13: EMI FILTER
B1: CRANKING MOTOR
BT1: BATTERY, 12V
BT2: BATTERY, 12V
C1–C3: EMI CAPACITOR
CB1: DC CONTROL POWER CIRCUIT BREAKER
CB3: GROUND FAULT CIRCUIT INTERRUPTER
CBGND: CONTROL BOX GROUND
CPGND: CONTROL PANEL GROUND
CR1: DIODE, REVERSE BATTERY
CR2: DIODE, FIELD FLASH
CR3–8: DIODE, BLOCKING
CT1: CURRENT TRANSFORMER
CT2: CURRENT TRANSFORMER
CT3: CURRENT TRANSFORMER
CT5: DROOP CURRENT TRANSFORMER
DS1: PANEL LIGHT
DS2: PANEL LIGHT
DS3: PANEL LIGHT
DS4: SYNCRONIZATION LIGHT
DS5: SYNCRONIZATION LIGHT
DS6: BATTLE SHORT LIGHT
DS7: AC CIRCUIT INTERRUPTER LIGHT
E1: AUXILLARY FUEL PUMP
FL1: LOW FUEL LEVEL FLOAT SWITCH
FL2: AUXILIARY FUEL PUMP FLOAT SWITCH
FU1: FUSE
G1: AC GENERATOR
G2: BATTERY CHARGING ALTERNATOR
GND: GROUND
HT: COOLANT HIGH TEMPERATURE SWITCH
J1: CONVENIENCE RECEPTACLE
J2: PARALLEL RECEPTACLE
J3: DIAGNOSTIC RECEPTACLE
J16: SWITCH BOX RECEPTACLE
K1: AC CIRCUIT INTERRUPTER
K2: CRANKING RELAY
K5: OVER / UNDER VOLTAGE RELAY
K8: OVERLOAD / SHORT CIRCUIT RELAY
K9: REVERSE POWER RELAY
K10: PERMISSIVE PARALLELING RELAY
K11: VOLTAGE SENSING RELAY
K12: ENGINE FAULT RELAY
K13: ELECTRICAL FAULT RELAY
K14: OVERSPEED RELAY
K15: START RELAY
K16: CRANK DISCONNECT RELAY
K17: K1 AUXILIARY RELAY
K19: FUEL LEVEL RELAY
K21: GOVERNOR, CONTROL, POWER
K22: GLOW PLUGS CONTACTOR
L0: OUTPUT TERMINAL
L1: OUTPUT TERMINAL
L2: OUTPUT TERMINAL
L3: OUTPUT TERMINAL
L4: STARTER SOLENOID
M1: AC VOLTMETER
M2: FREQUENCY METER
M3: TIME METER
M4: BATTERY CHARGING AMMETER
M5: FUEL LEVEL INDICATOR
M6: COOLANT TEMPERATURE INDICATOR
M7: OIL PRESSURE INDICATOR
M8: AC AMMETER
M9: KILOWATT METER
MPU: MAGNETIC PICKUP
MT4: BATTERY CHARGING AMMETER SHUNT
MT5: FUEL LEVEL SENDER
MT6: COOLANT TEMPERATURE SENDER
MT7: OIL PRESSURE SENDER
OP: LOW OIL PRESSURE SWITCH
P4: PLUG MALFUNCTION INDICATOR
R1: VOLTAGE ADJUST POTENTIOMETER
R2: FREQUENCY ADJUST POTENTIOMETER
R4: LOAD SHARING RHEOSTAT
R5: KVA SHARING RHEOSTAT
R6: SYNC LIGHTS DROPPING RESISTOR
R7: SYNC LIGHTS DROPPING RESISTOR
R8: SYNC LIGHTS DROPPING RESISTOR
R9: SYNC LIGHTS DROPPING RESISTOR
R10: BURDEN RESISTOR
R11: BURDEN RESISTOR
R12: BURDEN RESISTOR
R14: FIELD FLASH RESISTOR
R15: LED RESISTOR
⚠ R16: VOLTAGE ADJUST RESISTOR
S1: MASTER SWITCH
S2: PANEL LIGHT SWITCH
S5: AC CIRCUIT INTERRUPTER SWITCH
S6: AM / VM TRANSFER SWITCH
S7: BATTLE SHORT SWITCH
S10: DEAD CRANK SWITCH
S11: UNIT–PARALLEL SWITCH
⚠ S12: FREQUENCY SELECTOR SWITCH
S13: OVERSPEED SWITCH
S14: CRANK DISCONNECT SWITCH
S16: OVERSPEED RESET SWITCH
S17: EMERGENCY STOP SWITCH
SR1: SLAVE RECEPTACLE
T1: POTENTIAL TRANSFORMER
TB: TERMINAL BOARD (S)
TB1: VOLTAGE RECONNECTION TERMINAL BOARD
TB2: LOAD OUTPUT TERMINAL BOARD
V1–V4: VARISTOR AC LOAD LINES

INDICATES PIN ON DIAGNOSTIC RECEPTACLE (J3).

* LETTER DESIGNATION

⚠ NOTE: S12 AND R16 ARE NOT INCLUDED ON 400 HZ SETS.

FO-1. Electrical Schematic - MEP-804A/MEP-814A

A1: AC VOLTAGE REGULATOR
A2: MALFUNCTION INDICATOR
A3: KILOWATT TRANSDUCER
A4: LOAD MEASURING UNIT
A5: ELECTRONIC GOVERNOR CONTROL
A6: ELECTRONIC GOVERNOR ACTUATOR
A7: FREQUENCY TRANSDUCER
A8: RESISTOR ASSEMBLY
A9: FLOAT SWITCH MODULE
A10-A13: EMI FILTER
B1: CRANKING MOTOR
BT1: BATTERY, 12V
BT2: BATTERY, 12V
C1-C3: EMI CAPACITOR
CB1: DC CONTROL POWER CIRCUIT BREAKER
CB3: GROUND FAULT CIRCUIT INTERRUPTER
CBGND: CONTROL BOX GROUND
CPGND: CONTROL PANEL GROUND
CR1: DIODE, REVERSE BATTERY
CR2: DIODE, FIELD FLASH
CR3-9: DIODE, BLOCKING
CT1: CURRENT TRANSFORMER
CT2: CURRENT TRANSFORMER
CT3: CURRENT TRANSFORMER
CT5: DROOP CURRENT TRANSFORMER
DS1: PANEL LIGHT
DS2: PANEL LIGHT
DS3: PANEL LIGHT
DS4: SYNCRONIZATION LIGHT
DS5: SYNCRONIZATION LIGHT
DS6: BATTLE SHORT LIGHT
DS7: AC CIRCUIT INTERRUPTER LIGHT
E1: AUXILIARY FUEL PUMP
FL1: LOW FUEL LEVEL FLOAT SWITCH
FL2: AUXILIARY FUEL PUMP FLOAT SWITCH
FU1: FUSE
G1: AC GENERATOR
G2: BATTERY CHARGING ALTERNATOR
GRD: GROUND
HT: COOLANT HIGH TEMPERATURE SWITCH
J1: CONVENIENCE RECEPTACLE
J2: PARALLEL RECEPTACLE
J3: DIAGNOSTIC RECEPTACLE
J16: SWITCH BOX RECEPTACLE
K1: AC CIRCUIT INTERRUPTER
K2: CRANKING RELAY
K5: OVER/UNDER VOLTAGE RELAY
K8: OVERLOAD/SHORT CIRCUIT RELAY
K9: REVERSE POWER RELAY
K10: PERMISSIVE PARALLELING RELAY
K11: VOLTAGE SENSING RELAY
K12: ENGINE FAULT RELAY
K13: ELECTRICAL FAULT RELAY
K14: OVERSPEED RELAY
K15: START RELAY
K16: CRANK DISCONNECT RELAY
K17: K1 AUXILIARY RELAY
K19: FUEL LEVEL RELAY
K21: GOVERNOR, CONTROL, POWER
K22: GLOW PLUGS CONTACTOR
L0: OUTPUT TERMINAL
L1: OUTPUT TERMINAL
L2: OUTPUT TERMINAL
L3: OUTPUT TERMINAL
L4: STARTER SOLENOID
M1: AC VOLTMETER
M2: FREQUENCY METER
M3: TIME METER
M4: BATTERY CHARGING AMMETER
M5: FUEL LEVEL INDICATOR
M6: COOLANT TEMPERATURE INDICATOR
M7: OIL PRESSURE INDICATOR
M8: AC AMMETER
M9: KILOWATT METER
MPU: MAGNETIC PICKUP
MT4: BATTERY CHARGING AMMETER SHUNT
MT5: FUEL LEVEL SENDER
MT6: COOLANT TEMPERATURE SENDER
MT7: OIL PRESSURE SENDER
OP: LOW OIL PRESSURE SWITCH
P4: PLUG MALFUNCTION INDICATOR
R1: VOLTAGE ADJUST POTENTIOMETER
R2: FREQUENCY ADJUST POTENTIOMETER
R4: LOAD SHARING RHEOSTAT
R5: KVA SHARING RHEOSTAT
R6: SYNC LIGHTS DROPPING RESISTOR
R7: SYNC LIGHTS DROPPING RESISTOR
R8: SYNC LIGHTS DROPPING RESISTOR
R9: SYNC LIGHTS DROPPING RESISTOR
R10: BURDEN RESISTOR
R11: BURDEN RESISTOR
R12: BURDEN RESISTOR
R14: FIELD FLASH RESISTOR
R15: LED RESISTOR
R16: VOLTAGE ADJUST RESISTOR
S1: MASTER SWITCH
S2: PANEL LIGHT SWITCH
S5: AC CIRCUIT INTERRUPTER SWITCH
S6: AM/VM TRANSFER SWITCH
S7: BATTLE SHORT SWITCH
S10: DEAD CRANK SWITCH
S11: UNIT-PARALLEL SWITCH
S12: FREQUENCY SELECTOR SWITCH
S13: OVERSPEED SWITCH
S14: CRANK DISCONNECT SWITCH
S16: OVERSPEED RESET SWITCH
S17: EMERGENCY STOP SWITCH
SR1: SLAVE RECEPTACLE
T1: POTENTIAL TRANSFORMER
TB: TERMINAL BOARD (S)
TB1: VOLTAGE RECONNECTION TERMINAL BOARD
TB2: LOAD OUTPUT TERMINAL BOARD
V1-V4: VARISTOR AC LOAD LINES

NOTE: SIZE DESIGNATOR OMITTED FROM WIRE NUMBERS FOR CLARITY.

FO-2. Wiring Diagram - MEP-804A/MEP-814A (Sheet 1 of 4)

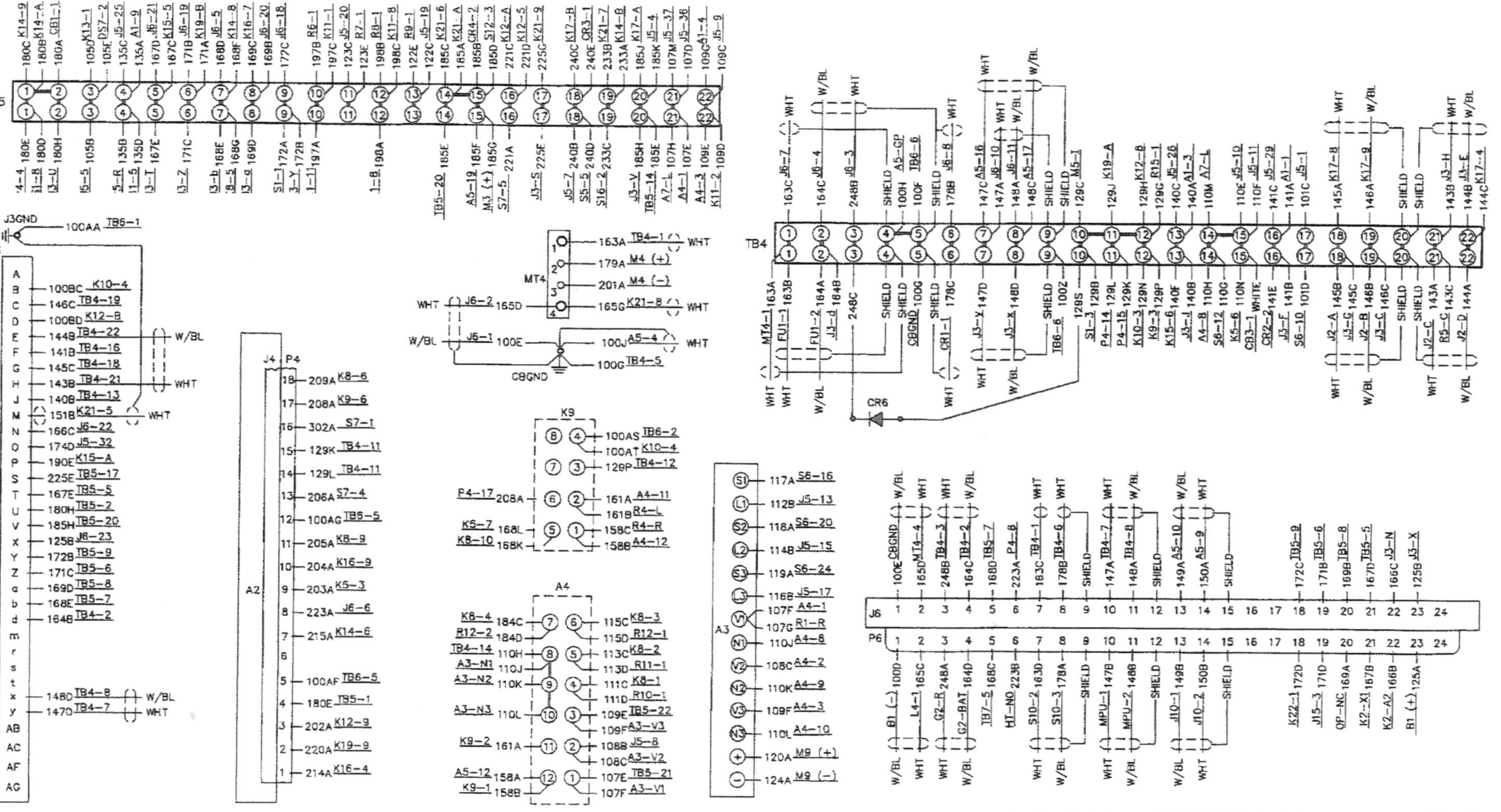

FO-2. Wiring Diagram - MEP-804A/MEP-814A (Sheet 2 of 4)

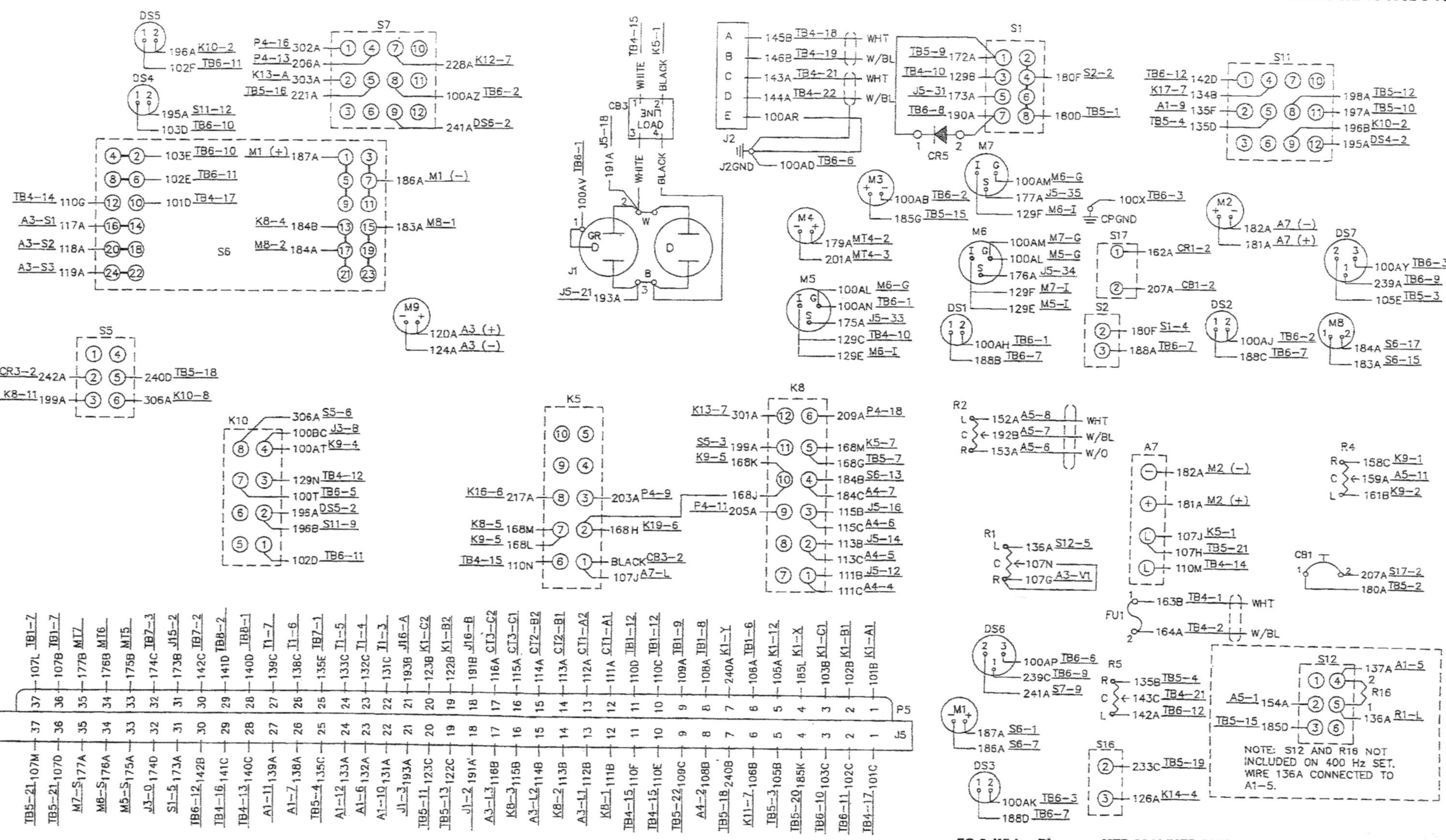

FO-2. Wiring Diagram - MEP-804A/MEP-814A (Sheet 3 of 4)

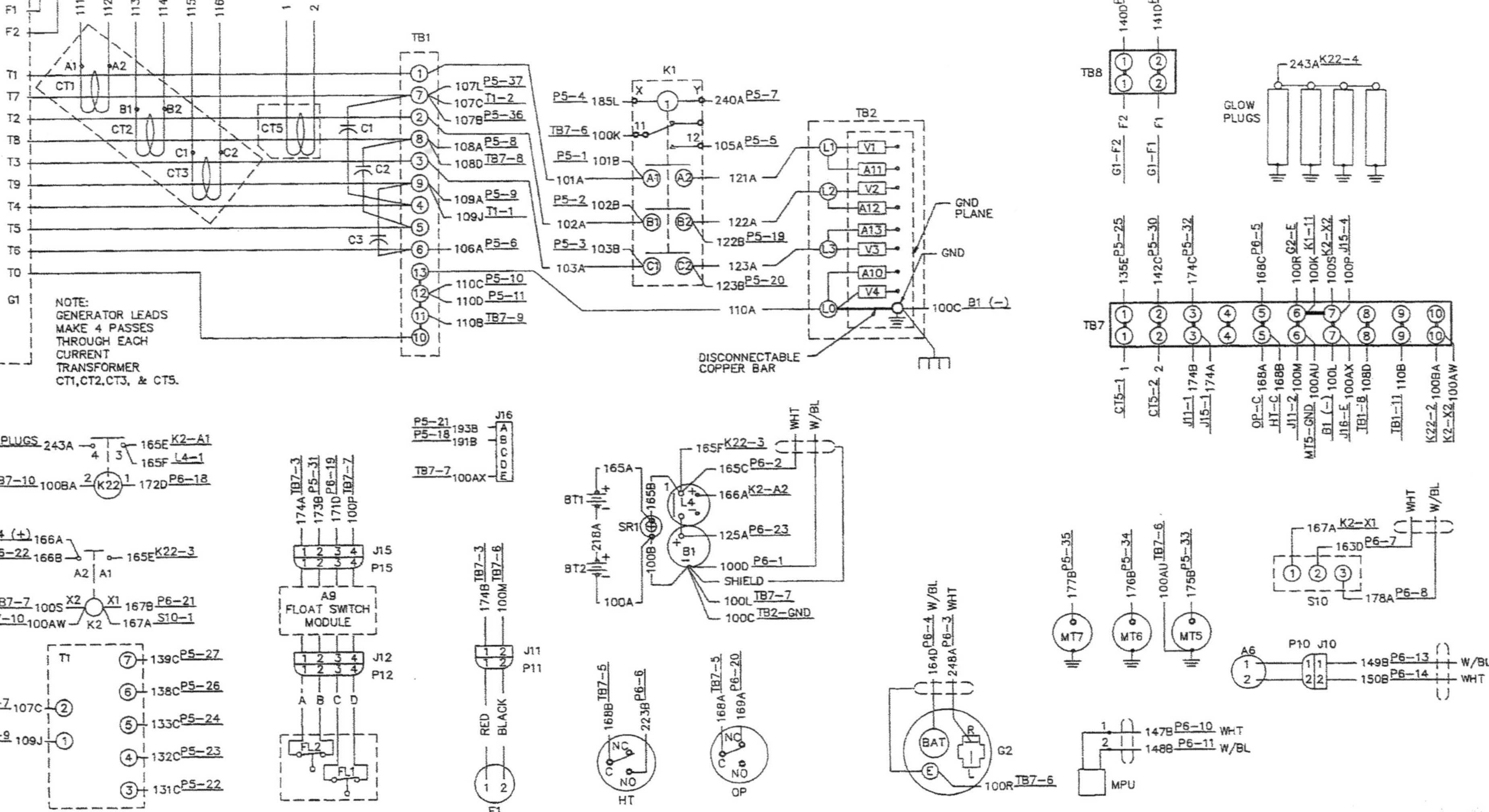

FO-2. Wiring Diagram - MEP-804A/MEP-814A (Sheet 4 of 4)

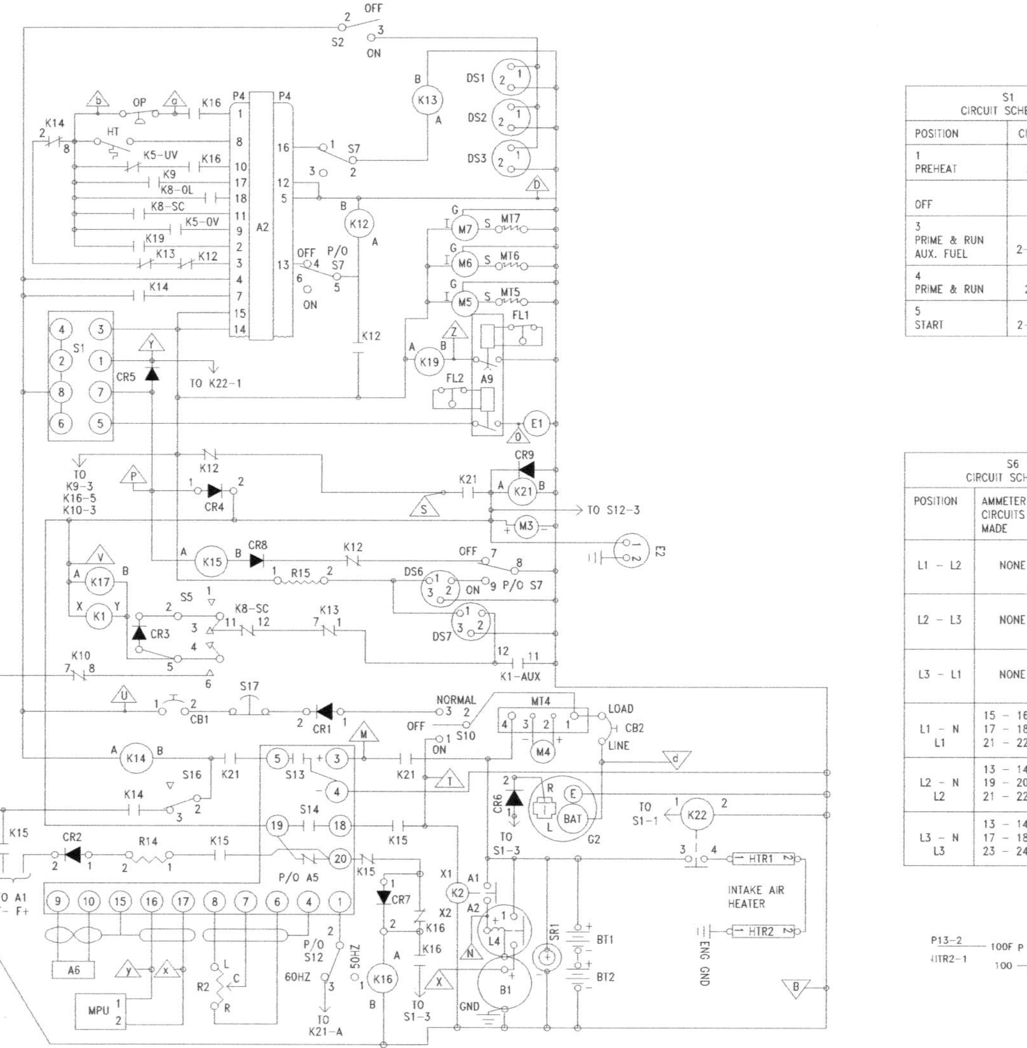

S1 CIRCUIT SCHEDULE	
POSITION	CIRCUITS MADE
1 PREHEAT	2-4-6-8-1
OFF	2-4-6-8 (JUMPERED)
3 PRIME & RUN AUX. FUEL	2-4-6-8-3-5
4 PRIME & RUN	2-4-6-8-3
5 START	2-4-6-8-3-7

S6 CIRCUIT SCHEDULE		
POSITION	AMMETER CIRCUITS MADE	VOLTMETER CIRCUITS MADE
L1 - L2	NONE	7 - 8 9 - 10
L2 - L3	NONE	3 - 4 5 - 6
L3 - L1	NONE	3 - 4 9 - 10
L1 - N L1	15 - 16 17 - 18 21 - 22	9 - 10 11 - 12
L2 - N L2	13 - 14 19 - 20 21 - 22	5 - 6 11 - 12
L3 - N L3	13 - 14 17 - 18 23 - 24	1 - 2 11 - 12

P13-2 —— 100F P —— ENG GND
4ITR2-1 —— 100

A1: AC VOLTAGE REGULATOR
A2: MALFUNCTION INDICATOR
A3: KILOWATT TRANSDUCER
A4: LOAD MEASURING UNIT
A5: ELECTRONIC GOVERNOR CONTROL
A6: ELECTRONIC GOVERNOR ACTUATOR
A7: FREQUENCY TRANSDUCER
A8: RESISTOR ASSEMBLY
A9: FLOAT SWITCH MODULE
A10–A13: EMI FILTER
B1: CRANKING MOTOR
BT1: BATTERY, 12V
BT2: BATTERY, 12V
C1–C3: EMI CAPACITOR
CB1: DC CONTROL POWER CIRCUIT BREAKER
CB2: BATTERY CHR CIRCUIT BREAKER
CB3: GROUND FAULT CIRCUIT INTERRUPTER
CBGND: CONTROL BOX GROUND
CPGND: CONTROL PANEL GROUND
CR1: DIODE, REVERSE BATTERY
CR2 DIODE, FIELD FLASH
CR3–9: DIODE, BLOCKING
CT1: CURRENT TRANSFORMER
CT2: CURRENT TRANSFORMER
CT3: CURRENT TRANSFORMER
CT5 DROOP CURRENT TRANSFORMER
DS1: PANEL LIGHT
DS2: PANEL LIGHT
DS3: PANEL LIGHT
DS4 SYNCRONIZATION LIGHT
DS5 SYNCRONIZATION LIGHT
DS6: BATTLE SHORT LIGHT
DS7: AC CIRCUIT INTERRUPTER LIGHT
E1: AUXILLARY FUEL PUMP
FL2: AUXILIARY FUEL PUMP FLOAT SWITCH
FU2: FUSE
G1: AC GENERATOR
G2: BATTERY CHARGING ALTERNATOR
GND: GROUND
H1: COOLANT HIGH TEMPERATURE SWITCH
HTR1: INTAKE AIR HEATER
HTR2: INTAKE AIR HEATER
J1: CONVENIENCE RECEPTACLE
J2: PARALLEL RECEPTACLE
J3: DIAGNOSTIC RECEPTACLE
J16: SWITCH BOX RECEPTACLE
K1: AC CIRCUIT INTERRUPTER
K2: CRANKING RELAY
K5: OVER / UNDER VOLTAGE RELAY
K8: OVERLOAD / SHORT CIRCUIT RELAY
K9: REVERSE POWER RELAY
K10: PERMISSIVE PARALLELING RELAY
K11: VOLTAGE SENSING RELAY
K12: ENGINE FAULT RELAY
K13: ELECTRICAL FAULT RELAY
K14: OVERSPEED RELAY
K15: START RELAY
K16: CRANK DISCONNECT RELAY
K17: K1 AUXILIARY RELAY

(x) INDICATES PIN ON DIAGNOSTIC RECEPTACLE (J3).

* LETTER DESIGNATION

K19: FUEL LEVEL RELAY
K21: GOVERNOR, CONTROL, POWER
K22: INTAKE HEATER CONTACTOR
L1: OUTPUT TERMINAL
L2: OUTPUT TERMINAL
L3: OUTPUT TERMINAL
L4: STARTER SOLENOID
M1: AC VOLTMETER
M2: FREQUENCY METER
M3: TIME METER
M4: BATTERY CHARGING AMMETER
M5: FUEL LEVEL INDICATOR
M6: COOLANT TEMPERATURE INDICATOR
M7: OIL PRESSURE INDICATOR
M8: AC AMMETER
M9: KILOWATT METER
MPU: MAGNETIC PICKUP
M14: BATTERY CHARGING AMMETER SHUNT
M15: FUEL LEVEL SENDER
M16: COOLANT TEMPERATURE SENDER
M17: OIL PRESSURE SENDER
N: OUTPUT TERMINAL
OP: LOW OIL PRESSURE SWITCH
P4: PLUG MALFUNCTION INDICATOR
R1: VOLTAGE ADJUST POTENTIOMETER
R2: FREQUENCY ADJUST POTENTIOMETER
R4: LOAD SHARING RHEOSTAT
R5: KVA SHARING RHEOSTAT
R6: SYNC LIGHTS DROPPING RESISTOR
R7: SYNC LIGHTS DROPPING RESISTOR
R8: SYNC LIGHTS DROPPING RESISTOR
R9: SYNC LIGHTS DROPPING RESISTOR
R10: BURDEN RESISTOR
R11: BURDEN RESISTOR
R12: BURDEN RESISTOR
R14: FIELD FLASH RESISTOR
R15: LED RESISTOR
(1) R16: VOLTAGE ADJUST RESISTOR
S1: MASTER SWITCH
S2: PANEL LIGHT SWITCH
S5: AC CIRCUIT INTERRUPTER SWITCH
S6: AM / VM TRANSFER SWITCH
S7: BATTLE SHORT SWITCH
S10: DEAD CRANK SWITCH
S11: UNIT-PARALLEL SWITCH
(1) S12: FREQUENCY SELECTOR SWITCH
S13: OVERSPEED SWITCH
S14: CRANK DISCONNECT SWITCH
S16: OVERSPEED RESET SWITCH
S17: EMERGENCY STOP SWITCH
SR1: SLAVE RECEPTACLE
T1: POTENTIAL TRANSFORMER
TB: TERMINAL BOARD (S)
TB1: VOLTAGE RECONNECTION TERMINAL BOARD
TB2: LOAD OUTPUT TERMINAL BOARD
V1–V4: VARISTOR AC LOAD LINES

(1) NOTE: S12 AND R16 ARE NOT INCLUDED ON 400 HZ SETS.

15KW-6115-24-FO-3-1B

FO-3. Electrical Schematic - MEP-804B/MEP-814B (Sheet 1 of 2)

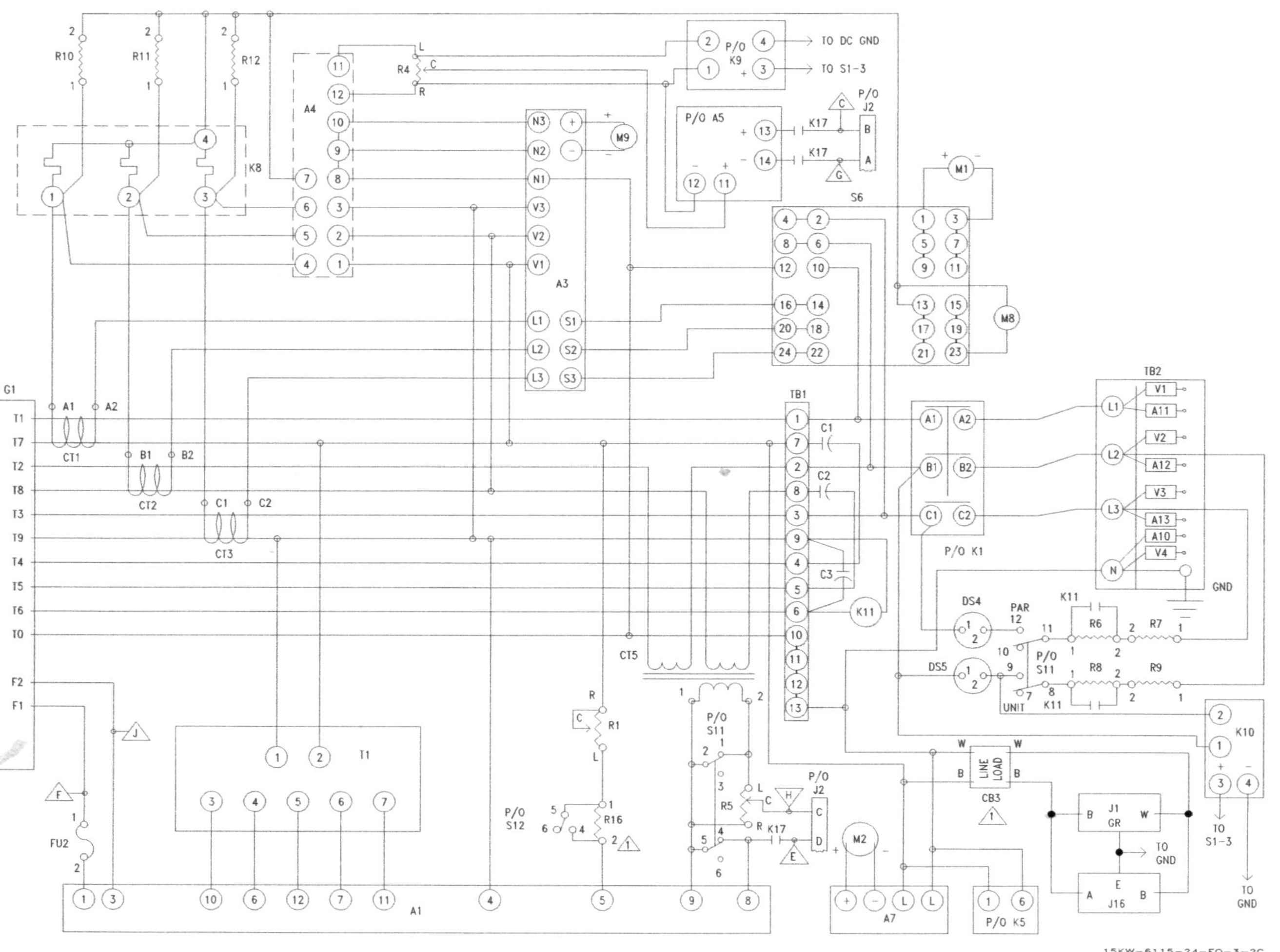

15KW-6115-24-FO-3-2C

FO-3. Electrical Schematic - MEP-804B/MEP-814B (Sheet 2 of 2)

A1: AC VOLTAGE REGULATOR
A2: MALFUNCTION INDICATOR
A3: KILOWATT TRANSDUCER
A4: LOAD MEASURING UNIT
A5: ELECTRONIC GOVERNOR CONTROL
A6: ELECTRONIC GOVERNOR ACTUATOR
A7: FREQUENCY TRANSDUCER
A8: RESISTOR ASSEMBLY
A9: FLOAT SWITCH MODULE
A10–A13: EMI FILTER
B1: CRANKING MOTOR
BT1: BATTERY, 12V
BT2: BATTERY, 12V
C1–C3: EMI CAPACITOR
CB1: DC CONTROL POWER CIRCUIT BREAKER
CB2: BATTERY CHG CIRCUIT BREAKER
CB3: GROUND FAULT CIRCUIT INTERRUPTER
CBGND: CONTROL BOX GROUND
CPGND: CONTROL PANEL GROUND
CR1: DIODE, REVERSE BATTERY
CR2: DIODE, FIELD FLASH
CR3–9: DIODE, BLOCKING
CT1: CURRENT TRANSFORMER
CT2: CURRENT TRANSFORMER
CT3: CURRENT TRANSFORMER
CT5: DROOP CURRENT TRANSFORMER
DS1: PANEL LIGHT
DS2: PANEL LIGHT
DS3: PANEL LIGHT
DS4: SYNCHRONIZATION LIGHT
DS5: SYNCRONIZATION LIGHT
DS6: BATTLE SHORT LIGHT
DS7: AC CIRCUIT INTERRUPTER LIGHT
E1: AUXILLARY FUEL PUMP
E2: PRIMARY FUEL PUMP
FL1: LOW FUEL LEVEL FLOAT SWITCH
FL2: AUXILIARY FUEL PUMP FLOAT SWITCH
FU2: FUSE
G1: AC GENERATOR
G2: BATTERY CHARGING ALTERNATOR
GND: GROUND
HT: COOLANT HIGH TEMPERATURE SWITCH
HTR1: INTAKE AIR HEATER
HTR2: INTAKE AIR HEATER
J1: CONVENIENCE RECEPTACLE
J2: PARALLEL RECEPTACLE
J3: DIAGNOSTIC RECEPTACLE
J16: SWITCH BOX RECEPTACLE

K1: AC CIRCUIT INTERRUPTER
K2: CRANKING RELAY
K5: OVER / UNDER VOLTAGE RELAY
K8: OVERLOAD / SHORT CIRCUIT RELAY
K9: REVERSE POWER RELAY
K10: PERMISSIVE PARALLELING RELAY
K11: VOLTAGE SENSING RELAY
K12: ENGINE FAULT RELAY
K13: ELECTRICAL FAULT RELAY
K14: OVERSPEED RELAY
K15: START RELAY
K16: CRANK DISCONNECT RELAY
K17: K1 AUXILIARY RELAY
K19: FUEL LEVEL RELAY
K21: GOVERNOR, CONTROL, POWER
K22: INTAKE AIR HEATER CONTACTOR
L1: OUTPUT TERMINAL
L2: OUTPUT TERMINAL
L3: OUTPUT TERMINAL
L4: STARTER SOLENOID
M1: AC VOLTMETER
M2: FREQUENCY METER
M3: TIME METER
M4: BATTERY CHARGING AMMETER
M5: FUEL LEVEL INDICATOR
M6: COOLANT TEMPERATURE INDICATOR
M7: OIL PRESSURE INDICATOR
M8: AC AMMETER
M9: KILOWATT METER
MPU: MAGNETIC PICKUP
MT4: BATTERY CHARGING AMMETER SHUNT
MT5: FUEL LEVEL SENDER
MT6: COOLANT TEMPERATURE SENDER
MT7: OIL PRESSURE SENDER
N: OUTPUT TERMINAL
OP: LOW OIL PRESSURE SWITCH
P4: PLUG MALFUNCTION INDICATOR
R1: VOLTAGE ADJUST POTENTIOMETER
R2: FREQUENCY ADJUST POTENTIOMETER
R4: LOAD SHARING RHEOSTAT
R5: KVA SHARING RHEOSTAT
R6: SYNC LIGHTS DROPPING RESISTOR
R7: SYNC LIGHTS DROPPING RESISTOR
R8: SYNC LIGHTS DROPPING RESISTOR
R9: SYNC LIGHTS DROPPING RESISTOR
R10: BURDEN RESISTOR
R11: BURDEN RESISTOR
R12: BURDEN RESISTOR
R14: FIELD FLASH RESISTOR
R15: LED RESISTOR
R16: VOLTAGE ADJUST RESISTOR
S1: MASTER SWITCH
S2: PANEL LIGHT SWITCH
S5: AC CIRCUIT INTERRUPTER SWITCH
S6: AM / VM TRANSFER SWITCH
S7: BATTLE SHORT SWITCH
S10: DEAD CRANK SWITCH
S11: UNIT-PARALLEL SWITCH
S12: FREQUENCY SELECTOR SWITCH
S13: OVERSPEED SWITCH
S14: CRANK DISCONNECT SWITCH
S16: OVERSPEED RESET SWITCH
S17: EMERGENCY STOP SWITCH
SR1: SLAVE RECEPTACLE
T1: POTENTIAL TRANSFORMER
TB: TERMINAL BOARD (S)
TB1: VOLTAGE RECONNECTION TERMINAL BOARD
TB2: LOAD OUTPUT TERMINAL BOARD
V1–V4: VARISTOR AC LOAD LINES

NOTE:
GENERATOR LEADS MAKE
4 PASSES THROUGH EACH
CURRENT TRANSFORMER
CT1, CT2, CT3 AND CT5.

NOTE:
SIZE DESIGNATOR OMITTED FROM
WIRE NUMBERS FOR CLARITY.

15KW-6115-24-FO-4-1B

FO-4. Wiring Diagram - MEP-804B/MEP-814B (Sheet 1 of 3)

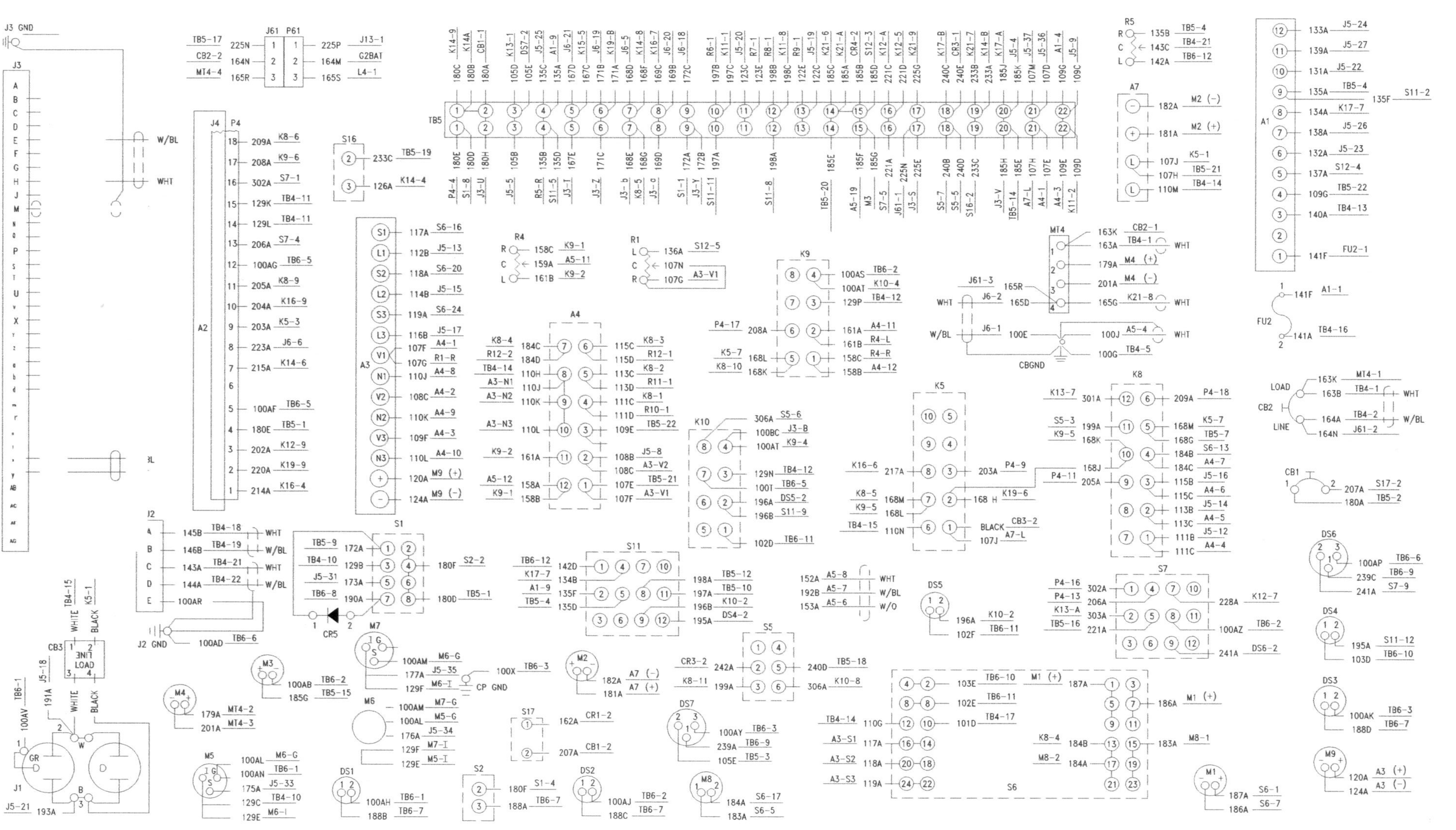

FO-4. Wiring Diagram - MEP-804B/MEP-814B (Sheet 2 of 3)

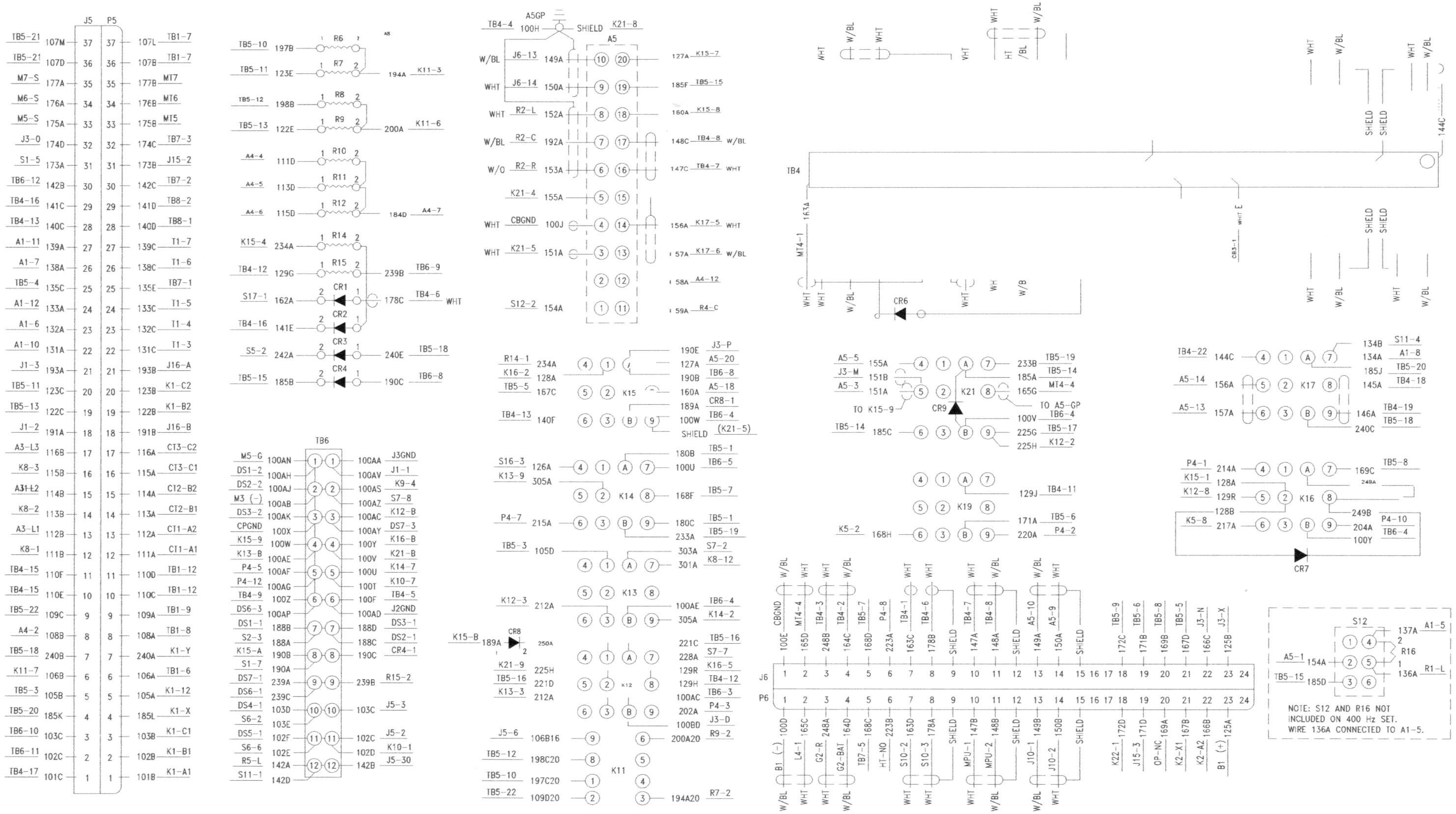

FO-4. Wiring Diagram - MEP-804B/MEP-814B (Sheet 3 of 3)

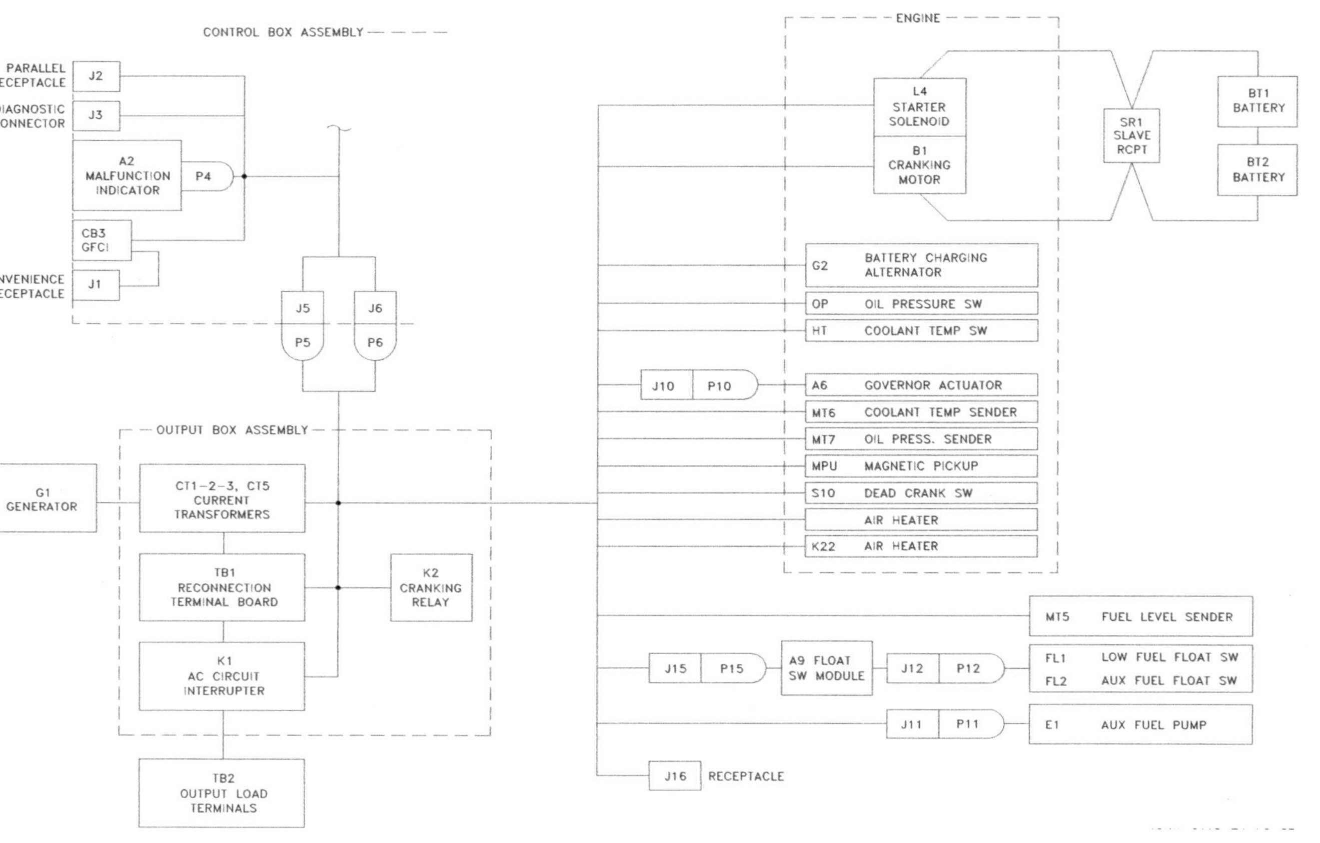

FO-5. Interconnect Diagram - MEP-804B/MEP-814B

The Metric System and Equivalents

Linear Measure

1 centimeter - 10 millimeters = .39 inch
1 decimeter = 10 centimeters = 3.94 inches
1 meter = 10 decimeters = 39.37 inches
1 dekameter = 10 meters = 32.8 feet
1 hectometer = 10 dekameters = 328.08 feet
1 kilometer = 10 hectometers = 3,280.8 feet

Weights

1 centigram - 10 milligrams = .15 grain
1 decigram = 10 centigrams = 1.54 grains
1 gram = 10 decigrams = .035 ounce
1 dekagram = 10 grams = .35 ounce
1 hectogram = 10 dekagrams = 3.52 ounces

1 kilogram = 10 hectograms = 2.2 pounds
1 quintal = 100 kilograms = 220.46 pounds
1 metric ton = 10 quintals = 1.1 short tons

Liquid Measure

1 centiliter = 10 milliliters = .34 fl. ounce
1 deciliter = 10 centiliters = 3.38 fl. ounces
1 liter = 10 deciliters = 33.81 fl. ounces
1 dekaliter = 10 liters = 2.64 gallons
1 hectoliter = 10 dekaliters = 26.42 gallons
1 kiloliter = 10 hectoliters = 264.28 gallons

Square Measure

1 sq. centimeter = 100 sq. millimeters = .155 sq. inch
1 sq. decimeter = 100 sq. centimeters = 15.5 sq. inches
1 sq. meter (centare) = 100 sq. decimeters = 10.76 sq. feet
1 sq. dekameter (are) = 100 sq. meters = 1,076.4 sq. feet
1 sq. hectometer (hectare) = 100 sq. dekameters = 2.47 acres
1 sq. kilometer = 100 sq. hectometers = .386 sq. mile

Cubic Measure

1 cu. centimeter = 1000 cu. millimeters = .06 cu. inch
1 cu. decimeter = 1000 cu. centimeters = 61.02 cu. inches
1 cu. meter = 1000 cu. decimeters = 35.31 cu. feet

Approximate Conversion Factors

To change	To	Multiply by
inches	centimeters	2.540
feet	meters	.305
yards	meters	.914
miles	kilometers	1.609
square inches	square centimeters	6.451
square feet	square meters	.093
square yards	square meters	.836
square miles	square kilometers	2.590
acres	square hectometers	.405
cubic feet	cubic meters	.028
cubic yards	cubic meters	.765
fluid ounces	milliliters	29.573
pints	liters	.473
quarts	liters	.946
gallons	liters	3.785
ounces	grams	28.349
pounds	kilograms	.454
short tons	metric tons	.907
pound-feet	newton-meters	1.356
pound-inches	newton-meters	.11296

To change	To	Multiply by
ounce-inches	newton-meters	.007062
centimeters	inches	.394
meters	feet	3.280
meters	yards	1.094
kilometers	miles	.621
square centimeters	square inches	.155
square meters	square feet	10.764
square meters	square yards	1.196
square kilometers	square miles	.386
square hectometers	acres	2.471
cubic meters	cubic feet	35.315
cubic meters	cubic yards	1.308
milliliters	fluid ounces	.034
liters	pints	2.113
liters	quarts	1.057
liters	gallons	.264
grams	ounces	.035
kilograms	pounds	2.205
metric tons	short tons	1.102

Temperature (Exact)

°F	Fahrenheit temperature	5/9 (after subtracting 32)	Celsius temperature	°C

PIN: 071749-000

www.ingramcontent.com/pod-product-compliance
Lightning Source LLC
Chambersburg PA
CBHW080413030426
42335CB00020B/2433
* 9 7 8 1 9 5 4 2 8 5 1 7 0 *